AF412737

Viral Gastroenteritis

Viral Gastroenteritis

Editors

U. Desselberger

Virologie Moléculaire et Structurale
UMR 2472, CNRS
1 Avenue de la Terrasse, Bât. 14B
91198 Gif-sur-Yvette Cedex
France

J. Gray

Enteric Virus Unit
Enteric, Respiratory and Neurological Virus Laboratory
Specialist and Reference Microbiology Division
Health Protection Agency
London NW9 5HT
United Kingdom

2003

ELSEVIER

Amsterdam – Boston – Heidelberg – London – New York – Oxford
Paris – San Diego – San Francisco – Singapore – Sydney – Tokyo

ELSEVIER SCIENCE B.V.
Sara Burgerhartstraat 25
P.O. Box 211, 1000 AE Amsterdam, The Netherlands

First edition 2003

Library of Congress Cataloging in Publication Data
A catalog record from the Library of Congress has been applied for.

British Library Cataloguing in Publication Data
A catalogue record from the British Library has been applied for.

ISBN: 0 444 51444 9

⊗ The paper used in this publication meets the requirements of ANSI/NISO Z39.48-1992 (Permanence of Paper). Printed in The Netherlands.

Cover image: Three-dimensional structures of rotavirus and Norwalk virus capsids determined by cryo-EM and X-ray crystallographic techniques, respectively. **Left:** Cut-away view of the rotavirus capsid: the structural proteins VP4, VP7, VP6 and VP2, which form the three concentric shells, are shown in red, yellow, blue and green, respectively. The flower-shaped transcription enzyme complex, in red, composed of VP1 and VP3, can be seen attached to the inside surface of VP2 (green) layer. **Right:** Depth-cued Cα-atom representation of the recombinant Norwalk virus capsid at 3.4 Å resolution. Photo by courtesy of Dr. B.V.V. Prasad, Baylor College of Medicine, Houston, TX77030.

Contents

SECTION III

ENTERIC ADENOVIRUSES

SECTION IV

NORWALK- AND SAPPORO-LIKE VIRUSES (HUMAN CALICIVIRUSES)

viii

Preface

Medical Virology has made rapid progress over the last 10 years, mainly due to the advent of highly sensitive and specific molecular detection techniques to analyse clinical materials. These methods have been rapidly introduced into most areas of clinical virology and have in many instances become the *gold standard* for diagnostic detection and analysis and also patient management. This development occurred in parallel with rapid progress in the basic science of virology which has lead to better understanding of virus morphology and structure function relationships, dissection of different steps of viral replication, manipulation of viral genomes for antigen expression and biological studies, better recognition of the true correlates of protection in many viral infections, and progress in vaccine development.

It is at the intersection of basic and applied clinical virology that this book wishes to make a contribution, directed towards the molecular analysis of viruses causing gastroenteritis in man as well as aspects of their pathogenesis, diagnosis, treatment, epidemiology and vaccine-related research. Those viruses comprise four major genera: rotaviruses (members of the *Reoviridae* family), entero-adenoviruses group F (members of the *Adenoviridae*), Norwalk- and Sapporo-like viruses, now termed novo- and sapoviruses (members of the *Caliciviridae*) and astroviruses (members of the *Astroviridae*). In addition viruses of the families *Picornaviridae*, *Parvoviridae*, *Coronaviridae* and some others have been implicated as causes of gastroenteritides in man. Viral causes are a significant component of infectious gastroenteritides which represent the most frequent infections worldwide.

Formally we have constructed the book in such a way that a general section (I) is followed by specific sections (II-VI) devoted to these differenct viruses. Basic data are presented in small introductory chapters which outline the present state-of-the-art and introduce the individual contributions. Those were written by leading experts in their fields on topics where basic and clinical sciences are at the forefront. Chapters have been cross-referenced. It is not claimed that completeness of coverage of all areas has been achieved but it is hoped that readers find the contributions interesting and stimulating.

The book is addressed to a large variety of people: physicians, paediatricians, virologists, immunologists, epidemiologists, molecular biologists, pathophysiologists and vaccinologists could all find chapters appealing to their specific interests as well as see their area of expertise in a larger context.

The editors have greatly enjoyed their interaction with the various authors and author groups whose enthusiasm and commitment have led to the production of great chapters and their delivery in time. We thank Mary Estes, Roger Glass, Harry Greenberg and Marion Koopmans for their comments on introductory chapters and for advice.

Special thanks are due to Lynne Bastow and Narguesse Stevens for their enthusiastic secretarial support.

The Publisher's staff have been excellent: in particular, thanks are due to Hendrik van Leusen, Ana Bela Sa Dias, Angelina Jokovic and the never tiring Joan Anuels.

The editors wish this book a fair reception and a wide distribution.

Cambridge, May 2002

U Desselberger
J Gray

Viral Gastroenteritis
U. Desselberger and J. Gray (editors)

SECTION I

Viral gastroenteritis: Causes, pathophysiology, immunology, treatment, and epidemiology

Introduction

Acute gastroenteritis is one of the most common illnesses affecting man and may be caused by a large variety of different microbes. The condition affects mainly children and the elderly, but increasingly also adults of all ages, often following foreign travel or the ingestion of contaminated food and water. Infection with gastroenteritis agents can be asymptomatic, or be followed by mild or severe disease including vomiting or diarrhoea or both, and can be fatal as a consequence of severe dehydration. In developing countries infants and young children (<5 years of age) may have repeated episodes of diarrhoea and vomiting amounting to a total of over a billion episodes per annum of which approximately 150 million are moderate to severe, and an estimated 2.4-2.8 million deaths occur; this represents one-quarter of all deaths in children under 5 years (Bern *et al.*, 1992; Murray and Lopez, 1997). In developed countries diarrhoea occurs less frequently, and the disease is fatal in only a very few cases; however, it leads to many hospitalizations and has considerable economic impact (Glass *et al.*, 2001).

Causative agents

The spectrum of causative agents differs in developed and developing countries. In developing countries about one quarter of diarrhoeal episodes are due to infection with enterotoxigenic *E. coli* (ETEC). Enteropathogenic *E. coli* (EPEC), *Shigella* species and *Campylobacter* are the cause of 15%, and rotavirus is the cause of 10-15% of episodes. In addition parasitic agents, e.g. *Entamoeba* and *Giardia,* occur in 5% of episodes. In developed countries, bacterial causes of diarrhoea in children are <5%, whilst rotaviruses and other viruses cause 32% and 15% of cases, respectively. In more than one third of the cases no causative agent is found.

Besides rotaviruses as the main etiologic agent, there are many other viral causes of diarrhoea. The main diarrhoeogenic agents comprise four virus families: rotaviruses, enteric adenoviruses, human caliciviruses (Norwalk- and Sapporo-like viruses, now termed noro- and sapoviruses) and astroviruses, and cause diarrhoea at frequencies of 20-30%, 5%, 5-10%, and 5%, respectively. Recently data from Vesikari's group have shown that the degree of undiagnosed diarrhoeal illness ('diagnostic gap'; Flewett *et al.*, 1987) depends critically on the severity of illness: in only 15% of moderately to

2

severely ill children was no microbiological causative agent found, whereas in children with mild diarrhoea this percentage amounted to about 40% (Pang *et al.*, 2000; Vesikari, 2001). In this study, the relative distribution of pathogens was: rotaviruses 24%, caliciviruses 19%, adenoviruses 4%, astroviruses 4%, indicating an improvement of diagnostic procedures and a more important role for caliciviruses as a cause of diarrhoea than was previously recognised. Data reported from outbreak investigations among children in Japan showed that human caliciviruses were an even more frequent cause than rotaviruses (Nakata *et al.*, 2000). In addition to the four main families of viruses, other viruses have been identified as causing human diarrhoea: toroviruses (Koopmans *et al.*, 1993), picobirnaviruses (Grohmann *et al.*, 1993) and enteroviruses (Yamashita *et al.*, 1991, 1993). The involvement of coronaviruses and parvoviruses that are well known causes of diarrhoea in animals, as causes of human disease is controversial. In the immunocompromised host (children with severe combined immunodeficiency (SCID), AIDS patients, tissue and organ transplant recipients etc.) viruses normally not causing disease in man are found as causes of chronic gastroenteritis (Grohmann *et al.*, 1993). Among those are mainly viruses of the herpes group (cytomegalovirus, herpes simplex virus), picobirnaviruses and adenoviruses other than the enteric adenoviruses. UD Parashar and R Glass have given a general overview of this area (Section I, Chapter 1).

Pathophysiology

The mucosa of the small intestine consists of arrays of long villi interspersed by crypts near their base. Villi and crypts are covered by a continuous monolayer of epithelial cells, those on the villi being highly differentiated for the purposes of absorption (particularly at the tip of the villi) and those in the crypts being undifferentiated and acting as a reservoir for proliferation and differentiation into absorptive cells. Crypts secrete chloride and immunoglobulin A (see below) into the gut lumen. Normally, villous absorption exceeds secretion (= net absorption).

Some viruses infect the mature enterocyte in the middle or upper villous epithelium of the small intestine (rotaviruses, adenoviruses, astroviruses); other viruses infect the crypt cells, e.g. parvoviruses (panleukopenia virus) and toroviruses (Breda virus) in animal models (Moon, 1994). Mainly the proximal small intestine is infected. After viral replication, epithelial cells become necrotic and are sloughed off, leading to a loss of enzymes breaking down carbohydrates and proteins (lactase, peptidases) and to primary malabsorption. This also leads to villous atrophy, followed by reactive crypt cell hyperplasia with increased numbers of mitoses in enteroblasts, cellular infiltration of the submucosa and hypersecretion which adds to the severity of diarrhoea. Recovery from the loss of the villi is relatively quick (7-10 days).

In the initial stages of diarrhoea, a marked ischemia is observed in the villi, preceding the death of enterocytes and suggesting that the observed changes are the result of localized systemic responses triggered by the infection of enterocytes and that diarrhoea is not simply the direct result of impaired enterocyte function (Starkey *et al.*, 1986; Osborne *et al.*, 1988; Stephen and Osborne, 1988). Recently, the first viral enterotoxin has been described in rotavirus infection (Tian *et al.*, 1995; Ball *et al.*, 1996; see Section II, Chapter 6). Furthermore, the autonomous nervous system seems to be affected in decreasing the mobility of the small intestine, thus contributing to the development of diarrhoea (Burrows and Merritt, 1984; Lundgren *et al.*, 2000). Pathophysiologically the diarrhoea is due to several factors: decreased glucose mediated sodium adsorption accompanied by increased undigested lactose in the gut lumen, and crypt hypersecretion as a consequence of electrocyte movements under the condition of increased cellular division (Stephen, 1988; Greenberg *et al.*, 1994). F Michelangeli and MR Ruiz have reviewed research on the pathophysiology of the gut in the context of viral diarrhoea (Section I, Chapter 2), and O Lundgren and L Svensson the role of the enteric nervous system which has recently been recognised as very important in this process (Section I, Chapter 3).

Immunology

The gastrointestinal immune system provides a significant proportion of the body's overall immune responses. The gut is the organ in which most immunoglobulins are produced. Gut associated lymphocytic tissue (GALT) occurs in Peyer's patches. scattered lymph node follicles, lymphoid cells in the epithelium, and submucosal lymphocytes. There are about 10^{10} immunoglobulin (Ig)-producing lymphocytes per metre of small bowel, compared to 2.5×10^{10} Ig-producing cells in bone marrow, spleen and lymph nodes together (Brandtzaeg, 1988). Most of the immunocytes in the gut produce dimers or polymers of immunoglobulins of subclass A (IgA) containing a small interconnecting peptide (J-chain). Those IgA dimers or polymers can be transported to the basal membrane of epithelium to a "secretory component" (SC) protein which acts as a specific receptor. IgA is then transcytosed through the epithelial cell and secreted into the gut lumen. The daily production in the gut is 40mg/kg of IgA which is more than the total daily production of IgG (30 mg/kg). The function of these IgAs is in bacterial neutralization, viral neutralization or antibody-dependent cytotoxicity (ADCC). IgA may also be involved in intracellular "neutralization" (Burns *et al.*, 1996; Desselberger, 1998). In rotavirus infections copro-IgAs were found to be the best correlate of protection (Coulson *et al.*, 1992; Feng *et al.*, 1994; Burns *et al.*, 1996; Feng *et al.*, 1997). By contrast, antibody directed against Norwalk-like viruses was not well correlated with subsequent protection (Parrino *et al.*, 1977; Graham *et al.*, 1994; Gray *et al.*, 1994). P Brandtzaeg and F E Johansen have produced a comprehensive update of this area of research (Section I, Chapter 4).

Treatment

Treatment of infantile diarrhoea is mainly by oral or intravenous rehydration. Several formulae of oral rehydration solution (ORS) have been devised, are recommended

4

by WHO and are widely used, mainly in developing countries (Bhan *et al.*, 1994). In severe cases intravenous fluid substitution with modified Ringer's solution is recommended (Santosham *et al.*, 1992). Oral immunoglobulins have been tried (Guarino *et al.*, 1994) but are not part of a standard treatment regime. The use of antimotility drugs in children is generally not advised due to severe side effects (Desselberger, 1999). However, enkephalinase inhibitors (racecadotril) which decrease hypersecretion but do not affect motility, have recently been demonstrated to be a safe and effective treatment of childhood diarrhoea (Salazar-Lindo *et al.*, 2000). The discussion of treatment options of viral diarrhoea has been dealt with by D Bass (Section I, Chapter 5).

Epidemiology

Despite the low mortality rate of infections with gastroenteritis viruses in developed countries, the disease burden is considerable. Thus, it has been calculated that children in the United States will experience 1.5-2.5 episodes of diarrhoea per year but an estimated 2 million children will visit a doctor or clinic, and 160,000 children will be hospitalized (Tucker *et al.*, 1998). A total of 400,000 adults in the United States is annually discharged from hospital with diarrhoea as their diagnosis, and 10-12% of all children hospitalized below the age of 5 years have diarrhoea (Glass *et al.*, 2001).

Viral gastroenteritis occurs in two distinct epidemiological patterns, endemic childhood diarrhoea and epidemic disease (Table 1; Glass *et al.*, 2001). The endemic childhood disease is mainly due to infection with rotaviruses, astroviruses and enteric adenoviruses. Transmission is mainly by person-to-person spread, but also possibly droplets, fomites or possibly zoonotic transmission. By the age of 5 years antibodies to these viruses are found in practically every child. By contrast, epidemic disease is found in all ages, mainly caused by Norwalk- and Sapporo-like viruses (human caliciviruses), rotaviruses of group B (in China), and sometimes astroviruses. They are mainly transmitted by food and water, but also by human-to-human contact (secondary cases). Human caliciviruses can infect repeatedly, even in the presence of virus-specific antibody, suggesting that the immunity is shortlived.

Whilst endemic childhood disease is likely to be controlled best by a vaccine programme, epidemic disease can be limited by outbreak control measures, improvements in food safety, and control of people handling or processing food. Risk factors to acquired diarrhoeal disease may be increased susceptibility (infants and the elderly), congenital and acquired immunodeficiencies (SCID, HIV-infected patients, transplant recipients), increased exposure by travel to developing countries, low socio-economic living conditions, ingestion of high risk foods or contaminated water, as well as concentration of susceptible people in hospital wards, extended care facilities for the elderly, day care centres, cruise ships and holiday camps.

An outline of epidemiological features of viral gastroenteritis is found in the chapter by UD Parashar and R Glass (Section I, Chapter 1).

Many of the research, management and surveillance aspects considered in this general section (I) will be returned to in the following, more specialised sections (II-VI).

The book updates, complements and actualises recent monographs on viral gastroenteritis (Kapikian, ed, 1994; Chiba *et al.*, eds, 1997; Chadwick and Goode, eds, 2001; Cohen *et al.*, eds, 2002).

Table 1

Epidemiological patterns of viruses causing acute gastroenteritis

	Endemic disease in childhood	Epidemic disease
Viruses	Group A rotaviruses Astroviruses Enteric adenoviruses Human caliciviruses Group C rotaviruses? Toroviruses? Coronaviruses??	Human caliciviruses Group B rotaviruses Astroviruses
Mode of transmission	person-to-person droplets and aerosols fomites	Food Water person-to-person droplets and aerosols
Reservoir	Humans	Humans Animals?
Immune states	High prevalence of specific antibody by 5 years of age	seroconversion in epidemic
Immunity	good	short term (human caliciviruses)
Virus variants	limited per site, variable between sites	Many genomic variants
Public health control measures	Vaccine (Group A rotaviruses)	outbreak control, improved safety of food supply and handling, water surveillance

* Slightly modified from Glass *et al.*, 2001

References

Ball JM, Tian P, Zeng CQY, Morris AP, Estes MK (1996). Age-dependent diarrhea induced by a rotaviral nonstructural glycoprotein. *Science* **272:** 101-104.

Bern C, Martines J, deZoysa I, Glass, RI (1992). The magnitude of the global problem of diarrhoeal disease: a ten-year update. *Bull. WHO* **70:** 705-714.

Bern C, Glass RI (1994). Impact of diarrhoeal disease worldwide. In: *Viral Infections of the Gastrointestinal Tract,* second edition, (Kapikian AZ, ed.), pp 1-26. M. Dekker, New York - Basel - Hong Kong.

Bhan MK, Mahalanabis D, Fontaine O *et al.* (1994). Clinical trials of improved oral rehydration salt formulations: a review. *Bull. WHO* **72:** 945-955.

Brandtzaeg P (1988). Immune responses to gut virus infections. In: *Viruses and the Gut* (MJG Farthing, ed.), pp 45-54, Swan Press, London.

Bresee JS, Glass RS, Ivanoff B and Gentsch JR (1999). Current status and future priorities for rotavirus vaccine development, evaluation and implementation in developing countries. *Vaccine* **17:** 2207-2222.

Burns JW, Siadat-Pajouh M, Krishnaney AA, Greenberg HB (1996). Protective effect of rotavirus VP6-specific IgA monoclonal antibodies that lack neutralizing activity. *Science* **272:** 104-107.

Burrows CF, Merritt AM (1984). Influence of coronavirus (transmissible gastroenteritis) infection on jejunal myoelectrical activity of the neonatal pig. *Gastroenterology* **87:** 386-391.

Chadwick D, Goode JA (Eds) (2001). *Gastroenteritis Viruses. Novartis Found. Symp.* 238, J Wiley and Sons, Chichester, pp 323.

Chiba S, Estes MK, Nakata S, Calisher CH (Eds) (1997). *Viral Gastroenteritis. Arch. Virol.* 142 (Suppl 12), 1-311.

Cohen J, Garbang-Chenon A, Pothier P (Eds) (2002). *Les Gastroentérites Virales.* Elsevier SAS, Paris, pp 325.

Coulson BS, Grimwood K, Hudson IL, Barnes GL, Bishop RF (1992). Role of antibody in clinical protection of children during infection with rotavirus. *J. Clin. Microbiol.* **30:** 1678-1684.

Desselberger U (1998). Prospects for vaccines against rotaviruses. *Rev. Med. Virol.* **8:** 43-52.

Desselberger U, (1999) Rotavirus infection: guidelines for treatment and prevention. *Drugs* **58:** 447-452.

Feng N, Burns JW, Bracey L, Greenberg HB (1994). Comparison of mucosal and systemic humoral immune responses and subsequent protection in mice orally inoculated with a homologous or heterologous rotavirus. *J. Virol.* **68:** 7766-7773.

Feng N, Vo PT, Chung D, Vo TVP, Hoshino Y, Greenberg H (1997). Heterotypic protection following oral immunization with live heterologous rotavirus in a mouse model. *J. Infect. Dis.* **175:** 330-341.

Flewett TH, Beards GH, Brown DW, Sanders RC (1987). The diagnostic gap in diarrhoeal aetiology. In: *Novel diarrhoea viruses. Ciba Found. Symp.* **128:** 238-249.

Glass R, Bresee J, Jiang BM, Gentsch J, Ando T, Fankhauser R, Noël J, Parashar U, Rosen B, Monroe SS (2001). Gastroenteritis viruses: an overview. *Novartis Found. Symp* **238:** 5-19.

Graham DY, Jiang X, Tanaka T, Opekun AR, Madore AP, Estes MK (1994). Norwalk virus infection in volunteers: new insights based on improved assays. *J. Infect. Dis.* **170:** 34-43.

Gray JJ, Cunliffe C, Ball J, Graham DY, Desselberger U, Estes MK (1994). Detection of immunoglobulin M (IgM), IgA and IgG Norwalk virus-specific antibodies by indirect enzyme-linked immunosorbent assay with baculovirus-expressed Norwalk virus capsid antigen in adult volunteers challenged with Norwalk virus. *J. Clin. Microbiol.* **32:** 3059-3063.

Greenberg HB, Clark HF, Offit PA (1994). Rotavirus pathology and pathophysiology. In: *Rotaviruses* (Ramig, RF ed), pp 256-283. Springer Verlag, Berlin-Heidelberg.

Grohmann GS, Glass RI, Pereira HG *et al.* (1993). Enteric viruses and diarrhea in HIV-infected patients. *N. Engl. J. Med.* **329:** 14-20.

Guarino A, Canani RB, Russo F, *et al.* (1994). Oral immunoglobulins for treatment of acute rotaviral infection. *Pediatrics* **93:** 12-16.

Kapikian AZ (Ed) (1994). *Viral Infections of the Gastrointestinal Tract.* Second Ed. M Dekker, New York, pp 785.

Kapikian AZ and Chanock RM (1996). Rotaviruses. In: *Fields Virology,* 3rd ed. (Fields BN, Knipe DM, Howley PM *et al.*, eds.), Lippincott-Raven, Philadelphia, pp 1657-1708.

Koopmans M, Petric M, Glass RI, Monroe SS (1993). Enzyme-linked immunosorbent assay reactivity of torovirus-like particles in faecal specimens for humans with diarrhoea. *J. Clin. Microbiol.* **31:** 2738-2744.

Lundgren O, Peregrin AT, Persson K, Kordasti S, Uhnoo I, Svensson L (2000). Role of the enteric nervous system in the fluid and electrolyte secretion of rotavirus diarrhea. *Science* **257:** 491-495.

Moon HW (1994). Pathophysiology of viral diarrhea. In: *Viral Infections of the Gastrointestinal Tract.* Second Edition. (Kapikian AZ, ed.), pp 27-52. M. Dekker, New York - Basel - Hong Kong.

Moser CA, Cookinham S, Coffin SE, Clark HF, Offit PA (1998). Relative importance of rotavirus-specific effector and memory B cells in protection against challenge. *J. Virol.* **72:** 1108-1114.

Murray CJ, Lopez AD (1997). Global mortality, disability, and the contribution of risk factors: Global burden of disease study. *Lancet* **349:** 1436-1442.

Nakata S, Honma S, Numata KK *et al.* (2000). Members of the family *Caliciviridae* (Norwalk virus and Sapporo virus) are the most prevalent cause of gastroenteritis outbreaks among infants in Japan. *J. Infect. Dis.* **181:** 2029-2032.

Offit PA (1994). Rotaviruses: immunological determinants of protection against infection and disease. *Adv. Virus Res.* **44:** 161-202.

Osborne MP, Haddon SJ, Spencer AJ at al (1988). An electron microscopic investigation of time-related changes of the intestine of neonatal mice infected with murrine rotavirus. *J. Pediatr. Gastroenterol.Nutr.* **7:** 236-298.

Parrino TA, Schreiber TS, Trier JS, Kapikian AZ, Blacklow NR (1977). Clinical immunity in acute gastroenteritis caused by Norwalk agent. *N. Engl. J. Med.* **297:** 86-89.

Pang XL, Honma S, Nakata S, Vesikari T (2000). Human caliciviruses in acute gastroenteritis of young children in the community. *J. Infect. Dis.* **181** (Suppl 2):

S288-S294.

Salazar-Lindo E, Santisteban-Ponce J, Chea-Woo E, Guiterrez M (2000). Racecadotril in the treatment of acute watery diarrhea in children. *N. Engl. J. Med.* **343:** 463-467.

Santosham M, Glass RI (1992). The management of acute diarrhea in children: oral rehydration maintenance and nutritional therapy. *Morb. Mort. Wkly Rep.* **41:** 1-20.

Starkey WG, Collins J, Wallis TS *et al.* (1986). Kinetics, tissue specificity and pathological changes in murine rotavirus infection of mice. *J. Gen. Virol.* **67:** 2625-2634.

Stephen J (1988). Functional abnormalities in the intestine. In: *Viruses and the Gut* (Farthing MJG, ed.), pp 41-44. Swan Press, London.

Stephen J, Osborne MP (1988). Pathophysiological mechanisms of diarrhoeal disease. In: *Bacterial infections of the respiratory and gastrointestinal mucosae.* (Donachie W, Griffiths E, Stephen J, eds), *SGM Symposium* **24:** pp 149-172. IRL Oxford University Press, Oxford.

Tian P, Estes MK, Hu Y, Ball JM, Zeng CQ, Schilling WP (1995). The rotavirus nonstructural glycoprotein NSP4 mobilizes Ca^{2+} from the endoplasmic reticulum. *J. Virol.* **69:** 5763-5772.

Tucker AW, Haddix AC, Bresee JS, Holman RC, Parashar UD, Glass RI (1998). Cost-effectiveness analysis of a rotavirus immunization programme for the United States. *J. Amer. Med. Ass.* **279:** 1371-1376.

Vesikari T (2001). Distribution of viral findings according to severity score of gastroenteritis episodes. Discussion. Novartis Found. Symp. **238:** 171-173.

Yamashita T, Kobayashi S, Sakae K *et al.* (1991). Isolation of cytopathic small round viruses with BSC-1 cells from patients with gastroenteritis. *J. Infect. Dis.* **164:** 954-957.

Yamashita T, Sakae K, Ishihara Y, Isomura S, Utagawa E (1993). Prevalence of newly isolated, cytopathic small round virus (Aichi strain) in Japan. *J. Clin. Microbiol.* **31:** 2938-2943.

Yuan LJ, Ward LA, Rosen BI, To TL and Saif LJ (1996). Systemic and intestinal antibody secreting cell responses and correlates of protective immunity to human rotaviruses in a gnotobiotic pigs model of disease. *J. Virol.* **70:** 3075-3083.

Viral Gastroenteritis
U. Desselberger and J. Gray (editors)

I, 1. Viral causes of gastroenteritis

Umesh D. Parashar and Roger I. Glass

Viral Gastroenteritis Section, Respiratory and Enteric Viruses Branch,
Division of Viral and Rickettsial Diseases, National Center for Infectious Diseases, Centers for Disease
Control and Prevention, Atlanta, GA 30333.

Introduction

Acute gastroenteritis is among the most common illnesses of humans and is caused by a variety of agents, including bacteria, viruses, parasites, toxins, and chemicals. The clinical spectrum ranges from asymptomatic or mild infection to severe, dehydrating illness with a fatal outcome; the latter occurs primarily in young children and in the elderly. In developing nations, children experience more than 3 episodes of gastroenteritis each year, leading to an estimated 2.5-3.2 million deaths [Bern, *et al.*, 1992; Murray and Lopez, 1997]. In industrialized nations, such as the United States, mortality from gastroenteritis is low, but the morbidity and health care costs associated with clinic visits and hospitalizations because of this disease are substantial [Tucker, *et al.*, 1998].

Causative viral agents

Before the early 1970s, no virus had been confirmed as a cause of acute gastroenteritis. The etiologic evaluation of patients with gastroenteritis was limited to investigations for a few bacterial and parasitic agents (e.g., *Salmonella, Shigella, Amoeba*); consequently, the causes of illness remained unidentified in a majority of cases. Clues that this "diagnostic void" [Flewett, *et al.*, 1987] might be filled by viruses were provided by studies in the 1940s and 1950s that demonstrated the transmission of infection to volunteers challenged with bacteria-free fecal filtrates from persons affected by gastroenteritis [Gordon, *et al.*, 1947]. However, attempts in the 1950s and 1960s to etiologically link a viral agent with gastroenteritis failed.

   In 1972, Norwalk virus became the first viral agent identified to cause gastroenteritis. Using immune electron microscopy, Kapikian et. al. identified the 27-nm Norwalk virus particle in stool filtrates of a volunteer challenged with fecal specimens from patients of an outbreak of gastroenteritis [Kapikian, *et al.*, 1972]. In the next few years, electron microscopy (EM) played a key role in the identification or confirmation of many other viruses causing gastroenteritis, such as rotaviruses [Bishop, *et al.*, 1973; Flewett *et al.*, 1973], astroviruses [Madeley and Cosgrove, 1975a, 1975b], enteric adenoviruses

10

[Wadell, *et al.*, 1987], and human caliciviruses of two different genogroups now called "Norwalk-like viruses" (NLVs) and "Sapporo-like viruses"(SLVs) [Chiba, *et al.*, 2000]. A number of other viruses whose link to human gastroenteritis is not well understood inhabit the gut, including picobirnaviruses [Grohmann, *et al.*, 1993; Rosen, *et al.*, 2000], toroviruses [Koopmans, *et al.*, 1993], coronaviruses, and Aichi virus [Yamashita, *et al.*, 1991; Yamashita, 1999]. The virologic properties of these agents and their EM appearances are presented in Table 1 and Fig. 1, respectively.

Table 1

Virologic properties of and means of detecting viruses associated with gastroenteritis among humans

Virus	Family	Size (nm)	Appearance using electron microscopy (EM)	Nucleic acid	Detection*
Rotavirus	*Reoviridae*	70	Wheel shaped, triple-layered capsid	dsRNA	EM, EIA, PAGE, RT-PCR, culture
Calicivirus	*Caliciviridae*	28-35	Small round structured viruses (SRSVs) with calices	ss(+)RNA	EM, EIA, RT-PCR
Astrovirus	*Astroviridae*	28-30	SRSV, star shaped	ss(+)RNA	EM, EIA, RT-PCR
Adenovirus	*Adenoviridae*	70-80	Icosahedral capsid	dsDNA	EM, EIA, RT-PCR, culture
Picobirnavirus	*Birnaviridae*	35	Small round virus	dsRNA	EM, PAGE, RT- PCR
Coronavirus	*Coronaviridae*	60-200	Pleomorphic with club-shaped projections	ss(+)RNA	EM
Torovirus	*Coronaviridae*	100-150	Pleomorphic with torus-shaped core	ss(+)RNA	EM
Aichi virus	*Picornaviridae*	30	Small round virus	ss(+)RNA	Culture, EIA

* EIA = Enzyme immunoassay, EM = Electron microscopy, PAGE = Polyacrylamide gel electrophoresis, RT-PCR = Reverse transcription- polymerase chain reaction.

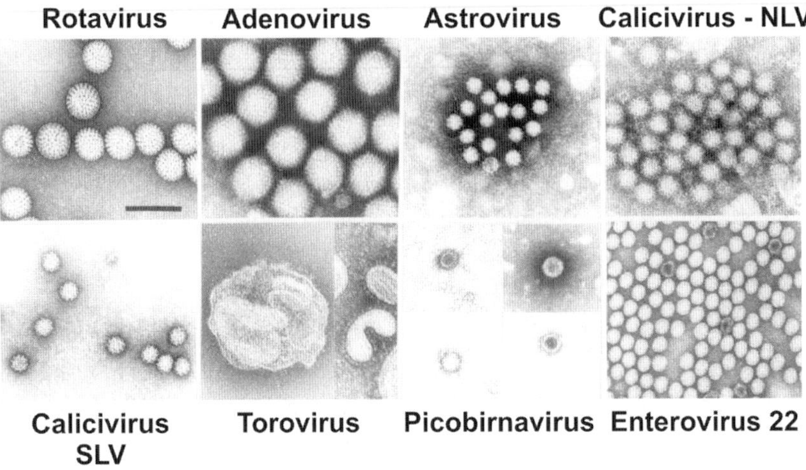

Fig. 1. Viral agents of gastroenteritis as seen by electron microscopy. Bar = 100 nm. (Reproduced with permission from Glass *et al.*, 2001)

Pathophysiology

The intestinal mucosa consists of arrays of long villi interspersed by crypts near their base. The epithelial cells covering the villi are highly differentiated for purposes of absorption, whereas those in the crypts are less differentiated and act as reservoirs for proliferation and differentiation into absorptive cells. The leading viral agents of gastroenteritis infect the mature enterocytes in the middle or upper villous epithelium of the small intestine. On pathologic examination, shortening and atrophy of the villi, round cell infiltration in the submucosa and lamina propria, and reactive hyperplasia of crypt cells are observed. Vacuolization and shedding of enterocytes from the villous tip may be seen early in infection. The infected epithelium ultimately becomes necrotic and sloughs off. The loss of absorptive villous epithelium coupled with proliferation of secretory crypt cells reverses the inherent absorptive state of the epithelium, resulting in secretory diarrhea with loss of fluids and electrolytes in the lumen. In addition, levels of brush border enzymes characteristic of differentiated cells (e.g., sucrase and lactase) are reduced, leading to accumulation of unmetabolized disaccharides in the gut lumen and consequent osmotic diarrhea [see also Section I, Chapter 2 of this book].

Other pathophysiologic mechanisms have been postulated for gastroenteritis caused by specific viruses. In the early stages of rotavirus infection before significant viral replication in epithelial cells, villous ischemia is induced that produces local changes mediated by endogenous, neuroactive, hormonal substances [Osborne, *et al.*, 1988; Greenberg, *et al.*, 1994]. A non-structural rotavirus protein, NSP4, has been identified as a viral enterotoxin. NSP4 increases the intracellular calcium levels and affects the permeability of plasma membranes, causing an efflux of chloride, sodium, and water, thereby inducing a secretory diarrhea [Ball, *et al.*, 1996; Estes, Section II, Chapter 6

12

of this book]. Furthermore, rotavirus appears to evoke intestinal fluid secretion through activation of the enteric nervous system, possibly through triggering by increased intracellular calcium the release of amines or peptides that stimulate dendrites or free nerve endings located beneath the epithelial layer [Lundgren, *et al.*, 2000; Lundgren and Svensson, Section I, Chapter 3 of this book]. In NLV disease, alterations in gastrointestinal motility are believed to play an important role in the pathogenesis of the characteristic nausea and vomiting [Meeroff, *et al.*, 1980].

Immunology

The mucosal immune system, quantitatively the largest immune system of the human body, plays a key role in protection against enteric viral infections [Brandtzaeg, 1998 and Section I, Chapter 4 of this book]. This system is characterized by several unique features, including the preferential production, transport, and secretion of IgA at mucosal surfaces. The precursors of mucosal IgA-producing plasma cells originate in organized lymphoepithelial structures (gut associated lymphocytic tissue) located in Peyer's patches, scattered lymph node follicles, lymphoid cells in the epithelium, and submucosal lymphocytes. Approximately 10^{10} immunoglobulin-producing cells are present per meter of bowel, compared with 2.5×10^{10} Ig producing cells in bone marrow, spleen, and lymph nodes together. Gut immunocytes produce dimers or polymers of IgA that are linked together by a J-chain. The IgA enters the cytoplasm of intestinal epithelial cells after attaching to receptors on the basolateral surface, acquires a secretory piece, and is secreted into the intestinal lumen.

Protection against rotavirus disease is correlated with presence of virus-specific secretory IgA antibodies in the feces and serum [Coulson, *et al.*, 1990; Coulson, *et al.*, 1992; Matson, *et al.*, 1993; Velazquez, *et al.*, 1996]. Since the presence of virus-specific IgA at the intestinal surface is short-lived, complete protection against natural rotavirus disease is only temporary. The presence of memory B and T cells in the lamina propria is believed to be important in the modification (i.e., reduction in severity) of disease from reinfection [Offit, 1996]. Activation of these memory cells to antibody-producing B cells and virus-specific cytotoxic T cells requires a few days; thus, these effector cells can shorten the duration of illness but cannot prevent it. Studies in mice have shown that both cell-mediated and humoral immune responses have a role in the resolution of rotavirus infection [Franco, *et al.*, 1995; McNeal, *et al.*, 1997; Gonzalez *et al.*, Section II, Chapter 11 of this book].

Studies of the immunity to NLVs have been hampered by the inability to cultivate these viruses in cell lines. Early studies indicated that approximately 50% of persons challenged with NLVs developed illness and as a result acquired short-term homologous immunity against the same strain that was correlated with serum antibody levels [Parrino, *et al.*, 1977]. Some of these studies paradoxically also demonstrated that persons with higher levels of preexisting antibody to NLVs were more likely to develop illness on challenge with virus [Johnson, *et al.*, 1990]. A more recent study using molecular assays confirmed that approximately 50% of volunteers challenged with NLVs are susceptible to illness, but it also demonstrated that almost 80% become infected, with some of these infections being asymptomatic [Graham, *et al.*, 1994; Gray, *et al.*, 1994].

Epidemiologic considerations

Age

The etiologic role of viruses is best defined for childhood gastroenteritis (Table 2). Rotaviruses are clearly the leading cause of severe gastroenteritis among children <5 years of age worldwide, causing an estimated 500,000-600,000 deaths each year [Miller, *et al.*, 2000]. With the improvement of detection assays, the role of both NLVs and SLVs in the etiology of childhood diarrhea is being increasingly recognized [Pang, *et al.*, 2000]. Astroviruses and enteric adenoviruses each cause from 2% to 9% of gastroenteritis episodes in children.

Limited data are available regarding the etiologic role of viruses in gastroenteritis among adults. In two recent community-based studies of diarrheal disease among adults in the Netherlands and England [Tompkins, *et al.*, 1999; de Wit, *et al.*, 2001], each of the enteric viruses were detected in 2%-9% of patients (Table 2). Despite the use in these studies of state-of-the-art assays for a variety of enteric pathogens, no organism was identified in 61% and 63% of the patients, respectively. Similarly, in a US study of more than 30,463 patients hospitalized with diarrhea [Slutsker, *et al.*, 1997], a major bacterial pathogen was identified in only 5.6% of cases, and no pathogen could be identified in 91.6% of patients. Ongoing studies may clarify what fraction of these gastroenteritis episodes of unidentified etiology among adults are caused by enteric viruses.

Table 2

Recent studies demonstrating the role of enteric viruses in the etiology of acute gastroenteritis among children and adults in various countries.

	% children infected			% adults infected	
	France [Bon, *et al.*, 1999]	China [Qiao, *et al.*, 1999]	Finland [Pang, *et al.*, 2000]	England [Tompkins, *et al.*, 1999]	The Netherlands [de Wit, *et al.*, 2001]
	inpatients / outpatients (N = 414)	inpatients (N = 186)	community (N = 832)	community (N = 761)	community (N = 857)
Any virus	72	NA	60	NA	15
Rotavirus	61	56	31	8	5
Caliciviruses	14	8	30	9	7
Astroviruses	6	9	9	3	2
Adenoviruses	3	3	6	3	2

NA = Not applicable.

14

Severity of gastroenteritis

In children, the etiologic fraction of gastroenteritis caused by viruses increases with increasing severity of illness. This is well illustrated in a study from Finland [Pang, *et al.*, 2000], in which viruses accounted for only 46% of mild cases of gastroenteritis but as many as 85% of moderate to severe cases. This pattern clearly reflected the above-average severity of rotavirus gastroenteritis, which caused 27% of mild cases, 50% of moderate cases, and 68% of severe cases of gastroenteritis. Based on a clinical scoring system, NLV gastroenteritis was second after rotavirus, whereas disease caused by astroviruses and SLVs was least severe.

Developing versus industrialized nations

Compared with children in industrialized countries, those in developing countries experience 2- 5 times as many diarrheal episodes [Glass, *et al.*, 2001]. The difference in the etiologic spectrum of illness in the two settings likely reflects difference in hygiene and sanitation. Thus, bacteria and parasites that are more often spread through contaminated food and water account for a substantial proportion of gastroenteritis episodes in developing countries, whereas enteric viruses that may be spread by airborne droplets or person-to-person contact are ubiquitous and therefore account for a greater proportion of diarrheal episodes in industrialized nations.

Endemic versus epidemic gastroenteritis

Endemic childhood gastroenteritis is caused primarily by rotaviruses and to a lesser extent by SLVs, astroviruses, and enteric adenoviruses. These viruses infect nearly all children in the first few years of life, resulting in long-lasting immunity. The universal nature of infection among children in both industrialized and developing nations suggests that improvements in hygiene and sanitation may not substantially reduce the incidence of disease, and vaccines may offer the best prevention strategy.

Epidemic gastroenteritis is most often caused by the NLVs [Fankhauser, *et al.*, 1998], although outbreaks associated with rotaviruses, astroviruses, and SLVs have been reported. NLV gastroenteritis affects persons of all ages, and volunteers become ill repeatedly on viral challenge. These observations indicate that either immunity to NLVs is short-lasting or the great antigenic diversity of NLVs precludes development of adequate immunity against all strains. Consequently, the development of effective NLV vaccines will be a challenging task, and prevention strategies are currently aimed at interrupting specific modes of transmission, including transmission through contaminated food and water.

HIV-Infected and other immunodeficient persons

Several studies have demonstrated an association between enteric viruses and gastroenteritis in adults infected with HIV while other studies have failed to do so (Table 3).

Table 3

Selected studies of detection of enteric viruses in adults with human immunodeficiency virus infection

Study	No of patients	Viruses	Method of detection	% detection with/without diarrhea	Association with diarrhea
Cunningham, *et al.*, 1988	123	Rotavirus	EIA	37/11	Yes
		Adenovirus	EM, biopsy, culture	22/5	Yes
Kaljot, *et al.*, 1989	153	Rotavirus	EIA	0	No
		Adenovirus	PAGE, EIA	5	No
Grohmann, *et al.*, 1993	110	Overall	As below	35/12	Yes
		Rotavirus	EIA	0/0	No
		Adenovirus	EIA	9/3	Yes
		Astrovirus	EIA/EM/PCR	12/2	Yes
		Calicivirus	EM	4/1	No
		Coronavirus	EM	3/2	No
		Picobirnavirus	PAGE	9/2	Yes
Schmidt, *et al.*, 1996	256	Overall	As below	24/10	Yes
		Adenovirus	EM, biopsy	9/3	Yes
		Coronavirus	EM, biopsy	15/7	Yes
Gonzalez, *et al.*, 1998	125	Overall	As below	2/10	No
		Rotavirus	EIA	0/0	No
		Adenovirus	EIA	2/5	No
		Astroviruses	PCR	0/0	No
		Norwalk virus	EIA	0/0	No
		Picobirnavirus	PAGE	0/4	No
Giordano, *et al.*, 1998	88	Rotavirus	EIA	0/0	No
		Picobirnavirus	PAGE	9/0	Yes
Giordano, *et al.*, 1999	120	Overall	As below	27/8	Yes
		Astrovirus	EIA	4/5	No
		Adenovirus	EIA	7/3	No
		Picobirnavirus	PAGE	15/0	Yes

In addition, HIV disease can modify the illness caused by the common enteric viruses and also increase susceptibility to infections by unconventional viral pathogens. In immunodeficient children, for example, rotavirus can cause a protracted diarrhea with prolonged viral excretion, and in rare instances can disseminate systemically and

16

cause hepatic infection [Gilger, *et al.*, 1992; Oshitani, *et al.*, 1994]. A recent study from Malawi showed that rotavirus was detected less frequently in HIV-infected than HIV-uninfected children, and the clinical outcome of disease and immune responses to infection were similar in both groups of children [Cunliffe, *et al.*, 2001]. In immunodeficient children, adenovirus infection can cause a disseminated infection of the lung, liver, bone marrow, heart, and brain and produce fulminant hepatitis [Krilov, *et al.*, 1990]. Cytomegalovirus (CMV) is a common and potentially serious opportunistic gastrointestinal pathogen in HIV-infected persons. Colitis is the most common manifestation, although CMV can infect any part of the gastrointestinal tract [Grant, Section IV, Chapter 4 of this book].

Prevention

The immense disease burden of viral gastroenteritis underscores the need for effective prevention strategies. Endemic rotavirus disease is clearly the most important target for prevention, and efforts to develop rotavirus vaccines were initiated many years ago when it became apparent that improvements in hygiene and sanitation were unlikely to interrupt viral transmission [Bresee, *et al.*, 1999]. A vaccine against rotavirus was licensed for the first time in the United States in 1998 but was withdrawn a year later following strong suspicion and evidence of its association with intussusception [Centers for Disease Control and Prevention, 1999; Murphy, *et al.*, 2001; Offit *et al.*, Section II, Chapter 13 of this book]. Despite this setback, efforts to develop other candidate vaccines are underway, and products currently in clinical trials may be available for use in the next few years. The role of other viruses, including caliciviruses, astroviruses, and adenoviruses in the etiology of severe endemic gastroenteritis needs to be better defined to determine whether these viruses should be targeted for prevention through vaccination.

For the prevention of epidemic viral gastroenteritis, efforts clearly need to be focused on caliciviruses. The epidemic spread of caliciviruses is facilitated by the low infectious dose (<100 viral particles), ability of the virus to survive at relatively high levels of chlorine and temperatures from freezing to 60°C, great genetic diversity, and lack of lasting immunity [Kapikian, *et al.*, 1996]. Efforts to prevent calicivirus outbreaks currently focus on identifying and eliminating sources of contamination of food and water. Person-to-person spread of caliciviruses can be reduced by good hygiene practices but is often difficult to interrupt; consequently, outbreaks spread by this mode in institutional settings (e.g., nursing homes) often run their natural course and terminate when susceptible persons are exhausted.

Treatment

No specific antiviral therapy is recommended for childhood viral gastroenteritis, emphasizing the importance of distinguishing it from the selected forms of bacterial and parasitic gastroenteritis that require treatment. Other than pertinent epidemiologic information (e.g., age of patient, history of travel, consumption of specific food items, immune status), certain clinical features of illness (e.g., incubation period, duration of illness) may provide etiologic clues but these features are not highly discriminating.

Standard therapy of viral enteric infections relies on maintenance of adequate hydration and electrolyte balance. Oral rehydration therapy (ORT) is the main treatment [Duggan, *et al.*, 1992], and its efficacy has been confirmed in numerous trials worldwide and demonstrated by a substantial decline in the global mortality from diarrheal disease. In patients with severe disease or those who are unable to tolerate ORT, intravenous fluid therapy is recommended [see also Bass, Section I, Chapter 5 of this book].

Antimotility agents are not recommended [Desselberger, *et al.*, 1999], but enkephalinase inhibitors that reduce hypersecretion but do not alter motility have been demonstrated to be effective in the management of childhood diarrhea [Salazar-Lindo, *et al.*, 2000]. The role of other experimental therapies, including oral administration of immunoglobulin or probiotics, is still under investigation [Szajewska and Mrukowicz, 2001].

References

Ball JM, Tian P, Zeng CQ-Y, Morris AP, Estes MK (1996). Age-dependent diarrhea induced by a rotaviral nonstructural glycoprotein. Science 272:101-4.

Bern C, Martines J, de Zoysa I, Glass RI (1992). The magnitude of the global problem of diarrhoeal disease: a ten year update. Bull World Health Organ 70:705-14.

Bishop RF, Davidson GP, Holmes IH, Ruck BJ (1973). Virus particles in epithelial cells of duodenal mucosa from children with viral gastroenteritis. Lancet i:1281-3.

Bon F, Fascia P, Dauvergne M, Tenenbaum D, Planson H, Petion AM, Pothier P, Kohli E (1999). Prevalence of group A rotavirus, human calicivirus, astrovirus, and adenovirus type 40 and 41 infection among children with acute gastroenteritis in Dijon, France. J Clin Microbiol 37:3055-8.

Brandtzaeg P (1998). Development and basic mechanisms of human gut immunity. Nutr Rev 56:S5-18

Bresee J, Glass RI, Ivanoff B, Gentsch J (1999). Current status and future priorities for rotavirus vaccine development, evaluation, and implementation in developing countries. Vaccine 17:2207-22.

Centers for Disease Control and Prevention (1999). Intussusception among recipients of rotavirus vaccine, United States, 1998-1999. Morb Mortal Wkly Rep 48:577-81.

Chiba S, Nakata S, Numata-Kinoshita K, Honma S (2000). Sapporo virus: history and recent findings. J Infect Dis 181 (suppl 2):S303-8.

Coulson BS, Grimwood K, Masendycz P, Lund J, Mermelstein N, Bishop R, Barnes G (1990). Comparison of rotavirus immunoglobulin A coproconversion with other indices of rotavirus infection in a longitudinal study in childhood. J Clin Microbiol 28:1367-74.

Coulson BS, Grimwood K, Hudson IL, Barnes GL, Bishop RF (1992). Role of copro-antibody in clinical protection of children during reinfection with rotavirus. J Clin Microbiol 30:1678-84.

Cunliffe NA, Gondwe JS, Kirkwood CD, Graham SM, Nhlane NM, Thindwa BDM, Dove W, Broadhead RL, Molyneux ME, Hart CA (2001). Effect of concomitant HIV infection on presentation and outcome of rotavirus gastroenteritis in Malawian children. Lancet 358:550-5.

Cunningham AL, Grohmann GS, Harkness J, Law C, Marriott D, Tindall B, Cooper DA (1988). Gastrointestinal viral infections in homosexual men who were symptomatic and seropositive for human immunodeficiency virus. J Infect Dis 158: 386-91.

Desselberger U (1999). Rotavirus infections: guidelines for treatment and prevention. Drugs 58:447-52.

de Wit MA, Koopmans MP, Kortbeek LM, van Leeuwen NJ, Bartelds AI, van Duynhoven YT (2001). Gastroenteritis in sentinel general practices, The Netherlands. Emerg Infect Dis 7:82-91.

Duggan C, Santosham M, Glass RI (1992). The management of acute diarrhea in children: oral rehydration, maintenance, and nutritional therapy. Morb Mortal Wkly Rep 41(RR-16):1-20.

Fankhauser RL, Noel JS, Monroe SS, Ando TA, Glass RI (1998). Molecular epidemiology of "Norwalk- like viruses" in outbreaks of gastroenteritis in the United States. J Infect Dis 178:1571-8.

Flewett TH, Beards GM, Brown DW, Sanders RC (1987). The diagnostic gap in diarrhoeal aetiology. Ciba Found Symp 128:238-49.

Flewett TH, Bryden AS, Davies H (1973). Virus particles in gastroenteritis. Lancet ii: 1497.

Franco AA, Greenberg HB. Role of B cells and cytotoxic T lymphocytes in clearance of and immunity to rotavirus infection in mice (1995). J Virol 69:7800-06.

Gilger MA, Matson DO, Conner ME, Rosenblatt HM, Finegold MJ, Estes MK (1992). Extraintestinal rotavirus infections in children with immunodeficiency. J Pediatr 120:912-7.

Giordano MO, Martinez LC, Rinaldi D, Guinard S, Naretto E, Casero R, Yacci MR, Depetris AR, Medeot SI, Nates SV (1998). Detection of picobirnavirus in HIV-infected patients with diarrhea in Argentina. J Acqu Imm Defic Syndr Hum Retrovirol 18:380-3.

Giordano MO, Martinez LC, Espul C, Martinez N, Isa MB, Depetris AR, Medeot SI, Nates SV (1999). Diarrhea and enteric emerging viruses in HIV-infected patients. AIDS Res Hum Retroviruses 15:1427-32.

Glass RI, Bresee J, Jiang B-M, Gentsch J, Ando T, Fankhauser R, Noel J, Parashar U, Rosen B, Monroe SS (2001). Gastroenteritis viruses: an overview. Novartis Found. Symp. 238:5-19.

Gonzalez GG, Pujol FH, Liprandi F, Deibis L, Ludert JE (1998). Prevalence of enteric viruses in human immunodeficiency virus seropositive patients in Venezuela. J Med Virol 55:288-92.

Gordon I, Ingraham HS, Korns RF (1947). Transmission of epidemic gastroenteritis to human volunteers by oral administration of fecal filtrates. J Exp Med 86:409-22.

Graham DY, Jiang X, Tanaka T, Opekun AR, Madore HP, Estes MK (1994). Norwalk virus infection of volunteers: new insights based on improved assays. J Infect Dis 170:34-43.

Gray JJ, Cunliffe C, Ball J, Graham DY, Desselberger U, Estes MK (1994). Detection of immunoglobulin M (IgM), IgA, and IgG Norwalk virus-specific antibodies by indirect enzyme-linked immunosorbent assay with baculovirus-expressed Norwalk virus capsid antigen in adult volunteers challenged with Norwalk virus. J Clin Microbiol 32:3059-63.

Greenberg HB, Clark HF, Offit PA (1994). Rotavirus pathology and pathophysiology. In: Ramig RF, ed. Rotaviruses. Berlin: Springer-Verlag, 255-83.

Grohmann GS, Glass RI, Pereira HG, Monroe SS, Hightower AW, Weber R, Bryan RT (1993). Enteric viruses and diarrhea in HIV-infected patients. N Engl J Med 329:14-20.

Johnson PC, Mathewson JJ, Dupont HL, Greenberg HB (1990). Multiple-challenge study of host susceptibility to Norwalk gatroenteritis in US adults. J Infect Dis 161:18-24.

Kaljot KT, Ling JP, Gold JWM, Laughon BE, Bartlett JG, Kotler DP, Oshiro LS, Greenberg HB (1989). Prevalence of acute enteric viral pathogens in acquired immunodeficiency syndrome patients with diarrhea. Gastroenterology 97:1031-2.

Kapikian AZ, Wyatt RG, Dolin R, Thornhill TS, Kalica AR, Chanock RM (1972). Visualization by immune electron microscopy of a 27 nm particle associated with acute infectious nonbacterial gastroenteritis. J Virol 10:1075-81.

Kapikian AZ, Estes MK, Chanock RM (1976). Norwalk group of viruses. In: Fields BN, Knipe DM, Howley PM, Chanock RM, Melnick JL, Monath TP, Roizman B, Straus SE, eds. Fields Virology. 3rd ed. Philadelphia: Lippincott-Raven, 1996: 783-810.

Koopmans M, Petric M, Glass RI, Monroe SS (1993). Enzyme-linked immunosorbent assay reactivity of torovirus-like particles in fecal specimens from humans with diarrhea. J Clin Microbiol 31:2738-44.

Krilov LR, Rubin LG, Frogel M, Gloster E, Ni K, Kaplan M, Lipson SM (1990). Disseminated adenovirus infection with hepatic necrosis in patients with human immunodeficiency virus infection and other immunodeficiency states. Rev Infect Dis 12:303-7.

Lundgren O, Peregrin AT, Persson K, Kordasti S, Uhnoo I, Svensson L (2000). Role of the enteric nervous system in the fluid and electrolyte secretion of rotavirus diarrhea. Science 287:491-5.

Madeley CR, Cosgrove BP (1975a). 28 nm particles in faeces in infantile gastroenteritis. Lancet ii:451-2.

Madeley CR, Cosgrove BP (1975b). Viruses in infantile gastroenteritis (letter). Lancet ii:124.

Matson DO, O'Ryan ML, Herrera I, Pickering LK, Estes MK (1993). Fecal antibody responses to symptomatic and asymptomatic rotavirus infections. J Infect Dis 167:577-83.

McNeal MM, Rae MN, Ward RL (1997). Evidence that resolution of rotavirus infection in mice is due to both CD4 and CD8 cell-dependant activities. J Virol 71: 8735-42.

Meeroff JC, Schreiber DS, Trier JS, Blacklow NR (1980). Abnormal gastric motor function in viral gastroenteritis. Ann Intern Med 92:370-3.

Miller MA, McCann L (2000). Policy analysis of the use of hepatitis B, *Haemophilus influenzae* type B-, *Streptococcus pneumoniae*-conjugate, and rotavirus vaccines in National Immunization Schedules. Health Econ 9:19-35.

Murphy TV, Gargiullo PM, Massoudi MS, Nelson DB, Jumaan AO, Okoro CA, Zanardi LR, Setia S, Fair EL, LeBaron CW, Schwartz B, Wharton M, Livingood JR (2001). Intussusception among infants given an oral rotavirus vaccine. N Engl J Med 344:564-72.

Murray CJ, Lopez AD (1997). Global mortality, disability, and the contribution of risk factors: global burden of disease study. Lancet 349:1436-42.

Offit PA (1996). Host factors associated with protection against rotavirus disease: the skies are clearing. J Infect Dis 174 (suppl 1):S59-64.

Osborne MP, Haddon SJ, Spencer AJ, Collins J, Starkey WG, Wallis TS, Clarke GJ, Worton KJ, Candy DCA, Stephen J (1988). An electron microscopic investigation of time-related changes in the intestine of neonatal mice infected with murine rotavirus. J Pediatr Gastroenterol Nutr 7:236-48.

Oshitani H, Kasolo FC, Mpabalwani M, Luo NP, Matsubayashi N, Bhat GH, Suzuki H, Numazaki Y, Zumla A, DuPont HL (1994). Association of rotavirus and human immunodeficiency virus infection in children hospitalized with acute diarrhea, Lusaka, Zambia. J Infect Dis 169:897-900.

Pang X-L, Honma S, Nakata S, Vesikari T (2000). Human caliciviruses in acute gastroenteritis of young children in the community. J Infect Dis 181 (suppl 2): S288-94.

Parrino TA, Schreiber DS, Trier JS, Kapikian AZ, Blacklow NR (1977). Clinical immunity in acute gastroenteritis caused by Norwalk agent. N Engl J Med 297: 86-9.

Qiao H, Nilsson M, Abreu ER, Hedlund K-O, Johansen K, Zaori G, Svensson L (1999). Viral diarrhea in children in Beijing, China. J Med Virol 57:390-6.

Rosen BI, Fang Z-Y, Glass RI, Monroe SS (2000). Cloning of human picorbirnavirus gene segments and development of an RT-PCR detection assay. Virology 277: 316-29.

Salazar-Lindo E, Santisteban-Ponce J, Chea-Woo E, Gutierrez M (2000). Racecadotril in the treatment of acute watery diarrhea in children. N Engl J Med 343:463-7.

Schmidt W, Schneider T, Heise W, Weinke T, Epple HJ, Stoffler-Meilicke M, Liesenfeld O, Ignatius R, Zeitz M, Riecken EO, Ullrich R (1996). Stool viruses, coinfections, and diarrhea in HIV-infected patients. J Acqu Imm Defic Syndr Hum Retrovirol 13:33-8.

Slutsker L, Ries AA, Greene KD, Wells JG, Hutwagner L, Griffin PM (1997).

Escherichia coli O157:H7 diarrhea in the United States: clinical and epidemiologic features. Ann Intern Med 126:505-13.

Szajewska H, Mrukowicz JZ (2001). Probiotics in the treatment and prevention of acute infectious diarrhea in infants and children: a systematic review of published randomized, double-blind, placebo-controlled trials. J Pediatr Gastroenterol Nutr 33:S17-25.

Tompkins DS, Hudson MJ, Smith HR, Eglin RP, Wheeler JG, Brett MM, Owen RJ, Brazier JS, Cumberland P, King V, Cook PE (1999). A study of infectious intestinal disease in England: microbiological findings in cases and controls. Commun Dis Public Health 2:108-13.

Tucker AW, Haddix AC, Bresee JS, Holman RC, Parashar UD, Glass RI (1998). Cost-effectiveness analysis of a rotavirus immunization program for the United States. J Am Med Ass 279:1371-6.

Velazquez FR, Matson DO, Calva JJ, Guerrero L, Morrow AL, Carter-Campbell S, Glass RI, Estes MK, Pickering LK, Ruiz-Palacios GM (1996). Rotavirus infections in infants as protection against subsequent infections. N Engl J Med 335: 1022-8.

Wadell G, Allard A, Johansson M, Svensson L, Uhnoo I (1987). Enteric adenoviruses. Ciba Found Symp 128:63-91.

Yamashita T, Kobayashi S, Sakae K, *et al.* (1991). Isolation of cytopathic small round viruses with BS-C-1 cells from patients with gastroenteritis. J Infect Dis 164: 954-7.

Yamashita T (1999). Biological and epidemiological characteristics of Aichi virus, as a new member of Picornaviridae [Japanese]. Uirusu 49:183-91.

I, 2.   Physiology and pathophysiology of the gut in relation to viral diarrhea

Fabián Michelangeli and Marie Christine Ruiz

Laboratorio de Fisiología Gastrointestinal, Centro de Biofísica y Bioquímica, Instituto Venezolano de Investigaciones Científicas (IVIC) Caracas, Venezuela

Introduction

The epithelium of the gastrointestinal tract forms a barrier between the outside world and the host. The intestine is lined by a folded continuous monolayer of columnar polarized epithelial cells joined together by tight junctions, dispersed enteroendocrine cells and M cells. The epithelium is permeant to small solutes and water which can cross via paracellular or transcellular pathways. Intestinal epithelial cells originate from anchored stem cells localized near the base of intestinal crypts and are subject to constant renewal, turning over every 4-7 days. Differentiation gives rise to columnar secretory and absorptive epithelial cells, as well as mucus-secreting goblet cells, Paneth cells and enteroendocrine cells (Madara, 1999). Epithelial proliferation may be enhanced under pathological conditions. This may have effects on the balance of functions by different epithelial cells along the crypt-villus axis (Barrett and Keely, 2000; Karam, 1999; Podolsky, 1993). The surface area of the intestinal epithelium is largely amplified (about 600-fold in the small intestine) by the presence of folds, villi, and cell microvillar structures. This enhances the capacity of the intestine to transport nutrients, electrolytes and water.

Many advances have been made in the understanding of intestinal electrolyte transport from the molecular to the whole-tissue level. This chapter emphasizes molecular mechanisms of intestinal epithelial ion transport processes, as well as the intra- and extracellular factors involved in their regulation, as a framework for the understanding of virus-induced gastroenteritis.

Abbreviations used in text and figures: 5HT: 5-hydroxytryptamine, serotonin; AC: adenylate cyclase; ACh: acetylcholine; AQP: water channels of the aquaporin family; CaCC: Ca^{2+}–activated chloride channel; cAMP: adenosine 3'-5'cyclic monophosphate; CF: cystic fibrosis; CFTR: cystic fibrosis transmembrane conductance regulator, chloride channel; cGMP: guanosine 3'-5'cyclic monophosphate; ENS: enteric nervous system; ER:endoplasmic reticulum; GC: guanylate cyclase; GLUT2: facilitated sugar transporter; GLUT5: facilitated sugar transporter; hIK1: a type of potassium channel; IP3: Inositol 1,4,5-triphosphate; Na^+/K^+ ATPase: Na^+/K^+ adenosine triphosphatase, Na^+/K^+ pump; NBD1, NBD2: nucleotide-binding domains of CFTR; NHE-3: Na^+/H^+ exchange; NO: Nitric oxide; NSP4: nonstructural protein 4 of rotavirus; PEDV: porcine epidemic diarrhea virus; PKA: protein kinase A; PKC: protein kinase C; PLC: phospholipase C; PMN: polymorphonuclear leucocytes; R: regulatory domain of CFTR; SGLT1: Na^+/glucose cotransporter; TGEV: transmissible gastroenteritis virus; TNF-α: tumor necrosis factor-α; VIP: vasoactive intestinal peptide; ZO-1: zonula occludens protein 1

Normal intestinal physiology

In the intestine, Na^+, K^+, $HCO3^-$, and Cl^- are transported actively determining fluid movements to maintain isotonicity between the intestinal lumen, the cell and the internal milieu. The intestine also contains ion-dependent mechanisms for the transport of peptides, amino acids, sugars, and bile acids (Lee *et al.*, 1994; Lentz, 1995; Mailliard *et al.*, 1995). In addition to the absorption of ions, nutrient uptake may collaborate in generating fluid movement by solvent drag (Pappenheimer *et al.*, 1994; Turner and Madara, 1995). The function of the intestine as a transporting machine is guaranteed by its structural and functional asymmetry at the cellular and crypt-villus axis levels and along the gut from duodenum to distal colon. The intestinal cells transport electrolytes and nutrients in a vectorial manner, generating electrochemical and osmotic gradients to drive absorption and/or secretion. This transport is made possible by the structural and functional polarity of the enterocyte. Transport proteins and receptors for regulatory mediators and toxins are asymmetrically distributed in the apical and basolateral cell membranes.

The diverse epithelial cell types are arranged so that there is a functional heterogeneity in electrolyte, fluid, and nutrient transport along the crypt-villus axis. It was a dogma for a long time that absorption and secretion took place in villus and crypt cells, respectively (Montrose *et al.*, 1999). This paradigm has been challenged since secretory mechanisms have been observed in some villus cells (Kockerling and Fromm, 1993; Stewart and Turnberg, 1989), and colonic crypt cells are able to absorb Na^+ and water in the absence of secretory stimuli (Naftalin and Pedley, 1999; Naftalin *et al.*, 1995; Singh *et al.*, 1995; Thiagarajah *et al.*, 2001). The small intestine is a leaky epithelium absorbing large volumes of fluid against small electrochemical gradients whereas the distal colon is a semi-tight epithelium capable of absorbing Na^+ against a large gradient.

Fluid balance in the intestine

The intestine handles large amounts of fluid containing water, nutrients and salts. Net fluid transport through and across the gut results from food intake, secretory and absorption processes and final excretion. On average, about two liters of water are ingested daily, together with nutrients and electrolytes. In addition, 7 liters of fluids are secreted by different parts of the gut to aid digestion (saliva, gastric and pancreatic juices, bile and intestinal secretion). In total, over 98% of these fluids are absorbed back into the body in the small and large intestine, accompanying nutrients and salts. Of these, about 5 liters are absorbed in the duodenum and jejunum, 2 liters in the ileum and 2 liters in the colon, remaining only 100 ml excreted in the stools in the healthy person. Normally, the ileo-cecal flow is approximately 2 liters but the colon may be able to absorb as much as 5-6 liters per day. Water movements in secretory and absorptive epithelia are secondary to electrolyte and nutrient transport and may occur through transcellular and/or paracellular pathways (Montrose *et al.*, 1999). Diarrhea occurs when secretory processes exceed the absorptive capacity of the intestine, from an impairment of nutrient and water absorption and/or from an increase in the gut motility inducing an accelerated transit.

Extracellular regulation of intestinal function

Intestinal transport is regulated by a number of extracellular stimuli, which may act on the luminal or serosal sides of the epithelium. These include endocrine, paracrine and neurocrine mediators (Fig. 1). The regulation of epithelial function is the result of the integrated response to all these systems. Enteroendocrine cells in the crypt and villus region release peptides and other substances such as serotonin (5HT), neurotensin, secretin, somatostatin, guanylin etc, across their basolateral or apical membranes [Madara and Trier, 1994]. The diffusion of these bioactive substances locally regulates the transport functions of neighboring cells (i.e., paracrine regulation) and signal to more distant cells along the crypt villous axis by stimulation of dendritic terminals of ENS neurons. These substances together with bloodborne peptides are involved in endocrine regulation of distant regions of the epithelium. Neurotransmitters such as VIP, 5HT, substance P and ACh, released from enteric nerve endings regulate epithelial functions of enterocytes (absorption and secretion) and enteroendocrine cells as well as motility and microcirculation. Adding to the complexity of regulatory mechanisms, the mast cells, neutrophils (PMN), macrophages and T lymphocytes from the immunological system and localized in the lamina propria, can also release substances (histamine, cytokines, adenosine, etc) that regulate the epithelium (immune regulation). On the other hand, molecules such as bile salts, fatty acids, and microbial toxins found in the luminal contents can also affect electrolyte and nutrient transport, acting directly on epithelial cells or indirectly by way of stimulation of enteroendocrine cells. The myofibroblasts form a continuous sheath underlying the basement membrane of the epithelial layer that release prostaglandins induced by serotonin, histamine, and TNF-α. Released prostaglandins may stimulate epithelial Cl$^-$ secretion (Berschneider and Powell, 1992; Kandil *et al.*, 1994; Lundgren, 1992; Sears and Kaper, 1996).

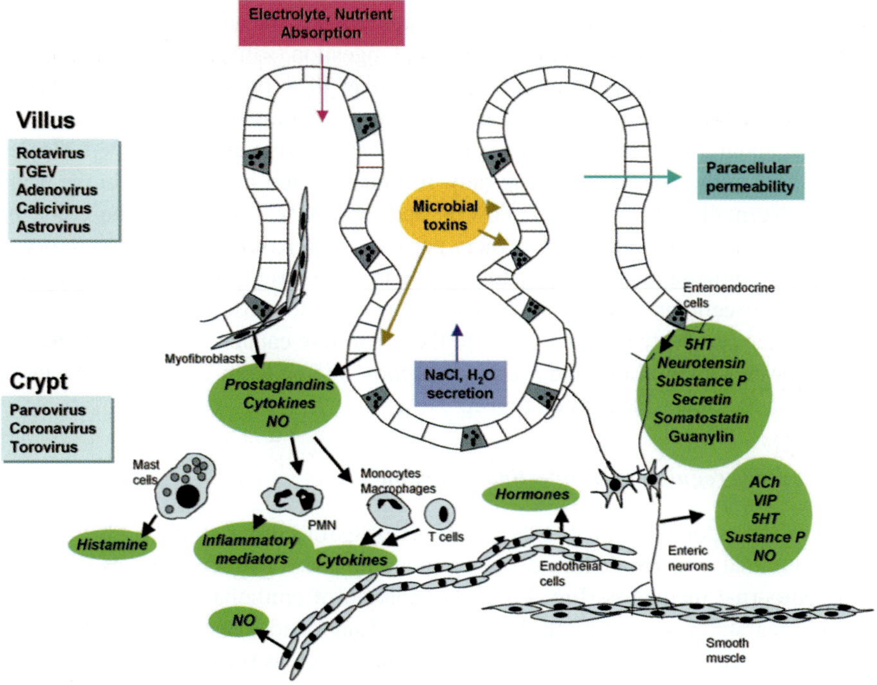

Fig. 1. Regulation of intestinal functions by the enteroendocrine, immune and enteric nervous systems. The intestinal epithelium is a continuous monolayer containing different cell types, mainly villus cells, crypt cells and enteroendocrine cells. Epithelial cells release cytokines, prostaglandins and nitric oxide (NO) in response to luminal stimuli. Enteroendocrine cells secrete mediators such as 5HT, neurotensin, substance P, secretin, somatostatin, and guanylin. The myofibroblast sheath releases prostaglandins, cytokines and NO. Other cells present in the lamina propria, such as mast cells, polymorphonuclear leucocytes (PMN) and T lymphocytes from the immunologic system let out a variety of chemical mediators (histamine, cytokines, prostaglandins, NO etc). The enteric nervous system releases VIP, 5HT, sustance P, ACh and NO. All three systems interact with each other at different levels to finally regulate transport function by enterocytes (secretion by crypt cells and absorption by villus cells), paracellular permeability, gut motility and microcirculation. Enteric viruses have a distinct predilection for different cell types in the intestinal epithelium. Rotavirus, TGEV, adenovirus, calicivirus and astrovirus infect the villus cells whereas parvovirus infects exclusively crypt cells. Coronavirus and torovirus infect basal villus cells and crypt cells. See text for discussion and references.

Intestinal absorption

Sugar transport

Nutrient absorption in the villus enterocyte depends to a large extent on the presence of a Na$^+$ gradient between the intra- and extracellular media generated by the Na$^+$/K$^+$

ATPase. This pump provides the driving force for the uphill transport of sugars and amino acids. On the other hand, both Na$^+$ and water absorption are largely dependent on the presence of nutrients in the apical side, especially glucose. Transport mechanisms are schematized in Fig. 2.

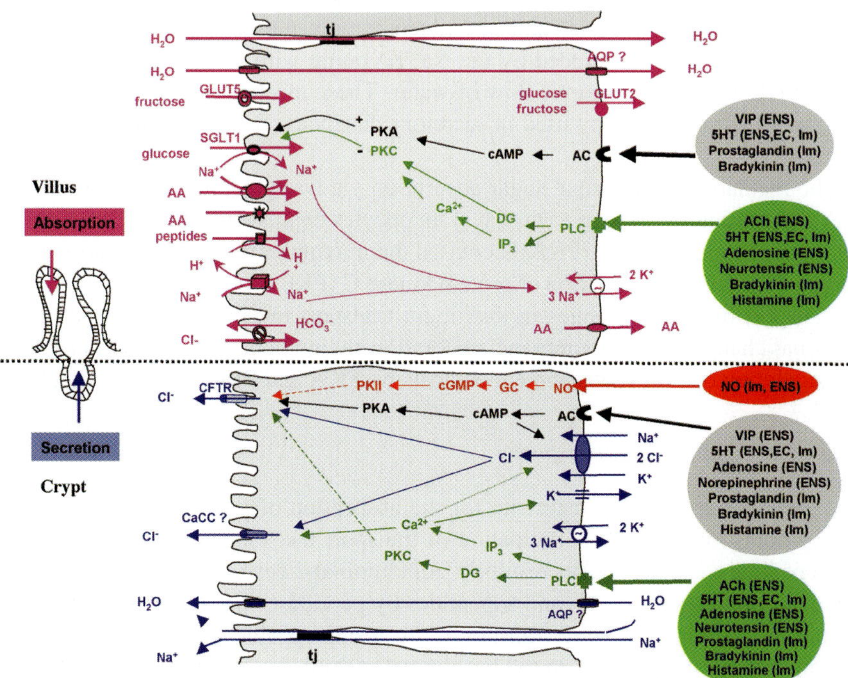

Fig. 2. Ion and nutrient transport mechanisms in the intestine. For the sake of clarity absorption mechanisms are represented in villus cells (upper part) and secretion mechanisms in crypt cells (lower part). *Villus cells*: A Na$^+$ gradient across the apical and basolateral membranes is generated by the operations of the basolateral Na$^+$/K$^+$ATPase. Glucose and Na$^+$ are taken up across the apical membrane of villus cells by the SGLT1 cotransporter driven by the Na$^+$ gradient. Fructose enters the cells driven by the chemical gradient through a facilitated diffusion pathway in the apical membrane (GLUT 5). Glucose and fructose exit through a facilitated diffusion pathway in the basolateral membrane (GLUT2). Amino acids (AA) enter the cell by Na$^+$-dependent or Na$^+$-independent transport mechanisms and exit across the basolateral membrane through another type of Na$^+$-independent mechanisms. Apical NaCl entry through Na$^+$/H$^+$ (NHE-3) and Cl$^-$/HCO$_3^-$ exchange drive electroneutral NaCl absorption in the small intestine and colon. Na$^+$ exits the cell through the basolateral Na$^+$/K$^+$ATPase, but the route of Cl$^-$ efflux is not known. The H$^+$ gradient created across the apical membrane by the Na$^+$/H$^+$ exchange drives the cotransport of peptides. Different chemical mediators coming from enteroendocrine (EC), immune (Im) and enteric nervous (ENS) systems act by binding to specific receptors in the basolateral membrane coupled to adenyl cyclase or phospholipase C which result in increases in the second messagers Ca^{2+}, diacylglycerol (DG) and cAMP. These regulate the activity of transporters, mainly SGLT1, by PKA and PKC mediated phosphorylation. *Crypt cells*: Chloride secretion in the small intestine and colon is depicted in the lower cell. Cl$^-$ enters cells at the basolateral membrane through Na$^+$/K$^+$/2Cl$^-$ cotransport and leaves cells

(cont. on p 28)

Sugars are transported across the apical membrane by different transport proteins. The Na^+/glucose cotransporter (SGLT1) accumulates glucose and galactose inside the enterocyte. The facilitated sugar transporter GLUT5 transports fructose down its chemical gradient. These sugars are then transported out of the cell by GLUT2, a facilitated transporter located in the basolateral membrane. The SGLT1 cotransporter has a stochiometry of 2 Na^+ and one glucose molecule per cycle. Sodium entering the cell by this way is extruded by the Na^+/K^+ pump with a concomitant transport of Cl^- and HCO_3^-, and the osmotic flow of water. These mechanisms provide the rational for oral rehydration therapy used in secretory diarrheas [Bass, Section I, Chapter 5 of this book].

The regulation of intestinal sugar absorption via Na^+/glucose cotransporters occurs over short (minutes) or long term (days). In oocytes expressing rabbit SGLT1, the activation of protein kinase A (PKA) increased the maximum rate of Na^+/glucose cotransport by 30%, and the activation of protein kinase C (PKC) decreased the maximum rate of transport by 60%. Changes in maximum transport rates are accompanied by proportional changes in the membrane area and in the number of cotransporters through the regulation of exocytosis and endocytosis by PKA and PKC (Wright *et al.*, 1997).

Amino acid and peptide transport

The small intestine is the main site for the absorption of protein digestion products. This process involves the participation of transport mechanisms for amino acids and small peptides. Large peptides resulting from luminal digestion by proteases are hydrolyzed to amino acids and di- and tripeptides by peptidases present in the brush border membrane. Amino acids are transported by group-specific transport systems localized in the apical as well as in the basolateral membrane. For most transport systems (B, $B^{o,+}$, Imino, b, X^-_{AG}, β) the driving force is the Na^+ gradient across the apical membrane. For the Na^+ independent transporters (systems y^+, $b^{o,+}$), the driving force is the electrical potential across the apical membrane. Absorbed amino acids cross

(Fig. 2 Legend cont.)   at the apical membrane through CFTR and perhaps Ca^{2+}-dependent Cl^- channels (CaCC). Operation of the cotransport is mainly driven by the Na^+ concentration gradient established by the Na^+/K^+ ATPase. Potassium taken up by the pump and the cotransporter exits the cell and recycles through the opening of basolateral K^+ channels. These three mechanisms operate to effectively accumulate Cl^- intracellularly above its electrochemical equilibrium. The opening of regulated rheogenic Cl^- channels in the apical membrane allows Cl^- to diffuse into the lumen, down its electrochemical gradient. The outflux of Cl^- and K^+ at the apical and basolateral membranes, respectively, contribute to generate a lumen negative transepithelial electrical potential, which drives the passage of Na^+ through the paracellular pathway. The accumulation of NaCl in the lumen in turn drives the osmotic flow of water for net secretion of fluid. As in the case of villus cells, different types of chemical mediators regulate transport functions by binding to receptors coupled to adenylate cyclase (AC) and phospholipase C (PLC), activating the respective signaling cascades. Nitric oxide activates guanylate cyclase (GC) to, in turn, modulate CFTR activity. Water crosses the epithelium through the tight junction (tj) and/or transcellular pathways through putative aquaporins (AQP). See text for discussion and references.

the basolateral membrane into the circulation through Na^+-independent transporters, different from those presents in the apical membrane (Ganapathy *et al.*, 1994).

Peptides are transported across the apical membranes by H^+/peptide cotransporters. The H^+ gradient across the apical membrane required to drive transport is assured by the existence of a Na^+/H^+ exchanger, partly converting the Na^+ gradient that exists across the brush border membrane into a H^+ gradient. Most of the peptides entering the enterocyte are digested to free amino acids by intracellular peptidases. A small fraction of peptides enters the blood crossing the basolateral membrane by H^+/peptide cotransport systems driven by the peptide concentration gradient. Short term regulation of amino acid transport is controlled by several hormones and neurotransmitters acting through the different second messengers (Ganapathy *et al.*, 1994).

Electroneutral NaCl absorption

A significant proportion of the Na^+ absorbed in the intestine occurs independently of nutrient transport. Part of the Na^+ is absorbed together with Cl^- without net charge transfer. Evidence supports a dual-exchanger model where the operation in parallel of Na^+/H^+ and Cl^-/HCO_3^- exchangers assures electroneutral NaCl transport (Montrose *et al.*, 1999). Supply of H^+ and HCO_3^- is dependent on the activity of intracellular carbonic anhydrase. The functioning of the system is driven by the Na^+ gradient across the apical membrane which in turn is maintained by the activity of the Na^+/K^+ ATPase located in the basolateral side. This mechanism is important in the small intestine and the proximal colon. In the distal colon, an electrogenic amiloride-sensitive Na^+ absorption regulated by aldosterone accounts for about 50% of the Na^+ reabsorbed. In this case Na^+ crosses the apical membrane through a rheogenic channel. Basolateral K^+ channels would recycle K^+ across the basolateral membrane. The paracellular flux of Cl^- driven by Na^+ movement would restore electroneutrality between apical and basolateral sides.

Intestinal secretion and its regulation

The small and large intestine are able to secrete a vast amount of fluid, mainly constituted by an isotonic secretion of NaCl. Chloride is transported from the plasma to the lumen through a transcellular pathway. The transport of Cl^- in mammalian secretory epithelia is not due to a Cl^- pump but results from the concourse of distinct transmembrane transport mechanisms asymmetrically located in polarized cells. The Cl^- secretory mechanism in crypt cells of the small and large intestine has been well characterized (Fig. 2; for review see Barrett and Keely, 2000; Montrose *et al.*, 1999; Morris, 1999). Chloride enters the cell across the basolateral membrane via an electroneutral $Na^+/K^+/2Cl^-$ cotransport mechanism. Operation of the cotransport is mainly driven by the Na^+ concentration gradient established by the ouabain-sensitive basolateral Na^+/K^+ ATPase. Potassium taken by the pump and by the cotransporter exits the cell and recycles through the opening of basolateral K^+ channels. These three mechanisms operate to effectively accumulate Cl^- intracellularly above its electrochemical equilibrium. The opening of regulated rheogenic Cl^- channels in the apical membrane allows Cl^- to diffuse into the lumen, down its electrochemical gradient. The outflux of Cl⁻ and K⁺ at the apical and basolateral membranes, respectively, contributes to generate a lumen negative transepithelial electrical potential. This voltage drives the passage of Na⁺ through the paracellular pathway. The accumulation of NaCl in the lumen in turn drives the osmotic flow of water for net secretion of fluid.

Chloride transport is mainly regulated by the opening of apical Cl⁻ and basolateral K⁺ channels, as a result of changes in intracellular second messengers such as cAMP and Ca^{2+}. More recent evidence suggests that the $Na^+/K^+/2Cl^-$ cotransporter is also subject to regulation (Haas and Forbush, 2000). The main apical chloride pathway in the intestine is the CFTR (cystic fibrosis transmembrane conductance regulator) chloride channel activated mainly by cAMP-dependent phosphorylation, resulting from adenylate cyclase activation by secretagogues such as VIP (Barrett and Keely, 2000). CFTR is an integral membrane protein with two membrane spanning domains, each including 6 putative transmembrane regions which form the channel pore (McCarty, 2000). On the cytoplasmic side, this protein presents two nucleotide-binding domains (NBD1, NBD2) that bind and hydrolyse ATP and a regulatory (R) domain. The latter is phosphorylated by PKA, but also by PKC. Both ATP hydrolysis by NBD and phosphorylation of the R domain are required for channel activation (Gadsby and Nairn, 1999). The CFTR channel can also be activated by cGMP dependent protein kinase II and, to a lesser extent, by Ca^{2+} and PKC (French *et al.*, 1995). Phosphorylation by PKC alone does not seem to open the channel but rather to sensitize the channel to subsequent phosphorylation by PKA, resulting in a potentiation of secretion (Jia *et al.*, 1997). Another level of possible control of CFTR activity is the regulation of the number of transporters at any time present in the membrane. This may occur by changing the rate of insertion or retrieval of channels containing vesicles, modulated by the concentrations of cAMP and/or Ca^{2+} (Morris, 1999).

In the intestine, Ca^{2+}-dependent secretory agonists such as ACh, histamine, bradykinin and neurotensin, evoke a small transient response even in the continuous presence of the agonist (Morris, 1999). This response contrasts with the sustained increase in chloride secretion induced by cAMP-dependent secretagogues. Secretion induced by the combination of the agonists acting via Ca^{2+} and cAMP is synergistic (Vajanaphanich *et al.*, 1995). On the other hand, patients with cystic fibrosis or mice with a targeted deletion of the CFTR gene do not have a chloride secretory response induced by cAMP- or Ca^{2+}-mediated agonists (Berschneider *et al.*, 1988; Grubb and Gabriel, 1997; Morris *et al.*, 1999; Taylor *et al.*, 1988). At this point, the presence of a Ca^{2+}–activated chloride channel (CaCC) involved in chloride secretion is somewhat controversial (Barrett and Keely, 2000). A Ca^{2+}-activated chloride conductance has been observed in intestine *in situ* and in cultured intestinal cell lines (Anderson *et al.*, 1992; McEwan *et al.*, 1994; Merlin *et al.*, 1998; Morris and Frizzell, 1993). Elevation of intracellular Ca^{2+} concentration does not appear to activate the CFTR channel directly. Instead, the regulatory effect of Ca^{2+} may be due to PKC-dependent phosphorylation of the channel and/or Ca^{2+} activation of the basolateral K⁺ channel. In this case, the chloride electrochemical gradient would be increased, and chloride would be secreted through the apical membrane via chloride channels (CFTR or not) open at any given time. Calcium-activated Cl⁻ channels may be a property of the developing intestine. Morris *et al.* (1992) have shown that HT29 cells exhibit Ca^{2+} activated Cl^- conductance that is lost as cells become polarized.

Functional and pharmacological characterization indicates the intervention of at least two types of basolateral K^+ channels involved in chloride secretion by intestinal cells. One channel can be activated by muscarinic agonists like carbachol, is activated by Ca^{2+}, insensitive to Ba^{2+} and blocked by clotrimazol and charibdotoxin (Devor and Duffey, 1992; Greger *et al.*, 1997; Jensen *et al.*, 1998; Lomax *et al.*, 1996; Vandorpe *et al.*, 1998). This channel may correspond to the recently described hIK1 potassium channel (Ishii *et al.*, 1997; Jensen *et al.*, 1998; Vandorpe *et al.*, 1998). The second one seems to be activated by cAMP (Greger *et al.*, 1997; McRoberts *et al.*, 1985).

The $Na^+/K^+/2Cl^-$ cotransporter appears to be under regulation by phosphorylation stimulated by a decrease in intracellular chloride. The nature of the protein kinase involved is not yet known. Thus, an apical chloride efflux would indirectly stimulate basolateral chloride uptake, providing a positive feedback for secretion (Lytle and Forbush, 1992; Lytle and Forbush, 1996; Torchia *et al.*, 1992). The activity of this cotransporter may be also regulated by insertion and retrieval of membrane transporter molecules modulated by the concentration of cAMP, and cytoskeletal dynamics (D'Andrea *et al.*, 1996; Hecht and Koutsouris, 1999; Karczewski and Groot, 2000; Matthews *et al.*, 1992; Matthews *et al.*, 1997; Matthews *et al.*, 1995).

Water transport

The intestine is capable of transporting large amounts of fluid even in the absence of any external hydrostatic or osmotic driving forces. The rate of water absorption is proportional to the rate of solute absorption (NaCl, amino acids and sugars); the fluid transported is isotonic to the saline bathing the epithelium. However, the molecular mechanisms of water transport (absorption or secretion) across the intestinal epithelia are not yet clear. It is a dogma that active salt and nutrient transport generates local osmotic gradients within the epithelium that drive water transport (Diamond, 1979, Diamond and Bossert, 1967; Fromter and Diamond, 1972). This transport may be carried out across paracellular and/or transcellular pathways. In the duodenum and proximal jejunum water movement across the epithelium is believed to occur by a paracellular pathway. In the colon the situation is different. It is a tight epithelium with a lower paracellular permeability, and water may be transported against an effective osmotic gradient. Naftalin *et al.* (1999) have proposed that the myofi-broblast-reticular sheath in the distal colon crypts serves as an additional diffusion barrier in series with the basal membrane, in order to generate a localized hyper-tonic Na^+ gradient to accomplish uphill water transport capable of dehydrating feces (Naftalin and Pedley, 1999; Naftalin *et al.*, 1999).

The paracellular flow of fluid and solutes is regulated by the permeability of tight junctions. These are characteristic of polarized absorptive and secretory epithelia. Tight junctions are dynamic structures that may adapt to a variety of situations, by mechanisms that begin to be elucidated. A number of enteric pathogens including rotavirus may affect tight-junction function (Sears, 2000). Numerous extracellular mediators including

32

hormones, neurotransmitters, biogenic amines and cytokines participate in a variety of physiological and pathological processes modulating tight junction permeability across epithelia (Nusrat *et al.*, 2000). These agents induce increases of intracellular calcium and adenosine 3'-5'cyclic monophosphate (cAMP) and may involve mechanisms such as the contraction of the cytoskeleton and the perijunctional ring. Other effectors include protein kinase C, myosin light-chain kinase, and phosphatidylinositol 3-kinase (Karczewski and Groot, 2000). Other factors such as glucose, when present in the apical side, appear to regulate tight junction permeability. It has been proposed that initiation of Na^+/glucose cotransport leads to activation of a Na^+/H^+ exchange, increased phosphorylation of myosin light chain, contraction of the perijunctional actomyosin ring, and, ultimately, increased permeability of intestinal tight junctions (Nusrat *et al.*, 2000).

The transcellular route for water movements involves the presence of molecular entities both at the brush border and the basolateral membranes of enterocytes. Water channels of the aquaporin family (AQP) have been identified in the small intestine (AQP2 and 7) and in the colon (AQP3, 4 and 8), located in villus and crypt cells (Ma and Verkman, 1999). A new aquaporin (AQP10) was identified in human small intestine and localized in absorptive duodenal and jejunal epithelial cells (Hatakeyama *et al.*, 2001).

Recently, based on experiments in *Xenopus* oocytes expressing SGLT1, Wright and Loo (2000) have proposed that this cotransporter may function as a water channel. Moreover, SGLT1 may couple the transport of Na^+ and sugar to water translocation, in this way functioning as a Na^+/glucose/H_2O cotransporter. The physiological role of these different transporters remains to be elucidated as well as their participation in pathological processes.

Pathophysiology of virus-induced diarrhea

Mechanisms of diarrhea

At least four mechanisms have been implicated in diarrheal syndromes: increased secretion, decreased absorption (ions and/or solutes and water), altered motility and increased permeability (Fig. 3). Viral infection induces the synthesis of viral proteins, cell lysis and inflammatory responses. Cell lysis permits the release of viral progeny, which in turn infect neighboring cells. The cytolytic infection leads to a decrease in the digestive and absorptive capacity of the intestine, generating a malabsorption component of diarrhea. Cell and viral proteins will act upon immune cells to release a variety of chemical mediators. The interaction between the immune enteroendocrine and enteric systems will lead to increases in intracellular second messengers Ca^{2+} and cAMP. Direct or mediated stimulation of enterocytes by viral protein (e.g., the rotaviral NSP4 enterotoxin) [see Estes, Section II, Chapter 6 of this book] could increase second messengers. These messengers may a) activate Cl^- channels inducing hypersecretion, b) act upon the cytoskeleton and the tight junction to increase paracellular permeability, c) derange intestinal motility and, d) activate transporter function or decrease enzyme expression to effectively impair nutrient and water absorption.

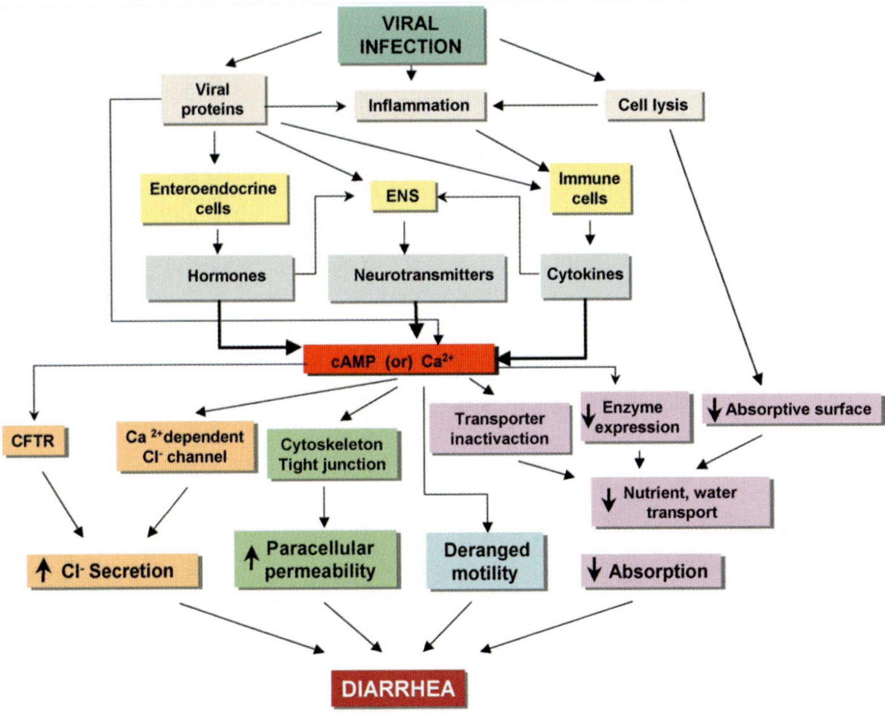

Fig. 3. Mechanisms of induction of diarrhea by viral pathogens. Four mechanisms have been implicated in diarrheal syndromes: increased secretion, decreased absorption (ions and/or solutes and water), altered motility and increased permeability. Usually, viruses cause diarrhea by more than one of these mechanisms. See text for details.

Usually, infectious agents, including viruses, cause diarrhea by more than one of these mechanisms. Viral pathogens have developed strategies to colonize different environments in the small intestine, disrupting the normal fluid balance of the gut and causing diarrhea. These enteric viruses are stable at low pH and resistant to digestive enzymes, important characteristics in their infective capacity. The viruses causing diarrhea in humans and animals may belong to different families (Table 1; for review see Glass *et al.*, 2001; Saif, 1999; Parashar and Glass Section I, Chapter 1 of this book). Among these viruses, rotaviruses, caliciviruses (including noro- and sapoviruses), enteric adenoviruses and astroviruses are the main agents inducing diarrhea in humans as well as in animals. In addition, picobirnaviruses and HIV have been associated with diarrheal syndromes in humans. Also, viruses belonging to the *Coronaviridae* (TGEV, PEDV, Torovirus), and *Parvoviridae* infect the intestine and are responsible for diarrhea in animals.

Table 1

Characteristics of viruses causing acute gastroenteritis

Virus	Family	Host	Site of infection	Mechanism of action	Type of diarrhea
Rotavirus	*Reoviridae*	human, animal	middle and tip villus enterocyte of jejunum and ileum	cytolytic infection, villus atrophy, crypt hyperplasia, NSP4 enterotoxin, activation of ENS, increase paracellular permeability, decreased digestive enzyme expression	mixed: primary secretory; enhanced permeability, increased motility, malabsorption at late stage; mild/severe
Calicivirus norovirus sapovirus	*Caliciviridae*	human, animal	villus enterocyte of duodenum and jejunum; patchy mucosal lesions.	cytolytic infection, villus atrophy, crypt hyperplasia, malabsorption of fat and xylose, diminished activity of disaccharidase enzymes, and delayed gastric emptying	mixed: primary malabsorption, secondary secretory; moderate
Enteric adenovirus	*Adenoviridae*	human, animal	base and side of villus and crypt enterocytes of jejunum, ileum and proximal colon	cytolytic infection, moderate villus atrophy	mixed: primary malabsorption, secondary secretory; moderate
Astrovirus	*Astroviridae*	human, animal	middle and tip villus enterocyte of duodenum, jejunum and ileum	cytolytic infection, mild or no villus atrophy and crypt hyperplasia	mixed: primary malabsorption, secondary secretory; mild/none
Enteric parvovirus	*Parvoviridae*	animal	primary systemic, basolateral pole of crypt enterocyte of jejunum and ileum	cytolytic infection of crypt enterocytes, severe villus atrophy, mucosal collapse, hemorrhagic diarrhea	malabsorption; severe, hemorragic.

Table 1

Continued

Virus	Family	Host	Site of infection	Mechanism of action	Type of diarrhea
Transmissible Gastroenteritis Virus (TGEV)	*Coronaviridae*	animal	entire villus enterocytes in multiple stages of differentiation, of duodenum, jejunum and ileum	cytolytic infection, pronounced villus atrophy, crypt hyperplasia	mixed: primary malabsorption, secondary secretory; severe
Porcine Epidemic Diarrhea Virus (PEDV)	*Coronaviridae*	animal	base and side of villus and crypt enterocytes of duodenum, jejunum, ileum and proximal colon	cytolytic infection, moderate villus atrophy	mixed: primary malabsorption, secondary secretory; moderate
Torovirus	*Coronaviridae*	animal, human	crypt and basal villus enterocytes of jejunum, ileum and colon	cytolytic infection, villus atrophy, crypt hyperplasia	role in human disease unknown
Picobirnavirus	*Picobirnaviridae*	animal, human	unknown	unknown	role in human disease unknown

The mechanisms by which viruses induce diarrhea are ill-defined. Most of them infect the small intestine enterocytes with a distinct tropism for different cell types (villus/crypt) and intestinal segment (duodenum, jejunum, ileum and colon). The majority of the viruses mentioned above penetrate the cell through the apical membrane, except for parvovirus where a systemic infection seems to precede penetration of crypt cells through the basolateral membrane (Saif, 1999). Once inside the cell, virus replication usually causes cell lysis and the release of viral progeny which in turn infects neighboring cells. The cytolytic infection often leads to villus atrophy and a decrease in the digestive and absorptive capacities of the intestine, generating a malabsorption component of diarrhea. As undigested and non-absorbed nutrients reach the colon, microbial fermentation leads to an accumulation of osmotically active solutes, adding an osmotic component to the syndrome. Secondary to villus enterocyte loss, a crypt hyperplasia follows, inducing a secretory contribution to diarrhea (Moon, 1994). This type of pathogenic mechanism may be responsible for the induction of diarrhea following infection by calicivirus, enteric adenovirus, TGEV and astrovirus. In the case of parvovirus, the lytic infection of the crypt enterocyte leads to a reduction of villus enterocyte production and induces a malabsorption type of diarrhea.

The extent of histopathological lesions often does not explain the intensity of diarrhea, suggesting that the mechanisms outlined might be a simplistic view of the process. In many cases, viral diarrhea may be a multifactorial disease where impaired digestion and absorption may be the major factors, with osmotic activity, secretion and hypermotility complicating the process. However, this sequence does not appear to operate for other viruses such as rotavirus, where a secreted viral protein (NSP4) seems to act as an enterotoxin triggering a simultaneous pleiades of complex effects (Estes *et al.*, 2001).

Rotavirus as a model for the genesis of viral diarrhea

Due to the clinical importance of rotavirus diarrhea and the fact that cell culture and animal models of the disease have been developed, our knowledge of the pathogenesis has greatly advanced during the past few years (for review see Estes *et al.*, 2001; Estes and Morris, 1999; Lundgren and Svensson, 2001; Morris and Estes, 2001; Ruiz *et al.*, 2000; Estes Section II, Chapter 6 of this book). Rotaviruses infect mature enterocytes of jejunum and ileum, inducing cell lysis, which will release the new rotavirus progeny and synthesized viral proteins into the lumen of the intestine. After several viral replication cycles, extensive cell death provokes shortening of microvilli and a reduction of the absorptive surface of the intestinal epithelium resulting in a late malabsorption component of diarrhea. Shortening of villi also induces crypt hyperplasia and hypersecretion, further increasing the syndrome (Burke and Desselberger, 1996; Greenberg *et al.*, 1994). However, these histopathological lesions appear at a late stage of infection when diarrhea is already installed. On the other hand, changes in ion homeostasis of the infected cell, in particular Ca^{2+}, have provided new insights into the pathogenic mechanisms that eventually lead to diarrhea.

Rotavirus infection of cultured cells induces changes in the homeostasis of both mono- and divalent cations that appear to be mediated by the synthesis of viral proteins (del Castillo *et al.*, 1991; Michelangeli *et al.*, 1991; Perez *et al.*, 1999; Ruiz *et al.*, 2000). Among these perturbances is a progressive increase in Ca^{2+} membrane permeability which leads to an elevation of cytosolic Ca^{2+} concentration and sequestered pools, without an apparent failure of regulatory mechanisms in virus-infected MA104, HT29 and Caco-2 cells (Brunet *et al.*, 2000a; Michelangeli *et al.*, 1991; Perez *et al.*, 1999). In parallel to the increase in plasma membrane permeability and cytosolic Ca^{2+} concentration, there is an enhancement of ER sequestered Ca^{2+} stores (Brunet *et al.*, 2000a; Michelangeli *et al.*, 1995; Perez *et al.*, 1999). This effect is likely to be due to the activation of ER Ca^{2+} pumps.

The nonstructural protein NSP4, expressed constitutively during rotavirus infection as well as added exogenously to noninfected cells, has been implicated in disturbances of Ca^{2+} homeostasis. The expression of recombinant NSP4 protein in insect cells induced an increase in cytosol Ca^{2+} concentration (Tian *et al.*, 1994). This effect was proposed to be due to the release of Ca^{2+} from intracellular stores without increases in plasma membrane Ca^{2+} permeability and linked to a membrane destabilizing activity of NSP4 (Tian *et al.*, 1995; Tian *et al.*, 1996).

Later during infection, another permeability pathway to Ca^{2+} is induced by a secreted or released viral product acting from the outside of the cell in Caco-2 cells (Brunet *et al.*, 2000a). This depends on the activation of the phospholipase C-IP$_3$ cascade, resulting in store depletion, which in turn may activate capacitative Ca^{2+} entry. It is reasonable to think that NSP4 is responsible for these effects. It is known that exogenously added recombinant-produced NSP4, or a cleaved form of NSP4 which is secreted from infected cells, increases $[Ca^{2+}]_i$ in HT29 cells (Zhang *et al.*, 2000).

These changes in $[Ca^{2+}]_i$ caused by infection or by exogenous NSP4 will have important consequences on the enterocyte physiology. These include actions on the cytoskeleton, ionic and nutrient transport, expression of digestive enzymes and eventually cell death which will be addressed below. These numerous modifications will lead to stimulation of secretion, diminished absorption, enhanced paracellular permeability and perhaps hypermotility which will participate in the pathogenesis of diarrhea from the early stages of the disease. We will analyse the different components of rotavirus diarrhea in the context of these four mechanisms.

Secretory component

That a primary secretory component of rotavirus diarrhea might exist was first proposed based on the effect of rotavirus on Ca^{2+} homeostasis of infected cells (Michelangeli *et al.*, 1991). The finding that the endogenous expression or the exogenous addition of the nonstructural protein NSP4 is capable of increasing $[Ca^{2+}]_i$ (Dong *et al.*, 1997; Tian *et al.*, 1994; Tian *et al.*, 1995) triggered a number of studies that lead to the hypothesis that this protein may act as a viral enterotoxin to induce secretion and diarrhea (Estes *et al.*, 2001; Estes and Morris, 1999; Morris and Estes, 2001). There is a series of evidence that supports this hypothesis: a) intraperitoneal or intraluminal injection of NSP4 or NSP4 peptide fragments cause age-dependent diarrhea in mice (Ball *et al.*, 1996; Sasaki *et al.*, 2001); b) NSP4 and/or a fragment of this protein are released from infected cell monolayers as a consequence of secretion and apparently not cell lysis (Zhang *et al.*, 2000); c) NSP4 or the synthetic NSP4$_{114-135}$ peptide potentiated cAMP-dependent Cl⁻ secretion in isolated mouse intestine mounted in Ussing chambers, similarly to carbachol (Ball *et al.*, 1996); d) exogenous addition of NSP4 or the NSP4$_{114-135}$ peptide to noninfected cells induced the release of Ca^{2+} in HT29 cells through the activation of the phospholipase C-IP$_3$ cascade and a secondary increase in plasma membrane Ca^{2+} permeability, similarly to cholinergic agents (Dong *et al.*, 1997). [For further details see Estes, Section II, Chapter 6 of this book].

The relationship between diarrhea, Ca^{2+} and NSP4 has been confirmed studying mutations in NSP4, which are associated with altered virus virulence. There is an association between the capacity of this protein to induce changes in intracellular Ca^{2+} in HT29 cells and the induction of diarrhea by the mutated NSP4 in mice. This mutation has been mapped between amino acids 131-140 (Zhang *et al.*, 1998). However, NSP4 may not be the only virulence factor as attenuation of rotavirus in culture cells appears to be unrelated to mutations in the NSP4 gene (Angel *et al.*, 1998; Ward *et al.*, 1997) and NSP4 sequences in both diarrheal and asymptomatically-infected kittens were closely related to each other (Oka *et al.*, 2001). Also, there was no apparent correlation in the deduced amino acid sequences corresponding to the proposed enterotoxic peptide region between rotaviruses recovered from children with and without diarrhea (Lee *et al.*, 2000). [For discussion of these questions see also Estes, Section II, Chapter 6 of this book].

The target cell for the NSP4 protein action is not yet known. NSP4 could act directly on uninfected crypt cells or mature enterocytes, or both, to induce secretion. In search of the chloride secretion pathway activated by NSP4, studies have been performed in CFTR knockout mice, which do not respond to Ca^{2+} and cAMP dependent secretagogues (Morris *et al.*, 1999). Rotavirus particles, NSP4, or its active NSP4$_{114-135}$ peptide can overcome secretory inhibition and elicit diarrhea when administered to CF mouse pups. The evidence was interpreted as showing that NSP4 induces diarrhea in neonatal mice through the activation of an age- and Ca^{2+}-dependent Cl⁻ secretory mechanism different from the CFTR (Morris *et al.*, 1999).

Within this attractive hypothesis of direct stimulation of crypt cells, several points remain to be clarified: a) the molecular identification of the putative Ca^{2+} dependent Cl⁻ channel; b) the reason why carbachol is not able to stimulate this Cl⁻ channel related to diarrhea in CFTR knockout mice; c) the identity and localization of the putative NSP4 receptor; d) the correlation between the Ca^{2+} dependent component of secretion and diarrhea. In this sense, the changes in short circuit current (Cl⁻ transport) in isolated mucosal sheets of mice intestine induced by NSP4$_{114-135}$ peptide or carbachol represent a small fraction of that induced by the cAMP dependent secretagogue, forskolin. Also, Cl⁻ influx measured in isolated crypts of CFTR⁻/⁻ pup mice induced by carbachol or NSP4 is small in comparison to forskolin as to explain the diarrheal syndrome. However, NSP4 and carbachol could potentiate CFTR-dependent secretion indirectly through the activation of the Ca^{2+}-stimulated basolateral K⁺ conductance, hyperpolarization and Cl⁻ uptake via the cotransport (Morris, 1999). Another experimental complication is that only a small percentage (20%) of the isolated crypts responded to either carbachol or NSP4 with increases in $[Ca^{2+}]_i$. Furthermore, it is not explained how NSP4 elicits these changes in isolated crypts if a luminal receptor is postulated and how NSP4 can reach deep into the lumen of a secreting crypt *in vitro* and in the *in vivo* intestine. The lumen of isolated crypt can be collapsed at the tip after isolation as it occurs in isolated gastric glands (Perez *et al.*, 2001), which would imply that NSP4 acted on basolateral receptors. Also, we should consider the sweeping out effect produced by the net flow of water accompanying crypt secretion. In this case, the water flow going out of the crypt would oppose the diffusional entry of NSP4. This effect has been shown for other secretory epithelia [Moody and Durbin, 1969]. These considerations do not at this point invalidate the hypothesis of NSP4 acting as an enterotoxin. However, they may cast doubts as to its role in the direct stimulation of crypt cells to induce Cl⁻ secretion.

The enteric nervous system is an integrative network of neurons that regulate gastrointestinal functions and might be the link between NSP4 and the secretion of water and electrolytes involved in rotavirus diarrhea. Pharmacological blockade of nervous transmission is able to inhibit secretion in the intestinal mucosa and to attenuate diarrhea induced by rotavirus infection (Lundgren *et al.*, 2000; Lundgren and Svensson, Section I, Chapter 3 of this book). Released or secreted NSP4 could be responsible for the activation of enteric neurons directly or indirectly through the stimulation of peptide and amine secretion from paracrine cells. In addition to NSP4, inflammatory mediators released from myofibroblasts and inflammatory cells during infection may be capable of regulating secretion and absorption directly or through the activation of the enteric nervous network and the paracrine system (Cooke, 2000; Montrose *et al.*, 1999; Moon, 1994; Powell *et al.*, 1999). Inflammation may contribute to fluid loss and diarrhea in other enteric viral infections for which an enterotoxin has not yet been described.

Malabsorption component

Rotavirus induces a lytic infection of the villus enterocyte leading to shortening of microvilli and a reduction of the absorptive surface of the intestinal epithelium resulting in a late malabsorption component of diarrhea (Burke and Desselberger, 1996; Greenberg *et al.*, 1994). Rotavirus infection provokes cell death of MA104 cells by necrosis in a Ca^{2+} dependent pathway (Perez *et al.*, 1998). The events triggered by the increase in cytosolic Ca^{2+} which lead to cell death in rotavirus infected cells are yet unknown. Calcium may participate in the generation of these alterations by activating Ca^{2+} dependent enzymes such as phospholipases A and C, proteases and endonucleases. On the other hand, the expression of NSP4 alone in MA104 cells induced cell death. Analysis of NSP4 deletion mutants indicates that a membrane-proximal region located within the cytoplasmic domain may mediate cytotoxicity (Newton *et al.*, 1997). Newly synthesized particles would be released into the medium after cell lysis and induce new infectious lytic cycles. However, we should note that an alternative model for the release of rotavirus particles before lysis of Caco-2 cells in a secretion-like process has been reported (Jourdan *et al.*, 1997). The importance of these findings in the *in vivo* infection remains to be further studied.

A reduction of electrolyte and nutrient absorption may be the consequence of other processes occurring earlier during infection before cell lysis. The Na^+ gradient across the

40

membrane is collapsed early during infection due to an increase in $[Na^+]_i$ (del Castillo *et al.*, 1991). If this occurred in the villus enterocyte, it would impair electroneutral NaCl absorption and Na^+ linked nutrient absorption with a consequent fluid loss and diarrhea.

The mechanism of rotavirus diarrhea may involve a generalized inhibition of Na^+/ solute cotransport systems. A strong inhibition of both Na^+/glucose (SGLT1) and Na^+/ leucine symport activities have been found in brush-border membrane vesicles isolated from rotavirus infected rabbits (Halaihel *et al.*, 2000a). The viral effect on this symport appears to be direct and perhaps linked to NSP4. The $NSP4_{114-135}$ peptide is a specific and noncompetitive inhibitor of SGLT1 (Halaihel *et al.*, 2000b).

Another mechanism that may be involved in a decrease of nutrient absorption during rotavirus infection is the reduction of digestive enzyme expression at the apical brush border membrane. During the course of infection of suckling mice and piglets with rotavirus, the activities of alkaline phosphatase, lactase, sucrase and malt-ase decreased (Collins *et al.*, 1988; Davidson *et al.*, 1977; Shepherd *et al.*, 1979). Furthermore, rotavirus infection of Caco-2 cells induces a decreased apical surface expression of sucrase-isomaltase which correlated with important alterations of the cytoskeleton (Jourdan *et al.*, 1998), perhaps due to an increase in $[Ca^{2+}]_i$ (Brunet *et al.*, 2000a; Brunet *et al.*, 2000b).

As pointed out above, a decrease in nutrient, electrolyte and water absorption due to reduction in the absorptive surface is the major and primary cause of diarrhea induced by other enteric viruses. Coronavirus (TGEV) has been used extensively to study enter-itis-associated changes in nutrient and electrolyte transport. A decreased Na^+/glucose transport has been shown in isolated piglet intestines mounted in Ussing chambers (Shepherd *et al.*, 1979; Telch *et al.*, 1981) as well as in brush border membrane vesicles (Keljo *et al.*, 1985). Also, Na^+ coupled alanine transport was reduced during TGEV infection in piglet intestine brush border membranes (Rhoads *et al.*, 1989).

Paracellular permeability component

The role of the paracellular pathway and the integrity of the tight junction in the bar-rier function and fluid balance in the intestine have been long recognized. A number of disease states, including inflammatory and diarrheal diseases, may be caused by primary defects in epithelial barrier function (Bjarnason *et al.*, 1995). Pathogens may act at dif-ferent levels such as direct cleavage of tight junctional structural proteins, modification of the actin cytoskeleton, activation of cellular signal transduction and, triggering trans-migration of polymorphonuclear cells across the epithelial cell barrier (Sears, 2000).

Profound changes in the permeability characteristics of the intestine of children during acute rotavirus diarrhea have been found (Stintzing *et al.*, 1986). Experiments performed in rotavirus infected Caco-2 and/or MDCK cell monolayers showed a dis-ruption of tight junctions and decrease of transepithelial resistance (Dickman *et al.*, 2000; Obert *et al.*, 2000; Svensson *et al.*, 1991). This was accompanied by increased transepithelial permeability to macromolecules (Dickman *et al.*, 2000; Obert *et al.*, 2000) and altered distribution of tight junction proteins claudin-1, occludin, and ZO-1 (Dickman *et al.*, 2000). These changes did not appear to be related to the observed changes in the cytoskeleton (Brunet *et al.*, 2000b; Obert *et al.*, 2000) but perhaps due to changes in cell metabolism caused by infection (Dickman *et al.*, 2000).

The modification of paracellular permeability might be induced by the endogenous expression of NSP4 in infected cells or released NSP4 acting as an enterotoxin on non-infected cells. In MDCK cell monolayers, apical addition of NSP4 induced a reduction in transepithelial electrical resistance, a redistribution of filamentous actin, an increase in paracellular permeability to extracellular markers and the targeting of the tight junction associated protein ZO-1 (Tafazoli *et al.*, 2001). These effects may be mediated by an increase in $[Ca^{2+}]_i$. Similar changes might be evoked by inflammatory mediators released during infection in intact tissue either directly or mediated by an action of the enteric nervous system (Lundgren and Svensson, 2001; Moon, 1994).

Altered motility

Deranged motility has been recognized as an important factor in the genesis of the diarrhea (Moon, 1994). The enteric nervous system is critical for the integration of intestinal electrolyte transport with motility. The enterotoxins produced by various bacteria induce effects on intestinal motility via an activation of the ENS. The effects of rotavirus enteritis on intestinal motility indicate that intestinal transit time is shortened. The possible involvement of the ENS in this response is not known (Lundgren and Svensson, 2001). However, again NSP4 may be participating in this response, either directly or indirectly, through the release of chemical mediators.

Changes in microcirculation have been also proposed as participating in the genesis of diarrhea induced by virus infection [Osborne *et al.*, 1991; Stephen, 1989; Stephen and Osborne, 1988]. According to this hypothesis, infection of villus cell by rotavirus would trigger the release of "neuroactive/hormonal substances" which would cause ischaemia by vasoconstriction and, in turn, villus shortening and a decrease in absorptive capacity. In addition, ischaemia induces hyperplasia [Stephen and Osborne, 1988] of immature cells and hypersecretion. These effects did not seem to be virus specific. The extent and participation of this component in virus-induced diarrhea remains to be assessed.

Conclusions

Based on the current knowledge of the effects of rotavirus infection on the physiology of the intestine at different levels of organization, a working model for the pathogenesis of rotavirus diarrhea is presented in Fig. 4. Rotavirus particles would bind to a functional receptor constituted by a complex of several molecules (sialic acid-dependent and -independent) on the apical membrane of the villus enterocyte (for review see Arias *et al.*, 2001; Lópet *et al.*, Section II, Chapter 2 of this book). For internalization into the cytoplasm, a Ca^{2+}-dependent endocytosis mechanism has been proposed (Ruiz *et al.*, 2000). During the first viral replication cycle, proteins and particles are synthesized. The enterotoxigenic viral protein NSP4 would be secreted. At the end of the cycle, a large number of newly synthesized viral particles are released after cell lysis, concomitantly with viral proteins.

42

The viral progeny will infect neighboring downstream cells, and the released NSP4 would trigger a series of events that would lead to diarrhea. Binding to putative apical receptors, NSP4 would induce an increase of $[Ca^{2+}]_i$ through a phospholipase C-IP3 pathway in villus and perhaps crypt enterocytes, and enteroendocrine cells. At the villus cell level, the enterotoxin would inhibit Na^+/nutrient cotransport and digestive enzyme activities decreasing water and solute transport. Concomitantly, the $[Ca^{2+}]_i$ increase would disrupt the cytoskeleton and the tight junction, leading to an enhanced paracellular permeability. Chloride secretion by crypt cells would also be stimulated. This would occur through a direct or indirect activation of the enteric nervous system and the release of chemical mediators.

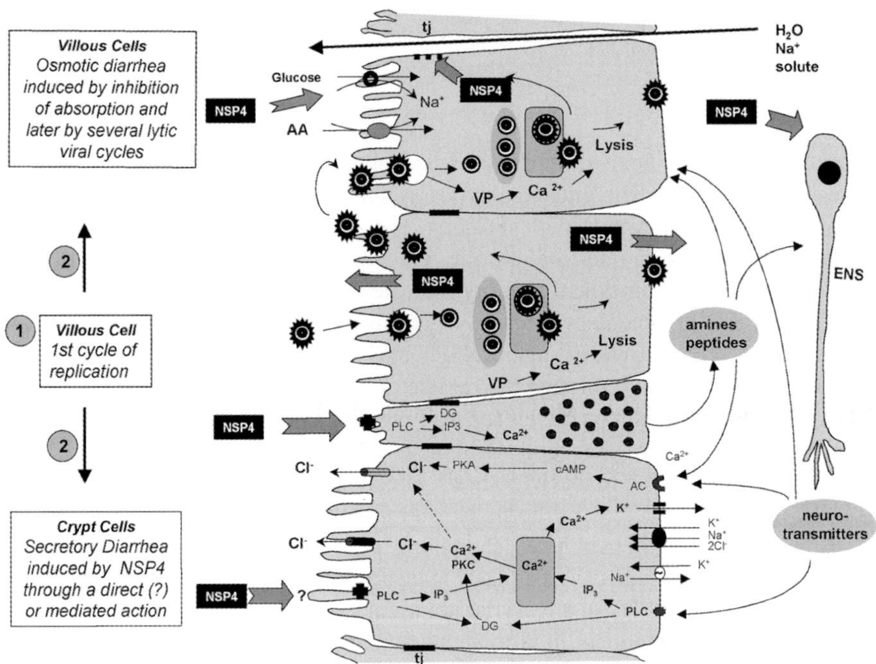

Fig. 4. Pathophysiological model of mixed type diarrhea induced by rotavirus infection. During the first cycle of infection the synthesis of rotaviral proteins in the infected cell leads to increase in plasma membrane permeability to Ca^{2+}, activation of regulatory mechanisms, and increase in Ca^{2+} content in the ER. The increase of cytosolic Ca^{2+} concentration in the infected cell provokes the activation of Ca^{2+} dependent enzymes, which in turn will induce cell lysis and release of viral proteins and viral progeny. NSP4 may act as a viral enterotoxin to induce secretory diarrhea by: a) Ca^{2+} dependent secretion of amines and peptides to induce stimulation of the enteric nervous system (ENS); b) activation of crypt cell Cl⁻ secretion by the release of neurotransmitters; c) Ca^{2+} dependent secretion by crypt cells or mature enterocytes by activation of Cl⁻ channels. NSP4 may also reduce nutrient and ion absorption. In parallel, released virus will infect downstream absorptive cells, leading to cell death and reduction of the absorptive surface generating an osmotic component of diarrhea. In addition, NSP4 will alter the integrity of the tight junction (tj), opening a paracellular permeability pathway and increasing secretion by solvent drag or exudation. Changes in motility may also occur in response to the activation of the ENS.

A direct stimulation of the crypt cell by NSP4, although unlikely, may also occur. As infected cells die in a calcium-mediated necrosis process, the absorptive surface is reduced further increasing the malabsorption component which is superimposed on the secretory one, resulting in a mixed type diarrhea. The understanding of the pathogenic processes of viral diarrheas may serve as the basis for a rational approach in the design of novel therapeutic strategies and the search for new antiviral drugs.

Acknowledgements

Work in the authors' laboratory is supported by Fundacyt, Venezuela (Grant N° S1-2001 000329). We thank Dr Jean Cohen for useful comments on the manuscript.

References

Anderson, M. P., Sheppard, D. N., Berger, H. A. and Welsh, M. J. (1992). Chloride channels in the apical membrane of normal and cystic fibrosis airway and intestinal epithelia. *Am J Physiol* **263**, L1-14.

Angel, J., Tang, B., Feng, N., Greenberg, H. B. and Bass, D. (1998). Studies of the role for NSP4 in the pathogenesis of homologous murine rotavirus diarrhea. *J Infect Dis* **177**, 455-8.

Arias, C. F., Guerrero, C. A., Mendez, E., Zárate, S., Isa, P., Espinosa, R., Romero, P. and López, S. (2001). Early events of rotavirus infection: the search for the receptor(s). *Novartis Found Symp* **238**, 47-60.

Ball, J. M., Tian, P., Zeng, C. Q., Morris, A. P. and Estes, M. K. (1996). Age-dependent diarrhea induced by a rotaviral nonstructural glycoprotein. *Science* **272**, 101-4.

Barrett, K. E. and Keely, S. J. (2000). Chloride secretion by the intestinal epithelium: molecular basis and regulatory aspects. *Annu Rev Physiol* **62**, 535-72.

Berschneider, H. M., Knowles, M. R., Azizkhan, R. G., Boucher, R. C., Tobey, N. A., Orlando, R. C. and Powell, D. W. (1988). Altered intestinal chloride transport in cystic fibrosis. *FASEB J* **2**, 2625-9.

Berschneider, H. M. and Powell, D. W. (1992). Fibroblasts modulate intestinal secretory responses to inflammatory mediators. *J Clin Invest* **89**, 484-9.

Bjarnason, I., MacPherson, A. and Hollander, D. (1995). Intestinal permeability: an overview. *Gastroenterology* **108**, 1566-81.

Brunet, J. P., Cotte-Laffitte, J., Linxe, C., Quero, A. M., Geniteau-Legendre, M. and Servin, A. (2000a). Rotavirus infection induces an increase in intracellular calcium concentration in human intestinal epithelial cells: role in microvillar actin alteration. *J Virol* **74**, 2323-32.

Brunet, J. P., Jourdan, N., Cotte-Laffitte, J., Linxe, C., Geniteau-Legendre, M., Servin, A. and Quero, A. M. (2000b). Rotavirus infection induces cytoskeleton disorganization in human intestinal epithelial cells: implication of an increase in intracellular calcium concentration. *J Virol* **74**, 10801-6.

Burke, B. and Desselberger, U. (1996). Rotavirus pathogenicity. *Virology* **218**, 299-305.

Collins, J., Starkey, W. G., Wallis, T. S., Clarke, G. J., Worton, K. J., Spencer, A. J., Haddon, S. J., Osborne, M. P., Candy, D. C. and Stephen, J. (1988). Intestinal enzyme profiles in normal and rotavirus-infected mice. *J Pediatr Gastroenterol Nutr* **7**, 264-72.

Cooke, H. J. (2000). Neurotransmitters in neuronal reflexes regulating intestinal secretion. *Ann N Y Acad Sci* **915**, 77-80.

D'Andrea, L., Lytle, C., Matthews, J. B., Hofman, P., Forbush, B., 3rd and Madara, J. L. (1996). Na:K:2Cl cotransporter (NKCC) of intestinal epithelial cells. Surface expression in response to cAMP. *J Biol Chem* **271**, 28969-76.

Davidson, G. P., Gall, D. G., Petric, M., Butler, D. G. and Hamilton, J. R. (1977). Human rotavirus enteritis induced in conventional piglets. Intestinal structure and transport. *J Clin Invest* **60**, 1402-9.

del Castillo, J. R., Ludert, J. E., Sanchez, A., Ruiz, M. C., Michelangeli, F. and Liprandi, F. (1991). Rotavirus infection alters Na+ and K+ homeostasis in MA-104 cells. *J.Gen.Virol.* **72**, 541-547.

Devor, D. C. and Duffey, M. E. (1992). Carbachol induces K+, Cl-, and nonselective cation conductances in T84 cells: a perforated patch-clamp study. *Am J Physiol* **263**, C780-7.

Diamond, J. M. (1979). Osmotic water flow in leaky epithelia. *J Membr Biol* **51**, 195-216.

Diamond, J. M. and Bossert, W. H. (1967). Standing-gradient osmotic flow. A mechanism for coupling of water and solute transport in epithelia. *J Gen Physiol* **50**, 2061-83.

Dickman, K. G., Hempson, S. J., Anderson, J., Lippe, S., Zhao, L., Burakoff, R. and Shaw, R. D. (2000). Rotavirus alters paracellular permeability and energy metabolism in Caco-2 cells. *Am J Physiol Gastrointest Liver Physiol* **279**, G757-66.

Dong, Y., Zeng, C. Q., Ball, J. M., Estes, M. K. and Morris, A. P. (1997). The rotavirus enterotoxin NSP4 mobilizes intracellular calcium in human intestinal cells by stimulating phospholipase C-mediated inositol-1,4,5- trisphosphate production. *Proc Natl Acad Sci USA* **94**, 3960-5.

Estes, M. K., Kang, G., Zeng, C. Q., Crawford, S. E. and Ciarlet, M. (2001). Pathogenesis of rotavirus gastroenteritis. *Novartis Found Symp* **238**, 82-96.

Estes, M. K. and Morris, A. P. (1999). A viral enterotoxin. A new mechanism of virus-induced pathogenesis. *Adv Exp Med Biol* **473**, 73-82.

French, P. J., Bijman, J., Edixhoven, M., Vaandrager, A. B., Scholte, B. J., Lohmann, S. M., Nairn, A. C. and de Jonge, H. R. (1995). Isotype-specific activation of cystic fibrosis transmembrane conductance regulator-chloride channels by cGMP-dependent protein kinase II. *J Biol Chem* **270**, 26626-31.

Fromter, E. and Diamond, J. (1972). Route of passive ion permeation in epithelia. *Nat New Biol* **235**, 9-13.

Gadsby, D. C. and Nairn, A. C. (1999). Control of CFTR channel gating by phosphorylation and nucleotide hydrolysis. *Physiol Rev* **79**, S77-107.

Ganapathy, V., Brandsch, M. and Leibach, F. H. (1994). Intestinal Transport of Amino Acid and Peptides. In *Physiology of the Gastrointestinal Tract*, edited by L. R.

Johnson. pp 1773-1794. Raven Press: New York.

Glass, R. I., Bresee, J., Jiang, B., Gentsch, J., Ando, T., Fankhauser, R., Noel, J., Parashar, U., Rosen, B. and Monroe, S. S. (2001). Gastroenteritis viruses: an overview. *Novartis Found Symp* **238**, 5-19.

Greenberg, H. B., Clark, H. F. and Offit, P. A. (1994). Rotavirus pathology and pathophysiology. *Curr Top Microbiol Immunol* **185**, 255-83.

Greger, R., Bleich, M. and Warth, R. (1997). New types of K+ channels in the colon. *Wien Klin Wochenschr* **109**, 497-8.

Grubb, B. R. and Gabriel, S. E. (1997). Intestinal physiology and pathology in gene-targeted mouse models of cystic fibrosis. *Am J Physiol* **273**, G258-66.

Haas, M. and Forbush, B., 3rd (2000). The Na-K-Cl cotransporter of secretory epithelia. *Annu Rev Physiol* **62**, 515-34.

Halaihel, N., Lievin, V., Alvarado, F. and Vasseur, M. (2000a). Rotavirus infection impairs intestinal brush-border membrane Na(+)- solute cotransport activities in young rabbits. *Am J Physiol Gastrointest Liver Physiol* **279**, G587-96.

Halaihel, N., Lievin, V., Ball, J. M., Estes, M. K., Alvarado, F. and Vasseur, M. (2000b). Direct inhibitory effect of rotavirus NSP4(114-135) peptide on the Na(+)-D-glucose symporter of rabbit intestinal brush border membrane. *J Virol* **74**, 9464-70.

Hatakeyama, S., Yoshida, Y., Tani, T., Koyama, Y., Nihei, K., Ohshiro, K., Kamiie, J. I., Yaoita, E., Suda, T., Hatakeyama, K. and Yamamoto, T. (2001). Cloning of a new aquaporin (aqp10) abundantly expressed in duodenum and jejunum. *Biochem Biophys Res Commun* **287**, 814-9.

Hecht, G. and Koutsouris, A. (1999). Myosin regulation of NKCC1: effects on cAMP-mediated Cl- secretion in intestinal epithelia. *Am J Physiol* **277**, C441-7.

Ishii, T. M., Silvia, C., Hirschberg, B., Bond, C. T., Adelman, J. P. and Maylie, J. (1997). A human intermediate conductance calcium-activated potassium channel. *Proc Natl Acad Sci USA* **94**, 11651-6.

Jensen, B. S., Strobaek, D., Christophersen, P., Jorgensen, T. D., Hansen, C., Silahtaroglu, A , Olesen, S. P. and Ahring, P. K. (1998). Characterization of the cloned human intermediate-conductance Ca2+- activated K+ channel. *Am J Physiol* **275**, C848-56.

Jia, Y., Mathews, C. J. and Hanrahan, J. W. (1997). Phosphorylation by protein kinase C is required for acute activation of cystic fibrosis transmembrane conductance regulator by protein kinase A. *J Biol Chem* **272**, 4978-84.

Jourdan, N., Brunet, J. P., Sapin, C., Blais, A., Cotte-Laffitte, J., Forestier, F., Quero, A. M., Trugnan, G. and Servin, A. L. (1998). Rotavirus infection reduces sucrase-isomaltase expression in human intestinal epithelial cells by perturbing protein targeting and organization of microvillar cytoskeleton. *J Virol* **72**, 7228-36.

Jourdan, N., Maurice, M., Delautier, D., Quero, A. M., Servin, A. L. and Trugnan, G. (1997). Rotavirus is released from the apical surface of cultured human intestinal cells through nonconventional vesicular transport that bypasses the Golgi apparatus. *J Virol* **71**, 8268-78.

Kandil, H. M., Berschneider, H. M. and Argenzio, R. A. (1994). Tumour necrosis factor alpha changes porcine intestinal ion transport through a paracrine mechanism involving prostaglandins. *Gut* **35**, 934-40.

Karam, S. M. (1999). Lineage commitment and maturation of epithelial cells in the gut. *Front Biosci* **4**, D286-98.

Karczewski, J. and Groot, J. (2000). Molecular physiology and pathophysiology of tight junctions III. Tight junction regulation by intracellular messengers: differences in response within and between epithelia. *Am J Physiol Gastrointest Liver Physiol* **279**, G660-5.

Keljo, D. J., MacLeod, R. J., Perdue, M. H., Butler, D. G. and Hamilton, J. R. (1985). D-Glucose transport in piglet jejunal brush-border membranes: insights from a disease model. *Am J Physiol* **249**, G751-60.

Kockerling, A. and Fromm, M. (1993). Origin of cAMP-dependent Cl- secretion from both crypts and surface epithelia of rat intestine. *Am J Physiol* **264**, C1294-301.

Lee, C. N., Wang, Y. L., Kao, C. L., Zao, C. L., Lee, C. Y. and Chen, H. N. (2000). NSP4 gene analysis of rotaviruses recovered from infected children with and without diarrhea. *J Clin Microbiol* **38**, 4471-7.

Lee, W. S., Kanai, Y., Wells, R. G. and Hediger, M. A. (1994). The high affinity Na+/ glucose cotransporter. Re-evaluation of function and distribution of expression. *J Biol Chem* **269**, 12032-9.

Lentz, M. J. (1995). Molecular aspects of hydrolysis and absorption. *Am J Clin Nutr* **61**, 946S.

Lomax, R. B., Warhurst, G. and Sandle, G. I. (1996). Characteristics of two basolateral potassium channel populations in human colonic crypts. *Gut* **38**, 243-7.

Lundgren, O. (1992). Neuroimmune modulation of epithelial function. An overview. *Ann N Y Acad Sci* **664**, 309-24.

Lundgren, O., Peregrin, A. T., Persson, K., Kordasti, S., Uhnoo, I. and Svensson, L. (2000). Role of the enteric nervous system in the fluid and electrolyte secretion of rotavirus diarrhea [see comments]. *Science* **287**, 491-5.

Lundgren, O. and Svensson, L. (2001). Pathogenesis of rotavirus diarrhea. *Microb Infect* **3**, 1145-56.

Lytle, C. and Forbush, B., 3rd (1992). The Na-K-Cl cotransport protein of shark rectal gland. II. Regulation by direct phosphorylation. *J Biol Chem* **267**, 25438-43.

Lytle, C. and Forbush, B., 3rd (1996). Regulatory phosphorylation of the secretory Na-K-Cl cotransporter: modulation by cytoplasmic Cl. *Am J Physiol* **270**, C437-48.

Ma, T. and Verkman, A. S. (1999). Aquaporin water channels in gastrointestinal physiology. *J Physiol* **517**, 317-26.

Madara, J. L. (1999). Epithelia: Biologic Principles of Organization. In: *Textbook of Gastroenterology*, edited by T. Yamada, D. H. Alpers, L. Laine, C. Owyang and D. W. Powell, pp 141-156, Lippincott Williams and Wilkins: Philadelphia.

Madara, J. L. and Trier J. S. (1994). The functional morphology of the mucosa of the small intestine. In: *Physiology of the Gastrointestinal Tract*, edited by L. R. Johnson. pp. 1209-1251. Raven Press: New York.

Mailliard, M. E., Stevens, B. R. and Mann, G. E. (1995). Amino acid transport by small intestinal, hepatic, and pancreatic epithelia. *Gastroenterology* **108**, 888-910.

Matthews, J. B., Awtrey, C. S. and Madara, J. L. (1992). Microfilament-dependent activation of Na+/K+/2Cl- cotransport by cAMP in intestinal epithelial monolayers. *J Clin Invest* **90**, 1608-13.

Matthews, J. B., Smith, J. A. and Hrnjez, B. J. (1997). Effects of F-actin stabilization or disassembly on epithelial Cl⁻ secretion and Na-K-2Cl cotransport. *Am J Physiol* **272**, C254-62.

Matthews, J. B., Smith, J. A. and Nguyen, H. (1995). Modulation of intestinal chloride secretion at basolateral transport sites: opposing effects of cyclic adenosine monophosphate and phorbol ester. *Surgery* **118**, 147-52; discussion 152-3.

McCarty, N. A. (2000). Permeation through the CFTR chloride channel. *J Exp Biol* **203**, 1947-62.

McEwan, G. T., Hirst, B. H. and Simmons, N. L. (1994). Carbachol stimulates Cl⁻ secretion via activation of two distinct apical Cl⁻ pathways in cultured human T84 intestinal epithelial monolayers. *Biochim Biophys Acta* **1220**, 241-7.

McRoberts, J. A., Beuerlein, G. and Dharmsathaphorn, K. (1985). Cyclic AMP and Ca^{2+}-activated K^+ transport in a human colonic epithelial cell line. *J Biol Chem* **260**, 14163-72.

Merlin, D., Jiang, L., Strohmeier, G. R., Nusrat, A., Alper, S. L., Lencer, W. I. and Madara, J. L. (1998). Distinct Ca^{2+}- and cAMP-dependent anion conductances in the apical membrane of polarized T84 cells. *Am J Physiol* **275**, C484-95.

Michelangeli, F., Liprandi, F., Chemello, M. E., Ciarlet, M. and Ruiz, M. C. (1995). Selective depletion of stored calcium by thapsigargin blocks rotavirus maturation but not the cytopathic effect. *J Virol* **69**, 3838-47.

Michelangeli, F., Ruiz, M. C., del Castillo, J. R., Ludert, J. E. and Liprandi, F. (1991). Effect of rotavirus infection on intracellular calcium homeostasis in cultured cells. *Virology* **181**, 520-527.

Montrose, M. H., Barrett, K. E. and Keely, S. J. (1999). Electrolyte secretion and absorption: small intestine and colon. In: *Textbook of Gastroenterology*, edited by T. Yamada, D. H. Alpers, L. Laine, C. Owyang and D. W. Powell. pp 320-354. Lippincott Williams and Wilkins: Philadelphia.

Moody, F. G. and Durbin, R. P. (1969). Water flow induced by osmotic and hydrostatic pressure in the stomach. *Am J Physiol* **207**, 255-261.

Moon, H. W. (1994). Pathophysiology of viral diarrhea. In *Viral Infections of the Gastrointestinal Tract*, 2nd ed., edited by A. Z. Kapikian. pp 27-52. Marcel Dekker, New York.

Morris, A. P. (1999). The regulation of epithelial cell cAMP- and calcium-dependent chloride channels. *Adv Pharmacol* **46**, 209-51.

Morris, A. P., Cunningham, S. A., Benos, D. J. and Frizzell, R. A. (1992). Cellular differentiation is required for cAMP but not Ca^{2+}-dependent Cl⁻ secretion in colonic epithelial cells expressing high levels of cystic fibrosis transmembrane conductance regulator. *J Biol Chem* **267**, 5575-83.

Morris, A. P. and Estes, M. K. (2001). Microbes and microbial toxins: paradigms for microbial-mucosal interactions. VIII. Pathological consequences of rotavirus infection and its enterotoxin. *Am J Physiol Gastrointest Liver Physiol* **281**, G303-10.

Morris, A. P. and Frizzell, R. A. (1993). Ca^{2+}-dependent Cl⁻ channels in undifferentiated human colonic cells (HT-29). II. Regulation and rundown. *Am J Physiol* **264**, C977-85.

Morris, A. P., Scott, J. K., Ball, J. M., Zeng, C. Q., O'Neal, W. K. and Estes, M. K. (1999). NSP4 elicits age-dependent diarrhea and Ca^{2+}-mediated I(-) influx into intestinal crypts of CF mice. *Am J Physiol* **277**, G431-G444.

Naftalin, R. J. and Pedley, K. C. (1999). Regional crypt function in rat large intestine in relation to fluid absorption and growth of the pericryptal sheath. *J Physiol* **514**, 211-27.

Naftalin, R. J., Zammit, P. S. and Pedley, K. C. (1995). Concentration polarization of fluorescent dyes in rat descending colonic crypts: evidence of crypt fluid absorption. *J Physiol* **487**, 479-95.

Naftalin, R. J., Zammit, P. S. and Pedley, K. C. (1999). Regional differences in rat large intestinal crypt function in relation to dehydrating capacity *in vivo*. *J Physiol* **514**, 201-10.

Newton, K., Meyer, J. C., Bellamy, A. R. and Taylor, J. A. (1997). Rotavirus nonstructural glycoprotein NSP4 alters plasma membrane permeability in mammalian cells. *J Virol* **71**, 9458-65.

Nusrat, A., Turner, J. R. and Madara, J. L. (2000). Molecular physiology and pathophysiology of tight junctions. IV. Regulation of tight junctions by extracellular stimuli: nutrients, cytokines, and immune cells. *Am J Physiol Gastrointest Liver Physiol* **279**, G851-7.

Obert, G., Peiffer, I. and Servin, A. L. (2000). Rotavirus-induced structural and functional alterations in tight junctions of polarized intestinal Caco-2 cell monolayers. *J Virol* **74**, 4645-51.

Oka, T., Nakagomi, T. and Nakagomi, O. (2001). A lack of consistent amino acid substitutions in NSP4 between rotaviruses derived from diarrheal and asymptomatically-infected kittens. *Microbiol Immunol* **45**, 173-7.

Osborne, M. P., Haddon, S. J., Worton, K. J., Spencer, A. J., Starkey, W. G., Thornber, D., Stephen, J. (1991). Rotavirus-induced changes in the microcirculation of intestinal villi of neonatal mice in relation to the induction and persistence of diarrhea. *J Pediatr Gastroenterol Nutr* **12**, 111-20.

Pappenheimer, J. R., Dahl, C. E., Karnovsky, M. L. and Maggio, J. E. (1994). Intestinal absorption and excretion of octapeptides composed of D amino acids. *Proc Natl Acad Sci USA* **91**, 1942-5.

Perez, J. F., Chemello, M. E., Liprandi, F., Ruiz, M. C. and Michelangeli, F. (1998). Oncosis in MA104 cells is induced by rotavirus infection through an increase in intracellular Ca^{2+} concentration. *Virology* **252**, 17-27.

Perez, J. F., Ruiz, M. C., Chemello, M. E. and Michelangeli, F. (1999). Characterization of a membrane calcium pathway induced by rotavirus infection in cultured cells. *J Virol* **73**, 2481-90.

Perez, J. F., Ruiz, M. C. and Michelangeli, F. (2001). Simultaneous measurement and imaging of intracellular Ca^{2+} and H^+ transport in isolated rabbit gastric glands. *J Physiol* **537**, 735-45.

Podolsky, D. K. (1993). Regulation of intestinal epithelial proliferation: a few answers, many questions. *Am J Physiol* **264**, G179-86.

Powell, D. W., Mifflin, R. C., Valentich, J. D., Crowe, S. E., Saada, J. I. and West, A. B. (1999). Myofibroblasts. I. Paracrine cells important in health and disease. *Am J Physiol* **277**, C1-9.

Rhoads, J. M., MacLeod, R. J. and Hamilton, J. R. (1989). Diminished brush border membrane Na-dependent L-alanine transport in acute viral enteritis in piglets. *J Pediatr Gastroenterol Nutr* **9**, 225-31.

Ruiz, M. C., Cohen, J. and Michelangeli, F. (2000). Role of Ca^{2+} in the replication and pathogenesis of rotavirus and other viral infections. *Cell Calcium* **28**, 137-149.

Saif, L. J. (1999). Comparative pathogenesis of enteric viral infections of swine. *Adv Exp Med Biol* **473**, 47-59.

Sasaki, S., Horie, Y., Nakagomi, T., Oseto, M. and Nakagomi, O. (2001). Group C rotavirus NSP4 induces diarrhea in neonatal mice. *Arch Virol* **146**, 801-6.

Sears, C. L. (2000). Molecular physiology and pathophysiology of tight junctions V. assault of the tight junction by enteric pathogens. *Am J Physiol Gastrointest Liver Physiol* **279**, G1129-34.

Sears, C. L. and Kaper, J. B. (1996). Enteric bacterial toxins: mechanisms of action and linkage to intestinal secretion. *Microbiol Rev* **60**, 167-215.

Shepherd, R. W., Gall, D. G., Butler, D. G. and Hamilton, J. R. (1979). Determinants of diarrhea in viral enteritis. The role of ion transport and epithelial changes in the ileum in transmissible gastroenteritis in piglets. *Gastroenterology* **76**, 20-4.

Singh, S. K., Binder, H. J., Boron, W. F. and Geibel, J. P. (1995). Fluid absorption in isolated perfused colonic crypts. *J Clin Invest* **96**, 2373-9.

Stephen, J. (1989). Functional abnormalities in the intestine. In: *Viruses and the Gut*, edited by M. J. G. Farthing. pp 19-44. Swan Press, London.

Stephen, J. and Osborne M. P. (1988). Pathophysiological mechanisms in diarrhoeal disease. In: *Bacterial Infections of Respiratory and Gastrointestinal Mucosae*, edited by W. Donachie, E. Griffiths and J. Stephen, pp 149-72. IRL Press, Oxford.

Stewart, C. P. and Turnberg, L. A. (1989). A microelectrode study of responses to secretagogues by epithelial cells on villus and crypt of rat small intestine. *Am J Physiol* **257**, G334-43.

Stintzing, G., Johansen, K., Magnusson, K. E., Svensson, L. and Sundqvist, T. (1986). Intestinal permeability in small children during and after rotavirus diarrhoea assessed with different-size polyethyleneglycols (PEG 400 and PEG 1000). *Acta Paediatr Scand* **75**, 1005-9.

Svensson, L., Finlay, B. B., Bass, D., von Bonsdorff, C. H. and Greenberg, H. B. (1991). Symmetric infection of rotavirus on polarized human intestinal epithelial (Caco-2) cells. *J Virol* **65**, 4190-7.

Tafazoli, F., Zeng, C. Q., Estes, M. K., Magnusson, K. E. and Svensson, L. (2001). NSP4 Enterotoxin of rotavirus induces paracellular leakage in polarized epithelial cells. *J Virol* **75**, 1540-6.

Taylor, C. J., Baxter, P. S., Hardcastle, J. and Hardcastle, P. T. (1988). Failure to induce secretion in jejunal biopsies from children with cystic fibrosis. *Gut* **29**, 957-62.

Telch, J., Shepherd, R. W., Butler, D. G., Perdue, M., Hamilton, J. R. and Gall, D.

G. (1981). Intestinal glucose transport in acute viral enteritis in piglets. *Clin Sci (Lond)* **61**, 29-34.

Thiagarajah, J. R., Pedley, K. C. and Naftalin, R. J. (2001). Evidence of amiloride-sensitive fluid absorption in rat descending colonic crypts from fluorescence recovery of FITC-labelled dextran after photobleaching. *J Physiol* **536**, 541-53.

Tian, P., Ball, J. M., Zeng, C. Q. and Estes, M. K. (1996). The rotavirus nonstructural glycoprotein NSP4 possesses membrane destabilization activity. *J Virol* **70**, 6973-81.

Tian, P., Estes, M. K., Hu, Y., Ball, J. M., Zeng, C. Q. and Schilling, W. P. (1995). The rotavirus nonstructural glycoprotein NSP4 mobilizes Ca^{2+} from the endoplasmic reticulum. *J Virol* **69**, 5763-72.

Tian, P., Hu, Y., Schilling, W. P., Lindsay, D. A., Eiden, J. and Estes, M. K. (1994). The nonstructural glycoprotein of rotavirus affects intracellular calcium levels. *J.Virol.* **68**, 251-257.

Torchia, J., Lytle, C., Pon, D. J., Forbush, B., 3rd and Sen, A. K. (1992). The Na-K-Cl cotransporter of avian salt gland. Phosphorylation in response to cAMP-dependent and calcium-dependent secretogogues. *J Biol Chem* **267**, 25444-50.

Turner, J. R. and Madara, J. L. (1995). Physiological regulation of intestinal epithelial tight junctions as a consequence of Na^+-coupled nutrient transport. *Gastroenterology* **109**, 1391-6.

Vajanaphanich, M., Schultz, C., Tsien, R. Y., Traynor-Kaplan, A. E., Pandol, S. J. and Barrett, K. E. (1995). Cross-talk between calcium and cAMP-dependent intracellular signaling pathways. Implications for synergistic secretion in T84 colonic epithelial cells and rat pancreatic acinar cells. *J Clin Invest* **96**, 386-93.

Vandorpe, D. H., Shmukler, B. E., Jiang, L., Lim, B., Maylie, J., Adelman, J. P., de Franceschi, L., Cappellini, M. D., Brugnara, C. and Alper, S. L. (1998). cDNA cloning and functional characterization of the mouse Ca^{2+}-gated K^+ channel, mIK1. Roles in regulatory volume decrease and erythroid differentiation. *J Biol Chem* **273**, 21542-53.

Ward, R. L., Mason, B. B., Bernstein, D. I., Sander, D. S., Smith, V. E., Zandle, G. A. and Rappaport, R. S. (1997). Attenuation of a human rotavirus vaccine candidate did not correlate with mutations in the NSP4 protein gene. *J Virol* **71**, 6267-70.

Wright, E. M., Hirsch, J. R., Loo, D. D. and Zampighi, G. A. (1997). Regulation of Na^+/glucose cotransporters. *J Exp Biol* **200**, 287-93.

Wright, E. M. and Loo, D. D. (2000). Coupling between Na^+, sugar, and water transport across the intestine. *Ann N Y Acad Sci* **915**, 54-66.

Zhang, M., Zeng, C. Q., Dong, Y., Ball, J. M., Saif, L. J., Morris, A. P. and Estes, M. K. (1998). Mutations in rotavirus nonstructural glycoprotein NSP4 are associated with altered virus virulence. *J Virol* **72**, 3666-72.

Zhang, M., Zeng, C. Q., Morris, A. P. and Estes, M. K. (2000). A functional NSP4 enterotoxin peptide secreted from rotavirus-infected cells. *J Virol* **74**, 11663-70.

Viral Gastroenteritis
U. Desselberger and J. Gray (editors)

51

I, 3.   The enteric nervous system and infectious diarrhea

Ove Lundgren[a] and Lennart Svensson[b]

[a]*Department of Physiology, Sahlgrenska Academy, Göteborg University, Box 432, S-405 30, Göteborg, Sweden*
[b]*Department of Virology, Swedish Institute for Infectious Disease Control, S-171 82 Solna, Sweden*

Introduction

It has been well established that autonomic nerve fibers control smooth muscles in blood vessels and in various hollow organs such as the gastrointestinal tract and the urinary bladder. The autonomic nervous control of epithelia is less well known and less well studied. However, a large number of histochemical reports on autonomic nerves have in great detail described autonomic fibres innervating epithelial cells. The mucosal lining of the gastrointestinal tract, for example, is provided with an extensive nervous supply from the enteric nervous system (ENS; Furness and Costa, 1987). During the last two decades experimental evidence for an involvement of these nerves in the pathophysiology of diarrhea has accumulated.

The present chapter is intended to provide background knowledge about the ENS and to discuss the role of ENS in secretory states of the small intestine. The chapter is organized in the following way: In the first part a general overview of the anatomy and physiology of the ENS is presented. This is followed by a description of the experimental evidence for the involvement of ENS in secretory states of the gut, primarily in cholera toxin-induced secretion which is the most thoroughly investigated secretory state. The involvement of ENS in rotavirus diarrhea is then discussed. Finally, the involvement of the ENS in diarrhea pathophysiology opens up new potential sites of action for drugs in the treatment of intestinal secretory states.

General aspects on the ENS

Morphological considerations

The ENS represents one of the major parts of the autonomic nervous system (ANS). This was recognized already by Langley (1921) in his classical monograph in which the ANS was divided into three major parts: the sympathetic, the parasympathetic and the enteric nervous systems. This view is again gaining acceptance since both morphological and functional studies of the ENS clearly suggest that the ENS can function

52

as an independent part of the ANS. The morphological evidence for such an independent function is rather clear-cut. There are almost as many neurons in the ENS as in the spinal cord (Furness and Costa, 1980). This enormous number of neurons in the gastrointestinal wall is controlled by very few efferent fibres from the central nervous system. In fact, there is on average one efferent nerve fibre for 300 enteric neurons (Agostoni *et al.*, 1956; Kuo *et al.*, 1982). The functional evidence for an independent role of the ENS is also well established. It has been known for some time that the intestinal peristaltic reflex functions in the absence of any connections with the central nervous system.

The ENS is mainly composed of two major nerve plexus (Fig. 1), the myenteric (between the two muscle layers) and the submucosal, and of their interconnections. In addition, a spare subserous plexus is found in the mesentery and on the outside of the muscle layer. Finally, nerves and cell bodies of the mucosa form a mucosal plexus. Most of the neurons of the ENS are confined to the gastrointestinal wall, but extrinsic (afferent or efferent) neurons are also found in the ENS.

The anatomical arrangement of the neurons of the ENS is, generally speaking, simple. Most myenteric neurons send projections to other myenteric ganglia or to the smooth muscles, while most submucous neurons project to other submucous neurons and the mucosa/submucosa. There is, however, both morphological and functional evidence that the two plexus are interconnected, myenteric neurons making contact with neurons of the submucous plexus and *vice versa*, as will be discussed later.

The individual nerve cells of the nerve plexus of the ENS were originally described by Dogiel (1895). He described three types of neurons. Type I cells are characterized by numerous short processes of irregular caliber which branch shortly after arising from the cell body ("cogwheel" appearance). One single, slender axon of uniform caliber leaves the cell body to enter one of the fasciculi of the plexus. Type II cells have a smooth soma and are provided with several long processes. They reach beyond the ganglion of origin and can be traced for considerable distances (up to 20 mm). Type III neurons, finally, have dendrites of intermediate length terminating within the same or an adjacent enteric ganglion.

As the neurons approach their effector cells they form a plexus (often called the ground plexus) together with other neurons, i.e. it is no longer possible to follow a specific nerve fibre. The ground plexus is easily demonstrated in the vascular wall where the innervation forms a plexus situated at the outer boundary between media and adventitia. Hence, the anatomical relationship between autonomic nerves and the cells controlled by the nerves is much less defined than the nervous control of skeletal muscle where one particular nerve fibre makes contact with a particular muscle fiber at the motor-end plate. In the gastrointestinal tract bundles of smooth muscle cells are controlled by a ground plexus. Similarly, the intestinal epithelium is controlled by a ground plexus situated just underneath it. The nerve fibres of the ground plexus have a beaded appearance. The beads contain varicosities in which the neurotransmitters are located and from which they are released into the so-called synaptic cleft.

The neurotransmitter released into the synaptic cleft reaches a comparatively high concentration of around 10^{-6} M (Ljung, 1970) which is decreased by several mechanisms.

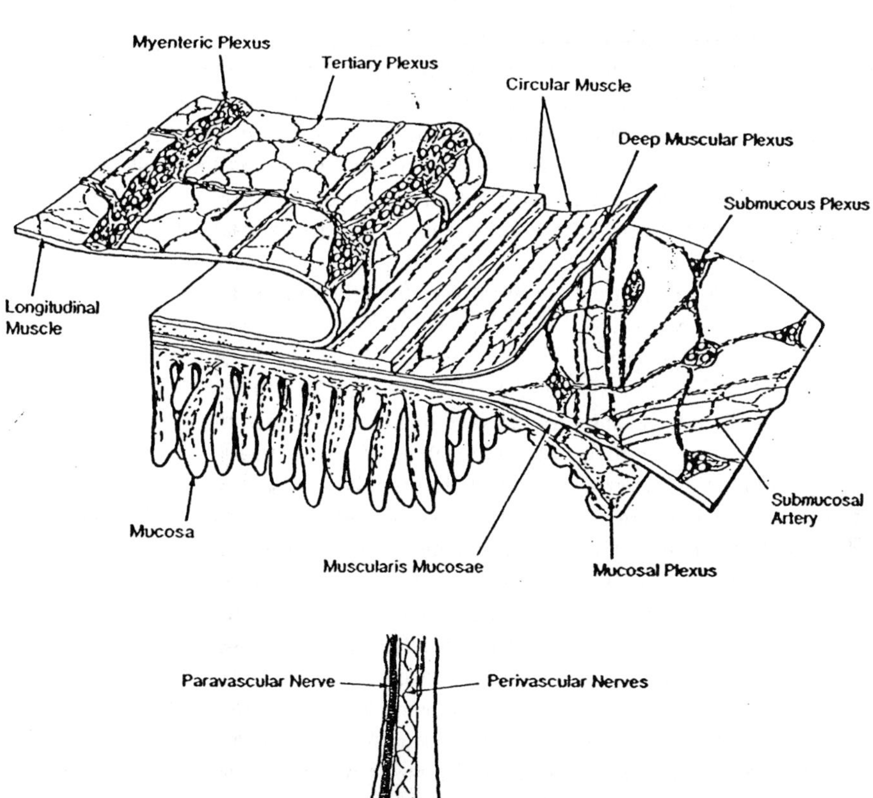

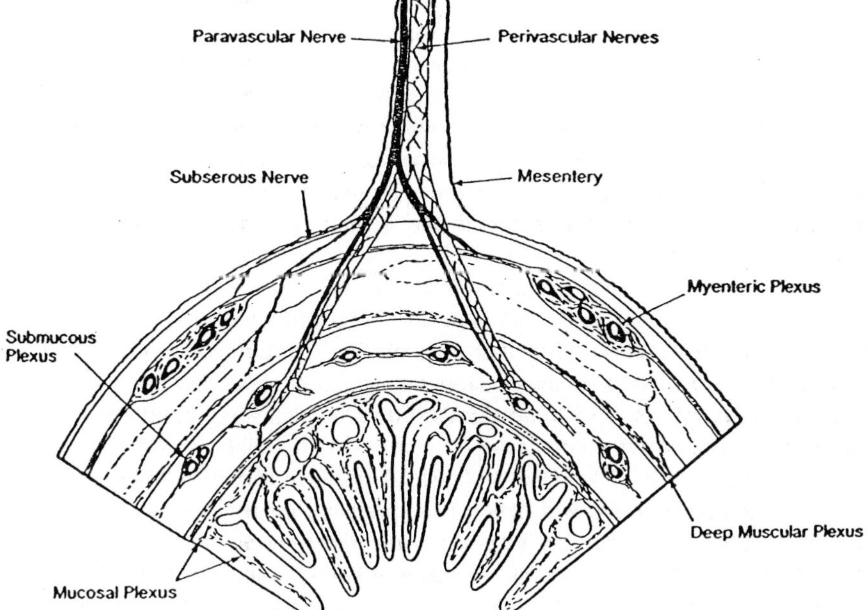

Fig. 1. The anatomical arrangement of the nerves of the small intestine. (From Furness and Costa, 1987; by permission).

54

The transmitter attaches to the appropriate receptor on the postsynaptic plasma membrane. The transmitter can also be degraded by enzymes in the synaptic cleft, as will be discussed below in relation to the treatment of rotavirus diarrhea. In the case of adrenergic (neuron with noradrenaline as neurotransmitter) and cholinergic (neuron with acetylcholine as neurotransmitter) neurons it is well established that there are re-uptake mechanisms that transport the transmitter "back" into the neuron from which it was previously released. Finally, there are receptors on the presynaptic membrane via which a negative feedback on the release of the transmitter is exerted ("autoregulation on neurotransmitter release"). This implies that an increased concentration of a transmitter in the synaptic cleft diminishes the release of that particular transmitter.

Histochemistry of the ENS

Up to the 1970`s it was generally believed that the neurons of the ANS including ENS either released acetylcholine or noradrenaline. With the advent of specific blocking agents for adrenergic and cholinergic receptors it became apparent that the ANS must contain neurons that release neurotransmitters other than those mentioned earlier. During the last 30 years, an increasing number of peptides has been isolated from the gastrointestinal tract and characterized biochemically. These peptides have been localized to the ENS and/or to certain epithelial cells of the gastrointestinal mucosa (see for example, Furness and Costa, 1987; Ekblad et al., 1991; McConalague and Furness, 1994). The polypeptides located in nerves have been ascribed a putative neurotransmitter function, which further underlines the degree of complexity of the ENS.

An overview of the neurotransmitters identified so far is given in Table 1. In the guinea pig, which is the most thoroughly investigated animal with regard to the histochemistry of ENS, it has been shown that the neurons innervating the intestinal mucosa can be divided into two major groups, cholinergic and non-cholinergic neurons.

It is now well established that neurons may contain more than one neurotransmitter (colocalization). In the ENS the number of putative transmitters in a single neuron is sometimes very high, some neurons containing up to seven putative transmitters. As pointed out above the ENS control of mucosal functions is to a large extent exerted by neurons with their somas in the submucous plexus. The most common among these neurons is the DYN/GAL/VIP neuron (for explanation of abbreviations, see Table 1) which constitutes 45% of all submucous neurons in the guinea pig small intestine. The corresponding figures for the other types of neurons are ChAT/DYN/SP: 10%; ChAT/CCK/CGRP/DYN/GAL/NPY/SOM: 30%; ChAT: 15% (Bornstein and Furness, 1988). ChAT (choline acetyltransferase) denotes the probable presence of acetylcholine in a neuron. It should be underlined that there may exist species differences (see e.g. Ekblad et al., 1991; McConalague and Furness, 1994). Morphologically, the neurons of the submucous plexus are, with one exception, monopolar, i.e. they have one long axon (Dogiel type 1). The ChAT/DYN/SP neuron, on the other hand, has two long axons (Dogiel type 2) connecting the mucosa with neurons in the myenteric plexus, the soma of the neuron being located in the submucosa. It is possible that this neuron is sensory, conveying afferent impulses from the mucosa to the myenteric plexus, a proposal also supported by electrophysiological observations (Bornstein *et al.*, 1989; Bertrand *et al.*, 1997).

Table 1

Established and putative neurtransmitters of ENS

Established transmitters
    Acetylcholine
    Noradrenaline

Transmitter candidates
    Adenosine triphosphate (ATP)
    Calcitonin gene-related peptide (CGRP)
    Cholecystokinin-8 (CCK-8)
    Dynorphin (DYN)
    Enkephalin (ENK)
    Galanin (GAL)
    Gamma aminobutyric acid (GABA)
    Gastrin releasing peptide (GRP)
    Neurokinin A (NKA; NK_2 receptor)
    Neurokinin B (NKB; NK_3 receptor)
    Neuromedin U
    Neuropeptide Y (NPY)
    Neurotensin (NT)
    Nitrous oxide (NO)
    Pituitary adenylyl cyclase activating peptide (PACAP)
    Serotonin
    Somatostatin (SOM)
    Substance P (SP; NK_1 receptor)
    Vasoactive intestinal polypeptide (VIP)

The functional implications of the colocalization of transmitters in the ENS are not well known. In many neurons one substance seems to have a major role in transmission while others have subsidiary or modulatory roles. Experimental observations indicate that the peptidergic transmitters are released from the neuron at higher rates of firing than the nonpeptidergic cotransmitter (Lundberg, 1981). Some neurons contain transmitters that seem to have opposite effects on the effector cell. This raises the question of whether some of the immunohistochemical findings are artifacts and/or whether the techniques used are so sensitive that they detect transmitter concentrations which are functionally unimportant. Bowers (1994) has proposed that large numbers of peptides in one and the same neuron can be explained by mechanisms of gene regulation implying that peptides of no functional importance may be produced by cells.

Electrophysiology of the ENS

The technical difficulties in studying the electrophysiology of the ENS are reflected in the fact that no method as yet exists to record electrical activity from single neurons (extra- or intracellular registration) of the ENS *in vivo*. Our current knowledge of the electrophysiology of ENS neurons is therefore mainly based on *in vitro* studies performed on isolated preparations of myenteric and/or submucosal plexus of the guinea pig.

Electrophysiological investigations of the neurons of the ENS using intracellular recording electrodes have shown that there are two major types of neurons, which according to Hirst (1979) are named S (cells with synaptic input) and AH (after-hyperpolarization) cells. Although this classification may be considered as too coarse, it suffices for the present discussion of reflex control of intestinal secretion. The S cells have a comparatively low resting potential and do not exhibit any hyperpolarizing after potential. The studies by Hirst (1979) suggested that the AH cells do not obtain any input from other neurons whereas the S cells do. Furthermore, the AH cells have a morphological appearance of the Dogiel type 2. These observations indicate that the AH cells may have a sensory function in the ENS as has also been experimentally verified (Bertrand *et al.*, 1997).

The neurons of the ENS "talk" with each other with the same "language" as the one used in the central nervous system. Postsynaptically, two types of changes of membrane potential may be recorded, namely excitatory or inhibitory potentials (excitatory postsynaptic potentials, EPSP, and inhibitory postsynaptic potentials, IPSP, respectively). EPSP denotes a depolarization of the membrane potential which, if large enough, may trigger an action potential. EPSPs can be fast or slow. The fast EPSP in the ENS is generally believed to be mediated by the action of acetylcholine on the postsynaptic membrane via a nicotinic receptor. IPSPs are associated with a hyperpolarization of the membrane potential of the postsynaptic neuron. The neurotransmitters causing IPSP in enteric neurons are not established.

Circuitry of the ENS

From a functional point of view, reflexes in the ENS can be divided into two major groups: axon reflexes or reflexes with neurons confined to the gastrointestinal wall (below named "intramural reflexes"). Axon reflexes are comprised of thin afferent nerve fibers connected to the central nervous system and often activated by mucosal noxious stimuli. The afferent fibre branches to make contact with an effector cell (epithelium, vascular smooth muscle) or another neuron. Figures 2 and 3 illustrate the two types of enteric nervous reflexes.

Combined morphological and electrophysiological studies have expanded our knowledge concerning the neuronal circuitry of the ENS (Bornstein and Furness, 1988; Bornstein *et al.*, 1984, 1986, 1988, 1989; Hodgkiss and Lees, 1983; Katayama *et al.*, 1986; Bertrand *et al.*, 1997). Interesting functional differences between the cholinergic and non-cholinergic submucosal neurons have then been revealed, the input to the two types of neurons being quite different. To exemplify, the extrinsic sympathetic adrenergic nerve fibres seem to make contact with the non-cholinergic but not with the cholinergic mucosal neurons. With regard to the enteric control of the epithelium, the non-adrenergic, non-cholinergic neurons in the submucosa (a DYN/GAL/VIP neuron) are of particular interest, since fluid and electrolyte transport may be controlled by such

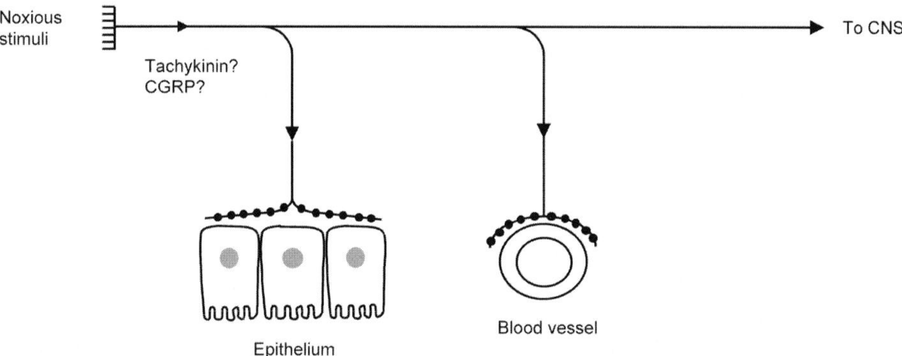

Fig. 2. Schematic illustration of a possible neural arrangement for axon reflexes in the small intestine. The axon branching off from the afferent nerve fibre makes direct contact with effector cells. CNS: central nervous system; CGRP: calcitonin gene related polypeptide.

neurons. These neurons are influenced by cholinergic, fast EPSPs induced by nerve cells located in both the myenteric and submucosal plexus, the major portion emanating from myenteric neurons.

Intestinal secretion and ENS

The discussion below regarding the pathophysiological mechanisms underlying the intestinal fluid losses caused by various secretagogues is based on the assumption that in the normal small intestine villi absorb and crypts secrete electrolytes and fluid. There are several observations to support this view. In particular, a hyperosmolar compartment, mainly accounted for by sodium chloride, is present in the lamina propria of the upper third of the villus (Jodal and Lundgren, 1986, 1996; Lundgren, 1988).

ENS and the intestinal secretion caused by cholera toxin

The textbook view of the pathophysiology of cholera secretion is that the fluid loss from the intestinal mucosa is explained by the toxin (via increases of intracellular cAMP), inhibiting the absorption of sodium chloride in the villus enterocytes by interfering with the Na^+/H^+ and/or Cl^-/HCO_3^- antiports. Furthermore, the toxin evokes a chloride secretion from the crypts by its direct action on the crypt cells. This model is mainly based on *in vitro* experiments, but observations made *in vivo* argue against this view. Firstly, using labelled cholera toxin it has been shown that the toxin does not reach the crypts when administered into the intestinal lumen (Weiser and Quill, 1975; Hansson *et al.*, 1984). Secondly, a marked hyperosmolality (of about 700 mOsm) has been demonstrated in cat villi in the face of net fluid secretion induced by exposing the mucosa to cholera toxin (Hallbäck *et al.*, 1979). The prerequisite for a fluid uptake

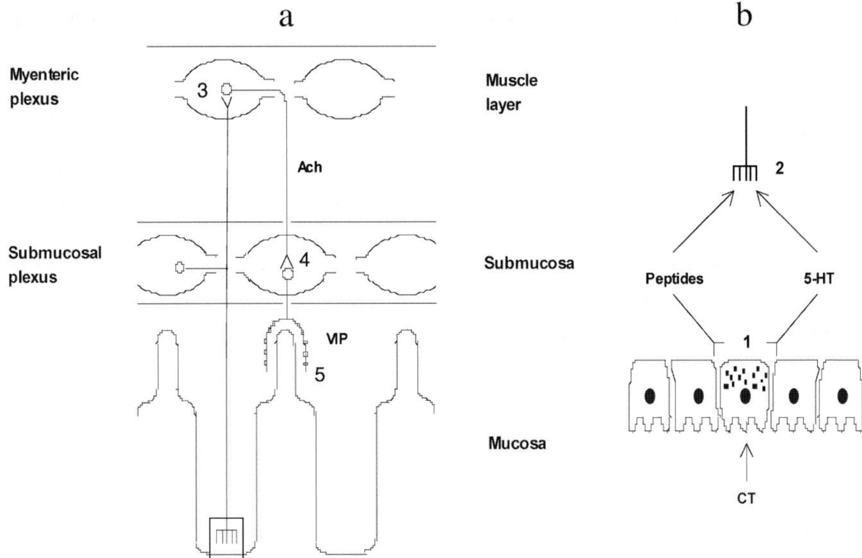

Fig. 3. Schematic illustration of the proposed model for the secretory nervous reflex in the ENS activated by exposing the intestinal mucosa to rotavirus or cholera toxin (CT). Most of the animal experimental work has been performed using CT. (a) The intramural reflex activated by the secretory agent is proposed to be made up by three neurons. The transmitter of the afferent neuron is not established. As discussed in the Text the interneuron is probably cholinergic and the efferent neuron releases vasoactive intestinal polypeptide (VIP) at the effector cell. (b). Illustration how an enterotoxin, such as CT may activate the enteric nervous system by releasing peptides and/or 5-hydroxytryptamine (5-HT) from the endocrine cells of the mucosa (indicated in Fig. as cell containing granules). It is possible that rotavirus also causes the release of amines/peptides via its effect on intracellular calcium concentration. It should be emphasized that the Fig. is highly schematic. The nerves activated by peptides and/or 5-HT are situated underneath the intestinal epithelium as depicted in more detail in Fig. 4. For a detailed discussion of the model, see Text. Numbers in the two panels indicate possible sites of pharmacological interventions discussed in Text. Ach: acetylcholine.

by villi is therefore present also in cholera. Hence, the toxin does not reach the crypts and villi are functioning more or less normally. Yet, the toxin evokes a net fluid secretion.

This paradox can be explained by the toxin activating the ENS. The experimental evidence for the involvement of ENS in the pathophysiology of cholera is extensive and has been obtained mainly from *in vivo* experiments on extrinsically denervated intestinal segments of cats and rats. The results of these experiments can be summarized as follows: (1) Tetrodotoxin (TTX; blocker of sodium channels in excitable cells) injected intraarterially to cats inhibits the intestinal fluid secretion evoked by cholera toxin (Cassuto *et al.*, 1981a). (2) Hexamethonium (nicotinic receptor antagonist) given intravenously to rats turns cholera secretion into absorption (Cassuto *et al.*, 1982a). Concomitantly, the increase of transepithelial potential difference (PD), caused by the toxin, is significantly attenuated (Tantisira *et al.*, 1990). (3) Lidocaine (local anesthetic) administered luminally or on the serosal surface reverses cholera secretion to fluid absorption and attenuates the increased PD (Cassuto *et al.*,1981a, 1983; Tantisira *et al.*, 1990). (4) Kirchgessner *et al.* (1992) monitored the activation of ENS neurons with a histochemical method using an antibody to the *fos* oncogene product the expression of which has been shown to be a marker of nervous activity. Exposing the intestinal mucosa to cholera toxin activated neurons in both the myenteric and submucosal plexuses. (5) Cholera toxin-induced secretion is accompanied by an augmented release of VIP, a neurotransmitter, into the venous effluent. Giving TTX attenuates both the VIP release and the fluid secretion (Cassuto *et al.*, 1981b). Atropine (blocker of acetylcholine receptors on effector cells) had no effect on cholera secretion (Cassuto *et al.*, 1982a), inferring that the transmitter at the effector cell probably is a peptide, presumably VIP. (6) Cholera toxin cannot elicit a secretory response in intestinal segments in which the myenteric plexus has been destroyed by exposing the serosal surface to benzalkonium chloride (Jodal *et al.*, 1993).

It has been pointed out above that there are at least two principally different types of reflexes in the ENS, axon reflexes and intramural reflexes. One way of differentiating between these two types is to perform a so-called "chronic" intestinal denervation. It consists of severing the periarterial nerves of the superior mesenteric artery. Two to four weeks later, when the nerves distal to the severing have degenerated, an acute experiment is performed on the animal. Sjöqvist (1991) showed that chronic denervation did not attenuate the effects of cholera toxin on net fluid transport in the small intestine of the rat, suggesting that the effect on fluid transport is not mediated via an axon reflex arrangement involving thin afferent nerve fibres.

Thus, the data summarized above clearly indicate that there exist intramural nervous reflexes in the small intestine which evoke fluid secretion when activated by cholera toxin. These observations made *in vivo*, together with those made on the ENS *in vitro* with *e.g.* electrophysiological techniques, are beginning to provide us with a coherent picture regarding the details of the intramural nervous reflex pathway(s) involved. A model for this is presented in Fig. 3a. It should be underlined that this model represents the simplest model that can be proposed on the basis of the current experimental observations.

According to the model of Fig. 3a the secretory reflex activated consists of three neurons. The ChAT/DYN/SP neuron mentioned earlier probably represents the sensory afferent neuron in the reflex although this is not firmly established. The effect of the nicotinic receptor antagonist hexamethonium suggests that there must be a cholinergic synapse in the secretory reflex. This is indicated by the cholinergic interneuron in the model of Fig. 3a. Finally, there are several observations to support the proposal that VIP is the neurotransmitter of the efferent neuron controlling the secretory enterocytes of the crypts.

In subsequent investigations a nervous involvement was also demonstrated for several other intestinal secretagogues including bile acid, an invasive strain of *Salmonella typhimurium* and the enterotoxins produced by *Escherichia coli*. In fact, all luminal secretagogues tested in our laboratory have been shown to activate the ENS in such a way that at least 60% of the fluid secretory response can be explained by a stimulation of the enteric nerves (Jodal and Lundgren, 1995).

The involvement of the ENS in intestinal secretory states may not only be of importance for epithelial functions but also for motility. This has been shown in experiments in which the enterotoxins produced by various bacteria have been investigated with regard to their effects on intestinal motility. For example, according to Mathias and Clench (1989) cholera toxin induces a particular motility pattern that they named "migrating action potential complexes". Functionally this pattern is very efficiently propelling the intestinal contents in an aboral direction. This motility effect of the toxin is also mediated via the ENS. Thus, the motility pattern can be attenuated by lidocaine and by nicotinic receptor blockade.

To summarize, bacterial enterotoxins produce an intestinal secretion and a propulsive motility response via an activation of the ENS. This response may be regarded as a defense mechanism against potentially harmful mucosal influence, the fluid secreted diluting the noxious agent and the increased motility propelling the intestinal contents in an aboral direction. The involvement of the ENS may also explain how enterotoxins which apparently do not reach the intestinal crypts (Weiser and Quill, 1975; Hansson *et al.*, 1984) can influence the secretory cells of the crypts.

ENS and the intestinal secretion evoked by rotavirus

The experiments briefly summarized above prompted a study to elucidate whether rotavirus-evoked fluid secretion in mice was also caused, at least in part, via an activation of the ENS. To test this, three types of experiments were performed (Lundgren *et al.*, 2000). Using an Ussing chamber [*in vitro* technique which makes it possible to measure the net potential difference (PD) across the intestinal wall established by epithelial electrolyte transport mechanisms] it was demonstrated that tetrodotoxin, lidocaine and mecamylamide (a nicotinic receptor blocker that is more lipophilic than hexamethonium) attenuated the increased PD observed in intestines exposed to virus in a dose dependent way. Similarly, in experiments in which the lumen of intact intestinal segments were perfused in an organ bath, tetrodotoxin, lidocaine and hexamethonium significantly lowered the monitored PD and often turned fluid secretion into fluid absorption in virus-infected intestines. Finally, giving lidocaine repeatedly intraperitoneally to awake mice inoculated with rotavirus significantly prevented the fecal losses of fluid. From the results obtained *in vitro* it was calculated that at least two thirds of the fluid and electrolyte secretion caused by the virus could be ascribed to an activation of the ENS.

It was pointed out above that there existed experimental evidence that enterotoxins influenced intestinal motility via the ENS. There are few studies of motility during virus diarrhea, and none of them has investigated the possible involvement of the ENS in the motility response. The reported studies suggest, however, that transit time for charcoal is increased in viral diarrhea in humans (Molla *et al.*, 1983). In line with this, Burrows and Merritt (1984) recorded an increasing number of motor activity fronts in the jejunum of neonatal pigs infected with the porcine coronavirus, transmissible gastroenteritis virus.

To summarize, several observations made during rotavirus enteritis in neonatal mice suggest that the secretory response is in part explained by an activation of the ENS. The involvement of the ENS may explain how the comparatively few cells at the villus tips infected by the virus can influence the intestinal crypts to augment their secretion of electrolytes and water. The possible involvement of the ENS in the motility response to rotavirus has not been tested.

Indirect evidence for the involvement of the ENS in human rotavirus infection comes from studies with enkephalinase inhibitors. Opiates have since long been known to inhibit intestinal fluid and electrolyte secretion as well as gut motility. The endogenous opiates, the enkephalins, are found in enteric nerves as neurotransmitters. It seems probable that there are enkephalin receptors on enteric neurons. Thus, there are several studies that demonstrate that there are enkephalin receptors on myenteric neurons the stimulation of which inhibits the nervous release of acetylcholine (see *e.g.* Nakayama *et al.*, 1990). Furthermore, indirect experimental evidence for a point of action for enkephalins in the ENS was provided by Eklund *et al.* (1988) who demonstrated that methionine-enkephalin given into the superior mesenteric artery caused a parallel reduction in cholera toxin induced net fluid secretion and VIP release into the intestinal venous effluent in cats. This nervous action of enkephalins can be enhanced by inhibitors of the enzymes degrading the enkephalins, the enkephalinases. In line with this the enkephalinase inhibitor acetorphan has been shown to attenuate acute diarrhea in children and adults (Hamaza *et al.*, 1999; Turck *et al.*, 1999; Lecomte, 2000). Finally, Salazar-Lindo *et al.* (2000) showed in a clinical trial that acetorphan (Racecadotril©) markedly inhibited stool output in young Peruvian children with rotavirus diarrhea (See also Bass, Section I, Chapter 5 of this book).

Many details regarding the ENS-linked hypothesis of rotavirus-induced fluid secretion remain to be elucidated. For example, the pharmacology of the secretory reflex involved has not been studied. The reflex of Fig. 3a is mainly derived from studies of enterotoxins and, in particular, cholera enterotoxin. Although it seems possible that rotavirus evokes intestinal fluid and electrolyte secretion via a similar nervous reflex this has to yet to be demonstrated. Another major question is how virus can activate enteric nerves. In the case of bacterial enterotoxin-evoked fluid secretion it has been proposed that enterotoxins induce the release of amines/peptides from the endocrine cells of the intestinal epithelium via their effects on intracellular second messengers. To exemplify, several lines of evidence indicate that cholera toxin causes the release of 5-hydroxytryptamine (5-HT) from the enterochromaffin cells (Jodal and Lundgren, 1995; see Fig. 3b). Alone or together, the secreted amines/peptides activate nervous dendrites located underneath the intestinal epithelium. It seems possible that rotavirus *per se* and/or the recently discovered rotavirus enterotoxin NSP4 may function in a similar manner by increasing the intracellular calcium concentration (Tian *et al.*, 1994; Ball *et al.*, 1996; Fig. 4).

Fig. 4 also illustrates another mechanism that may explain how rotavirus activates the ENS. It is now recognized that epithelial cells function as "sensors" for microorganisms. For example, when exposed to bacteria the cells release a wide range of biologically active compounds, such as cytokines, prostaglandins and nitrous oxide

(Kagnoff and Eckmann, 1997). These compounds also participate in the inflammatory response. It is established that receptors located on enteric neurons exist for some of those substances and that they, alone or together, may cause a membrane depolarization of dendrites to induce action potentials (Kirkup *et al.*, 2001). For a detailed discussion of all the mechanisms underlying rotavirus diarrhea, the reader is referred to a review by Lundgren and Svensson (2001).

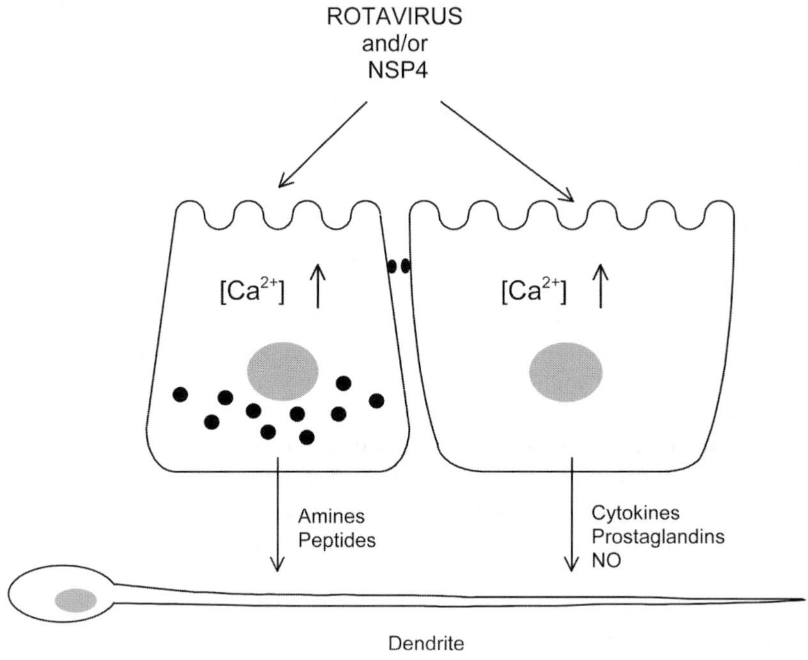

Fig. 4. Schematic illustration of how rotavirus may activate ENS. The left cell depicts an endocrine cell which under the influence of rotavirus or the rotavirus enterotoxin NSP4 may release amines and/or peptides. The right cell illustrates an enterocyte which when exposed rotavirus and/or NSP4 will to rotavirus or NSP4 may produce cytokines, prostaglandins and NO. In all cases rotavirus and/or NSP4 will effect an increase of intracellular calcium levels. It is proposed that such an increased concentration of calcium is of functional importance for the release of the different compounds. The details of the stimulated secretory nervous reflex is depicted on Fig. 3a.

Sites of action for the pharmacological treatment of diarrhea

The replacement of fluid losses in diarrhea with an oral solution containing glucose and sodium chloride represented, when introduced, a major therapeutic breakthrough (Hirschhorn *et al.*, 1968). It relies on an intact absorptive capacity of the intestinal epithelium. It would be advantageous if the oral glucose-salt solution could be combined with a drug that attenuated the intestinal secretion of fluid.

The involvement of the ENS in the pathophysiology of intestinal secretory states opens up new potential sites of actions for drugs in the treatment of diarrhea. In Fig. 3 a number of possible sites of intervention in the secretory reflex(es) are indicated by numbers. Intestinal secretion evoked by diarrheal agents that activate the ENS via the intestinal endocrine cells can be influenced by drugs that decrease the amine/peptide release from those cells (site number 1; see also Fig. 4). Thus, it has been demonstrated that fluid secretion evoked by cholera toxin or bile salt can be significantly attenuated by calcium channel blockers of the L-type via an effect on release of 5-HT from the EC cells (Timar Peregrin *et al.*,1997a,b, 1999)

Other sites for pharmacological interventions are receptor(s) on the nerve dendrites for the substances released from the endocrine cells and/or enterocytes (site number 2 in Figures 3 and 4). In line with this proposal it has been shown that 5-HT_3 receptor antagonists may attenuate cholera toxin-induced secretion (Cassuto *et al.*, 1982b; Beubler *et al.*, 1989; Sjöqvist *et al.*, 1992). According to the hypothesis illustrated in Figures 3b and 4 it may also be possible to attenuate cholera toxin and rotavirus fluid secretion using blockers of various peptide receptors as well as by blocking those biologically active compounds released from the enterocytes. This has not yet been tested experimentally.

The sites numbered 3 and 4 in Fig. 3 illustrate the possibility of interfering with synaptic transmission in a secretory reflex. The effect of nicotinic cholinergic receptor blockers on intestinal fluid secretion may be explained by such an action. The advantage of such a drug is that a nicotinic receptor seems to be involved in all secretory states studied so far. One disadvantage is that a nicotinic receptor antagonist probably inhibits intestinal motility to such an extent that the elimination of the noxious agent from the gut is prolonged. Furthermore, nicotinic receptor blockade may influence several other organ functions in the body. To exemplify, arterial blood pressure may be lowered dramatically.

It seems also possible to influence the release of synaptic transmitters via presynaptic receptors. This was discussed earlier in some detail in connection with the effect of enkephalinase inhibitors on rotavirus evoked intestinal secretion. Finally, blocking the neurotransmitter(s) at the enterocytes (number 5 in Fig. 3) is another possible way of attenuating the fluid secretion in diarrhea. In line with this VIP receptor blockers have been shown to attenuate enterotoxin-induced fluid secretion in experimental animals (Mourad and Nassar, 2000). Their clinical significance remains to be demonstrated.

The research performed in the authors' laboratories was supported by the Swedish Medical Research Council (grants 2855 and 10392). The excellent secretarial help of Ms Eva Magnusson is gratefully acknowledged.

References

Agostoni, E., Chinnok, J.E., De Burg Daly, M., Murray, J.G. (1956). Functional and histological studies to the heart, lungs and abdominal viscera in the cat. *J. Physiol.* (Lond.) **135**, 182-205.

Ball J.M., Tian P., Zeng C.Q.-Y., Morris A.P., Estes M.K. (1996). Age-dependent diarrhea induced by a rotaviral nonstructural glycoprotein. *Science* **272**, 101-104.

Bertrand, P.P., Kunze, W.A.A., Bornstein, J.C., Furness, J.B., Smith, M.L. (1997). Analysis of the responses of myenteric neurons in the small intestine to chemical stimulation of the mucosa. *Am. J. Physiol.* **273**, G422-G435.

Beubler, E., Kollar F., Saria A., Bukhave K., Rask-Madsen J. (1989): Involvement of 5-hydroxytryptamine, prostaglandin-E$_2$, and cyclic adenosine monophosphate in cholera- toxin-induced fluid secretion in the small intestine of the rat *in vivo*. *Gastroenterology* **96**, 368-376.

Bornstein, J.C., Costa, M., Furness, J.B., Lees, G.M. (1984). Electrophysiology and enkephalin immunoreactivity of identified myenteric plexus neurones of guinea-pig small intestine. *J. Physiol.* (Lond.) **351**, 313-325.

Bornstein, J.C., Costa, M., Furness, J.B. (1986). Synaptic inputs to immunohistochemically identified neurones in the submucous plexus of the guinea-pig small intestine. *J. Physiol.* (Lond.) **381**, 465-482.

Bornstein, J.C., M. Costa, Furness, J.B. (1988). Intrinsic and extrinsic inhibitory synaptic inputs to submucosal neurones of the guinea-pig small intestine. *J. Physiol.* (Lond.), **398**, 371-390.

Bornstein, J.C., Furness, J.B. (1988). Correlated electrophysiological and histochemical studies of submucosal neurons and their contribution to understanding enteric neural circuits. *J. Auton. Nerv. Syst.* **25**, 1-13.

Bornstein, J.C., Furness, J.B., Costa, M. (1989). An electrophysiological comparison of substance P-immunoreactive neurons and other neurons in the guinea-pig submucosal plexus. *J. Auton. Nerv. Syst.* **26**, 113-120.

Bowers, C.W. (1994). Superfluous neurotransmitters? *Trends Neurosci.* **17**, 315-320.

Burrows, C.F., Merritt, A.M. (1984). Influence of coronavirus (transmissible gastroenteritis) infection on jejunal myoelectrical activity of the neonatal pig. *Gastroenterology* **87**, 386-391.

Cassuto, J., Fahrenkrug, J., Jodal, M., Tuttle, R., Lundgren, O. (1981a). Release of vasoactive intestinal polypeptide from the cat small intestine exposed to cholera toxin. *Gut* **22**, 958-963.

Cassuto, J., Jodal, M., Tuttle, R., Lundgren, O. (1981b). On the role of intramural nerves in the pathogenesis of cholera toxin-induced intestinal secretion. *Scand. J. Gastroenterol.* **16**, 377-384.

Cassuto, J., Jodal, M., Lundgren, O. (1982a). The effect of nicotinic and muscarinic receptor blockade on cholera toxin induced intestinal secretion in rats and cats. *Acta Physiol. Scand.* **114**, 573-577.

Cassuto, J., Jodal, M., Tuttle R. Lundgren, O. (1982b) 5-Hydroxytryptamine and cholera secretion. *Scand. J. Gastroenterol.* **17**, 695-703

Cassuto, J., Siewert, A., Jodal, M., Lundgren, O. (1983). The involvement of intramural nerves in cholera toxin induced intestinal secretion. *Acta Physiol. Scand.* **117**, 195-202.

Dogiel, A.S. (1895). Zur Frage über die Ganglien der Darmgeflechte bei den Säugetieren. *Anat. Anzeiger* **10**, 517-528.

Ekblad, E., Håkanson, R., Sundler, F. (1991). Microanatomy and chemical coding of peptide-containing neurons in the digestive tract. In: Daniel, E.E. (ed.). *Neuropeptide function in the gastrointestinal tract.* CRC Press, Boca Raton, FA.

Eklund, S., Sjöqvist, A., Fahrenkrug, J., Jodal, M., Lundgren, O. (1988). Somatostatin and methionine-enkephalin inhibit cholera toxin-induced jejunal net fluid secretion and release of vasoactive intestinal polypeptide in the cat *in vivo.* *Acta Physiol. Scand.* **133**, 551-557.

Furness, J.B., Costa, M. (1980). Types of nerves in the enteric nervous system. *Neuroscience* **5**, 1-20.

Furness, J.B., Costa, M. (1987). *The enteric nervous system,* Churchill Livingstone, New York.

Hallbäck, D.-A., Jodal, M., Lundgren, O. (1979). Effects of cholera toxin on villous tissue osmolality and fluid and electrolyte transport in the small intestine of the cat. *Acta Physiol. Scand.* **107**, 239-249.

Hamaza, H., Khalifa, H.B., Baumer, P. Berard, H., Lecomte, J.M. (1999). Racecadotril versus placebo in the treatment of acute diarrhoea in adults. *Aliment. Pharmacol. Ther.* **13**, Suppl **6**,15-19.

Hansson, H.A., Lange, S., Lonnroth, I. (1984). Internalization *in vivo* of cholera toxin in the small intestinal epithelium of the rat. *Acta Pathol. Microbiol. Immunol. Scand.* [A] **92**, 15-21.

Hirschhorn, N., Kinzie, J.L., Sachar, D.B., Northrup, R.S., Taylor, J.O., Ahmad, S.Z., Phillips, R.A. (1968). Decrease in net stool output in cholera during intestinal perfusion with glucose-containing solutions. *N Engl J Med.* **279**, 176-181.

Hirst, G.D.S. (1979). Mechanisms of peristalsis. *Br. Med. Bull.* 35, 263-268.

Hodgkiss, J.P., Lees, G.M. (1983). Morphological studies of electrophysiologically identified myenteric plexus neurons of the guinea-pig ileum. *Neuroscience* **8**, 593-608.

Jodal, M., Holmgren, S., Lundgren, O., Sjöqvist, A. (1993). Involvement of the myenteric plexus in the cholera toxin-induced net fluid secretion in the rat small intestine. *Gastroenterology* **105**, 1286-1293.

Jodal, M., Lundgren, O. (1986). Countercurrent mechanisms in the mammalian gastrointestinal tract. *Gastroenterology* **91**, 225-241.

Jodal, M. and Lundgren, O. (1995). Neural reflex modulation of intestinal epithelial transport. In: Gaginella, T.S. (ed.) *Regulatory mechanisms in gastrointestinal function.* pp. 99-144. CRC Press, Boca Raton, FA.

Jodal, M., Lundgren, O. (1996). Do intestinal villi secrete? *Acta Physiol. Scand.* **158**, 115-118.

Kagnoff, M.F., Eckmann, L. (1997). Epithelial cells as sensors for microbial infection. *J. Clin. Invest.* **100**, 6-10.

Katayama, Y., Lees, G.M., Pearson, G.T. (1986). Electrophysiology and morphology of vasoactive-intestinal-peptide-immunoreactive neurones of the guinea-pig ileum. *J. Physiol.* (Lond.) **378**, 1-11.

Kirchgessner, A.L., Tamir, H., Gershon, M.D. (1992). Identification and stimulation by serotonin of intrinsic sensory neurons of the submucosal plexus of the guinea pig gut: Activity-induced expression of Fos immunoreactivity. *J. Neurosci.* **12**, 235-248.

Kirkup, A.J., Brundsen, A.M., Grundby, D. (2001). Receptors and transmission in the brain-gut axis: potential for novel therapies. I. Recptors on visceral afferents. *Am. J. Physiol.* **280**, G787-G794.

Kuo, D.C., Yang, G.C.H., Yamasaki, D.S., Krauthamer, G.H. (1982). A wide-field electron microscopic analysis of the fiber constituents of the major splanchnic nerve in the cat. *J. Comp. Neurol.* **210**, 49-58.

Langley, J.N. (1921). *The autonomic nervous system*, Heffner, London.

Lecomte, J.M. (2000). An overview of clinical studies with racecadotril in adults. *Int. J. Antimicrob. Agents* **14**, 81-87.

Ljung, B. (1970). Nervous and myogenic mechanisms in the control of a vascular neuroeffector system. *Acta Physiol. Scand. Suppl.* **349**, 1-68.

Lundberg, J.M. (1981). Evidence for coexistence of vasoactive intestinal polypeptide (VIP) and acetylcholine in neurons of cat exocrine glands. *Acta Physiol. Scand. Suppl.* **496**, 1-57.

Lundgren, O. (1988). Factors controlling absorption and secretion in the small intestine. In: Donachie, W., Griffiths, E., Stephen, J. (eds.). *Bacterial infections of respiratory and gastrointestinal mucosae.* pp. 97-112. IRL Press, Oxford.

Lundgren, O., Svensson, L. (2001). Pathogenesis of rotavirus diarrhea. *Microb. Infect.* **3**, 1145-1156.

Lundgren, O., Timar Peregrin, A., Persson, K., Kordasti, S., Uhnoo, I., Svensson, L. (2000). Role of the enteric nervous system in the fluid and electrolyte secretion of rotavirus diarrhea. *Science* **287**, 491-495.

Mathias, J.R., Clench, M.H. (1989). Alterations of small intestine motility by bacteria and their enterotoxins. In: Wood, J.D. (ed.) *Handbook of Physiology, sect. 6. The gastrointestinal system, volume 1.* pp. 1301-1334. American Physiology Society, Bethesda, MD.

McConalague, K., Furness, J.B. (1994). Gastrointestinal neurotransmitters. In: Fuller, P., Shulkes, A. (eds.) *The gut as an endocrine organ.* pp. 51-76. Baillière's Clin. Endocrinol. Metabol. Baillière Tindall, London.

Molla, A., Molla, A.M., Sarker, S.A., Khatun, M. (1983). Whole-gut transit time and its relationship to absorption of macronutrients during diarrhoea and after recovery. *Scand. J. Gastroenterol.* **18**, 537-543.

Mourad, F.H., Nassar C.F. (2000). Effect of vasoactive intestinal polypeptide (VIP) antagonism on rat jejunal fluid and electrolyte secretion induced by cholera and *Escherichia coli* enterotoxins. *Gut* **47**, 382-386.

Nakayama S., Taniyama, K., Matsuyama, S., Ohgushi, N., Tsunekawa, K., Tanaka, C. (1990). Regulatory role of enteric mu and kappa opioid receptors in the release of acetylcholine and norepinephrine from guinea pig ileum. *J. Pharmacol. Exp. Ther.* **254**, 792-798.

Salazar-Lindo, E., Santisteban-Ponce, J., Chea-Woo, E., Gutierrez, M. (2000). Racecadotril in the treatment of acute watery diarrhea in children. *N. Engl. J. Med.* **343**, 463-467.

Sjöqvist, A. (1991). Interaction between antisecretory opioid and sympathetic mechanisms in the rat small intestine. *Acta Physiol. Scand.* **142**, 127-132.

Sjöqvist, A., Cassuto J., Jodal M., Lundgren O. (1992): Actions of serotonin antagonists on cholera-toxin-induced intestinal fluid secretion. *Acta Physiol. Scand.* **145**, 229-237.

Tantisira, M.H., Fändriks, L., Jönson, C., Jodal, M., Lundgren, O. (1990). Studies of cholera toxin-induced changes of alkaline secretion and transepithelial potential difference in the rat intestine *in vivo*. Acta Physiol. Scand. 138, 75-84.

Tian P., Hu Y., Schilling W.P., Lindsay D.A., Eiden J., Estes M.K. (1994) The nonstructural glycoprotein of rotavirus affects intracellular calcium levels. *J. Virol.* **68**, 251-257.

Timar Peregrin, A., Ahlman, H. Jodal, M., Lundgren. O. (1997b). Effects of calcium channel blockade on intestinal fluid secretion: sites of action. *Acta Physiol. Scand.* **160**, 379-386.

Timar Peregrin A., Ahlman, H., Jodal M., Lundgren O. (1999). Involvement of serotonin and calcium channels in the intestinal fluid secretion evoked by bile salt and cholera toxin. *Br. J. Pharmacol.* **127**, 887-894.

Timar Peregrin, A., Svensson, M., Jodal, M., Lundgren, O. (1997a). Calcium channels and intestinal fluid secretion: an experimental study *in vivo* in rats. *Acta Physiol. Scand.* **160**, 371-378.

Turck, D., Berard, H., Fretault, N., Lecomte, JM. (1999). Comparison of racecadotril and loperamide in children with acute diarrhoea. *Aliment. Pharmacol. Ther.* **13** Suppl **6**:27-32.

Weiser, M.M., Quill, H. (1975). Intestinal villus and crypt cell responses to cholera toxin. *Gastroenterology* **69**, 472-482.

I, 4.   Immunology of the gut

Per Brandtzaeg and Finn-Eirik Johansen

*Laboratory for Immunohistochemistry and Immunopathology (LIIPAT), Institute of Pathology,
University of Oslo, Rikshospitalet, N-0027 Oslo, Norway*

Introduction

An increasing variety of viral agents are suspected in the aetiology of acute as well as
protracted diarrhoea (Desselberger, 2000). Both DNA and RNA viruses are obligate
intracellular microorganisms, so those infecting via the gut must in one way or another
affect the intestinal mucosal lining. However, the inconsistently observed damage of
enterocytes, villous atrophy, and crypt hyperplasia should not necessarily be ascribed
to cytopathic effects of the actual viruses but may reflect immunopathology resulting
from an immunological attack on the infected host cell or from immune-mediated
inflammation. Due to ethical considerations, there are only few studies of the pathogenesis of mucosal injury caused by gastroenteritis in humans (Davidson and Barnes,
1979; Phillips, 1989). Overt or obscure cellular damage could result from comple-
ment- or T cell-mediated cytotoxicity, antibody-dependent cell-mediated cytotoxicity
(ADCC), complement-induced inflammation, or activation of Th1 cells that secrete
proinflammatory cytokines. However, similar immune reactions are involved in virus
elimination and mucosal healing; the clinical outcome most likely reflects the balance
that is continuously evolving in the fine-tuning of the respective defence mechanisms
of the pathogen and its host. This inherent biological problem, that host defence
against pathogenic microorganisms and immunopathology basically go hand-in-hand,
is opposed in the mucosae by a complex secretory immune system protecting the epithelial lining principally without causing damage.

The main focus of this chapter is the adaptive defence mechanisms exerted by the
human mucosal immune system in the gut. Secretory immunity is of special interest because asymptomatic virus infections, or prevention of infections, seems to be
related to a relatively high local antibody level – at least for poliovirus (Ogra and
Karzon, 1969) and apparently also for rotavirus (Coulson *et al.*, 1992; Matson *et al.*,
1993). Experiments in mice and gnotobiotic pigs support this notion by showing that
improved protection correlates with an enhanced specific B-cell response in the gut
after oral rotavirus inoculation (Feng *et al.*, 1994; Moser and Offit, 2001; Yuan *et al.*,
1996, 1998).

Neonatal specific mucosal immunity

The enterocytes play a vital role in the defence of the neonate, not only by forming a mechanical barrier but also by transferring breast milk-derived maternal antibodies from the gut lumen, thus providing passive systemic immunity in the newborn period. This enterocytic immunoglobulin (Ig) transmission differs remarkably among species. In the ungulate (horse, cattle, sheep, pig) the whole length of the intestine is involved in a non-selective protein uptake, including all Ig isotypes in a poorly defined pinocytotic process. Because colostrum of these animals is particularly rich in IgG, this antibody class will preferentially reach the circulation of the neonate via its gut epithelium during the two first postnatal days, after which so-called 'gut closure' takes place (Mackenzie, 1990). Rodents, on the other hand, express an Fc receptor specific for IgG apically on neonatal enterocytes in the proximal small intestine. This receptor (FcRn), which disappears at weaning, has been particularly well characterized on enterocytes of the neonatal rat; it is a major histocompatibility complex (MHC) class I-related molecule associated with β2-microglobulin (Simister and Mostov, 1989). Complexes of FcRn and IgG are internalized in clathrin-coated pits at the base of the microvilli; binding of the ligand takes place in the acidic luminal environment, and IgG release occurs at physiological pH on the basolateral face of the enterocyte, after which the receptor is recycled.

Contrary to the above mentioned species, the human fetus acquires maternal IgG via the placenta (Mackenzie, 1990), and perhaps to some extent from swallowed amniotic fluid via FcRn expressed by fetal enterocytes (Israel *et al.*, 1993). Indeed, a bi-directional transport mechanism for IgG has been demonstrated in a human intestinal epithelial cell line (Dickinson *et al.*, 1999), but the functional significance of FcRn on enterocytes in the human newborn remains unknown. Intestinal uptake of secretory IgA (SIgA) antibodies after breast-feeding appears of little or no importance in the support of systemic immunity (Ogra *et al.*, 1977; Klemola *et al.*, 1986), except perhaps in the preterm infant (Weaver *et al.*, 1991). Although gut closure in humans normally seems to occur mainly before birth, a patent mucosal barrier function may not be established until after 2 years of age; the different variables involved in this process are poorly defined (van Elburg *et al.*, 1992).

Postnatal specific mucosal immunity

The gastrointestinal epithelial barrier is monolayered and therefore quite vulnerable. Nevertheless, most babies growing up under privileged conditions show remarkably good resistance to infections if their innate nonspecific mucosal defence mechanisms are adequately developed. This can be explained by the fact that immune protection of their mucosae is additionally provided by maternal IgG antibodies which are distributed in interstitial tissue fluid at a concentration 50-60% of the intravascular level (Offit *et al.*, 2002). In the first postnatal period, only occasional traces of SIgA and secretory IgM (SIgM) normally occur in the intestinal juice, whereas some IgG is more often present – either as a result of external FcRn-mediated transmission or, perhaps more likely, passive epithelial 'leakage' from the highly vascularized lamina propria (Persson *et al.*, 1998) which particularly after 34 weeks of gestation contains readily detectable maternal IgG (Brandtzaeg *et al.*, 1991). However, an optimal mucosal barrier function in the neonatal period unquestionably depends on an appropriate supply of breast milk as highlighted in relation to mucosal infections, especially in the developing countries. Exclusively breast-fed infants are better protected against a variety of infections (Pisacane *et al.*, 1994; Wold and Hanson, 1994; Newman, 1995; Wright *et al.*, 1998), and epidemiological data suggest that the risk of dying from diarrhoea is reduced 14-24 times in nursed children (Hanson *et al.*, 1993; Anonymous, 1994). In the westernized part of the world, the protective value of breast-feeding is clinically most apparent in preterm infants (Hylander *et al.*, 1998). Experiments in neonatal rabbits have clearly indicated that SIgA is a crucial anti-microbial component of breast milk (Dickinson *et al.*, 1998). The role of secretory antibodies for mucosal homeostasis is furthermore supported by the fact that knock-out mice lacking SIgA and SIgM show increased mucosal leakiness and overstimulation of systemic (IgG) immunity against *Escherichia coli* (Johansen *et al.*, 1999).

After the peak of passive immunity mediated by maternal IgG and secretory antibodies from breast milk, the survival of the infant will to an increasing extent depend on its own adaptive immune responses (Offit *et al.*, 2002). At mucosal surfaces such responses are largely expressed by IgA antibody production (Brandtzaeg *et al.*, 1999). The cellular basis for this first-line humoral defence is the fact that exocrine glands and secretory mucosae contain most of the body's activated B cells – terminally differentiated to Ig-producing blasts and plasma cells (collectively called immunocytes). Most of these immunocytes (70-90%) produce dimers and some larger polymers of IgA (collectively called pIgA) which, along with pentameric IgM, can be transported through serous type of secretory epithelia (Brandtzaeg, 1974a, 1975; Brandtzaeg *et al.*, 1968) to act as SIgA and SIgM in mucosal surface defence (Fig. 1). This function depends on the polymeric Ig receptor (pIgR) – a 110-kD epithelial glycoprotein also known as membrane secretory component (SC).

IgA-producing immunocytes are normally undetectable in human intestinal mucosa before 10 days of age but thereafter a rapid increase takes place. IgM immunocytes usually remain predominant up to 1 month after birth (Brandtzaeg *et al.*, 1991; Brandtzaeg, 1998). Adult salivary IgA levels are reached quite late in childhood, but only a small increase of IgA-producing cells has been reported in intestinal mucosa after the first year. These observations have notably been made in industrialized countries; a faster development of the mucosal IgA system is usually seen in children from developing countries, reflecting the adaptability of local immunity according to the environmental antigenic load (Hoque *et al.*, 2000).

72

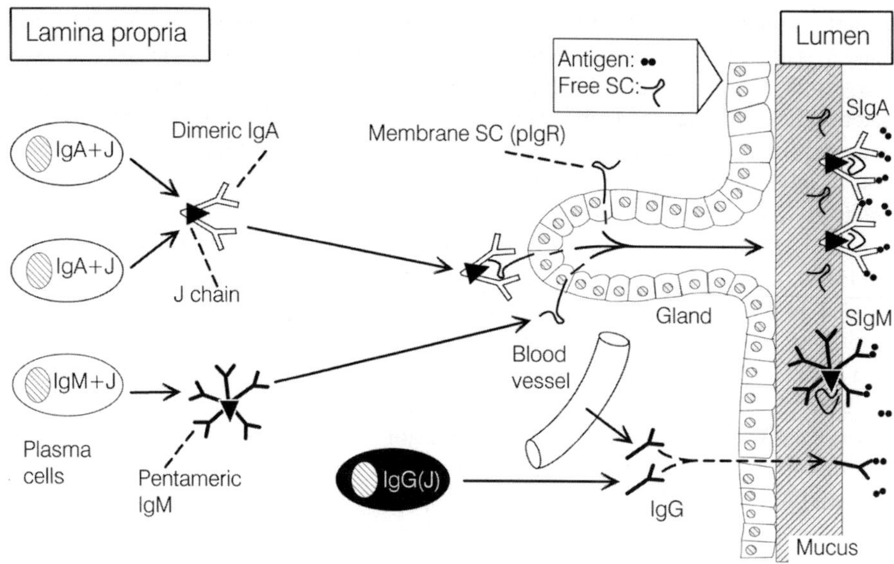

Fig. 1. Model for external transport of J chain-containing dimeric IgA and pentameric IgM by membrane secretory component (SC), or the polymeric Ig receptor (pIgR), expressed basolaterally on glandular epithelial cells. The polymeric Ig molecules are produced with incorporated J chain (IgA + J and IgM + J) by mucosal plasma cells. The resulting secretory Ig molecules (SIgA and SIgM) act in a first line of defence by performing immune exclusion of antigens in the mucus layer on the epithelial surface. Although J chain is often (70-90%) produced by mucosal IgG plasma cells (Brandtzaeg *et al.*, 1999), it does not combine with this Ig class but is degraded intracellularly as denoted by (J) in the figure. Locally produced and serum-derived IgG is not subjected to active external transport, but can be transmitted paracellularly to the lumen as indicated. Free SC (depicted in mucus) is generated when unoccupied pIgR (top symbol) is cleaved at the apical face of the epithelial cell in the same manner as bound SC in SIgA and SIgM.

Mucosa-Associated Lymphoid Tissue (MALT)

Induction of gastrointestinal immunity

Lymphoid cells are located in three distinct tissue compartments in the gut: organized gut-associated lymphoid tissue (GALT), the lamina propria, and the surface epithelium (Fig. 2). GALT comprises the Peyer's patches, the appendix and numerous solitary lymphoid follicles, especially in the large bowel. All these lymphoid structures are believed to represent inductive sites for gastrointestinal immune responses (Brandtzaeg *et al.*, 1999). The lamina propria and the epithelial compartment constitute effector sites but are nevertheless important in terms of cellular expansion and differentiation within the mucosal immune system. GALT and other MALT structures (see below) are covered by a characteristic follicle-associated epithelium (FAE), which contains membrane (M) cells (Fig. 2). These specialized thin epithelial cells are especially effective in the uptake of live and dead particulate antigens from the gut lumen, and many enteropathogenic infectious bacterial (e.g., *Salmonella* spp., *Vibrio cholerae*) and viral (e.g., poliovirus, reovirus, HIV) agents use the M cells as portals of entry (Frey and Neutra, 1997; Hathaway and Kraehenbuhl, 2000).

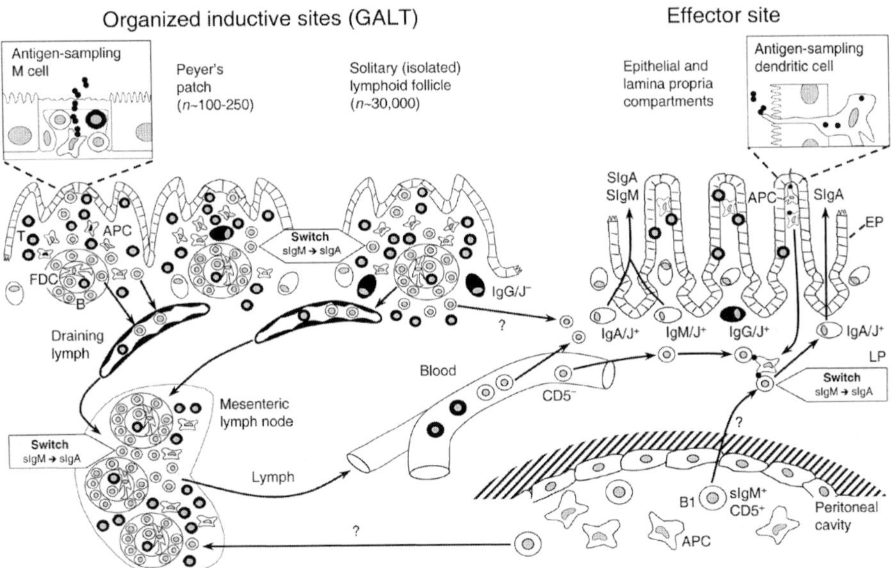

Fig. 2. Antigen-sampling and B cell-switching sites for induction of intestinal IgA responses. The classical inductive sites are constituted by gut-associated lymphoid tissue (GALT) which is equipped with antigen-sampling M cells, T-cell areas (T), B-cell follicles (B), and antigen-presenting cells (APCs). Class switching from surface (s)IgM to sIgA occurs in GALT and mesenteric lymph nodes; from here primed B and T cells home to the lamina propria (LP) via lymph and blood. T cells mainly end up in the epithelium (EP). Primed B cells may also migrate from solitary follicles directly into LP. The sIgA⁺ cells differentiate to plasma cells that produce dimeric IgA with J chain (IgA/J⁺) which becomes secretory IgA (SIgA). Bone marrow-derived sIgM⁺ B2 cells (CD5⁻) may give rise to pentameric IgM (IgM/J⁺) and secretory IgM (SIgM). B1 cells (CD5⁺) from the peritoneal cavity reach the LP by an unknown route (?), perhaps via mesenteric lymph nodes. These sIgM⁺ cells are particularly abundant in mice and may switch to sIgA within the LP under the influence of APCs that have sampled luminal antigen as dendritic cells within the epithelium. The sIgA⁺ cells then differentiate to plasma cells that provide SIgA mainly directed against the commensal gut flora. Red dots denote antigen. Modified from Brandtzaeg *et al.* (2001).

MALT structures resemble lymph nodes with B-cell follicles, intervening T-cell areas, and a variety of antigen-presenting cell (APC) subsets, but there are no afferent lymphatics supplying antigens for immunological stimulation. Therefore, the exogenous stimuli must come directly from the gut lumen, apparently in the main via the M cells. Among the T cells, the CD4[+] helper subset predominates – the ratio between CD4 and CD8 cells being similar to that of other peripheral T-cell populations (Brandtzaeg *et al.*, 1999). In addition, B cells aggregate together with T cells in the M-cell pockets, which thus represent the first contact site between immune cells and luminal antigens (Yamanaka *et al.*, 2001). The B cells may carry out important antigen-presenting functions in this compartment, perhaps promoting antibody diversification and immunological memory. Other types of professional APCs, macrophages and dendritic cells (DCs), are located below the FAE and between the follicles.

Pioneer studies, performed in animals almost 30 years ago, demonstrated that immune cells primed in GALT are functionally linked to mucosal effector sites by an integrated migration or 'homing' pathway (Brandtzaeg, 1996). T cells activated by microbial and other antigens in GALT, preferentially differentiate to CD4[+] helper cells which – aided by DCs and secretion of cytokines such as transforming growth factor (TGF)-β and interleukin (IL)-10 – induce the differentiation of antigen-specific B cells to predominantly IgA-committed plasma blasts. These blasts proliferate and differentiate further on their route through mesenteric lymph nodes and the thoracic duct into the blood stream (Fig. 2). They thereafter home preferentially to the gut mucosa and complete their terminal differentiation to IgA-producing plasma cells locally – most likely under the influence of DCs which may pick up luminal antigens at the secretory effector site (Rescigno *et al.*, 2001). As reviewed elsewhere (Brandtzaeg *et al.*, 1999), the migration of lymphoid cells into the mucosal lamina propria is facilitated by 'homing receptors' that interact with ligands on the microvascular endothelium at the effector site ('addressins') – with an additional fine-tuned level of navigation conducted by local chemoattractant cytokines (chemokines). Under normal conditions, therefore, the lamina propria microvasculature exerts a 'gatekeeper' function to allow selective extravasation of primed lymphoid cells belonging to the mucosal immune system, but this restriction is not maintained during inflammation.

As a reflection of the temporary immunological naivety seen in the newborn period (Offit *et al.*, 2002), very few B cells with IgA-producing capacity (presumably GALT-derived) are present in peripheral blood of newborns ($<8/10^6$ lymphocytes), but after 1 month of age this number increases remarkably ($\sim600/10^6$ lymphocytes) in parallel with the progressive environmental stimulation of GALT (Stoll *et al.*, 1993). An initial early elevation of positive cells can be seen in pre-term infants, especially in those with intrauterine infections, although IgM production dominates in these cases. Later on in childhood, B-cell trafficking between inductive and effector sites is reflected by a significant correlation between the number of IgA antibody-producing cells in the circulation and in the lamina propria, as shown for the small intestine of patients with rotavirus infection (Brown *et al.*, 2000).

Additional sources of gastrointestinal B cells

Although GALT constitutes the major part of human MALT, induction of mucosal immune responses apparently takes place also in the palatine tonsils and other lymphoepithelial structures of Waldeyer's pharyngeal ring, including nasopharynx-associated lymphoid tissue such as the adenoids in humans, and probably also bronchus-associated lymphoid tissue. Accumulating evidence suggests that a certain regionalization exists in the mucosal immune system, especially a dichotomy between the gut and the upper aerodigestive tract with regard to homing properties and terminal differentiation of B cells (Brandtzaeg *et al.*, 1999, 2002). This disparity may be explained by micro-environmental differences in the antigenic repertoire as well as adhesion molecules and chemokines involved in preferential local leucocyte extravasation. It appears that primed immune cells selectively home to effector sites corresponding to the inductive sites where they initially were triggered by antigens. Such compartmentalization within the 'common' or integrated mucosal immune system has to be taken into account in the development of local vaccines. Nevertheless, as reviewed elsewhere (Brandtzaeg *et al.*, 1987a; Brandtzaeg, 1989), there is interaction among the various immune compartments, and even the secretory and the systemic lymphoid cell systems are not completely segregated. Thus, it is possible to obtain a detectable SIgA response also by the parenteral route of immunization.

In the mouse, the peritoneal cavity is an additional source of intestinal lamina propria B cells ($CD5^+$ B1 cells) giving rise to so-called natural SIgA antibodies in the gut lumen (Fig. 2), particularly directed against commensal bacteria (Macpherson *et al.*, 2000). Nevertheless, there is no evidence to suggest that B1 cells are significantly involved in intestinal IgA production in humans (Brandtzaeg *et al.*, 2001), although human secretions contain considerable levels of polyreactive SIgA antibodies recognizing both self and microbial antigens (Bouvet and Fischetti, 1999). Interestingly, however, even in mice the traditional B2 cells, but not the B1 cells, contribute along with $CD4^+$ T cells to intestinal clearance of rotavirus infection by a specific intestinal IgA response (Kushnir *et al.*, 2001).

Immune-effector compartments in the gut

Lamina propria B cells and local immunity

The adult normal gut mucosa contains approximately 10^{10} immunocytes per metre of intestine, or at least 80% of all Ig-producing cells in the body. Some 80-90% are IgA immunocytes, and a relatively large fraction consists of the IgA2 subclass (17-64%) compared with the proportion (7-25%) seen in peripheral lymphoid tissue, tonsils, and airway mucosae (Brandtzaeg *et al.*, 1999). However, IgA2 immunocytes predominate only in the large bowel mucosa. The concentration ratios of the two SIgA subclasses in various exocrine secretions are quite similar to the relative distribution of IgA1 and IgA2 immunocytes at the corresponding effector sites, attesting to the fact that pIgA of both isotypes are equally well transported externally (Jonard *et al.*, 1984; Müller *et al.*, 1991; Feltelius *et al.*, 1994). The relative increase of the IgA2 subclass in secretions compared with serum may be important for the stability of secretory antibodies because SIgA2, in contrast to SIgA1, is resistant to several IgA1-specific proteases which are produced by a variety of potentially pathogenic bacterial species (Kilian *et al.*, 1996).

More than 90% of the mucosal IgA immunocytes synthesize a small polypeptide called the joining (J) chain (Brandtzaeg, 1974b). The J chain is essential for correct polymerization of pIgA (and also pentameric IgM) and for the subsequent binding of Ig polymers to the pIgR that is expressed basolaterally on the intestinal crypt cells (Brandtzaeg and Prydz, 1984; Johansen *et al.*, 2000, 2001). The ligand-receptor complexes are transported through the secretory epithelial cells and, by apical pIgR cleavage, released into the gut lumen (Fig. 1). The extracellular portion of the receptor (~80 kD) remains as so-called bound SC in SIgA and SIgM, thereby stabilizing the secretory antibodies. Particularly the covalent bonding between SC and the α-chain of pIgA makes SIgA the most stable antibody operating in external secretions (Brandtzaeg *et al.*, 1999; Norderhaug *et al.*, 1999; Johansen *et al.*, 2000).

The secretory antibodies are generated principally to perform immune exclusion by complexing with soluble antigens and by binding to the surface of microorganisms, thus preventing their adherence to epithelial cells and penetration of the surface barrier (Fig. 3, left panel). Interestingly, the transported pIgA and pentameric IgM antibodies may even inactivate infectious agents – e.g., rotavirus, influenza virus and HIV – inside of secretory epithelial cells and carry the pathogens and their products back to the lumen (Fig. 3, middle panel), thus avoiding cytolytic damage to the epithelium (Mazanec *et al.*, 1992; Burns *et al.*, 1996; Bomsel *et al.*, 1998; Norderhaug *et al.*, 1999). Both the agglutinating and virus-neutralizing antibody effect of pIgA is superior compared with monomeric antibodies (Brandtzaeg *et al.*, 1987a; Renegar *et al.*, 1998), and SIgA antibodies may block microbial invasion quite efficiently. This has been particularly well documented in relation to HIV (Mazzoli *et al.*, 1997); specific SIgA antibodies isolated from human colostrum were shown to be more efficient in this respect than comparable IgG antibodies (Hocini and Bomsel, 1999). Moreover, as discussed above, serum IgG antibodies mainly depend on a passive paracellular transfer to reach mucosal surfaces (Fig. 1), and the same is true for the locally produced IgG that is detectable in various secretions (Bouvet and Fischetti, 1999). Also, notably, in human gut mucosa less than 5% of the immunocytes normally produce IgG – a figure that in IgA deficiency, however, is increased to ~25% together with ~75% IgM immunocytes (Brandtzaeg *et al.*, 1999). The proportion of IgG immunocytes is also strikingly increased in inflamed mucosae (Brandtzaeg *et al.*, 1987a).

The protective role of SIgA has been questioned by observations in IgA knockout (IgA$^{-/-}$) mice (Mbawuike *et al.*, 1999). These mice remain healthy under ordinary laboratory conditions; and when challenged with influenza virus, they show similar pulmonary virus levels and mortality as control wild-type (IgA$^{+/+}$) mice. However, leakage of serum IgG antibodies through an irritated mucosal surface lining (Fig. 1), according to the principle of 'pathotopic potentiation of local immunity' (Fazekas de St. Groth, 1951), has probably a greater protective role in the respiratory tract than in the gut (Persson *et al.*, 1998). In addition, the IgA$^{-/-}$ mice usually show a compensatory SIgM response. Interestingly, IgA$^{-/-}$ mice on exclusive parenteral nutrition have reduced IgA anti-influenza virus titres in the upper respiratory tract and no compen-

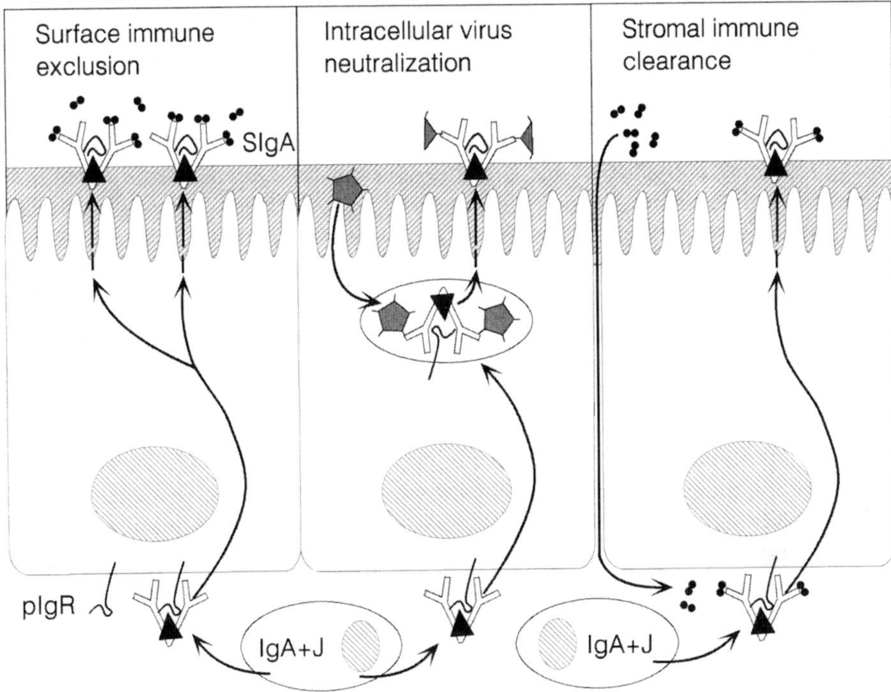

Fig. 3. Schematic representation of three levels at which dimeric IgA or secretory IgA (SIgA) may provide immune protection in the gut after being produced with J chain (IgA + J) by plasma cells in the lamina propria. Left: dimeric IgA is transported by the polymeric Ig receptor (pIgR) across epithelial cells and released into the lumen as SIgA antibodies that perform immune exclusion by interaction with luminal antigens (••). Middle: dimeric IgA antibodies interact with viral antigens within apical epithelial endosomes during pIgR-mediated transport, thereby performing intracellular virus neutralization and removal of viral products. Right: dimeric IgA antibodies interact with penetrating antigens in the lamina propria and shuttle them back to the lumen by pIgR-mediated transport.

satory SIgM which most likely explains that they show impaired mucosal immunity (Renager *et al.*, 2001). The fact that individuals with selective IgA deficiency do not suffer significantly more than others from intestinal virus infections, may largely be ascribed to their consistently enhanced SIgM and IgG1 response in the gut (Brandtzaeg and Nilssen, 1995). Altogether, a secretory antibody response appears to be essential for adequate mucosal protection. This notion is supported by the finding that systemic IgG antibody production against *E. coli* is triggered in pIgR knockout mice lacking both SIgA and SIgM (Johansen *et al.*, 1999).

The complexity of local immunity is emphasized by the finding that clearance of rotavirus after infection, as well as long-term resistance to reinfection, was not impaired in IgA$^{-/-}$ mice (O'Neal *et al.*, 2000). Careful analysis of different variables suggested that in the absence of IgA, the protective effect was primarily mediated by IgG antibodies. Notably, however, both clinical data and animal experiments have suggested that when IgA is present, rotavirus infection is primarily cleared and subsequently prevented by intestinal SIgA antibodies (O'Neal et al., 2000). In a similar study, the role of IgA versus IgG in gut immunity to reovirus was recently investigated (Silvey et al., 2001). Although this virus does not normally cause disease in humans, it constitutes a useful rodent model for enteric virus infections. Both wild-type and IgA$^{-/-}$ mice were competent at clearing a primary infection, but only wild-type mice were fully protected against entry of reovirus through Peyer's patch M cells upon subsequent infection; partial protection shown by the IgA$^{-/-}$ mice was ascribed to a compensatory SIgM response (Silvey et al., 2001). Although IgG or IgA antibodies to the σ1 protein of reovirus could protect when preincubated with the virus prior to oral viral challenge (Silvey et al., 2001), only the corresponding pIgA secreted as SIgA antibodies protected against virus entry when administered systemically via the hybridoma 'back-pack' model (Marian R. Neutra; personal communication). Together these data imply that secretory antibodies are superior at providing protection against initial infection of Peyer's patches by this enteric virus. In addition to mucosal antibodies, local cytotoxic T-lymphocyte responses may be important to prevent systemic seeding of virus, particularly when the pathogen persistently tends to replicate in gut mucosa, which is the case for HIV (Berzofsky et al., 2001).

Non-inflammatory mucosal antigen clearance

Compared with IgG antibodies, IgA is a poor activator of complement, and external transport of pIgA-containing immune complexes has been suggested as an efficient, non-inflammatory antigen clearance mechanism (Fig. 3, right panel). This notion has recently been supported by experiments performed in vivo (Robinson et al., 2001). Pentameric IgM (in contrast to hexameric IgM without J chain) also appears to have poor complement-activating properties and can therefore most likely support the non-inflammatory functions of pIgA in competition with corresponding pro-inflammatory IgG antibodies (Brandtzaeg et al., 1999). Interestingly, monomeric IgA or IgG antibodies, when cross-linked to pIgA via the same antigen, may contribute to such pIgR-mediated epithelial trancystosis of foreign material (Mazanec et al., 1993). Conversely, IgG antibodies against infectious agents and dietary proteins may by themselves adversely affect mucosal penetrability for a variety of exogenous proteins while contributing to local protection. This possibility has been suggested by experiments ex vivo (Brandtzaeg and Tolo, 1977) and in vivo (Lim and Rowley, 1982); IgG antibodies against one soluble antigen were shown to enhance mucosal penetration of bystander molecules. The mucosal integrity can apparently be damaged by lysosomal enzymes released from polymorphonuclear granulocytes which are attracted when complement-activating immune complexes are formed locally.

The proinflammatory potential of IgG is probably less important in the gut of infants who are breast-fed because milk SIgA antibodies will exert a non-inflammatory blocking effect. Moreover, breast milk contains large amounts of the soluble complement inhibitor protectin (Bjørge *et al.*, 1993). Also, this factor and other complement regulatory proteins are expressed by the gastrointestinal epithelium (Berstad and Brandtzaeg, 1998) which most likely will counteract immune complex-mediated damage (type III hypersensitivity) of the epithelial lining (Andoh *et al.*, 2001).

There is further experimental evidence to suggest that IgA in various ways may influence mucosal homeostasis through its binding to the Fcα receptor (CD89) when present on lamina propria leucocytes, although in the normal state CD89 expression is not detectable on human intestinal macrophages (Smith *et al.*, 2001; Hamre *et al.*, 2002). However, this situation is probably different in the presence of inflammation with extravasation of circulating leucocytes. Interestingly, IgA can downregulate the secretion of the proinflammatory cytokine tumour necrosis factor (TNF)-α from activated monocytes and inhibit activation-dependent generation of reactive oxygen intermediates in neutrophils and monocytes (Wolf *et al.*, 1994a,b). On the other hand, pIgA or aggregated monomeric IgA can trigger monocytes to show increased activity such as TNF-α secretion (Devière *et al.*, 1991), and can also cause upregulation of the costimulatory molecule B7 on APCs (Geissmann *et al.*, 2001) and induce eosinophil degranulation (Abu-Ghazaleh *et al.*, 1989). This proinflammatory potential of IgA probably reflects the need for reinforcement of mucosal antigen elimination mechanisms when immune exclusion fails.

Difficulties in evaluating the impact of intestinal secretory immunity on virus infections

As discussed above and elsewhere (Brandtzaeg, 1989), it is difficult to evaluate the protective effect of SIgA and SIgM antibodies during intestinal virus infection because there is always a concurrent induction of systemic humoral immunity both after natural infection (or enteric live vaccines) and when non-proliferating virus-like particles are mucosally applied together with an adjuvant (Velazquez *et al.*, 1996; O'Neal *et al.*, 1998). The effect of serum antibodies both locally in the gut and peripherally may be to inhibit further spread of the infectious agent by neutralization and immune elimination. In the gut lumen, however, IgG (and probably also monomeric IgA) is less stable than SIgA and may thus be of little protective value, although binding of antibody fragments to the biliary protein Fv (Fv fragment-binding protein) has been suggested to reinforce immune exclusion (Bouvet and Fischetti, 1999).

There are also methodological problems rendering evaluation of intestinal antibody responses rather difficult. To analyze intestinal fluid or fecal extracts for secretory immunity may be misleading due to external leakage of serum IgA antibodies, and because even SIgA may be considerably degraded in such samples (Johansen *et al.*, 1999). Thus, it is often possible to measure intestinal IgA antibodies to for instance rotavirus, whereas a corresponding SIgA activity may be undetectable – depending on whether the employed immunoenzyme assay reveals the α-chain of IgA or bound SC. Another problem is the fact that an unknown fraction of the most avid SIgA antibodies may remain adsorbed to viruses in feces after the extraction procedure.

Mucosal T cells and their putative roles

Although the lymphoid cells in the human lamina propria comprise some B lymphocytes that can be induced to proliferate (Farstad *et al.*, 2000), they are mainly constituted by terminally differentiated IgA-producing immunocytes and T cells (Brandtzaeg *et al.*, 1999). The latter resemble other peripheral T lymphocytes in that they virtually always express the T-cell receptor (TCR)-α/β and show similar predominance of CD4$^+$ over CD8$^+$ cells (Brandtzaeg *et al.*, 1998). Phenotypically they appear as memory/effector cells, but they show a low level of proliferation and little expression of CD25, the high-affinity receptor for IL-2. Also, they are refractory to CD3-triggered activation *in vitro*, but can be activated through CD2-CD28 costimulation (Abreu-Martin and Taragan, 1996). Thus, lamina propria T cells are apparently in a state of restricted reactivation.

The intraepithelial lymphocytes (IELs) comprise mostly T cells, normally amounting to 5-10 per 100 epithelial cells (Fig. 2). B lymphocytes are virtually absent from the surface epithelium outside GALT structures (Farstad *et al.*, 2000). As reviewed elsewhere (Brandtzaeg *et al.*, 1998; Helgeland and Brandtzaeg, 2000), the surface epithelium contains in addition a small population of non-T non-B lymphocytes of obscure origin and function. Thus, the composition of intraepithelial T cells differs markedly from that of lamina propria T cells; CD8$^+$ IELs dominate (80-90%) over CD4$^+$ IELs, and unusual cell subpopulations exist in the intestinal surface epithelium. Such peculiar subsets consist of T cells expressing the alternative TCR-γ/δ, the $\alpha\alpha$ homodimer of the CD8 molecule (both the TCR-α/β^+ and TCR-γ/δ^+ subsets contain CD8$\alpha\alpha^+$ subpopulations), as well as CD4$^+$CD8$^+$ 'double positive' and CD4$^-$CD8$^-$ 'double negative' T cells (Hayday *et al.*, 2001). These subsets may contribute substantially to the IEL pool in mice and rats, which also contain unique T-cell subpopulations that coexpress natural killer (NK)-cell markers (Guy-Grand *et al.*, 1996; Helgeland *et al.*, 1997). Consistent with expression of the CD8 coreceptor and their content of cytoplasmic granules, human and murine IELs show vivid cytotoxicity *in vitro*, and this can be elicited through both the CD3 complex, TCR-α/β, and TCR-γ/δ (Sydora *et al.*, 1993; Lundqvist *et al.*, 1996). In addition, IELs are capable of spontaneous cytotoxicity against certain targets (Roberts *et al.*, 1993; Guy-Grand *et al.*, 1996). Interestingly, IELs are potent producers of IFN-γ after *in vitro* stimulation (Lundqvist *et al.*, 1996), and this capacity may be important for protection of neighbouring epithelial cells against intracellular infectious agents (Chardes *et al.*, 1994). In mice, there is also evidence suggesting that intraepithelial T cells can provide help for B-cell maturation and IgA production in the lamina propria (Fujihashi *et al.*, 1993).

Since the discovery that a major IEL subpopulation in chicken and mice belongs to the γ/δ T-cell lineage, considerable information has accumulated over the last ten years describing various putative developmental pathways for intestinal T cells. The balance of evidence suggests that most TCR-γ/δ^+ and TCR-α/β^+CD8$\alpha\alpha^+$ IELs mature independently of the thymus (Hayday *et al.*, 2001), but these subpopulations are quite small in the human gut (Brandtzaeg *et al.*, 1998; Helgeland and Brandtzaeg, 2000). Far more controversial are the developmental pathways suggested for the TCR-α/β^+CD8$\alpha\beta^+$ and

TCR-α/β⁺CD4⁺ IEL subsets. In experimental mouse models, all IEL subpopulations can be generated independently of the thymus. Some investigators have proposed that intraepithelial T cells differentiate from bone marrow-derived precursors in the intestinal surface epithelium, and that TCR-α/β⁺CD8αα⁺ IELs represent an immature intermediate developmental stage (Helgeland and Brandtzaeg, 2000).

Possible antimicrobial functions of intraepithelial T cells

It is generally believed that IELs play a role in a first line of microbial defence but only a few studies have directly demonstrated such a function (Hayday *et al.*, 2001). Experiments in mice have suggested that certain infectious intracellular pathogens elicit a dynamic disseminated IEL response which may be quite persistent. Thus, enteric infection with reovirus, rotavirus and *Toxoplasma gondii* resulted in rapid accumulation of cytotoxic TCR-α/β⁺ IELs that lyzed infected intestinal epithelial cells. The effector cells belonged to the CD8αβ⁺ subpopulation and recognized intracellular peptides presented on classical MHC class I molecules, thus being similar to the conventional CD8αβ⁺ cells of systemic lymphoid organs (Helgeland and Brandtzaeg, 2000).

Transfer of germ-free animals to a specific pathogen-free environment converts the composition and functional profile of IELs to that of conventional animals, suggesting that the response of these cells is primarily directed against the indigenous microbial flora. However, attempts to identify the stimulatory bacteria have been inconclusive. To elucidate the nature of the antigens recognized by IELs, their TCR repertoire has been subjected to detailed investigation. Surprisingly, it has been found in humans, mice and rats that TCR-α/β⁺ IELs express an oligoclonal antigen receptor repertoire and the same limited number of T-cell clones can be identified over long distances in the gut (Helgeland and Brandtzaeg, 2000). This result is difficult to reconcile with the possibility that IELs recognize diverse microbial antigens. Furthermore, inbred mice and rats express individually distinct oligoclonal repertoires, indicating that TCR-α/β⁺ IELs recognize different antigens in different animals. It has been claimed that this observation excludes the possibility of specific recognition of indigenous microorganisms, which are believed to be quite comparable but not phenotypically identical among animals of the same species (Regnault *et al.*, 1994). Moreover, the same investigators reported that TCR-α/β⁺ IELs in germ-free mice also express oligoclonal repertoires, further speaking against a role of the intestinal microflora in the clonal T-cell selection process (Regnault *et al.*, 1996).

The possibility remains that different T-cell clones recognize different antigenic epitopes of the same microbial species, and certain features of the TCR repertoire suggests that it is shaped by the intestinal flora. Thus, the widely disseminated clones were found to vary markedly in size at different levels of the gut and some clones occurred only locally (Regnault *et al.*, 1994; Dogan *et al.*, 1996). Also, we have shown that microbial colonization of rats leads to selective expansion of TCR-α/β⁺ clones bearing certain variable β gene segments (Helgeland *et al.*, 1996). Finally, data from chicken suggest that the first T cells to arrive in the intestinal epithelium during ontogeny express a polyclonal repertoire (Dunon *et al.*, 1994), which then might be subjected to oligoclonal selection by naturally occurring microbial antigens in postnatal life. In our rat study, the TCR Vβ profile in germ-free *versus* conventionalized animals could reflect transition from a polyclonal to an oligoclonal response. Altogether, it is tempting to speculate that IELs recognize conserved microbial antigens. Thus, rat CD4[+]CD8[+] double-positive IELs have been reported to respond *in vitro* to mycobacterial (Nakamura *et al.*, 1995) and *E. coli* heat-shock proteins (Kimura *et al.*, 1996-97), and the intraepithelial number of T cells expressing such specificities was related to the luminal presence of the actual bacteria. Furthermore, γ/δ T cells may recognize unprocessed antigen directly without any restriction element involved, in a manner similar to antibodies (Guehler *et al.*, 1999).

Another characteristic of the γ/δ IEL subset is the association of certain TCR Vγ and Vδ gene segments with defined anatomical sites. Thus, in the human gut, Vδ1 in expressed by some 70% of the TCR-γ/δ[+] IELs, apparently in the main together with Vγ8 (Brandtzaeg *et al.*, 1998). This may reflect a preferential specificity directed against selected gut antigens, and molecular analyses of Vδ1 sequences have suggested an individually imprinted oligoclonality along extensive gut segments for considerable time periods. Interestingly, Groh *et al.* (1998) showed that Vδ1 TCR-γ/δ[+] IELs could recognize stress-induced nonclassical MHC class I (or class Ib) molecules on epithelial cells, mainly MICA and MICB. This observation might hint to the existence of an immune surveillance mechanism for detection of damaged, infected or transformed intestinal epithelium, and/or the possibility for stimulated secretion of immunoregulatory cytokines from the TCR-γ/δ[+] IELs. Several studies have suggested that this subset is important in mucosal homeostasis, particularly with regard to low-dose tolerance mediated by IL-10-secreting CD4[+] T cells, as well as maintenance of IgA production (Fujihashi *et al.*, 1992, 1999).

Intraepithelial T cells as part of compensatory mucosal responses in immunodeficiency

As mentioned above, there is regularly a compensatory overproduction of SIgM and IgG1 in the intestinal mucosa of subjects with IgA deficiency; this probably explains why they have relatively few problems with gut infections. In contrast, infections are much more common in the respiratory tract of such patients, especially in those with little or no SIgM compensation in the airways (Brandtzaeg *et al.*, 1987b). In patients with hypogammaglobulinaemia that virtually lack antibody production, a compensatory enhancement of T cell-mediated defence by cytotoxic IELs of the TCR-α/β[+]CD8[+] phenotype apparently takes place (Brandtzaeg and Nilssen, 1995). In IgA deficiency, such enhancement of cellular mucosal immunity may include both TCR-α/β[+] and TCR-γ/δ[+] IELs.

Interestingly, the number of intestinal TCR-γ/δ[+] IELs is generally increased in AIDS patients, although with a wide range probably reflecting individual responses to opportunistic infections (Nilssen *et al.*, 1996). Conversely, the almost total depletion of mucosal CD4[+] cells is not compensated for by a numerical increase of duodenal CD8[+] cells. In AIDS patients with particularly short life expectancy (<7 months), γ/δ

IELs are decreased to virtually normal levels. Our longitudinal studies in a few patients supported this observation, suggesting that γ/δ IELs might be involved in prolonging the life of patients with AIDS. More recently we have also observed in AIDS that opportunistic infections appear to stimulate intestinal IgA and IgG1 responses, which are reduced by highly active antiretroviral combination therapy (Nilssen *et al.*, 2002).

Summary

Secretory immunity is desirable in the defence against intestinal virus infections because it can operate both at the luminal face of the epithelium and intracellularly in infected epithelial cells without causing mucosal damage. In addition, IELs may act in front-line defence by eliminating infected epithelial cells. Several virological studies have shown that natural gut infection and enteric vaccination are more efficient in giving rise to SIgA antibodies than parenteral vaccination. Also, the intraluminal vaccines have been much more efficient when consisting of live than killed viruses. Like natural infections, live vaccines give rise not only to SIgA antibodies, but also to longstanding serum IgG and IgA responses. SIgM and serum IgM antibodies often appear in the acute infection phase but decline after covalescence. In infants and subjects with selective IgA deficiency, however, SIgM may play a major protective role against infectious agents (Ogra *et al.*, 1974; Mellander *et al.*, 1986; Brandtzaeg *et al.*, 1987b).

Despite our advanced mechanistic understanding of the protective effects of mucosal immune responses against gut viruses, it is generally difficult to determine unequivocally the importance of SIgA versus serum antibodies (Glass *et al.*, 1996). The degree of immunological protection may range from complete inhibition of reinfection (poliovirus and murine reovirus) to reduction of symptoms (rotavirus). It is furthermore not clear to what extent T cell-mediated mechanisms or ADCC are involved in this effect. Nevertheless, although the importance of numerous innate intestinal defence mechanisms should not be underestimated, the paramount protective role of specific immunity is emphasised by the numerous gut infections often seen in patients afflicted with AIDS.

Acknowledgements

Studies in the authors' laboratory are supported by the Research Council of Norway and the Norwegian Cancer Society. Hege Eliassen Bryne and Erik K. Hagen are thanked for their excellent assistance with the manuscript.

References

Abreu-Martin, M.T., Targan, S.R. (1996). Regulation of immune responses of the intestinal mucosa. Crit. Rev. Immunol. 16, 277-309.

Abu-Ghazaleh, R.I., Fujisawa, T., Mestecky, J., Kyle, R.A., Gleich, G.J. (1989). IgA-induced eosinophil degranulation. J. Immunol. 142, 2393-2400.

Andoh, A., Kinoshita, K., Rosenberg, I., Podolsky, D.K. (2001). Intestinal trefoil factor induces decay-accelerating factor expression and enhances the protective activities against complement activation in intestinal epithelial cells. J. Immunol. 167, 3887-93.

Anonymous (1994). A warm chain for breastfeeding. Lancet 344, 1239-41.

Berzofsky, J.A., Ahlers, J.D. Belyakov, I.M. (2001). Strategies for designing and optimizing new generation vaccines. Nature Rev. Immunol. 1, 209-19.

Berstad, A.E., Brandtzaeg, P. (1998). Expression of cell-membrane complement regulatory glycoproteins along the normal and diseased human gastrointestinal tract. Gut 42, 522-9.

Bjørge, L., Jensen, T.S., Vedeler, C.A., Ulvestad, E., Kristoffersen, E.K., Matre, R. (1993). Soluble CD59 in pregnancy and infancy. Immunol. Lett. 36, 233-4.

Bomsel, M., Heyman, M., Hocini, H., Lagaye, S., Belec, L., Dupont, C., Desgranges, C. (1998). Intracellular neutralization of HIV transcytosis across tight epithelial barriers by anti-HIV envelope protein dIgA or IgM. Immunity 9, 277-87.

Bouvet, J.-P. and Fischetti, V.A. (1999). Diversity of antibody-mediated immunity at the mucosal barrier. Infect. Immun. 67, 2687-91.

Brandtzaeg, P. (1974a). Mucosal and glandular distribution of immunoglobulin components: differential localization of free and bound SC in secretory epithelial cells. J. Immunol. 112, 1553-9.

Brandtzaeg, P. (1974b). Presence of J chain in human immunocytes containing various immunoglobulin classes. Nature 252, 418-20.

Brandtzaeg, P. (1975). Human secretory immunoglobulin M. An immunochemical and immunohistochemical study. Immunology 29, 559-70.

Brandtzaeg, P. (1989). Immune responses to gut virus infections. pp. 45-54. In: Viruses and the Gut (Ed.: Farthing MJG), Smith Kline and French Laboratories, Swan Press, London.

Brandtzaeg, P. (1996). History of oral tolerance and mucosal immunity. Ann. N.Y. Acad. Sci. 778, 1-27.

Brandtzaeg, P. (1998). Development and basic mechanisms of human gut immunity. Nutr. Rev. 56, S5-18.

Brandtzaeg, P., Nilssen, D.E. (1995). Mucosal aspects of primary B-cell deficiency and gastrointestinal infections. Curr. Opin. Gastroenterol. 11, 532-540.

Brandtzaeg, P., Prydz, H. (1984). Direct evidence for an integrated function of J chain and secretory component in epithelial transport of immunoglobulins. Nature 311, 71-73.

Brandtzaeg, P., Tolo, K. (1977). Mucosal penetrability enhanced by serum-derived antibodies. Nature 266, 262-3.

Brandtzaeg, P., Fjellanger, I., Gjeruldsen, S.T. (1968). Immunoglobulin M: local synthesis and selective secretion in patients with immunoglobulin A deficiency. Science 160, 789-91.

Brandtzaeg, P., Baklien, K., Bjerke, K., Rognum, T.O., Scott, H., Valnes, K. (1987a). Nature and properties of the human gastrointestinal immune system. In:

Immunology of the Gastrointestinal Tract (Ed. by K. Miller and S. Nicklin), pp 1-85. CRC Press, Boca Raton, Florida.

Brandtzaeg, P., Karlsson, G., Hansson, G., Petruson, B., Bjørkander, J., Hanson, L.Å. (1987b). The clinical condition of IgA-deficient patients is related to the proportion of IgD- and IgM-producing cells in their nasal mucosa. Clin. Exp. Immunol. 67, 626-36.

Brandtzaeg, P., Nilssen, D.E., Rognum, T.O., Thrane, P.S. (1991). Ontogeny of the mucosal immune system and IgA deficiency. Gastroenterol. Clin. North. Am. 20, 397-439.

Brandtzaeg, P., Farstad, I.N., Helgeland, L. (1998). Phenotypes of T cells in the gut. Chem. Immunol. 71, 1-26.

Brandtzaeg, P., Farstad, I.N., Johansen, F.E., Morton, H.C., Norderhaug, I.N., Yamanaka, T. (1999). The B-cell system of human mucosae and exocrine glands. Immunol. Rev. 171, 45-87.

Brandtzaeg, P., Baekkevold, E.S., Morton, H.C. (2001). From B to A the mucosal way. Nature Immunol. 2, 1093-4.

Brandtzaeg, P., Johansen, F.-E., Baekkevold, E.S., Farstad, I.N. (2002). Direct tracking of mucosal B-cell migration in humans reveals distinct effector-site differences. Abstract 2476, 11th International Congress of Mucosal Immunology, Orlando. Mucosal Immunol Update 10 (Nos. 2 and 3).

Brown, K.A., Kriss, J.A., Moser, C.A., Wenner, W.J., Offit, P.A. (2000). Circulating rotavirus-specific antibody-secreting cells (ASCs) predict the presence of rotavirus-specific ASCs in the human small intestinal lamina propria. J. Infect. Dis. 182, 1039-43.

Burns, J.W., Siadat-Pajouh, M., Krishnaney, A.A., Greenberg, H.B. (1996). Protective effect of rotavirus VP6-specific IgA monoclonal antibodies that lack neutralizing activity. Science 272, 104-7.

Chardes, T., Buzoni-Gatel, D., Lepage, A., Bernard, F., Bout, D. (1994). Toxoplasma gondii oral infection induces specific cytotoxic CD8 α/β^+ Thy-1$^+$ gut intraepithelial lymphocytes, lytic for parasite-infected enterocytes. J. Immunol. 153, 4596-603.

Coulson, B.S., Grimwood, K., Hudson, I.L., Barnes, G.L., Bishop, R.F. (1992). Role of coproantibody in clinical protection of children during reinfection with rotavirus. J. Clin. Microbiol. 30, 1678-84.

Davidson, G.P., Barnes, G.L. (1979). Structural and functional abnormalities of the small intestine in infants and young children with rotavirus enteritis. Acta Paediatr. Scand. 68, 181-6.

Desselberger, U. (2000). Gastroenteritis viruses: research update and perspectives. Gastroenteritis viruses, Novartis Foundation Symposium 238, London, UK, 16-18 May 2000. Mol. Med. Today 6, 383-4.

Devière, J., Vaerman, J.-P., Content, J., Denys, C., Schandene, L., Vandenbussche, P., Sibille, Y., Dupont, E. (1991). IgA triggers tumor necrosis factor α secretion by monocytes: a study in normal subjects and patients with alcoholic cirrhosis. Hepatology 13, 670-75.

Dickinson, B. L., Badizadegan, K., Wu, Z., Ahouse, J. C., Zhu, X., Simister, N. E., Blumberg, R. S., Lencer, W. I. (1999). Bidirectional FcRn-dependent IgG transport in a polarized human intestinal epithelial cell line. J. Clin. Invest. 104, 903-11.

Dickinson, E. C., Gorga, J. C., Garrett, M. Tuncer, R., Boyle, P., Watkins, S. C., Alber, S. M., Parizhskaya, M., Trucco, M., Rowe, M. I., Ford, H. R. (1998). Immunoglobulin A supplementation abrogates bacterial translocation and preserves the architecture of the intestinal epithelium. Surgery 124, 284-90.

Dogan, A., Dunn-Walters, D.K., MacDonald, T.T., Spencer, J. (1996). Demonstration of local clonality of mucosal T cells in human colon using DNA obtained by microdissection of immunohistochemically stained tissue sections. Eur. J. Immunol. 26, 1240-5.

Dunon, D., Schwager, J., Dangy, J.P., Cooper, M.D., Imhof, B.A. (1994). T cell migration during development: homing is not related to TCR V β 1 repertoire selection. EMBO J. 13, 808-15.

Farstad, I.N., Carlsen, H., Morton, H.C., Brandtzaeg, P. (2000). Immunoglobulin A cell distribution in the human small intestine: phenotypic and functional characteristics. Immunology 101, 354-363.

Fazekas de St. Groth, S. (1951). Studies in experimental immunology of influenza. IX. The mode of action of pathotopic adjuvants. Aust. J. Exp. Biol. Med. Sci. 29, 339-52.

Feltelius, N., Hvatum, M., Brandtzaeg, P., Knutson, L., Hällgren, R. (1994). Increased jejunal secretory IgA and IgM in ankylosing spondylitis: normalization after treatment with sulfasalazine. J. Rheumatol. 21, 2076-81.

Feng, N.G., Burns, J.W., Bracy, L., Greenberg, H.B. (1994). Comparison of mucosal and systemic humoral immune responses and subsequent protection in mice orally inoculated with a homologous or a heterologous rotavirus. J. Virol. 68, 7766-73.

Frey, A. and Neutra, M.R. (1997). Targeting of mucosal vaccines to Peyer's patch M cells. Behring Inst. Mitt. 98, 376-89.

Fujihashi, K., Dohi, T., Kweon, M.N., McGhee, J.R., Koga, T., Cooper, M.D., Tonegawa, S., Kiyono, H. (1999). $\gamma\delta$ T cells regulate mucosally induced tolerance in a dose-dependent fashion. Int. Immunol. 11, 1907-16.

Fujihashi, K., Taguchi, T., Aicher, W.K., McGhee, J.R., Bluestone, J.A., Eldridge, J.H., Kiyono, H. (1992). Immunoregulatory functions for murine intraepithelial lymphocytes: γ/δ T cell receptor-positive (TCR$^+$) T cells abrogate oral tolerance, while α/β TCR$^+$ T cells provide B cell help. J. Exp. Med. 175, 695-707.

Fujihashi, K., Yamamoto, M., McGhee, J.R., Kiyono, H. (1993). $\alpha\beta$ T cell receptor-positive intraepithelial lymphocytes with CD4$^+$, CD8- and CD4$^+$, CD8$^+$ phenotypes from orally immunized mice provide Th2-like function for B cell responses. J. Immunol. 151, 6681-91.

Geissmann, F., Launay, P., Pasquier, B., Lepelletier, Y., Leborgne, M., Lehuen, A., Brousse, N., Monteiro, R. C. (2001). A subset of human dendritic cells expresses IgA Fc receptor (CD89), which mediates internalization and activation upon cross-linking by IgA complexes. J. Immunol. 166, 346-52.

Glass, R.I., Gentsch, J.R., Ivanoff, B. (1996). New lessons for rotavirus vaccines. Science 272, 46-8.

Groh, V., Steinle, A., Bauer, S., Spies, T. (1998). Recognition of stress-induced MHC molecules by intestinal epithelial gammadelta T cells. Science 279, 1737-40.

Guehler, S.R., Bluestone, J.A., Barrett, T.A. (1999). Activation and peripheral expansion of murine T-cell receptor gamma delta intraepithelial lymphocytes. Gastroenterology 116, 327-34.

Guy-Grand, D., Cuenod-Jabri, B., Malassis-Seris, M., Selz, F., Vassalli, P. (1996). Complexity of the mouse gut T cell immune system: identification of two distinct natural killer T cell intraepithelial lineages. Eur. J. Immunol. 26, 2248-56.

Hamre, R., Farstad, I.N., Brandtzaeg, P., Morton, H.C. (2002). Expression and modulation of CD89 IgA Fc receptor and the Fc receptor γ chain on myeloid cells in human blood and tissues: implications for the biological role of CD89. Scand. J. Immunol., in press.

Hanson, L. Å., Ashraf, R., Carlsson, B., Jalil, F., Karlberg, J., Lindblad, B. S., Khan, S. R., Zaman, S. (1993). Child health and the population increase. In: Peace, Health and Development (Ed. by L. Å. Hanson and L. Köhler), pp. 31-8. University of Göteborg and the Nordic School of Public Health, NHV Report 4, Göteborg.

Hathaway, L. J., Kraehenbuhl, J. P. (2000). The role of M cells in mucosal immunity. Cell Mol. Life Sci. 57, 323-32.

Hayday A., Theodoridis, E., Ramsburg, E., Shires, J. (2001). Intraepithelial lymphocytes: exploring the Third Way in immunology. Nature Immunol. 2, 997-1003.

Helgeland, L., Brandtzaeg, P. (2000). Development and function of intestinal B and T cells. Microbiol. Ecol. Health Dis. 12 (Suppl. 2), 110-27.

Helgeland, L., Vaage, J.T., Rolstad, B., Halstensen, T.S., Midtvedt, T., Brandtzaeg, P. (1997). Regional phenotypic specialization of intraepithelial lymphocytes in the rat intestine does not depend on microbial colonization. Scand. J. Immunol. 46, 349-57.

Helgeland, L., Vaage, J.T., Rolstad, B., Midtvedt, T., Brandtzaeg, P. (1996). Microbial colonization influences composition and T-cell receptor Vβ repertoire of intraepithelial lymphocytes in rat intestine. Immunology 89, 494-501.

Hocini, H., Bomsel, M. (1999). Infectious human immunodeficiency virus can rapidly penetrate a tight human epithelial barrier by transcytosis in a process impaired by mucosal immunoglobulins. J. Infect. Dis. 179, S448-53.

Hoque, S.S., Ghosh, S., Poxton, I.R. (2000). Differences in intestinal humoral immunity between healthy volunteers from UK and Bangladesh. Eur. J. Gastroenterol. Hepatol. 12, 1185-93.

Hylander, M.A., Strobino, D.M., Dhanireddy, R. (1998). Human milk feedings and infection among very low birth weight infants. Pediatrics 102, E38 (pp. 6, electronic paper).

Israel, E.J., Simister, N., Freiberg, E., Caplan, A., Walker, W.A. (1993). Immunoglobulin G binding sites on the human foetal intestine: a possible mechanism for the passive transfer of immunity from mother to infant. Immunology 79, 77-81.

Johansen, F.-E., Pekna, M., Norderhaug, I.N., Haneberg, B., Hietala, M.A., Krajci, P., Betsholtz, C., Brandtzaeg, P. (1999). Absence of epithelial immunoglobulin A transport, with increased mucosal leakiness, in polymeric immunoglobulin receptor/secretory component-deficient mice. J. Exp. Med. 190, 915-22.

Johansen, F.-E., Braathen, R., Brandtzaeg, P. (2000). Role of J chain in secretory immunoglobulin formation. Scand. J. Immunol. 52, 240-248.

Johansen, F.-E., Braathen, R., Brandtzaeg, P. (2001). The J chain is essential for polymeric Ig receptor-mediated epithelial transport of IgA. J. Immunol. 167, 5185-92.

Jonard, P.P., Rambaud, J.C., Dive, C., Vaerman, J.P., Galian, A., Delacroix, D.L. (1984). Secretion of immunoglobulins and plasma proteins from the jejunal mucosa. Transport rate and origin of polymeric immunoglobulin A. J. Clin. Invest. 74, 525-35.

Kilian, M., Reinholdt, J., Lomholt, H., Poulsen, K., Frandsen, E.V. (1996). Biological significance of IgA1 proteases in bacterial colonization and pathogenesis: critical evaluation of experimental evidence. APMIS 104, 321-38.

Kimura, Y., Sakai, T., Takeuchi, M., Matsumoto, Y., Watanabe, K., Yuuki, M., Takada, T., Yoshikai, Y. (1996-97). An unique CD4[+]CD8[+] intestinal intraepithelial lymphocyte specific for DnaK (Escherichia coli HSP70) may be selected by intestinal microflora of rats. Immunobiology 196, 550-66.

Klemola, T., Savilahti, E., Leinikki, P. (1986). Mumps IgA antibodies are not absorbed from human milk. Acta. Pediatr. Scand. 75, 230-2.

Kushnir, N., Bos, N.A., Zuercher, A.W., Coffin, S.E., Moser, C.A., Offit, P.A., Cebra, J.J. (2001). B2 but not B1 cells can contribute to CD4[+] T-cell-mediated clearance of rotavirus in SCID mice. J. Virol. 75, 5482-90.

Lim, P.L., Rowley, D. (1982). The effect of antibody on the intestinal absorption of macromolecules and on intestinal permeability in adult mice. Int. Arch. Allergy Appl. Immunol. 68, 41-6.

Lundqvist, C., Melgar, S., Yeung, M.M., Hammarström, S., Hammarström, M.L. (1996). Intraepithelial lymphocytes in human gut have lytic potential and a cytokine profile that suggest T helper 1 and cytotoxic functions. J. Immunol. 157, 1926-34.

Mackenzie, N.M. (1990). Transport of maternally derived immunoglobulin across the intestinal epithelium. In: Ontogeny of the Immune System of the Gut (Ed. by T. T. MacDonald), pp. 69-81. CRC Press, Boca Raton, Florida.

Macpherson, A.J., Gatto, D., Sainsbury, E., Harriman, G.R., Hengartner, H., Zinkernagel, R.M. (2000). A primitive T cell-independent mechanism of intestinal mucosal IgA responses to commensal bacteria. Science 288, 2222-6.

Matson, D.O., O'Ryan, M.L., Herrera, I., Pickering, L.K., Estes, M.K. (1993). Fecal antibody responses to symptomatic and asymptomatic rotavirus infections. J. Infect. Dis. 167, 577-83.

Mazanec, M.B., Kaetzel, C.S., Lamm, M.E., Fletcher, D., Nedrud, J.G. (1992). Intracellular neutralization of virus by immunoglobulin A antibodies. Proc. Natl. Acad. Sci. USA 89, 6901-5.

Mazanec, M.B., Nedrud, J.G., Kaetzel, C.S., Lamm, M.E. (1993). A three-tiered view of the role of IgA in mucosal defense. Immunol. Today 14, 430-5.

Mazzoli, S., Trabattoni, D., Lo Caputo, S., Piconi, S., Ble, C., Meacci, F., Ruzzante, S., Salvi, A., Semplici, F., Longhi, R., Fusi, M. L., Tofani, N., Biasin, M., Villa, M. L., Mazzotta, F., Clerici, M. (1997). HIV-specific mucosal and cellular immunity in HIV-seronegative partners of HIV-seropositive individuals. Nat. Med. 3, 1250-7.

Mbawuike, I.N., Pacheco, S., Acuna, C.L., Switzer, K.C., Zhang, Y., Harriman, G.R. (1999). Mucosal immunity to influenza without IgA: an IgA knockout mouse model. J. Immunol. 162, 2530-7.

Mellander, L., Carlsson, B., Hanson, L.A. (1986). Secretory IgA and IgM antibodies to *E. coli* O and poliovirus type I antigens occur in amniotic fluid, meconium and saliva from newborns. A neonatal immune response without antigenic exposure: a result of anti-idiotypic induction? Clin. Exp. Immunol. 63, 555-61.

Moser, C.A., Offit, P.A. (2001). Distribution of rotavirus-specific memory B cells in gut-associated lymphoid tissue after primary immunization. J. Gen Virol. 82, 2271-4.

Müller, F., Frøland, S.S., Hvatum, M., Radl, J., Brandtzaeg, P. (1991). Both IgA subclasses are reduced in parotid saliva from patients with AIDS. Clin. Exp. Immunol. 83, 203-9.

Nakamura, T., Matsuzaki, G., Takimoto, H., Nomoto, K. (1995). Age-associated changes in the proliferative response of rat intestinal intraepithelial leukocytes to bacterial antigens. Gastroenterology 109, 748-54.

Newman, J. (1995). How breast milk protects newborns. Sci. Am. 273, 58-61.

Nilssen, D.E., Müller, F., Øktedalen, O., Frøland, S.S., Fausa, O., Halstensen, T.S., Brandtzaeg, P. (1996). Intraepithelial γ/δ T cells in duodenal mucosa are related to the immune state and survival time in AIDS. J. Virol. 70, 3545-50.

Nilssen, D.E., Øktedalen, O., Brandtzaeg, P. (2002). Duodenal IgA and IgG1 production is increased in advanced AIDS but tends to be normalized by triple combination therapy. Submitted

Norderhaug, I.N., Johansen, F.-E., Schjerven, H., Brandtzaeg, P. (1999). Regulation of the formation and external transport of secretory immunoglobulins. Crit. Rev. Immunol. 19, 481-508.

Offit, P.A., Quarles, J., Gerber, M.A., Hackett, C.J., Marcuse, E.K., Kollman, T.R., Gellin, B.G., Landry, S. (2002). Addressing parents' concerns: do multiple vaccines overwhelm or weaken the infant's immune system? Pediatrics 109, 124-9.

Ogra, P.L., Karzon, D.T. (1969). Distribution of poliovirus antibody in serum, nasopharynx and alimentary tract following segmental immunization of lower alimentary tract with poliovaccine. J. Immunol. 102, 1423-30.

Ogra, S.S., Weintraub, D., Ogra, P.L. (1977). Immunological aspects of human colostrum and milk. III. Fate and absorption of cellular and soluble components in the gastrointestinal tract of the newborn. J. Immunol. 119, 245-8.

Ogra, P.L., Coppola, P.R., MacGillivray, M.H., Dzierba, J.L. (1974). Mechanism of mucosal immunity to viral infections in γA immunoglobulin-deficiency syndromes. Proc. Soc. Exp. Biol. Med. 145, 811-6.

O'Neal, C.M., Clements, J.D., Estes, M.K., Conner, M.E. (1998). Rotavirus 2/6 virus-like particles administered intranasally with cholera toxin, *Escherichia coli* heat-labile toxin (LT), and LT-R192G induce protection from rotavirus challenge. J. Virol. 72, 3390-3.

O'Neal, C.M., Harriman, G.R., Conner, M.E. (2000). Protection of the villus epithelial cells of the small intestine from rotavirus infection does not require immunoglobulin A. J. Virol. 74, 4102-9.

Persson, C.G., Erjefält, J.S., Greiff, L., Erjefält, I., Korsgren, M., Linden, M., Sundler, F., Andersson, M., Svensson, C. (1998). Contribution of plasma-derived molecules to mucosal immune defence, disease and repair in the airways. Scand. J. Immunol. 47, 302-13.

Phillips, A.D. (1989). Mechanisms of mucosal injury: human studies. pp. 30-40. In: Viruses and the Gut (Ed.: Farthing MJG), Smith Kline and French Laboratories, Swan Press, London.

Pisacane, A., Graziano, L., Zona, G., Granata, G., Dolezalova, H., Cafiero, M., Coppola, A., Scarpellino, B., Ummarino, M., Mazzarella, G. (1994). Breast feeding and acute lower respiratory infection. Acta Paediatr. 83, 714-8.

Regnault, A., Cumano, A., Vassalli, P., Guy-Grand, D., Kourilsky, P. (1994). Oligoclonal repertoire of the CD8αα and the CD8αβ TCR-α/β murine intestinal intraepithelial T lymphocytes: evidence for the random emergence of T cells. J. Exp. Med. 180, 1345-58.

Regnault, A., Levraud, J.P., Lim, A., Six, A., Moreau, C., Cumano, A., Kourilsky, P. (1996). The expansion and selection of T cell receptor αβ intestinal intraepithelial T cell clones. Eur. J. Immunol. 26, 914-21.

Renegar, K.B., Jackson, G.D., Mestecky, J. (1998). *In vitro* comparison of the biologic activities of monoclonal monomeric IgA, polymeric IgA, and secretory IgA. J. Immunol. 160, 1219-23.

Renegar, K.B., Johnson, C.D., Dewitt, R.C., King, B.K., Li, J., Fukatsu, K., Kudsk, K.A (2001). Impairment of mucosal immunity by total parenteral nutrition: requirement for IgA in murine nasotracheal anti-influenza immunity. J. Immunol. 166, 819-25.

Rescigno, M., Urbano, M., Valzasina, B., Francolini, M., Rotta, G., Bonasio, R., Granucci, F., Kraehenbuhl, J.P., Ricciardi-Castagnoli, P. (2001). Dendritic cells express tight junction proteins and penetrate gut epithelial monolayers to sample bacteria. Nat. Immunol. 2, 361-7.

Roberts, A.I., O'Connell, S.M., Biancone, L., Brolin, R.E., Ebert, E.C. (1993). Spontaneous cytotoxicity of intestinal intraepithelial lymphocytes: clues to the mechanism. Clin. Exp. Immunol. 94, 527-32.

Robinson, J.K., Blanchard, T.G., Levine, A.D., Emancipator, S.N., Lamm, M.E. (2001). A mucosal IgA-mediated excretory immune system *in vivo*. J. Immunol. 166, 3688-92.

Silvey, K.J., Hutchings, A.B., Vajdy, M., Petzke, M.M., Neutra, M.R. (2001). Role of immunoglobulin A in protection against reovirus entry into murine Peyer's patches. J. Virol. 75, 10870-9.

Simister, N.E., Mostov, K.E. (1989). An Fc receptor structurally related to MHC class I antigens. Nature 337, 184-7.

Smith, P.D., Smythies, L.E., Mosteller-Barnum, M., Sibley, D.A., Russell, M.W., Merger, M., Sellers, M.T., Orenstein, J.M., Shimada, T., Graham, M.F., Kubagawa, H. (2001). Intestinal macrophages lack CD14 and CD89 and consequently are down-regulated for LPS- and IgA-mediated activities. J. Immunol. 167, 2651-6.

Stoll, B.J., Lee, F.K., Hale, E., Schwartz, D., Holmes, R., Ashby, R., Czerkinsky, C., Nahmias, A.J. (1993). Immunoglobulin secretion by the normal and the infected newborn infant. J. Pediatr. 122, 780-6

Sydora, B.C., Mixter, P.F., Holcombe, H.R., Eghtesady, P., Williams, K., Amaral, M.C., Nel, A., Kronenberg, M. (1993). Intestinal intraepithelial lymphocytes are activated and cytolytic but do not proliferate as well as other T cells in response to mitogenic signals. J. Immunol. 150, 2179-91.

van Elburg, R.M., Uil, J.J., de Monchy, J.G., Heymans, H.S. (1992). Intestinal permeability in pediatric gastroenterology. Scand. J. Gastroenterol. 194 Suppl., S19-S24.

Velazquez, F.R., Matson, D.O., Calva, J.J., Guerrero, L., Morrow, A.L., Carter-Campbell, S., Glass, R.I., Estes, M.K., Pickering, L.K., Ruiz-Palacios, G.M. (1996). Rotavirus infections in infants as protection against subsequent infections. N. Engl. J. Med. 335, 1022-8.

Weaver, L.T., Wadd, N., Taylor, C.E., Greenwell, J., Toms, G.L. (1991). The ontogeny of serum IgA in the newborn. Pediatr. Allergy Immunol. 2, 72-5.

Wold, A.E., Hanson, L.Å. (1994). Defense factors in human milk. Curr. Opin. Gastroenterol. 10, 652-8.

Wolf, H.M., Fischer, M.B., Pühringer, H., Samstag, A., Vogel, E., Eibl, M.M. (1994a). Human serum IgA downregulates the release of inflammatory cytokines (tumor necrosis factor-α, interleukin-6) in human monocytes. Blood 83, 1278-88.

Wolf, H.M., Vogel, E., Fischer, M.B., Rengs, H., Schwarz, H.P., Eibl, M.M. (1994b). Inhibition of receptor-dependent and receptor-independent generation of the respiratory burst in human neutrophils and monocytes by human serum IgA. Pediatr. Res. 36, 235-43.

Wright, A.L., Bauer, M., Naylor, A., Sutcliffe, E., Clark, L. (1998). Increasing breastfeeding rates to reduce infant illness at the community level. Pediatrics 101, 837-44.

Yamanaka, T., Straumfors, A., Morton, H., Fausa, O., Brandtzaeg, P., Farstad, I. (2001). M cell pockets of human Peyer's patches are specialized extensions of germinal centers. Eur. J. Immunol. 31, 107-17.

Yuan, L., Ward, L.A., Rosen, B.I., To, T.L., Saif, L.J. (1996). Systemic and intestinal antibody-secreting cell responses and correlates of protective immunity to human rotavirus in a gnotobiotic pig model of disease. J. Virol. 70, 3075-83.

Yuan, L., Kang, S.Y., Ward, L.A., To, T.L., Saif, L.J. (1998). Antibody-secreting cell responses and protective immunity assessed in gnotobiotic pigs inoculated orally or intramuscularly with inactivated human rotavirus. J. Virol. 72, 330-8.

I, 5.   Treatment of viral gastroenteritis

Dorsey Bass

*Packard Children's Hospital at Stanford, Stanford University Hospital,
Stanford CA 94305-5208*

Introduction

Every child in the world is infected with gastroenteritis viruses during their first years
of life. In the developed world, the result is substantial morbidity (Glass *et al.*, 1996)
and expense while in developing countries hundreds of thousands of children die
(Medicine, 1986). A good deal of the difference in outcomes is due to differences
in the availability of treatment. Viral gastroenteritis (mainly after rotavirus infection)
directly causes death through dehydration. Thus, treatment and prevention of dehydration is the first goal in caring for infants with gastroenteritis. Viral gastroenteritis can
also inflict considerable nutritional insult on children due to anorexia, vomiting, malabsorption, and traditional therapies in which nutritionally poor diets are offered (Jelliffe,
1987). Finally, a bout of viral gastroenteritis can serve as trigger point for an extended
downward spiral of chronic diarrhea and malnutrition (Haffejee, 1990b). The optimal
solution to the disease burden of viral gastroenteritis would be prevention via effective
vaccines but that option remains several years away. In this section, I will discuss the
treatment of the deficiencies of fluid, electrolytes and nutrients. While there are already
very good cost-effective strategies for correcting the sequelae of viral gastroenteritis,
there is little yet available to effectively treat the acute symptoms of vomiting and
diarrhea. Several potential modes of symptomatic therapy for viral gastroenteritis are
discussed below including probiotics, antivirals, passive immunotherapy, and antidiarrheals.

Traditional therapy of diarrheal disease has often been based on the empiric observation that withholding feeds and fluids results in less stooling (Jelliffe, 1987). Obviously
this may lead to disastrous consequences in severe gastroenteritis with dehydration.
The first major advance in treatment of dehydration was the development of effective
intravenous rehydration therapy in the 1950's. Intravenous therapy requires sophisticated and relatively expensive equipment and expertise; thus, effective rehydration
therapy remained unavailable to most of the world's children. Initial attempts at producing solutions for oral rehydration therapy (ORT) contained excess glucose and led
to frequent osmotic diarrhea and hypernatremic dehydration (Paneth, 1980). The legacy
of these early solutions lives on as concern about hypernatremia with ORT, even with

its improved formulations. The development of effective ORT in the late 1960's was an offshoot of increasing knowledge of intestinal absorptive physiology — essentially the idea of glucose coupled sodium transport as a fundamental mechanism of sodium and water absorption (Hirschhorn, 1968). Further advances in rehydration include the realization that rapid rehydration and prompt refeeding improve outcomes.

Rehydration

Although much debate and speculation continue about the relative contribution of various pathophysiologic mechanisms in viral gastroenteritis, the net result of moderate to severe disease is a deficit of body water and sodium known as dehydration. The relative deficits of sodium and water determine whether the child has hypo-, normo-, or hypernatremic dehydration. The first step in treatment is assessing the degree of dehydration. Table 1 lists some of the signs and symptoms used in the assessment of mild, moderate, and severe dehydration. Unlike adults, children tend to maintain blood pressure until very advanced states of hypovolemia are reached. If the child is in or nearly in shock or is otherwise obtunded, rehydration commences with rapid intravenous administration of 20-40 ml/kg of isotonic fluid such as normal saline or lactated Ringers solution (130 mEq/L Na^+, 4 mEq/L K^+, 109 mEq/L Cl^-, 28 mEq/L $HCO3^-$). The great majority of infants are not in shock and may commence ORT.

Table 1

Signs and symptoms of dehydration in young children

Signs and Symptoms	Mild Dehydration	Moderate Dehydration	Severe Dehydration
Body weight loss (%)	3-5%	6-9%	10% or greater
Appearance	Alert, thirsty	Irritable, lethargic,thirsty	Limp, lethargic
Pulse	Normal	Rapid, weak	Rapid, feeble
Blood pressure	Normal	Normal	Decreased to undetectable
Capillary refill	Normal	+/-2 Seconds	>3 Seconds
Respiration	Normal	Deep, +/-rapid	Deep and rapid
Skin turgor	Normal	Decreased	Poor
Mucous membranes	Moist	Dry	Very dry
Fontanel/eyes	Normal	Sunken	Very sunken
Urine output	Normal	Decreased	Absent

Various ORT formulations have been studied and are in use (Table 2). Modern ORT solutions generally contain 45-90 mEq of sodium per liter and approximately equimolar carbohydrate as glucose or glucose polymer. The World Health Organization recommends ORT with a glucose to sodium molar ratio of no greater than 1.4. Glucose polymers have the advantage of lowering the solution osmolarity although a meta-analysis of treatment of non cholera (mostly rotavirus) diarrhea suggested only a slight advantage for rice-based polymer solutions (Gore *et al.*, 1992). Some ORT solutions utilize traditional complex carbohydrate sources such as those found in rice water to produce a low osmolarity solution which may be more culturally acceptable. Most solutions contain approximately 20 mEq of potassium to make up for enteral losses in form of acetate, bicarbonate or citrate. At least one authority suggests the base content is not critical for successful rehydration (Walker-Smith, 1992). Solutions not designed as ORT such as sports drinks, soft drinks, and broths are inappropriate and dangerous due to either excessive carbohydrates or salt. The compositions of some appropriate and inappropriate solutions are shown in Table 2. Additional substances such as amino acids have been added to ORT in attempts to capitalized on other intestinal co-transporter systems, but these have not proven clinically superior to glucose based ORT (Bhan, 1994).

Table 2

Comparison of oral rehydration solutions and other clear liquids

Solution/drink	Glucose (mEq/L)	Sodium (mEq/L)	Glucose/ Sodium ratio	Potassium (mEq/L)	Base (mEq/L)	Osmolarity (mMol/L)	*Appropriate for ORT*
WHO ORT	111	90	1.2	20	30	310	YES
Pedialyte	140	45	3.1	20	30	250	YES
Rehydrolyte	140	75	1.9	20	30	310	YES
Infalyte	70	45	1.4	25	30	200	YES
Gatorade	255	20	13	3	3	330	NO
Apple juice	690	3	230	32	0	730	NO
Cola	700	2	350	0	13	750	NO
Chicken broth	0	250	—	8	0	500	NO

ORT can be administered by bottle, cup, spoon, or nasogastric tube. Vomiting can be minimized by administering ORT slowly. The goal is to achieve rehydration within the first four hours of therapy. The fluid deficit is calculated as 10ccORT/kg body weight for each one percent of estimated dehydration or body weight loss (Table 1). One simple but practical recommendation is to administer one teaspoon (5 ml) of ORT each minute. For a typical 10 kg infant this would provide 120cc/kg over the first 4 hours of therapy. Properly administered ORT has been shown repeatedly to be effective in hypo-and hypernatremic dehydration with lower incidence of seizures observed than with intravenous therapy (Pizarro, 1983; Pizarro, 1984). After rehydration has been achieved, regular feedings are resumed with ORT offered after each stool to compensate for ongoing fluid and salt losses until the diarrhea episode resolves.

In both developing and developed worlds, slow refeeding is an established tradition. This practice was derived from the observation that stooling is increased immediately after a feeding. Recently, a number of studies in developing and developed countries have shown that the duration and nutritional impact of acute gastroenteritis is reduced by offering a reasonable age appropriate diet as soon as rehydration is complete (Hjelt *et al.*, 1989; Quak *et al.*, 1989). Breastfed infants are nursed throughout the rehydration period. To the surprise of older clinicians, a number of studies have shown that most infants with gastroenteritis tolerate lactose-containing diets quite well although the occasional infant may develop severe osmotic diarrhea due to acquired lactase deficiency (Armitstead *et al.*, 1989; Haffejee, 1990b; Szajewska *et al.*, 1997).

There has been considerable interest in whether specific nutritional interventions could accelerate recovery from diarrheal illness. Recent studies in Bangladesh have shown that cooked green banana or pectin added to a normal diet can hasten resolution of chronic diarrhea (Rabbani, 2001). Micronutrients have also been examined as possible therapies for acute diarrhea. Vitamin A when given prophylactically in areas of the world where children are at risk for deficiency, seems to decrease childhood mortality/morbidity from diarrhea and respiratory illness (Rahman, 2001). Results of vitamin A therapy during acute diarrhea have been less impressive (Dewan, 1995; Yurdakok, 2000; Khatun, 2001). Zinc supplements have been shown to hasten the resolution of acute watery diarrhea in the developing world (Bhutta, 2000). Whether such supplements would be effective in developed nations where deficiencies are much less common is not known.

Probiotics are live cultures of nonpathogenic bacteria which are administered as therapy or prophylaxis. As a rule, the strains do not stably colonize the gut, thus continuous dosing is required. Some strains are not easily recovered in stools. A variety of strains of bacteria and some yeast have been used to treat or prevent diarrhea. Many strains are lactobacilli such as are found in yogurts, others (bifidobacteria) are derived from the flora of breast fed infants, and *Saccharomyces bouladi* is a non-pathogenic yeast. Excellent safety profiles have been observed. The mechanisms by which these probiotics protect the gut remain unclear. Hypotheses include; change in luminal pH due to local fermentation, competition with pathogens for mucosal receptors, and stimulation of the local immune system by activating enterocytes to secrete specific cytokines (Marteau, 2001). There is evidence that probiotic therapy does enhance the IgA response to rotavirus (Majamaa *et al.*, 1995). *Lactobacillus* GG has been shown to reduce the duration of rotavirus excretion. Probiotics have been studied extensively in rotavirus gastroenteritis with *Lactobacillus* spp. most widely used. In one large

(300 subjects), double-blinded, European, multicenter study, *Lactobacillus* GG or placebo were administered in ORT to children with acute diarrhea (Guandalini *et al.*, 2000). Most of the patients had relatively mild symptoms. Rotavirus-infected children who were treated experienced about 20 hours less diarrhea than those who received placebo. Ten percent of placebo-treated infants had symptoms longer than one week compared to 2 percent of the *Lactobacillus*-treated infants. A recent meta-analysis of *Lactobacillus* as therapy for childhood diarrhea concluded that it is an effective treatment but the clinical significance of reducing duration of stooling by less than one day remains questionable (Van Niel *et al.*, 2002). Perhaps new, well powered trials with early intervention might examine more relevant endpoints such as determining whether probiotic therapy can prevent hospitalization or the need for intravenous rehydration. Most studies of probiotics for acute infantile diarrhea have taken place in developed countries. There is a need to extend these studies to the developing world where the need is most acute and the baseline intestinal microflora is quite different.

Antiviral drug therapy for viral gastroenteritis has been explored mainly in the laboratory. A number of compounds have been shown to be inhibitory to rotavirus replication *in vitro*, including ribavirin, isoflavines, and other adenosine analogues (Smee, 1982; De *et al.*, 1984; Kitaoka *et al.*, 1986). They probably owe their anti-rotavirus activity to inhibition of S-adenosylhomocysteine hydrolase, a key enzyme in regulating methylations including those that are required for the maturation of viral mRNA. Results using these drugs in animal experiments have been largely unimpressive (Schoub, 1977; Smee, 1982; Kitaoka *et al.*, 1986). A number of herbal and tea extracts have also been reported to inhibit rotavirus *in vitro* (Husson *et al.*, 1991; Clark *et al.*, 1998; Kudi and Myint, 1999). An essential question for antiviral therapy in gastroenteritis is whether much clinical benefit can be expected if rotavirus has already done most of its damage prior to the onset of symptoms in a disease with a natural duration of only a few days.

Several other approaches to inhibiting rotavirus infection have been examined. Protease inhibitors administered enterally can prevent the trypsin-mediated cleavage of VP4 which is essential for rotavirus replication and increases its infectivity (Vonderfecht *et al.*, 1988). Unfortunately, severe inhibition of gut luminal proteases leads to severe protein maldigestion, and thus this is not a practical therapy. Another approach has been the administration of glycoproteins or lipids which may function as pseudo-receptors or receptor competitors for the virus. Such entities have been identified in human breast milk and intestinal mucins (Yolken *et al.*, 1992; Yolken *et al.*, 1994; Kiefel *et al.*, 1996; Reading *et al.*, 1998). None of these approaches have been reported in clinical trials.

Passive immunotherapy has been examined for both prevention and treatment of viral gastroenteritis. Antibodies of human, bovine or egg yolk derivation have been administered enterally to infants with rotavirus diarrhea. Many of these studies have shown some decrease in stooling in treated infants. Human serum immunoglobulin containing rotavirus-specific antibodies administered orally to immunocompromised children with chronic rotavirus infection was able to associate with rotavirus in the gut, resulting in the formation of large immune complexes rather than free virus. Antigen shedding was significantly reduced after antibody therapy, although the virus was not eliminated in all cases (Losonsky *et al.*, 1985). Enteric administration of antibody has shown some efficacy in normal children. A single dose of human immunoglobulin preparation that had a high neutralization titer against rotavirus (800-3,200) significantly accelerated the rate of virus clearance and recovery from diarrhea in children infected with rotavirus (Guarino *et al.*, 1994). However, another similar study showed no significant effect of orally administered immunoglobulin (Ventura *et al.*, 1993). Bovine immunoglobulin against rotavirus has been produced by hyperimmunizing cattle with human rotaviruses (Brüssow *et al.*, 1987). Either the milk of hyperimmunized cattle or serum immunoglobulins resuspended in milk formula significantly reduced the incidence of rotavirus illness in 3 to 7 months old children when orally administered. Results of treatment of active disease with bovine antibodies has also produced conflicting results. In some studies no effect was observed (Hilpert *et al.*, 1987) while in others diarrhea was reduced (Sarker *et al.*, 1998).

An alternative means of producing anti-rotavirus immunoglobulins is to harvest specific immunoglobulins from the yolks of eggs laid by chickens that had been immunized with rotavirus. Egg yolks contain large quantities of specific immunoglobulin following intramuscular immunization of hens, and the antibody remains stable for long periods of time. Oral dosing of mice and calves with these immunoglobulins significantly reduces overall disease (Ebina *et al.*, 1990; Kuroki *et al.*, 1994; Kuroki *et al.*, 1997). In mice, this protection was directly associated with reduced rotavirus antigen distribution within the intestinal tract, suggesting that replication of the virus has been inhibited. Recent human trials in Bangladesh suggested that such egg yolk preparations may have potential therapeutic value in human infants as well (Sarker, 2001).

Cytokines and hormones play an important role in host defense and self regulation and have been examined as potential therapeutic agents for viral gastroenteritis. Interferons are cytokines which were discovered and named for their antiviral properties. Rotavirus and astrovirus are sensitive *in vivo* to the effects of class I and class II interferons (IFN) as well as Interleukin 1 (Bass, 1997). However in mice, treatment with IFN had no effect on rotavirus replication or disease (Angel *et al.*, 1999). Knockout mice lacking receptors for class I IFN and IFN gamma negative mice cleared virus as easily as their normal counterparts. These results hint at mucosal host defense systems with multiple levels of redundancy. In calves IFN has been reported as efficacious in suppressing rotavirus disease (Schwers *et al.*, 1985). Cytokines and hormones which affect intestinal epithelial integrity such as epithelial growth factor, TGF 1, hydrocortisone, and thyroid hormone have been tested experimentally as therapeutic aids for viral enteritis. Despite improvement in functional assays of the infected epithelium, epidermal growth factor and TGF failed to decrease diarrhea or increase weight gain in infected swine (Zijlstra, 1994; Rhoads *et al.*, 1995). Corticosteroids lead to early maturation of the suckling mouse gut and prevented symptomatic murine rotavirus (EDIM) infection (Wolf *et al.*, 1981). Corticosteroid treatment of suckling pigs infected with transmissable gastroenteritis virus lead to improved villous structure and improved sodium transport in intestinal mucosa (Rhoads *et al.*, 1988). *In vivo* experiments have suggested that prostaglandins may exert an antiviral effect on rotaviruses (Superti *et al.*, 1998).

Therapies aimed against symptoms of viral gastroenteritis have not been very useful in young children who typically have the most severe disease. Some compounds such as kaolin, smectite, and pysillium fiber change the consistency of stools by binding the free water but do not really alter the net fluid loss. Antiemetics are ineffective in gastroenteritis in young children and cause frequent dystonic reactions in dehydrated infants. Bismuth subsalicylate has some effect on clinical symptoms, probably due to antisecretory effects (Soriano *et al.*, 1991), but has not been recommended routinely largely because of concerns about potential metabolic toxicity, such as concern for triggering Reye's syndrome. Cholestyramine, an anion exchange resin which binds bile salts, appeared to decrease clinical symptoms in a small study (Vesikari and Isolauri, 1985). Currently available antidiarrheal medications have been disappointing in efficacy for viral enteritis in young children. Most of the antidiarrheals are based on narcotics or narcotic analogues which act mainly by altering intestinal motility. Narcotics such as tincture of opium are inappropriate for young children with acute gastroenteritis. Loperamide is a synthetic opioid which is relatively free of central effects and has virtually no abuse potential. Loperamide has been found ineffective for the treatment of early childhood diarrhea in several studies (Owens *et al.*, 1981; Kassaem *et al.*, 1983; Vesikari and Isolauri, 1985) and has been known to cause life threatening ileus and central nervous system intoxication in children with gastroenteritis (Bhutta and Tahir, 1990; Minton and Henry, 1990; Schwartz and Rodriquez, 1991)

Racecadotril is a new class of antidiarrheal medication which inhibits the degradation of endogenous opioid-like enkephalins. Animal studies showed that racecadotril had much less tendency to cause small intestinal bacterial overgrowth and far less CNS toxicity than loperamide (Duval-Iflah, 1999). In preliminary studies in children and adults it is remarkably well tolerated with no CNS effects, no increase in vomiting, and only one pediatric ileus reported to date. Racecadotril has been reported to decrease stool output in young children with rotavirus diarrhea by approximately 50% and duration of diarrhea from 72 hours to 23 hours (Salazar-Lindo *et al.*, 2000; Cezard, 2001). Thus far, racecadotril appears to be a very promising adjunctive therapy to ORT in viral gastroenteritis (See also Lundgren and Svensson, Section I, Chapter 3 of this book).

In summary, current recommended therapy for viral gastroenteritis in children is limited to rehydration and nutritional support. Current and potential therapies are summarized in Table 3. It is hoped that safe, effective and affordable treatments will be developed in coming years.

Table 3

Therapies for Viral Gastroenteritis

Treatment	Efficacy	Safety	cost
ORT	Excellent	Excellent	Low
Early Feeding	Good	Good	Low
Probiotics	Fair-good	Good	Moderate
Zinc	Fair (developing world)	Good	Low
Immunoglobulin	Fair	Good	High
Antiviral	Poor	Unknown	High
Antiemetics	Poor	Poor	Moderate-high
Antidiarrheal-opiate agonists (loperamide, diphenoxylate)	Poor	Poor	High
Cytokines/hormones	Not known	?????	High
Racecadotril	Possibly good	Possibly good	Not yet available

References

Angel J, Franco MA, Greenberg HB, Bass D. Lack of a role for type I and type II interferons in the resolution of rotavirus-induced diarrhea and infection in mice. *J Interferon Cytokine Res* 1999; **19**: 655-9.

Armitstead J, Kelly D, Walker-Smith J. Evaluation of infant feeding in acute gastroenteritis. *J Pediatr Gastroenterol Nutr* 1989; **8**: 240-4.

Bass DM. Interferon gamma and interleukin 1, but not interferon alfa, inhibit rotavirus entry into human intestinal cell lines. *Gastroenterology* 1997; **113**: 81-9.

Bhan MK. Clinical trials of improved oral rehydration salt formulations: a review. *Bull WHO* 1994; **72**: 945-55.

Bhutta TI, Tahir KI. Loperamide poisoning in children [letter] [see comments]. *Lancet* 1990; **335**: 363.

Bhutta ZA. Therapeutic effects of oral zinc in acute and persistent diarrhea in children in developing countries: pooled analysis of randomized controlled trials. *Am J of Clin Nutr* 2000; **72**: 1516-22.

Brüssow H, Hilpert H, Walther I, Sidoti J, Mietens C, Bachmann P. Bovine milk immunoglobulins for passive immunity to infantile rotavirus gastroenteritis. *J Clin Microbiol* 1987; **25**: 982-6.

Cezard JP. Efficacy and tolerability of racecadotril in acute diarrhea in children. *Gastroenterology* 2001; **120**: 799-805.

Clark KJ, Grant PG, Sarr AB, Belakere JR, Swaggerty CL, Phillips TD, Woode GN. An *in vitro* study of theaflavins extracted from black tea to neutralize bovine rotavirus and bovine coronavirus infections. *Vet Microbiol* 1998; **63**: 147-57.

De CE, Bergstrom DE, Holy A, Montgomery JA. Broad-spectrum antiviral activity of adenosine analogues. *Antiviral Res* 1984; **4**: 119-33.

Dewan V. A randomized controlled trial of vitamin A supplementation in acute diarrhea. *Indian Pediatrics* 1995; **32**: 21-5.

Duval-Iflah Y. Effects of racecadotril and loperamide on bacterial proliferation and on the central nervous system of the newborn gnotobiotic piglet. *Aliment Pharmacol Ther* 1999; **13 Suppl 6**: 9-14.

Ebina T, Tsukada K, Umezu K, Nose M, Tsuda K, Hatta H, Kim M, Yamamoto T. Gastroenteritis in suckling mice caused by human rotavirus can be prevented with egg yolk immunoglobulin (IgY) and treated with a protein-bound polysaccharide preparation (PSK). *Microbiol Immunol* 1990; **34**: 617-29.

Glass RI, Kilgore PE, Holman RC, Jin S, Smith JC, Woods PA, Clarke MJ, Ho MS, Gentsch JR. The epidemiology of rotavirus diarrhea in the United States: surveillance and estimates of disease burden. *J Infect Dis* 1996; **174 Suppl 1**: S5-11.

Gore SM, Fontaine O, Pierce NF. Impact of rice based oral rehydration solution on stool output and duration of diarrhoea: meta-analysis of 13 clinical trials [see comments]. *Br Med J* 1992; **304**: 287-91.

Guandalini S, Pensabene L, Zikri MA, Dias JA, Casali LG, Hoekstra H, Kolacek S, Massar K, Micetic-Turk D, Papadopoulou A, de Sousa JS, Sandhu B, Szajewska H, Weizman Z. Lactobacillus GG administered in oral rehydration solution to children with acute diarrhea: a multicenter European trial. *J Pediatr Gastroenterol Nutr* 2000; **30**: 54-60.

Guarino A, Canani RB, Russo S, Albano F, Canani MB, Ruggeri FM, Donelli G, Rubino A. Oral immunoglobulins for treatment of acute rotaviral gastroenteritis. *Pediatrics* 1994; **93**: 12-6.

Haffejee IE. Cow's milk-based formula, human milk, and soya feeds in acute infantile diarrhea: a therapeutic trial. *J Pediatr Gastroenterol Nutr* 1990; **10**: 193-8.

Haffejee IE. Persistent diarrhoea following gastroenteritis. *J Diarrhoeal Dis Res* 1990; **8**: 143-6.

Hilpert H, Brüssow H, Mietens C, Sidoti J, Lerner L, Werchau H. Use of bovine milk concentrate containing antibody to rotavirus to treat rotavirus gastroenteritis in infants. *J Infect Dis* 1987; **156**: 158-66.

Hirschhorn N. Decrease in net stool output in cholera during intestinal perfusion with glucose-containing solutions. *N Engl J Med* 1968; **279**: 176-81.

Hjelt K, Paerregaard A, Petersen W, Christiansen L, Krasilnikoff PA. Rapid versus gradual refeeding in acute gastroenteritis in childhood: energy intake and weight

102

gain. *J Pediatr Gastroenterol Nutr* 1989; **8**: 75-80.

Husson GP, Vilagines P, Sarrette B, Vilagines R. Antiviral effect of Haemanthus albiflos leaves extract on herpes virus, adenovirus, vesicular stomatitis virus, rotavirus and poliovirus. *Ann Pharm Fr* 1991; **49**: 40-8.

Jelliffe EF. Traditional practices concerning feeding during and after diarrhoea (with special reference to acute dehydrating diarrhoea in young children). *World Rev Nutr Dietet* 1987; **53**: 218-95.

Kassaem AS, Madkour AB, Massoub BS, Mehanna ZB. Loperamide in acute childhood diarrhea. *J Diarrh Dis Res* 1983; **1**: 10-16.

Khatun UH. A randomized controlled clinical trial of zinc, vitamin A or both in undernourished children with persistent diarrhea in Bangladesh. *Acta Paediatr* 2001; **90**: 376-80.

Kiefel MJ, Beisner B, Bennett S, Holmes ID, von Itzstein M. Synthesis and biological evaluation of N-acetylneuraminic acid-based rotavirus inhibitors. *J Med Chem* 1996; **39**: 1314-20.

Kitaoka S, Konno T, De CE. Comparative efficacy of broad-spectrum antiviral agents as inhibitors of rotavirus replication *in vitro*. *Antiviral Res* 1986; **6**: 57-65.

Kudi AC, Myint SH. Antiviral activity of some Nigerian medicinal plant extracts. *J Ethnopharmacol* 1999; **68**: 289-94.

Kuroki M, Ohta M, Ikemori Y, Icatlo Jr. F, Kobayashi C, Yokoyama H, Kodama Y. Field evaluation of chicken egg yolk immunoglobulins specific for bovine rotavirus in neonatal calves. *Arch Virol* 1997; **142**: 843-51.

Kuroki M, Ohta M, Ikemori Y, Peralta RC, Yokoyama H, Kodama Y. Passive protection against bovine rotavirus in calves by specific immunoglobulins from chicken egg yolk. *Arch Virol* 1994; **138**: 143-8.

Losonsky GA, Johnson JP, Winkelstein JA, Yolken RH. Oral administration of human serum immunoglobulin in immunodeficient patients with viral gastroenteritis. A pharmacokinetic and functional analysis. *J Clin Invest* 1985; **76**: 2362-7.

Majamaa H, Isolauri E, Saxelin M, Vesikari T. Lactic acid bacteria in the treatment of acute rotavirus gastroenteritis. *J Pediatr Gastroenterol Nutr* 1995; **20**: 333-8.

Marteau PR. Protection from gastrointestinal diseases with the use of probiotics. *Am J Clin Nutr* 2001; **73**: 430S-6S.

Medicine Io The prospects for immunizing against rotavirus disease. *New vaccine development. Establishing priorities. Diseases of importance in developing countries.* Washington, D. C., National Academy Press. 1986; **II:** 308-318.

Minton NA, Henry JA. Loperamide poisoning in children [letter; comment]. *Lancet* 1990; **335**: 788.

Owens JR, Broadhead R, Hendrickse RG, Jaswal OP, Gangal RN. Loperamide in the treatment of acute gastroenteritis in early childhood. Report of a two centre, double-blind, controlled clinical trial. *Ann Trop Paediatr* 1981; **1**: 135-41.

Paneth N. Hypernatremic dehydration of infancy: an epidemiologic review. *Am J Dis Child* 1980; **134**: 785-92.

Pizarro D. Oral rehydration in hypernatremic and hyponatremic diarrheal dehydration. *Am J Dis Child* 1983; **137**: 730-4.

Pizarro D. Hypernatremic diarrheal dehydration treated with "slow" (12-hour) oral rehydration therapy: a preliminary report. *J Pediatr* 1984; **104**: 316-9.

Quak SH, Low PS, Quah TC, Teo J. Oral refeeding following acute gastro-enteritis: a clinical trial using four refeeding regimes. *Ann Trop Paediatr* 1989; **9**: 152-5.

Rabbani GH. Clinical studies in persistent diarrhea: dietary management with green banana or pectin in Bangladeshi children. *Gastroenterology* 2001; **121**: 554-60.

Rahman MM. Simultaneous zinc and vitamin A supplementation in Bangladeshi children: randomised double blind controlled trial. *Br Med J* 2001; **323**: 314-8.

Reading PC, Holmskov U, Anders EM. Antiviral activity of bovine collectins against rotaviruses. *J Gen Virol* 1998; **79**: 2255-63.

Rhoads JM, Macleod RJ, Hamilton JR. Effect of glucocorticoid on piglet jejunal mucosa during acute viral enteritis. *Pediatr Res* 1988; **23**: 279-82.

Rhoads JM, Ulshen MH, Keku EO, Chen W, Kandil HM, Woodard JP, Liu SC, Fuller CR, Leary Jr. H, Lecce JG. Oral transforming growth factor-alpha enhances jejunal mucosal recovery and electrical resistance in piglet rotavirus enteritis. *Pediatr Res* 1995; **38**: 173-81.

Salazar-Lindo E, Santisteban-Ponce J, Chea-Woo E, Guitierrez M. Racecadotril in the treatment of acute watery diarrhea in children. *N Engl J Med* 2000; **343**: 463-7.

Sarker SA. Randomized, placebo-controlled, clinical trial of hyperimmunized chicken egg yolk immunoglobulin in children with rotavirus diarrhea. *J Pediatr Gastroenterol Nutr* 2001; **32**: 19-25.

Sarker SA, Casswall TH, Mahalanabis D, Alam NH, Albert MJ, Brüssow H, Fuchs GJ, Hammerstrom L. Successful treatment of rotavirus diarrhea in children with immunoglobulin from immunized bovine colostrum. *Pediatr Infect Dis J* 1998; **17**: 1149-54.

Schoub BD. Antiviral activity of ribavirin in rotavirus gastroenteritis of mice. *Antimicrob Ag Chemother* 1977; **12**: 543-4.

Schwartz RH, Rodriquez WJ. Toxic delirium possibly caused by loperamide [letter; comment]. *J Pediatr* 1991; **118**: 656-7.

Schwers A, Vanden BC, Maenhoudt M, Beduin JM, Werenne J, Pastoret PP. Experimental rotavirus diarrhoea in colostrum-deprived newborn calves: assay of treatment by administration of bacterially produced human interferon (Hu-IFN alpha 2). *Ann Rech Vet* 1985; **16**: 213-8.

Smee DF. Inhibition of rotaviruses by selected antiviral substances: mechanisms of viral inhibition and *in vivo* activity. *Antimicrob Ag Chemother* 1982; **21**: 66-73.

Soriano BH, Avendano P, O'Ryan M, Braun SD, Manhart MD, Balm TK Soriano HA. Bismuth subsalicylate in the treatment of acute diarrhea in children: a clinical study. *Pediatrics* 1991; **87**: 18-27.

Superti F, Amici C, Tinari A, Donelli G, Santoro MG. Inhibition of rotavirus replication by prostaglandin A: evidence for a block of virus maturation. *J Infect Dis* 1998; **178**: 564-8.

Szajewska H, Kantecki M, Albrecht P, Antoniewicz J. Carbohydrate intolerance after acute gastroenteritis--a disappearing problem in Polish children. *Acta Paediatr* 1997; **86**: 347-50.

Van Niel CW, Feudtner C, Garrison MM, Christakis DA. Lactobacillus therapy for acute infectious diarrhea in children: A meta-analysis. *Pediatrics* 2002; **109**: 678-84.

Ventura A, Nassimbeni G, Martelossi S, Bohm P, D'Agaro PL. [Experience with gamma globulins per os in the therapy and prevention of infectious diarrhea]. *Pediatr Med Chir* 1993; **15**: 343-6.

Vesikari T, Isolauri E. A comparative trial of cholestyramine and loperamide for acute diarrhoea in infants treated as outpatients. *Acta Paediatr Scand* 1985; **74**: 650-4.

Vonderfecht, SL, Miskuff RL, Wee SB, Sato S, Tidwell RR, Geratz JD, Yolken RH. Protease inhibitors suppress the *in vitro* and *in vivo* replication of rotavirus. *J Clin Invest* 1988; **82**: 2011-6.

Walker-Smith JA. Advances in the management of gastroenteritis in children. *Br J Hosp Med* 1992; **48**: 582-5.

Wolf JL, Cukor G, Blacklow NR, Dambrauskas R, Trier JS. Susceptibility of mice to rotavirus infection: effects of age and corticosteroid administration. *Infect. Immun.* 1981; **33**: 565-574.

Yolken RH, Ojeh C, Khatri IA, Sajjan U, Forstner JF. Intestinal mucins inhibit rotavirus replication in an oligosaccharide-dependent manner. *J Infect Dis* 1994; **169**: 1002-6.

Yolken RH, Peterson JA, Vonderfecht SL, Fouts ET, Midthun K, Newburg DS. Human milk mucin inhibits rotavirus replication and prevents experimental gastroenteritis. *J Clin Invest* 1992; **90**: 1984-91.

Yurdakok K. Vitamin A supplementation in acute diarrhea. *J Pediatr Gastroenterol Nutr* 2000; **31**: 234-7.

Zijlstra RT. Effect of orally administered epidermal growth factor on intestinal recovery of neonatal pigs infected with rotavirus. *J Pediatr Gastroenterol Nutr* 1994; **19**: 382-90.

SECTION II

Rotaviruses

Introduction

Rotaviruses (RVs) are the main aetiologic agent of viral gastroenteritis in infants and young children, and in the young of a large variety of animal species. Since the discovery of human RVs 28 years ago much has been learnt about their basic virology, but also their pathogenesis, immunology and epidemiology. Finally great efforts have been made to develop rotavirus vaccines.

Based on several recent reviews (Ramig, 1994; Estes, 2001; Kapikian *et al.*, 2001; Iqbal and Shaw, 1997; Desselberger, 1998a; Gray and Desselberger, 2000) the basic facts are reviewed on the following pages.

Rotaviruses are a genus of the *Reoviridae* family and possess a genome of 11 segments of double-stranded RNA (dsRNA), ranging in size from approximately 660 (segment 11) to 3,300 (segment 1) base pairs (bp). The total genomic size is approximately 18,550 bps. The genome codes for 6 structural viral proteins (VPs; termed VP1, VP2, VP3, VP4, VP6 and VP7) and 6 non-structural proteins (NSPs; termed NSP1 - NSP6). Gene-protein coding assignments and also protein-function assignments have been made for all RNA segments and are summarised in a Table of the chapter by D Poncet (Section II, Chapter 5).

Rotaviruses are triple-layered icosahedral particles measuring 75nm in diameter. The core layer is formed by VP2 and contains the viral genome and the proteins VP1 (the RNA-dependent RNA polymerase) and VP3 (several capping enzyme functions). The inner capsid (intermediate layer) contains 260 VP6 trimers which are interrupted by 132 aqueous channels of 3 different types in relation to capsid symmetry. The outer capsid (third layer) consists of 260 VP7 trimers and 60 spike-like VP4 dimers. Including the VP4 projections, the particles measure about 100nm in diameter. Much of the detailed structure has been revealed by the groups of Prasad and Yeager (Prasad *et al.*, 1988, 1990; Lawton *et al.*, 1997; Yeager *et al.*, 1990, 1994). Prasad's group has recently taken the morphological investigation from a static to a dynamic approach by analysing the structure of actively transcribing double-layered particles. From this it could be concluded that the nascent transcripts of mRNA exit through the 12 type 1 channels in the VP6/VP7 layers. This is enabled by the accumulation of transcription-active complexes consisting of VP1 and VP3 (and possibly VP2) at the inside of the core opposite the type 1 channels (Lawton *et al.*, 1997; Lawton *et al.*, 1999, 2000; Prasad and Estes, 2000). Two chapters of this section describe recent work in this area; Pesavento *et al.* (BVV Prasad's group) report new data on the genome organisation of

106

rotavirus (Section II, Chapter 1), and Rey *et al.* review the data which have allowed to establish the three-dimensional structure of VP6, the inner (and major) capsid protein of rotavirus (Section II, Chapter 2).

Rotaviruses have been classified according to the structure and immunological reactivities of VP6 (defining at least 5, possibly 7 groups, A-G; and within group A, at least 4 subgroups: I, II, I+II, nonI, nonII) and of the outer capsid proteins VP4 and VP7. As the latter both elicit neutralising antibodies, a dual typing and classification system has emerged (Estes, 2001) which, when applied to a large number of strains, has shown that there are at least 14 different VP7-specific types, termed G types (G derived from *glyco*protein) and at least 20 different VP4-specific types, termed P types (P derived from *protease*-sensitive protein). All P and G types have been unambiguously distinguished by sequencing of the relevant genes (genotypes). Genotypes are distinguished by at least 10-12% differences in the nucleotide sequences of the relevant genes. Whilst all G genotypes have also been characterised as serotypes, this is not the case for all P genotypes. Thus, the following nomenclature has been agreed: each virus is designated a P genotype [in square brackets] and a P serotype (as far as available), and a G genotype/serotype (numbers coinciding); thus, the human Wa strain is defined as P1A[8]G1 type, etc. Lists of G types and P types known at present are provided by Estes (2001); the lists are likely to grow as more isolates are characterised.

The major steps of RV replication are as follows:
- adsorption to a cellular receptor (likely to involve coreceptors), or direct penetration into the cytoplasm;
- removal of the outer capsid layer (VP7 and VP4);
- mRNA production in the cytoplasm from single-shelled (bi-layered) subviral particles;
- mRNA translation into six structural and six nonstructural proteins;
- involvement of NSP2 and NSP5 in the assembly of single-shelled particles containing VP2, VP1, VP3 and VP6 and a full complement of 11 segments of single-stranded RNAs (ssRNAs);
- synthesis of dsRNA (replication) in precore particles;
- formation of aggregates of bilayered particles (pseudocrystals termed viroplasm) in the cytoplasm;
- particle maturation into double-shelled (triple-layered) particles by NSP4 acting as an intracellular receptor for bilayered particles and by apposition of VP7 and VP4 in the rough endoplasmic reticulum (RER) after glycosylation of VP7 (occurrence of transiently enveloped particles in the RER containing VP4 and VP7);
- liberation of triple-layered, non-enveloped infectious particles by cell lysis.

The cycle of viral replication is completed in approximately 12 hours. Aspects of rotavirus replication are addressed in the chapters by López and Arias (receptor and coreceptor characterisation; Section II, Chapter 3), Patton *et al.* (the role of virus-specific structural and nonstructural gene products in viral replication, Section II, Chapter 4), Poncet (detailed update on translation mechanisms, Section II, Chapter 5) and Taylor and Bellamy (interaction of NSP4 with viral and intracellular proteins; Section II, Chapter 7). The activation/suppression of cellular proteins by rotavirus infection can be measured in a very comprehensive way by the microarray technique, and first results are reported by Feigelstock *et al.* (Greenberg's group, Section II, Chapter 9). Chapters 1 and 2 of Section II also touch on replication aspects. Not enclosed in this volume are interesting new data from G. Trugnan's and J Cohen's groups who found that in both MA104 and CaCo-2 cells cellular membrane microdomains enriched in cholesterol and sphingolipids (rafts) are involved in the assembly of rotavirus by direct early interaction of VP4 with rafts, triggering atypical targeting of the nascent virus (Jourdan *et al.*, 1997; Nejmeddine *et al.*, 2000; Sapin *et al.*, 2002).

The pathogenesis of rotaviruses has been extensively studied as good animal models are available, due to the vast host reservoir of rotaviruses. Most work has been carried out in mice, gnotobiotic piglets and calves and rabbits. Rotavirus infection of rats has recently emerged as a promising animal model and is characterised and its potential described by Ciarlet *et al.* (Section II, Chapter 10).

Rotaviruses infect the apical cells of the villi of the small intestine, causing cell death and desquamation (Greenberg *et al.*, 1994). The necrosis of the apical villi reduces digestion, followed by diarrhoea due to primary malabsorption. Villous atrophy is compensated for by a reactive crypt cell hyperplasia, accompanied by increased secretion which also contributes to the diarrhoea. Recovery is by replacement of the desquamated villous epithelium with enteroblasts ascending from crypts. By then the diarrhoea stops. The pathogenic process takes 5 to 7 days. It is remarkable that diarrhoea starts before villous lesions become morphologically obvious (see below). A L Servin describes in detail pathophysiological processes in the cell following rotavirus infection (Section II, Chapter 8).

Viral factors determining pathogenicity have been investigated in several animal models. The results are not uniform. In various virus-host models different viral proteins have been identified as correlated with pathogenicity (VP3, VP4, NSP1, NSP2, VP7, NSP4; Burke and Desselberger, 1996).

Recent studies have shown that NSP4 or a C terminal peptide derived thereof, has a diarrhoeogenic effect, i.e. acts as a viral enterotoxin (Ball *et al.*, 1996; Morris *et al.*, 1999). The protein seems to be secreted in a truncated form, interacts with specific (still unidentified) receptors on the surfaces of uninfected cells and thereby triggers intracellular messenger cascades leading to an increase of the intracellular calcium level and an efflux of chloride and water (Estes *et al.*, 2001). This is an early event, occurring before cellular lysis due to rotavirus replication is seen. M Estes has reviewed the present status of research into NSP4, checking the experimental data against the predictions of the model mechanisms of action (Section II, Chapter 6)

Host-specific factors may also play a role in the generation of diarrhoea. It was shown that the autonomous nervous system of the gut may be involved as drugs affecting autonomous nerves can influence the extent of diarrhoea (Lundgren *et al.*, 2000; Lundgren and Svensson, Section I, Chapter 3 of this book).

After a short incubation period of 24 to 48 hours the illness has a sudden onset, with watery diarrhoea, vomiting and often rapid dehydration. In developing countries rotavirus diarrhoea represents a major cause of infantile death (Bern *et al.*, 1992 and see below). It should be noted though that clinical symptoms after RV infection vary widely and that asymptomatic infections are not rare.

As large numbers of virus particles are shed during the acute phase of the illness, diagnosis using electron microscopy, passive particle agglutination test, or enzyme-linked immunosorbent assays is relatively easy and straightforward. Prolonged presence of viral RNA in the gut has been observed using RT-PCR techniques (Richardson *et al.*, 1998).

Treatment is mainly with oral rehydration solution (ORS) formulae which have been approved by the WHO and are applied worldwide. The use of antimotility drugs is discouraged (Desselberger, 1999), but may change with the use of encephalinase inhibitors drug treatment (Salazar-Lindo *et al.*, 2000; Bass, Section I, Chapter 5 of this book).

A lot of work has gone into the analysis of the immune response to rotavirus infections and trying to identify the true correlates of protection. Most of this work has been done in animal models. Whilst there is a complex humoral and also a cell-mediated immune response, it has become clear that rotavirus-specific local secretory IgA antibodies (copro-IgA) represent the best correlate of protection (Offit, 1994; Yuan *et al.*, 1996; Moser *et al.*, 1998). Gonzalez *et al.* (Franco's group, Section II, Chapter 11) have started to study the memory B and T cell responses of children after rotavirus infection, including mechanisms of immune cells to locate to the gut (see also the review by Brandtzaeg and Johansen; Section I, Chapter 4).

The epidemiology of rotavirus infections is complex. In developing countries, approximately 125 million cases of rotavirus infection occur annually in children under 5 years of age of which 18 million are moderately severe to severe; at least 600,000 children die annually from rotavirus infection in these countries. In developed countries the morbidity is high (several million cases of severe diarrhoea per annum), but the mortality is very low, mainly due to the universal availability of rehydration therapy (Bern *et al.*, 1992; Bresee *et al.*, 1999).

Rotaviruses are transmitted mostly via the feco-oral route. Group A RVs are the major cause of human infections. RVs are universally endemic. In temperate climates outbreaks with a strict seasonal winter/spring pattern occur whilst in tropical regions infections are spread more evenly throughout the year. At any one time and site there is cocirculation of RVs of different G/P combinations. As VP4 and VP7 are encoded by different RNA segments (see Table 1 in Section II, Chapter 6), various combinations of G and P types can arise after reassortment in doubly-infected cells and have been observed *in vivo* and *in vitro,* both in man and in animals (Desselberger, 1998a; Iturriza-Gómara *et al.*, 2000). Viruses carrying G1P[8], G2P[4], G3P[8] and G4P[8] represent over 95% of human strains corcirculating in most countries, and G1P[8] strains occur most frequently. However, viruses of other G and P types have been isolated in increasing numbers and are found at high frequencies in countries of South America, Africa and South/East Asia (Desselberger *et al.*, 2001; Cunliffe *et al.*, 2002). Iturriza-Gómara *et al.* (Section II, Chapter 12) have reviewed the molecular epidemiology of rotaviruses and considered the genetic mechanisms underlying their diversity.

Group B rotaviruses have been found to be the cause of waterborne outbreaks of gastroenteritis in China (Hung, 1988) and of sporadic cases in India (Krishnan *et al.*, 1999; Sen *et al.*, 2000). Group C rotaviruses were confirmed as the cause of small family outbreaks of gastroenteritis and of sporadic cases (Matsumoto *et al.*, 1989; Caul *et al.*, 1990; Jiang *et al.*, 1995) but may cause frequent inapparent infections (James *et al.*, 1997; Steele and James, 1999, Kuzuya *et al.*, 2001; Iturriza-Gómara *et al.*, in preparation).

As RV infections are the major cause of viral gastroenteritis in infancy and early childhood, from the onset of the recognition of RVs as causative agents efforts have been made to develop a vaccine (for review see Vesikari, 1997; Desselberger, 1998b; Bresee *et al.*, 1999). After many years of endeavour, a live-attenuated rhesus rotavirus (RRV)-based, tetravalent (TV) human reassortant vaccine was licensed by the FDA in August 1998 for universal use in infancy, applying three doses at 2, 4 and 6 months of age (CDC, 1999a). The vaccine contains VP7 molecules of human G1, G2 and G4 strains on RRV monoreassortants and G3 VP7 on RRV; it thus covers potentially more than 95% of rotaviruses circulating in infants and young children in most countries in temperate climates. The vaccine has only a moderate effect in preventing infection, but has efficacy of 80% or higher in preventing severe disease. Between September 1998 and July 1999 approximately 1.5 million doses of the vaccine were administered in the United States. The emergence of gut intussusception was observed in a small number of infants after the first dose and lead the US Centers for Disease Control (CDC) to recommend postponing the administration of the RRV-TV to infants scheduled to receive the vaccine during the following months (CDC, 1999b). In October 1999 the ACIP recommendation was withdrawn although the FDA licence has not been revoked (CDC; 1999c). There are at present discussions on the true attributable risk to develop intussusception after vaccination, and a risk benefit analysis on the use of the vaccine in developing countries is ongoing (For review and further details see Kapikian, 2001; Murphy *et al.*, 2001). Offit *et al.* have reviewed the current status of the development of human rotavirus vaccines (Section II, Chapter 13).

Other vaccine candidates are being investigated (e.g. Offit *et al.*, 1994; Treanor *et al.*, 1995; Clark *et al.*, 1996; Conner *et al.*, 1996; Chen *et al.*, 1998; Clements-Mann *et al.*, 1999; Herrmann *et al.*, 1999; Jiang *et al.*, 1999; Bernstein *et al.*, 1999; Coste *et al.*, 2000). Yuan and Saif (Section II, Chapter 14) and Herrmann (Section II, Chapter 15) report on animal model work in which rotavirus-like particles (VLPs) and rotavirus-specific DNA molecules, respectively, are explored as possible candidate vaccines.

For a number of single-stranded RNA viruses of positive and negative polarity of their genomic RNA, reverse genetics systems have been developed by constructing cDNA clones which are infectious in the presence or absence of supporter functions or helper viruses. These systems allow exact manipulation of cDNAs (deletion/insertion, rearrangements, site-directed mutagenesis) and thus to dissect structure-function relationships as well as to study biological characteristics at a molecular level. The establishment of such a system for rotaviruses and other genera of the *Reoviridae* is one of the major goals of ongoing research (Desselberger and Estes, 2000).

References

Ball JM, Tian P, Zeng CQY, Morris AP, Estes MK (1996). Age-dependent diarrhea induced by a rotaviral nonstructural glycoprotein. *Science* **272:** 101-104.

Bern C, Martinez J, deZoysa I, Glass, RI (1992). The magnitude of the global problem of diarrhoeal disease: a ten-year update. *Bull. WHO* **70:** 705-714.

Bernstein DI, Sack DA, Rothstein E *et al.* (1999). Efficacy of live, attenuated, human rotavirus vaccine 89-12 in infants: a randomized, placebo controlled trial. *Lancet* **354:** 287-290.

Bresee JS, Glass RS, Ivanoff B, Gentsch JR (1999). Current status and future priorities for rotavirus vaccine development, evaluation and implementation in developing countries. *Vaccine* **17:** 2207-2222.

Burke B, Desselberger U, (1996). Rotavirus pathogenicity. *Virology* **218;** 299-305.

Caul EO, Ashley CR, Darville JM, Bridger JC (1990). Group C rotavirus associated with fatal enteritis in a family outbreak. *J. Med. Virol.* **30:** 201-205.

CDC (1999a). Rotavirus vaccine for the prevention of rotavirus gastroenteritis among children. Recommendations of the Advisory Committee on Immunization Practices (ACIP). *Morb. Mortal. Wkly. Rep.* **48:** 1-23.

CDC (1999b). Intussusception among recipients of rotavirus vaccine in United States, 1998-1999. *Morb. Mortal. Wkly. Rep.* **48:** 577-581.

CDC (1999c). Withdrawal of rotavirus vaccine recommendation. *Morb. Mortal. Wkly. Rep.* **48:** 1007.

Chen SC, Jones DH, Fignan FF *et al.* (1998). Protective immunity induced by oral immunization with a rotavirus DNA vaccine encapsulated in microparticles. *J. Virol.* **72:** 5757-5761.

Clark HF, Offit PA, Ellis RCV *et al.* (1996). The development of multivalent bovine rotavirus (strain WC3) reassortant vaccines in infants. *J. Infect. Dis.* **174 (Suppl.1):** S73-S80.

Clements-Mann ML, Makhene MK, Mrukowicz J *et al.* (1999). Safety and immunogenicity of live attenuated human-bovine (UK) reassortant rotavirus vaccines with VP7-specificity for serotypes 1, 2, 3 or 4 in adults, children and infants. *Vaccine* **17:** 2715-2725.

Conner ME, Zarley CD, Hu B *et al.* (1996). Virus-like particles as a rotavirus subunit vaccine. *J. Infect. Dis.* **174 (Suppl. 1):** S88-S92

Coste A, Sirard JL, Johansen K, Cohen J, Kraehenbuehl JP (2000). Nasal immunization of mice with virus-like particles protects offspring against rotavirus diarrhea. *J. Virol.* **74:** 8966-8971.

Cunliffe NA, Bresee JS, Gentsch JR, Glass R, Hart CA (2002). The expanding diversity of rotaviruses. *Lancet* **359:** 640-642.

Desselberger U, (1998a). Reoviruses. In: *Topley and Wilson's Microbiology and Microbial Infections.* 9th ed., Vol.1: *Virology* (Mahy, BWJ and Collier, L eds.), E. Arnold, London-Sydney-Auckland, pp 537-550.

Desselberger U, (1998b). Prospect for vaccines against rotaviruses. *Rev. Med. Virol.* **8:** 43-52.

Desselberger U, (1999) Rotavirus infection: guidelines for treatment and prevention. *Drugs* **58:** 447-452.

Desselberger U, Estes MK (2000). Future rotavirus research. In: *Methods in Molecular Medicine. Rotaviruses.* (J Gray, U Desselberger, eds). Vol **34,** pp 239-258. Humana Press, Totowa, NJ.

Desselberger U, Iturriza-Gómara M, Gray JJ (2001). Rotavirus epidemiology and surveillance. *Novartis Found. Symp.* **238:**125-147 (Discussion 147-152).

Estes MK (2001). Rotaviruses and their replication. In: *Fields Virology,* 4th ed., (Knipe DM, Howley PM, *et al.,* eds.), Lippincott Williams & Wilkins, Philadelphia, pp 1747-1785.

Estes MK, Kang GD, Zeng CQY, Crawford SE, Ciarlet M (2001). Pathogenesis of rotavirus gastroenteritis. *Novartis Found. Symp.* **238:** 82-96 (Discussion 96-100).

Gray JJ, Desselberger, U (eds) (2000). *Rotaviruses. Methods and Protocols.* Methods in Molecular Medicine, Humana Press, Totowa, NJ.

Greenberg HB, Clark HF, Offit PA (1994). Rotavirus pathology and pathophysiology. In: *Rotaviruses* (Ramig, RF ed), Springer Verlag, Berlin-Heidelberg, pp 256-283.

Herrmann JE, Chen SC, Jones DH *et al.* (1999). Immune responses and protection obtained by oral immunization with rotavirus VP4 and VP7 DNA vaccines encapsulated in microparticles. *Virology* **259:** 148-153.

Hung T (1988). Rotavirus and adult diarrhoea. *Adv. Virus. Res.* **35:** 193-218.

Iqbal N, Shaw RD (1997). Rotaviruses. In: *Clinical Virology* (Richman DD, Whitley RJ, Hayden RG, eds), pp 765-785. Churchill Livingstone, New York-Edinburgh-London.

Iturizza-Gómara M, Isherwood B, Desselberger U, Gray JJ (2000). Reassortment *in vivo:* a driving force for diversity of rotavirus strains isolated in the UK between 1995 and 1999. *J. Virol.* **75:** 3696-3705.

James VL, Lambden PR, Carl EO, Booke SJ, Clarke IN (1997). Seroepidemiology of human group C rotavirus in the UK. *J. Med. Virol.* **52:** 86-91.

Jiang B, Dennehy PH, Spangenberger S, Gentsch JR, Glass RI (1995). First detedtion of group C rotavirus in fecal specimens of children with diarrhoea in the United States. *J. Infect. Dis.* **172:** 45-50.

Jiang B, Estes MK, Barone C *et al.* (1999). Heterotypic protection from rotavirus infection in mice vaccinated with virus-like particles. *Vaccine* **17:** 1005-1013.

Jourdan N, Maurice M, Delautier D, Quero AM, Servin AL, Trugnan G (1997). Rotavirus is released from the apical surface of cultured human intestinal cells through nonconventional vesicular transport that bypasses the Golgi apparatus. *J. Virol.* **71:** 8268-8278.

Kapikian AZ (2001). A rotavirus vaccine for prevention of severe diarrhoea in infants and young children: development, utilization and withdrawal. *Novartis Found. Symp.* **238:**153-171 (Discussion 171-179).

Kapikian AZ, Hóshino Y, Chanock RM (2001). Rotaviruses. In: *Fields Virology,* 4th ed. (Knipe DM, Howley PM *et al.,* eds.), Lippincott Williams & Wilkins, Philadelphia, pp 1787-1833.

Krishnan T, Sen A, Choudhury JS, Das S, Naik TN, Bhattacharya SK (1999).

Emergence of adult diarrhoea virus in Calcutta, India. *Lancet* **353:** 380-381.

Kuzuya M, Fujii R, Hamano J, Ohota R, Ogura H, Yamada M (2001). Seroepidemiology of human group C rotavirus in Japan based on a blocking enzyme-linked immunosorbent assay. *Clin. Diagn. Lab. Immunol.* **8:**161-165.

Lawton JA, Estes MK, Prasad BVV (1997). Three-dimensional visualization of mRNA release from actively transcribing rotavirus particles. *Nature Struct. Biol.* **4:** 118-121.

Lawton JA, Estes MK, Prasad BV (1999). Comparative structural analysis of transcriptionally competent and incompetent rotavirus-antibody complexes. *Proc. Natl. Acad. Sci. USA* **96:** 5428-5433.

Lawton JA, Estes MK, Prasad BV (2000). Mechanism of genome transcription in segmented dsRNA viruses. *Adv. Virus Res.* **55:** 185-229.

Lundgren O, Peregrin AT, Persson K, Kordasti S, Uhnoo I, Svensson L (2000). Role of the enteric nervous system in the fluid and electrolyte secretion of rotavirus diarrhea. *Science* **257:** 491-495.

Matsumoto K, Hatano J, Kobayashi K, Hasegawa A *et al.* (1989). An outbreak of gastroenteritis associated with acute rotaviral infection in school children. *J. Infect. Dis.* **160:** 611-615.

Morris AP, Scott JK, Ball JM, Zeng CQ, O'Neal WK, Estes MK (1999). NSP4 elicits age-dependent diarrhoea and Ca^{2+} mediated ion influx into intestinal crypts of CF mice. *Amer.J. Physiol.* **277:** G431-444.

Moser CA, Cookinham S, Coffin SE, Clark HF, Offit PA (1998). Relative importance of rotavirus-specific effector and memory B cells in protection against challenge. *J. Virol.* **72:** 1108-1114.

Murphy TV, Gargiullo PM, Massoudi MS, Nelson DB, Jumaan AO, Okoro CA, Zanardi LR, Setias Fair E, LeBaron CW, Wharton M, Livingood JK (2001). Intussusception among infants given an oral rotavirus vaccine. *N. Engl. J. Med.* **344:** 564-572.

Nejmeddine M, Trugnan G, Sapin C, Kohli L, Svensson L, López S, Cohen J (2000). Rotavirus spike protein VP4 is present at the plasma membrane and is associated with microtubules in infected cells. *J. Virol.* **74:** 3313-3320.

Offit PA (1994). Rotaviruses: immunological determinants of protection against infection and disease. *Adv. Virus Res.* **44:** 161-202.

Offit PA, Khoury CA, Moser CA *et al.* (1994). Enhancement of rotavirus immunogenicity by microencapsidation. *Virology* **203:** 134-143.

Prasad BVV, Burns JW, Marietta Z, Estes MK, Chiu, W (1990). Localization of VP4 neutralization sites in rotavirus by three-dimensional cryo-electron microscopy. *Nature* **343:** 476-479.

Prasad BVV, Estes MK (2000). Electron cryo-microscopy and computer image processing techniques. In: *Rotaviruses. Methods and Protocols.* Methods in Molecular Medicine, (Gray JJ, Desselberger U Eds). Humana Press, Totowa, NJ, pp 9-31.

Prasad BVV, Wang GJ, Clerx JP, Chiu W (1988). Three-dimensional structure of rotavirus. *J. Mol. Biol.* **199:** 269-275.

Ramig F (Ed) (1994). *Rotaviruses*. Pp 380. Springer Verlag, Heidelberg.

Richardson S, Grimwood K, Gorrell R, Palombo E, Barnes G, Bishop R (1998). Extended excretion of rotavirus after severe diarrhoea in young children. *Lancet* **351:** 1844-1848.

Salazar-Lindo E, Santisteban-Ponce J, Chea-Woo E and Guiterrez M. (2000). Racecadotril in the treatment of acute watery diarrhea in children. *N. Engl. J. Med.* **343;** 463-467.

Sapin C, Colard O, Delmas O *et al.* (2002). Rafts promote assembly and atypical targeting of a nonenveloped virus, rotavirus, in CaCo-2 cells. *J. Virol.* **76:** 4591-4602.

Sen A, Kobayashi N, Das S, Krishnan T, Bhattacharya SK, Urasawa S, Naik TN (2000). Amplification of various genes of human Group B rotavirus from stool by RT-PCR. *J. Clin. Virol.* **17:** 177-181.

Steele AD, James VL (1999) Seroepidemiology of group C rotavirus in South Africa. *J. Clin. Microbiol.* **37:** 4142-4144.

Treanor JJ, Clark HF, Pichichero M *et al.* (1995). Evaluation of the protective efficacy of a serotype 1 bovine-human rotavirus reassortant vaccine in infants. *Pediatr. Infect. Dis. J.* **14:** 301-307.

Vesikari T (1997). Rotavirus vaccines against diarrhoeal disease. *Lancet* **350:** 1538-1541.

Yeager M, Berriman JA, Baker, TS, Bellamy AR (1994). Three-dimensional structure of the rotavirus haemagglutinin VP4 by cryo-electron microscopy and difference map analysis. *EMBO J.* **13:** 1011-1018.

Yeager M, Dryden KA, Olson NH, Greenberg HB, Baker TS (1990). Three-dimensional structure of rhesus rotavirus by cryo-electron microscopy and image reconstruction. *J. Cell. Biol.* **110:** 2133-2144.

Yuan LJ, Ward LA, Rosen BI, To TL, Saif LJ (1996). Systemic and intestinal antibody secreting cell responses and correlates of protective immunity to human rotaviruses in a gnotobiotic pigs model of disease. *J. Virol.* **70:** 3075-3083.

Viral Gastroenteritis
U. Desselberger and J. Gray (editors)

II, 1. Structural organization of the genome in rotavirus

Joseph B. Pesavento[1], Mary K. Estes[3] and B.V. Venkataram Prasad[1,2]

[1] *Verna and Marrs McLean Department of Biochemistry and Molecular Biology*
[2] *W.M. Keck Center for Computational Biology*
[3] *Department of Molecular Virology and Microbiology, Baylor College of Medicine, Houston, TX 77030*

Introduction

Rotavirus is a well-characterized genus of the *Reoviridae* family, which consists of 11 genera (Fields, 1996). Other well-studied genera in this family include *orthoreovirus* (Reinisch *et al.*, 2000), *orbivirus* (Grimes *et al.*, 1997; Gouet *et al.*, 1999), *aquareovirus* (Nason *et al.*, 2000), *cypovirus* (Hill *et al.*, 1999; Zhang *et al.*, 1999), and *phytoreovirus* (Zhou *et al.*, 2001). One of the distinguishing features of these viruses is that their genomes consist of multiple segments of dsRNA often enclosed within multiple capsid layers. During the replication of these viruses, the viral genome is converted into messenger RNA for subsequent progeny template generation and synthesis of structural and non-structural proteins. Because host cells do not possess the capability of transcribing the dsRNA genome, these viruses have evolved to carry within themselves all the enzymatic machinery required for transcription. These viruses have the unique capability of transcribing the multiple segments of genomic dsRNA, simultaneously and repeatedly within the intact capsid layers in a process known as endogenous transcription. Structural knowledge of how the genome is organized inside these viruses and how it interacts with the transcription enzymes is important to understand the molecular mechanisms of endogenous transcription.

Structural description of rotavirus

The genome of rotavirus consists of eleven segments of dsRNA ranging in size from 667 to 3302 nucleotides, and its genes code for six structural proteins and six non-structural proteins (Fig. 1a) (Estes, 2001). Rotavirus is a triple layered icosahedral particle of ~1000 Å diameter including the spikes (Fig. 1b and 1c). The genome is packaged entirely within the innermost capsid layer. This innermost layer is composed of 120 copies of VP2 (Fig. 1d), a known RNA binding protein (Labbé *et al.*, 1991; Lawton *et al.*, 1997b). The transcription enzymes VP1 and VP3 are attached as a heterodimeric complex (flower-shaped features in Fig. 1c) to the inside surface of VP2 at the 5-fold symmetry positions (Prasad *et al.*, 1996). Such a conclusion is consistent with biochemical studies which indicated that VP1, the RNA-dependent RNA polymerase (Valenzuela *et al.*, 1991), interacts with VP3, the guanylyl and methyl transferase (Liu *et al.*, 1992), to form a transcription enzyme complex.

The addition of 780 copies of VP6 on top of the VP2 layer with T=13 icosahedral symmetry produces the double layered particle (DLP). The outermost layer of the triple layered particle (TLP) consists of 780 copies of the glycoprotein VP7 which has T=13 symmetry, and 120 copies of VP4, the cell attachment protein, arranged as 60 dimers emerging from the surface of this shell (Fig. 1b) (Prasad *et al.*, 1988; Prasad *et al.*, 1990; Shaw *et al.*, 1993; Yeager *et al.*, 1994). During cell entry, the TLP loses the VP7 and VP4 proteins, and the resulting DLP becomes transcriptionally active within the cytoplasm (Estes, 2001).

The rotavirus TLPs and DLPs contain 132 porous channels which allow for the exchange of compounds in aqueous solution to the inside of the capsid (Fig. 1b). There are 12 type I channels, each located at the icosahedral five-fold vertices of the TLP and DLP. Immediately surrounding each of the type I channels there are five type II channels. VP4 is attached to the edge of each one of the 60 type II channels. Finally, there are 60 type III channels, each at one of the hexavalent positions immediately neighboring the icosahedral three-fold axis.

Questions related to genome organization

What is known about the structure of the genome?

Structural comparisons of native and recombinant empty particles showed that the RNA genome exhibits significant icosahedral ordering, and the outermost layer of the rotavirus genome actually has a dodecahedral cage-like organization, with the dsRNA helices localized around the VP1/VP3 transcription complexes at the five-fold vertices (Fig. 1c and 1e) (Prasad *et al.*, 1996). This icosahedrally ordered outer layer of RNA consists of nearly 25% of the total mass of the genome. Such an orderly organization of the genome is thought to be important for endogenous transcription, as it is conducive for the orchestrated movement of RNA relative to the VP1/VP3 enzyme complex during mRNA production. New mRNA transcripts are translocated from the particle interior through the type I channels at the 5-fold axes (Lawton *et al.*, 1997a: Fig. 1f and 1g, see below). Further inward radially within the particle, the dsRNA appears as regularly spaced concentric layers of density (Fig. 2a; Pesavento *et al.*, 2001). These concentric layers appear as peaks in the radially averaged density profiles (Fig. 2c). The presence of these concentric layers of RNA has also been observed in other members of the Reoviridae such as bluetongue virus (BTV; Gouet *et al.*, 1999), cypovirus (CPV; Hill *et al.*, 1999), and orthoreovirus (Reinisch *et al.*, 2000),

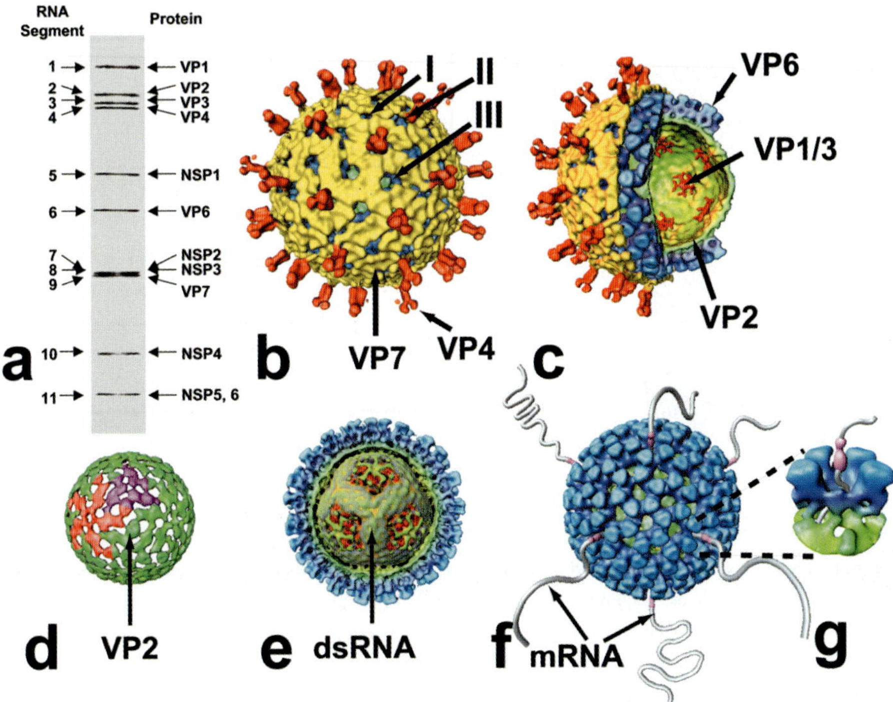

Fig. 1. Structural organization of rotavirus. a) RNA gel showing rotavirus genes and encoded products. b) Surface representation of the rotavirus structure. Three types of channels, the VP4 spikes (red) and VP7 capsid layer (in yellow) are indicated by arrows. c) Cut-away of the rotavirus structure showing the internal VP6 (blue) and VP2 (green) layers. The flower-shaped VP1/VP3 complex, attached to the inside of VP2 layer complex is shown in red. d) Structural organization of the VP2 layer. In one of the 60 dimers that constitute this layer, the two VP2 subunits are shown in red and purple. e) Genomic RNA in the rotavirus structure. The VP6 and VP2 layers are "peeled off" half way to expose the genomic organization. The outer layer of RNA has a dodecahedral appearance. f) Structure of the actively transcribing DLPs, with proposed exit pathway out of the type I channel at the five-fold vertex. g) A close-up cutaway view depicting the exit pathway in one of the channels. The transcripts are shown as gray strands in panels (f) and (g). The bowling-pin shaped density due to the exiting transcript as seen in the reconstructions of actively transcribing DLPs (Lawton *et al.*, 1997a) is shown in pink.

What are the conditions for endogenous transcription?

Once the virus penetrates the host cell membrane and the outer capsid protein layer comes off, the DLP becomes transcriptionally competent. The DLP transcribes its gene segments, and released mRNA transcripts are free to be translated in the cytoplasm by the host cell machinery. Such translation and subsequent replication of the genome ultimately leads to the assembly of new virus progeny (Mason *et al.*, 1980). None of

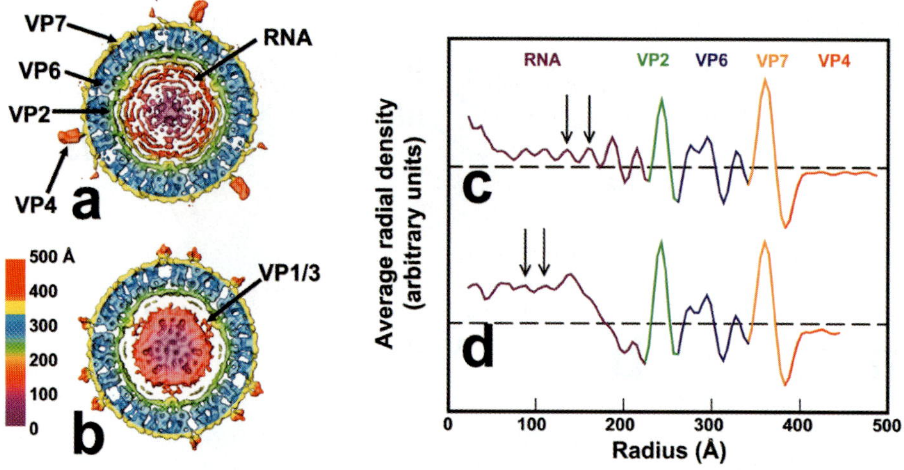

Fig. 2. Structural features of genomic RNA density. a) Equatorial cross-sectional slice from the rotavirus reconstruction under native pH conditions. In this native expanded state, the concentric layers of RNA density are easily visible as touching the inside of the VP2 layer (green). b) Similar equatorial section as in (a) showing the RNA in the condensed state after ammonium high pH treatment. Color bar indicates protein composition by radial distance as in Fig. 1. c) Radial density profile of c) native and d) condensed conditions. Arrows indicate spacing between RNA layers.

the rotaviral proteins, by itself, is capable of transcription; thus, the structural integrity of the DLP is an important requirement for transcription to occur. The removal of the outer VP7 layer, which sits on the VP6 layer, is another necessary requirement for transcription to occur. *In vitro* studies have shown that mature virions with an intact VP7 layer are transcriptionally incompetent. Some anti-VP6 monoclonal antibodies are also known to inhibit transcription (Ginn *et al.*, 1992; Lawton *et al.*, 1999). The site of transcription, which is the location of the VP1/VP3 enzyme complex, is about 150 Å away from the tip of VP6 where VP7 interacts. How then do the outer shell of VP7 or certain VP6-specific antibodies inhibit transcription? At what stage is the transcription inhibited – is it at the stage of transcriptional initiation, or elongation, or translocation?

How does rotavirus undergo continuous and simultaneous transcription?

Given a sufficient excess of the necessary precursors, it has been shown that members of the *Reoviridae* continuously transcribe their genome segments until all of the biochemical precursors are depleted (Cohen, 1977). This continuous and repeated nature of genome transcription clearly indicates a high efficiency for this process. It also suggests that some type of internal ordering of the genome segments and transcription enzymes is required for a gene segment to be transcribed repeatedly. Others have shown that all of the genome segments are transcribed both independently of one another and yet simultaneously (Gillies *et al.*, 1971; Smith and Furuichi 1982). Since the shorter gene segments are created more quickly, they tend to accumulate in a linear fashion much more rapidly, but, nevertheless, all of the mRNA transcripts are present from the earliest time points. These findings indicate that rotavirus does not undergo selective processing of certain genes in any particular time order.

How fast does endogenous transcription occur?

Orthoreovirus endogenously transcribes its 10 gene segments from a single layered core particle. It was shown some time ago that reovirus RNA is transcribed by its dsRNA polymerase at a maximum rate of ~50 nucleotides per second (Bartlett *et al.*, 1974). Using this value as an upper estimate in rotavirus, with a genome of ~18,550 base pairs, it should only take slightly more than six minutes for the entire genome to be transcribed by a single enzyme complex. If we assume full occupancy of twelve transcription complexes with genome segments, it would take the DLP on average 30 seconds to replicate one complete copy of the genome. This high rate of transcription suggests that some type of ordering of the dsRNA is required if it is to be efficiently transcribed and rapidly cycled through the enzyme complex for the subsequent round of transcription.

How do the mRNA transcripts exit the virus and substrates for transcription enter particles?

Structural studies using electron cryomicroscopy (cryo-EM) techniques on the actively transcribing DLPs have shown that the nascent mRNA transcripts exit the DLP through the type I channels at the five-fold vertices (Fig. 1f and 1g; Lawton *et al.*, 1997a). In some of the cryo-EM images of the transcribing DLPs, three or four transcripts emerging from single particles can be visualized. Such an observation is consistent with the notion of simultaneous and independent transcription of the segments. In these images, some of the transcripts extend outwards from the particle by as much as 600 Å. In reconstructions of these particles, the nascent transcripts appear as an area of density slightly above the virus surface at the type I channels. Based on these studies, an exit pathway for the transcripts has been proposed. The mRNA transcript from the enzyme complex exits through a channel in the VP2 layer, in the immediate vicinity of the 5-fold axis, and exits the DLP through the type I channels in the VP6 layer (Fig. 1g). Exit of transcripts through the channels at the 5-fold axes has also been shown for the transcriptionally competent cores of bluetongue virus (BTV) (Diprose *et al.*, 2001). From the X-ray crystallographic analysis of BTV core crystals soaked with various ligands, Diprose *et al.* (2001) have shown that cores expand slightly in the presence of Mg^{2+} ions and that such an expansion results in opening of small pores at the 5-fold axis in the innermost VP3 layer (equivalent of VP2 in rotavirus), providing exit points for the transcripts. These studies have also shown that small pores adjacent to the 5-fold axis may provide conduits for the entry of substrates during transcription.

Does the initiation of transcription proceed immediately to continuous elongation?

To further understand the process of endogenous transcription and to examine how VP7, and certain anti-VP6 antibodies affect transcription Lawton *et al.* (2001), carried out studies to analyze the kinetics and products of transcription, using high-resolution polyacrylamide gel electrophoresis. These studies provided further insights into the early events of transcription, transcript initiation and cap formation, in the context of the virus architecture. These studies showed that during transcription in DLPs, two types of shorter particle-bound capped oligonucleotides are formed in addition to full-length transcripts. The shorter oligonucleotides (5-7 nucleotides in length) result from a brief pause in the transcription, before they proceed to become full-length transcripts. Interestingly, under transcribing conditions, in both mature TLPs and in DLPs decorated with anti-VP6 F(ab)$_2$ that inhibit transcription (Lawton *et al.*, 1999), similar shorter capped oligonucleotides are formed although full-length transcripts are not produced. The shorter oligonucleotides that are produced in the structural context of the TLP are functional products that become full-length transcripts, based on pulse-chase experiments, once the TLPs are converted to DLPs. This indicates that in TLPs, both the polymerase and the capping enzymes are fully functional and that the presence of the VP7 layer interferes with processive elongation. Similar results are seen with Fab-decorated DLPs.

To examine how VP7 affects the enzymatic activity of the polymerase, Lawton *et al.* (2001) carried out transcription studies in the absence of the capping reaction co-factor, S-adenosyl–methionine (SAM). SAM, a component of the transcription reaction mixture, is known to significantly increase the efficiency of the polymerase by increasing the rate of nucleotide incorporation (Spencer and Garcia, 1984). In the absence of SAM, DLPs generate full-length transcripts, with about 50% efficiency and without any pause, whereas the TLPs become transcriptionally silent. When SAM is added back to transcriptionally inert TLPs, shorter oligonucleotides are formed. These results suggest that the presence of VP7 has an inhibitory effect of the enzymatic activity of the polymerase, and that this activity can be partially resurrected by SAM. Since DLPs are able to synthesize full-length transcripts in the absence of SAM, removal of VP7 may influence the architectural properties in the vicinity of the polymerase complex, perhaps acting like a switch, to allow processive elongation.

Mechanistically it is not clear how VP7 and other ligands, such as the binding of certain Fabs to the surface of VP6, can have such a profound influence on the molecular events that take place within the VP2 layer. It is possible that removal of VP7 confers subtle conformational changes in the VP2 layer or at the interface between the VP2 and VP6 layers (Lawton *et al.*, 1999). This idea is supported by recent X-ray crystallographic studies on BTV cores in the presence of nucleotides, which indicate subtle structural changes (Diprose *et al.*, 2001). Higher resolution structural studies of the TLPs, DLPs, and actively transcribing DLPs may provide better insight into the mechanism of transcriptional inhibition by ligands which bind to VP6. Taken together, these data suggest that the observed pause during transcription in DLPs could represent a kinetic barrier, in the form of specific molecular events taking place before processive elongation, which is overcome by conformational alterations brought about by the removal of VP7. In the presence of the VP7 layer, however, the path is blocked. It is possible that the pause site serves to function as a checkpoint mechanism to keep transcription physically halted until the VP7 protein layer has been removed and the DLP becomes transcriptionally competent once it is in the cytoplasm. Another important conclusion from these studies is that, since the short oligonucleotides (5-7 nucleotides) are capped, the sites of transcription initiation and capping are in close proximity.

Are proteins responsible for maintaining the genome organization in rotavirus?

In addition to VP1 and VP3, the only other structural protein that is in close proximity of the genomic RNA is VP2, which forms the innermost shell of the rotavirus structure. VP2 binds to RNA through a positively charged amino terminus that interacts with the outermost layer of the dsRNA (Prasad *et al.*, 1996; Lawton *et al.*, 1997b; Zeng *et al.*, 1998). Biochemical studies of baculovirus-expressed virus-like particles (VLPs) constructed with truncated VP2 proteins revealed that the nucleic acid binding domain is localized between amino acid residues 1 to 132 (Labbé *et al.*, 1991). Examination by cryo-EM of mutants with deletions in this region and comparison with the structure of VLPs carrying full-length VP2 showed that the N-terminus of VP2 is localized near the five-fold axis where the transcription complexes are located (Lawton *et al.*, 1997b). X-ray crystallographic studies of BTV cores (Grimes *et al.*, 1998) and orthoreovirus (Reinisch *et al.*, 2000) also showed that the N-termini of VP2 counterparts in these viruses, VP3 and σ2, respectively, are located near the 5-fold axis facing the interior of the virus.

The innermost shell is formed by 60 dimers of VP2 on a T=1 icosahedral lattice (Fig. 1d; Lawton *et al.*, 1997b). This kind of icosahedral organization with 120 subunits, also referred to as T=2 icosahedral assembly (Grimes *et al.*, 1998), is rather unique and has so far only been found in dsRNA viruses. It was first observed in the aquareovirus structure (Shaw *et al.*, 1996). Later, the X-ray crystallographic analysis of BTV cores (Grimes *et al.*, 1998) and orthoreovirus cores (Reinisch *et al.*, 2000) provided an atomic-resolution description of such an organization. This "T=2" icosahedral assembly is observed not only in the members of the *Reoviridae* (Hill *et al.*, 1999; Zhou *et al.*, 2001) but also in other dsRNA viruses such as the bacteriophage phi6 (Butcher *et al.*, 1997) and the L-A virus of yeast origin (Caston *et al.*, 1997). Such an organization may be a fundamental requirement for the endogenous transcription of dsRNA in these viruses. The "T=2" organization with two molecules in the icosahedral asymmetric unit may have evolved to serve the dual purpose to properly position the transcription enzyme complex and to organize the genome to facilitate endogenous transcription.

How does the rotavirus genome undergo reversible condensation and expansion?

Recent cryo-EM analysis of rotavirus examined under various chemical conditions revealed a remarkable ability of the rotavirus genome to undergo reversible condensation and expansion (Pesavento *et al.*, 2001). These studies have provided further insights into the genome organization and nature of interactions between the genome and the internal proteins. At high pH in the presence of ammonium ions, the genome condenses to a radius of 180 Å from the original radius of 220 Å (Fig. 2a and 2b). Apart from VP4, which undergoes a structural change, the rest of the capsid remains intact under these conditions. Assuming that dsRNA behaves like dsDNA (Livolant and LeForestier, 1996), at the concentrations of RNA calculated from the volume corresponding to these radii, dsRNA is likely to exhibit local hexagonal packing with an inter-strand separation of ~31 Å and ~25 Å for the expanded and condensed states, respectively. These values translate to ~28 Å and ~25 Å separations between the layers as seen in the radial density plots of the normal and pH-treated structures respectively (Fig. 2c and 2d). When the ammonium, high pH-treated particles are brought back to physiological pH, the genome structure returns to its normal radius. The re-expanded genome in DLPs treated similarly is transcriptionally functional indicating that the genome has not undergone any covalent modifications and that the transformation is merely structural. These studies illustrate the remarkable stability of the capsid and resilience of the genome, which may be attributes required to carry out the continuous transcription of multiple segments within the capsid.

An important observation of the studies by Pesavento *et al.* (2001), which has direct relevance to protein-genome interactions, is that the condensation is concentric with respect to the particle center and a dark mass of density is consistently seen in the center of each of the particles in the cryo-EM images of ammonium, high pH-treated virus. The reconstruction of these particles shows density linking the condensed core with the VP2 layer at all the 5-fold axes clearly indicating a strong interaction between the genome core and the VP1/VP3 complexes, which are anchored to the inner surface of the VP2 layer at the 5-fold axes. It is unlikely that the condensed core would have remained at the center without these symmetrically disposed interactions.

In addition to the interactions with VP1/VP3, the interactions between VP2 and the genome may play an important role in the structural organization of the genome. As mentioned earlier, VP2 is an RNA binding protein. Upon returning to the normal expanded condition, all the interactions between VP2 and the genome are restored. The observation that the condensation occurs radially outwardly instead of inwardly indicates that the interaction between VP2 and the genome is perhaps pH-dependent, and that at high pH VP2 is unable to interact with the genome. Genome interaction with VP2 may be important for maintaining the appropriate spacing between the dsRNA strands in the native expanded state, and disruption of these interactions at high pH, possibly due to deprotonation of critical amino acid residues, may initiate the observed condensation. Once disassociated from VP2, primarily an effect of high pH, the dsRNA subsequently is free to be condensed by ammonium ions. The ammonium ions may bridge the adjacent dsRNA strands by hydrogen bonding to their phosphate groups and reduce the inter-strand spacing to cause condensation (Fig. 2b). Thus the condensation is a synergistic effect of high pH and ammonium ions. High pH alone is not sufficient to cause condensation. On any model for the structural organization of the genome, these studies impose further constraint of its ability to reversibly condense and expand.

Model for the dsRNA genome organization

A plausible model for the structural organization of the genome that emerges from the above biochemical and structural information on rotavirus and on other members of the *Reoviridae*, is that each dsRNA segment is spooled around a transcription enzyme complex at the 5-fold axes inside the inner most capsid layer (Prasad *et al.*, 1996; Gouet *et al.*, 1999; Pesavento *et al.*, 2001). This model allows up to 12 independent transcription complexes, each with an individual dsRNA segment attached for concurrent transcription. Such a model is also consistent with the observation that no more than 12 segments are found in any of the members of *Reoviridae* (Mertens, 2000; Lawton *et al.*, 1997a). A stylized version of this model is shown in Fig. 3a. Each dsRNA segment is depicted as an inverted cone at the 5-fold vertex surrounding a transcription enzyme complex. The matching of areas of RNA density from this model with that in the reconstruction suggests that the model is generally consistent with the reconstruction (Fig. 3b). In addition to allowing for simultaneous and independent transcription of the dsRNA segments, this model also provides a simple mechanistic explanation for the ability of the genome to undergo reversible expansion and condensation. The isometric and concentric condensation is achieved simply by reducing the inter-strand separation in each of these cones.

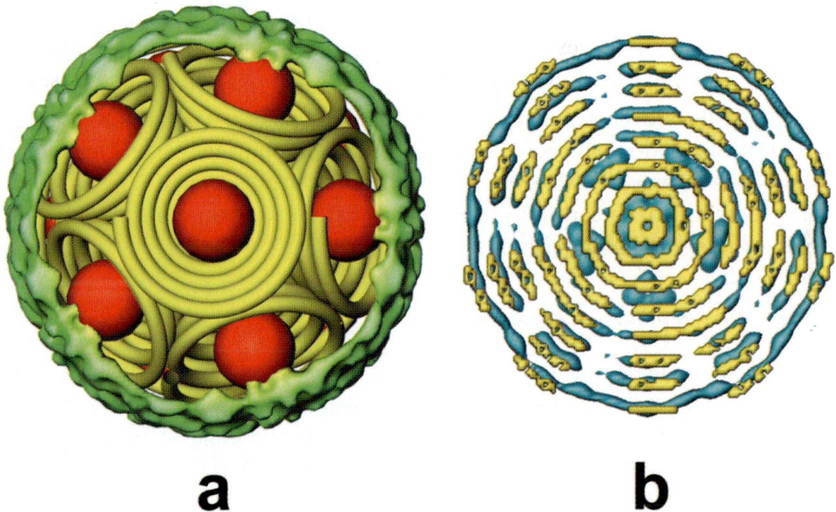

Fig. 3. Model for the organization of dsRNA. a) A model of the structural organization of the genome shown inside the VP2 layer (green). Red balls indicate transcription complexes of VP1/VP3 attached to the inside of the VP2 layer at the five-fold vertices. Segments of dsRNA (yellow) are shown as inverted conical spirals around each VP1/VP3 complex. b) Equatorial slice of the concentric RNA layers from the rotavirus reconstruction (in yellow), at native pH condition, overlaid with a similar equatorial slice from the RNA model (in cyan) to indicate that model corresponds generally well with the reconstruction in both the observed density and the spacing between the layers.

124

Several questions remain with regard to the actual organization of the genomic RNA, and answers to these questions would help to validate the model. First, are all eleven segments contained within a single capsid particle? Based on the equimolarity of the genome segments extracted from purified particles, and also low particle to pfu ratio of ~10:1 with some strains, one can argue that most likely each particle has a full complement of 11 RNA segments (Hundley *et al.*, 1985). Second, are each of the twelve enzyme complexes associated with their own individual dsRNA segment, or are there eleven complexes with RNA attached and one empty complex? Or is there an extra copy of one of the genome segments at this twelfth complex? Finally, how are the 3' and 5' ends of a single dsRNA segment connected in order to allow continuous and repeated cycles of transcription? Does a linker molecule such as a short oligonucleotide stretch or a peptide connect them, or is there some mechanism where the transcription complex finds the 3´end and processively transcribes the segment?

It is clearly of interest to answer the question of how many RNA segments are packaged inside each capsid. Present structural studies suffer from the use of icosahedral averaging, which cannot give precise structural information on the genome organization, unless the genome obeys icosahedral symmetry (Crawford *et al.*, 2001). An alternative method is to conduct single particle analysis on just the condensed core of each particle, as the boundary between the genome and the capsid interior is easily discernible. Using a non-symmetrical reconstruction method on just the condensed core, it should be possible to see in the reconstruction both the numbers and relative sizes of each genome segment. For example, if it were the case that only one of each of the eleven segments is housed in the capsid, the reconstruction should show a clear distinction between the density and location of the largest and smallest segments. This proposal assumes of course that all eleven segments are packaged in the same locations from capsid to capsid, and that the structural methods would be discerning enough to distinguish this.

A model for the genome encapsidation

Subsequent to endogenous transcription and release of the transcripts, another critical step in the life cycle of the virus is genome encapsidation. The question of how the correct set of multiple segments of the genome is packaged inside the virus particle remains a mystery. The current model for structural organization of the genome perhaps provides a clue. Based on this model we could envisage that assembly and genome encapsidation proceed concurrently. In this model, twelve units, each composed of pentamers of dimers of VP2, a transcription enzyme (VP1/VP3) complex, and an individual dsRNA segment, interact to form the single layered particle to provide a scaffold for the subsequent assembly of the VP6 layer. Proteinaceous parts of each of these units may represent the replicase complex in which a positive strand of the RNA, brought in with the aid of non-structural proteins (NSP2/NSP5), is fed into the enzyme complex for the synthesis of the negative strand and the formation of the duplex RNA that is spooled around the enzyme complex. This model raises the important question as to how a correct set of 11 distinct segments (as in rotavirus) is brought together. It is possible that specific RNA-RNA interactions will coordinate this process.

Conclusion

In summary, biochemical and recent structural studies have provided enormous insight into the fascinating phenomenon of endogenous transcription in rotavirus and other members of the *Reoviridae*. Simultaneous and repeated processing of viral dsRNA segments to capped transcripts within the structural boundaries of the virus particle raises several other mechanistic questions such as to how each dsRNA segment is transcribed by the polymerase, how it unwinds before being processed by the polymerase, how it anneals back to recycle, and how the transcript is capped. Future biochemical studies together with X-ray crystallographic and cryo-EM studies of individual proteins or complexes of proteins involved in transcription may provide answers to these questions and as well lead to a better understanding of how the segments are encapsidated.

Acknowledgements

This work is supported by grants from the National Institutes of Health AI 36040 to BVVP and DK 31044 to MKE. J.B.P. acknowledges the support of NSF training grant BIR-9256580.

References

Bartlett, N. M., Gillies, S. C. , Bullivant, S. and Bellamy, A. R. (1974). Electron microscopy study of reovirus reaction cores. *J Virol* **14**: 315-26.

Butcher, S. J., Dokland, T., Ojala, P. M., Bamford, D. H. and Fuller, S. D. (1997). Intermediates in the assembly pathway of the double-stranded RNA virus phi6. *EMBO J* **16**: 4477-87.

Caston, J. R., Trus, B. L. , Booy, F. P. , Wickner, R. B. , Wall, J. S. and Steven, A. C. (1997). Structure of L-A virus: a specialized compartment for the transcription and replication of double-stranded RNA. *J Cell Biol* **138**: 975-85.

Cohen, J. (1977). Ribonucleic acid polymerase activity associated with purified calf rotavirus. *J Gen Virol* **36**: 395-402.

Crawford, S. E., Mukherjee, S. K., Estes, M. K., Lawton, J. A., Shaw, A. L., Ramig, R. F., and Prasad, B. V. (2001). Trypsin cleavage stabilizes the rotavirus VP4 spike. *J Virol* **75**: 6052-61.

Diprose, J. M., Burroughs, J. N., Sutton, G. C., Goldsmith, A., Gouet, P., Malby, R., Overton, I., Zientara, S., Mertens, P. P. C., Stuart, D. I., and Grimes, J. M. (2001). Translocation of portals for the substrates and products of viral transcription complex: the bluetongue virus core. *EMBO J* **20**: 7229-39.

Estes, M. K. (2001). Rotaviruses and their replication. In: *Fields Virology*, 4th Edition, edited by D. M. Knipe, P.M. Howley, *et al.*, pp 1747-85, Lippincott Williams and Wilkins, Philadelphia.

Fields, B. N. (1996). The Reoviridae. In: *Fields Virology*, 3rd Edition, edited by B.N. Fields, D. M. Knipe, *et al.*, pp 1553-6, Lippincott-Raven, Philadelphia.

Gillies, S., Bullivant, S. and Bellamy, A. R. (1971). Viral RNA polymerases: electron microscopy of reovirus reaction cores. *Science* **174**: 694-6.

Ginn, D. I., Ward, R. L., Hamparian, V. V. and Hughes, J. H. (1992). Inhibition of rotavirus *in vitro* transcription by optimal concentrations of monoclonal antibodies specific for rotavirus VP6. *J. Gen Virol* **73**: 3017-22.

Gouet, P., Diprose, J. M., Grimes, J. Malby, M., R., Burroughs, J. N., Zientara, S., Stuart, D. I., and Mertens, P. P. (1999). The highly ordered double-stranded RNA genome of bluetongue virus revealed by crystallography. *Cell* **97**: 481-90.

Grimes, J. M., Burroughs, J. N., Gouet, P., Diprose, J. M., Malby, R., Zientara, S., Mertens, P. C. P. and Stuart, D. I. (1998). The atomic structure of the bluetongue virus core. *Nature* **395**: 470-478.

Grimes, J. M., Jakana, J., Ghosh, M., Basak, A. K., Roy, P., Chiu, W., Stuart, D. I. and Prasad, B. V. (1997). An atomic model of the outer layer of the bluetongue virus core derived from X-ray crystallography and electron cryomicroscopy. *Structure* **5**: 885-93.

Hill, C. L., Booth, T. F., Prasad, B. V., Grimes, J. M., Mertens, P. P., Sutton, G. C. and Stuart D. I. (1999). The structure of a cypovirus and the functional organization of dsRNA viruses. *Nat Struct Biol* **6**: 565-8.

Hundley, F., Biryahwaho, B., Gow, M. and Desselberger, U. (1985). Genome rearrangements of bovine rotavirus after serial passage at high multiplicity of infection. *Virology*. **143:** 88-103.

Labbé, M., Charpilienne, A., Crawford, S. E., Estes, M. K. and Cohen, J. (1991). Expression of rotavirus VP2 produces empty corelike particles. *J Virol* **65**: 2946-52.

Lawton, J. A., Estes, M. K. and Prasad, B. V. (1997a). Three-dimensional visualization of mRNA release from actively transcribing rotavirus particles. *Nat Struct Biol* **4**: 118-21.

Lawton, J. A., Estes, M. K. and Prasad, B. V. (1999). Comparative structural analysis of transcriptionally competent and incompetent rotavirus-antibody complexes. *Proc Natl Acad Sci USA* **96**: 5428-33.

Lawton, J. A., Zeng, C. Q., Mukherjee, S. K., Cohen, J., Estes, M. K. and Prasad, B. V. (1997b). Three-dimensional structural analysis of recombinant rotavirus-like particles with intact and amino-terminal-deleted VP2: implications for the architecture of the VP2 capsid layer. *J Virol* **71**: 7353-60.

Liu, M., Mattion, N. M. and Estes, M. K. (1992). Rotavirus VP3 expressed in insect cells possesses guanylyltransferase activity. *Virology* **188**: 77-84.

Livolant, F. and LeForestier, A. (1996). Condensed phases of DNA: structures and phase transitions. *Prog. Polymer Sci.* **21**: 1115-64.

Mason, B. B., Graham, D. Y. and Estes, M. K. (1980). *In vitro* transcription and translation of simian rotavirus SA11 gene products. *J Virol* **33**: 1111-21.

Mertens, P. P. C. (2000). Orbiviruses and Coltiviruses--General features. In: *Encyclopedia of Virology,* edited by R. G. Webster and A. Granoff pp 941-956. Academic Press, London.

Nason, E. L., Samal, S. K. and Venkataram Prasad, B. V. (2000). Trypsin-induced structural transformation in aquareovirus. *J Virol* **74**: 6546-55.

Pesavento, J. B., Lawton, J. A., Estes, M. K. and Prasad, B. V. V. (2001). The reversible condensation and expansion of the rotavirus genome. *Proc Natl Acad Sci USA* **98**: 1381-6.

Prasad, B.V.V., Wang, G. J., Clerx, J. P. and Chiu, W. (1988). Three-dimensional structure of rotavirus. *J. Mol. Biol.* **199: 269-75.**

Prasad, B.V.V., J.W. Burns, E. Marrietta, M. K. Estes and Chiu, W. (1990). Localization of VP4 neutralization sites in rotavirus by three-dimensional cryo-electron microscopy. *Nature* **343: 476-9.**

Prasad, B. V. V., Rothnagel, R., Zeng, C. Q., Jakana, J., Lawton, J. A., Chiu, W. and Estes, M. K. (1996). Visualization of ordered genomic RNA and localization of transcriptional complexes in rotavirus. *Nature* **382**: 471-3.

Reinisch, K. M., Nibert, M. L. and Harrison, S.C. (2000). Structure of the reovirus core at 3.6 Å resolution. *Nature* **404**: 960-7.

Shaw, A. L., Rothnagel, R., Chen, D., Ramig, R. F., Chiu, W. and Prasad, B. V. (1993). Three-dimensional visualization of the rotavirus hemagglutinin structure. *Cell* **74**: 693-701.

Shaw, A. L., Samal, S. K., Subramanian, K. and Prasad, B. V. (1996). The structure of aquareovirus shows how the different geometries of the two layers of the capsid are reconciled to provide symmetrical interactions and stabilization. *Structure* **4**: 957-67.

Smith, R. E. and Furuichi, Y. (1982). The double-stranded RNA genome segments of cytoplasmic polyhedrosis virus are independently transcribed. *J Virol* **41**: 326-9.

Spencer, E. and Garcia, B. I. (1984). The effcect of S-adenosylmethionine on human rotavirus RNA synthesis. *J. Virol* **52**. 188-97.

Valenzuela, S., Pizarro, J., Sandino, A. M., Vasquez, M., Fernandez, J., Hernandez, O., Patton, J. and Spencer, E. (1991). Photoaffinity labeling of rotavirus VP1 with 8-azido-ATP: identification of the viral RNA polymerase. *J Virol* **65**: 3964-7.

Yeager, M., Berriman, J. A., Baker, T. S. and Bella, A. R. (1994). Three-dimensional structure of the rotavirus haemagglutinin VP4 by cryo-electron microscopy and difference map analysis. *EMBO J* **13**: 1011-8.

Zeng, C. Q., Estes, M. K., Charpilienne, A. and Cohen, J. (1998). The N terminus of rotavirus VP2 is necessary for encapsidation of VP1 and VP3. *J Virol* **72**: 201-8.

Zhang, H., Zhang, J., Yu, X., Lu, X., Zhang, Q., Jakana, J., Chen, D. H., Zhang, X. and Zhou, Z. H. (1999). Visualization of protein-RNA interactions in cytoplasmic polyhedrosis virus. *J Virol* **73**: 1624-9.

Zhou, Z. H., Baker, M. L., Jiang, W., Dougherty, M., Jakana, J., Dong, G., Lu, G. and Chiu, W. (2001). Electron cryomicroscopy and bioinformatics suggest protein fold models for rice dwarf virus. *Nat Struct Biol* **8**: 868-73.

II, 2. The three-dimensional structure of rotavirus VP6

F.A. Rey[1], J. Lepault[1] and J. Cohen[1,2]

[1] *Virologie Moléculaire et Structurale – UMR 2472 CNRS – INRA 1, Avenue de la Terrasse, Bât. 14C, 91198, Gif-sur-Yvette, Cedex, France*
[2] *Present adress: INRA, Domaine de Vilvert, 78352 Jouy-en-Josas, Cedex, France*

Overview of rotavirus capsid

Rotaviruses are non-enveloped viruses with a complex icosahedral architecture composed of three concentric protein shells and 60 spikes. The overall diameter of the complete particle is about 1000Å. The two outer shells are both organized in a T=13 *levo* icosahedral lattice implicating a single type of interactions between proteins VP7 and VP6 which form, respectively, the outer and middle layer of the viral capsid (Lawton *et al.*, 2000). This symmetry dictates the presence of 780 (i.e., 13 × 60) subunits of both VP6 and VP7, organized as 260 trimers, in the virion. The spike protein VP4 is present as 60 dimers inserted into channels that open radially across the two outer layers of the virion. Recent structural studies of the spike have shown that the sialic acid binding domain, at the head of the spike has a galectin-like fold (Dormitzer *et al.*, 2002). The inner layer is made of 120 copies of protein VP2 which form asymmetric dimers arranged on a T=1 lattice. This protein interacts directly with the viral genome together with replication enzymes VP1 and VP3 (Lawton *et al.*, 1997b). The lower symmetry of the VP2 shell introduces a symmetry mismatch between the inner and the middle layers, just as the 120 VP4 subunits in their interaction with the VP6/VP7 shell.

The principal antigenic determinants of rotaviruses, giving rise to neutralizing antibodies, are contained in the outer layer proteins VP7 and VP4. The dominant immunogen of rotaviruses is, however, protein VP6, perhaps because it is the major capsid protein, accounting for more than 50% of the total protein mass of the virion (Estes, 1996). Covered in the extra-cellular particle by the VP7 layer, VP6 is not easily recognized by the primary humoral immune response. There is therefore no noticeable antigenic drift, resulting in a more conserved polypeptide sequence. Antibodies directed against VP6 are non-neutralizing and in general are group- or subgroup-reactive. It has nevertheless been shown that immunization with recombinant virus-like particles (VLP2/6) containing only the inner two protein layers (VP2 and VP6) results in some protection against further challenge by the parent virus (Choi *et al.*, 1999; Choi *et al.*, 2000; Ciarlet *et al.*, 1998). In addition, it was found that non-neutralizing immunoglobulin A (IgA) directed against VP6 has a protective effect in the mouse model, suggesting that this protein could be exposed to the mucosal immune response, perhaps during transcytosis of the IgA through the epithelial intestinal cells (Burns *et al.*, 1996). In agreement with this hypothesis, mice genetically knocked out for the J piece of immunoglobulins and therefore unable to transcytose were found to be not protected when immunized with VLP2/6 (Schwartz-Cornil *et al.*, 2002).

During the viral life cycle, the spikes and the outer layer of the virion are lost upon penetration into the target cell. The process of rotavirus entry (see also López and Arias, Section II, Chapter 3 of this book), leads to the delivery of double lay-ered particles (DLPs, virus particles depleted of VP7 and VP4) into the cytoplasm. The DLPs do not disassemble in the infected cell and constitute the viral transcription units. Messenger RNA molecules have been visualized emerging from channels along the 5-fold icosahedral axes by electron cryo-microscopy and image reconstruction of *in vitro* transcribing DLPs (Lawton *et al.*, 1997a). Replication begins on mRNA tem-plates after the cellular machinery has translated the viral proteins. Immature particles bud into the ER acquiring a transient lipid envelope, presumably via interactions of the VP6 layer with the viral nonstructural protein NSP4, which is present at the ER mem-brane (Taylor *et al.*, 1992, Xu *et al.*, 2000; Taylor and Bellamy, Section II, Chapter 7 of this book).

Transcription by rotavirus DLPs has been studied in some detail *in vitro*. It was found that particles devoid of the VP6 layer are transcription-incompetent, indicating that VP6 plays some structural role as a necessary component in this process (Bican *et al.*, 1982). The presence of VP7 bound to VP6 in the TLP inhibits *in vitro* transcription. Certain anti-VP6 antibodies have also been shown to inhibit transcription (Lawton *et al.*, 1999; Thouvenin *et al.*, 2001). Although some experiments suggest a direct physical link between VP6 and the viral polymerase VP1 (Zeng *et al.*, 1998), structural studies of the DLPs point to an interaction via protein VP2, which lies in between the internal VP1 enzyme and the VP6 layer (Lawton *et al.*, 1997b). Crystallographic stud-ies of bluetongue virus, a member of another genus of the *Reoviridae* family, have sug-gested a limited conformational rearrangement of the middle layer concomitant with the opening of a pore in the inner layer at the 5-fold axes for extrusion of transcripts (Diprose *et al.*, 2001). A similar VP2-VP6 conformational change limited to the region surrounding the icosahedral 5-fold axes can also be proposed for rotaviruses. The block of transcription by VP7 and by some anti-VP6 antibodies could thus be explained by an inhibition of this necessary rearrangement of the capsid (Thouvenin *et al.*, 2001).

Molecular architecture of protein VP6

VP6 forms a pear-shaped trimer that is 95Å long with a triangular cross-section of about 60 Å per side at the bottom (the "base") and a roughly hexagonal outline of 45 Å in diameter at the top (the "head") (Mathieu *et al.*, 2001). The polypeptide chain folds into two distinct domains that constitute the base and the head of the trimer (Fig. 1). Each subunit buries a surface of about 4500 Å^2 in the trimer, accounting for its high stability. A predominantly α-helical domain (domain B), containing amino acid (aa) residues 1 to 150 and 335 to 397, forms the base. This domain consists of a bundle of eight α-helices and a β-hairpin packing vertically against the side of the molecule. The top of the trimer is formed by a jelly-roll β-barrel, which forms the outward facing "head" of the molecule (termed domain H). It contains amino acids 151 through 334, which constitute an insertion into the B domain. It contains two antiparallel β sheets, one with four ("CHEF" sheet) and one with six (A'A"BIDG) β-strands. There is also a β-hairpin (D'D") insertion into the jelly-roll motif. The D'D" hairpin is responsible for most of the interactions between H domains at the top of the trimer. The A'A"BIDG β-sheet lies at the center of the domain, flanked at either side by the exposed CHEF sheet and the internal D'D" β-hairpin. The H domain is oriented with its β-strands roughly parallel to the radial direction in the viral particle, similar to the jelly-rolls found in other members of the *Reoviridae* (Grimes *et al.*, 1995; Liemann *et al.*, 2002). This orientation is also found in the major capsid proteins of double-stranded DNA viruses, like the adenovirus hexon protein (Athappilly *et al.*, 1994) or in papovavirus major capsid proteins (SV-40 (Stehle *et al.*, 1996), polyomavirus (Stehle *et al.*, 1997) and papillomavirus (Chen *et al.*, 2000)). This is in contrast to most T=3 or T=4 single stranded RNA viruses, in which the jelly-roll structure is found in an orientation such that the β-strands are perpendicular to the virus radial direction.

There is an aqueous channel along the central 3-fold axes of the trimer. This channel is blocked at the top of the molecule by a hydrophobic plug formed by Pro 251 from each subunit. This plug is reminiscent of the "β-constriction" present on the molecular axis of the adenovirus hexon trimer (Athappilly *et al.*, 1994). It was indeed shown that a point mutation from proline to aspartic acid at position 251 results in a molecule unable to trimerize (López *et al.*, 1994), suggesting an important contribution of this residue to the stability of the folded molecule. The central channel is also constricted at the center of the molecule by a Zn^{2+} atom, coordinated by His 153 from each subunit. Site-directed mutagenesis of His 153 leads to a less stable molecule as assayed by protease sensitivity (Erk *et al.*, 2002). At the base, the central channel opens into a fairly big water filled cavity. In contrast to the base, the trimer interface at the head is extensive and resembles the interior of globular proteins, with tight packing of mainly hydrophobic side-chains. In particular, the D'D" β-hairpin (aa residues 233-253) and the HI loop (aa residues 297-313) from the jelly-roll of the clockwise neighbor, exhibit an intricate network of contacts. A single amino acid change from proline to glutamine at position 309 in the HI loop was indeed reported to reduce the stability of VP6 trimers on SDS-PAGE and at low pH values compared with the normal gene product (Shen *et al.*, 1994).

Below the D'D" β-hairpin there are hydrophobic contacts between the middle β-sheets (A'A"BIDG) of the three subunits. The packing is such that side chains from the lateral strands A'A" contact residues from strands DG of the clockwise next subunit, thereby extending the hydrophobic core resulting from the packing between the middle sheet and the D'D" β-hairpin into the inter-chain contact. The two central strands B and I exhibit charged residues in this face of the β-sheet which are directed towards the central aqueous channel.

Site-directed mutagenesis studies have shown that a VP6 mutant with His153 changed to serine exhibited an increased susceptibility to proteolytic digestion (see above).

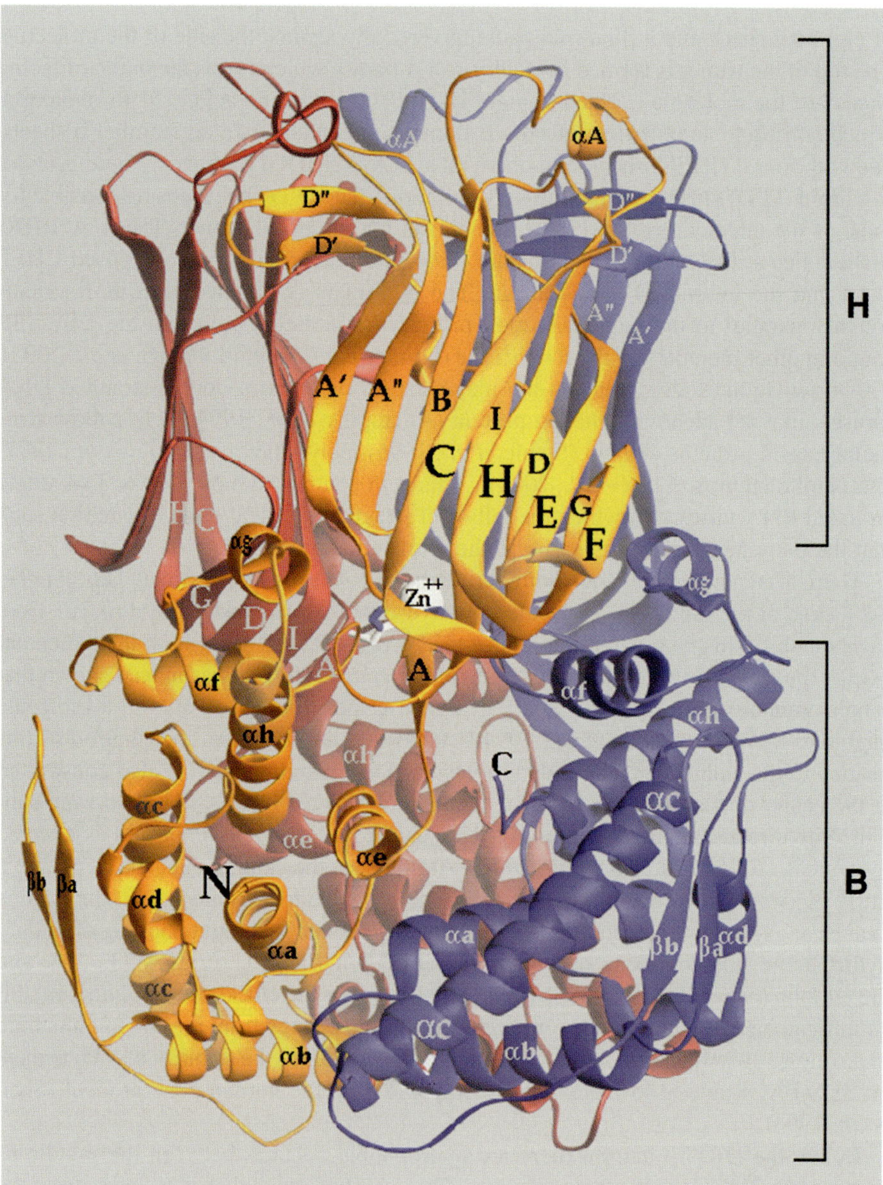

Fig. 1. Diagram showing the folded structure of the VP6 polypeptide chain in the trimer. Beta strands are represented by ribbons with arrowheads, alpha helices by coiled ribbons and connecting loops by thin tubes. The three subunits in the trimer are displayed in yellow, pink and blue. Lower- and upper-case letters label secondary structure elements in domains B and H, respectively. Beta strands in domain H are labeled according to the standard nomenclature for β-rolls (Rossmann *et al.*, 1989). White ball indicates the position of a Zn^{2+} ion on the molecular three-fold axis. Letters A and C label the amino- and carboxy-termini of the polypeptide chain, respectively.

However, this mutant was found to assemble correctly on cores as demonstrated by both electron microscopy and by rescuing the transcriptase activity of reconstituted DLPs (Erk *et al.*, 2002). The susceptibility to the protease was highest in the isolated trimer, whereas upon assembly no proteolytic cleavage was detected. This feature was accentuated in the H153S mutant, but was also observed for the wildtype protein. These observations suggest that, in solution, the basal domain of the VP6 trimer is more flexible. The flexibility depends upon both the presence of the zinc ion and on pH.

Comparison of VP6 with corresponding proteins of other viruses of the *Reoviridae* family

A trimer with a head domain consisting of a jelly-roll seems to be the hallmark of proteins forming the second layer of viruses in the *Reoviridae* family. Furthermore, in all cases for which the structure is known — protein VP7 from orbiviruses, protein $\mu 1$ of reovirus and protein VP6 of rotavirus — the jelly-roll domain appears as an insertion within a mainly α-helical protein. The right-handed twist of the subunits along the 3-fold molecular axis is an additional common feature. Although there is no detectable sequence conservation, it is clear from the structures that these three proteins are indeed homologous, and it is therefore reasonable to predict that the corresponding protein from other members of the *Reoviridae* for which no structure is (yet) known will display similar features.

What is the biological reason for this conservation? One possibility is membrane penetration, since all viruses must be capable of membrane translocation. In the case of reoviruses and orbiviruses, there is evidence pointing to the middle layer protein which can undergo cleavage and display alternative trimeric conformations as being responsible for membrane interactions leading to entry. Studies of enveloped virus proteins have shown that proteolytic cleavages can play a major role in priming viral fusogenic proteins which then remain in a metastable state at the viral surface. The energy necessary to merge the viral and cellular membranes is released through a refolding event that is triggered by interactions with the target cell (Eckert *et al.*, 2001; Skehel *et al.*, 2001). Viruses of the family *Reoviridae* are non-enveloped, and so the membrane fusion step cannot occur, yet the large, 70 nm diameter transcription units must cross the cellular membrane to reach the cytoplasm of the target cell. Orbivirus double-layered particles are infectious in insect cells, in the absence of the external layer, and are therefore capable of interacting with membranes (Mertens *et al.*, 1996). Orbivirus VP7 is indeed susceptible to cleavage at the interface between head and base (Basak *et al.*, 1996). The structure of an alternative conformation of the VP7 trimer has also been reported (Basak *et al.*, 1997). This alternative conformation could be due to a cleavage at the junction of the head and base domains, although the resolution of the structure determination (6 Å) did not allow reaching a definitive conclusion (Basak *et al.*, 1997). In the case of reovirus, an autolytic event separates the myristylated N terminus of $\mu 1$ ($\mu 1N$) from the rest of the protein ($\mu 1C$). This cleavage has been proposed to prime the protein for membrane penetration, and a major structural rearrangement of the $\mu 1$

134

trimer has been proposed to take place during the membrane disruption step (Liemann *et al.*, 2002). There is no evidence that group A rotavirus VP6 is capable of interacting with membranes. However, it will be important to determine whether the structure of VP6 was maintained only as a remnant of an ancestral membrane interaction function, or as reflecting some as yet unidentified function during entry. *In vitro* studies have shown that isolated VP6 from group A and C rotaviruses is indeed susceptible to cleavage, and this cleavage also occurs at the interface between head and base [J. Lepault, unpublished data].

Molecular polymorphism of VP6

VP6 has the capacity of auto-assembly and gives rise to polymorphic cylindrical or spherical structures of high molecular weight. VP6 tubes have been observed in infected cells, in the stool from rotavirus-infected patients, and in cells expressing recombinant VP6 (Estes *et al.*, 1987). The biological function of these assemblies is not known. It could be a way of storage of VP6 during viral replication, since experiments suggest an inhibitory effect of VP6 on the viral replicase as tested *in vitro* (J. Cohen, unpublished data). The *in vitro* physico-chemical conditions leading to the different assemblies of VP6 were characterized, and it was shown that pH and ionic strength were the most important parameters (Lepault *et al.*, 2001) (Fig. 2). In essence, at medium ionic strength (NaCl concentrations between 0.2 to 0.6 M) and at a pH of around 4.5, a heterogeneous population of spherical particles of different sizes was observed, displaying most probably icosahedral symmetry. The smallest spherical particles display a diameter of about 750 Å, which is bigger than that of the DLP (700 Å), suggesting a role of the rotavirus core in the compact assembly of the DLP. In addition, many of the spherical VP6 particles were not complete. At neutral pH, tubes of about 450 Å diameter were observed (Fig. 2), whereas at intermediate pH values (of around 6) the tubes display roughly the same radius as the smallest spherical particles (750 Å). All of these VP6 assemblies are destabilized by high Ca^{2+} concentrations, and it was shown that above 100 mM concentrations of this ion, the only observed species is the isolated trimer. This observation was used to obtain crystals of VP6 by exploring crystallisation conditions in the presence of Ca^{2+} (Petitpas *et al.*, 1998). Because the spheres made by VP6 alone are heterogeneous, no 3D reconstruction was attempted. Only the very regular tubular assemblies of VP6 were analyzed further by cryo-EM, and the atomic model of VP6 was fitted into the resulting 20 Å resolution image reconstructions. It was found that in the small tubes each VP6 trimer exhibits three different types of lateral, 2-fold related contacts. In the 750 Å diameter tubes, two of the contacts are identical. Because radial 3-fold symmetry is incompatible with helical symmetry, the symmetry of the VP6 trimer has to be broken in order to build such assemblies. In contrast, icosahedral symmetry implies 3-fold symmetry, and assemblies of triangulation T=13 and bigger (corresponding to sizes of the particles that are equal or bigger than the DLP) can preserve the quasi 3-fold contacts very well. The fact that the spherical (likely icosahedral) particles were obtained only at acidic pH were

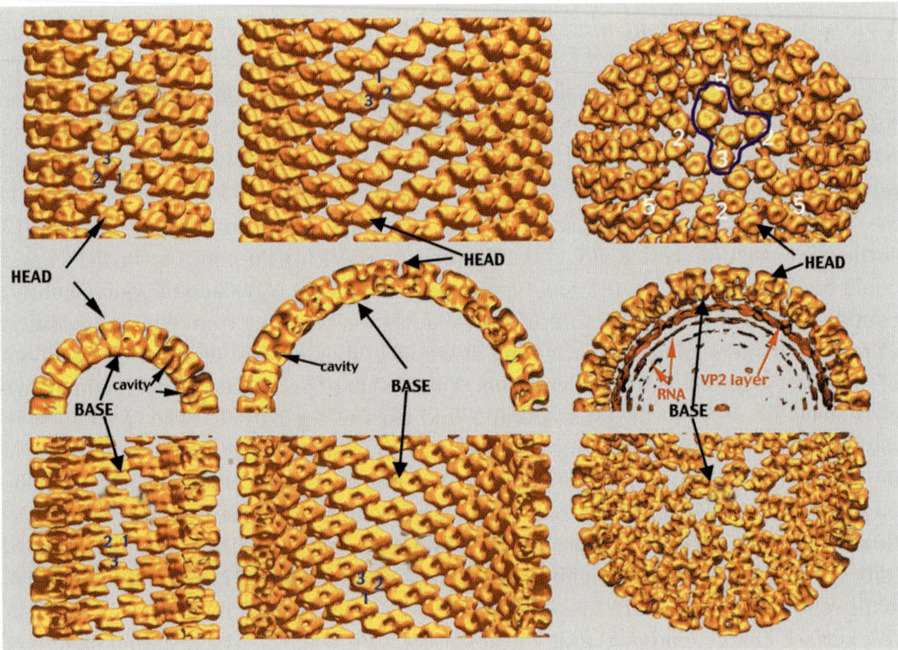

Figure 2. Cryo-EM reconstructions of different assemblies made by protein VP6. Left, 450 Å diameter tubes, center, 750 Å diameter tubes, right, DLP reconstruction extended to 17Å resolution. The top row shows a view from the outside, and the bottom row a view from the inside. The middle view shows an orthogonal slice about 80Å thick. Indicated on the tubes are the cavities at the center of the VP6 base, and the three non-equivalent contacts 1,2 and 3. In the DLP, white numbers mark the different icosahedral symmetry axes, and a blue contour encircles 5 icosahedrally independent trimers. The side-view (middle-right panel) of the DLP contains the internal VP2 layer as well as density corresponding to double-stranded RNA and core proteins (labeled in red). For clarity, this density has been removed in the other two views. The reconstruction of DLP has been kindly provided by Dr V Prasad (Mathieu *et al.*, 2001).

interpreted as evidence that the VP6 trimers (in absence of VP2) are capable of making assemblies displaying three identical contacts (for instance, on a 3-fold icosahedral axis) only when they are fully protonated. This interpretation is consistent with the crystal structure of VP6 which indicates that, at neutral pH, the faces of the VP6 trimers that come into contact during assembly are negatively charged. At intermediate pH, the trimers are capable of making two identical contacts only (large tubes), and at neutral pH only one of the three low pH contacts remains (small tubes). At alkaline pH no regular assembly can be made by VP6, suggesting that the conserved contact, present in spheres, big and small tubes, is important for all of the observed regular particles.

136

VP6 in the viral particle

Cryo-EM image reconstructions of both the intact viral particle and the DLP were used to fit the atomic model of VP6 and examine lateral contacts between neighboring trimers. This work also allowed the identification of its contacts with the other capsid proteins of the virion (see below). As in the case of orbivirus VP7 and reovirus $\mu 1$, the VP6 trimers interact via contacts at the 2-fold and quasi 2-fold axes of the T=13 surface icosahedral lattice (Fig. 3). The observed VP6/VP6 contacts in the capsid could be classified as two different types: quasi-equivalent (Q) and non-equivalent (N) contacts, according to the root mean square deviation resulting from the superposition of pairs of trimers related to each other at the local dyads. Most of the contacts (4 out of a total of 6 icosahedrally independent contacts) are Q contacts. The remaining two N contacts lie close to the icosahedral 5-fold axes of the particle. The Q contacts in the DLPs or TLPs were indistinguishable from the conserved contact, found twice in the big tubes and once in the small tubes of VP6. This observation suggests that the presence of the VP2 layer not only causes compaction of the particle through distortions at the two N contacts close to the 5-fold axes, but also allows the formation of three identical contacts by neighboring VP6 trimers at neutral pH, overcoming electrostatic repulsion. This in turn suggests that the VP2 layer very likely displays a net positive surface charge, compensating the acidity of VP6 (Fig. 2). The resulting pattern of lateral interactions between adjacent trimers leads to a model for an assembly pathway for the VP6 layer on top of the inner layer. Because of the symmetry mismatch, the only place where the symmetries of the two layers match is at the icosahedral 3-fold axes. A strong VP6/VP2 contact at the center of each of the 20 triangular faces of the icosahedron could thus initiate polymerization of the second layer via the most favorable Q contacts, identical to those found in the tubes. The final double layer is then a mosaic of 20 faces, connected by N contacts at the vertices of the icosahedron, in order to fit the smaller size of the DLP dictated by the inner layer.

The resulting arrangement of the VP6 trimers in the viral particle is such that the vertical hairpins βa-βb in the B domain (Fig. 1), are directed towards solvent channels. The amino-acid sequence in the βa-βb hairpin is highly conserved. In particular, the walls of the channels at the 5-fold axes which have a smaller diameter are formed exclusively by side-chains from this β-hairpin. The conserved distribution of side chains is probably due to functional reasons, given that the side chains do not participate in contacts stabilizing the folded VP6 molecule.

Contacts with proteins from the other layers of the viral capsid

The pseudo-atomic model of the rotavirus middle layer obtained by fitting the crystal structure into the cryo-EM reconstruction identifies the surfaces of VP6 that interact with the other viral proteins (see also Pesamento *et al.*, Section II, Chapter 1 of this book).

Figure 3. Diagram illustrating the different types of contacts made by each VP6 trimer within the different regular assemblies. Contacts are indicated by bars, blue for the strongest contact, grey for an intermediate contact, and whit for weak contacts. The right panel shows the contacts on the subviral particle (and in the sphere obtained at low pH), where most of the trimers exhibit three blue contacts. The middle panel corresponds to the large tubes (intermediate pH) where the strong contact is made twice per trimer, and the left panel corresponds to the small tubes (neutral pH) in which each trimer makes three different contacts and where the strong blue contact is present only once per trimer. No regular assemblies are obtained at more alkaline pH. Thus, in the absence of the inner VP2 layer, the ability to make repeated strong lateral contacts depends on the protonation state of VP6.

Inner layer

The VP6 side chains that contact the inner layer were analyzed by site-directed mutagenesis for their importance in assembly of DLP (Charpilienne *et al.*, 2002). The residues participating in the interactions with VP2 come essentially from the bottom side of helix αb which runs horizontally at the base of the trimer, and from the βb-αc loop. This conserved region of polypeptide chain (aa residues 60 to 74) is quite hydrophobic, and the crystal structure shows several conserved leucine residues with their side chains exposed to solvent. Although no reverse genetics system is available for rotaviruses, it is possible to recoat rotavirus cores with mutagenized VP6 molecules and to show that leucines 65, 70 and 71 are important for the interaction with VP2. Triple and double mutants of these three residues (but not single mutants) were affected in their VP2 interactions. Comparison of the phenotypes of 13 mutants indicated that all three positions interact with VP2, but that eliminating only one of them is not enough to destabilize the interaction, the highest significance being attributed to the leucine residue at aa position 71. Another interesting result was obtained by mutagenizing aa position 32, which is close to the 3-fold axis at the bottom of the trimer. This position is a strictly conserved glutamine in group A and C rotaviruses. Changing this glutamine into glutamic acid does not affect assembly of VP6 onto the inner layer, but the resulting DLPs are not transcription-competent (Charpilienne *et al.*, 2002), suggesting that this additional negative charge may affect a putative concerted VP2/VP6 conformational change that is necessary for rotavirus transcription.

138

Outer layer and spike

VP7 interacts in a one-to-one fashion with VP6: trimers of VP7 lie directly on top, contacting aa residues of loops A'A", BC and HI at the top of the jelly-roll head of the molecule. Several of the amino acids involved in these contacts are highly conserved, like the stretch between amino acids Arg 296 and Pro 302 just before helix αA. These contacts leave a cavity around the 3-fold axis, which extends the VP6 central solvent channel into the VP6/VP7 complex.

VP4 contacts simultaneously four VP6 trimers inside channels made at the quasi 6-folds axes directly surrounding the 5-fold axes (type II channels; Prasad *et al.*, 1988). The VP6 surface of interaction is essentially the CHEF β-sheet (Fig. 1), with the spike reaching the upper part of the VP6 base. A bulge at the beginning of the F strand in VP6 seems to be an important feature of the contact with the spike protein (see also Pesamento *et al.*, Section II, Chapter 1 of this book)

Concluding remarks

The structural studies of VP6 by X-ray crystallography and by electron microscopy have led to a pseudo-atomic model of its helical assemblies and of the viral middle layer. Calculation of the surface electrostatic potential shows that VP6 is an acidic molecule, indicating that repulsion between capsomers is likely to be responsible for the different assemblies obtained at different pH. It also suggests that the inner shell of the viral particle plays an essential role in compensating the negative electrostatic charges of VP6. If we think of the evolution of rotaviruses, we may hypothesize that at the origin the protein that now forms the middle layer was a coat protein of a simpler, single shelled virus. With the advent of an additional internal protein layer capable of enclosing the genome, capsid closure did not remain essential for assembly of the second layer. The ability of rotavirus VP6 to make closed icosahedral particles might reflect a remnant of such an ancestral function. By analogy, the acquisition by rotaviruses of an external layer and spikes, for which membrane destabilizing properties have been demonstrated (Charpilienne *et al.*, 1997; Dowling *et al.*, 2000), may have led to the loss of an ancient membrane penetration function for VP6.

In summary, protein VP6 plays a central role in the organization of the virion; acting as an adaptor between two distinct viral functions, i.e. cell entry (outer layer) and genomic RNA packaging (inner layer). It also seems to perform additional functions related to a possible concerted conformational switch of the viral particle necessary for transcription.

References

Athappilly, F. K., Murali, R., Rux, J. J., Cai, Z. and Burnett, R. M. (1994). The refined crystal structure of hexon, the major coat protein of adenovirus type 2, at 2.9 Å resolution. *J Mol Biol* **242**, 430-455.

Basak, A. K., Gouet, P., Grimes, J., Roy, P. and Stuart, D. (1996). Crystal structure of the top domain of African horse sickness virus VP7: comparisons with bluetongue virus VP7. *J Virol* **70**, 3797-3806.

Basak, A. K., Grimes, J. M., Gouet, P., Roy, P. and Stuart, D. I. (1997). Structures of orbivirus VP7: implications for the role of this protein in the viral life cycle. *Structure* **5**, 871-883.

Bican, P., Cohen, J., Charpilienne, A. and Scherrer, R. (1982). Purification and characterization of bovine rotavirus cores. *J Virol* **43**, 1113-1117.

Burns, J. W., Siadat-Pajouh, M., Krishnaney, A. A. and Greenberg, H. B. (1996). Protective effect of rotavirus VP6-specific IgA monoclonal antibodies that lack neutralizing activity. *Science* **272**, 104-107.

Charpilienne, A., Abad, M. J., Michelangeli, F., Alvarado, F., Vasseur, M., Cohen, J. and Ruiz, M. C. (1997). Solubilized and cleaved VP7, the outer glycoprotein of rotavirus, induces permeabilization of cell membrane vesicles. *J Gen Virol* **78**, 1367-1371.

Charpilienne, A., Lepault, J., Rey, F. and Cohen, J. (2002). Identification of rotavirus VP6 residues located at the interface with VP2 that are essential for capsid assembly and transcriptase activity. *J Virol* **76**, 7822-7831.

Chen, X. S., Garcea, R. L., Goldberg, I., Casini, G. and Harrison, S. C. (2000). Structure of small virus-like particles assembled from the L1 protein of human papillomavirus 16. *Mol Cell* **5**, 557-567.

Choi, A. H., Basu, M., McNeal, M. M., Clements, J. D. and Ward, R. L. (1999). Antibody-independent protection against rotavirus infection of mice stimulated by intranasal immunization with chimeric VP4 or VP6 protein. *J Virol* **73**, 7574-7581.

Choi, A. H., Basu, M., McNeal, M. M., Flint, J., VanCott, J. L., Clements, J. D. and Ward, R. L. (2000). Functional mapping of protective domains and epitopes in the rotavirus VP6 protein. *J Virol* **74**, 11574-11580.

Ciarlet, M., Crawford, S. E., Barone, C., Bertolotti-Ciarlet, A., Ramig, R. F., Estes, M. K. and Conner, M. E. (1998). Subunit rotavirus vaccine administered parenterally to rabbits induces active protective immunity. *J Virol* **72**, 9233-9246.

Diprose, J. M., Burroughs, J. N., Sutton, G. C., Goldsmith, A., Gouet, P., Malby, R., Overton, I., Zientara, S., Mertens, P. P., Stuart, D. I. and Grimes, J. M. (2001). Translocation portals for the substrates and products of a viral transcription complex: the bluetongue virus core. *EMBO J* **20**, 7229-7239.

Dormitzer, P. R., Sun, Z. Y., Wagner, G. and Harrison, S. C. (2002). The rhesus rotavirus VP4 sialic acid binding domain has a galectin fold with a novel carbohydrate binding site. *EMBO J* **21**, 885-897.

Dowling, W., Denisova, E., LaMonica, R. and Mackow, E. R. (2000). Selective membrane permeabilization by the rotavirus VP5* protein is abrogated by mutations in an internal hydrophobic domain. *J Virol* **74**, 6368-6376.

Eckert, D. M. and Kim, P. S. (2001). Mechanisms of viral membrane fusion and its inhibition. *Annu Rev Biochem* **70**, 777-810.

Erk, I., Huet, J. C., Duarte, M., Duquerroy, S., Rey, F., Cohen, J. and Lepault, J. (2003). A zinc ion controls assembly and stability of the major capsid protein of rotavirus, *J Virol* **77**, 3596-3601.

Estes, M. K. (1996). Rotaviruses and their replication. In *Fields Virology, Third Edition*. Fields, B.N., Howley, P.M. *et al.*, eds., pp. 1625-1655. Lippincott-Raven, Philadelphia.

Estes, M. K., Crawford, S. E., Peñaranda, M. E., Petrie, B. L., Burns, J. W., Chan, W. K., Ericson, B., Smith, G. E. and Summers, M. D. (1987). Synthesis and immunogenicity of the rotavirus major capsid antigen using a baculovirus expression system. *J Virol* **61**, 1488-1494.

Grimes, J., Basak, A., Roy, P. and Stuart, D. (1995). The crystal structure of bluetongue virus VP7. *Nature* **373**, 167-170.

Lawton, J. A., Estes, M. K. and Prasad, B. V. (1997a). Three-dimensional visualization of mRNA release from actively transcribing rotavirus particles. *Nat Struct Biol* **4**, 118-121.

Lawton, J. A., Estes, M. K. and Prasad, B. V. (1999). Comparative structural analysis of transcriptionally competent and incompetent rotavirus-antibody complexes. *Proc Natl Acad Sci U S A* **96**, 5428-5433.

Lawton, J. A., Estes, M. K. and Prasad, B. V. (2000). Mechanism of genome transcription in segmented dsRNA viruses. *Adv Virus Res* **55**, 185-229.

Lawton, J. A., Zeng, C. Q., Mukherjee, S. K., Cohen, J., Estes, M. K. and Prasad, B. V. (1997b). Three-dimensional structural analysis of recombinant rotavirus-like particles with intact and amino-terminal-deleted VP2: implications for the architecture of the VP2 capsid layer. *J Virol* **71**, 7353-7360.

Lepault, J., Petitpas, I., Erk, I., Navaza, J., Bigot, D., Dona, M., Vachette, P., Cohen, J. and Rey, F. A. (2001). Structural polymorphism of the major capsid protein of rotavirus. *EMBO J* **20**, 1498-1507.

Liemann, S., Chandran, K., Baker, T. S., Nibert, M. L. and Harrison, S. C. (2002). Structure of the reovirus membrane-penetration protein, Mu1, in a complex with is protector protein, Sigma3. *Cell* **108**, 283-295.

López, S., Espinosa, R., Greenberg, H. B. and Arias, C. F. (1994). Mapping the subgroup epitopes of rotavirus protein VP6. *Virology* **204**, 153-162.

Mathieu, M., Petitpas, I., Navaza, J., Lepault, J., Kohli, E., Pothier, P., Prasad, B. V., Cohen, J. and Rey, F. A. (2001). Atomic structure of the major capsid protein of rotavirus: implications for the architecture of the virion. *EMBO J* **20**, 1485-1497.

Mertens, P. P. C., Burroughs, J. N., Walton, A., Wellby, M. P., Fu, H., Ohara, R. S., Brookes, S. M. and Mellor, P. S. (1996). Enhanced infectivity of modified bluetongue virus particles for two insect cell lines and for two culicoides vector species. *Virology* **217**, 582-593.

Petitpas, I., Lepault, J., Vachette, P., Charpilienne, A., Mathieu, M., Kohli, E., Pothier, P., Cohen, J. and Rey, F. A. (1998). Crystallization and preliminary X-ray analysis of rotavirus protein VP6. *J Virol* **72**, 7615-7619.

Prasad, B. V., Wang, G. J., Clerx, J. P. and Chiu, W. (1988). Three-dimensional structure of rotavirus. *J Mol Biol* **199**, 269-275.

Rossmann, M. G. and Johnson, J. E. (1989). Icosahedral RNA virus structure. *Annu*

Rev Biochem **58**, 533-573.

Schwartz-Cornil, I., Benureau, Y., Greenberg, H., Hendrickson, B. and Cohen, J. (2002). Heterologous protection induced by the inner capsid structural proteins of rotavirus requires transcytosis of mucosal immunoglobins. *J Virol* **76**, 8110-8117.

Shen, S., Burke, B. and Desselberger, U. (1994). Rearrangement of the VP6 gene of a group A rotavirus in combination with a point mutation affecting trimer stability. *J Virol* **68**, 1682-1688.

Skehel, J. J., Cross, K., Steinhauer, D. and Wiley, D. C. (2001). Influenza fusion peptides. *Biochem Soc Trans* **29**, 623-626.

Stehle, T., Gamblin, S. J., Yan, Y. and Harrison, S. C. (1996). The structure of simian virus 40 refined at 3.1 Å resolution. *Structure* **4**, 165-182.

Stehle, T. and Harrison, S. C. (1997). High-resolution structure of a polyomavirus VP1-oligosaccharide complex: implications for assembly and receptor binding. *EMBO J* **16**, 5139-5148.

Taylor, J. A., Meyer, J. C., Legge, M. A., O'Brien, J. A., Street, J. E., Lord, V. J., Bergmann, C. C. and Bellamy, A. R. (1992). Transient expression and mutational analysis of the rotavirus intracellular receptor: the C-terminal methionine residue is essential for ligand binding. *J Virol* **66**, 3566-3572.

Thouvenin, E., Schoehn, G., Rey, F., Petitpas, I., Mathieu, M., Vaney, M. C., Cohen, J., Kohli, E., Pothier, P. and Hewat, E. (2001). Antibody inhibition of the transcriptase activity of the rotavirus DLP: a structural view. *J Mol Biol* **307**, 161-172.

Xu, A., Bellamy, A. R. and Taylor, J. A. (2000). Immobilization of the early secretory pathway by a virus glycoprotein that binds to microtubules. *EMBO J* **19**, 6465-74.

Zeng, C. Q., Estes, M. K., Charpilienne, A. and Cohen, J. (1998). The N terminus of rotavirus VP2 is necessary for encapsidation of VP1 and VP3. *J Virol* **72**, 201-208.

U. Desselberger and J. Gray (editors)

II, 3. Attachment and post-attachment receptors for rotavirus

Susana López and Carlos F. Arias

Departamento de Genética y Fisiología Molecular, Instituto de Biotecnología, Universidad Nacional Autónoma de México, Cuernavaca, Morelos 62250, Mexico.

Introduction

Group A rotaviruses are nonenveloped viruses that have a genome of 11 segments of double-stranded RNA contained in a triple-layered protein capsid. The outermost layer is composed of two proteins, VP4 and VP7. The smooth external surface of the virus is made up of 780 copies of the glycoprotein VP7, while 60 spike-like structures, formed by dimers of VP4, extend about 12 nm from the VP7 surface (Estes, 1996; Prasad *et al.*, 1990). These two proteins are involved in the initial interactions of the virus with the host cell which lead to a productive cell infection; accordingly, both proteins are targets of neutralizing and protective antibodies against the virus (reviewed in Kapikian and Chanock, 1996). Given their key role in the viral replication cycle, and as targets of the protective host immune response, these two proteins have been studied extensively.

VP7 is a calcium-binding glycoprotein of 326 amino acids that forms a thin layer on the surface of the viral particle. Reconstructions from cryoelectron microscopy images have shown that this protein forms trimers on the surface of the virion (Prasad and Chiu, 1994; Yeager *et al.*, 1990); recent biochemical studies have confirmed these observations and have suggested that the rotavirus surface protein layer assembles from calcium-dependent VP7 trimers, and that the dissociation of these trimers might be the biochemical basis for the EDTA-induced rotavirus uncoating (Dormitzer *et al.*, 2000a). Also, it has recently been described that the treatment of the viral particle with trypsin, besides cleaving VP4, causes a conformational change in VP7 (Crawford *et al.*, 2001). Despite earlier work implicating this protein in cell attachment (Fukuhara *et al.*, 1988; Sabara *et al.*, 1985), recent studies indicate that VP4 is the main player in this event (Crawford *et al.*, 1994; Kirkwood *et al.*, 1998; Zárate *et al.*, 2000b). Thus, the role of VP7 during the early interactions of the virus with the cell is not clear. However, it has been shown that VP7 can modulate some of the VP4-mediated virus phenotypes, including receptor binding (Beisner *et al.*, 1998; Méndez *et al.*, 1996), and it has been suggested that it might interact with cell surface molecules after the

144

initial attachment of the virus through the spike protein (Coulson *et al.*, 1997; Estes, 1996; Méndez *et al.*, 1999). It has been observed that the VP7 sequences from different rotavirus strains contain the LDV (at aa 237 to 239) and GPR (at aa 253 to 255) tripeptide sequence binding motifs for integrins αxβ2 and α4β1, respectively (Coulson *et al.*, 1997). However, the functionality of these sites has not been proven directly. The virus neutralizing antibodies directed to VP7 have been mapped to three major antigenic domains named A, B and C, which are located at amino acids 86 to 101, 142 to 152, and 208 to 221, respectively (Fig. 1).

Rotavirus VP7 protein

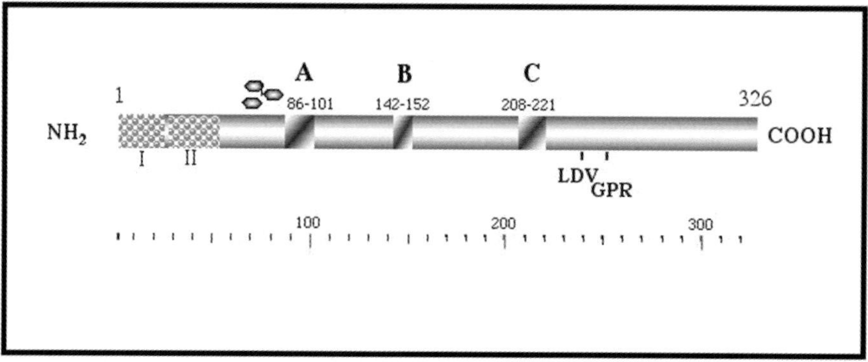

Fig. 1. Structural features of the outer capsid protein VP7. I and II denote the two amino terminal hydrophobic signals. The glycosylation site known to be used by SA11 at aa 76 is indicated by . A, B and C depict the three major antigenic sites described for group A rotaviruses (located at amino acids 86 to 101, 142 to 152, and 208 to 221, respectively). The position of the LDV and GPR tripeptide sequence binding motifs for integrins αxβ2 and α4β1, located at amino acids 237 to 239 and 253 to 255, respectively, are shown.

VP4 has essential functions in the early virus-cell interactions, including receptor binding and cell penetration (Crawford *et al.*, 1994; Ludert *et al.*, 1996; Zárate *et al.*, 2000b). The properties of this protein are therefore important determinants of host range, virulence, and induction of protective immunity. Several discrete functional domains involved in the early interactions of rotaviruses with the host cell have been described on VP4 (Fig. 2):

i) *In vitro* treatment of virions with trypsin results in the specific cleavage of VP4, of 776 amino acids, at three closely-spaced arginines located at amino acid positions 231, 241, and 247, to yield polypeptides VP8 (aa 1-231) and VP5 (aa 248-776), both of which remain associated to the virion (Arias *et al.*, 1996; López *et al.*, 1985). This cleavage results in the enhancement of viral infectivity (Arias *et al.*, 1996; Clark *et al.*, 1981; Espejo *et al.*, 1981; Estes *et al.*, 1981; López *et al.*, 1985), and although not formally proven, it is believed that a productive infection

Rotavirus VP4 protein

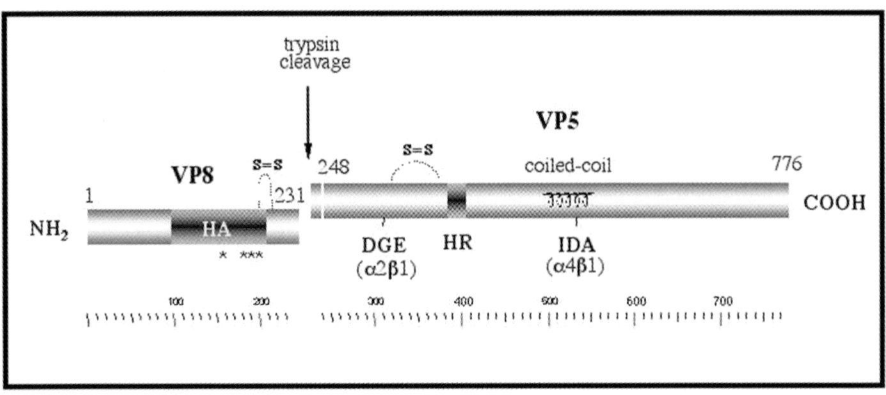

Fig. 2. Structural features of the outer capsid protein VP4. An arrow indicates the trypsin cleavage regions, and the numbers indicate the boundaries between VP8 and VP5. In VP8, the hemagglutination domain (HA) (aa 93 to·208) is shadowed; the asterisks below this domain indicate the amino acids that have been shown to be important in the SA binding activity of this protein (aa 155, and 188 to 190). The disulfide bridges between Cys 203 and Cys 216, and between Cys 318 and Cys 380, are indicated by S=S. In VP5, the position of the DGE and IDA tripeptide sequence binding motifs for integrins α2β1 and α4β1, located at aa 308-310 and 538-540, respectively, are shown. The hydrophobic region (HR), which has been proposed to be a putative fusion domain (aa 385-404), and a predicted heptad repeat (aa 494-554) which might form part of a coiled-coil structure are also depicted.

is absolutely dependent on it. It has been shown that the ability of rotaviruses to induce the formation of syncytia in MA104 cells supplemented with cholesterol depends on the cleavage of VP4 at arginine 247 (Gilbert and Greenberg, 1998). In general, the cleavage of VP4 does not affect cell binding, and it rather seems to be required for the entry of the virus into the cell's cytoplasm and/or for the uncoating of the virus particle;

ii) Many rotavirus strains are able to agglutinate red blood cells, and this agglutination has been shown to be mediated by the interaction of VP4 with sialic acid (SA) on the surface of erythrocytes. It has been shown that the domain responsible for this interaction resides on the VP8 subunit of VP4, between amino acids 93 and 208 (Fiore *et al.*, 1991; Fuentes Panana *et al.*, 1995), with tyrosine residue 155, and the tripeptide YYS, at positions 188 to 190, having an essential role in the SA-binding activity of the protein (Isa *et al.*, 1997).

iii) The prediction of the secondary structure of VP4 suggests that it has at least two different structural domains. One, comprising the amino terminal 60% of the protein, is predicted to be a globular domain rich in β-strands, while the second domain, represented by the remaining carboxy-terminal part of VP4, is rich in long stretches of α-helix (Estes, 1996; Isa *et al.*, 1997; López *et al.*, 1991). One of these stretches, of 63 residues, has a predicted coiled-coil structure, which might be the structural motif responsible for the dimerization of VP4. Recent characterisation of a monomeric recombinant VP4 protein produced in insect cells, and of its *in vitro* proteolytic fragments, has confirmed these structural predictions (Dormitzer *et al.*, 2001).

iv) The VP5 subunit of VP4 contains a hydrophobic region between amino acids 385 and 404 that shares sequence similarity with an internal fusogenic domain of the E1 glycoprotein of some alphaviruses (Mackow *et al.*, 1988). It has been shown that this domain on VP4 is capable of permeabilizing model, and bacterial membranes (Denisova *et al.*, 1999; Dowling *et al.*, 2000), and it has been suggested that it might facilitate rotavirus penetration into cells (Dowling *et al.*, 2000). Although this seems likely, direct evidence of its involvement on virus entry is lacking.

v) VP4 contains the integrin ligand site DGE, at amino acids 308 to 310 (Coulson *et al.*, 1997; Hewish *et al.*, 2000), and it has been shown that this tripeptide is used by rotaviruses to interact with integrin α2β1 during either their initial attachment to the cell surface, or as a second interaction site (Zárate *et al.*, 2000a). This protein also contains the tripeptide IDA at amino acids 538 to 540, which is a ligand motif for integin α4β1 (Coulson *et al.*, 1997), however, the functionality of this site has not yet been demonstrated directly.

vi) Two major antigenic domains for neutralizing antibodies have been mapped on VP4. The first is coincident with the SA-binding domain in VP8, while the second is around amino acids 305 and 393 in VP5. These two regions in VP5 are probably very close in the folded protein, since it has been shown that cysteine residues 318 and 380 form one of the two disulphide bridges that have been characterized in some rotavirus strains (Patton *et al.*, 1993).

The cryoelectron microscopy image reconstructions of trypsinized rotavirus particles have shown that the spikes formed by VP4 are structured as dimers with lobed heads (Shaw *et al.*, 1993; Yeager *et al.*, 1994). Image reconstructions of rotavirus particles bound by Fab fragments from monoclonal antibodies directed to VP8 or VP5, showed that VP8 is located in the heads of the spikes, while an epitope which maps to the hydrophobic region of VP5 described above is located just proximal to the heads (Dryden *et al.*, 2000). Recently, the structure of a protease-resistant core of a recombinant VP8 protein was obtained by NMR spectroscopy. As previously suggested by the analysis of the secondary structure of VP8 (Isa *et al.*, 1997), this core was found to consist of a monomeric, compactly folded 12-stranded anti-parallel β-sandwich, which fits into the lobed heads of the spikes seen by cryoelectron microscopy. The VP8 core was found to have the same basic fold as the carbohydrate-binding domains in members of the galnectin family of lectins (Dormitzer *et al.*, 2000b). Trypsin cleavage of a purified recombinant VP4 protein has shown that trypsin removes 3.3 kDa from the N-terminus of VP8, and about 30 kDa from the C-terminus of VP5, suggesting that the N- and C- termini of the virion-associated VP4 are sequestered from the protease, or folded into protease-resistant structures as a direct or indirect consequence of inter-actions between VP4 and VP6 or VP7 in the viral particle (Dormitzer, *et al.*, 2001).

Putative rotavirus receptors

The recognition of a specific receptor on the cell surface determines, at least in part, the tissue, host, and age-dependent susceptibility to a virus infection (Haywood, 1994). Rotaviruses have a specific cell tropism, infecting primarily the mature enterocytes at the tip of intestinal villi, and the susceptibility of these cells seems to be limited to a narrow age window (Kapikian and Chanock, 1996). Also, rotaviruses exhibit a marked host range restriction, since strains isolated from a given animal species do not replicate productively, or do so to a very limited extent, in a different species (Ciarlet *et al.*, 2000). The genetic basis for this restriction is not clear, and several viral genes have been implicated, which vary depending on the animal model studied; however, in addition to the nonstructural viral protein NSP1, VP4 and VP7 have frequently been associated with host range restriction (Broome *et al.*, 1993; Ciarlet *et al.*, 2000; Hoshino *et al.*, 1995). These observations suggest that differentiated enterocytes must bear specific receptors for these viruses. The strict tropism shown by rotaviruses *in vivo,* is also displayed *in vitro,* since they are able to bind to a variety of cell lines, while infecting efficiently only those cells of renal or intestinal epithelium origin (Estes and Cohen, 1989).

Different rotavirus strains display different requirements to bind, and thus infect, susceptible cells. The cell attachment of some strains isolated from animals (other than humans) is greatly diminished by treatment of cells with neuraminidase (NA), indicat-ing the need for sialic acid on the cell surface (Ciarlet and Estes, 1999; Fukudome *et al.*, 1989; Keljo and Smith, 1988; Méndez *et al.*, 1993). The interaction with a SA-containing receptor, however, does not seem to be essential, since variants which no longer need SA to infect the cells can be isolated from SA-dependent strains (Ludert *et al.*, 1996; Méndez *et al.*, 1993). In addition, the infectivity of many animal rotavirus strains and most, if not all, strains isolated from humans is not affected by treatment of cells with neuraminidase (Ciarlet and Estes, 1999; Fukudome *et al.*, 1989; Méndez *et al.*, 1999). Thus, there is a great interest in identifying the NA-resistant cellular receptor(s) for rotavirus, and to determine the role it (they) may have on the narrow tropism observed for these viruses.

The identification of the cellular receptor(s) is a topical area of rotavirus research, but despite the effort of many groups for more than a decade the advances in this field have been slow. However, the recent description of several receptor candidates allow us to foresee a speedier characterization of the functional cellular machinery involved in the attachment and cell entry of rotaviruses (Table 1), and hopefully will also help in the understanding of the mechanism used by these viruses to enter the cell, a key step in the virus replication cycle.

148

Table 1

Cellular molecules proposed to play a role as rotavirus receptors

Cell molecule	Rotavirus strain[*]	Cell type	Reference[#]
Gangliosides, GM1	SA11 (Si)	HT29, LLC-MK2	Superti and Donelli, 1991
GM1 and GM3	KU and MO (Hu)	MA104	Guo *et al.*, 1999
NeuGc-GM3	OSU (Po)	piglet enterocytes	Rolsma *et al.*, 1998
300 and 330 kDa glycoproteins	RRV (Si)	murine enterocytes	Bass *et al.*, 1991
Galactose in 66 to 97 kDa glycoproteins	CR48 (Po), SA11(Si), and NCDV(Bo)	MA104	Jolly *et al.*, 2000
Integrins $\alpha2\beta1$, $\alpha4\beta1$, and $\alpha x\beta2$	SA11 (Si), RV5 (Hu)	MA104, Caco-2, K562	Coulson *et al.*, 1997 Hewish *et al.*, 2000
Integrin $\alpha v\beta3$	RRV (Si), nar3 (Si), Wa (Hu)	MA104	Guerrero *et al.*, 2000a
Hsc70	RRV (Si), nar3 (Si), Wa (Hu)	MA104, Caco-2	Guerrero *et al.*, 2002

[*] Abbreviations: Si, simian; Po, porcine; Bo, bovine; Hu, Human.
[#] As listed in the reference section.

A number of glycoconjugates have been shown to bind to, and to block the infectivity of NA-sensitive animal rotavirus strains, and some of them have been suggested to play a role as possible receptors, like the GM3 ganglioside containing N-glycolyl neuraminic acid as the sialic acid moiety (NeuGc-GM3) in newborn piglet intestine (Rolsma *et al.*, 1998), GM1 in LLC-MK2 cells (Superti and Donelli, 1991), and 300-330 kDa glycoproteins in murine enterocytes (Bass *et al.*, 1991). It has also been suggested that the NA-resistant ganglioside GM1 may act as a receptor for some human rotavirus strains in MA104 cells (Guo *et al.*, 1999). Recently, it has been suggested that rotaviruses of human, simian, and porcine origin specifically recognize galactose as an important component of glycoprotein receptors in MA104 cells (Jolly *et al.*, 2000).

In addition to glycolipids, several integrins have been proposed to facilitate the initial steps of virus infection. VP4 contains the DGE and IDA tripeptide sequence motifs known to interact with integrins $\alpha2\beta1$ and $\alpha4\beta1$, respectively, while VP7 contains the $\alpha x\beta2$ integrin ligand site GPR, and the $\alpha4\beta1$ binding motif LDV (Coulson *et al.*, 1997; Hewish *et al.*, 2000). Antibodies to the integrin subunits $\alpha2$, $\beta2$, and $\alpha4$, as well as peptides that mimic their ligand sites, were shown to block the infectivity of NA-resistant and -sensitive rotavirus strains (Coulson *et al.*, 1997; Guerrero *et al.*, 2000a), and it was also shown that integrins α2β1 and α4β1 can mediate the attachment and entry of rotavirus SA11 into the human myelogenous leukemic cell line K562 (Hewish *et al.*, 2000). Furthermore, Guerrero *et al.* (Guerrero *et al.*, 2000a) demonstrated that in MA104 cells integrin αvβ3 interacts with NA-sensitive and -resistant strains at a post-attachment step, and it was shown that this integrin is capable of promoting rotavirus infection of the poorly permissive Chinese hamster ovary cells. More recently, it has been suggested that, similar to integrin αvβ3, the heat shock cognate protein, hsc70, may also function as a post-attachment receptor for rotaviruses (Guerrero *et al.*, 2002).

The possibility that several cell surface molecules may be involved in the early interactions of rotaviruses with the host cell which ultimately lead to the entry of the virus particle into the cytoplasm of the cell, is not far-fetched. In fact, the need for multiple receptor binding events to achieve an efficient cell entry is becoming a frequent observation in virus-host cell interactions. Viruses from different families [e.g. adenovirus (Nemerow, 2000), human immunodefficiency virus (Ugolini *et al.*, 1999), herpes simplex virus (Spear, 1993), HHV-6 (Campadelli-Fiume, 2000), reovirus (Barton *et al.*, 2001)], utilize at least two different receptors to interact with their host cells: the attachment receptors which in general allow the virus to rapidly bind to the cell surface; and the entry receptors, also known as co-receptors, or post-attachment receptors, which facilitate the entry of virus into the cell. It has been observed that this second interaction frequently induces conformational changes of the viral surface proteins that are essential for the penetration of the virus into the cellular cytoplasm.

In the case of rotaviruses, several lines of evidence suggest that these viruses need to interact with more than one cell surface molecule to enter the cell, using during this process different domains of the virus surface protein VP4 (Fig. 2), and probably of VP7. In this chapter, we will summarize the work of our group carried out during the past few years along these lines, and will place our findings in the context of the present knowledge in the field, to propose a working model for the early interactions of the virus with the host cell.

To understand the early events of rotavirus infection, we have undertaken the comparative characterization of three rotavirus strains which have different requirements to initially bind to, and thus infect, the host cell: the NA-sensitive simian rotavirus strain RRV (G3P5[3]), its NA-resistant variant nar3, and the human rotavirus strain Wa (G1P1A[8]), which is naturally resistant to the NA-treatment of cells.

Rotavirus interaction with the host cell is a multistep process

To characterize if the NA-resistant and -sensitive strains share at least one interaction with cell surface molecules, we developed an infection assay designed to detect competition for cell surface molecules at both attachment and post-attachment steps (Méndez *et al.*, 1996). In this assay we found that the human strain Wa efficiently competed with the infectivity of the NA-sensitive porcine rotavirus strain YM, and that of the variant nar3 both in untreated, as well as in NA-treated cells. This competition was nonreciprocal since YM and nar3 did not compete the infectivity of Wa. In contrast, a two-direction competition between the variant nar3 and the NA-sensitive strain was found. The fact that the competition between the two NA-resistant strains, nar3 and Wa, was not reciprocal strongly suggested that they bind to different molecules. In addition, the NA-sensitive phenotype clearly differentiated strains, like RRV or YM, from nar3 and Wa. Altogether, these findings suggested the existence of at least three cellular structures involved in rotavirus cell infection, with at least one being shared by human, NA-sensitive animal, and NA-resistant animal variant strains.

The comparison of the binding characteristics of wtRRV and nar3 to MA104 cells showed that both, the NA-sensitive and NA-resistant interactions of these viruses with the cell are mediated through two different domains of VP4. It was shown that RRV attaches to the cell through VP8, while nar3 does so through the VP5 domain of VP4 (Zárate *et al.*, 2000a, b). This observation was supported by the fact that neutralizing antibodies to VP8 (1A9, 7A12, M11 and M14) blocked the attachment of RRV to cells, but not of the variant nar3, while a MAb to VP5 (2G4) inhibited the binding of nar3, but not that of RRV. In addition, recombinant VP8 and VP5 proteins produced in bacteria as fusion products with GST, were capable of inhibiting the binding and infection of wt and variant viruses, respectively, when pre-incubated with the cell (Zárate *et al.*, 2000a). While nar3 interacts directly (through VP5) with the NA-resistant receptor, wtRRV seems to engage in the two interactions described in a sequential manner, since MAb 2G4, despite selectively blocking the binding of nar3, efficiently neutralized the infectivity of both viruses (Méndez *et al.*, 1993).

The sequential interaction of RRV with two molecules on the surface of MA104 cells was further supported by the observation that a monoclonal antibody (MAb 2D9) directed to a cell surface antigen specifically blocked the infectivity of both wtRRV and nar3, but competed only the attachment of the variant virus, indicating that wtRRV is blocked at a post-binding step (López *et al.*, 2000). Since MAb 2D9 also blocked the infectivity of nar3 in NA-treated cells, and prevented the cell attachment of the recombinant GST-VP5 protein, but did not affect the binding of GST-VP8, it seems to be directed to the NA-resistant receptor used by nar3 to attach to the cell, or to a molecule closely associated to it.

As an approach to identify potential cell molecules that could serve as rotavirus receptors, MA104 cells were treated with the non-ionic detergent octyl-β-glucoside (OG) under non-lytic conditions, a treatment that rendered the cells largely refractory to infection (Guerrero *et al.*, 2000b; Jolly *et al.*, 2001). The OG extract was shown to inhibit rotavirus infection when incubated with the viruses in solution, and the molecules responsible for at least part of this inhibitory activity were isolated and characterized. Two of these proteins were identified as the heat shock protein hsc70, and the β3 integrin subunit (Guerrero *et al.*, 2000b, 2002).

Taking together the findings gathered from the binding characterization of different rotavirus strains, the cellular molecules that have been suggested to play a role during these interactions, and the fact that the entry of rotaviruses seems to be a multistep process, we have proposed a working model for the early interactions of rotavirus with the host cell (Fig. 3). In this model we have divided the interactions of rotavirus with the cell surface in two phases; the initial phase is represented by the interaction of the virus with an attachment receptor(s), while the second phase represents later interactions of the virus particle with post-attachment receptor(s). The data that support this model will be discussed in the following sections.

MODEL FOR THE EARLY INTERACTIONS OF ROTAVIRUSES WITH MA104 CELLS

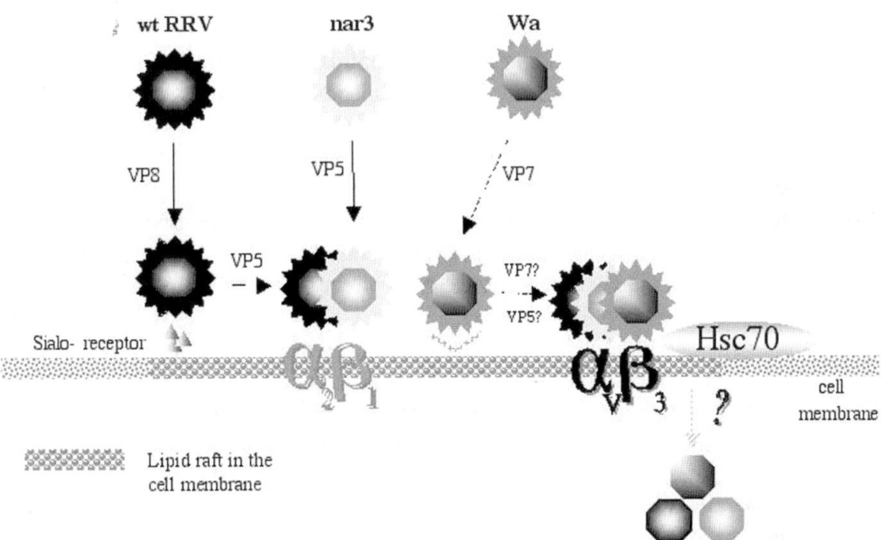

Fig. 3. Working model for the early interactions of rotavirus with MA104 cells. Wild type RRV interacts initially with a SA-containing cell receptor through the VP8 domain of VP4. Subsequent to this interaction, which might induce a conformational change in VP4, the virus interacts with a second, NA-resistant cell receptor, here proposed to be the α2β1 integrin. This interaction takes place through the DGE binding motif of VP5, present at aa 308-310. This second virus-cell interaction might facilitate a third interaction of the virus with β3 integrin and/or hsc70. The SA-independent variant nar3 is proposed to interact directly, through VP5, with the α2β1 integrin. The nature of the SA-containing, and NA-resistant cellular receptors to which wt RRV and human rotavirus Wa initially bind have not been determined, however, gangliosides NeuGc-GM3 and GM1 represent good candidates, respectively (see text). After their initial contact with the cell, all three rotavirus strains are proposed to interact with the β3 integrin subunit and/or with hsc70, interactions that might mediate the penetration of the viruses to the interior of the cell by a mechanism which is poorly understood, but which finally ends up with the uncoating of the virus. In this model most, if not all, of the molecules involved in rotavirus binding and entry are proposed to form a complex, probably embedded in glycosphingolipid-enriched lipid microdomains on the cell surface (see text).

152

Attachment receptors

I)   Gangliosides

There seems to be agreement on the fact that rotaviruses can be classified as NA-resistant or NA-sensitive strains, based on their ability or inability to infect cells treated with NA. Apparently, the NA-sensitive strains, like the simian strain RRV, initially interact with a SA-containing cell receptor through the VP8 SA-binding domain. The identity of the SA-containing molecule has not been determined, although good candidates are ganglioside NeuGc-GM3 (Delorme *et al.*, 2001; Rolsma *et al.*, 1998), and integrin $\alpha 2\beta 1$ (see below).

In the NA-resistant strains, such as the human strain Wa, the infection is not blocked by pretreatment of the target cells with sialidases. However, the infectivity of the human rotaviruses KUN and MO, which are also NA-resistant, was reported to be inhibited by ganglioside GM1 (Guo *et al.*, 1999), whose sialic acid is not cleaved by NA, although direct binding of the viruses to this ganglioside was not demonstrated.

More recently, using a thin-layer chromatogram binding assay, Delorme *et al.* (2001) determined the glycosphingolipid binding specificities of both NA-sensitive and NA-resistant animal rotavirus strains. They found that the simian SA11 and the bovine NCDV rotaviruses, which require SA to infect the cells, bound to external SA residues in gangliosides, while the bovine strain UK, which is NA-resistant, recognized gangliosides containing internal SA residues, which are resistant to NA treatment. Although the effect of these gangliosides on the infectivity of rotaviruses was not determined, they concluded that the widespread concept that the NA-resistant strains bind in a SA-independent manner is probably misleading. Instead, they proposed that the NA-resistant strains bind to gangliosides with internal sialic acids that are not removed by NA treatment (Delorme *et al.*, 2001). Altogether, these results suggest that the initial interaction of rotaviruses with the cell is most probably through glycosphingolipids present in the plasma membrane; this first interaction might allow the viral particle to dock in the cell surface, and then to search for more specific receptor(s) in two dimensions, which can be more efficient than searching in three dimensions (Adam and Delbruck, 1968).

II)   Integrins

Integrins are a family of α/β heterodimers of cell adhesion receptors that mediate cell-extracellular matrix and cell to cell interactions, and are known to function as signalling receptors for a variety of cellular processes including spreading, migration, proliferation, differentiation and survival (Hynes, 1992; Juliano and Haskill, 1993; Ruoslahti and Reed, 1994). These cell molecules are commonly used as receptors for several different viruses (see Table 2), with integrin $\alpha v\beta 3$ being, so far, the most frequently employed as virus receptor (Gavrilovskaya *et al.*, 1998; Nemerow and Stewartt, 2001; Nelsen-Salz *et al.*, 1999; Pulli *et al.*, 1997; Wickham *et al.*, 1993).

Table 2

Viruses that Use Integrins as Receptors

Virus	Integrin	Cell type [*]	Reference[#]
Echovirus 1, and 8	$\alpha2\beta1$	Hela	Bergelson *et al.*, 1992, 1993.
Echovirus 9	$\alpha v\beta3$	GMK	Nelsen-Salz *et al.*, 1999.
Parechovirus 1	$\alpha v\beta1$, $\alpha v\beta3$	A549	Pulli *et al.*, 1997, Triantafilou *et al.*, 2000.
Coxsackie A9	$\alpha v\beta3$	GMK	Roivainen *et al.*, 1994.
Foot and mouth disease virus	$\alpha v\beta3$,$\alpha v\beta6$	LLC-MK2, SW80/$\beta3$	Berinstein *et al.*, 1995, Jackson *et al.*, 2000.
Papillomavirus	$\alpha6\beta1$, $\alpha v\beta5$	CV1	Evander *et al.*, 1997.
Adenovirus 2, 5, 37, 19p.	$\alpha v\beta3$, $\alpha v\beta5$	Hela, M21, A549	Wickham *et al.*, 1993, Nemerow and Stewart, 2001.
Adeno-associated virus	$\alpha v\beta5$	Hela	Summerford *et al.*, 1999.
Hantavirus	$\alpha v\beta3$	Vero E6	Gavrilovskaya *et al.*, 1998.
Rotavirus	$\alpha4\beta1$, $\alpha x\beta2$, $\alpha2\beta1$, $\alpha v\beta3$	K562, Caco-2, MA104	Coulson *et al.*, 1997, Haywood *et al.*, 2000, Zárate *et al.*, 2000b, Guerrero *et al.*, 2000a.

* Cell origin: Hela = human cervix adenocarcinoma, GMK,Vero E6, and CV-1 = african green monkey kidney; LLC-MK2, and MA104 = rhesus monkey kidney; SW80/$\beta3$ = human colon carcinoma cells transfected with $\beta3$ integrin; M21 = human melanoma; A549 = human lung carcinoma; K562 human chronic myelogenous leukemia; Caco-2 = human colon adenocarcinoma. # As listed in the reference section.

As previously mentioned, it has been described that several integrins are involved in the interaction of rotavirus with its host cell. The finding that integrin ligand-binding motifs are present in the VP4 of most rotavirus strains (Coulson *et al.*, 1997) and the characterization of the binding and infectivity properties of the NA-resistant variant nar3, allowed us to show that the initial interaction of nar3 with the cell surface is with integrin $\alpha2\beta1$, through the DGE integrin binding domain present in VP5; this conclusion is supported by the following data: antibodies to the $\alpha2$ subunit reduce the infectivity of both wt RRV and nar3, but only block the cell attachment of the latter virus; MAbs to $\alpha2$ block the attachment of the recombinant GST-VP5 fusion protein but not that of GST-VP8; GST-VP5 specifically displaces the cell binding of nar3, while a GST-VP5 mutant polypeptide in which the $\alpha2$ integrin binding motif DGE was mutagenized to AGE no longer displaces it, and a synthetic VP4 peptide containing the $\alpha2\beta1$ integrin binding motif DGE efficiently inhibits the attachment of nar3, but not that of RRV (Zárate *et al.*, 2000a, b).

Accordingly, we have proposed that subsequent to the initial interaction with SA, RRV interacts with a second cell receptor, most probably integrin α2β1, through the DGE binding motif in VP5 (Zárate *et al.*, 2000a). The ability of the NA-resistant variant nar3 to interact directly with this integrin is likely to be the result of a slight different conformation of its VP4 protein, compared to that of the wtRRV protein (Méndez *et al.*, 1993; Méndez *et al.*, 1996).

Integrins α4β1 and αxβ2 have also been implicated in rotavirus cell infection (Coulson *et al.*, 1997; Hewish *et al.*, 2000). The role of these integrins has not been determined yet, however, given that no additivity was observed when mixtures of antibodies to these and other integrins were tested (Guerrero *et al.*, 2000a), they may represent alternative interaction sites for rotaviruses.

The results obtained in the infection competition assay described previously suggest that the human rotavirus Wa initially attaches to a cell surface molecule that is used by RRV and nar3, in a subsequent step after their interaction with α2β1 integrin. It can not be ruled out, however, that the Wa attachment receptor is not actually used by RRV and nar3, but that human rotavirus Wa interferes with the infectivity of these viruses by binding to a molecule that might be located in close proximity to either α2β1, or to post-attachment receptors αVβ3 and hsc70 (see below). The cellular molecule initially used by Wa to bind to cells and the viral protein responsible for this interaction have not been characterized, although recent data from our laboratory suggest that this virus binds initially to the cell surface through VP7, rather than through VP5 or VP8 (Villatoro A., *et al.* manuscript in preparation).

Post-attachment receptors

I)   *Integrin* αvβ3

The relevance of this integrin for rotavirus infection was demonstrated by the fact that antibodies to the β3 subunit reduce the infectivity of rotavirus strains RRV, nar3, and Wa; accordingly, when vitronectin, a β3 integrin ligand, is pre-incubated with the cells, it specifically blocks rotavirus infectivity by about 70% (Guerrero *et al.*, 2000a). In addition, it has been shown that the expression of β3 integrin in poorly permissive CHO cells facilitates the infectivity of rotaviruses; CHO cells stably transfected with the αv and β3 integrin subunit genes were three to four times more susceptible to rotavirus infection than the parental CHO cell line. This increase in infectivity was blocked by incubation of the cells with either MAbs to β3 or vitronectin (Guerrero *et al.*, 2000a). Furthermore, antibodies to αvβ3, as well as ligands to this integrin, reduced the infectivity of rotaviruses, but did not block their attachment to cells, indicating that rotaviruses interact with αvβ3 at a post-binding step. Also, it was found that the interaction of rotaviruses with αvβ3 integrin was RGD-independent, as expected from the fact that neither VP4 nor VP7 have this integrin binding motif (Guerrero *et al.*, 2000a).

Since integrins α2β1, α4β1, and αxβ2 have been suggested to play a role during rotavirus infection (Coulson *et al.*, 1997; Zárate *et al.*, 2000a), blocking experiments, using mixtures of antibodies directed to these integrins and to αvβ3 were performed. An additive blocking effect was found when mixtures of antibodies to integrins α2β1 and αvβ3 were used, suggesting that these two integrins are involved in different stages of rotavirus infection (Guerrero *et al.*, 2000a).

II)   *The heat shock protein hsc70*

Hsc70, an abundant cytoplasmic and nuclear protein, is a constitutive member of the stress-inducible hsp70 protein family that functions in normal cellular physiology. The proteins in this family are highly conserved ATPases which have been associated with a number of functions, including protein folding, translocation across biological membranes, and assembly and disassembly of oligomeric complexes. In response to different stress conditions these proteins prevent the formation of protein aggregates by stabilizing unfolded intermediates which are subsequently refolded to the native state or degraded (Hartl, 1996; Mayer and Bukau, 1998; Morimoto *et al.*, 1997). In particular, hsc70 has been shown to favour protein transport across organellar membranes, bind nascent polypeptides, and dissociate clathrin from clathrin coats (Oglesbee *et al.*, 1993).

Despite the fact that hsc70 has no export signal sequence, it has been shown that it can be exposed on the surface of various cell types (Multhoff and Hightower, 1996). The presence of this protein on the surface of MA104 cells was demonstrated by flow cytometry and immunofluorescence of non-permeabilized cells (Guerrero *et al.*, 2002).

The role of hsc70 during rotavirus infection was evaluated by testing the ability of MAbs to hsc70 to block infectivity when pre-incubated with the cell. It was found that these MAbs specifically blocked the infectivity of rotaviruses RRV, nar3, and Wa to about 80%. Despite their efficient blocking activity, these MAbs did not prevent the binding of the viruses to the surface of the cells, indicating that they interfere with the virus infectivity at a post-binding step. These data suggest that hsc70 is involved in rotavirus cell entry and/or uncoating during MA104 cell infection, although the precise role of this protein has yet to be determined. Although the viral protein involved in the interaction with hsc70 has not been defined, we found that infectious, triple layered particles, but not non-infectious, double layered particles, bind to hsc70 in an ELISA assay, and this binding can be prevented by preincubation of the viral particles with monoclonal antibodies to both virus surface proteins, but not by antibodies to VP6, suggesting that either VP4 or VP7 is responsible for this interaction. Finally, it has been shown that hsc70 is present on the surface of both rotavirus-susceptible and non-susceptible cells, indicating that this protein alone is not responsible for determining the cell tropism of these viruses.

Lipid microdomains seem to be important for rotavirus infection

We have recently shown that the infectivity of rotaviruses is partially blocked by metabolic inhibitors of N-glycosylation and glycolipid synthesis, while it is not affected by the inhibition of the cellular O-glycosylation. In addition, we have also shown that sequestration of cholesterol from the cell membrane with methyl-β-cyclodextrin reduces the infectivity of the three viruses, RRV, nar3, and Wa, by more than 90%, while not affecting their binding to the cell (Guerrero *et al.*, 2000b). These observations suggest that the virus receptor(s) might be forming part of cell membrane glycosphingolipid-enriched lipid microdomains, termed rafts (Simons and Ikonen, 1997). These rafts are thought to function as specialized platforms for apical cell sorting of proteins and signal transduction, and have also been reported to serve as the entry portal for the SV40 virus (Stang *et al.*, 1997), and as the site of assembly for HIV (Nguyen and Hildreth, 2000), and measles virus (Vincent *et al.*, 2000). Our findings indicate that although the receptor(s) retain their ability to bind rotavirus particles in cholesterol-deficient cells, in order to fully promote virus entry they must probably be organized in a lipid microdomain. Finally, the observation that integrin αvβ3 (Green *et al.*, 1999) and protein hsc70 (Isa *et al.*, unpublished data) associate to rafts is also consistent with the possibility that the functional rotavirus receptor might be forming part of these microdomains.

In summary, the data presented here are consistent with the existence of several rotavirus receptors, which might be tightly organized, maybe forming a complex in glycosphingolipid-enriched rafts. The requirement for several cell molecules which need to be present and organized in a precise fashion, might explain the exquisite cell and tissue tropism of these viruses. It remains to be established which, if any, of the receptor molecules described so far is indeed non replaceable, and whether in fact there exists a unique pathway of infectivity for rotaviruses, with distinct entry points for different strains.

Future directions

Despite the information that has been accumulated about the functional domains on VP4 and VP7, the viral proteins (and their particular domains) involved in the interaction with the post-attachment receptors αvβ3 and hsc70 are yet to be defined. In the particular case of human rotaviruses, it is also important to define if the initial contact with the cell surface is through VP7 rather than through VP4. The initial interaction of both NA-resistant and -sensitive viruses with the attachment receptors (gangliosides and/or integrins) might induce conformational changes in the viral particle that could facilitate more specific contacts with the post-attachment receptors (αvβ3 and Hsc70) and expose domains in the viral proteins that may otherwise be buried within hydrophobic pockets, and which are important for the entry of the virus into the cell. One of such domains could be the fusion peptide of VP4, that has been shown to have lipophilic activity in model membranes, as well as on bacterial membranes (Denisova *et al.*, 1999; Dowling *et al.*, 2000). A rearrangement of the coiled-coil region present in the carboxy terminal half of VP5 might potentially expose the fusion region, as has been observed for the fusion regions of other viruses such as influenza virus, HIV, Ebola virus, and Newcastle disease virus, among others [For a review see (Eckert and Kim, 2001)].

The two cellular proteins that have been identified as post-attachment receptors for rotaviruses, $\alpha v \beta 3$ and hsc70, are known to have important functions during normal cell physiology. It remains to be determined if these proteins serve only as anchors on the membrane for the virus during its transit into the cellular cytoplasm, or if some of the specific functions of these cellular proteins are relevant for this event.

Integrins can signal through the cell membrane in either direction: The extracelular binding activity of integrins is regulated from the inside of the cell (inside-out signalling), while the binding of their ligands elicits signals that are transmitted into the cell (outside-in signalling) (Giancotti and Ruoslahti, 1999). Fundamental in the integrin signalling is, in most cases, the cytoplasmic tail of the beta integrin subunit. The activation of tyrosine kinases triggered by integrins is well documented, and it has been shown for adenovirus and the adeno-associated virus that the integrin-dependent activation of PI3 kinase and its association with p13OCAS, is required for the viruses to enter the cell (Li *et al.*, 2000). It remains to be determined if the signal transduction induced by integrins is important for rotavirus entry.

In addition to integrins $\alpha 2 \beta 1$ and $\alpha v \beta 3$, two other integrins have been implicated in the cell entry of rotaviruses, $\alpha 4 \beta 1$ and $\alpha x \beta 2$. The role of these molecules will have to be determined. It should be explored whether they represent alternative pathways of infection, or are accessory molecules relevant for virus infectivity.

So far not a single putative receptor has been associated with the full susceptibility of a cell to rotavirus infection. Thus, it is important to determine if there are still some other nonidentified cell molecules involved in rotavirus cell infection. It remains to be determined which is the contribution of each of the proposed cell molecules involved in rotavirus infection to the rotavirus-permissive status of a cell.

Given that several cell molecules have been proposed to play a role at the attachment or post-attachment steps, and given the evidence that lipid microdomains are involved in rotavirus infection, it is important to explore whether the functional receptor for rotaviruses is a multiprotein complex organized in these lipid microdomains in the cell membrane, and to determine if rotaviruses enter the cell through these microdomains.

It is also of the upmost importance to determine if the receptors identified in cultured cells play this role during the infection of mature enterocytes in the gut. If not, it is essential to identify the virus cell receptors in the gut.

Much remains to be learned about the process of binding and penetration of rotaviruses. The characterization of the nature of the interactions that occur between the cellular and viral partners, and the signal transduction pathways potentially triggered by the early virus-cell contacts, should give insight into the elaborate mechanism used by these viruses to enter cells.

Acknowledgments

We would like to emphasize that the work presented here is the result of the excellent work, enthusiasm, and friendship of all the students, technicians, and investigators that have been, or currently are, part of our team, we would like to acknowledge the

158

efforts of all of them. This work was partially supported by grants 75197-527106 and 55000613 from the Howard Hughes Medical Institute, G0012-N9607 from the National Council for Science and Technology-Mexico, and IN201399 from DGAPA-UNAM.

References

Adam, G., and Delbruck, M. (1968) Reduction od dimentionality in biological diffusion processes. In: A. Rich and N. Davidson (eds). Structural chemistry and molecular biology. W. H. Freeman, San Francisco, p 198-215.

Arias, C. F., Romero, P., Alvarez, V., and López, S. (1996). Trypsin activation pathway of rotavirus infectivity. *J Virol* **70,** 5832-5839.

Barton, E. S., Forrest, J. C., Connolly, J. L., Chappell, J. D., Liu, Y., Schnell, F. J., Nusrat, A., Parkos, C. A., and Dermody, T. S. (2001). Junction adhesion molecule is a receptor for reovirus. *Cell* **104,** 441-451.

Bass, D. M., Mackow, E. R., and Greenberg, H. B. (1991). Identification and partial characterization of a rhesus rotavirus binding glycoprotein on murine entero-cytes. *Virology* **183,** 602-610.

Beisner, B., Kool, D., Marich, A., and Holmes, I. H. (1998). Characterisation of G serotype dependent non-antibody inhibitors of rotavirus in normal mouse serum. *Arch Virol* **143,** 1277-1294.

Berinstein, A., Roivainen, M., Hovi, T., Mason, W., and Baxt, B. (1995). Antibodies to the vitronectin receptor (integrin $\alpha v\beta 3$) inhibit binding and infection of Foot-and-mouth disease virus to cultured cells. *J Virol* **69,** 2664-2666.

Bergelson, J. M., Shepley, M. P., Chan, B. M. C., Hemler, M. E., and Finberg, R. W. (1992). Identification of the integrin VLA-2 as a receptor for echovirus I. *Science* **255,** 1718-1720.

Bergelson, J. M., St. John, N., Kawaguchi, S., Chan, M., Stubdal, H., Modlin, J., and Finberg, R. W. (1993). Infection by echoviruses 1 and 8 depends on the alpha 2 subunit of human VLA-2. *J Virol* **67,** 6847-6852.

Broome, R. L., Vo, P. T., Ward, R. L., Clark, H. F., and Greenberg, H. B. (1993). Murine rotavirus genes encoding outer capsid proteins VP4 and VP7 are not major determinants of host range restriction and virulence. *J Virol* **67,** 2448-2455.

Campadelli-Fiume, R. (2000). Virus receptor arrays, CD46 and human herpesvirus 6. *Trends Microbiol* **8,** 436-438.

Ciarlet, M., and Estes, M. K. (1999). Human and most animal rotavirus strains do not require the presence of sialic acid on the cell surface for efficient infectivity. *J Gen Virol* **80,** 943-948.

Ciarlet, M., Estes, M. K., and Conner, M. E. (2000). Simian rhesus rotavirus is a unique heterologous (non-lapine) rotavirus strain capable of productive replica-tion and horizontal transmission in rabbits. *J Gen Virol* **81,** 1237-1249.

Clark, S. M., Roth, J. R., Clark, M. L., Barnett, B. B., and Spendlove, R. S. (1981). Trypsin enhancement of rotavirus infectivity: mechanism of enhancement. *J Virol*

39, 816-822.

Coulson, B. S., Londrigan, S. H., and Lee, D. J. (1997). Rotavirus contains integrin ligand sequences and a disintegrin-like domain implicated in virus entry into cells. *Proc Natl Acad Sci USA* **94,** 5389-5394.

Crawford, S., Mukherjee, S. K., Estes, M. K., Lawton, J. A., Shaw, A. L., Ramig, R. F., and Prasad, B. V. V. (2001). Trypsin cleavage stabilizes rotavirus VP4 spike. *J Virol* **75,** 6052-6061.

Crawford, S. E., Labbé, M., Cohen, J., Burroughs, M. H., Zhou, Y. J., and Estes, M. K. (1994). Characterization of virus-like particles produced by the expression of rotavirus capsid proteins in insect cells. *J Virol* **68,** 5945-5952.

Delorme, C., Brüssow, H., Sidoti, J., Roche, N., Karlsson, K. A., Neeser, J. R., and Teneberg, S. (2001). Glycosphingolipid binding specificities of rotavirus: identification of a sialic acid-binding epitope. *J Virol* **75,** 2276-2287.

Denisova, E., Dowling, W., LaMonica, R., Shaw, R., Scarlata, S., Ruggeri, F., and Mackow, E. R. (1999). Rotavirus capsid protein VP5* permeabilizes membranes. *J Virol* **73,** 3147-3153.

Dormitzer, P., Greenberg, H., and Harrison, S. C. (2000a). Purified recombinant rotavirus VP7 forms soluble, calcium-dependent trimers. *Virology* **277,** 420 - 428.

Dormitzer, P. R., Sun, Z.-Y. J., Wagner, G., and Harrison, S. C. (2000b). *VII International Symposium on Double-Stranded RNA Viruses, Aruba. Abstract* W2-2

Dormitzer, P., Greenberg, H., and Harrison, S. C. (2001). Proteolysis of monomeric recombinant rotavirus VP4 yield an oligomeric VP5* core. *J Virol* **75,** 7339-7350.

Dowling, W., Denisova, E., Lamonica, R., and Mackow, E. R. (2000). Selective membrane permeabilization by the rotavirus VP5* protein is abrogated by mutations in an internal hydrophobic domain. *J Virol* **74,** 6368-6376.

Dryden, K. A., Tihova, M., Bellamy, A. R., Greenberg, H. B., and Yeager, M. (2000) *VII International Symposium on Double-Stranded RNA Viruses, Aruba. Abstract* P2-15

Eckert, D. M., and Kim, P. S. (2001). Mechanisms of viral membrane fusion and its inhibition. *Annu Rev Biochem* **70,** 777-810.

Espejo, R. T., López, S., and Arias, C. (1981). Structural polypeptides of simian rotavirus SA11 and the effect of trypsin. *J Virol* **37,** 156-160.

Estes, M. K. (1996). Rotaviruses and their replication. *In:* Fields, B.M., Knipe, D. M., Howley, P. M. *et al.* (eds) Fields Virology. Lippincott-Raven Press, Philadelphia, p.1625-1655.

Estes, M. K., and Cohen, J. (1989). Rotavirus gene structure and function. *Microbiol Rev* **53,** 410-449.

Estes, M. K., Graham, D. Y., and Mason, B. B. (1981). Proteolytic enhancement of rotavirus infectivity: molecular mechanisms. *J Virol* **39,** 879-888.

Evander, M., Fraze, r. I. H., Payne, E., Qi, Y. M., Hengst, K., and McMillan, N. A. (1997). Identification of the alpha6 integrin as a candidate receptor for papillomaviruses. *J Virol* **71,** 2449-2456.

Fiore, L., Greenberg, H. B., and Mackow, E. R. (1991). The VP8 fragment of VP4 is the rhesus rotavirus hemagglutinin. *Virology* **181,** 553-563.

Fuentes Panana, E. M., López, S., Gorziglia, M., and Arias, C. F. (1995). Mapping the hemagglutination domain of rotaviruses. *J Virol* **69,** 2629-2632.

Fukudome, K., Yoshie, O., and Konno, T. (1989). Comparison of human, simian, and bovine rotaviruses for requirement of sialic acid in hemagglutination and cell adsorption. *Virology* **172,** 196-205.

Fukuhara, N., Yoshie, O., Kitaoka, S., and Konno, T. (1988). Role of VP3 in human rotavirus internalization after target cell attachment via VP7. *J Virol* **62,** 2209-2218.

Gavrilovskaya, I. N., Shepley, M., Shaw, R., Ginsberg, M. H., and Mackow, E. R. (1998). beta3 Integrins mediate the cellular entry of hantaviruses that cause respiratory failure. *Proc Natl Acad Sci USA* **95,** 7074-7079.

Giancotti, F. G., and Ruoslathi, E. (1999). Integrin signaling. *Science* **285,** 1028-1032.

Gilbert, J. M., and Greenberg, H. B. (1998). Cleavage of rhesus rotavirus VP4 after arginine 247 is essential for rotavirus-like particle-induced fusion from without. *J Virol* **72,** 5323-5327.

Green, J. M., Zhelesnyak, A., Frazier, W. A., and Brown, E. J. (1999). Role of cholesterol in formation and function of a signaling complex involving $\alpha v \beta 3$, integrin-associated protein (CD47), and heterotrimeric G proteins. *J Cell Biol* **146,** 673-682.

Guerrero, C. A., Bouyssounade, D., Zárate, S., Espinosa, R., Romero, P., Méndez, E., López, S., and Arias, C. F. (2002). The heat shock cognate protein 70 is involved in rotavirus cell entry. *J Virol* **76,** 4096-4102.

Guerrero, C. A., Méndez, E., Zárate, S., Isa, P., López, S., and Arias, C. F. (2000a). Integrin $\beta 3$ mediates rotavirus cell entry. *Proc Natl Acad Sci USA* **97,** 14644-14649.

Guerrero, C. A., Zárate, S., Corkidi, G., López, S., and Arias, C. F. (2000b). Biochemical characterization of rotavirus receptors in MA104 cells. *J Virol* **74,** 9362-9371.

Guo, C., Nakagomi, O., Mochizuki, M., Ishida, H., Kiso, M., Ohta, Y., Suzuki, T., Miyamoto, D., Jwa Hidari, K. I., and Suzuki, Y. (1999). Ganglioside GM1a on the cell surface is involved in the infection by human rotavirus KUN and MO strains1. *J Biochem (Tokyo)* **126,** 683-688.

Hartl, F.U. (1996). Molecular chaperones in cellular protein folding. *Nature* **381,** 571-579.

Haywood, A. M. (1994). Virus receptors: Binding, adhesion strengthening, and changes in viral structure. *J Virol* **68,** 1-5.

Hewish, M. J., Takada, Y., and Coulson, B. S. (2000). Integrins $\alpha 2 \beta 1$ and $\alpha 4 \beta 1$ can mediate SA11 rotavirus attachment and entry into cells. *J Virol* **74,** 228-236.

Hoshino, Y., Saif, L. J., Kang, S. Y., Sereno, M. M., Chen, W. K., and Kapikian, A. Z. (1995). Identification of group A rotavirus genes associated with virulence of a porcine rotavirus and host range restriction of a human rotavirus in the gnotobiotic piglet model. *Virology* **209,** 274-280.

Hynes, R. O. (1992). Integrins: versatility, modulation, and signaling in cell adhesion. *Cell* **69,** 11-25.

Isa, P., López, S., Segovia, L., and Arias, C. F. (1997). Functional and structural analysis of the sialic acid-binding domain of rotaviruses. *J Virol* **71,** 6749-6756.

Jackson, T., Sheppard, D., Denyer, M., Blakemore, W., and King, A. M. (2000). The epithelial integrin $\alpha v \beta 6$ is a receptor for foot-and-mouth-disease virus. *J Virol* **74,** 4949-4956.

Jolly, C. L., Beisner, B., Ozser, E., and Holmes, I. H. (2001). Non-lytic extraction and characterisation of receptors for multiple strains of rotavirus. *Arch Virol* **146,** 1307-1323.

Jolly, C. L., Beisner, B. M., and Holmes, I. H. (2000). Rotavirus infection of MA104 cells is inhibited by Ricinus lectin and separately expressed single domains. *Virology* **275,** 89-97.

Juliano, R. L., and Haskill, S. (1993). Signal transduction from the extracellular matrix. *J Cell Biol* **120,** 577-585.

Kapikian, A. Z., and Chanock, R. M. (1996). In: Fields, B.M., Knipe, D. M., Howley, P. M. *et al.* (eds) *Fields Virology*, 3rd ed., Lippincott-Raven, Philadelphia, p. 1657-1708.

Keljo, D. J., and Smith, A. K. (1988). Characterization of binding of simian rotavirus SA-11 to cultured epithelial cells. *J Pediatr Gastroenterol Nutr* **7,** 249-256.

Kirkwood, C. D., Bishop, R. F., and Coulson, B. S. (1998). Attachment and growth of human rotaviruses RV-3 and S12/85 in Caco-2 cells depend on VP4. *J Virol* **72,** 9348-9352.

Li E, Stupack D. G., Brown S. L., Klemke R., Shlaepfer D.D., and Nemerow G. R. (2000) Association of p130CAS with phosphatidylinositol-3-OH kinase mediates adenovirus cell entry. *J Biol Chem* **275**, 14729-14735

López, S., Arias, C. F., Bell, J. R., Strauss, J. H., and Espejo, R. T. (1985). Primary structure of the cleavage site associated with trypsin enhancement of rotavirus SA11 infectivity. *Virology* **144,** 11-19.

López, S., Espinosa, R., Isa, P., Zárate, S., Méndez, E., and Arias, C. F. (2000). Characterization of a monoclonal antibody directed to the surface of MA104 cells that blocks the infectivity of rotaviruses. *Virology* **273,** 160-168

López, S., López, I., Romero, P., Méndez, E., Soberón, X., and Arias, C. F. (1991). Rotavirus YM gene 4: analysis of its deduced amino acid sequence and prediction of the secondary structure of the VP4 protein. *J Virol* **65,** 3738-3745.

Ludert, J. E., Feng, N. G., Yu, J. H., Broome, R. L., Hoshino, Y., and Greenberg, H. B. (1996). Genetic mapping indicates that VP4 is the rotavirus cell attachment protein *in vitro* and *in vivo*. *J Virol* **70,** 487-493.

Mackow, E. R., Shaw, R. D., Matsui, S. M., Vo, P. T., Dang, M. N., and Greenberg, H. B. (1988). The rhesus rotavirus gene encoding protein VP3: location of amino acids involved in homologous and heterologous rotavirus neutralization and identification of a putative fusion region. *Proc Natl Acad Sci USA* **85,** 645-649.

Mayer, M. P., and Bukau, B. (1998). Hsp70 chaperone systems: diversity of cellular functions and mechanism of action. *Biol Chem* **379,** 261-268.

Méndez, E., Arias, C. F., and López, S. (1993). Binding to sialic acids is not an essential step for the entry of animal rotaviruses to epithelial cells in culture. *J Virol* **67,** 5253-5259.

Méndez, E., Arias, C. F., and López, S. (1996). Interactions between the two surface proteins of rotavirus may alter the receptor-binding specificity of the virus. *J Virol* **70,** 1218-1222.

Méndez, E., López, S., Cuadras, M. A., Romero, P., and Arias, C. F. (1999). Entry of rotaviruses is a multistep process. *Virology* **263,** 450-459.

Morimoto, R. I., Kline, M. P., Bimston, D. N., and Cotto, J. J. (1997). The heat-shock response: regulation and function of heat-shock proteins and molecular chaperones. *Essays Biochem* **32,** 17-29.

Multhoff, G., and Hightower, L. E. (1996). Cell surface expression of heat shock proteins and the immune response. *Cell Stress Chaperones* **1,** 167-176.

Nelsen-Salz, B., Eggers, H. J., and Zimmermann, H. (1999). Integrin $\alpha v \beta 3$ (vitronectin receptor) is a candidate receptor for the virulent echovirus 9 strain Barty. *J Gen Virol* **80,** 2311-2313.

Nemerow, G. R. (2000). Cell receptors involved in adenovirus entry. *Virology* **274,** 1-4.

Nemerow, G. R., and Stewartt, P. L. (2001). Antibody neutralization epitopes and integrin binding sites on nonenveloped viruses. *Virology* **288,** 189-191.

Nguyen, D. H., and Hildreth, J. E. K. (2000). Evidence for budding of HIV-1 selectively from glycolipid-enriched membrane lipid rafts. *J Virol* **74,** 3264-3272.

Oglesbee, M. J., Kenney, H., Kenney, T., and Krakowka, S. (1993). Enhanced production of morbillivirus gene-specific RNAs following induction of the cellular stress response in stable persistent infection. *Virology* **192,** 556-567.

Patton, J. T., Hua, J., and Mansell, E. A. (1993). Location of intrachain disulfide bonds in the VP5* and VP8* trypsin cleavage fragments of the rhesus rotavirus spike protein VP4. *J Virol* **67,** 4848-4855.

Prasad, B. V., Burns, J. W., Marietta, E., Estes, M. K., and Chiu, W. (1990). Localization of VP4 neutralization sites in rotavirus by three-dimensional cryo-electron microscopy. *Nature* **343,** 476-479.

Prasad, B. V., and Chiu, W. (1994). Structure of rotavirus. *Curr Top Microbiol Immunol* **185,** 9-29.

Pulli, T., Koivunen, E., and Hyypia, T. (1997). Cell-surface interactions of echovirus 22. *J Biol Chem* **272,** 21176-21180.

Roivainen, M., Piirainen, L., Hovi, T., Virtanen, I., Riikonen, T., Heino, J., and Hyypia, T. (1994). Entry of coxsackievirus A9 into host cells: specific interactions with alpha v beta 3 integrin, the vitronectin receptor. *Virology* **203,** 357-365.

Rolsma, M. D., Kuhlenschmidt, T. B., Gelberg, H. B., and Kuhlenschmidt, M. S. (1998). Structure and function of a ganglioside receptor for porcine rotavirus. *J Virol* **72,** 9079-9091.

Ruoslahti, E., and Reed, J. C. (1994). Anchorage dependence, integrins, and apoptosis. *Cell* **77,** 477-478.

Sabara, M., Gilchrist, J. E., Hudson, G. R., and Babiuk, L. A. (1985). Preliminary characterization of an epitope involved in neutralization and cell attachment that is located on the major bovine rotavirus glycoprotein. *J Virol* **53,** 58-66.

Shaw, A. L., Rothnagel, R., Chen, D., Ramig, R. F., Chiu, W., and Prasad, B. V. (1993).

Three-dimensional visualization of the rotavirus hemagglutinin structure. *Cell* **74,** 693-701.

Simons, K., and Ikonen, E. (1997). Functional rafts in cell membranes. *Nature* **387,** 569-572.

Spear, P. G. (1993). Entry of alphaherpesviruses into cells. *Semin Virol* **4,**167-180.

Stang, E., Kartenbeck, J., and Parton, R. G. (1997). Major histocompatibility complex class I molecules mediate association of SV40 with caveolae. *Mol Biol Cell* **8,** 47-57.

Summerford, C., Bartlett, J., and Samulski, R. J. (1999). $\alpha v\beta 5$ integrin: a co-receptor for adeno-associated virus type 2 infection. *Nature Biotechnol* **5,** 78-82.

Superti, F., and Donelli, G. (1991). Gangliosides as binding sites in SA-11 rotavirus infection of LLC-MK2 cells. *J Gen Virol* **72,** 2467-2474.

Triantafilou, K., Triantafilou, M., Yakada, Y., and Fernadez, N., (2000). Human parechovirus 1 utilizes integrins $\alpha v\beta 3$ and $\alpha v\beta 1$ as receptors. *J Virol* **74,** 5856-5862.

Ugolini, S., Mondor, I., and Sattentau, Q. J. (1999). HIV-1 attachment: another look. *Trends Microbiol* **7,** 144-149.

Vincent, S., Gerlier, D., and Manie, S. N. (2000). Measles virus assembly within membrane rafts. *J Virol* **74,** 9911-9915.

Wickham, T.J., Mathias, P., Cheresh, D.A., and Nemerow, G.R. (1993). Integrins $\alpha v\beta 3$ and $\alpha v\beta 5$ promote adenovirus internalization but no virus attachment. *Cell* **73,** 309-319.

Yeager, M., Berriman, J. A., Baker, T. S., and Bellamy, A. R. (1994). Three-dimensional structure of the rotavirus haemagglutinin VP4 by cryo-electron microscopy and difference map analysis. *EMBO J* **13,** 1011-1018.

Yeager, M., Dryden, K. A., Olson, N. H., Greenberg, H. B., and Baker, T. S. (1990). Three-dimensional structure of rhesus rotavirus by cryoelectron microscopy and image reconstruction. *J Cell Biol* **110,** 2133-2144.

Zárate, S., Espinosa, R., Romero, P., Guerrero, C. A, Arias, C. F., and López, S. (2000a). Integrin alpha2 beta1 mediates the cell attachment of the rotavirus neuraminidase-resistant variant nar3. *Virology* **278,** 50-54.

Zárate, S., Espinosa,R., Romero,P., Méndez, E., Arias, C.F., and López, S. (2000b). The VP5 domain of VP4 can mediate attachment of rotaviruses to cells. *J Virol* **74,** 593-599.

Viral Gastroenteritis
U. Desselberger and J. Gray (editors)

165

II, 4. Rotavirus genome replication: role of the RNA-binding proteins

John T. Patton*, Karen Kearney and Zenobia Taraporewala

Laboratory of Infectious Diseases, National Institutes of Allergy and Infectious Diseases, National Institutes of Health, 50 South Drive, MSC 8026, Room 6314

Introduction

The rotavirus virion is a triple-layered icosahedron containing eleven segments of double-stranded (ds)RNA. The innermost layer consists of the core lattice protein, VP2, arranged with T=1 symmetry (Lawton *et al.*, 1997a). Projecting inward from each of the twelve vertices formed by the VP2 layer is one copy each of the viral RNA polymerase, VP1 (Valenzuela *et al.*, 1991), and the mRNA-capping enzyme, VP3 (Chen *et al.*, 1999; Lawton *et al.*, 1997b; Lui *et al.*, 1992). Together, VP1, VP2, VP3, and the dsRNA genome constitute the core of the virion. The genome segments are believed to be organized within the core in such a way that each is associated with one of the vertices (Gouet *et al.*, 1999; Pesavento *et al.*, 2001). The middle protein layer of the virion is formed by VP6, and the outer protein layer is formed by the glycoprotein, VP7, and the spike protein, VP4 (Prasad *et al.*, 1988).

The RNA polymerase activity of double-layered particles (cores surrounded by VP6) catalyzes the synthesis of the eleven viral mRNAs (Prasad *et al.*, 1996). The nascent transcripts are extruded through channels present at the 5-fold axes of the double-layered particles (Lawton *et al.*, 1997a). The viral mRNAs possess 5'-caps but are not polyadenylated. Translation of the mRNAs results in the synthesis of six structural (VP1-VP4, VP6, VP7) and six nonstructural (NSP1-NSP6) proteins (Estes, 2001). Of the twelve proteins encoded by the virus, seven possess RNA-binding activity. As might be expected, most of the RNA-binding proteins are directly involved in the replication of the viral genome. However, at least one of the RNA-binding proteins also plays a critical role in the efficient translation of viral mRNAs (Piron *et al.*, 1998).

During genome replication, the viral mRNAs serve as templates for the synthesis of minus-strand RNA, to form dsRNA (Chen *et al.*, 1994). The synthesis of viral dsRNA and the assembly of cores and double-layered particles occur in electron-dense inclusions (viroplasms) that form in the cytoplasm soon after infection (Estes, 2001). Characterization of intracellular subviral particles has indicated that three species of RNA-containing replication intermediates (RIs) are present in the infected cell: (i) the pre-core RI (which contains the structural proteins VP1 and VP3), (ii) the core

RI (VP1, VP2, and VP3), and (iii) the double-layered RI (VP1, VP2, VP3 and VP6) (Gallegos and Patton, 1989) (Fig. 1). Pulse-chase experiments have suggested that the pre-core RI is the precursor of the core RI, which in turn is the precursor of the double-layered RI. In addition to structural proteins, the components of the RIs also include nonstructural proteins (Aponte *et al.*, 1996). As reviewed below, the nonstructural proteins, NSP2 and NSP5, of the core and double-layered RIs are probably essential for packaging and replication of viral RNA.

The fact that the eleven genome segments are produced and packaged in equimolar amounts in rotavirus-infected cells demonstrates that RNA packaging and replication are highly coordinated processes (Patton, 1990). The absence of naked dsRNA in infected cells indicates that packaging occurs prior to replication. Hence, it must be the mRNAs, and not the dsRNAs, that contain the cis-acting signals that operate during packaging. Enzymatic analyses have shown that core RIs and double-layered RIs, but not pre-core RIs, have associated RNA polymerase (replicase) activity that directs the synthesis of dsRNA *in vitro* (Gallegos and Patton, 1989). This finding and the results of studies addressing the impact of RNase on the replicase activity and size of RIs have suggested that genome replication occurs concurrently with the packaging of viral mRNAs into the core and double-layered RIs (Patton, 1986; Patton and Gallegos, 1990). VP6 associates with core RIs, even as the synthesis of dsRNA still proceeds, yielding double-layered RIs with associated replicase activity. It is tempting to speculate that during replication, the mRNA templates move into the RI through the same channels at the 5-fold axes, but in the reverse direction, from which nascent transcripts will be extruded from double-layered particles later in the replication process.

Properties of viral RNA

Since all eleven viral mRNAs are replicated by the same polymerase machinery, the viral mRNAs almost certainly must share common cis-acting signals recognized by the polymerase. On the other hand, packaging of the complete complement of eleven viral mRNAs into RIs, *vis-a-vis*, assortment, requires each mRNA to contain a unique cis-acting signal that allows it to be distinguished from the other viral mRNAs.

The eleven mRNAs of the group A rotaviruses range in size from ~0.7 to 3.3 kb, and with the exception of gene 11, are monocistronic (Estes, 2001). Typically, the 5'- and 3'-untranslated regions (UTRs) of the mRNAs are 9 to 49 and 17 to 182 nucleotides in length, respectively. The only primary sequences that are conserved among the rotavirus mRNAs are present at the 5'- and 3'-ends and these are located entirely within the UTRs. The absence of more extensive stretches of conserved sequences raises the possibility that the common cis-acting signals in the mRNAs recognized by the viral RNA polymerase are of limited size. Alternatively, the common cis-acting signals may consist partly or entirely of secondary structures and, therefore, may be of greater complexity than the limited nature of conserved sequences at the termini of the mRNAs might suggest. The 5'-ends of the mRNAs of all group A rotaviruses initiate with the sequence 5'-GGC(A/U)$_{5-7}$-3', and most reports have indicated that the mRNAs end

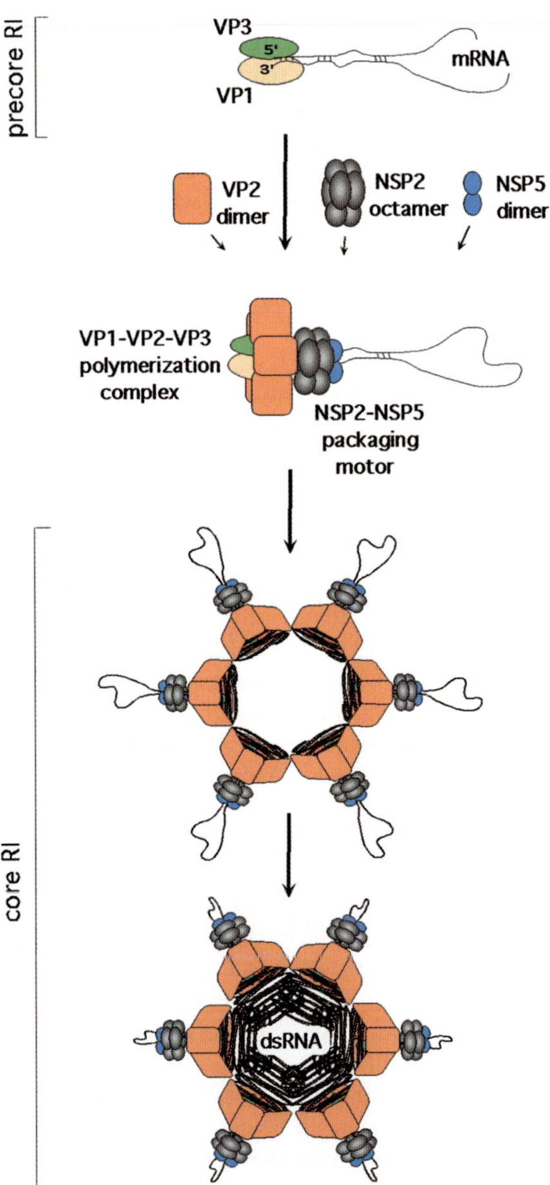

Fig. 1. Intermediates formed in rotavirus genome replication. The precore RI is formed by the association of viral mRNA, the RNA polymerase VP1, and the capping enzyme VP3. Interaction of precore RIs with the core lattice protein VP2 and the NTPase and helix-destabilizing protein NSP2 and protein kinase NSP5 produces core RIs. The VP2-mediated assembly of precore RIs into core-like structures activates the viral RNA polymerase, leading to the synthesis of dsRNA from the mRNAs. NSP2 and NSP5 complexes function as packaging motors facilitating the translocation of viral mRNA into the core as dsRNA synthesis proceeds (Adopted from Patton, 2001, and Patton and Spencer, 2001).

with the heptanucleotide sequence, 5'-UGUGACC-3' (Mitchell and Both, 1990b). The entire 3'-heptanucleotide consensus sequence cannot be considered essential for RNA packaging and replication since viable rotaviruses have been isolated which have gene-specific variations in this portion of one of their genome segments (Table 1). Not only have site-specific substitutions been found in the 3'-consensus sequence, but in addition, viral RNAs have been identified which contain nucleotide insertions or deletions within the consensus sequence (Patton *et al.*, 2001). The only residues of the consensus sequence that appear to be invariant are the last two nucleotides, CC.

Table 1

Variations in the 3'-consensus sequence of rotavirus mRNAs

Genome Segment (protein product)	Virus strain	3'-Sequence		Reference
—	—	5'-...UGUGACC-3'	(consensus)	Desselberger and McCrae, 1994
gene 2 (VP2)	simian SA11	5'-...U<u>A</u>UGACC-3'*		Mitchell and Both, 1990b
gene 5 (NSP1)	bovine UK	5'-...UGUGACC-3'	(consensus)	unpublished results
gene 5 (NSP1)	canine K9	5'-...UGUGA<u>A</u>CC-3'	(insertion)	Okada *et al.*, 1999
gene 5 (NSP1)	human KU	5'-...UGUGA<u>A</u>CC-3'	(insertion)	Okada *et al.*, 1999
gene 5 (NSP1)	rhesus RRV variant	5'-...UGU<u>UU</u>CC-3'		unpublished results
gene 5 (NSP1)	rhesus RRV variant	5'-...UGUUCC-3'	(deletion)	unpublished results
gene 5 (NSP1)	simian SA11	5'-...UGUGA<u>A</u>CC-3'	(insertion)	Mitchell and Both, 1990a Patton *et al.*, 2001
gene 6 (VP6)	simian SA11	5'-...UGUGACC-3'	(consensus)	Both *et al.*, 1984b
gene 7 (NSP3)	simian SA11	5'-...UGUG<u>G</u>CC-3'		Both *et al.*, 1984a

* Underlined residues represent variations from the consensus sequence.

Rotavirus group A variants have been identified which have atypical electropherotypes due to the duplication or deletion of sequences within one or more of the genome segments (Desselberger, 1996; Tian *et al.*, 1993). The fact that these viruses are non-defective and capable of high titer growth in cell culture indicates that size is not a major factor affecting the packaging or replication of the viral mRNAs (Desselberger, 1996; Patton *et al.*, 2001). The identification of rotavirus variants with deletions in the open reading frame (ORF) of a genome segment has provided evidence that the packaging and replication signals are located at the ends of the mRNAs. Studies of variants containing sequence deletions and rearrangements within the gene 5 dsRNA have indicated that NSP1, an RNA-binding protein containing a conserved zinc finger (Hua *et al.*, 1994; Mitchell and Both, 1990a), is not required for replication of the virus in cell culture (Hua and Patton, 1994; Taniguchi *et al.*, 1996). (See also Pesavento *et al.*, Section II, Chapter 1 of this book).

Cis-acting replication signals

The development of a cell-free system that supports the synthesis of dsRNA from exogenous mRNA represents an important milestone in studies on rotavirus replication, providing a mechanism by which it has been possible to define elements in viral mRNAs that promote minus-strand synthesis (Chen *et al.*, 1994). The source of replicase activity in the system is virion-derived cores which have been disrupted (or 'opened') by incubation in hypotonic buffer. Addition of mutant viral mRNAs to the open-core replication system has shown that deletion of the 3'-consensus sequence prevents the mRNAs from serving as templates for minus-strand synthesis (Patton *et al.*, 1996; Wentz *et al.*, 1996). The effectiveness of the 3'-consensus sequence in promoting minus-strand synthesis is abrogated if not positioned precisely at the 3'-end of the viral mRNA (Patton *et al.*, 1996; Wentz *et al.*, 1996). Site-specific mutagenesis has revealed that it is the 3'-CC of the 3'-consensus sequence that is most critical for minus-strand synthesis (Chen *et al.*, 2001). Studies using the open-core system to explore the role of the 3'-CC in dsRNA synthesis have indicated that the 3'-CC interacts with the viral RNA polymerase and the G-dinucleotide, ppGpG or pGpG, to form a stable ternary complex that operates to promote minus-strand initiation (Chen and Patton, 2000). The pathway by which the open-core system produces the G-dinucleotide is not known, but it may be by a template-independent process, since the compound is made even in the absence of exogenous mRNA or endogenous dsRNA (Chen and Patton, 2000).

Studies using the open-core replication system have also provided evidence that regions of viral mRNAs, besides the 3'-consensus sequence, are important for efficient minus-strand synthesis. In particular, deletion of the 5'-UTR or the 3'-UTR upstream of the 3'-consensus sequence has been shown to produce mutant mRNAs that are replicated much less efficiently than full-length mRNAs *in vitro* (Patton *et al.*, 1996; Wentz *et al.*, 1996). Even deletion of sequences within the ORF of rotavirus mRNAs has been shown to yield mutant mRNAs that are poorly replicated (Patton *et al.*, 1999). Notably, with the exception of the RNA termini (see above), sequences that make up the 5'-UTR, the ORF, and upstream region of the 3'-UTR are not conserved among the eleven viral mRNAs. Hence, it seems likely that it is not the primary sequences of these non-conserved regions that are important for replication, but rather the contribution of these regions to the intramolecular folding of the RNA. Indeed, results of replication assays performed with mutant mRNAs combined with computer modeling of the higher order structure of rotavirus mRNAs suggest that the promoter for minus-strand synthesis is formed by basepairing in cis of complementary sequences proximal to the 5'- and 3'-ends of the viral mRNAs (Fig. 2) (Chen and Patton, 1998).

170

The 3'-terminal consensus sequence is predicted to extend as a single-stranded tail from the panhandle formed by the interaction of the ends of the mRNA (Chen and Patton, 1998). If the 3'-terminal sequence becomes basepaired either by mutagenesis of the mRNA or by annealing of complementary oligonucleotides to the 3'-terminal sequence, the ability of the mRNA to promote dsRNA synthesis is inhibited (Barros *et al.*, 2001; Chen *et al.*, 2001).

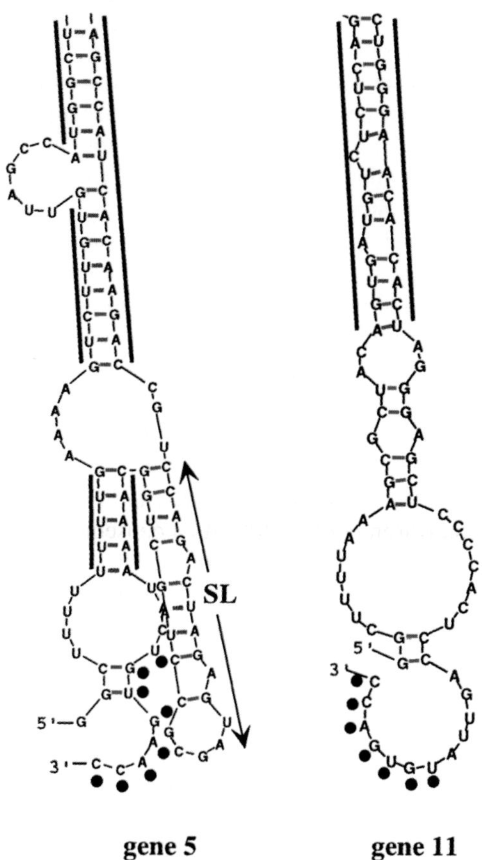

Fig. 2. Putative promoter structure for minus-strand synthesis produced by basepairing between the ends of rotavirus mRNAs. The mfold program version 3.1 (Zuker *et al.*, 1999; Mathews *et al.*, 1999) was used to predict the most stable folded structure of the gene 5 and 11 mRNAs of simian rotavirus SA11. Regions of the 5' and 3' ends predicted to anneal, forming the 'panhandle', are overlined and nucleotides that make up the 3'-consensus sequence are indicated with dots. Like most of the putative structures determined for rotavirus mRNAs, all or most of the 3'-consensus sequence exists in an unbasepaired form and the 5'- and 3'-termini are in close proximity to one another. Also note that extensive unbasepaired regions, forming bubbles, are located between the 3'-consensus sequence and the panhandle and that a stem-loop (SL) motif can be located near the termini.

The lack of either a cell-free gene-specific packaging system or a reverse genetics system has prevented identification of the cis-acting signals involved in the assortment of viral mRNAs into virus particles. Given that none of the rotavirus proteins have been found to possess gene-specific RNA-binding activity, it seems unlikely that assortment of the eleven viral mRNAs is mediated solely by protein-RNA interactions. Instead, it seems more probable that interactions between the eleven mRNAs in trans provide the basis for assortment (Patton and Spencer, 2000). Proteins may facilitate the necessary RNA-RNA interactions by (i) concentrating the viral mRNAs to a single area of the cell (i.e., the viroplasm), (ii) causing the folding of mRNAs in a manner that favors productive RNA-RNA interactions, or (iii) stabilizing interactions that occur between molecules of RNA.

Cis-acting translation signals

In addition to its role in minus-strand synthesis, the 3'-consensus sequence contains a cis-acting signal that operates as a translation enhancer (Chizhikov and Patton, 2000; Vende *et al.*, 2000). The activity of the translation enhancer is mediated by NSP3, a rotavirus RNA-binding protein that specifically recognizes the last four to five nucleotides of the 3'-consensus sequence (Poncet *et al.*, 1994). The simultaneous interaction of NSP3 with the 3'-consensus sequence and the initiation factor eIF4GI facilitates the circularization of viral mRNAs in polysomes, thereby increasing the efficiency of viral gene expression (Piron *et al.*, 1998). As a consequence of its dual function in replication and translation, there is considerable selective pressure on the 3'-consensus sequence to be maintained such that it can efficiently promote both processes. (See also Poncet, Section II, Chapter 5 of this book).

RNA-binding proteins involved in genome replication

The common protein components of the core and double-layered RIs are VP1, VP2, VP3, NSP2 and NSP5. Remarkably, all of these proteins have affinity for RNA, although the nature of their specificity varies significantly (Table 2). As summarized below, studies on the structure and biochemical activities of the RNA-binding proteins have contributed importantly to our understanding of the mechanism of rotavirus RNA packaging and replication.

VP1

Analyses of the sequence and biochemical activities of VP1 indicate that this protein is the viral RNA-dependent RNA polymerase (RDRP), responsible for both transcribing the dsRNA genome to produce mRNA and replicating the mRNA template to produce dsRNA. The analyses have shown that: (i) VP1 contains the GDD and other sequence motifs characteristic of RNA polymerases (Mitchell and Both, 1990b), (ii) VP1 has affinity for nucleotides, and the crosslinking of the photoreactable nucleotide azido-ATP to VP1 inhibits transcription (Valenzuela *et al.*, 1991), (iii) VP1 is the only protein of double-layered particles or RIs with specific affinity for the 3'-end of viral mRNAs (Patton, 1996), and (iv) VP1 is a common component of RIs and recombinant virus-like particles with replicase activity (Gallegos and Patton, 1989; Zeng *et al.*, 1996).

Table 2

Properties of the rotavirus RNA-binding proteins involved in RNA replication

Protein	RNA specificity	Enzymatic or other activity	Structural features
VP1	3'-end of viral mRNA	RNA-dependent RNA polymerase, requires VP2 for activity	Located at the vertices of the core (~12 copies per virion)
VP2	ssRNA, weak affinity for dsRNA, nonspecific	Assembly and RNA-binding activity essential for RNA replication	Core matrix protein, forms T=1 icosahedron from 60 dimers of VP2
VP3	ssRNA, nonspecific	Multifunctional capping enzyme: guanylyltransferase, methyltransferase	Located at the vertices of the core (~12 copies per virion)
NSP2	ssRNA, nonspecific	NTPase, helix-destabilizing activity; possible molecular motor for packaging; induces hyper-phosphorylation of NSP5	Forms doughnut-shaped octamers, interacts with VP1 and NSP5
NSP5	ssRNA and dsRNA	Low autophosphokinase activity, possible modulator of NSP2 function	O-glycosylated phosphoprotein, forms dimers, interacts with NSP2 and NSP6

VP2

Despite compelling circumstantial evidence that VP1 functions as the viral RDRP, *in vitro* assays performed with purified recombinant VP1 have shown that the protein alone lacks enzymatic activity. However, when recombinant VP2 is included with VP1 in such assays, the polymerase exhibits activity that allows it to synthesize dsRNA from viral mRNA (Patton *et al.*, 1997). Thus, although VP1 alone can specifically recognize and form a complex with the 3'-end of viral mRNA, the core lattice protein must be present for activation of the polymerase. From these results, it can be speculated that it is minus-strand elongation, and not the formation of the initiation complex for the minus strand, that requires VP2. Additional insight into the role that VP2 plays in RNA synthesis comes from studies of tsF, a mutant rotavirus with a temperature-sensitive (ts) lesion in the VP2 gene (Gombold and Ramig, 1987). At the nonpermissive temperature, tsF-infected cells contain reduced levels of dsRNA and replicase activity, and the mutant VP2 is defective in its ability to assemble into core RIs (Mansell and Patton, 1990). These findings suggest that not only is VP2 required for polymerase activity, but the protein must also undergo assembly if it is to support the synthesis of RNA. The requirement for VP2 assembly to achieve dsRNA synthesis provides a possible mechanism by which the processes of viral replication and packaging are coordinated, and an explanation as to why naked dsRNA is not produced in rotavirus-infected cells.

Although VP2 binds both single-stranded (ss) and dsRNA nonspecifically, the affinity of the protein is much greater for ssRNA (Boyle and Holmes, 1986). The RNA-binding domain of VP2 has been mapped by deletion mutagenesis to its N-terminus (Labbé *et al.*, 1994). Structural analysis of virus-like particles has indicated that the N-terminus is also required for the interaction of VP1 with the VP2 pentamers that form the vertices of the core (Lawton *et al.*, 1997b). Although co-incubation of VP1 and VP2 will support the synthesis of dsRNA *in vitro*, co-incubation of VP1 and N-terminally deleted VP2 will not. Thus, the ability of VP2 to bind to RNA and/or interact with VP1 is crucial for RNA replication (Patton *et al.*, 1997; Zeng *et al.*, 1998). While the precise role of VP2 in dsRNA synthesis remains unresolved, it is possible that the pentamers formed by VP2 serve as platforms to which the mRNA template binds and on which VP1 catalyzes minus-strand synthesis.

VP3

Electrophoretic mobility shift assays have shown that VP3 has sequence-independent affinity for ssRNA. These experiments have also provided evidence that the protein has higher affinity for uncapped than capped RNA (Patton and Chen, 1999). Consistent with this property are the results of several studies which have indicated that VP3 is the viral guanylyltransferase and, therefore, is responsible for the capping of viral mRNAs. These results include the observation that VP3 shares sequence motifs common to guanylyltransferases and that, in the presence of GTP, VP3 will form a reversible covalent adduct with GMP (Liu *et al.*, 1992; Mitchell and Both, 1990b; Pizarro *et al.*, 1991). UV-crosslinking studies have shown that VP3 also has affinity for S-adenosyl-methionine (SAM), the substrate for methylation of eukaryotic cap structures (Chen *et al.*, 1999). Thus, VP3 is likely to have methyltransferase activity in addition to guanylyltransferase activity (Chen *et al.*, 1999). As a consequence, VP3 can be considered to be a multifunctional capping enzyme. The capping of rotavirus mRNAs also requires an RNA 5'-triphosphate phosphohydrolase (RNA 5'-triphosphatase), an enzyme that removes the γ-phosphate from the 5'-end of a nascent RNA transcript prior to the addition of GMP by the guanylyltransferase. Since purified double-layered particles are capable of producing capped mRNAs *in vitro* that lack the γ-phosphate, the triphosphatase activity must be associated with one of the protein components of the particles (i.e., VP1, VP2, VP3 or VP6). However, attempts to demonstrate that this activity is associated with VP3 or any other protein of double-layered particles have so far not been successful (Chen *et al.*, 1999).

The fact that recombinant VP1 and VP2 can support the synthesis of dsRNA *in vitro* in the absence of VP3 is *prima facie* evidence that VP3 has no function in RNA replication. However, the level of replicase activity associated with open cores is significantly greater than that achieved by co-incubating VP1 and VP2, suggesting that VP3 may increase the efficiency of RNA replication. Similar to findings for VP1, VP3 is located at the vertices formed by VP2, and the interaction of VP3 with the vertices is abrogated if the N-terminus of VP2 is deleted (Lawton *et al.*, 1997b). Not only does their physical proximity at the vertices raise the possibility that VP3 could influence the polymerase activity of VP1, but recent studies showing that the methylation substrate, SAM, has an impact on the rate of nucleotide incorporation by the RNA polymerase provides strong evidence for it (Lawton *et al.*, 2001). Formation of the initiation complex for minus-strand synthesis requires one or more core proteins, the mRNA template, and GTP or the GTP derivative, pGpG or ppGpG (Chen and Patton, 2000). Because VP3 has affinity for GTP, it may be speculated that the protein plays some role in RNA replication, perhaps recruiting GTP to the initiation complex or generating the G-dinucleotide substrate for minus-strand elongation (see above). The affinity of VP3 for one end of the mRNA and VP1 for the other, combined with the interaction of these proteins at the vertices of the core, raises the possibility that VP1 and VP3 could facilitate interactions between the ends of the mRNAs that lead to the formation of the 5'-3' panhandles that promote minus-strand synthesis.

NSP2

NSP2 is a highly conserved, basic protein that localizes to viroplasms and that binds to ssRNA nonspecifically with high affinity (Kattoura *et al.*, 1992; Taraporewala *et al.*, 1999). The protein has only weak affinity for dsRNA. NSP2 self assembles into stable octamers, possibly by the interaction of NSP2 tetramers (Schuck *et al.*, 2001). The octamers are both the predominant and the functional form of NSP2 in the infected cell (Taraporewala *et al.*, 2002). Incubation with rNTPs and Mg^{2+} causes a conformational shift in NSP2 octamers to a more condensed form (Schuck *et al.*, 2001).

NSP2 octamers bind to ssRNA cooperatively, yielding higher order complexes consisting of a single RNA molecule and multiple copies of the octamer (Taraporewala *et al.*, 1999). NSP2 octamers also possess an rNTP- and Mg^{2+}-independent helix-destabilization activity that can disrupt RNA-DNA and RNA-RNA duplexes (Taraporewala and Patton, 2001). In addition to its other activities, NSP2 octamers have an associated Mg^{2+}-dependent NTPase activity that can hydrolyze any of the four rNTPs to rNDPs and phosphates (Taraporewala *et al.*, 1999). *In vitro* studies have indicated that the γ-phosphates released by hydrolysis are covalently linked to NSP2, thereby resulting in the phosphorylation of the protein. NSP2 that is transiently expressed alone in uninfected cells is also phosphorylated, but a phosphorylated form of NSP2 has not been detected in rotavirus-infected cells (Taraporewala *et al.*, 1999). The fact that the octamer is associated with RIs, binds to ssRNA, generates energy by NTP hydrolysis, and undergoes ligand-induced conformational changes has led to the proposal that the octamer operates as a molecular motor, facilitating the packaging of the mRNA template into the core RI (Schuck *et al.*, 2001). The properties of the NSP2 octamer are remarkably similar to those of P4, the RNA-translocating protein of the dsRNA bacteriophage φ6, which also possesses non-specific NTPase activity and possibly helicase activity (Juuti *et al.*, 1998). Although the P4 protein self-assembles into a hexamer and not an octamer, incubation with NTPs induce a conformational change in the P4 hexamer, a result that is reminiscent of the effect that NTPs have on the conformation of the NSP2 octamer (Juuti *et al.*, 1998; Schuck *et al.*, 2001).

Studies of *tsE(1400)*, a simian SA11 rotavirus mutant with *ts* lesion in the NSP2 gene (Gombold *et al.*, 1985), have also supported the hypothesis that NSP2 plays a critical role in packaging. Most notably, *ts*E-infected cells produce unusually high numbers of empty virus particles at the nonpermissive temperature, suggesting that RNA packaging is an NSP2-dependent process. The lack of viroplasm formation in *ts*E-infected cells at the nonpermissive temperature also provides evidence that NSP2 plays an essential role in the formation of these inclusions (Ramig and Petrie, 1984).

In addition to NSP2, several other viral proteins including NSP5 accumulate in viroplasms of infected cells. The importance of NSP2 and NSP5 in the formation of these inclusions is demonstrated by the observation that the co-expression of these two proteins in uninfected cells generates viroplasm-like structures (Fabbretti *et al.*, 1999). Not only does this suggest that NSP2 and NSP5 are essential and sufficient for viroplasm formation, but these results also provide evidence of physical interaction between these proteins *in vivo*. Indeed, co-immunoprecipitation and two-hybrid assays have indicated that these proteins specifically interact with one another (Afrikanova *et al.*, 1998; Gonzalez *et al.*, 1998; Poncet *et al.*, 1997). Of undefined importance are the results of protein-protein crosslinking studies, which suggest that NSP2 interacts with the viral polymerase VP1 in addition to NSP5 (Kattoura *et al.*, 1994).

NSP5

NSP5 is a dimeric, serine/threonine-rich protein that is slightly acidic and that has non-specific affinity for both ss and dsRNA (Poncet *et al.*, 1997; Vende *et al.*, 2002). The protein undergoes O-linked glycosylation (González and Burrone, 1991) and phosphorylation in infected cells (Afrikanova *et al.*, 1996; Blackhall *et al.*, 1997). Because of variations in the extent of phosphorylation, several phosphorylated isomers of NSP5 have been detected (Blackhall *et al.*, 1997; Poncet *et al.*, 1997). The sources of the enzymatic activities involved in the phosphorylation of NSP5 have not been fully resolved. More than one study has concluded that NSP5 possesses an intrinsic low level of protein kinase activity that results in a basal level of NSP5 phosphorylation (Blackhall *et al.*, 1997; Poncet *et al.*, 1997; Vende *et al.*, 2002). Additional phosphorylation of NSP5, or hyperphosphorylation, requires its interaction with cellular or viral proteins. The hyperphosphorylation of NSP5 may result from the action of cellular protein kinases or from the interaction of the viral protein with cellular proteins that cause the upregulation of the autokinase activity of NSP5 (Afrikanova *et al.*, 1996; Blackhall *et al.*, 1998; Poncet *et al.*, 1997). However, the following results suggest that it is more likely that NSP2 mediates the hyperphosphorylation of NSP5: (i) Incubation of purified NSP2 with NSP5 *in vitro* yields hyperphosphorylated NSP5 (Afrikanova *et al.*, 1998; Vende *et al.*, 2002), and (ii) Co-expression of NSP2 and NSP5 in uninfected cells leads to the hyperphosphorylation of NSP5, which is not the case if NSP5 is expressed alone (Poncet *et al.*, 1997). Remarkably, when NSP2 and NSP5 are co-expressed in uninfected cells maintained in the presence of kinase inhibitors, NSP5 does not undergo hyperphosphorylation, and perhaps more importantly, NSP5 does not accumulate in viroplasms (Poncet *et al.*, 1997). Hence, not only are both NSP2 and NSP5 required for viroplasm formation, but the NSP2-induced hyperphosphorylation of NSP5 appears critical for the assembly of these inclusions.

The mechanism by which NSP2 causes the hyperphosphorylation of NSP5 is unclear, but could involve the induction of a dormant kinase activity associated with either NSP2 or NSP5 that is activated upon interaction of the two proteins. An alternative possibility is that the NTPase activity of NSP2 generates phosphate moieties that are transferred from NSP2 to NSP5, resulting in the hyperphosphorylation of NSP5. If this hypothesis is correct, then the absence of detectable phosphorylated NSP2 in the infected cell may result from the rapid and efficient transfer of phosphate groups from NSP2 to NSP5. The observation that the level of hyperphosphorylated NSP5 increases in infected cells which are maintained in the presence of phosphatase inhibitors, suggests that NSP5 is transiently hyperphosphorylated *in vivo* and that cellular phosphatases catalyze the removal of phosphate groups from the protein (Blackhall *et al.*, 1998). Thus, it can be speculated that NSP5 and cellular phosphatases play critical roles in the cycling of NSP2 between phosphorylated and non-phosphorylated forms and provide a means to dispense of the phosphate residues generated by the NTPase activity of NSP2. Since binding of an rNTP induces a conformational shift in the NSP2 octamer, hydrolysis of that rNTP by the NTPase activity of NSP2 and transfer of the cleaved phosphate moiety from NSP2 to NSP5, may be essential for the octamer to return to its less condensed form, rendering it competent to once again bind an NTP molecule (Schuck *et al.*, 2001).

As a consequence of the linkage between packaging and replication, all rotavirus genomic dsRNA in the infected cell exists within one or more layers of capsid protein (Patton and Spencer, 2000). The presence of the layers of capsid protein prevents the interaction of the nonstructural RNA-binding proteins with genomic dsRNA. As a consequence, even though NSP5 has affinity for dsRNA, it is unlikely that NSP5 is bound to genomic dsRNA in the infected cell. Then what is the importance of the dsRNA-binding activity of NSP5? The answer to this question is suggested by studies on $\phi6$, a bacteriophage with a segmented dsRNA genome, which coordinates RNA replication and the packaging of its mRNA templates into procapsids in a manner analogous to that observed for the rotaviruses (Mindich, 1999). Importantly, Qiao *et al.* (1995) showed that the presence of higher order structures in $\phi6$ mRNA templates impedes the packaging of the RNA into procapsids. Extrapolation of this finding to the rotavirus system suggests that it is possible that secondary and tertiary structures in rotavirus mRNAs can similarly interfere with packaging and that the function of the dsRNA-binding activity of NSP5 is to target these structures for disruption. Since NSP5 has no detectable helicase activity (unpublished results), binding of NSP5 alone to the higher order structures would not be expected to be sufficient to cause their removal. However, due to its affinity for NSP2, a protein with helix-destabilization activity, it is possible that NSP5 recruits NSP2 to the higher order structures, and thereby, indirectly causes their disruption. Once destabilized, the extensive quantities of NSP2 and NSP5 in viroplasms, and the affinity of these proteins for ssRNA, may prevent higher order structures from reforming in the mRNAs. The site of initial interaction between NSP2 and NSP5 and viral mRNA in the infected cell is not known. But the fact that these proteins have affinity for RNA and localize to viroplasms, suggests that they could be responsible for the intracellular transport of viral mRNAs to these inclusions. Besides possibly transporting mRNAs to viroplasms and preparing mRNAs for packaging and replication, the interaction of NSP2 and NSP5 with viral mRNAs may be sufficiently extensive to have a secondary effect, which is to impede the interferon-induced RNA-dependent activation of the protein kinase PKR (Vende *et al.*, 2002). If this is the case, then NSP2 and NSP5 may suppress immediate antiviral responses that could inhibit a productive rotavirus infection.

Besides interacting with NSP2, evidence exists that NSP5 interacts with NSP6, a small protein of unknown function that localizes with NSP2 and NSP5 to viroplasms in infected cells (Torres-Vega *et al.*, 2000). While many rotaviruses encode NSP6, the existence of some that do not (group A strains Mc323 and Alabama and all group C strains) suggests that NSP6 is not required for virus replication (Torres-Vega *et al.*, 2000).

Future directions

Our insight into the mechanism of rotavirus RNA replication has advanced considerably in the last several years due to the development of the open-core replication system and to the structural and functional analysis of proteins associated with the RIs. Unfortunately, our understanding of even the most rudimentary events that govern genome assortment remains poor, largely because we lack the necessary assay systems required to explore this process. For example, without an *in vitro* packaging system, it is not possible to fully understand the role of nonstructural RNA-binding proteins such as NSP2 and NSP5 in the formation of cores containing the complete complement of viral RNAs. And, while the open-core system has been instrumental in defining the nature of the *cis*-acting replication signals in viral mRNAs, the lack of both an *in vitro* packaging system and a reverse genetics system has prevented us from identifying *cis*-acting packaging and/or assortment signals in viral mRNAs.

Although we and others have made numerous attempts to develop a reverse genetics system for the rotaviruses, to date these have been unsuccessful. More than a decade ago, Roner *et al.* (1990) described protocols wherein reovirus could be generated in cells via transfection of combinations of *in vitro* expressed protein and purified viral RNA. More recently, Roner and Joklik (2001) reported success in applying these protocols to the development of a reverse genetics system for the reoviruses. Indeed, with this system, the investigators indicate that they have succeeded in replacing the S2 gene of the reovirus with a chimeric RNA containing a CAT ORF and the S2 termini (Roner and Joklik, 2001). If this reverse genetics system can be substantiated as a viable approach for manipulating the reovirus genome, not only will it prove useful for elucidating the molecular biology and pathogenesis of reovirus, but it also may provide direction in the development of a reverse genetics system for the rotaviruses. If for no other reason than providing a tool for the engineering of new vaccine strains, the development of a reverse genetics system for the rotaviruses remains a worthwhile pursuit.

References

Afrikanova, I., Miozzo, M.C., Giambiago, S., Burrone, O. (1996). Phosphorylation generates different forms of rotavirus NSP5. J. Gen. Virol. 77, 2059-2065.

Afrikanova, I., Fabbretti, E., Miozzo, M.C., Burrone, O.R. (1998). Rotavirus NSP5 phosphorylation is up-regulated by interaction with NSP2. J. Gen. Virol. 79, 2679-2686.

Aponte, C., Poncet, D., Cohen, J. (1996). Recovery and characterization of a replicase complex in rotavirus-infected cells using a monoclonal antibody against NSP2. J. Virol. 70, 985-991.

Barros, M., Chen, D., Patton, J.T., Spencer, E. (2001). Identification of sequences in rotavirus mRNAs important for minus strand synthesis using antisense oligonucleotides. Virology 288, 71-80.

Blackhall, J., Fuentes, A., Hansen, K., Magnusson, G. (1997). Serine protein kinase activity associated with rotavirus phosphoprotein NSP5. J. Virol. 71, 138-144.

Blackhall, J., Muñoz, M., Fuentes, A., Magnusson, G. (1998). Analysis of rotavirus nonstructural protein NSP5 phosphorylation. J. Virol. 72, 6398-6405.

Both, G.W., Bellamy, A.R., Siegman, L.J. (1984a). Nucleotide sequence of the dsRNA genome segment 7 of simian 11 rotavirus. Nucleic Acids Res. 12, 1621-1626.

Both, G.W., Siegman, L.J., Bellamy, A.R., Ikegami, N., Shatkin, A.J., Furuichi, Y. (1984b). Comparative sequence analysis of rotavirus genomic segment 6 - the gene specifying viral subgroups 1 and 2. J. Virol. 51, 97-101.

Boyle, J.F., Holmes, K.V. (1986). RNA-binding proteins of bovine rotavirus. J. Virol. 58, 561-568.

Chen, D., Patton, J.T. (1998). Rotavirus RNA replication requires a single-stranded 3'-end for efficient minus-strand synthesis. J. Virol. 72, 7387-7396.

Chen, D., Patton, J.T. (2000). *De novo* synthesis of minus-strand RNA by the rotavirus RNA polymerase in a cell-free system involves a novel mechanism of initiation. RNA 6, 1455-1467.

Chen, D.Y., Zeng, C.Q.-Y., Wentz, M.J., Gorziglia, M., Estes, M.K., Ramig, R.F. (1994). Template-dependent, *in vitro* replication of rotavirus RNA. J. Virol. 68, 7030-7039.

Chen, D., Luongo, C.L., Nibert, M.L., Patton, J.T. (1999). Rotavirus open cores catalyze 5'-capping and methylation of exogenous RNA: Evidence that VP3 is a methyltransferase. Virology 265, 120-130.

Chen, D., Barros, M., Spencer, E., Patton, J.T. (2001). Features of the 3'-consensus sequence of rotavirus mRNAs critical to minus strand synthesis. Virology 281, 221-229.

Chizhikov, V., Patton, J.T. (2000). A four-nucleotide translation enhancer in the 3'-terminal consensus sequence of the nonpolyadenylated mRNAs of rotavirus. RNA 6, 814-825.

Desselberger, U. (1996). Genome rearrangements of rotaviruses. Adv. Virus Res. 46, 69-95.

Desselberger, U., McCrae, M. A. (1994). The rotavirus genome. Curr. Top. Microbiol. Immunol. 185, 31-66.

Estes, M.K. (2001). Rotaviruses and their replication, p. 1747-1785. In D. Knipe, M. Howley, *et al.* (eds.), Fields Virology, 4th ed., Lippincott Williams and Wilkins, Philadelphia, PA.

Fabbretti, E., Afrikanova, I., Vascotto, F., Burrone, O.E. (1999). Two non-structural rotavirus proteins, NSP2 and NSP5, form viroplasm-like structures *in vivo*. J. Gen. Virol. 80, 333-339.

Gallegos, C.O., Patton, J.T. (1989). Characterization of rotavirus replication intermediates: A model for the assembly of single-shelled particles. Virology 172, 616-627.

Gombold, J.L., Ramig, R.F. (1987). Assignment of simian rotavirus SA11 temperature-sensitive mutant groups A, C, F, and G to genome segments. Virology 161, 463-473.

Gombold, J.L., Estes, M.K., Ramig, R.F. (1985). Assignment of simian rotavirus SA11 temperature-sensitive mutant groups B and E to genome segments. Virology 143, 309-320.

González, S.A., Burrone, O.R. (1991). Rotavirus NS26 is modified by addition of single O-linked residues of N-acetylglucosamine. Virology 182, 8-16.

González, R.A., Torres-Vega, M.A., López, S., Arias, C.F. (1998). *In vivo* interactions among rotavirus nonstructural proteins. Arch. Virol. 143: 981-996; Erratum in: Arch. Virol. 143, 2064.

Gouet, P., Grimes, J.M., Malby, R., Burroughs, J.N., Zientara, S., Stuart, D.I., Mertens, P.P.C. (1999). The highly ordered double-stranded RNA genome of bluetongue virus revealed by crystallography. Cell 97, 481-490.

Hua, J., Patton, J.T. (1994). The carboxyl-half of the rotavirus nonstructural protein NS53 (NSP1) is not required for virus replication. Virology 198, 567-576.

Hua, J., Chen, X., Patton, J.T. (1994). Deletion mapping of the rotavirus metalloprotein NS53 (NSP1): The conserved cysteine-rich region is essential for virus-specific RNA binding. J. Virol. 68, 3990-4000.

Juuti, J.T., Bamford, D.H., Tuma, R., Thomas Jr, G.J. (1998). Structure and NTPase activity of the translocating protein (P4) of the bacteriophage ϕ6. J. Mol. Biol. 279, 347-359.

Kattoura, M.D., Chen, X., Patton, J.T. (1994). The rotavirus RNA-binding protein NS35 (NSP2) forms 10S multimers and interacts with the viral RNA polymerase.

Virology 202, 803-813.

Kattoura, M., Clapp, L.L., Patton, J.T. (1992). The rotavirus non-structural protein, NS35, is a nonspecific RNA-binding protein. Virology 191, 698-708.

Labbé, M., Baudoux, P., Charpilienne, A., Poncet, D., Cohen, J. (1994). Identification of the nucleic acid binding domain of the rotavirus VP2 protein. J. Gen. Virol. 75, 3423-3430.

Lawton, J.A., Estes, M.K., Prasad, B.V. (1997a). Three-dimensional visualization of mRNA release from actively transcribing rotavirus particles. Nat. Struct. Biol. 4, 118-121.

Lawton, J.A., Zeng, C.Q., Mukherjee, S.K., Cohen, J., Estes, M.K., Prasad, B.V. (1997b). Three-dimensional structural analysis of recombinant rotavirus-like particles with intact and amino-terminal-deleted VP2: implications for the architecture of the VP2 capsid layer. J. Virol. 71, 7353-7360.

Lawton, J.A., Estes, M.K., Prasad, B.V. (2001). Identification and characterization of a transcription pause site in rotavirus. J. Virol. 75, 1632-1642.

Liu, M., Mattion, N.M., Estes, M.K. (1992). Rotavirus VP3 expressed in insect cells possesses guanylyltransferase activity. Virology 188, 77-84.

Mansell, E. A., Patton, J.T. (1990). Rotavirus RNA replication: VP2, but not VP6, is necessary for viral replicase activity. J. Virol. 64, 4988-4996.

Mansell, E.A., Ramig, R.F., Patton, J.T. (1994). Temperature-sensitive lesions in the capsid proteins of the rotavirus mutants *ts*F and *ts*G that affect virion assembly. Virology 204, 69-81.

Mathews, D.H., Sabina, J., Zuker, M., Turner, D.H. (1999). Expanded sequence dependence of thermodynamic parameters improves prediction of RNA secondary structure. J. Mol. Biol. 288, 911-940.

Mindich, L. (1999). Reverse genetics of dsRNA bacteriophage phi6. Adv. Virus Res. 53, 341-353.

Mitchell, D.B., Both, G.W. (1990a). Conservation of a potential metal binding motif despite extensive sequence diversity in the rotavirus nonstructural protein NS53. Virology 174, 618-621.

Mitchell, D.B., Both, G.W. (1990b). Completion of the genomic sequence of the simian rotavirus SA11: nucleotide sequences of segments 1, 2, and 3. Virology 177, 324-331.

Okada, J., Kobayashi, N., Taniguchi, K., Shiomi, H. (1999). Functional analysis of heterologous NSP1 genes in the genetic background of simian rotavirus SA11. Arch. Virol. 144, 1439-1449.

Patton, J.T. (1986). Synthesis of simian rotavirus SA11 double-stranded RNA in a cell free system. Virus Res. 6, 217-233.

Patton, J.T. (1990). Evidence for equimolar synthesis of double-strand RNA and minus-strand RNA in rotavirus-infected cells. Virus Res. 17, 199-208.

Patton, J.T. (1996). Rotavirus VP1 alone specifically binds to the 3'-end of viral mRNA but the interaction is not sufficient to initiate minus-strand synthesis. J. Virol. 70, 7940-7947.

Patton, J.T. (2001). Rotavirus RNA replication and gene expression. Novartis Found.

Symp. 238, 64-77.

Patton, J.T., Chen, D. (1999). RNA-binding and capping activities of proteins in rotavirus open cores. J. Virol. 73, 1382-1391.

Patton, J.T., Gallegos, C.O. (1990). Rotavirus RNA replication: single-strand RNA extends from the replicase particle. J. Gen. Virol. 71, 1087-1094.

Patton, J.T., Spencer, E. (2000). Genome replication and packaging of segmented double-stranded RNA viruses. Virology 277, 217-225.

Patton, J.T., Spencer, E. (2001) RNA structure and the replication of the rotavirus segmented double-stranded RNA genome. Recent Res. Devel. Virol. 3, 529-539.

Patton, J.T., Wentz, M., Xiaobo, J., Ramig, R.F. (1996). *Cis*-acting signals that promote genome replication in rotavirus mRNA. J. Virol. 70, 3961-3971.

Patton, J.T., Jones, M.T., Kalbach, A.N., He, Y.-W., Xiaobo, J. (1997). Rotavirus RNA polymerase requires the core shell protein to synthesize the double-stranded RNA genome. J. Virol. 71, 9618-9626.

Patton, J.T., Chnaiderman, J., Spencer, E. (1999). Open reading frame in rotavirus mRNA specifically promotes synthesis of double-stranded RNA: Template size also affects replication efficiency. Virology 264, 167-180.

Patton, J.T., Taraporewala, Z., Chen, D., Chizhikov, V., Jones, M., Elhelu, A., Collins, M., Kearney, K., Wagner, M., Hoshino, Y., Gouvea, V. (2001). Effect of intragenic rearrangement and changes in the 3'-consensus sequence on NSP1 expression and rotavirus replication. J. Virol. 75, 2076-2086.

Pesavento, J.B., Lawton, J.A., Estes, M.K., Prasad, B.V.V. (2001). The reversible condensation and expansion of the rotavirus genome. Proc. Natl. Acad. Sci. USA 98, 1381-1386.

Piron, M., Vende, P., Cohen, J., Poncet, D. (1998). Rotavirus RNA-binding protein NSP3 interacts with eIF4GI and evicts the poly(A) binding protein from eIF4F. EMBO J. 17, 5811-5821.

Pizarro, J.L., Sandino, A.M., Pizarro, J.M., Fernandez, J., Spencer, E. (1991). Characterization of rotavirus guanylyltransferase activity associated with polypeptide VP3. J. Gen. Virol. 72, 325-332.

Poncet, D., Laurent, S., Cohen, J. (1994). Four nucleotides are the minimal requirement for RNA recognition by rotavirus non-structural protein NSP3. EMBO J. 13, 4165-4173.

Poncet, D., Lindenbaum, P., Haridon, R.L., Cohen, J. (1997). *In vivo* and *in vitro* phosphorylation of rotavirus NSP5 correlates with its localization in viroplasms. J. Virol. 71, 34-41.

Prasad, B.V.V., Wang, G.J., Clerx, J.P.M., Chiu, W. (1988). Three-dimensional structure of rotavirus. J. Mol. Biol. 199, 269-275.

Prasad, B.V.V., Rothnagel, R., Zeng, C.Q.-Y., Jakana, J., Lawton, J.A., Chiu, W., Estes, M.K. (1996). Visualization of ordered genomic RNA and localization of transcriptional complexes in rotavirus. Nature 382, 471-473.

Qiao, X., Qiao, J., Mindich, L. (1995). Interference with bacteriophage phi6 genomic RNA packaging by hairpin structures. J. Virol. 69, 5502-5505.

Ramig, R.F., Petrie, B.L. (1984). Characterization of temperature-sensitive mutants of simian rotavirus SA11: protein synthesis and morphogenesis. J. Virol. 49, 665-673.

Roner, M.R., Sutphin, L.A., Joklik, W.K. (1990). Reovirus RNA is infectious. Virology 179, 845-852.

Roner, M.R., Joklik, W.K. (2001). Reovirus reverse genetics: Incorporation of the CAT gene into the reovirus genome. Proc. Natl. Acad. Sci. USA 98, 8036-8041.

Schuck, P., Taraporewala, Z., McPhie, P., Patton, J.T. (2001). Rotavirus nonstructural protein NSP2 self-assembles into octamers that undergo ligand-induced conformational changes. J. Biol. Chem. 276, 9679-9687.

Taniguchi, K., Kojima, K., Urasawa, S. (1996). Nondefective rotavirus mutants with an NSP1 gene which has a deletion of 500 nucleotides, including a cysteine-rich zinc finger motif-encoding region (nucleotides 156-248), or which has a nonsense codon at nucleotides 153-155. J. Virol. 70, 125-130.

Taraporewala, Z.F., Patton, J.T. (2001). Identification and characterization of the helix-destabilizing activity of rotavirus nonstructural protein NSP2. J. Virol. 75, 4519-4527.

Taraporewala, Z., Chen, D., Patton, J.T. (1999). Multimers formed by the rotavirus nonstructural protein NSP2 bind to RNA and have nucleoside triphosphatase activity. J. Virol. 73, 9934-9943.

Taraporewala, Z.F., Schuck, P., Ramig, R.F., Patton, J.T. (2002). Analysis of a rotavirus temperature-sensitive mutant indicates that NSP2 octamers are the functional form of the protein. J. Virol. 76, 7082-7093.

Tian, Y., Tarlow, O., Ballard, A., Desselberger, U., McCrae, M.A. (1993). Genomic concatemerization / deletion in rotaviruses: a new mechanism for generating rapid genetic change of potential epidemiological importance. J. Virol. 67, 6625-6632.

Torres-Vega, M.A., González, R.A., Duarte, M., Poncet, D., López, S., Arias, C.F. (2000). The C-terminal domain of rotavirus NSP5 is essential for its multimerization, hyperphosphorylation and interaction with NSP6. J. Gen. Virol. 81, 821-830.

Valenzuela, S., Pizarro, J., Sandino, A.M., Vasquez, M., Fernandez, J., Hernandez, O., Patton, J., Spencer, E. (1991). Photoaffinity labeling of rotavirus VP1 with 8-azido-ATP: Identification of the viral RNA polymerase. J. Virol. 65, 3964-3967.

Vende, P., Piron, M., Castagne, N., Poncet, D. (2000). Efficient translation of rotavirus mRNA requires simultaneous interaction of NSP3 with the eukaryotic translation initiation factor eIF4G and the mRNA 3' end. J. Virol. 74, 7064-7071.

Vende, P., Taraporewala, Z.F., Patton, J.T. (2002). RNA-binding activity of the rotavirus phosphoprotein NSP5 includes affinity for double-stranded RNA. J. Virol. 76, 5291-5299.

Wentz, M.J., Patton, J.T., Ramig, R.F. (1996). The 3'-terminal consensus sequence of rotavirus mRNA is the minimal promoter of negative-strand RNA synthesis. J. Virol. 70: 7833-7841.

Zeng, C.Q.-Y., Wentz, M.J., Estes, M.K., Ramig, R.F. (1996). Characterization and replicase activity of double-layered and single-layered rotavirus-like particles expressed from baculovirus recombinants. J. Virol. 70, 2736-2742.

Zeng, C.Q.-Y., Estes, M.K., Charpilienne, A., Cohen, J. (1998). The N terminus of rotavirus VP2 is necessary for encapsidation of VP1 and VP3. J. Virol. 72, 201-208.

Zuker, M., Mathews, D.H., Turner, D.H. (1999). Algorithms and thermodynamics for RNA secondary structure prediction: A practical guide, p. 11-43. In: J. Barciszewski and B.F.C. Clark (eds.), RNA Biochemistry and Biotechnology, NATO ASI Series, Kluwer Academic Publishers, Dordrecht, NL.

II, 5. Translation of rotavirus mRNAs in the infected cell

Didier Poncet

Unité Mixte de Recherche Centre National de la Recherche Scientifique
Institut National de la Recherche Agronomique, Virologie Moléculaire et Structurale,
Centre de Recherche INRA, 78352 Jouy-en-Josas

Abbreviations used

eIF, eukaryotic Initiation Factor; eEF, eukaryotic Elongation Factor; IF,EF, initiation, elongation Factor (procaryote); mRNA, messenger RNA; UTR, UnTranslated Region; Met-tRNAi, initiator methionine tRNA; PABP, Poly(A) Binding Protein; IRES, Internal Ribosome Entry Site; ORF, Open Reading Frame; upORF: upstream ORF

Introduction

Protein synthesis is a complex process requiring a very sophisticated biological machinery. Approximatively one fourth of the genes of a microorganism like *Mycoplasma genitalium* encodes proteins or RNA involved in translation, and in eukaryotic cells the complete set of proteins or RNA involved in translation has not been identified yet. No virus genome is large enough to encode the whole set of genes necessary to build the translation apparatus; instead, viruses entirely rely on their ability to detour the cellular translation machinery to synthesise their proteins. Thus, the study of virus replication has given penetrating insights into this cellular function (Gale *et al.*, 2000). The *Reoviridae* are no exception; the 'cap' structure of the messanger (m) RNA was first identified at the 5' end of reovirus mRNA many years ago (Furuichi *et al.*, 1976). In this chapter, I will focus on the basic mechanisms of translation and try to show how they fit (or not) with what is known about rotavirus gene expression. In this review I have analysed the 18492 bp of the genome of the RF strain; whilst I have not conducted an extensive statistical analysis of all the rotavirus genes published I have checked whether what has been found to be relevant for the RF strain is true for other rotaviruses.

Three steps for the synthesis of protein (Fig. 1)

The process of protein synthesis is usually divided into three steps. The first step, translation initiation, includes the recruitment of the ribosome onto the mRNA up to

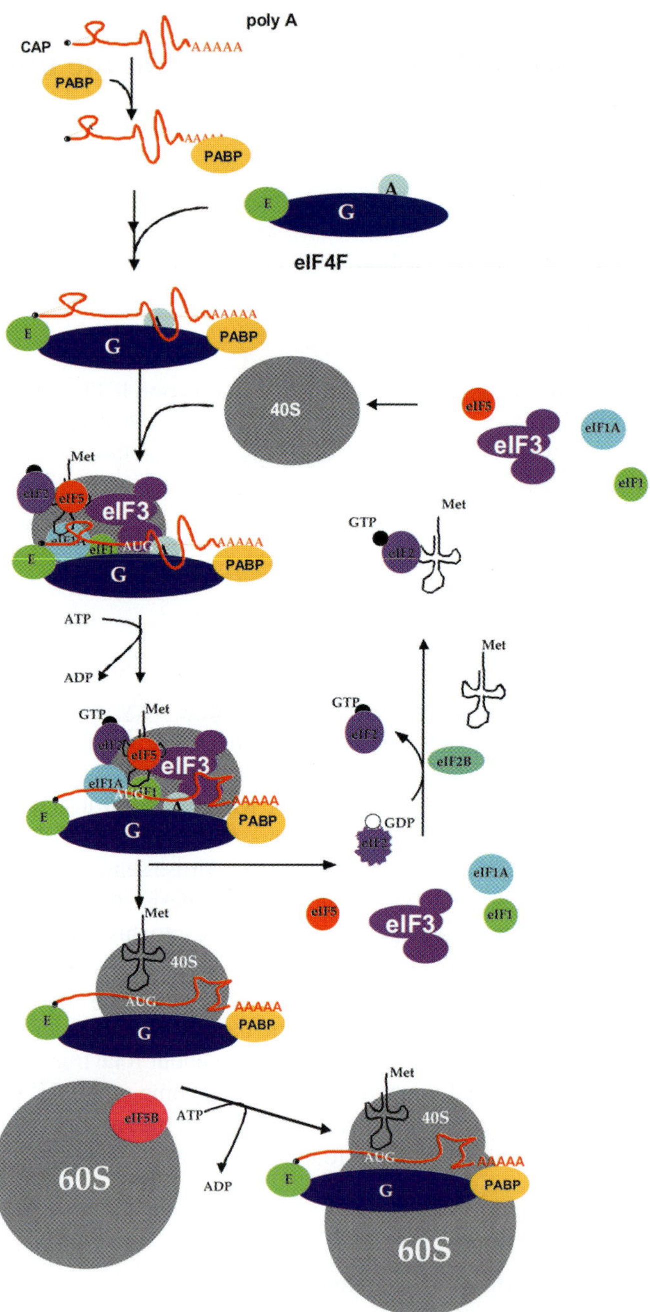

Fig. 1. Schematic representation of translation initiation in eukaryotes. See text for details.

Table 1

Characteristics of the genome of bovine rotavirus strain RF

SEGMENT		5' UNTRANSLATED REGION		CODING REGION			3' UNTRANSLATED REGION		
number (nt length)	GenBank accession	5' end sequence	nt.length/AUG environment	length (nt)	No. of codons (Mw of protein)	protein, functions	stop codon environment	length (nt)	3'end sequence
1 (3302)	J04346	GGCUAUUAAA	19/UACA**AUG**G	3267	1089 (124911)	VP1, RNA polymerase	**UAG**AGC	17	UGUGACC
2 (2687)	X14057	GGCUAUUAAA	16/UUCA**AUG**G	2643	881 (102431)	VP2, RNA binding, core	**UAA**AC	28	UAUGACC
3 (2592)	AY116592	GGCUUUUUAAA	39/UCUG**AUG**G	9	3 (248)	?	**UAA**ACA	2544	CGUGACC
			50/AAAC**AUG**A	2508	836 (97959)	VP3, Guanylyl and methyl transferase	**UGA**GCU	34	
4 (2362)	U65924	GGCUAUAAA	9/AAA**AUG**G	2331	777 (86728)	VP4, spike, cell attachment	**UAA**GCA	22	UGUGACC
			67/ UCAG**AUG**A	177	59 (7017)	?	**UAA**GUU	2118	
5 (1581)	M22308	GGCUUUUAUUUAU	11/UUU**AUG**A	18	6 (576)	?	**UGA**GCC		UGUGACC
			32/AGCC**AUG**G	1476	492 (58690)	NSP1, RNA binding	**UAA**AUA	73	
6 (1356)	K02254	GGCUUUUAAA	23/CAAC**AUG**G	1194	398 (44814)	VP6, structural	**UGA**GGA	139	UGUGACC
7 (1074)	Z21639	GGCUUUUAAU	25/GUUG**AUG**C	942	314 (36324)	NSP3, RNA binding viral translation factor	**UAG**UCA	107	UGUGACC

Table 1

Continued

SEGMENT		5' UNTRANSLATED REGION		CODING REGION			3' UNTRANSLATED REGION		
number (nt length)	GenBank accession	5' end sequence	nt.length/AUG environment	length (nt)	No. of codon (Mw of protein)	protein, functions	stop codon environment	length (nt)	3'end sequence
8 (1059)	Z21640	GGCUUUUAAA	46/AGCCAUGG	954	318 (36680)	NSP2 RNA helicase	UAAUUC	59	UGUGACC
9 (1062)	X65940	GGCUUUAAAA	48/UUUUAAUGU	981	327 (37060)	VP7, outer capsid	UAGGUG	33	UGUGACC
			111/UAUAUGU	918	306 (34597)	glycoprotein	UAGGUG	33	
			234/UCAAUGG	894	298 (33654)		UAGGUG	33	
10 (751)	AY116593	GGCUUUUAAAA	41/AAAGAUGG	528	176 (20366)	NSP4, intracellular receptor, viral enterotoxin	UAAGAG	182	UGUGACC
			20/AGUGAUGU	597	199 (21630)	NSP5, phosphoprotein, RNA binding	UAGGUC	49	
11 (666)	AF188126	GGCUUUAAA	78/AAAAUGA	297	99 (11487)	NSP6, ?	UGAACG	291	UGUGACC

the formation of the first peptide bond between the initiator methionine and the second amino acid. Reiterative addition of a new amino acid to the growing peptide chain constitutes the elongation step, and the last step, termination, includes the release of the peptide from the ribosome and the release of the ribosome from the mRNA. These three steps also relate to the three regions of a mRNA. The 5' UnTranslated Region (5'UTR) extends between the 5' cap structure present on most cellular mRNAs and the initiation codon (generally "AUG"). The Open Reading Frame (ORF) corresponds to the coding part of the mRNA and ends with one of three possible stop codons (UAG, UGA, or UAA). The 3' UnTranslated Region (3'UTR) starts after the stop codon and ends in most cellular mRNA with a poly(A) sequence which is not encoded by the gene but added co-transcriptionally after maturation of the 3' end of the pre-mRNA.

Translation initiation *(Hershey and Merrick, 2000; Pestova et al., 2001)*

The cap structure on mRNA was discovered on reovirus RNA in 1974 (Furuichi and Shatkin, 2000); it consists of a methyl-7-GTP connected to the 5' end of the first base of the mRNA (by two pyrophosphoryl bonds in a 5'-5' configuration). In the cytoplasm, this structure is recognised by eIF4E (eukaryotic Initiation Factor 4E), the "cap binding protein" (Marcotrigiano *et al.*, 1997). eIF4E interacts with eIF4G and, together with the ATP-dependent helicase eIF4A, forms the heterotrimeric complex eIF4F. Binding of the mRNA to the eIF4F complex is thought to precede the binding of the small ribosomal subunit 40S to the 5' end of the mRNA.The 40S ribosomal sub-unit binds the mRNA in the form of a 43S preinitiation complex made of the ribosome small unit 40S loaded with the ternary complex eIF2/GTP/Met-tRNAi and initiation factors eIF3, eIF5, eIF1 and eIF1A. eIF3 is a 600kd multiprotein complex made of about ten non-identical subunits; it has an anti 60S joining activity, interacts with the central part of eIF4G and probably makes the link between the eIF4G-mRNA complex and the 43S preinitiation complex.

Recently it has been shown that scanning of the 5' UTR by the 43S complex requires eIF1, eIF1A and ATP hydrolysis. Once the ribosome has reached the AUG codon, the stable codon-anticodon base pairing stimulates (via eIF5) the GTP hydrolysis activity of eIF2 and the subsequent release of eIF2-GDP, eIF3 and eIF5. The optimal nucleotide sequence context for initiation of translation of vertebrate mRNA ("Kozak consensus sequence" as it is frequently called) is defined by a purine at position -3 **(or at least a purine)** and a G at position +4 A/GxxAUGG (Kozak, 1991) (position 1 being the A in the initiation codon). Initiation codons which do not match these two positions are subjected to "leaky scanning", and initiation at a downstream ATG could occur.

The joining of ribosomal subunits to form an active 80S ribosome requires the hydrolysis of GTP by eIF5B, a homologue of prokaryotic IF2, whose GTPase activity is specifically activated by the two ribosomal subunits. Joining of the 60S subunit onto the 40S subunit and subsequent release of some of the initiation factors ends with a mRNA loaded with an 80S ribosome ready for the formation of the first peptide bond between the initiation methionine and the next amino acid.

Elongation *(Merrick and Nyborg, 2000)*

Elongation is the decoding step, one of the most beautiful processes in biology. It has been extensively studied and got a new start with the recent determination of the ribosome 3D structure (Dahlberg, 2001; Ramakrishnan, 2002; Spahn *et al.*, 2001). Elongation is not fundamentally different in eukaryotes and prokaryotes, and two (eEF1A and eEF2) of the three factors required are evolutionary related. Elongation consists of the addition of a new amino acid to the growing polypeptide chain according to the next codon presented at the A site of the ribosome. Three elongation factors are required in eukaryotes (eEF1A, eEF2 and the GTP exchange factor eEF1B). eEF1A forms a ternary complex with the amino acetylated-tRNA (aa-tRNA) and GTP and brings the aa-tRNA to the A site of the ribosome. After formation of the peptidyl bond which is ensured by the ribozyme activity (Cech, 2000; Noller *et al.*, 1992) of the ribosomal RNA, eEF2 moves the peptidyl tRNA to the P site of the ribosome and allows the precise pulling through of three nucleotideportions of the mRNA.

Termination *(Welch et al., 2000)*

Termination of translation requires the intervention of two termination factors, eRF1 and eRF3. The presence of a termination codon in the A site of the ribosome is recognized by the polypeptide chain release factor eRF1 the structure of which mimics a tRNA. Binding of eRF1 induces the cleavage of the ester bound between the 3' end of the tRNA and the nascent peptide present at the P site. The GTPase activity of eRF3 is involved in the release of the termination factors from the ribosome. It is not known for eukaryotes how the mRNA is freed from the ribosome and how the ribosome subunits are recycled for a new round of translation initiation.

The translation of rotavirus proteins

Upstream ORF and Kozak rule *(Geballe and Sachs, 2000)*

Recently Kozak's model for translation initiation (first AUG-first served – scanning model) has been challenged: a large proportion of AUGs that are effectively used for translation does not fit the "Kozak environment", and many mRNAs contain an upstream open reading frame (upORF) which is not used (Hellen and Sarnow, 2001; Jackson, 2000).

Exceptions to this rule were first observed in virus genomes in which large and frequently GC-rich 5' UTR sequences with several short upstream ORFs were observed. This has led to the concept of "IRES", Internal Ribosome Entry Sites. Using IRES, the ribosome binds directly onto the AUG without scanning. Cellular mRNAs from tightly regulated genes with the same characteristics have since been described and, in spite of some resistance (Kozak, 2001b), cellular IRES are now well documented (Schneider *et al.*, 2001).

It should be underlined that rotavirus mRNAs contain very short 5' UTRs (Table 1). The shortest is found in the 5' UTR of gene 4 that in many strains is only 9 nucleotides long. In the RF strain, the next in frame AUG is 243 nucleotides downstream, which, if used, would encode a 78 kD VP4 (much shorter than 87kD full length VP4) which has not been observed. Taking into account that the 5' cap is held by eIF4E and that a ribosome covers around 30 nucleotides, there is no possibility for the tRNA placed in the middle of the ribosome to meet the first AUG. It is required that the ribosome is precisely directed to the first AUG probably by a mechanism similar to IRES, or that it scans the mRNA backwards. Such reverse scanning has been proposed (Kozak, 2001a) but seems to be limited to the 40S subunit. Interestingly, between the two gene 4 AUGs, a short out-of-frame ORF (58aa) exists which is conserved in some rotavirus genes 4 although in shorter forms (20 to 40 aa long). Whether or not these ORFs are used in the infected cell should receive attention.

Some cellular genes contain short ORFs upstream the main ORF. The major constraint on reinitiation is the size of the upORF; reinitiation declines as the length of upORF increases, and reinitiation has not been observed with upORFs exceeding 55 codons. Gene 3 of rotavirus strain RF presents a first AUG within a moderate Kozak environment, encoding a short 9 nucleotides long ORF. Its position 39 nucleotides from the 5' end of the mRNA makes it a good candidate for a upORF regulating the expression of the downstream VP3-ORF, which starts with an AUG in a moderate Kozak environment 50 nucleotides from the 5' end. Gene 5 of the RF rotavirus strain (encoding NSP1), also presents an upORF with a weak AUG 11 nucleotides downstream of the 5' end. If this AUG is used, the upORF potentially encodes an 18-aa peptide that ends only 6 nucleotides before the "strong" AUG for the NSP1-ORF. This short distance could allow reinitiation to occur at the NSP1-ORF after translation of the upORF-NSP1. This short upORF is found on all the genes 5 analysed. In some cases, the stop codon is mutated, resulting in the fusion of the upORF to the NSP1-ORF. Nevertheless, it should also be noted that this upORF is positionned very close to the 5' end of the viral gene, in a region where oligonucleotides are chosen for RT-PCR cloning. Thus its conservation could, at least in part, result from the choice of oligonucleotides and does not necessarily reflect biological importance.

Gene 7 presents two in frame AUGs 25 and 34 nucleotides from the 5' end. Only the second AUG is in a strong Kozak environment. Furthermore, the first AUG is mutated in an avian (Ito *et al.*, 2001) and a human (Gault *et al.*, 2001) rotavirus; a functional NSP3 protein is produced when translation starts at the second AUG (Piron *et al.*, 1999; Poncet *et al.*, 1994).

Glycoproteins are synthesised on the endoplasmic reticulum (ER) membrane from a precursor carrying a signal peptide which is cleaved during the translocation of the glycoprotein into the ER. Rotavirus gene 9 encoding the glycoprotein VP7 is remarkable for possessing two signal peptides (H1 and H2) (Stirzaker *et al.*, 1990). In the case of the RF rotavirus strain, three in frame AUGs exist but only the first and third (which precede H1 and H2) are conserved amongst all rotaviruses. Only the third AUG, positioned 135 nucleotides downstream of the 5' end, is in a good Kozak environment, and, therefore, H2 is the likely favourite signal peptide for the processing of VP7.

A consequence of the first AUG/first serve concept is that a mRNA could not encode two overlapping proteins. In some rotavirus strains, gene 11 encodes two proteins, NSP5 and NSP6 (Mattion *et al.*, 1991; Torres-Vega *et al.*, 2000). Both proteins are synthesised *in vivo* during rotavirus infection and possibly interact with each other. NSP6 is encoded in an ORF starting 58 nucleotides downstream of the initiation codon for NSP5. How the synthesis of these two proteins is regulated is not known. However, it is interesting to note that the NSP5 ORF has a poor AUG environment that could allow leaky scanning and downstream initiation for the synthesis of NSP6.

ORF and STOP codon

The fidelity of translation is assured by the specificity of the tRNA amino acylation, and by the specificity of the codon-anticodon recognition. This last point explains the importance of the codon choice: amongst the codons that code for the same amino acid, viruses and host could have different strategies. Rotavirus prefers codons with less G or C content (http://www.kazusa.or.jp/codon/); for example, the CCC codon for proline is used 15 times less by rotavirus than by its host, and the UUA codon for leucine is used 5 times more frequently. However, how the abundancy of use of one codon affects the production of a protein is far from clear. It seems that the codon environment is more important than the codon itself and that codon pairs are important (Fedorov *et al.*, 2002). The effect of a codon pair can be paradoxical, since doublets of frequent codons can slow down the translation of a protein (Irwin *et al.*, 1995). It is reasonable to think that for rotavirus the selective pressure required by the melting of the two RNA strands in the subviral particle during transcription is more important than the codon usage and thus induces a high primary A+U content.

Rotavirus genes can rearrange during replication. Rearrangement generally induces a duplication of the gene which then evolves by deletions and point mutations. The duplicated part of rearranged genes normally does not form a new ORF and is devoid of its own initiation codon (Desselberger, 1996). Two mutants of gene 5 that encode a much reduced NSP1 (40 and 50 aa) have been characterised (Taniguchi *et al.*, 1996). In the case of mutant A10, a stop codon shortens NSP1 to a peptide of 40 aa length but the remainder of the protein is encoded by a second ORF in frame with the first one. Hence, a stop codon read-through might allow the synthesis of the full-length protein. For the A16 mutant the situation is more complicated as the remainder of the protein is encoded in a second ORF which is not in the same frame as the first one. Hence, a frameshift would only allow the synthesis of an incomplete protein. No product corresponding to NSP1 of normal size has been observed with the two mutants, but a few percent of read-through or frameshift could not be excluded. Interestingly, the quantity of NSP3 produced by the NSP1 deficient mutants is significantly increased whereas the quantity of the other viral proteins is not. How a higher expression of NSP3 could compensate the deletion of NSP1 or how the deletion of NSP1 specifically stimulates NSP3 translation remains to be explored. Recently a human rotavirus with a rearrangement in gene 7 has been described (Gault *et al.*, 2001). The rearrangement deeply modified the encoded protein and the NSP3 produce by the mutant was twice the size of a normal protein.

Analysis of the nucleotide context of the translation termination codon has shown that the nucleotide following the termination codon is not random and that the sequence UGAC is very poorly used in eukaryotes, inducing 2 to 5% readthrough (Brown *et al.*, 1990; Li and Rice, 1993). It should be underlined that this combination is not used in the rotavirus RF genome (Table I) nor by the A10 and A16 gene 5 mutants above. UAA(A/G) and UGA(A/G) are the preferred termination codons in eukaryotes and in the RF genome.

Release of the ribosome from the mRNA occurs somewhere between the stop codon and the end of the mRNA. How the ribosome subunits are separated and how they are recycled for a new round of translation are still open questions.

The 3' end: A major role for NSP3 (Fig. 2)

A nonstructural viral protein is bound to the 3' end of viral mRNA

The most remarkable feature of the 3' ends of the mRNAs of rotavirus and other member of the *Reoviridae* is the absence of a poly(A) tail. Instead, all the rotavirus genes and thus rotavirus mRNAs end with the same short sequence. In the case of group A rotaviruses, the last five nucleotides UGACC are conserved amongst all the segments. The conservation of the 3' terminal sequences prompted us to look at nonstructural proteins susceptible to bind specifically to these conserved sequences. We developed monoclonal antibodies against NSP2, NSP3, NSP5 and NSP1 proteins (Aponte *et al.*, 1993; Poncet *et al.*, 1997; Poncet, unpublished results) and used them to immunoprecipitate the covalent protein-RNA complexes formed by UV irradiation of rotavirus- infected cells. The cross-linked RNAs were shortened with RNAse T1 to produce 3' OH ends which then can be labelled with gamma ^{32}P-ATP and polynucleotide kinase. If RNA has been crosslinked to the protein, then the ribonucleoprotein complex will be labelled with ^{32}P and detected by autoradiography after SDS-PAGE. When applied to the above NSPs, this technique labelled all of them (Aponte *et al.*, 1996; Poncet *et al.*, 1993; Poncet *et al.*, 1997) except for NSP1 (Poncet, D. unpublished results) which, paradoxically, has been the first rotavirus RNA binding protein identified (Brottier *et al.*, 1992).

When we analysed the RNA cross-linked to these proteins, it appeared that the product cross-linked to NSP2 was resolved as a ladder, suggesting an absence of specificity, whereas those bound to NSP3 were resolved in 5 bands of 9 to 15 nucleotides long. Hybridisation of the NSP3-linked RNAs (not RNase T1 treated) to rotavirus dsRNA, mRNA and cDNA, showed that the RNAs bound to NSP3 *in vivo* were of viral origin, of positive polarity, derived from the eleven genes and localised on the 3' end of the genes. Enzymatic sequencing of some of the RNA fragments definitively showed that the very 3' end UGACC sequence of the viral mRNAs was bound to NSP3 (Poncet *et al.*, 1993). Production of recombinant protein in the baculovirus system and then in *E. coli* allowed to investigate further the RNA sequence requirement for binding to NSP3 (Piron *et al.*, 1999; Poncet *et al.*, 1994). First, it demonstrated that NSP3 is able to bind RNA in the absence of other viral proteins. Use of synthetic ribonucleotides showed the exquisite specificity of binding of the viral protein. An RNA as short

194

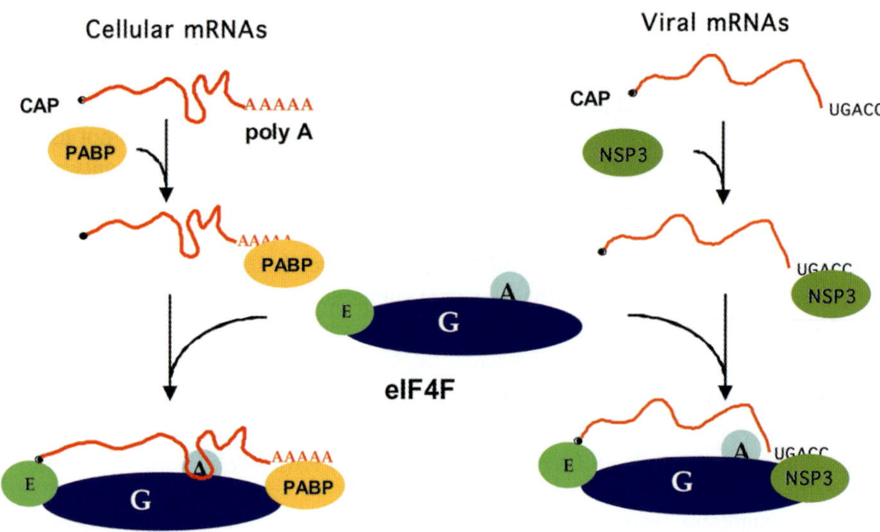

Fig. 2. Comparison of translation initiation of cellular poly-adenylated mRNAs with translation initiation of rotavirus mRNAs. Only the first steps are shown (see Fig. 1). See text for details.

as four nucleotides corresponding to the 3' terminal sequence of rotavirus mRNA (GACC) was sufficient to be recognized by NSP3. Moreover, addition or deletion of only one base to this sequence totally abolished the binding of the protein. To map the RNA binding domain of NSP3 we turned to the expression of deletion mutants in *E. coli*. This work led to the identification of the RNA-binding domain in the N terminal half of NSP3 in a region well conserved between NSP3s of group A and group C rotaviruses. Surprisingly, we showed that a dimer of NSP3 was bound to one molecule of RNA. The recent determination of the crystal structure of the NSP3 RNA-binding domain (Deo *et al.*, 2002) has confirmed all the above results and allowed a better understanding of the sequence specificity of the viral protein. It was shown that NSP3 forms an asymmetric dimer presenting in the shape of a heart. The asymmetric homodimerisation creates a highly basic tunnel in which the last four bases of the RNA are stacked. The tunnel is closed at one end, precluding binding to an internal GACC RNA sequence. Extensive contacts of the two monomers with each other and with the RNA explain the high affinity and stability of the RNA-protein complex. Despite being well conserved, the three bases just upstream of the consensus sequence do not make contact with NSP3 and are dispensable for binding to NSP3, the last four nucleotides providing the entire binding energy.

NSP3 is a viral translation initiation factor

The role of NSP3 became evident when this protein was used in a yeast two-hybrid screening system to identify a cellular protein with which it interacts (Piron *et al.*, 1998).

The most frequent partner thus identified was a new form of eIF4G. The encoding cDNA isolated contained an additional N-terminal sequence with which NSP3 interacts. Simultaneously, N. Sonenberg's laboratory identified the same cDNA, and it became evident that the previously published eIF4G cDNA was incomplete (Imataka *et al.*, 1998).

In vitro (Vende *et al.*, 2000) and *in vivo* (Chizhikov and Patton, 2000; Vende *et al.*, 2000) studies have shown that addition or expression of NSP3 enhances translation of a reporter gene which ends with the UGACC sequence. This enhancement requires the simultaneous presence of the N-terminal RNA- and C-terminal eIF4G-binding domains of NSP3 and the UGACC sequence at the 3' end of the target mRNA (Vende *et al.*, 2000; Fig. 2). Deletion of the RNA binding domain or the last 20 amino acids (eIF4G binding domain) of NSP3 precluded translation of the target RNA as did the addition of extranucleotides or poly(A) at the 3'end of the mRNA. Thus, NSP3 behaves like the poly(A) binding protein (PABP) in recruiting the viral mRNA for eIF4G. As expression of non poly-adenylated mRNA is very low in eukaryotic cells, NSP3 plays a pivotal role in enabling rotavirus gene expression.

Exceptions to the 3' end UGACC sequence were first thought to be sequencing errors, incomplete cDNAs or possibly non-viable viral mutants of the quasi-species. Then, RT-PCR with specific primers did not allow taking into account all the published sequences as their 3' end sequences correspond to those of the primer used for RT-PCR and not the natural sequence. Recently it has been clearly shown that some rotavirus strains bear segments with a "non-canonical" 3' end. Remarkably, sequence determination performed by three different laboratories on three isolates of the most commonly used rotavirus strain, SA11, showed that gene 5 of this simian strain ends with the "non-canonical" 3' GAACC (Mitchell and Both, 1990; Patton *et al.*, 2001). *In vitro*, the GAACC sequence is bound by recombinant NSP3 of the SA11 strain, but with less affinity than the UGACC sequence (Patton *et al.*, 2001). The recombinant NSP3 of the RF strain does not bind the UAACC sequence (Poncet *et al.*, 1994). This difference could arise from subtle differences in the amino acid sequence of these two proteins. The low ability of NSP3 to bind to the GAACC sequence has been interpreted as a way to reduce the translation of an otherwise toxic protein (NSP1) (Patton *et al.*, 2001). Instead of lowering the importance of NSP3 in translation regulation, these observations emphasise the role of NSP3 in allowing translational control of rotavirus gene expression.

Control of cellular mRNA translation by rotavirus

For a long time, it has been suspected that eIF4G could interact physically with the PABP (Sachs, 2000). *In vivo*, mRNAs which are both capped and polyadenylated are translated more efficiently than mRNAs that are either capped or polyadenylated. The molecular basis for synergy between cap and poly(A) is not established.

Interaction between PABP and eIF4G has been first demonstrated in yeast in which interaction between the two proteins requires the presence of a poly(A) sequence bound to the second RNA recognition motif (RRM2) of the PABP (Tarun and Sachs,

196

1996). Despite numerous attempts, the PABP-eIF4G interaction remained elusive in mammalian cells. Thus, it rapidly became evident that the newly identified N-terminal sequence of eIF4G could be the target for PABP. We first showed that PABP is part of the eIF4F complex (made of eIF4G, E and A, see above) and that addition of the C-terminal part of NSP3 evicts PABP from the complex (Piron *et al.*, 1998). The same phenomenon was observed in rotavirus-infected cells; as infection progresses, less and less PABP could be co-immunoprecipitated with eIF4G whereas more and more NSP3 was detected. The quantity of the other components of eIF4F remained unchanged, indicating that NSP3 was taking the place of PABP in the course of rotavirus infection. Simultaneously, Dr N. Sonenberg's group showed that RRM2 of the human PABP interacts directly with the same region of eIF4G but, contrary to yeast, this interaction does not depend upon the presence of a poly(A) bound to PABP (Imataka *et al.*, 1998). Removing the PABP from eIF4F complexes is probably the molecular basis for the lower synthesis of cellular protein observed during rotavirus infection (Heath and Birch, 1988; Piron *et al.*, 1998). Indeed, *in vitro* and *in vivo* experiments showed that addition of NSP3 lowered the translation of poly(A) mRNA (Borman *et al.*, 2000; Michel *et al.*, 2000; Vende *et al.*, 2000).

How could the removal of PABP from eIF4G lower translation or why should RNA be circularised?

Some time ago it has been shown *in vivo* that the cap structure and the poly(A) tail act synergistically to enhance translation (Gallie, 1991; Gallie, 1998; Otero *et al.*, 1999). Synergy is measured (Otero *et al.*, 1999) by the ratio of the amount of translation of a capped and polyadenylated RNA over the sum of translations of an only capped mRNA and of an only polyadenylated mRNA: it depends strongly on the cellular system used and amounts to 10 to 27 in CHO cells (Gallie *et al.*, 1996), to 15 in protoplasts (Gallie, 1991) and the *Drosophila* embryo (Gebauer *et al.*, 1999) and to 7 in yeast (Otero *et al.*, 1999).

The replication-dependent histone mRNAs are the only known messenger RNAs that do not have poly(A) tails. Formation of the 3' end of replication-dependent histones requires the recognition by U7snRNPs of a purine rich sequence 10-15 nucleotides upstream of a highly conserved stem-loop structure. The stem-loop sequence (six nucleotides stem, four nucleotides loop) is bound by an RNA-binding protein SLBP, and cleavage occurs at its 3' end. The stem-loop structure when placed at the 3' end of the mRNA is sufficient and necessary for the efficient translation of a reporter gene introduced into the cell cytoplasm (Gallie *et al.*, 1996). Despite the absence of poly(A), the histone mRNA shows a similar degree of synergy with the 5' cap structure. These observations show that the three kinds of 3' end of mRNA (rotavirus, poly(A) and histone stem-loop) act cooperatively with the cap to enhance translation and probably reflect a fundamental mechanism for translation. Synergy probably reflects the competition between mRNAs for access to some limiting translation factors. *In vitro*, synergy is observed only when mRNA is in competition with endogenous poly(A) mRNA (Preiss and Hentze, 1998) or when ribosomes and/or translation factors are scarce (Borman *et al.*, 2000; Michel *et al.*, 2000). The high value of synergy for rotavirus mRNA in ST- NSP3 cells (35 versus a value of 20 for cap-poly(A) in the same cell line) could reflect a tougher competition between poly(A) mRNA for translation factors (probably eIF4G) induced by expression of NSP3.

Otero *et al.* (Otero *et al.*, 1999) suggested that part of the synergy between cap and poly(A) was due to an unknown factor interacting with PABP, because synergy can be obtained in "trans" with a PABP disabled for mRNA or eIF4G binding (Gray *et al.*, 2000; Otero *et al.*, 1999).

However, it should be underlined that the eIF4G- and poly(A)-binding domains of yeast PABP are overlapping on the primary sequence, and it could be difficult to totally disentangle one function from the other. The residual synergy observed with PABP mutants could originate from a low RNA- or eIF4G-binding activity not detectable by gel retardation or pull-down assays but still efficient enough to stimulate translation. It should be noted that if NSP3 is disabled for eIF4G- or 3' end RNA-binding, then there is no stimulation of translation of any mRNA (Vende *et al.*, 2000). Furthermore, experiments using depleted rabbit reticulocyte lysate (in which cap-poly(A) synergy is observed), show that addition of NSP3 abolishes the cap-poly(A) synergy (Borman *et al.*, 2000). The mechanism underlying synergy is not fully established. It has been proposed that a circularization of the mRNA is induced by interactions of PABP with the 3' end of the mRNA, of eIF4E with the 5' cap structure and of both factors with eIF4G. Circularization of the mRNA would facilitate the recycling of the ribosomes or enhance formation of complexes of mRNA and translation factors. Circular RNA has been observed when the three yeast factors (eIF4G, 4E and PABP) are allowed to interact with RNA (Wells *et al.*, 1998). Circular polysomes have been observed by EM (Christensen *et al.*, 1987; Yazaki *et al.*, 2000). However, there is no evidence that ribosomes could be reused immediately for another round of translation. On the other hand, synergy is increased when extracts are depleted of ribosomes and of ribosome-associated translation initiation factors, suggesting that synergy may come from a higher affinity of circularised RNA for these factors. Moreover, synergy is insensitive to the distance between termination codon and the poly(A) tail or to secondary structure present in the 3' UTR (Michel *et al.*, 2000).

Open questions

Replication-translation balance

It is remarkable that the replicase activity of rotavirus, thought to involve VP1, also requires the presence of a cytosine doublet near the end of the mRNA. VP1 has been shown to bind to the GACC 3' sequence although with less affinity than NSP3 (Chen *et al.*, 2001). Replication of mRNA proceeds 3' to 5' while ribosomes progress on the same substrate from the 5' to the 3'end. Thus, it seems reasonable to conclude that a mRNA cannot be replicated *and* translated at the same time. However, it has not been established that a viral mRNA is irrevocably committed to replication *or* translation as soon as it emerges from the viral particle. It could be a great advantage for the virus to have developed a "quality control" procedure for its mRNA allowing the replication of mRNA only when it has already been efficiently translated (i.e. bound by NSP3). Regulation of the accessibility of the 3' terminal sequence, by modification of the mRNA secondary structure, could control the balance between replication and translation. Other RNA-binding viral or cellular proteins could probably be involved in this process. It should be noted that on average the 3' UTRs of rotavirus genes are much smaller than cellular 3' UTRs; this could reflect the absence of signals for cellular regulation like intracellular localisation, nucleo-cytoplasmic transport or mRNA turnover. In rotavirus segments 1-4, and 8,9 and 11, the 3' UTR is less than 60 nucleotides long. This length would allow only one or two ribosomes to make contact with the end of the mRNA. If we consider that a ribosome covers 30 nucleotides and that NSP3 covers the last five nucleotides (Deo *et al.*, 2002), then the ribosomes certainly come into contact with NSP3 at the end of the translation of the gene. The ribosomes could remove NSP3 from the mRNA by scanning the 3' UTR downstream the termination codon and thus free the 3' end for replication.

RoXaN: another cellular factor interacting with NSP3

If cap-poly(A) synergy is due to an additional factor interacting with PABP, as proposed by Otero *et al.* (1999), then, to completely fulfil the role of PABP, NSP3 should interact with the same factor.

Screens by the yeast two-hybrid system using PABP as bait have identified translation termination factor eRF3 as PABP partner (Cosson *et al.*, 2002; Hoshino *et al.*, 1999). Whereas it has been shown that eRF3 interacts with PABP, the role of this interaction in recycling a post-termination ribosome to the 5' end of the mRNA or positioning ribosomes for a subsequent round of initiation has not been established. Recently, a genetic interaction has been observed between PABP and eIF5 in yeast (Searfoss *et al.*, 2001; Valentini *et al.*, 2002). Two PABP binding proteins have also been identified: whereas PAIP1 stimulates translation (Craig *et al.*, 1998), PAIP2 which competes with PAIP1 for binding to PABP, is a translation repressor (Khaleghpour *et al.*, 2001).

We did not find interactions between eRF3 and NSP3 nor were eIF5, PAIP1 or PAIP2 obtained in our NSP3 two-hybrid screen. However, in the same two-hybrid screen which led us to the identification of eIF4G, another cDNA encoding an unknown ORF has been isolated repeatedly (Poncet, in preparation). The protein encoded by the cDNA has been called "RoXaN I" (X protein associated with Rotavirus NSP3). Through screening of a HeLa cell cDNA library, EST sequences and the human genome draft, we have been able to reconstitute its full-length cDNA which is more than 5.5 kb long and encodes a protein of 977 amino acids (Fig. 3). It is the N-terminal part of RoXaN which strongly interacts with NSP3. Recently another cDNA encoding a homologous protein RoXaN-II has been obtained. *In silico* analysis of these two proteins showed the presence of a TPR (tetratricopeptide repeat, sequence; a protein-protein interaction motif at their N terminus (Andrade *et al.*, 2001; Blatch and Lassle, 1999), and the presence of at least six zinc fingers in the remaining sequence. NSP3 interacts with RoXaN-I between

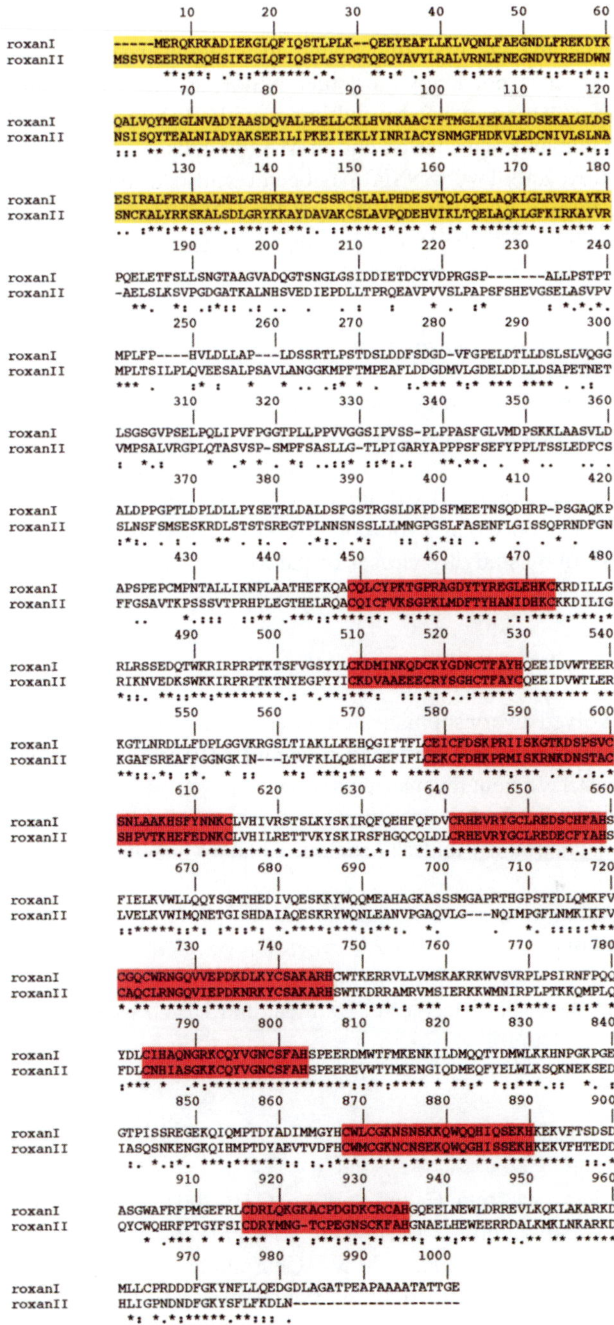

Fig. 3. Comparison of RoXaN I and II amino acid sequences. Amino acid sequence was deduced from DNA sequence of AK023849 and AF188530 for RoXaN I and from AK000325 and AW964076 for RoxanII. Stars indicate identical residues, double points indicate strongly similar residues and simple point weakly similar residues. Dashes indicate deletions or gaps. The tetratricopeptide regions are highlighted in yellow and the zinc fingers in red. The Clustal program by default was used (http://npsa-pbil.ibcp.fr/cgi-bin/npsa automat.pl?page=npsa clustalw.html).

the TPR sequence and the first zinc finger, in a region weakly conserved between RoXaN-I and -II. It is not known whether NSP3 also interacts with RoXaN-II or whether RoXaN-I can form hetero-multimers with RoXaN-II. RoXaN-I mRNA is found in all the human tissues tested and in MA104 cells.

Interaction of RoXaN-I with NSP3 has been confirmed in rotavirus infected cells, and a complex formed by eIF4G/NSP3, and RoXaN-I has been detected (Poncet, in preparation). Nevertheless, the exact function of RoXaN in cells and during rotavirus infection will still have to be established.

A family problem: what about the Orthoreovirus genus?

Orthoreoviruses also have non-polyadenylated mRNAs but the functional equivalent of NSP3 has not yet been identified (Poncet, unpublished results). Has reovirus developed another strategy of translational control (for example through the antiviral PKR response which targets eIF2 (Samuel, 1998), or have all the traces between the two viral genera been erased by evolution? Certainly, there is still much to learn about translational control by studying viruses of the *Reoviridae* familly.

Conclusion

Rotavirus has evolved a very sophisticated way to ensure the expression of its genome. However, many questions still remain to be answered. Some of these questions would get a faster answer if efficient reverse genetics systems for the highly segmented ds RNA viruses were available. Such a procedure is much awaited by many investigators of the dsRNA virus community and will remain a major challenge in the years to come.

Acknowledgement

Thanks to Maria, Jean, Patrice, Pierre, Damien, and Ulrich for critical reading of the manuscript.

References

Andrade, M.A., Perez-Iratxeta, C. and Ponting, C.P. (2001) Protein repeats: structures, functions, and evolution. *J Struct Biol.* **134**, 117-131.

Aponte, C., Mattion, N.M., Estes, M.K., Charpilienne, A. and Cohen, J. (1993) Expression of two bovine rotavirus non- structural proteins (NSP2, NSP3) in the baculovirus system and production of monoclonal antibodies directed against the expressed proteins. *Arch Virol.* **133**, 85-95.

Aponte, C., Poncet, D. and Cohen, J. (1996) Recovery and characterization of a replicase complex in rotavirus-infected cells by using a monoclonal antibody against

NSP2. *J Virol.* **70**, 985-991.

Blatch, G.L. and Lassle, M. (1999) The tetratricopeptide repeat: a structural motif mediating protein-protein interactions. *Bioassays* **21**, 932-939.

Borman, A.M., Michel, Y.M. and Kean, K.M. (2000) Biochemical characterisation of cap-poly(A) synergy in rabbit reticulocyte lysates: the eIF4G-PABP interaction increases the functional affinity of eIF4E for the capped mRNA 5'- end. *Nucleic Acids Res.* **28**, 4068-4075.

Brottier, P., Nandi, P., Bremont, M. and Cohen, J. (1992) Bovine rotavirus segment 5 protein expressed in the baculovirus system interacts with zinc and RNA. *J Gen Virol.* **73**, 1931- 1938.

Brown, C.M., Stockwell, P.A., Trotman, C.N. and Tate, W.P. (1990) Sequence analysis suggests that tetranucleotides signal the termination of protein synthesis in eukaryotes. *Nucleic Acids Res.* **18**, 6339-6345.

Cech, T.R. (2000) Structural biology. The ribosome is a ribozyme. *Science* **289**, 878-879.

Chen, D., Barros, M., Spencer, E. and Patton, J.T. (2001) Features of the 3'-consensus sequence of rotavirus mRNAs critical to minus strand RNA synthesis. *Virology* **282**, 221-229.

Chizhikov, V. and Patton, J.T. (2000) A four-nucleotide translation enhancer in the 3'-terminal consensus sequence of the nonpolyadenylated mRNAs of rotavirus. *RNA* **6**, 814-825.

Christensen, A.K., Kahn, L.E. and Bourne, C.M. (1987) Circular polysomes predominate on the rough endoplasmic reticulum of somatotropes and mammotropes in the rat anterior pituitary. *Am J Anat.* **178**, 1-10.

Cosson, B., Couturier, A., Chabelskaya, S., Kiktev, D., Inge-Vechtomov, S., Philippe, M. and Zhouravleva, G. (2002) Poly(A)-binding protein acts in translation termination via Eukaryotic Release Factor 3 interaction and does not influence [PSI(+)] propagation. *Mol Cell Biol* **22**, 3301- 3315.

Craig, A.W., Haghighat, A., Yu, A.T. and Sonenberg, N. (1998) Interaction of poly-adenylate-binding protein with the eIF4G homologue PAIP enhances translation. *Nature* **392**, 520-523.

Dahlberg, A.E. (2001) Ribosome structure. The ribosome in action. *Science* **292**, 868-869.

Deo, R.C., Groft, C.M., Rajashankar, K.R. and Burley, S.K. (2002) Recognition of the rotavirus mRNA 3' consensus by an asymmetric NSP3 homodimer. *Cell* **108**, 71-81.

Desselberger, U. (1996) Genome rearrangements of rotaviruses. *Adv Virus Res.* **46**, 69-95.

Fedorov, A., Saxonov, S. and Gilbert, W. (2002) Regularities of context-dependent codon bias in eukaryotic genes. *Nucleic Acids Res.* **30**, 1192-1197.

Furuichi, Y., Muthukrishnan, S., Tomasz, J. and Shatkin, A.J. (1976) Mechanism of formation of reovirus mRNA 5'-terminal blocked and methylated sequence, m7GpppGmpC. *J. Biol. Chem.* **251**, 5043-5053.

Furuichi, Y. and Shatkin, A.J. (2000) Viral and cellular mRNA capping: past and prospects. *Adv Virus Res.* **55**, 135-184.

Gale, M., Jr., Tan, S.L. and Katze, M.G. (2000) Translational control of viral gene expression in eukaryotes. *Microbiol Mol Biol Rev.* **64**, 239-280.

Gallie, D.R. (1991) The cap and poly(A) tail function synergistically to regulate mRNA translational efficiency. *Genes Dev.* **5**, 2108-2116.

Gallie, D.R. (1998) A tale of two termini: A functional interaction between the termini of an mRNA is a prerequisite for efficient translation initiation. *Gene* **216**, 1-11.

Gallie, D.R., Lewis, N.J. and Marzluff, W.F. (1996) The histone 3'-terminal stem-loop is necessary for translation in Chinese hamster ovary cells. *Nucleic Acids Res.* **24**, 1954-1962.

Gault, E., Schnepf, N., Poncet, D., Servant, A., Teran, S. and Garbarg-Chenon, A. (2001) A human rotavirus with rearranged genes 7 and 11 encodes a modified NSP3 protein and suggests an additional mechanism for gene rearrangement. *J Virol.* **75**, 7305-7314.

Geballe, A.P. and Sachs, M.S. (2000) Translational control by upstream open reading frame. In: Hershey, J.W.B., Mathews, M.B. and Sonenberg, N. (eds.), *Translational control of gene expression.* pp. 595-614. Cold Spring Harbor Press, Cold Spring Harbor, NY.

Gebauer, F., Corona, D.F., Preiss, T., Becker, P.B. and Hentze, M.W. (1999) Translational control of dosage compensation in *Drosophila* by Sex-lethal: cooperative silencing via the 5' and 3' UTRs of msl-2 mRNA is independent of the poly(A) tail. *EMBO J.* **18**, 6146-6154.

Gray, N.K., Coller, J.M., Dickson, K.S. and Wickens, M. (2000) Multiple portions of poly(A)-binding protein stimulate translation *in vivo. EMBO J.* 19, 4723-4733.

Heath, R.L. and Birch, C.J. (1988) Synthesis of human rotavirus polypeptides in cell culture. *J Med Virol.* **25**, 91-103.

Hellen, C.U. and Sarnow, P. (2001) Internal ribosome entry sites in eukaryotic mRNA molecules. *Genes Dev.* **15**, 1593-1612.

Hershey, J.W.B. and Merrick, W.C. (2000) Pathway and mechanism of initiation of protein synthesis. In: Hershey, J.W.B., Mathews, M.B. and Sonenberg, N. (eds.), *Translational control of gene expression.* pp. 33-88. Cold Spring Harbor Press, Cold Spring Harbor, NY.

Hoshino, S., Imai, M., Kobayashi, T., Uchida, N. and Katada, T. (1999) The eukaryotic polypeptide chain releasing factor (eRF3/GSPT) carrying the translation termination signal to the 3'-poly(A) tail of mRNA. Direct association of erf3/GSPT with polyadenylate-binding protein. *J Biol Chem.* **274**, 16677-16680.

Imataka, H., Gradi, A. and Sonenberg, N. (1998) A newly identified N-terminal amino acid sequence of human eIF4G binds poly(A)-binding protein and functions in poly(A)-dependent translation. *EMBO J.* **17**, 7480-7489.

Irwin, B., Heck, J.D. and Hatfield, G.W. (1995) Codon pair utilization biases influence translational elongation step times. *J Biol Chem.* **270**, 22801-22806.

Ito H, Sugiyama M, Masubuchi K, Mori Y and Minamoto, N. (2001) Complete nucleotide sequence of a group A avian rotavirus genome and a comparison with its counterparts of mammalian rotaviruses. *Virus Res.* **75**, 123-138.

Jackson, R.J. (2000) A comparative view of initiation site selection mechanisms.

In: Hershey, J.W.B., Mathews, M.B. and Sonenberg, N. (eds.), *Translational control of gene expression.* pp 127-183. Cold Spring Harbor Press, Cold Spring Harbor, NY.

Khaleghpour, K., Svitkin, Y.V., Craig, A.W., DeMaria, C.T., Deo, R.C., Burley, S.K. and Sonenberg, N. (2001) Translational repression by a novel partner of human poly(A) binding protein, Paip2. *Mol Cell* **7**, 205-216.

Kozak, M. (1991) Structural features in eukaryotic mRNAs that modulate the initiation of translation. *J Biol Chem.* **266**, 19867-19870.

Kozak, M. (2001a) Constraints on reinitiation of translation in mammals. *Nucleic Acids Res.* **29**, 5226-5232.

Kozak, M. (2001b) New ways of initiating translation in eukaryotes? *Mol Cell Biol.* **21**, 1899-1907.

Li, G. and Rice, C.M. (1993) The signal for translational readthrough of a UGA codon in Sindbis virus RNA involves a single cytidine residue immediately downstream of the termination codon. *J Virol.* **67**, 5062-5067.

Marcotrigiano, J., Gingras, A.C., Sonenberg, N. and Burley, S.K. (1997) Cocrystal structure of the messenger RNA 5' cap-binding protein (eIF4E) bound to 7-methyl-GDP. *Cell* **89**, 951-961.

Mattion, N.M., Mitchell, D.B., Both, G.W. and Estes, M.K. (1991) Expression of rotavirus proteins encoded by alternative open reading frames of genome segment 11. *Virology* **181**, 295-304.

Merrick, W.C. and Nyborg, J. (eds.). (2000) The protein biosynthesis elongation cycle. Cold Spring Harbor Press, Cold Spring Harbor, NY.

Michel, Y.M., Poncet, D., Piron, M., Kean, K.M. and Borman, A.M. (2000) Cap-poly(A) synergy in mammalian cell-free extracts. Investigation of the requirements for poly(A)-mediated stimulation of translation initiation. *J Biol Chem.* **275**, 32268-32276.

Mitchell, D.B. and Both, G.W. (1990) Conservation of a potential metal binding motif despite extensive sequence diversity in the rotavirus nonstructural protein NS53. *Virology* **174**, 618-621.

Noller, H.F., Hoffarth, V. and Zimniak, L. (1992) Unusual resistance of peptidyl transferase to protein extraction procedures. *Science* **256**, 1416-1419.

Otero, L.J., Ashe, M.P. and Sachs, A.B. (1999) The yeast poly(A)-binding protein Pab1p stimulates *in vitro* poly(A)-dependent and cap-dependent translation by distinct mechanisms. *EMBO J.* **18**, 3153-3163.

Patton, J.T., Taraporewala, Z., Chen, D., Chizhikov, V., Jones, M., Elhelu, A., Collins, M., Kearney, K., Wagner, M., Hoshino, Y. and Gouvea, V. (2001) Effect of intragenic rearrangement and changes in the 3' consensus sequence on NSP1 expression and rotavirus replication. *J Virol.* **75**, 2076-2086.

Pestova, T.V., Kolupaeva, V.G., Lomakin, I.B., Pilipenko, E.V., Shatsky, I.N., Agol, V.I. and Hellen, C.U. (2001) Molecular mechanisms of translation initiation in eukaryotes. *Proc Natl Acad Sci USA* **98**, 7029-7036.

Piron, M., Delaunay, T., Grosclaude, J. and Poncet, D. (1999) Identification of the RNA-binding, dimerization and eIF4GI-binding domains of rotavirus NSP3.

J. Virol. **73**, 5411-5421.

Piron, M., Vende, P., Cohen, J. and Poncet, D. (1998) Rotavirus RNA-binding protein NSP3 interacts with eIF4GI and evicts the poly(A) binding protein from eIF4F. *EMBO J.* **17**, 5811- 5821.

Poncet, D., Aponte, C. and Cohen, J. (1993) Rotavirus protein NSP3 (NS34) is bound to the 3' end consensus sequence of viral mRNAs in infected cells. *J. Virol.* **67**, 3159-3165.

Poncet, D., Laurent, S. and Cohen, J. (1994) Four nucleotides are the minimal requirement for RNA recognition by rotavirus non-structural protein NSP3. *EMBO J.* **13**, 4165- 4173.

Poncet, D., Lindenbaum, P., L'Haridon, R. and Cohen, J. (1997) *In vivo* and *in vitro* phosphorylation of rotavirus NSP5 correlates with its localization in viroplasms. *J Virol.* **71**, 34-41.

Preiss, T. and Hentze, M.W. (1998) Dual function of the messenger RNA cap structure in poly(A)-tail- promoted translation in yeast. *Nature.* **392**, 516-520.

Ramakrishnan, V. (2002) Ribosome structure and the mechanism of translation. *Cell* **108**, 557-572.

Sachs, A. (2000) Physical and functionnal interactions between the mRNA cap structure and the poly(A) tail. In: Hershey, J.W.B., Mathews, M.B. and Sonenberg, N. (eds.), *Translational control of gene expression.* pp. 447-465. Cold Spring Harbor Press, Cold Spring Harbor, NY.

Samuel, C.E. (1998) Reoviruses and the interferon system. *Curr Top Microbiol Immunol.* **233**, 125-145.

Schneider, R., Agol, V.I., Andino, R., Bayard, F., Cavener, D.R., Chappell, S.A., Chen, J.J., Darlix, J.L., Dasgupta, A., Donze, O., Duncan, R., Elroy-Stein, O., Farabaugh, P.J., Filipowicz, W., Gale, M., Jr., Gehrke, L., Goldman, E., Groner, Y., Harford, J.B., Hatzglou, M., He, B., Hellen, C.U., Hentze, M.W., Hershey, J., Hershey, P., Hohn, T., Holcik, M., Hunter, C.P., Igarashi, K., Jackson, R., Jagus, R., Jefferson, L.S., Joshi, B., Kaempfer, R., Katze, M., Kaufman, R.J., Kiledjian, M., Kimball, S.R., Kimchi, A., Kirkegaard, K., Koromilas, A.E., Krug, R.M., Kruys, V., Lamphear, B.J., Lemon, S., Lloyd, R.E., Maquat, L.E., Martinez-Salas, E., Mathews, M.B., Mauro, V.P., Miyamoto, S., Mohr, I., Morris, D.R., Moss, E.G., Nakashima, N., Palmenberg, A., Parkin, N.T., Pe'ery, T., Pelletier, J., Peltz, S., Pestova, T.V., Pilipenko, E.V., Prats, A.C., Racaniello, V., Read, G.S., Rhoads, R.E., Richter, J.D., Rivera-Pomar, R., Rouault, T., Sachs, A., Sarnow, P., Scheper, G.C., Schiff, L., Schoenberg, D.R., Semler, B.L., Siddiqui, A., Skern, T., Sonenberg, N., Tahara, S.M., Thomas, A.A., Toulme, J.J., Wilusz, J., Wimmer, E., Witherell, G. and Wormington, M. (2001) New ways of initiating translation in eukaryotes. *Mol Cell Biol.* **21**, 8238-8246.

Searfoss, A., Dever, T.E. and Wickner, R. (2001) Linking the 3' poly(A) tail to the subunit joining step of translation initiation: relations of Pab1p, eukaryotic translation initiation factor 5b (Fun12p), and Ski2p-Slh1p. *Mol Cell Biol.* **21**, 4900-4908.

Spahn, C.M., Beckmann, R., Eswar, N., Penczek, P.A., Sali, A., Blobel, G. and Frank,

J. (2001) Structure of the 80S ribosome from *Saccharomyces cerevisiae*-tRNA-ribosome and subunit-subunit interactions. *Cell* **107**, 373-386.

Stirzaker, S.C., Poncet, D. and Both, G.W. (1990) Sequences in rotavirus glycoprotein VP7 that mediate delayed translocation and retention of the protein in the endoplasmic reticulum. *J Cell Biol.* **111**, 1343-1350.

Taniguchi, K., Kojima, K. and Urasawa, S. (1996) Nondefective rotavirus mutants with an NSP1 gene which has a deletion of 500 nucleotides, including a cysteine-rich zinc finger motif- encoding region (nucleotides 156 to 248), or which has a nonsense codon at nucleotides 153-155. *J Virol.* **70**, 4125-4130.

Tarun, S.Z., Jr. and Sachs, A.B. (1996) Association of the yeast poly(A) tail binding protein with translation initiation factor eIF-4G. *EMBO J.* **15**, 7168-7177.

Torres-Vega, M.A., Gonzalez, R.A., Duarte, M., Poncet, D., Lopez, S. and Arias, C.F. (2000) The C-terminal domain of rotavirus NSP5 is essential for its multimerization, hyperphosphorylation and interaction with NSP6. *J Gen Virol.* **81**, 821-830.

Valentini, S.R., Casolari, J.M., Oliveira, C.C., Silver, P.A. and McBride, A.E. (2002) Genetic interactions of yeast eukaryotic translation initiation factor 5A (eIF5A) reveal connections to poly(A)-binding protein and protein kinase C signaling. *Genetics* **160**, 393-405.

Vende, P., Piron, M., Castagne, N. and Poncet, D. (2000) Efficient translation of rotavirus mRNA requires simultaneous interaction of NSP3 with the eukaryotic translation initiation factor eIF4G and the mRNA 3' end. *J Virol.* **74**, 7064-7071.

Welch, E.M., Wang, W. and Peltz, S.W. (2000) Translation termination:it's not the end of the story. In: Hershey, J.W.B., Mathews, M.B. and Sonenberg, N. (eds.), *Translational control of gene expression.* pp. 467-485. Cold Spring Harbor Press, Cold Spring Harbor, NY.

Wells, S.E., Hillner, P.E., Vale, R.D. and Sachs, A.B. (1998) Circularization of mRNA by eukaryotic translation initiation factors. *Mol Cell.* **2**, 135-140.

Yazaki, K., Yoshida, T., Wakiyama, M. and Miura, K. (2000) Polysomes of eukaryotic cells observed by electron microscopy. *J Electron Microsc.* **49**, 663-668.

Viral Gastroenteritis
U. Desselberger and J. Gray (editors)
© 2003 Elsevier Science B.V. All rights reserved

II, 6.   The rotavirus NSP4 enterotoxin: Current status and challenges

Mary K. Estes

*Department of Molecular Virology and Microbiology, Texas Gulf Coast Digestive Diseases Center,
Baylor College of Medicine, Houston, Texas 77030-3498*

Introduction

The clinical significance of rotavirus-induced diarrheal disease, which causes high morbidity in children in developed countries and high mortality in children in developing countries, makes disease prevention a global priority. Vaccination is the most effective method of disease prevention, and effective treatments or vaccines for any microbial infection generally are targeted to interfere with key molecular steps in pathogenesis. Rotavirus-induced diarrheal disease is the outcome of a complex interplay of host and viral factors. This review summarizes new information discovered since 1996 showing that rotaviruses produce an enterotoxin that has pleiotropic properties. Increasing evidence indicates that this enterotoxin is a key virulence factor that functions early after infection to initiate cell signaling, resulting in chloride secretion and diarrhea. Information supporting this new idea of how rotaviruses cause disease and arguments that have refuted this idea are discussed from the author's perspective.

Rotavirus virulence genes

Virulence genes are most directly defined for viruses and other microorganisms by genetic analyses. For viruses such as rotaviruses with segmented genomes, where recombination between markers on the same segment either does not occur or is extremely rare and where a reverse genetics system is not available, the genetic analysis of choice is to study reassortant viruses. In this approach, intertypic crosses between two virus types, which differ with regard to their ability to induce disease and exhibit different electrophoretic mobilities of each RNA segment, are carried out, and progeny are analyzed to determine if the disease phenotype maps to an individual genome segment. Such reassortant analyses for rotaviruses in several animal models indicate that multiple viral genes can determine the pathogenicity of rotaviruses (Desselberger, 1997). Not all genes implicated in pathogenicity are associated with diarrhea. Depending on the animal model used, pathogenicity is measured in different ways, being based on induction of disease, the ability of a virus to simply infect an animal, or the ability of a virus to replicate in the inoculated animal and then spread to uninoculated littermates. Virulence genes identified in any model code for both structural (VP3, VP4, and VP7) and nonstructural (NSP1, NSP2, NSP3, NSP4) proteins. The virulence genes associated with *diarrhea induction* are more limited and include those that code for VP3, VP4, VP7 and NSP4 (Hoshino *et al.*, 1995; Tauscher *et al.*, 1997; Bridger *et al.*, 1998). The surface proteins (VP4 and VP7) on the virus are likely to be involved in virus stability, virus attachment and penetration into cells. If these functions are lacking and an infection is not initiated, disease clearly will not occur. VP4 has also been shown to affect growth in intestinal cells (Kirkwood *et al.*, 1998). The precise role for the inner capsid protein, VP3, which is part of an enzyme complex needed for the synthesis and capping of viral mRNA remains unclear, but this protein may affect mRNA synthesis and replication efficiency. The role for NSP4 as a virulence gene responsible for diarrhea in the piglet model (Hoshino *et al.*, 1995) was initially not understood. This protein was known to be involved in viral morphogenesis suggesting that it might affect the synthesis of new virus particles, and thus affect pathogenesis. In 1996, a new function of NSP4 as an enterotoxin was discovered (Ball *et al.*, 1996), and this and other properties of this protein more clearly explain its role in pathogenesis.

A viral enterotoxin: a new molecular mechanism of action for a viral protein helps understand previously unexplained aspects of rotavirus pathogenesis

Prior to the reassortant studies that identified NSP4 as a gene that could cause diarrhea in piglets, many studies had examined the mechanisms of rotavirus-induced diarrhea. Several studies indicated that diarrhea occurs secondary to the destruction of enterocytes and is associated with villus blunting and malabsorption (Conner and Ramig, 1997). However, many other investigators observed that infected animals exhibit profuse diarrhea *prior* to the detection of histologic lesions and villus blunting in the intestine, suggesting that other mechanisms could be responsible for the induction of diarrhea (Bohl *et al.*, 1978; McAdaragh *et al.*, 1980; Mebus, 1976; Theil *et al.*, 1978; Osborne *et al.*, 1988). Such investigators suggested that rotavirus infection must induce changes (functional impairment) in the epithelial cells of the small intestine sufficient to induce diarrhea *before* histopathologic lesions become apparent, and this was clearly seen if animals were euthanized and studied within a few hours after the onset of diarrhea. Conversely, in adult rabbits infected with rotavirus, villus blunting is seen yet diarrhea does not occur (Ciarlet *et al.*, 1998), again suggesting that mechanisms in addition to simple histologic changes leading to malabsorption play a role in rotavirus-induced diarrhea.

A series of subsequent studies in rotavirus-infected mice concluded that lactose malabsorption is one of several factors which contribute to the loss of fluid at the peak of diarrhea, but malabsorption could not be responsible for diarrheal induction or perpetuation (Stephen, 1989). The small intestine showed a significant change from absorption to frank secretion, and this was thought to be due to the induction of hypersecretion rather than inhibition of absorption. Rotavirus-induced diarrhea then was proposed to result from complex mechanisms that involve the localized responses of the gut to injury triggered by the infection of enterocytes rather than by simple malabsorption (Stephen, 1989). Further studies suggested that a soluble neuroactive/hormonal factor(s) released from infected enterocytes might initiate and perpetuate disease (Spencer *et al.*, 1990). However, no specific factors were identified at that time.

In 1996, the rotavirus NSP4 was reported to be the first viral enterotoxin (Ball *et al.*, 1996). NSP4, and an active 22 amino acid (aa) synthetic peptide from NSP4 (aa residues 114-135), were found to induce dose-dependent and age-dependent diarrhea in the absence of histological alterations, that resembles virus-induced disease in neonatal suckling mice. The peptide also promotes calcium-dependent chloride secretory currents when added to ileal mucosa of young mice (Ball *et al.*, 1996). The diarrhea dose$_{50}$ (DD$_{50}$) of the NSP4 114-135 peptide in outbred CD1 mice is more than ten fold higher than the DD$_{50}$ of the protein, indicating that this peptide does not contain the entire enterotoxin domain.

The NSP4 protein from several other group A rotaviruses (murine EW and porcine OSU and Gottfried), and from a group C rotavirus (human Ehime 9301) have since been shown to induce diarrhea in suckling mice (Horie *et al.*, 1999; Zhang *et al.*, 1998; Sasaki *et al.*, 2001). Although the group C and group A NSP4 proteins have virtually no sequence identity (Deng *et al.*, 1995; Horie *et al.*, 1997b), they share conserved structural features including the presence of hydrophobic domains in the amino terminal domain and a large, more hydrophilic cytoplasmic domain that contains a predicted amphipathic alpha helix and enterotoxin activity (Fig. 1). These results confirm that structure, rather than a primary amino acid sequence, is important in enterotoxin function (Zhang *et al.*, 1998; and see below).

The discovery that NSP4 is an enterotoxin resulted from a series of basic studies that were focused on understanding the molecular mechanism of how NSP4 (called NS28 at that time) regulates the unique process of rotavirus morphogenesis in which newly made subviral particles "bud" into the endoplasmic reticulum (ER). In this unusual process, NSP4, which is initially synthesized as a transmembrane ER-specific protein with its C-terminus localized in the cytoplasm, had been shown to function as an intracellular receptor that binds to VP6, the protein on the surface of newly made subviral particles (Au *et al.*, 1989; Taylor *et al.*, 1993). The intracellular binding of VP6-coated particles to NSP4 was proposed to be an event that initiates the budding process, in which rotavirus particles acquire a transient envelope while transversing the ER membrane; in this process, particles also acquire the two proteins, VP4 and VP7, that make up the outer capsid shell. It was further proposed that this morphogenetic process might involve regulation of calcium fluxes from the ER because calcium is an important factor for the oligomerization of NSP4 in the ER (Poruchynsky *et al.*, 1991) and for maintaining the stability of VP7 on the outer surface of particles (Estes *et al.*, 1979). In addition, calcium fluxes in some cells regulate normal cellular vesicular budding.

The idea that NSP4 might alter calcium regulation in cells initially was tested in insect cells. Intracellular expression of NSP4 was found to be associated with elevated

Features of NSP4: An Intracellular Receptor and Enterotoxin

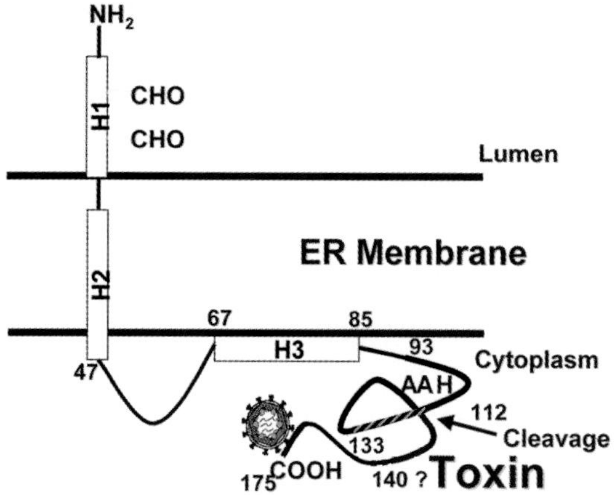

Fig. 1. Features of NSP4: An intracellular receptor and enterotoxin.

A linear representation of the topology of a monomer of NSP4 as a transmembrane ER-glycoprotein. NSP4 is co-translationally glycosylated (-CHO) and its N-terminus is localized in the lumen of the ER. There are two N-terminal hydrophobic domains (H1 and H2). A large cytoplasmic domain contains several domains including a third hydrophobic domain (H3), an amphiphatic alpha helix (AAH), the enterotoxin domain and a C-terminal domain that binds nascent double-layered subviral particles. A site of cleavage that releases a peptide of aa 112-175 into the medium of virus-infected cells has also been identified (Zhang *et al.*, 2000). A high resolution structure of a peptide representing aa 95-137 indicates this self-associating peptide is a parallel, tetrameric coiled-coil that contains a divalent metal binding site that is a putative binding site for calcium ions (Bowman *et al.*, 2000).

intracellular calcium $[Ca^{2+}]i$ while no other rotavirus protein, or wild-type baculovirus protein, affected calcium regulation (Tian *et al.*, 1994) (Fig. 2A). Further studies showed that NSP4 (or the NSP4 114-135 peptide) exogenously applied to insect cells also causes an increase in intracellular calcium by a phospholipase C (PLC)-mediated process. Increases in $[Ca^{2+}]i$ associated with NSP4 expressed endogenously or added exogenously were found to result from the release of intracellular thapsigargin-sensitive stores in the ER but they are mediated by different upstream pathways as intracellularly expressed NSP4 does not stimulate PLC (Tian *et al.*, 1995).

These findings led to an initial model for how NSP4 induces diarrhea (Fig. 3). This model hypothesized two ways by which rotavirus infection results in expression of viral proteins, including NSP4, that could lead to inducing calcium-dependent chloride secretion in host cells. First, NSP4 synthesized in virus-infected cells may release $[Ca^{2+}]i$ by interacting with ER Ca^{2+} pools. Second, after being released from virus infected cells, NSP4 may bind to an unidentified molecule, on or within the plasma

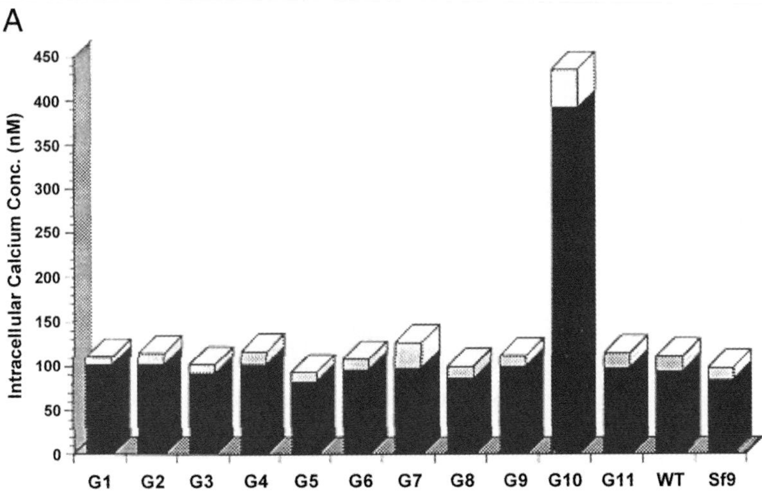

Fig. 2. Calcium mobilization by different forms of NSP4. Intracellular calcium concentration ($[Ca^{2+}]i$) in Sf9 cells infected with wild-type (wt) baculovirus or baculovirus recombinants expressing each of the 11 genes of a group A rotavirus individually.

(A). G1 to G11 represent baculovirus recombinants expressing rotavirus gene 1 to gene 11. Sf9 cells were infected with each baculovirus at a high multiplicity (MO1 of 10). $[Ca^{2+}]i$ was measured between 32 to 36 hours p.i. using the fluorescent indicator fura-2. Each column shows the mean (top of shaded box) and the standard deviation (open box) of three to five independent experiments. (N = 3 in G1, G3, G7, G11; N = 4 in G2, G4, G5, G6, G8, G9; and N = 5 in G10, wt and uninfected Sf9 cells). Three measurements were made in each independent experiment. The SA11 gene 10 NSP4 protein causes diarrhea in suckling mice (see Text). From Tian *et al.* (1994).

membrane of uninfected neighboring cells, to mobilize $[Ca^{2+}]i$. The $[Ca^{2+}]i$ increase generated by either mechanism could stimulate endogenous fluid secretory pathways within the intestinal mucosa and cause the early onset of secretory diarrhea observed in animal models. This model resulted in several testable predictions (Fig. 3), which have been addressed in subsequent studies and substantiate the model. A brief summary of these studies follows.

NSP4 is released from cells

The first hypothesis suggested that NSP4 may be released from virus-infected cells. Studies of the proteins in the media of rotavirus-infected monkey kidney cells or human intestinal cells showed that a cleavage product of NSP4 (NSP4 112-175) is secreted from cells by a Golgi-independent pathway early after infection and prior to the appearance of virus particles (Zhang *et al.*, 2000). This secreted form of NSP4 con-

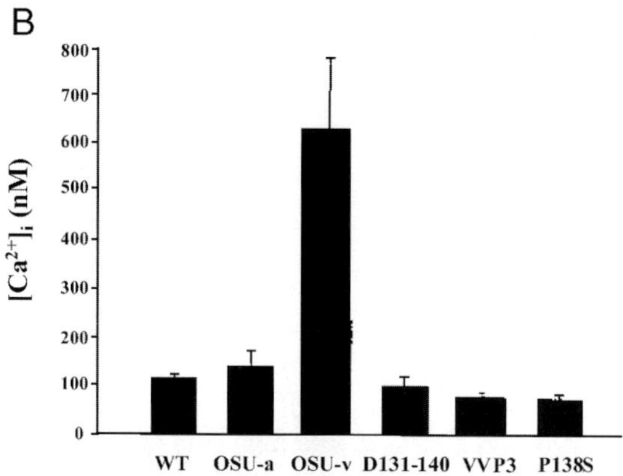

Fig. 2. (B). $[Ca^{2+}]i$ in Sf9 cells expressing NSP4 and NSP4 mutants. Fura-2/AM-loaded cells were superfused continuously with Na-HEPES buffer, and intracellular calcium was measured by ratio imaging. The averaged ratio signal obtained from each cell was digitally saved as a log file. The collected values from cells imaged within a single experiment (6 to 10 cells) were then averaged to give an experimental observation of 1($n = 1$). WT, cells infected with wild-type baculovirus ($n = 3$); OSU-a, cells infected with recombinant baculovirus expressing the attenuated OSU-a NSP4 ($n = 5$); OSU-v, cells infected with recombinant baculovirus expressing the virulent OSU-v NSP4 ($n = 3$); D131-140, cells infected with recombinant baculovirus expressing a deletion mutant (lacking aa 131-140 of NSP4); VVP3, cells infected with recombinant baculovirus expressing a substitution mutant; P138S, cells infected with recombinant baculovirus expressing a point mutant. Standard errors of the means are shown by the bars. From Zhang *et al.* (1998).

tains the enterotoxin domain and, as expected, has the ability to mobilize $[Ca^{2+}]i$ and to induce diarrhea in suckling mice. Ongoing studies are characterizing the form of NSP4 that is released from infected polarized epithelial cells. This is important because virus infection of polarized cells has been shown to contain a factor released into the medium that can mobilize calcium in neighboring non-infected cells as a result of a PLC-dependent efflux of Ca^{2+} from the ER and by extracellular Ca^{2+} influx (Brunet *et al.*, 2000). It seems likely that this factor will be NSP4. This idea is supported by the new finding that rotaviruses and NSP4 associate with lipid rafts in the polarized Caco-2 cells and that such rafts promote apical targeting and virus assembly (Sapin *et al.*, 2002).

Avirulent rotaviruses possess mutant NSP4 genes

A second prediction was that some avirulent rotaviruses would have mutant NSP4 genes. This idea was tested by sequencing pairs of parental virulent and tissue-culture derived avirulent virus strains of two independent porcine rotavirus strains (Zhang *et al.*, 1998). Specific mutations in the genes for NSP4 were found in both attenuated viruses at aa positions 135 and 138, suggesting that these residues may be important in virulence. Identification of the mutation at aa 138 extended the toxin domain past aa 135.

Expression in insect cells of the NSP4 protein from the attenuated OSU virus and of other mutant forms of NSP4, including a deletion mutant that lacked aa 131-140 failed to mobilize [Ca^{2+}]i, while the NSP4 from the virulent porcine OSU virus mobilized [Ca^{2+}]i, and the levels of expression of each of these proteins were similar (Fig. 2B; Zhang *et al.*, 1998). The mutant proteins also failed to mobilize [Ca^{2+}]i in human intestinal epithelial HT-29 cells although the cells were responsive to the protein from the virulent virus (Zhang *et al.*, 1998). Finally, the mutant proteins failed to induce diarrhea or were attenuated in their ability to induce diarrhea in suckling mice compared to the virulent form of NSP4 (Table 1). These data are consistent with the idea that attenuation of at least *some* strains of rotavirus by passage in tissue culture is associated with mutations in the enterotoxin. The mutation at aa 138 (from proline to serine) could result in an altered conformation of the protein and this, in addition to the fact that a peptide containing aa 114-135 but composed of D-amino acids instead of L-amino acids does not induce diarrhea, suggests that structure and not the primary sequence of this protein is critical for enterotoxin activity. The idea that the NSP4 structure is important for enterotoxin function is supported by additional studies that have shown that the NSP4 protein from a group C rotavirus, which shares little sequence identity with the NSP4s from group A rotaviruses, also causes diarrhea in sucking mice (Horie *et al.*, 1999).

Table 1

Summary of biological properties of NSP4 and NSP4 mutants

NSP4	[Ca^{2+}]$_i$ mobilization in insect cells	[Ca^{2+}]$_i$ mobilization in human intestinal epithelial cells	Diarrhea induction in neonatal mice	
OSU-v NSP4	6-fold ↑	10-fold ↑	57%*	(13/23)
OSU-a NSP4	1.4-fold ↑	1.2-fold ↑	16%	(4/25)
P138S	No increase	No increase	0%	(0/12)
D131-140	No increase	ND	0%	(0/12)
P138A	3.3-fold ↑	3.2-fold ↑	18%*	(2/11)
S138P	3.1-fold ↑	2.3-fold ↑	18%*	(2/11)

*P < 0.05 (From Zhang *et al.*, 1998 and unpublished data).
D131-140: deletion mutant lacking aa 131-140.

Intestinal cells possess a receptor for NSP4

The third prediction is that intestinal cells have a receptor for NSP4. Several lines of evidence support this idea although the receptor has not yet been identified. For example, the effective dose 50 for mobilization of [Ca^{2+}]i in human intestinal cells is low (4.6 nM), and mobilization of [Ca^{2+}]i by the addition of exogenous NSP4 is ablated by treatment of cells with proteases (Dong *et al.*, 1997). In addition, a D-aa NSP4 114-135 peptide does not induce diarrhea in contrast to some "lytic toxins" that function by directly perturbing cellular membranes and that function in both L- and

Model of NSP4–Induced Diarrhea

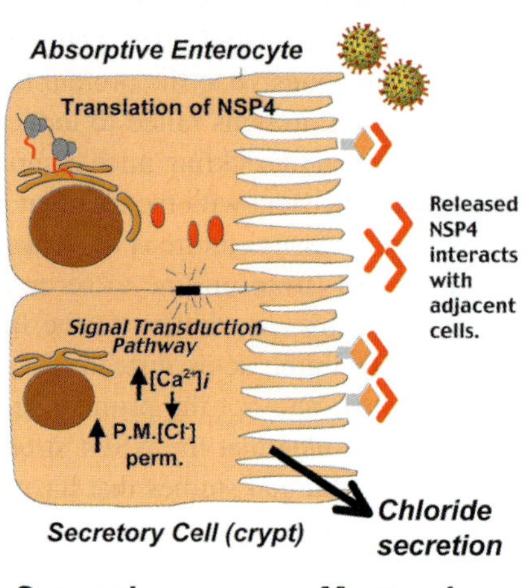

Fig. 3. Model of NSP4-induced diarrhea. Rotavirus-infected enterocytes produce virus-specific proteins including NSP4. NSP4 is released from cells by a nonclassical vesicular secretion pathway, and extracellular NSP4 acts in a paracrine fashion by interacting with a receptor for NSP4 on adjacent cells. These secretory cells may be crypt cells or villus enterocytes. Binding of NSP4 triggers a signal transduction pathway that mobilizes intracellular Ca^{2+} and results in increased plasma membrane Cl^- permeability. The plasma membrane permeability and resulting diarrhea are age-dependent based on studies in CFTR mice (Morris *et al.*, 1999). Extracellular NSP4 can also cause a disruption of the organization of filamentous actin and transepithelial cell resistance (not shown). Model modified from Estes and Morris (1999)

D-aa forms. Polarized kidney cells respond to treatment with exogenous NSP4 only when the protein is added to the apical surface, suggesting an apical receptor (Tafazoli *et al.*, 2001). Finally, direct binding studies indicate that intestinal epithelial cells contain multiple receptors that exhibit different binding affinities (Zeng and Estes, unpublished data). Identification of the receptor has been more challenging than anticipated because of the lack of a good method to label NSP4 to high specific activity, and due to the fact that, unfortunately, the biologic activity of NSP4 is lost by iodination. Additional work is needed to identify the cellular receptor for NSP4.

Age-dependent disease results from changes in the signaling pathway (receptor/intracellular signaling/channel)

Studies in mice lacking the cystic fibrosis transmembrane regulator (CFTR) were initiated because these animals lack the cAMP-dependent chloride channel, and so we predicted that these mice would still exhibit rotavirus- and NSP4-induced diarrhea if disease involves the activity of a calcium-regulated chloride channel. Such mice are indeed susceptible to diarrhea caused by rotavirus, as are children with cystic fibrosis (Angel *et al.*, 1998; Morris *et al.*, 1999). Rotavirus NSP4 also induces diarrhea in these mice, demonstrating that the CFTR- Cl channel is not necessary for diarrhea elicited by rotavirus (Morris *et al.*, 1999). Further analyses of the age-dependent responses in these mice have shown that Ca^{2+} mobilization is age-independent but halide (e.g., I or Cl) permeability is age-dependent (Morris *et al.*, 1999). This result indicates that the age-restricted rotavirus-induced diarrhea is not due to an age-dependent receptor for NSP4 but is due to an age-dependent event downstream in the signaling pathway from Ca^{2+} mobilization. Thus, there appears to be an age-dependent regulation or expression of a Ca^{2+}-dependent Cl channel in the intestinal mucosa of CFTR knockout mice that was previously not recognized, because young animals had not been studied previously.

The studies in CFTR knockout mice also illustrated that NSP4 is a novel secretory agonist because such mice do not respond to a variety of other classical secretagogues such as cholera toxin, the $[Ca^{2+}]i$-mobilizing carbachol and the cAMP-mobilizing forskolin. Thus, rotavirus- and NSP4-induced secretory diarrhea is not mediated by CFTR, and the molecular identity of the responsible channel remains to be determined.

Antibody to NSP4 will reduce rotavirus-induced diarrheal disease

A final prediction of our initial model was that antibody to NSP4 should reduce rotavirus-induced diarrheal disease. This would be analogous to protection elicited against bacterial-induced diarrheas by antibody to their enterotoxins. Antibody to NSP4 can reduce both the incidence and severity of rotavirus-induced diarrheal disease in suckling mice (Ball *et al.*, 1996; Yu and Langridge., 2001; Ball *et al.*, 2002; Zeng *et al.*, 2002). This protection can be demonstrated by immunizing adult female mice with purified NSP4 protein or peptide (Ball *et al.*, 1996; Ball *et al.*, 2002) or by feeding such mice with potatoes engineered to express the NSP4 114-135 peptide fused to cholera toxin B (Yu *et al.*, 2001) and monitoring protection from rotavirus-induced disease in pups born to such dams. Alternatively, diarrhea induced by rotavirus infection can be treated by oral administration of antibody to pups with diarrhea (Ball *et al.*, 2002).

These studies indicate that NSP4 is a new candidate rotavirus vaccine. It could be used alone, or in conjunction with other rotavirus proteins known to induce protective immunity, as a nonreplicating vaccine. Such a vaccine offers a new approach to rotavirus vaccination that may avoid the unexpected adverse effects of intussusception associated with oral administration of the first licensed, live attenuated, RRV-reassortant based rotavirus vaccine (Murphy *et al.*, 2001).

Does NSP4 have other functions?

As summarized above, NSP4 has been shown to have previously unexpected properties, and it has become increasingly clear that the NSP4 function is complex. One of the difficulties in studying specific functions of NSP4, or of any rotavirus gene product, is the lack of a reverse genetics system to knock out a specific rotavirus gene. Therefore, rotavirus researchers remain challenged by the need to design other, often indirect, methods to study protein function. While this approach has some limitations, the NSP4 story has clearly been confirmed by investigating predicted functions with a variety of methods. The results are of interest because they answer questions raised more than 10 years ago when many investigators questioned how rotavirus-induced diarrhea was initiated *prior* to histologic changes, and they also have triggered new interest in the pathogenesis of viral-induced diarrhea in general to address questions regarding host cell responses to infection. Finally, these results suggest the involvement of previously unappreciated common mechanisms of pathogenesis of enteric bacterial and viral agents.

It is clear that NSP4 has unique properties depending on whether it is functioning as an intracellular or an extracellular molecule, either in an autocrine or paracrine fashion. While we have some understanding of new properties for this protein, more can be expected. Details of the signaling pathways elicited by the different forms of NSP4 remain to be dissected, and the full complement of cellular proteins, which function in the signaling pathways remains to be identified.

Evidence already exists that indicates that NSP4 has other functions that may help sustain the diarrheal response. Intracellular NSP4 has been shown to immobilize protein secretion in epithelial cells (Xu *et al.*, 2000). It remains unknown if this reduces epithelial cell function such as absorption by reducing sucrase-isomaltase expression in intestinal cells (Jourdan *et al.*, 1998), or if this might also alter the host immune response to rotavirus infection by reducing the surface expression of MHC molecules. Exogenously added NSP4 114-135 peptide is able to function as a full competitive inhibitor of the Na^+-D-glucose symporter SGLT1 in young rabbit brush border membranes (Halaihel *et al.*, 2000) suggesting it might potentiate malabsorption. Exogenous treatment of polarized kidney cells with NSP4 can result in a reduction of transepithelial cell resistance, a redistribution of the tight junction associated ZO-1 protein, and an increase in paracellular permeability (Tafazoli *et al.*, 2001). These effects are delayed (requiring 24-30 hours) relative to similar effects seen in rotavirus-infected polarized cells (Ciarlet *et al.*, 2001; Dickman *et al.*, 2000; Obert *et al.*, 2000); thus, different pathways may be involved. Alternatively, intracellular expression of NSP4 may alter transepithelial resistance by a different and faster mechanism. The enteric nervous system also plays a role in sustaining rotavirus-induced diarrhea, and it remains unknown if NSP4 plays a role in activating secondary mediators of this system (Lundgren *et al.*, 2000; Lundgren and Svensson, Section I, Chapter 3 of this book).

NSP4 has been expressed in a variety of prokaryotic and eukaryotic systems to analyze the different functions of NSP4 (Browne *et al.*, 2000; Mohan *et al.*, 2000; Newton *et al.*, 1997; Taylor *et al.*, 1998; Taylor and Bellamy, Section II, Chapter 7 of this book); sometimes the results from these studies do not seem to be consistent. Because our knowledge of the pleiotropic properties of this protein is expanding, one must remain open-minded about the strengths and weaknesses of individual experimental systems, and apparently discrepant results should be used to design new approaches to more rigorously test reported new functions. Whether expression in certain systems affects the structure of NSP4 that may alter function is a overriding question that needs to be understood more clearly. Information from the crystal structure of a portion of NSP4 (NSP4 aa 95-137) is of interest because this portion of the molecule that encompasses *part* of the enterotoxin domain, is predicted to contain a metal (Ca $^{2+}$)-binding region, and it may form a homotetrameric pore (Bowman *et al.*, 2000).

Sequence analyses indicate that there are at least four genotypes (A-D) of NSP4 genes of group A rotaviruses (Ciarlet *et al.*, 2000; Horie *et al.*, 1997a), and NSP4 genotypes A and B may cluster according to species of origin of virus strains. It is possible that the function of specific types of NSP4 may differ according to its interaction with other viral or cellular proteins. A big challenge is to understand the full effects of intracellular NSP4. Understanding such interactions will be difficult in the absence of a reverse genetics system. Further studies are clearly needed before the full range of properties of NSP4, alone or in conjunction with other viral or cellular proteins that may play a role in pathogenesis or viral morphogenesis is completely understood.

Do challenges to the NSP4 model hold up?

Several papers have reported results that seem to challenge the NSP4 model summarized above. Do these papers represent formidable challenges? The answer seems to be *no* as there are explanations for the results in these papers. It seems worthwhile to present the arguments.

Ward *et al.* (1997) reported that attenuation of a human rotavirus vaccine candidate did not correlate with mutations in the NSP4 protein gene. This result is not surprising because there are many mechanisms of virus attenuation, and we and others have never proposed that mutation of NSP4 is the only mechanism of attenuation of a rotavirus strain by passage in tissue culture. In the absence of detecting a change in NSP4, it is important to know the basis of attenuation of this human rotavirus vaccine candidate, 89-12, because presumably its NSP4 could still cause diarrhea. This idea could be tested by production and characterization of a monoreassortant containing the 89-12 NSP4 gene on the genetic background of an apathogenic virus.

Other studies have compared the sequences of NSP4 genes from viruses isolated from children or animals with diarrhea, and from subjects without diarrhea, and failed to find sequence changes in the gene for NSP4 that correlate with disease phenotype (Oka *et al.*, 2000; Lee *et al.*, 2000). Such studies are generally not analyzing parental stool- and tissue culture-derived virus pairs, but rather strains associated with disease outcome in individuals that may reflect host immunity or host factors active at the time of infection. Such studies would be stronger if the asymptomatic and symptomatic phenotype of the viruses were confirmed in rotavirus-naïve animals and if tissue culture-adapted avirulent viruses derived from the symptomatic virus were studied.

Two studies have reported changes in the "toxin domain" of NSP4 in human rotaviruses associated with symptomatic compared to asymptomatic disease in neonatal infections. Such infections may be unusual cases because these "neonatal" viruses seem to become endogenous in specific neonatal wards. The changes found in NSP4 were a substitution at aa 135 or aa 138, the two sites implicated in the studies of parental virulent-tissue culture derived-avirulent porcine rotavirus pairs (Kirkwood *et al.*, 1996; Pager *et al.*, 2000; Zhang *et al.*, 1998). While these changes are of interest because they support the studies of virulent parental-avirulent tissue culture-derived virus pairs, it should be noted that there were other sequence changes in these human viruses in the genes that encode VP4 and VP7. Thus, no single gene was associated with changes in virulence, and it may be that a combination of changes is most frequently responsible for alterations in virulence.

It is important to remember that if an attenuated rotavirus *contains an altered VP4 or VP7 and fails to initiate an infection*, it will not cause disease even if it carries a fully virulent NSP4 gene. Thus, it is possible to find virulent NSP4 sequences in attenuated rotaviruses that are attenuated because of other genetic changes preventing initiation of infection. This scenario is a possible explanation for the report that NSP4 may not be critical in the pathogenesis of murine RV diarrhea (Angel *et al.*, 1998). These authors reported that pairs of virulent and tissue culture-adapted murine rotaviruses did not have changes in the toxin region of their NSP4 genes, but the sequences of the other genes important for virulence and possible growth in intestinal cells were not reported. Others have shown that the NSP4 protein of at least one of the same murine strains can cause diarrhea in suckling mice (Horie *et al.*, 1999).

Sequence changes in regions of NSP4 other than the enterotoxin domain have been reported in some studies of virulent/avirulent viruses (Angel *et al.*, 1998; Chang *et al.*, 1999). In these studies, the sequences of the other genes potentially associated with virulence were not studied. Since NSP4 is a protein with pleiotropic properties, it is possible that changes in other regions of the protein may alter the functions of NSP4 in ways that currently we do not fully understand. For example, changes in other domains of NSP4 could affect its folding or oligomerization thereby changing the interactions with cellular signaling molecules, or newly made viral particles, or affecting intercellular transport. Alterations in these other NSP4 functions could affect pathogenesis indirectly.

Does the NSP4 enterotoxin have clinical significance for disease in children or animals besides rodents?

There is currently no direct proof that NSP4 is involved in disease induction in children but there is reason to believe it will be involved. NSP4 has been shown to trigger a signaling cascade that results in mobilization of $[Ca^{2+}]i$ in human intestinal cells (Dong *et al.*, 1997), and children who have recovered from rotavirus infections make antibodies to NSP4 (Johansen *et al.*, 1999). Thus, NSP4 is expressed during infections in humans. For bacterial toxins, the author is unaware of any example of a bacterial toxin that functions only in rodents and is not relevant to other species. Although sometimes there are initial difficulties in confirming toxin function in all animal models, technical difficulties such as protease degradation of toxin can be overcome and have been shown to have yielded initial negative results with bacterial toxin testing in some animal models (Debreuil, 1997). It is not clear either whether testing of purified baculovirus- or bacterially-expressed NSP4 fully mimics the native enterotoxin released from virus-infected cells. Thus, the full host range of NSP4 biologic activities remains to be determined.

Finally, it is of interest to know if antibodies to NSP4 will be a useful correlate of protection from rotavirus-induced disease. This is an important question because a correlate of protection for rotavirus remains unknown in spite of many vaccine trials in children and studies of pathogenesis in animals. While intestinal or fecal antibodies of IgA or IgG subclasses seem to be necessary for protection, the protein or protein(s) targeted by this antibody are not known. Initial studies using fixed insect cells expressing NSP4 have reported weak antibody responses to NSP4 (Chang *et al.*, 2001; Ishida *et al.*, 1997). New ELISA-based systems to detect antibody to NSP4 may yield more meaningful data (Johansen *et al.*, 1999; Kang *et al.*, 2002).

New questions and future studies

The discovery of an enterotoxin in rotaviruses was unexpected, but additional studies by several laboratories have now confirmed the initial observations for the simian rotavirus SA11 NSP4 with homologous proteins of other rotavirus strains. Many questions remain to be answered such as: What is the cellular receptor for NSP4? What is the nature of the age-dependent calcium-regulated chloride channel? Is it possible that NSP4 itself may function as a channel? What cellular molecules are involved in the signaling pathways? Are distinct pathways elicited downstream of calcium mobilization, as seen upstream of calcium mobilization, depending on whether NSP4 is functioning intracellularly or extracellularly? Are the cellular pathways elicited by rotavirus NSP4 unique to this virus or are similar pathways elicited by other viral or bacterial pathogens? If intestinal epithelial cells respond in a unique way to rotavirus infection, does co-infection with a bacterial pathogen exacerbate the severity of disease or complications? And finally, will vaccines that incorporate NSP4, or treatments that target NSP4, significantly improve the outcome of rotavirus infections in young children who are the least able to withstand the consequences of rotavirus-induced diarrheal disease?

Acknowledgements

The work on the rotavirus enterotoxin NSP4 summarized in this article includes that of former students, postdoctoral fellows, and collaborators. The author is grateful for having the opportunity to work with these talented investigators and to the National Institutes of Health for partial funding of the research in her laboratory.

220

References

Angel J, Tang B, Feng N, Greenberg HB, Bass D (1998). Studies on the role for NSP4 in the pathogenesis of homologous murine rotavirus diarrhea. *J. Infect. Dis.* **177**: 455-458.

Au KS, Chan W-K, Burns JW, Estes MK (1989). Receptor activity of rotavirus nonstructural glycoprotein NS28. *J. Virol.* **63**: 4553-4562.

Ball JM, Tian P, Zeng CQ, Morris AP, Estes MK (1996). Age-dependent diarrhea induced by a rotaviral nonstructural glycoprotein. *Science* **272**: 101-104.

Ball JM, Zeng CQ-Y, Estes MK (2002). Rotavirus-induced diarrhea in neonatal mice is significantly reduced by maternal antibodies to the NSP4 enterotoxic peptide. Submitted for publication.

Bohl EH, Kohler EM, Saif LJ, Cross RF, Agnes AG, Theil KW (1978). Rotavirus as a cause of diarrhea in pigs. *J. Am. Vet. Med. Assoc.* **172**: 458-463.

Bowman G, Nodelman I, Levy O, Lin S, Tian P, Zamb T, Udem S, Venkataraghavan B, Schutt C (2000). Crystal structure of the oligomerization domain of NSP4 from rotavirus reveals a core metal-binding site. *J. Mol. Biol.* **304**: 861-871.

Bridger JC, Tauscher GI, Desselberger U (1998). Viral determinants of rotavirus pathogenicity in pigs: evidence that the fourth gene of a porcine rotavirus confers diarrhea in the homologous host. *J. Virol.* **72**: 6929-6931.

Browne EP, Bellamy AR, Taylor JA (2000). Membrane-destabilizing activity of rotavirus NSP4 is mediated by a membrane-proximal amphipathic domain. *J. Gen. Virol.* **81**: 1955-1959.

Brunet JP, Cotte-Laffitte J, Linxe C, Quero AM, Geniteau-Legendre M, Servin A (2000). Rotavirus infection induces an increase in intracellular calcium concentration in human intestinal epithelial cells: role in microvillar actin alteration. *J. Virol.* **74**: 2323-2332.

Chang KO, Yandal OH, Yuan L, Hodgins DC, Saif LF (2001). Antibody-secreting cell responses to rotavirus proteins in gnotobiotic pigs inoculated with attenuated or virulent human rotavirus. *J. Clin. Microbiol.* **39**: 2807-2813.

Chang KO, Kim YJ, Saif LJ (1999). Comparisons of nucleotide and deduced amino acid sequences of NSP4 genes of virulent and attenuated pairs of group A and C rotaviruses. *Virus Genes* **18**: 229-233.

Ciarlet M, Crawford SE, Estes MK (2001). Differential infection of polarized epithelial cell lines by sialic acid-dependent and sialic acid-independent rotavirus strains. *J. Virol.* **75**:11834-11850.

Ciarlet M, Gilger MA, Barone C, McArthur M, Estes MK, Conner ME (1998). Rotavirus disease, but not infection and development of intestinal histopathological lesions, is age-restricted in rabbits. *Virology* **251**: 343-360.

Ciarlet M, Liprandi F, Conner ME, Estes MK (2000). Species specificity, interspecies relatedness of NSP4 genetic groups by comparative NSP4 sequence analyses of animal rotaviruses. *Arch. Virol.* **145**: 371-383.

Conner ME, Ramig RF (1997). Viral Enteric Diseases. In: Nathanson N, Ahmed R, Gonzalez-Scarano F, Griffin DE, Holmes KV, Murphy FA, Robinson HL (eds).

Viral Pathogenesis, pp 713-743. Lippincott-Raven Publ., Philadelphia.

Deng Y, Fielding PA, Lambden PR, Caul EO, Clarke IN (1995). Molecular characterization of the eleventh RNA segment from human group C rotavirus. *Virus Genes* **10**: 239-243.

Debreuil J (1997). Escherichia coli STb enterotoxin. *Microbiology* **143**: 1783-1795.

Desselberger U (1997). Viral factors determining rotavirus pathogenicity. *Arch. Virol. Suppl.* **13**: 131-139.

Dickman K, Hempson SJ, Anderson J, Lippe S, Zhao LBR, Shaw R (2000). Rotavirus alters paracellular permeability and energy metabolism in Caco-2 cells. *Am. J. Physiol. Gastrointest. Liver. Physiol.* **279**: G757-G766.

Dong Y, Zeng CQ, Ball JM, Estes MK, Morris AP (1997). The rotavirus enterotoxin NSP4 mobilizes intracellular calcium in human intestinal cells by stimulating phospholipase C-mediated inositol 1,4,5-trisphosphate production. *Proc. Natl. Acad. Sci. USA* **94**: 3960-3965.

Estes MK, Graham DY, Smith EM, Gerba CP (1979). Rotavirus stability and inactivation. *J. Gen. Virol.* **43**: 403-409.

Estes MK, Morris AP (1999). A viral enterotoxin: A new mechanism of virus-induced pathogenesis. In: Paul P, Francis DH (eds). *Mechanisms in the Pathogenesis of Enteric Diseases 2*, pp 73-82. Plenum Press, New York..

Halaihel N, Lievin V, Ball JM, Estes MK, Alvarado F, Vasseur M (2000). Direct inhibitory effect of rotavirus NSP4 (114-135) peptide on the Na^+-D-glucose symporter of rabbit intestinal brush border membrane. *J. Virol.* **74**: 9464-9470.

Horie Y, Masamune O, Nakagomi O (1997a). Three major alleles of rotavirus NSP4 proteins identified by sequence analysis. *J. Gen. Virol.* **78**: 2341-2346.

Horie Y, Nakagomi O, Oseto M, Masamune O, Nagagomi O (1997b). Conserved structural features of nonstructural glycoprotein NSP4 between group A and group C rotaviruses. *Arch. Virol.* **142**: 1865-1872.

Horie Y, Nakagomi O, Koshimura Y, Nakagomi T, Suzuki Y, Oka T, Sasaki S, Matsuda Y, Watanabe S (1999). Diarrhea induction by rotavirus NSP4 in the homologous mouse model system. *Virology* **262**: 398-407.

Hoshino Y, Saif LJ, Kang SY, Sereno MM, Chen WK, Kapikian AZ (1995). Identification of group A rotavirus genes associated with virulence of a porcine rotavirus, host range restriction of a human rotavirus in the gnotobiotic piglet model. *Virology* **209**: 274-280.

Ishida SI, Feng N, Gilbert JM, Tang B, Greenberg HB (1997). Immune responses to individual rotavirus proteins following heterologous and homologous rotavirus infection in mice. *J. Infect. Dis.* **175**: 1317-1323.

Johansen K, Hinkula J, Espinoza F, Levi M, Zeng CQY, Vesikari T, Estes MK, Svensson L (1999). Humoral and cell-mediated immune responses in humans to the NSP4 enterotoxin of rotavirus. *J.Med. Virol.* **59**: 369-377.

Jourdan N, Brunet JP, Sapin C, Blais A, Cotte-Laffitte J, Forestier F, Quero AM, Trugnan G, Servin AL (1998). Rotavirus infection reduces sucrase-isomaltase expression in human intestinal epithelial cells by perturbing protein targeting and organization of microvillar cytoskeleton. *J. Virol.* **72**: 7228-7236.

Kang G, Ciarlet M, Zeng CQ-Y, Crawford SE, Cheng E, Jayaram H, Conner M, Nakata S, Chiba M, Estes MK (2002). Development and evaluation of an enzyme immunoassay for antibodies to the rotavirus enterotoxin non-structural protein 4. Submitted for publication.

Kirkwood CD, Bishop RF, Coulson BS (1998). Attachment and growth of human rotaviruses RV-3, S12/85 in Caco-2 cells depend on VP4. *J. Virol.* **72**: 9348-9352.

Lee C, Wang Y, Kao C, Zao CLC, Chen H (2000). NSP4 gene analysis of rotaviruses recovered from infected children with and without diarrhea. *J. Clin. Microbiol.* **38**: 4471-4477.

Lundgren O, Peregrin AT, Persson K, Kordasti S, Uhnoo I, Svensson L (2000). Role of the enteric nervous system in the fluid and electrolyte secretion of rotavirus diarrhea [see comments]. *Science* **287**: 491-495.

McAdaragh JP, Bergeland ME, Meyer RC, Johnshoy MW, Benefield DA, Hammer R (1980). Pathogenesis of rotaviral enteritis in gnotobiotic pigs: a microscopic study. *Am. J. Vet. Res.* **41**: 1572-1581.

Mebus CA (1976). Reovirus-like calf enteritis. *Am. J. Dig. Dis.* **21**: 592-599.

Mohan K, Dermody TS, Atreya C (2000). Mutations selected in rotavirus enterotoxin NSP4 depend on the context of its expression. Virology **275**: 125-132.

Morris AP, Scott JK, Ball JM, Zeng CQ, O'Neal WK, Estes MK (1999). NSP4 elicits age-dependent diarrhea and Ca^{++}.mediated I(-). influx into intestinal crypts of CF mice. *Am. J. Physiol.* **277**: G431-G444.

Murphy TV, Gargiullo PM, Massoudi MS, Nelson DB, Jumaan AO, Okoro CA, Zanardi LR, Setia S, Fair E, LeBaron CW, Wharton M, Livingood JR (2001). Intussusception among infants given an oral rotavirus vaccine. *N. Engl. J. Med.* **344**: 564-572.

Newton K, Meyer JC, Bellamy AR, Taylor JA (1997). Rotavirus nonstructural glycoprotein NSP4 alters plasma membrane permeability in mammalian cells. *J. Virol.* **71**: 9458-9465.

Obert G, Peiffer I, Servin AL (2000). Rotavirus-induced structural, and functional alterations in tight junctions of polarized intestinal Caco-2 cell monolayers. *J. Virol.***74**: 4645-4651.

Oka T, Nakagomi T, Nakagomi O (2001). A lack of consistent amino acid substitutions in NSP4 between rotaviruses derived from diarrheal and asymptomatically-infected kittens. *Microbiol. Immunol.* **45**: 173-177.

Osborne MP, Haddon SJ, Spencer AJ, Collins J, Starkey WG, Wallis TS, Clarke GJ, Worton KJ, Candy DC, Stephen J (1988). An electron microscopic investigation of time-related changes in the intestine of neonatal mice infected with murine rotavirus. *J. Pediatr. Gastroenterol. Nutr.* **7**:236-248.

Pager C, Alexander JJ, Steele AD (2000). South African G4P[6] asymptomatic and symptomatic neonatal rotavirus strains differ in their NSP4, VP8*, and VP7 genes. *J. Med. Virol.* **62**: 208-216.

Poruchynsky MS, Maass DR, Atkinson PH (1991). Calcium depletion blocks the maturation of rotavirus by altering the oligomerization of virus-encoded proteins in the ER. *J. Cell. Biol.* **114**: 651-661.

Sapin C, Colard O, Delmas O, Tessier C, Breton M, Enouf V, Chwetzoff S, Quanich J, Cohen J, Wolf C, Trugnan G (2002). Rafts promote assembly and atypical targeting of a nonenveloped virus in Caco-2 cells. *J. Virol.* **76**:4591-4602.

Sasaki S, Horie Y, Nakagomi T, Oseto M, Nakagomi O (2001). Group C rotavirus NSP4 induces diarrhea in neonatal mice. *Arch. Virol.* **146**:801-806.

Spencer AJ, Osborne MP, Haddon SJ, Collins J, Starkey WG, Cy DC, Stephen J (1990). X-ray microanalysis of rotavirus-infected mouse intestine: a new concept of diarrheal secretion. *J. Pediatr. Gastroenterol. Nutr.* **10**: 516-529.

Stephen J. Functional abnormalities in the intestine (1989). In: Farthing MJG (ed). *Viruses and the Gut.* pp 41-44. Swan Press, London.

Tafazoli F, Zeng CQ-Y, Estes MK, Magnusson KE, Svensson L (2001). The NSP4 enterotoxin of rotavirus induces paracellular leakage in polarized epithelial cells. *J.Virol.* **75**:1540-1546.

Taylor JA, O'Brien JA, Lord VJ, Meyer JC, Bellamy AR (1993). The RER-localized rotavirus intracellular receptor: A truncated purified soluble form is multivalent and binds virus particles. *Virology* **194**:807-814.

Tauscher GI, Desselberger U (1997). Viral determinants of rotavirus pathogenicity in pigs: production of reassortants by asynchronous coinfection. *J. Virol.* **71**: 853-857.

Taylor JA, O'Brien JA, Yeager M (1998). The cytoplasmic tail of NSP4, the endoplasmic reticulum-localized non-structural glycoprotein of rotavirus, contains distinct virus binding and coiled coil domains. *EMBO J.* **15**:4469-4476.

Theil KW, Bohl EH, Cross RF, Kohler EM, Agnes AG (1978). Pathogenesis of porcine rotaviral infection in experimentally inoculated gnotobiotic pigs. *Am. J. Vet. Res.* **39**: 213-220.

Tian P, Estes MK, Hu Y, Ball JM, Zeng C Q-Y, Schilling WP (1995). The rotavirus nonstructural glycoprotein NSP4 mobilizes Ca^{2+} from the endoplasmic reticulum. *J. Virol.* **69**: 5763-5772.

Tian P, Hu Y, Schilling WP, Lindsay DA, Eiden J, Estes MK (1994). The nonstructural glycoprotein of rotavirus affects intracellular calcium levels. *J. Virol.* **68**: 251-257.

Ward RL, Mason BB, Bernstein DI, Sander DS, Smith VE, Zandle GA, Rappaport RS (1997). Attenuation of a human rotavirus vaccine candidate did not correlate with mutations in the NSP4 protein gene. *J. Virol.* **71**:6267-6270.

Xu A, Bellamy AR, Taylor JA (2000). Immobilization of the early secretory pathway by a virus glycoprotein that binds to microtubules. *EMBO J.* **19**: 6465-6474.

Yu J, Langridge WH (2001). A plant-based multicomponent vaccine protects mice from enteric diseases. *Nat. Biotechnol.* **19**: 548-552.

Zeng CQ-Y, Zhang M, Conner ME, Estes MK (2002). Homotypic and heterotypic protection against rotavirus diarrhea by immunization with the rotavirus enterotoxin. In preparation.

Zhang M, Zeng CQ-Y, Morris AP, Estes MK (2000). A functional NSP4 enterotoxin peptide secreted from rotavirus-infected cells. *J. Virol.* **74**:11663-11670.

Zhang M, Zeng CQ-Y, Dong Y, Ball JM, Saif LJ, Morris AP, Estes MK (1998). Mutations in rotavirus nonstructural glycoprotein NSP4 are associated with altered virus virulence. *J. Virol.* **72**: 3666-3672.

Viral Gastroenteritis
U. Desselberger and J. Gray (editors)

II, 7. Interaction of the rotavirus nonstructural glycoprotein NSP4 with viral and cellular components

John A. Taylor and A. Richard Bellamy

School of Biological Sciences, University of Auckland, Private Bag 92019, Auckland, New Zealand

Introduction

The assembly and pathogenesis of rotavirus exhibit a number of unique features. Virion morphogenesis occurs by an unusual process in which double-layered subviral particles (DLPs), which are assembled in cytoplasmic inclusions, bud across the membrane of the endoplasmic reticulum (ER) and acquire a transient lipid envelope. The budding of these particles into the ER lumen is driven by the interaction between the particle and the non-structural glycoprotein NSP4 – which is embedded in the ER with a major domain exposed on the cytoplasmic side of the membrane. In this respect, rotavirus replication mirrors the assembly of some enveloped viruses that bud into intracellular membrane compartments following the interaction of the viral nucleocapsid with the cytoplasmic tail of a surface glycoprotein. During the final steps of rotavirus maturation, the lipid envelope and NSP4 are removed from the particle prior to release of the virion from the infected cell. Concurrent with, or shortly after budding, the outer capsid proteins VP4 and VP7 are assembled on the surface of the particle. Non-enveloped triple-layered virions are then released, either by direct cell lysis or, in polarised epithelial cells, after transport along a vesicular pathway from the ER to the plasma membrane (Jourdan *et al.*, 1997).

Recently, following discovery of the enterotoxic properties of NSP4 (Ball *et al.*, 1996), attention has focussed on the role of this non-structural rotavirus glycoprotein in the pathogenesis of viral infection. Since enterotoxic properties of NSP4 are discussed extensively elsewhere in this volume (Estes, Section II, Chapter 6), this brief review will address the structure and function of NSP4, the intracellular interactions which occur between NSP4 and viral and host components and the potential consequences of these interactions for the rotavirus-infected cell.

Structure and membrane topology of NSP4

NSP4 is encoded by genomic segment 10 of group A rotaviruses. The major features of the polypeptide (Both *et al.*, 1983) were first inferred from the nucleic acid sequence of

226

this segment for the SA11 strain of the virus. Subsequent analysis of NSP4 sequences from a large number of human and animal virus strains indicated the existence of distinct NSP4 genotypes. Nevertheless there is a high degree of sequence homology between strains, indicating that the major structural features of the protein are conserved amongst all group A rotaviruses (Kirkwood and Palombo, 1997; Horie *et al.*, 1997a; Ciarlet *et al.*, 2000). Analysis of group C rotavirus strains has indicated that the major structural features may also be conserved between groups - despite the lack of primary sequence homology overall (Horie *et al.*, 1997b).

The inferred amino acid sequence of NSP4 revealed a polypeptide of 175 residues with 3 distinct hydrophobic segments contained within the amino terminal half of the molecule (Fig. 1). The protein contains two N-linked high-mannose oligosaccharide residues (at Asn 8 and 18), indicating that the amino terminus of the molecule is oriented toward the lumen of the ER. Analysis of the membrane topology of full-length NSP4 and of truncated variants synthesised *in vitro* confirmed this distribution and revealed that the protein spans the membrane only once *via* a single transmembrane domain (H2), exhibiting a topology typical of type II membrane proteins. For NSP4, the carboxy-terminal 128 residues are presented on the cytoplasmic side of the membrane (Bergmann *et al.*, 1989; Fig. 1). Within this cytoplasmic region, there is a further short hydrophobic segment which lies between residues 67 and 80 (H3). This region of the protein is thought to be associated with the cytoplasmic face of the ER membrane, leaving the remainder of the C-terminal residues projecting into the cytoplasm.

More detailed analysis of the NSP4 sequence has indicated the presence of motifs that are characteristic of secondary and tertiary structural features. A repeating heptad pattern between residues 95 and 136 is predictive of the α-helical coiled coil structures that are commonly found in the fusion glycoproteins of animal viruses (Skehel and Wiley, 1998). This prediction has been verified following spectroscopic analysis of a soluble polypeptide fragment which comprises residues 86-175 (Taylor *et al.*, 1996). More recently, determination of the crystal structure of a synthetic peptide comprising residues 95-137 (2.3Å resolution) has confirmed the presence of a parallel tetrameric coiled coil that is likely to mediate the oligomerisation of the full length glycoprotein (Bowman *et al.*, 2000). Crystallographic analysis also yielded the surprising discovery that a single calcium ion is bound in the hydrophilic core of the coiled coil domain. A second region of the protein (residues 55 – 71) may also adopt a helical secondary structure. This motif is notable for the extreme amphipathic character of the predicted α-helix, a feature of many proteins and peptides that exhibit membrane destabilising activity. This notion is supported by the observation that the membrane destabilising activity of NSP4 maps to this region of the protein (Browne *et al.*, 2000).

The lack of amino acid sequence homology between NSP4 and any other protein sequence in current databases, even over relatively short stretches of the protein, means that few clues to the intracellular functions of the molecule are available. This difficulty is compounded by the lack of a reverse genetic system for rotaviruses through which the function of variant viral gene products could be evaluated. Insight into the function of NSP4 during virus infection therefore has come mainly from

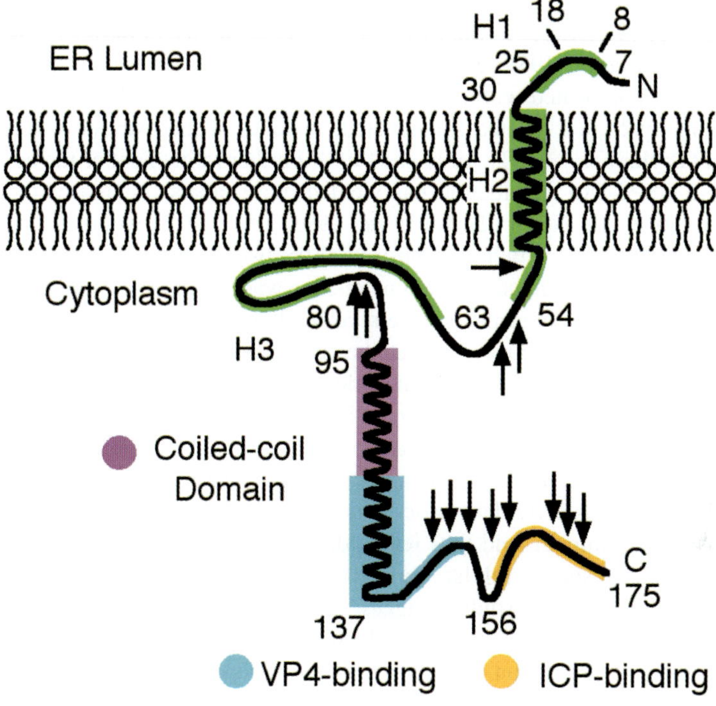

Fig. 1. Schematic diagram of NSP4 showing the topology defined by Bergmann *et al.* (1989). The hydrophobic domains H1, H2 and H3 are outlined in green. Within H1, Asn 8 and 18 are glycosylated with high mannose carbohydrate. H2 spans the ER membrane and H3 interacts with the cytoplasmic face of the membrane. Structural and functional regions of the cytoplasmic domain are colour coded (note the DLP binding domain is here referred to as ICP binding), and arrows indicate sites sensitive to added protease. Note that only a single subunit is represented, although the protein is thought to exist as a homotetramer. Diagram reproduced with permission from Bowman *et al.* (2000). The microtube binding domain (Xu, *et al.*, 2000) localises to the C-terminal 50 amino acids.

in vitro studies using purified peptides and from expression of mutant forms of NSP4 in mammalian and insect cell systems.

Interaction of NSP4 with other rotavirus proteins

Studies into the mechanism of rotavirus replication revealed that purified DLPs (isolated by CsCl gradient centrifugation from infected MA104 cells) bound to crude membranes prepared from rotavirus-infected but not mock-infected cells. Expression of individual rotavirus genes in both mammalian and insect cells, using recombinant vaccinia or baculoviruses, demonstrated that NSP4 was the viral component that mediated the binding of particles to the membrane, thus acting as an intracellular receptor for the DLP (Au *et al.,* 1989; Meyer *et al.,* 1989). Confirmation that a binding site for the particle was located within the cytoplasmic domain of NSP4 was provided by mutational analyses that deleted (or mutated) residues at the extreme C-terminus of the protein (Taylor *et al.,* 1992; Au *et al.,* 1993). Further refinement of the assays used to study the NSP4-DLP interaction has yielded more precise details of the binding domain present within NSP4 (Taylor *et al.,* 1993, 1996). This work has identified a minimal binding domain which is comprised of the twenty carboxy-terminal residues. The integrity of the C-terminal methionine residue was found to be crucial for binding of the particle. Further experiments have revealed that the DLP binding site on NSP4 lies within a protease-sensitive 'tail' at the terminus of a protease-resistant 'stalk' formed by the remainder of the cytoplasmic region (O'Brien *et al.,* 2000). Further details of the interaction of the receptor with the DLP await the relevant structural details of both NSP4 and the DLP at atomic resolution.

While the binding of DLPs to membranes containing NSP4 occurs with high affinity ($K_d \sim 10^{-10}$ M), attempts in our laboratory to decorate particles with soluble forms of the NSP4 cytoplasmic domain, and thus reveal details of the binding site on the particle surface, have been unsuccessful, largely due to difficulties encountered in the isolation of stable complexes. These studies suggest that soluble NSP4, even in a tetrameric form, binds only weakly to the particle and emphasise the likely importance of a cooperative, multivalent interaction between the two components in driving the transfer of particles across the ER membrane *in vivo*. Soaking crystals of the DLP in the relevant peptide and difference analysis may prove to be a more successful approach to elucidating details of the interaction.

In addition to binding to DLPs, NSP4 can also associate with the outer capsid proteins VP4 and VP7. A complex of all three viral proteins can be isolated from lysates of infected MA104 cells using sucrose density gradient centrifugation (Maass and Atkinson, 1990; Poruchynsky and Atkinson, 1991). Analysis of the NSP4-VP4 interaction using protein synthesised *in vitro* has also suggested that interactions occur between these proteins, and a binding site for VP4 has been identified within residues 128-154 of NSP4 (Au *et al.,* 1993).

While these studies point to the existence of multiple interactions between NSP4 and other viral components within rotavirus-infected cells, several questions remain to be answered before the precise role of this glycoprotein in virus assembly is understood. For example, it is not clear whether the role of NSP4 is solely to direct the budding of DLPs across the ER membrane, or whether the interaction between these components may 'prime' the particle for assembly of the outer capsid. Evidence in support of this latter hypothesis is provided by the observation that DLPs incubated with soluble NSP4 aggregate and change their appearance when analysed by negative staining in the electron microscope (Taylor *et al.,* 1993). One possible interpretation of this finding is that particles may undergo a temporary conformational change induced by NSP4 which could facilitate the correct assembly of outer capsid components. However precise knowledge of the timing, intracellular location and mechanism of removal of the transient envelope and the dissociation of NSP4 from virus particles prior to release of mature virions from infected cells is lacking. Reconstitution of purified virus components within *in vitro* systems probably now offers the best prospect of addressing these and other related issues.

Interaction of NSP4 with host cellular components

A central issue in contemporary virology is the need to understand better how viral proteins interact with components of the host cell during infection on the one hand, and how these interactions impact on virus replication and pathogenesis on the other. Viral glycoproteins have been studied intensively, and these studies have provided many insights into the basic cellular processes that are involved in the targeting and retention of membrane proteins in mammalian cells. The remainder of this review therefore focuses on research that has addressed the interaction of NSP4 with cellular components and how the endogenous expression of NSP4 affects the biology of the host cell.

Interactions of NSP4 with cellular membranes

Immuno-electron microscopy localises NSP4 to the ER membrane of the infected cell (Petrie *et al.*, 1982). Removal of the oligosaccharide moiety from the protein by endoglycosidase H indicates that the carbohydrate is not modified by medial Golgi enzymes and that the distribution of NSP4 is restricted to a pre-Golgi compartment (Kabcenell and Atkinson, 1985). More recently, the intracellular distribution of NSP4 has been examined using fluorescent confocal microscopy. Gonzalez *et al.* (2000) reported that NSP4 was organized mostly as ring-like or semicircular structures found in close association with viroplasms in rotavirus-infected MA104 cells. Results from our own laboratory have indicated that NSP4 exhibits a punctate staining pattern in transfected Cos-7 cells, suggesting that the protein is found in discrete membranous structures which are concentrated in the perinuclear region, but extend outward along radial trajectories toward the cell periphery (Fig. 2; Xu *et al.*, 2000).

Several studies have addressed the ability of NSP4, or peptides derived from NSP4, to interact with lipid membranes. Tian *et al.* (1996) showed that the enterotoxic peptide ($NSP4_{114-135}$) exhibits membrane destabilising activity (MDA) that was selective for anionic phospholipids. A similar activity was reported for NSP4 and certain truncated derivatives following their expression in a bacterial system (Browne *et al.*, 2000). However in the latter study, the membrane destabilising region mapped to a different region of the protein that contained the third hydrophobic domain (H3) and an adjacent sequence rich in basic residues. Deletion of this region, or expression of only those sequences containing the enterotoxic peptide and oligomerisation domain, did not affect membrane integrity, but enhanced the potency of membrane destabilisation when fused to H3.

While the above studies demonstrate the potential of NSP4 to disrupt the structure of lipid bilayers though the action of one or more specific regions of the protein, the

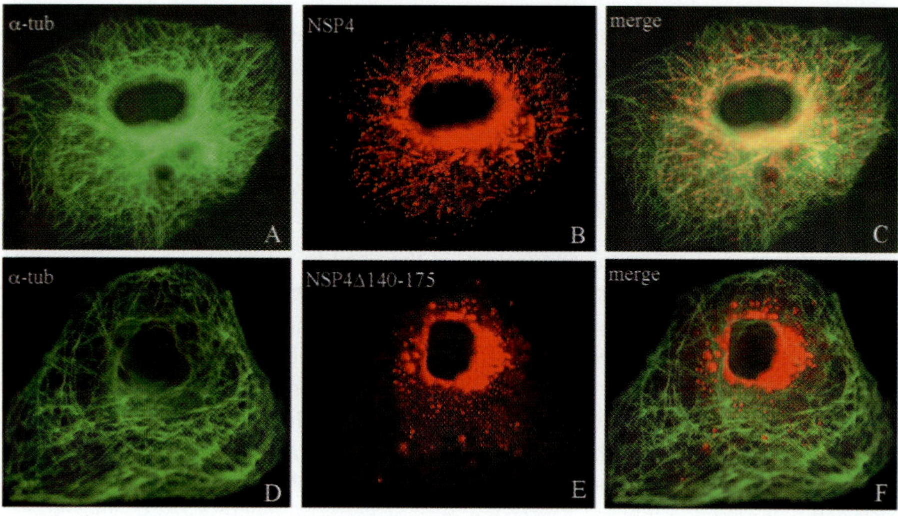

Fig. 2. Co-localisation of NSP4 with microtubules in Cos-7 cells. Cells were transfected with a construct encoding either NSP4 or a deletion mutant (NSP4 Δ140-175) that lacks the tubulin-binding domain. After 48 hrs, cells were fixed, permeabilised and labelled with antibody against α-tubulin and NSP4, and analysed by confocal fluorescent microscopy. Note the co-localisation of NSP4 (red) with the microtubule network (green). Reproduced with permission from Xu *et al.* (2000).

relevance of this activity to the virus replication cycle and/or viral pathogenesis is less obvious. The membrane destabilisation activity by NSP4 has been suggested to be a possible mechanism by which the transient envelope that is acquired by particles that bud into the ER might be removed (Tian *et al.*, 1996). The elevation of intracellular calcium ion concentration, which is observed in *Spodoptera frugiperda* (Sf9) insect cells infected with a recombinant baculovirus that expresses NSP4, could also result from destabilisation of the ER membrane, leading to a leakage of stored Ca^{2+} (Tian *et al.*, 1994;1995). This activity may also account for the cytopathic effects observed following expression of NSP4 in mammalian cells (Newton *et al.*, 1997). However interpretation of results from these types of studies should be made with caution, given the possibility that synthetic peptides or fragments of NSP4 expressed in heterologous systems may not behave in a manner that reflects the activity of the intact protein in the rotavirus-infected cell.

Recently, Huang *et al.* (2001) have used circular dichroism spectroscopy to examine the alteration in protein secondary structure that accompanies the interaction of NSP4 with model membranes. These studies showed that full-length NSP4 undergoes a modest increase in α-helical content in the presence of small unilamellar vesicles containing anionic phospholipids. A region of NSP4 (amino acids $_{114-135}$) also preferentially adopts an α-helical conformation in the presence of negatively charged membranes, this transition being further enhanced in the presence of cholesterol. This observation led the authors to speculate that a cytoplasmic fragment of NSP4 containing the enterotoxic peptide may partition into lipid microdomains within the cell. Given the observation that a 63 amino acid peptide fragment (presumed to be a cleavage product derived from the cytoplasmic domain of NSP4) is secreted from both rotavirus-infected MA104 cells and insect cells infected with a recombinant baculovirus (Zhang *et al.*, 2000), different forms of NSP4 may exist in rotavirus-infected cells; each of these forms might exhibit different functions and different distributions within the infected cell.

Interaction of NSP4 with molecular chaperones in the ER

Correct folding and assembly of membrane glycoproteins in the ER is essential to prevent protein aggregation and to facilitate the maturation of viruses that assemble in this compartment. Several molecular chaperones that mediate protein folding in the ER have been identified and their association with rotavirus proteins, including NSP4, has been examined. Mirazimi *et al.* (1998) demonstrated that NSP4 interacts with calnexin, a membrane bound and lectin-like chaperone. This interaction is transient and is critically dependent on the correct glycosylation of NSP4. Furthermore enzymatic removal of glucose residues on the N-linked glycan units was necessary for disruption of the NSP4-calnexin complex. In contrast, protein disulphide isomerase (PDI) binds selectively to unglycosylated NSP4 (Mirazimi and Svensson, 1998). These results may reflect sequential interactions between NSP4 and different chaperones in the ER during protein folding. In contrast to the calnexin and PDI findings, NSP4 does not interact with the chaperones BiP and endoplasmin, both of which were found to associate with the outer capsid proteins VP4 and VP7 (Xu *et al.*, 1998).

Interaction of NSP4 with microtubules

NSP4 lacks a recognisable KDEL or K[X]KXX retention motif at its extreme C-terminus. These sequences have been shown to mediate the retention of some viral glycoproteins in the RER compartment (Jackson *et al.*, 1990). While an N-terminal tri-peptide sequence has been implicated as constituting an ER retention signal in VP7 (Maass and Atkinson, 1994), details of how NSP4 is retained in this compartment are unknown. To investigate whether retention of NSP4 in the ER could be mediated through interaction with a cellular component, we have recently examined whether NSP4 might be bound by a cellular protein. A variety of biochemical approaches demonstrated that NSP4 bound a cellular protein with an apparent M.W. of 50 kDa; this protein was subsequently identified as tubulin (Xu *et al.*, 2000). The tubulin binding domain was localised to the C-terminal 50 amino acid residues of NSP4. Fluorescent confocal microscopy confirmed that NSP4 binds tubulin and that the viral glycoprotein does indeed co-localise with the microtubule filaments of the transfected cell (Fig. 2).

Whether microtubule binding represents a potential mechanism for the retention of NSP4 in the ER is unclear, because truncated forms which lack an intact microtubule binding domain localise to juxtanuclear aggregates and not, as might be expected, to the plasma membrane. However, a recent report suggests that larger deletions of the cytoplasmic domain may be sufficient to target NSP4 to the plasma membrane (Mirazimi and Svensson, 2000). The C-terminal ER retention motif present in adenovirus E3 19K has also been demonstrated to bind microtubules, suggesting that retention in the ER and microtubule binding may be a feature of proteins of this type (Dahllof *et al.*, 1991).

A further outcome of microtubule binding by NSP4 in transfected Cos-7 cells is disruption of ER-to-Golgi vesicular trafficking. NSP4 expression causes the retention of the spike glycoprotein of Vesicular Stomatitis Virus in a pre-Golgi membrane compartment as well as causing a redistribution of the endogenous protein β-COP, a marker for the intermediate compartment (Xu *et al.*, 2000). These results suggest that microtubule binding by NSP4 may contribute to the pathogenic effects of rotavirus infection by inhibiting the cellular trafficking of plasma membrane components. Indeed the trafficking of sucrose isomaltase to the plasma membrane is known to be inhibited in rotavirus-infected Caco-2 cells (Jourdan *et al.*, 1998), although the block in trafficking of this protein was thought to occur at a post-Golgi step. Further analysis of membrane protein trafficking in rotavirus-infected cells will be required to confirm whether the effect of NSP4 on microtubule-mediated vesicle transport observed in transfected cells represents an authentic feature of the virus-infected cell. The binding of NSP4 to microtubules is intriguing in the light of evidence that VP4 also is found in association with microtubules in both transfected and rotavirus-infected cells (Nejmeddine *et al.*, 2000). Given that these two rotavirus proteins are known to interact, it is tempting to speculate that the microtubule cytoskeleton may play a role in the localisation of viral components during assembly.

Microtubules have also been implicated in the release of a peptide cleavage product of NSP4 (NSP4$_{112-175}$), during virus infection (Zhang *et al.*, 2000). Although this peptide contains the microtubule binding domain, both the mechanistic details of its secretion and whether direct binding to microtubules is necessary for intracellular transport of NSP4$_{112-175}$ remain unclear. VP22, a microtubule binding protein expressed by herpes simplex virus, is also released from the infected cells *via* a non-classical secretary pathway (Elliott and O'Hare, 1997).

Summary and perspectives

The studies we have described above provide some preliminary insight not only into the potential pathogenic roles played by NSP4 in rotavirus replication, but also into the effect(s) of NSP4 expression on animal cells and the protein-protein and protein-lipid interactions that involve this protein. There are several important aspects which require further research. High resolution structural analysis of the cytoplasmic region is required to reveal molecular details of its interaction with VP4, the DLP and microtubules. The identification of a Ca^{2+} binding site in the hydrophilic core of the oligomerisation domain has prompted speculation that NSP4 may function as a transmembrane Ca^{2+} channel. Biochemical and electrophysiological approaches are now required – as have been applied successfully to other virus-encoded ion channels. The mechanism by which the cleaved cytoplasmic fragment (NSP4$_{112-175}$) is secreted also requires clarification because the release of an enterotoxic peptide from an ER-anchored type II glycoprotein suggests that novel mechanisms may be involved. It will also be of interest to examine the fate of any NSP4 cleavage products that may be retained in the cell. Finally, there is the need for caution in attributing certain functions to NSP4 on the basis of cell-based expression work, the results of which may not reflect the behaviour of the protein in rotavirus-infected cells (Mohan *et al.*, 2000). For these and other reasons, there remains much interesting additional work to be carried out on this unusual non-structural viral glycoprotein.

Acknowledgements

We acknowledge the helpful comments of Dr Judith O'Brien. Some of the work reported upon here was supported by grants to JT and ARB from the Marsden Fund and the Health Research Council of New Zealand.

References

Au KS, Chan WK, Burns JW, Estes MK. (1989). Receptor activity of rotavirus nonstructural glycoprotein NS28. *J. Virol.* **63:** 4553-4562.

Au KS, Mattion NM, Estes MK. (1993). A subviral particle binding domain on the rotavirus nonstructural glycoprotein NS28. *Virology* **194:** 665-673.

Ball JM, Tian P, Zeng CQ, Morris AP, Estes MK (1996). Age-dependent diarrhoea induced by a rotaviral nonstructural glycoprotein. *Science* **272:** 101-104.

Bergmann CC, Maass D, Poruchynsky MS, Atkinson PH, Bellamy AR (1989). Topology of the non-structural rotavirus receptor glycoprotein NS28 in the rough endoplasmic reticulum. *EMBO J.*, **8:** 1695-1703.

Both GW, Siegman LJ, Bellamy AR, Atkinson PH (1983). Coding assignment and nucleotide sequence of simian rotavirus SA11 gene segment 10: location of glycosylation sites suggests that the signal peptide is not cleaved. *J. Virol.*, **48:** 335-339.

Bowman GD, Nodelman IM, Levy O, Lin SL, Tian P, Zamb TJ, Udem SA, Venkataraghavan B, Schutt CE. (2000). Crystal structure of the oligomerization domain of NSP4 from rotavirus reveals a core metal-binding site. *J. Mol. Biol.* **304:** 861-871.

Browne EP, Bellamy AR, Taylor JA (2000). Membrane-destabilizing activity of rotavirus NSP4 is mediated by a membrane-proximal amphipathic domain. *J. Gen. Virol.* **81:** 1955-1959.

Ciarlet M, Liprandi F, Conner ME, Estes MK (2000) Species specificity and interspecies relatedness of NSP4 genetic groups by comparative NSP4 sequence analyses of animal rotaviruses. *Arch. Virol.* **145:** 371-383.

Dahllof B, Wallin M, Kvist S (1991). The endoplasmic reticulum retention signal of the E3/19K protein of adenovirus-2 is microtubule binding. *J. Biol. Chem.* **266:**

234

1804-1808.

Elliott G, O'Hare P (1997). Intercellular trafficking and protein delivery by a herpesvirus structural protein. *Cell* **88:** 223-233.

Gonzalez RA, Espinosa R, Romero P, López S, Arias CF (2000). Relative localization of viroplasmic and endoplasmic reticulum-resident rotavirus proteins in infected cells. *Arch Virol.* **145:** 1963-1973.

Horie Y, Masamune O, Nakagomi O. (1997a). Three major alleles of rotavirus NSP4 proteins identified by sequence analysis. *J Gen Virol.* **78:** 2341-2346.

Horie Y, Nakagomi T, Oseto M, Masamune O, Nakagomi O (1997b). Conserved structural features of nonstructural glycoprotein NSP4 between group A and group C rotaviruses. *Arch.Virol.* **142:** 1865-1872.

Huang H, Schroeder F, Zeng CQ, Estes MK, Schoer JK, Ball, JM (2001) Membrane interactions of a novel viral enterotoxin: rotavirus nonstructural glycoprotein NSP4. *Biochemistry* **40:** 4169-4180.

Jackson MR, Nilsson T, and Peterson PA (1990). Identification of a consensus motif for retention of transmembrane proteins in the endoplasmic reticulum. *EMBO J.* **10:** 3153-3162.

Jourdan N, Brunet JP, Sapin C, Blais A, Cotte-Laffitte J, Forestier F, Quero AM, Trugnan G, Servin AL (1998). Rotavirus infection reduces sucrase-isomaltase expression in human intestinal epithelial cells by perturbing protein targeting and organization of the cytoskeleton. *J. Virol.,* **72:** 7228-7236.

Jourdan N, Maurice M., Delautier D, Quero AM, Servin AL, Trugnan G (1997). Rotavirus is released from the apical surface of cultured human intestinal cells through nonconventional vesicular transport that bypasses the Golgi apparatus. *J. Virol.,* **71:** 8268-8278.

Kabcenell AK, Atkinson PH (1985). Processing of the rough endoplasmic reticulum membrane glycoproteins of rotavirus SA11. *J. Cell Biol.* **101:** 1270-1280.

Kirkwood CD, Palombo EA (1997) Genetic characterization of the rotavirus nonstructural protein, NSP4. *Virology* **236:** 258-265.

Maass DR, Atkinson PH (1994). Retention by the endoplasmic reticulum of rotavirus VP7 is controlled by three adjacent amino-terminal residues. *J. Virol.* **68:** 366-378.

Maass DR, Atkinson PH. (1990) Rotavirus proteins VP7, NS28, and VP4 form oligomeric structures. *J. Virol.* **64:** 2632-2641.

Meyer JC, Bergmann CC, Bellamy AR (1989). Interaction of rotavirus cores with the nonstructural glycoprotein NS28. *Virology* **171:** 98-107.

Mirazimi A, Nilsson M, Svensson L (1998). The molecular chaperone calnexin interacts with the NSP4 enterotoxin of rotavirus *in vivo* and *in vitro*. *J. Virol.* **72:** 8705-8709.

Mirazimi A, Svensson L. (1998). Carbohydrates facilitate correct disulfide bond formation and folding of rotavirus VP7. *J. Virol.* **72:** 3887-3892.

Mirazimi A, Svensson L (2000). A C-terminal region of the NSP4 enterotoxin of rotavirus is involved in ER retention. *Poster Presentation, 7th International Symposium on Double Stranded RNA Viruses,* Aruba, 2nd-7th December 2000.

Mohan KV, Dermody TS, Atreya CD (2000). Mutations selected in rotavirus enterotoxin NSP4 depend on the context of its expression. *Virology* **275:** 125-32.

Nejmeddine M, Trugnan G, Sapin C, Kohli E, Svensson L, López S, Cohen J (2000). Rotavirus spike protein VP4 is present at the plasma membrane and is associated with microtubules in infected cells. *J. Virol.* **74:** 3313-3320.

Newton K, Meyer JC, Bellamy AR, Taylor JA (1997). Rotavirus nonstructural glycoprotein NSP4 alters plasma membrane permeability in mammalian cells. *J. Virol.* **71:** 9458-9465.

O'Brien JA, Taylor JA, Bellamy AR (2000). Probing the structure of rotavirus NSP4: a short sequence at the extreme C terminus mediates binding to the inner capsid particle. *J. Virol.* **74:** 5388-5394.

Petrie BL, Graham DY, Hanssen H, Estes MK (1982). Localization of rotavirus antigens in infected cells by ultrastructural immunocytochemistry. *J. Gen. Virol.* **63:** 457-467.

Poruchynsky MS Atkinson PH (1991). Rotavirus protein rearrangements in purified membrane-enveloped intermediate particles. *J. Virol.* **65:** 4720-4727.

Skehel JJ, Wiley DC (1998). Coiled coils in both intracellular vesicle and viral membrane fusion. *Cell* **95:** 871-874.

Taylor JA, O'Brien JA, Yeager M (1996). The cytoplasmic tail of NSP4, the endoplasmic reticulum-localized non-structural glycoprotein of rotavirus, contains distinct virus binding and coiled coil domains. *EMBO J.* **15:** 4469-4476.

Taylor JA, O'Brien JA, Lord VJ, Meyer JC, Bellamy AR (1993). The RER-localised rotavirus intracellular receptor: a truncated purified soluble form is multivalent and binds virus. *Virology* **194:** 807-814.

Taylor JA, Meyer JC, Legge MA, O'Brien JA, Street JE, Lord VJ, Bergmann CC, Bellamy AR (1992). Transient expression and mutational analysis of the rotavirus intracellular receptor: the C-terminal methionine residue is essential for ligand binding. *J. Virol.* **66:** 3566-3572.

Tian P, Yanfang H, Schilling WP, Lindsay DH, Estes MK (1994). Rotavirus NSP4 affects intracellular calcium levels. *J. Virol.,* **68:** 251-257.

Tian P, Estes MK. Hu Y, Ball JM, Zeng CQ, Schilling WP (1995). The rotavirus nonstructural glycoprotein NSP4 mobilizes Ca^{2+} from the endoplasmic reticulum. *J. Virol.* **69:** 5763-5772.

Tian P, Ball JM, Zeng CQ, Estes MK (1996). The rotavirus nonstructural glycoprotein NSP4 possesses membrane destabilization activity. *J. Virol.* **70:** 6973-6981.

Xu A, Bellamy AR, Taylor JA (2000). Immobilization of the early secretory pathway by a virus glycoprotein that binds to microtubules. *EMBO J.* **19:** 6465-6474.

Xu A, Bellamy AR, Taylor JA (1998). BiP (GRP78) and Endoplasmin (GRP94) are induced following rotavirus infection and bind transiently to an endoplasmic reticulum-localized virion component. *J. Virol.* **72:** 9865-9872.

Zhang M, Zeng CQ, Morris AP, Estes MK (2000). A functional NSP4 enterotoxin peptide secreted from rotavirus-infected cells. *J Virol.* **74:** 11663-11670.

II, 8. Effects of rotavirus infection on the structure and functions of intestinal cells

Alain L. Servin

Institut National de la Santé et de la Recherche Médicale (INSERM), Unité 510, Pathogènes et Fonctions des Cellules Epithéliales Polarisées, Faculté de Pharmacie Paris XI, F-92296 Châtenay-Malabry, France

Introduction

Viral gastroenteritis is an important cause of pediatric morbidity and mortality throughout the world, especially among infants and young children in developing countries. Research has recently focused on obtaining a better understanding of the mechanisms of pathogenicity of rotavirus.

Experimental studies of rotavirus pathogenicity have often been carried out using the simian kidney epithelial MA104 cells which are routinely used to produce progeny virus. However, the target of the rotavirus is known to be the mature enterocyte of the small intestine. The fact that information has been obtained using unpolarized kidney MA104 cells, which do not express specific intestinal functions, has limited our understanding of the intestinal pathogenicity of rotavirus. New investigations into the pathophysiological mechanisms of rotavirus infection are now being carried out using appropriate *in vitro* polarized and fully differentiated human intestinal cell models. The pioneering studies by Superti *et al.* (1991) and Swensson *et al.* (1996) have revealed that the cultured human intestinal HT-29 and Caco-2 cells are permissive towards rotavirus infection *in vitro*. During the past decade, these cultured human polarized intestinal cells have been extensively used in the study of human specific intestinal cell functions. Moreover, these cell lines are now widely used in investigations of bacterial pathogenicity.

The polarity of the epithelial cells lining the intestinal epithelium means that the plasma membrane has three domains; the apical surface facing the lumen, the lateral surface facing the adjacent cells and the basal surface interacting with the underlying connective tissue (for a review see Louvard *et al.*, 1993). The cytoskeleton is essential for establishing and maintaining the structural organization of polarized epithelial intestinal cells. Several cytoskeleton-associated proteins play a pivotal role in the architectural organization of the polarized cells (Fig. 1). It is well documented that several gastrointestinal epithelial functions are influenced by the establishment and the maintenance of the polarized organization of the epithelial intestinal cells. Moreover, the junctional complex regulates paracellular solute flow and lateral diffusion between the apical and basolateral plasma membrane domains (for a review see Denker and

Nigam, 1998) (Fig. 2). At the molecular level, the "gate" function results from the organization and activity of specific proteins localized in the basolateral domain which function as cell-to-cell adhesion and/or closing molecules. The "fence" function allows the separation between the apical and the basolateral domains, thereby segregating cell surface proteins and lipids in each domain.

The cultured polarized intestinal cell lines express common and/or specific characteristics of several of the cell phenotypes that line the intestinal epithelium (for a review see Zweibaum *et al.*, 1991). T84 cells are known to exhibit crypt cell morphology and form a tight monolayer with a high level of transepithelial resistance (TER), especially when grown on collagen-coated permeable supports. Depending on the culture conditions, the parental human intestinal epithelial cell lines HT-29 and Caco-2, established from a human colonic adenocarcinoma by Fogh, have been shown to undergo morphological and functional differentiation *in vitro* which is characteristic of the mature enterocytes of the small intestine. The differentiation of Caco-2 cells which form tight monolayers occurs spontaneously in culture whereas the parental HT-29 cell line remains undifferentiated when cultured in glucose-containing medium. Different HT-29 cell subpopulations, such as the absorptive HT-29 glc$^-$ subpopulation and muco-secreting HT29-MTX and HT29-FU subpopulations, have been obtained in differents culture conditions. Moreover, clones of the HT-29 and Caco-2 cell lines have been established, such as the absorptive Caco-2$_{BB2}$, Caco-2/TC7 cells and HT29-19A cells, and the mucin secreting HT29-Cl.16E cells. These cell lines are now being used as models for studying enterovirus pathogenicity, and in particular for investigating the impact of viral infection on the structure and specific functions of the absorptive and mucin secreting intestinal cells. During the last five years, a set of studies using polarized and fully-differentiated cell models have identified several pivotal steps in rotavirus pathogenicity and revealed the molecular mechanisms of action of several rotavirus proteins acting as virulence factors.

Attachment, entry and trafficking of rotavirus in polarized intestinal cells

The attachment of rotavirus to cells followed by cell-entry is a complex and multistep process that involves the recognition of cell receptor(s) and co-receptor(s), protease cleavage and cell-entry (Mendez *et al.*, 1999; López and Arias, Section II, Chapter 3 of this book). Like for many other enterovirulent microorganisms, it may be necessary for rotavirus to bind to a membrane-bound receptor(s) before the virus can enter the cells. Rotavirus VP4, and/or its VP5* and VP8* cleavage products, mediate virus attachment in undifferentiated, non-intestinal MA104 cells and in polarized, fully-differentiated Caco-2 cells (Kirkwood *et al.*, 1998). VP7 protein may also play a role in rotavirus cell entry but its function remains unclear. Two mechanisms of cell entry have been proposed: direct penetration by fusion with the cell membrane (Gilbert and Greenberg, 1997; 1998) and endocytosis-mediated entry. Experiments conducted with MA104 cells have gained evidence for the role of Ca^{2+} in endocytosis-mediated rotavirus cell entry (for a review see Ruiz *et al.*, 2000).

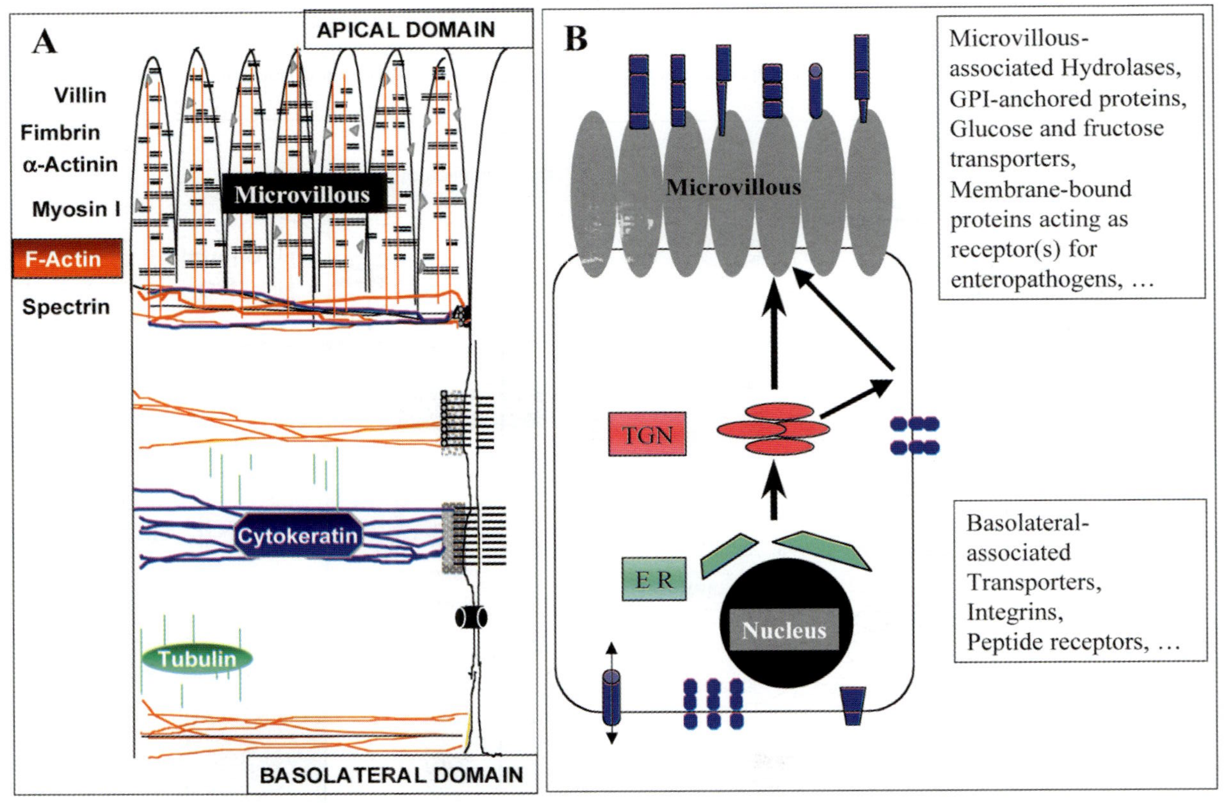

Fig. 1. Schematic ultrastructural and functional organization of polarized, fully-differentiated intestinal cells. In right, arrows indicate the two pathways of endogenous plasma membrane proteins cell-sorting in polarized epithelial cells. A, structural organisation of cell; B, functional organisation of cell.

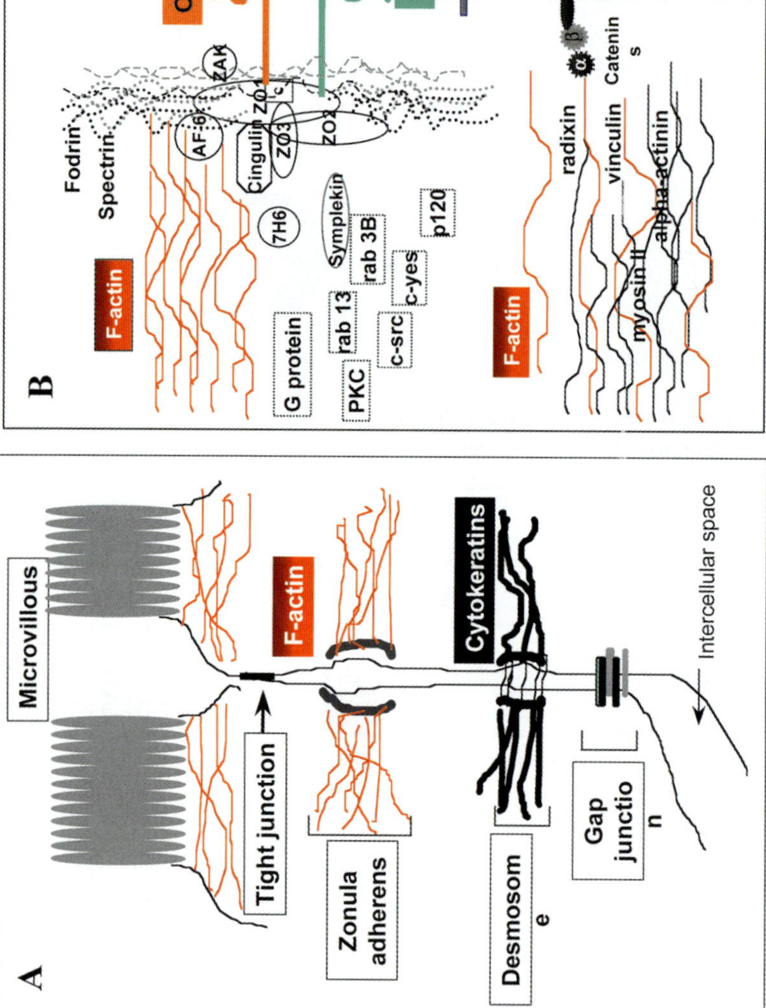

Fig. 2. Molecular organization of the junctional domain in polarized, fully-differentiated intestinal cells. The tight junction (TJ) which contains occludin and proteins of the ZO and claudin families, functions as a permeability barrier and separates apical from basolateral membrane domains. TJ-associated proteins and numerous signaling molecules identified in or near the TJ provide anchoring sites for cytoskeleton proteins and/or regulate the TJ assembly/disassembly events. The zonula adherens contains E-cadherins and functions as a cell anchor site. The desmosomes provide anchoring sites for cytoskeleton proteins.

The critical Ca^{2+} concentration $[Ca^{2+}]$ governing the triple-layered particle (TLP)-to-double-layered particle (DLP) transformation has been shown to be strain-dependent. The proposed model of endocytosis-mediated rotavirus cell entry hypothesized that equilibration of endosomal $[Ca^{2+}]$ to intracellular Ca^{2+} $[Ca^{2+}]_i$ favored full virion uncoating and that the free outer capsid proteins were able to permeabilize and then to lyse the endosomal membranes. It has been also proposed that VP5*, by selectively permeabilizing the early endosomal membrane, could accelerate the lowering of endosomal $[Ca^{2+}]$ (Dowling et al., 2000). An alternative explanation has recently been proposed by Martin et al. (2001). After infectious rotavirus enters into early endosomal vesicles, the progressive decrease of endosomal $[Ca^{2+}]$ favors progressive virion decapsidation and subsequent permeabilization of endosomal membrane. This permeabilization would in turn accelerate a three-step process: lowering of endosomal $[Ca^{2+}]$, uncoating, and permeabilization, which would be reiterated until the transcriptional active double-layered particles (DLPs) were delivered into the cytoplasm of infected cells. The recent, novel observation that, under physiological conditions, the Ca^{2+} dissociation constant values governing the TLP to DLP transformation of both RF and SA11 rotaviruses were well above the intracytoplasmic Ca^{2+} concentration of various cells, may explain why TLP uncoating takes place within vesicles (possibly endosomes) during the entry process.

Competition studies currently focus on identifying the brush border-associated molecule(s) that act as receptor(s) and/or co-receptor(s) for rotavirus in intestinal cells. Lipids rafts and cholesterol-glycosphingolipid-rich microdomains have recently attracted much attention (for a review see Brown and London, 1998). They are associated with cell functions that include potocytosis, non-clathrin-dependent transcytosis, endocytosis, calcium regulation, and signal transduction. Interestingly, it has been reported that caveolae could serve as concentration devices which permit the internalization of bacterial toxins (Orlandi and Fishman, 1998; Coconnier et al., 2001) and bacteria (Shin et al., 2000; Guignot et al., 2001). Caveolae have also been found to be involved in the uptake of respiratory syncytial virus antigen by dendritic cells (Werling et al., 1997). Recent reports have documented the localization of animal and human rotavirus receptor(s) and co-receptor(s), many of which are probably immersed in membrane functional lipid microdomains and/or rafts (Guo et al., 1999; Guerrero et al., 2000; López et al., 2000; Delorme et al., 2001). Most of these data have been obtained using MA104 cells which are non-intestinal and undifferentiated. However, the data can be extrapolated to polarized differentiated intestinal cells, since these lipid microdomains and/or rafts have been detected at the apical surface, accumulating in small regions of the cell membrane characterized by isolated or grapes of clusters. Sialic acid (SA) residues localized on the surface of target host cells have been shown to be involved in rotavirus infectivity, although several rotaviruses infect cells by means of a SA-independent mechanism(s) (for a review see Ciarlet and Estes, 1999). According to Zárate et al. (2000), the cleavage products of VP4 protein, i.e., the VP5* polypeptide of the nar3 variant and VP8* polypeptide of wild-type RRV, bind to the asialo receptor and SA-containing receptors, respectively. The KUN and MO strains recognize the GM1a ganglioside (Guo et al., 1999), and the SA-dependent OSU strain recognize the GM3 ganglioside (Rolsma *et al.*, 1998). Moreover, the neuraminidase-sensitive simian SA11 and bovine NCDV, and the neuraminidase-insensitive bovine UK rotavirus strains show distinct ganglioside binding specificities since they recognized gangliotetraosylceramide (GA1; also known as asialo-GM1), gangliotriaosylceramide (GA2; also know as asialo-GM2), NeuGc-GM3, and GD1a gangliosides differently (Delorme *et al.*, 2001).

Observations that the spike protein, VP4, and the major capsid protein, VP7, contain the alphaxbeta2 integrin ligand site GPR and the alpha4beta1 integrin ligand site LDV (Coulson *et al.*, 1997), respectively, have focused studies on the role of integrins as receptor(s) and/or co-receptors for rotavirus cell entry. Recent reports have demonstrated that cellular alpha2beta1, alpha4beta1 and alphaXbeta2 integrins act as receptors and/or co-receptors for rotavirus cell attachment and cell entry (Hewish *et al.*, 1999; Zárate *et al.*, 2000; Guerrero *et al.*, 2000; Londrigan *et al.*, 2000). However, these data were obtained with MA104 cells, detached undifferentiated HT-29 cells and detached differentiated Caco-2 cells. The choice of non-intestinal MA104 cells, undifferentiated intestinal HT-29 cells or detached differentiated Caco-2 cells raises several questions about the role of integrin as a rotavirus receptor and/or co-receptor in polarized, fully-differentiated intestinal cells. The undifferentiated MA104 and HT-29 cells could be infected by rotavirus by means of a integrin-dependent mechanism since these cells effectively express integrins over their entire surface. Detached, polarized, fully-differentiated Caco-2 cells can be infected by rotavirus by means of an integrin-dependent mechanism, since detached cells artificially expose the basolateral domain in which the integrins are located. Indeed, it has been clearly demonstrated that integrins are strikingly segregated at the basolateral domain of polarized, fully-differentiated intestinal cells (Vachon *et al.*, 1993). In consequence, the role of integrins in rotavirus infection remains unclear, since these receptors and/or co-receptor(s) were never found at the brush border of the polarized differentiated intestinal cells, which is the natural route of infection. However, the role of integrins as receptors and/or co-receptor(s) in intestinal rotavirus infection could be of interest, since M cells express integrins at their luminal membrane (Schulte *et al.*, 2000). M cells are distinctive epithelial cells that occur only in the follicle-associated epithelia that overlie organized mucosa-associated lymphoid tissues (for a review see Kraehenbuhl and Neutra, 2000). They are structurally and functionally specialized for transepithelial transport, delivering foreign antigens and microorganisms to organized lymphoid tissues within the mucosae of the small and large intestines. Moreover, M cells play an important part in delivering vaccines to the mucosal immune system. Viral transmission *via* M cells has been documented (for a review see Neutra *et al.*, 1996), but rotavirus infection via M cells has not yet been demonstrated. It would be interesting to investigate this possibilty in the future with regard to the route of delivery for rotavirus vaccines.

Rotavirus infection causes cell death by modifying Ca^{2+} homeostasis (Perez *et al.*, 1998; 1999) and/or by apoptosis (Superti *et al.*, 1996) in undifferentiated MA104 and HT-29 cells, respectively. In polarized fully differentiated rotavirus-infected Caco-2 cells (Dickman *et al.*, 2000), observation of increased production of lactase, decreased mitochondrial oxygen consumption, and reduced cellular ATP are indicative of mitochondrial alteration and apoptosis. Enhanced condensed and marginated nuclear chromatin, exposure of phosphatidylserine in the plasma membrane, appearance of cytoplasmic cytochrome C and dissipation of the mitochondrial transmembrane potential, and large-scale fragmentation of DNA have been observed in a subpopulation of rotavirus-infected Caco-2 cells (Chaïbi C, Cotte-Laffitte J, Geniteau-Legendre M, unpublished results). Based on the observed cell lysis in MA104 cells, it has been postulated that, as a result of cell desquamation, high levels of progeny viruses are delivered into the lumen, leading to the infection of adjacent cells in the intestinal epithelium. The transport and apical delivery of rotaviruses in infected mature enterocytes at the tip of the microvilli of the small intestine has been poorly investigated. The use of polarized fully differentiated intestinal Caco-2 cells have led to new insights into the transport and release of rotavirus in human intestinal cells prior to cell lysis (Jourdan *et al.*, 1997). A non-conventional, vesicular transport of newly synthesized viruses from the endoplasmic reticulum to the apical plasma membrane that by-passes the Golgi complex has been identified (Fig. 3). This demonstrates that vesicular transport permits the release of progeny viruses from the apical domain, expressing brush border, to the lumen without affecting the integrity of the polarized intestinal cells.

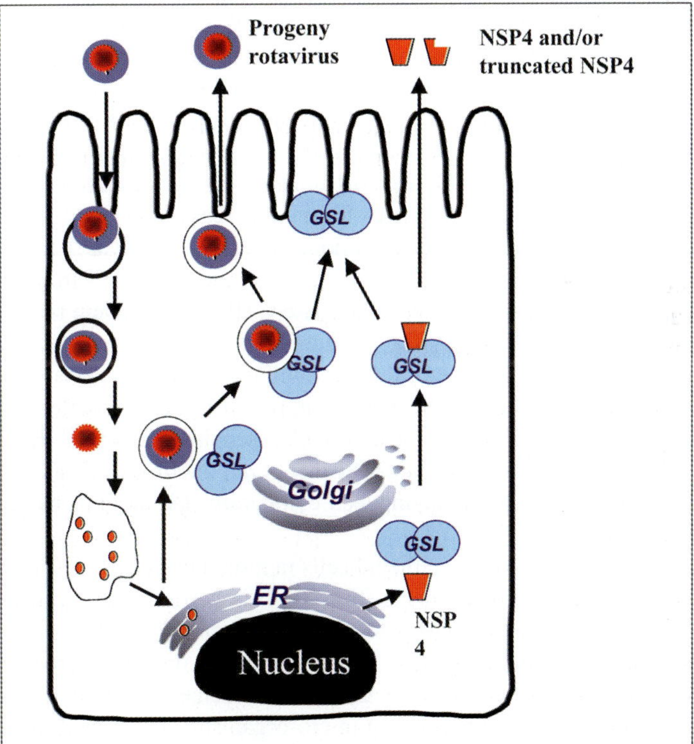

Fig. 3. Proposed model for rotavirus entry and intracellular transport of progeny rotaviruses and non-structural NSP4 protein in polarized, fully-differentiated intestinal cells. GSL, glycosphingolipid-enriched microdomains; ER, endoplesmic reticulum.

Rotavirus-induced structural lesions in polarized intestinal cells

Investigations of rotavirus pathogenicity using the cultured human polarized fully differentiated intestinal cells have recently produced an alternative explanation of the mechanism(s) by which rotavirus alters the enterocytes of the small intestine. Microbial pathogens have developed many sophisticated ways of infecting their hosts and causing disease. For example, enterovirulent bacteria subvert functional membrane-bound proteins as receptors for colonizing epithelia and exploit the cell-signalling pathways to interact with the host cells (for a review see Finlay and Falkow, 1997). For example, enteropathogenic *Escherichia coli* promotes changes in the intestinal cell cytoskeleton by piracy of host-cell Ca^{2+}-dependent signalization cascades influencing the distribution of cytoskeleton proteins (for a review see Vallance and Finlay, 2000). Recent reports have documented that rotavirus induces specific structural lesions in polarized, fully-differentiated intestinal cells without affecting the polarized organization of the cells or integrity of the cell monolayer forming a barrier (Fig. 4).

Organization of the apical domain of intestinal cells is under the control of several cytoskeletal-interacting, Ca^{2+}-sensitive proteins (for a review see Louvard *et al.*, 1992) (Fig. 1). In the presence of high concentrations of Ca^{2+}, a mechanism involving the breakdown of F-actin is known to be mediated by Ca^{2+}-dependent, actin-severing proteins such as villin. Rotavirus infection in undifferentiated MA104 and HT-29 cells is followed by changes in Ca^{2+} homeostasis (for a review see Ruiz *et al.*, 2000). Brunet *et al.* (2000a) have reported that rotavirus infection of polarized, fully differentiated intestinal Caco-2 cells promotes an increase in the intracellular concentration of Ca^{2+} ($[Ca^{2+}]_i$) without modifying the polarized organization of the cells. At the primary stage of infection, the increase in $[Ca^{2+}]_i$ results from an increase in the Ca^{2+} permeability of the plasmalemma. At a late stage of infection, the increase in $[Ca^{2+}]_i$ results from an increase in both the extracellular Ca^{2+} influx and the release of Ca^{2+} from the intracellular organelles, mainly the ER. A phospholipase C (PLC)-dependent mechanism is in part responsible for the increase in $[Ca^{2+}]_i$. Moreover, viral replication is necessary for the increase in $[Ca^{2+}]_i$ and damage of the microvillar cytoskeleton to occur. In turn, at a late stage of rotavirus infection, the increase in $[Ca^{2+}]_i$ has a direct effect on the brush border-associated, cytoskeleton interacting, Ca^{2+}-sensitive proteins, F-actin, villin (Brunet *et al.*, 2000a), and microtubules (Brunet *et al.*, 2000b) which are dramatically disassembled. In contrast, rotavirus-induced cytokeratin-18 rearrangements are Ca^{2+}-independent (Brunet *et al.*, 2000b).

The organization of polarized epithelial cells in monolayers provides a permeability barrier between different environments (Fig. 2). This barrier function means that the junctional complex acts as a « gate » to provide a permeability barrier between the mucosal and serosal environments and enables vectorial transport across the cellular layer (for a review see Denker and Nigam, 1998). Mechanistic studies have revealed four major ways by which enterovirulent micro-organisms may assault the tight junctional complex (for review see Sears, 2000). It was shown that rotavirus infection promotes structural and functional injuries at the tight junctions (TJ) in the cell-cell junctional complex of cultured polarized human intestinal Caco-2 cell monolayers

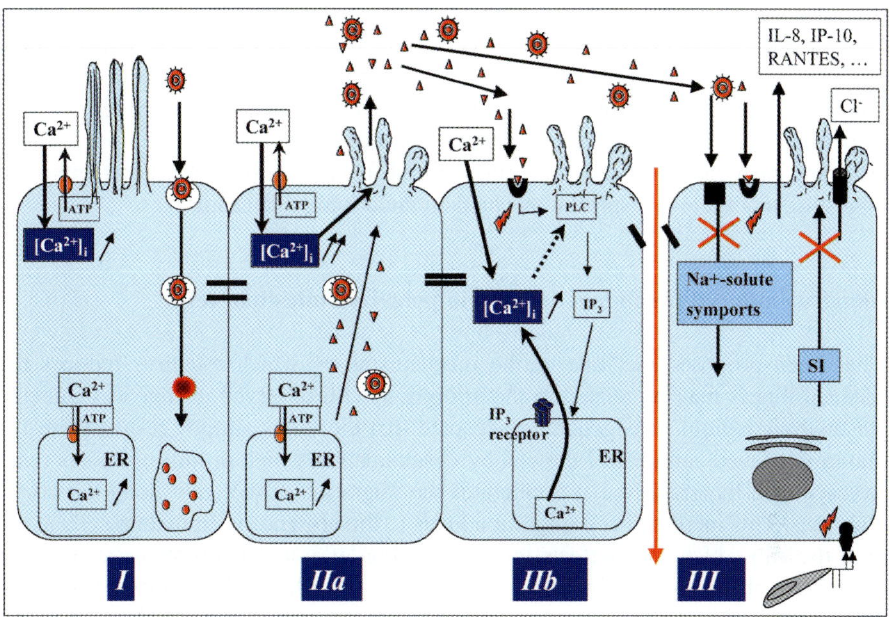

Fig. 4. Proposed schematic model of rotavirus pathogenicity in polarized, fully-differentiated intestinal cells. Adapted from Brunet *et al.* (2000a). *In I*: Rotavirus enters the cell through a Ca^{2+}-dependent endocytosis pathway. At an early stage of infection, increase in $[Ca^{2+}]_i$ results only from extracellular Ca^{2+} entry, partially compensated by an activation of the Ca^{2+}-ATPase pump of the ER. *In IIa*: At a late stage of infection, a strong increase in $[Ca^{2+}]_i$ results from Ca^{2+} mobilization from the intracellular organelles, mainly the ER, probably through a PLC-dependent mechanism. The $[Ca^{2+}]_i$ rise induces microvillar cytoskeleton proteins disassembly. Progeny rotaviruses and rotaviral proteins (red triangle), such as NSP4 are apically transported and released in the luminal space. *In IIb*: Apically released rotaviral proteins including NSP4 (or fragments thereof) would interact with membrane-bound receptor(s) to promote an IP_3-dependent mobilization of Ca^{2+} from the ER and a plasmalemma Ca^{2+} influx which in turn induces a transient microvillar F-actin disassembly. *In III*: Rotavirus infection is accompanied by an increased paracellular permeability, resulting in part from the disassembly of TJ-associated proteins playing a pivotal role in the sealing of TJ. Rotavirus infection is followed by the secretion of IL-8, GRO-α, IP-10, and RANTES and activation of nuclear transcription factors NF-κB, STAT1 and ISGF3. Functional defects develop in rotavirus-infected cells. Rotavirus infection decreases brush border expression of sucrase-isomaltase (SI) which is coupled to the inhibition of SI intracellular transport. Rotavirus infection strongly inhibits both Na^+-D-glucose (SGLT1) and Na^+-L-leucine transports. A direct or mediated action of NSP4 inhibits SGLT1, and by mobilizing intracellular Ca^{2+} would induces chloride secretion. Stimulation of enteric nervous system by virus or rotaviral proteins may also induce chloride secretion. Red lightning indicate activated cell signaling. All these cell events develop without that the polarized organization, integrity and viability of the cells are affected. PLC, phospholipase C; IP_3, inositol-1,4,5-triphosphate.

without damaging the cell or the integrity of the monolayer (Dickman *et al.*, 2000; Obert *et al.*, 2000). TJ-associated ZO-1, claudin-1 and occludin were redistributed, whereas in contrast the zonula adherens-associated E-cadherin was not affected. This change in TJ-associated protein distribution is accompanied by increased transepithelial permeability. Interestingly, the rotavirus-induced structural and functional changes in TJ are independent of the rotavirus-induced apical cytoskeleton rearrangements, suggesting that there is a specific mechanism underlying this change.

Rotavirus-induced functional defaults in polarized intestinal cells

It has been proposed that one of the mechanisms by which rotavirus induces the diarrheal illness may be related to the atrophy of villi observed in rotavirus-infected intestinal epithelium. It is generally accepted that the villus atrophy results from the rotavirus-induced cell death followed by desquamation which in turn promotes reactive crypt cell hyperplasia that accelerates the migration of the cells along the crypt-villus axis. This increased cell migration leads to the presence of immature cells at the tip of the villi which have defects in the brush border-associated proteins responsible for specific intestinal functions. However, regardless of the severity of the histopathologic lesions and in the absence of enterocytes destruction, a decrease in disaccharidase activities has been observed. It has now been demonstrated that several gastrointestinal functions are influenced by the establishment and maintenance of the polarization of epithelial intestinal cells (for a review see Louvard *et al.*, 1992) (Fig. 1). Recent data obtained in cultured rotavirus-infected polarized, fully-differentiated intestinal cells demonstrate that the infecting virus induces specific defects in brush border-associated functional proteins, without affecting the viability or the integrity of the cells.

Rotavirus infection in polarized, fully-differentiated Caco-2 cells is followed by a defect in brush border-associated hydrolase expression (Jourdan *et al.*, 1998) (Fig. 5). Indeed, sucrase-isomaltase (SI) activity and apical expression are specifically and selectively decreased by rotavirus infection without any apparent cell destruction. Rotavirus infection does not affect SI biosynthesis, maturation or stability, but does induce a blockade of SI transport to the brush-border. Rotavirus infection in Caco-2 cells is accompanied by a decrease in the expression of lactase at the brush border, but the mechanism by which the virus promotes this effect is unknown (Martin S, Cotte-Laffitte J, Geniteau-Legendre M, unpublished results). The impact of rotavirus infection on intestinal transport functions, such as Na^+-D-glucose (SGLT1) and Na^+-L-leucine symport activities, has been investigated in brush border membranes (BBM) of young rabbit enterocytes (Halaihel *et al.*, 2000a). Rotavirus infection strongly inhibited both Na^+-D-glucose and Na^+-L-leucine transport into BBM vesicles, although the effect on amino acid transport occurred more slowly. Interestingly, the decrease in the activities of symporters took place in the absence of any apparent histological damage. Moreover, the observation that the level of SGLT1 protein in BBM of rotavirus-infected enterocytes remained unchanged, indicated that the virus has a direct effect on the symporter. Taken together, these reports provided fresh evidence that

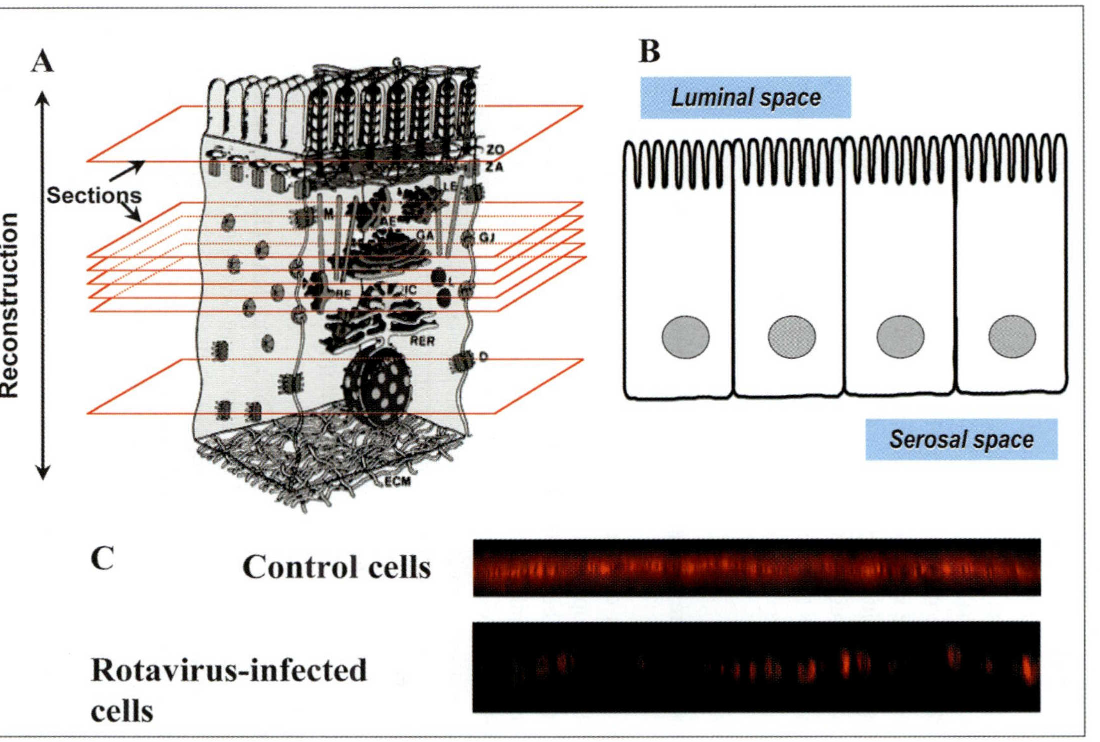

Fig. 5. Confocal laser scanning microscopy (CLSM) showing a decrease in sucrase-isomaltase expression at the brush border of rotavirus-infected human polarized, fully-differentiated intestinal Caco-2 cells. Schematic representation of polarized, fully-differentiated intestinal cell is from Louvard *et al.* (1992). A, scheme of CLSM analysis; B, schematic representation of cultured intestinal cell monolayer; C, CLSM scanning of brush border-associated sucrase-isomaltase (SI), labelled by indirect immunofluorescence test (mAb 8A9 directed against human SI, rhodamine-conjugated secondary Ab). Control cells show homogeneous distribution of staining; in rotavirus-infected cells the SI labelling is disrupted.

an alternative model of rotavirus pathophysiology does exist, independently and/or before the histological lesions develop, and that it participates in the onset of rotavirus-induced diarrheal illness.

In parallel to the rotavirus-induced defects in the activities of intestine-specific proteins as those implicated in the maintenance of the absorption/secretion balance, it was established that rotavirus infection is accompanied by the increased paracellular movements of Na^+ and H_2O. The cellular and molecular mechanism(s) by which the rotavirus induces the accumulation of fluid in the lumen are still under investigation. One promising way of investigation is related to the recent observation by Lundgren *et al.* (2000) that the enteric nervous system (ENS) is probably involved in rotavirus-induced fluid and electrolyte secretion. This exciting finding, which was obtained by a pharmacological approach, indicates that it may be possible to treat rotaviral diarrhea by an innovative drug strategy that specifically targets the signaling pathways activated by the neuromediators released after rotavirus infection (See also Lundgren and Svensson, Section I, Chapter 3 of this book).

Rotavirus-induced cytokine responses in polarized intestinal cells

The potential part played by cytokines in host responses to intestinal microbial infections has highlighted the importance of this aspect of host defense. Increased levels of inflammatory cells have been observed in the lamina propria of rotavirus-infected intestinal epithelium. This inflammatory response generally results from the production of chemokines which could act as mediators of immune/inflammatory responses, such as the recruitment and activation of different populations of leukocytes. In agreement with clinical observations, the production of cytokines in response to rotavirus infection in cultured human intestinal epithelial polarized, fully-differentiated cells has been recently reported (Sheth *et al.*, 1996; Casola *et al.*, 1998; Rollo *et al.*, 1999). The production of the chemotactic and cell-activating cytokine, interleukin 8 (IL-8), growth-related peptide α (GRO-α), IFN-stimulated protein-10 (IP-10), granulocyte-macrophage colony-stimulating factor (GM-CSF), and RANTES (regulated upon activation, normal T cell expressed and secreted) were stimulated, whereas production of TNFα, IL-1α, IFNα, IFNγ, macrophage inflammatory protein 1α (MIP-α), monocyte chemotactic protein 1 (MCP-1) or IL-6 remained unchanged. Rotavirus infection increased IL-8-, GRO-α-, and RANTES-specific RNA and chemokine secretion in undifferentiated HT-29 and polarized, fully-differentiated HT29-19A cells (IL-8, GRO-α, and RANTES), T84 cells (IL-8), and Caco-2 cells (IL-8, but not GRO-α and RANTES). Consistent with the observed rotavirus-induced stimulation of cytokine mRNAs, Rollo *et al.* (1999) reported that the nuclear transcription factors NF-κB, STAT1 and IFN-stimulated gene factor-3 (ISGF3) were activated following rotavirus infection in undifferentiated HT-29 cells. Although infectious rotavirus more effectively stimulated NF-κB activation and IL-8 production, the fact that genetically inactivated rotavirus was capable of activating NF-κB and STAT1 suggests that rotavirus may activate NF-κB by different mechanisms. LaMonica *et al.* (2001) found that rotaviruses primarily activated NF-κB through a TRAF2-NF-κB-inducing kinase signaling pathway and that rotavirus capsid protein VP4 and its N-terminal VP8* cleavage product, having TNRF-associated factor (TRAF) binding motifs, directed pathway activation through interactions with cellular TRAFs.

Rotavirus proteins participate independently in rotavirus pathogenicity in polarized intestinal cells

The pioneering report by Ball *et al.* (1996) has shown that the non-structural protein NSP4 of the simian rotavirus strain SA11 promotes diarrhea in a mouse model system. Moreover, it has been recently demonstrated that group A NSP4s are able to induce diarrhea (Horie *et al.*, 1999). These important findings lead to the conclusion that rotavirus protein(s) are able to trigger a rotavirus-like illness, independently and/or together with the viruses. Following this demonstration, the main focus of investigations has been NSP4 activities (for reviews Estes and Morris, 2000; Morris and Estes, 2001; Estes, Section II, Chapter 6 of this book), but the activities of other structural proteins, some of which may display specific intracellular targeting (Nejmedine *et al.*, 2000), may be investigated in the future.

Adding NSP4 to polarized, fully-differentiated HT29-19A cells induced $[Ca^{2+}]_I$ mobilization by stimulating both the intracellular release of Ca^{2+} and the extracellular influx of Ca^{2+}, controlled by a mechanism involving PLC activation and inositol 1,4,5-triphosphate production (Dong *et al.*, 1997), suggesting that a receptor-mediated signalling pathway is involved in the NSP4-induced changes in $[Ca^{2+}]_I$. The activity of NSP4 at the junctional domain of polarized Madin-Darby canine kidney (MDCK-1) cells has been investigated by Tafazoli *et al.* (2001). The apical application of NSP4 was followed by a reduction in the TER, a redistribution of the TJ-associated ZO-1 protein, and an increase in paracellular permeability, as previously observed in rotavirus-infected Caco-2 cells (Dickman *et al.*, 2000; Obert *et al.*, 2000). However, the NSP4 effect developed after prolonged exposure (24-30 hours) and was reversed after removing the NSP4. This differs from the short-term effect observed 12-24 hours post-infection in Caco-2 cells.

The NSP4 (112-175) fragment was found to promote an increase in $[Ca^{2+}]_I$ in HT29-19A cells and diarrhea in neonatal mice, as does full-length NSP4 (Zhang *et al.*, 2000). Moreover, the NSP4 (114-135) peptide mobilized intracellular calcium and induced secretory chloride in the same way as NSP4. The NSP4 (114-135) peptide was also found to be a specific, fully non-competitive inhibitor of the Na^+-D-glucose symporter SGLT1 in young rabbit BBM (Halaihel *et al.*, 2000b). A secreted cleavage product of NSP4, the NSP4 (112-175) fragment, has been detected in the media of undifferentiated HT-29 cells infected with rotavirus strain SA11 (Zhang *et al.*, 2000). The secretion of the NSP4 (112-175) fragment in HT29-19A cells apparently occurred by a nonclassical, Golgi-independent pathway that involves both the microtubule and actin microfilament networks (Zhang *et al.*, 2000). Thus, it has been proposed that endogenously produced NSP4 in rotavirus-infected cells could be transported from the

ER to the apical domain in association with lipids rafts (Fig. 3). This hypothesis fits in with a recent observation showing that NSP4 and NSP4 (114-135) interact directly with membrane lipids, mimicking caveolae or caveolae-like microdomains or lipid rafts (Huang *et al.*, 2001). Interestingly, Jourdan *et al.* (1997) after demonstrating the non-conventional vesicular transport of newly synthesized viruses from the ER to the apical plasma membrane that bypass the Golgi complex, suggested that virus particles could be transported by lipid rafts. Activity of rotavirus-secreted protein(s) against the brush border cytoskeleton of the polarized, fully-differentiated Caco-2 cells has been reported by Brunet *et al.* (2000a). Culture media of rotavirus-infected cells were able to promote a transient increase in $[Ca^{2+}]_I$ in naïve cells as a result of a PLC-dependent efflux of Ca^{2+} from the ER and by extracellular Ca^{2+} influx, accompanied by transient microvillar F-actin disassembly. SDS-PAGE electrophoresis-Western blotting has revealed the presence of a protein having the apparent molecular mass of NSP4 in the culture media of rotavirus-infected Caco-2 cells (Martin S, Linxe C, and Geniteau-Legendre M, unpublished results).

References

Ball JM, Tian P, Zeng CQY, Morris AP, Estes MK (1996) Age-dependent diarrhea induced by a rotraviral nonstructural glycoprotein. *Nature* **272**: 101-107.

Brown DA, London E (1998) Functions of lipids rafts in biological membranes. *Annu. Rev. Cell Dev. Biol.* **14**:111-136.

Brunet JP, Cotte-Laffitte J, Linxe C, Quéro AM, Géniteau-Legendre M, Servin AL (2000a) Rotavirus infection induces an increase in intracellular calcium concentration in human intestinal epithelial cells: role in microvillar actin alteration. *J. Virol.* **74**: 2323-2332.

Brunet JP, Jourdan N, Cotte-Laffitte J, Linxe C, Géniteau-Legendre M, Servin AL, Quéro AM (2000b) Rotavirus infection induces cytoskeleton disorganization in human intestinal epithelial cells: implication of an increase in intracellular calcium concentration. *J. Virol.* **74**:10801-10806.

Casola A, Estes MK, Crawford SE, Ogra PL, Ernst PB, Garofalo RP, Crowe SE (1998) Rotavirus infection of cultured intestinal epithelial cells induces secretion of CXC and CC chemokines. *Gastroenterology* **114**;947-955.

Ciarlet M, Estes MK (1999) Human and most animal rotavirus strains do not require the presence of sialic acid on the cell surface for efficient infectivity. *J. Gen. Virol.* **80**: 943-948.

Ciarlet M, Estes MK (2001) Interactions between rotavirus and gastrointestinal cells. *Curr. Opin. Microbiol.* **4**: 435-441.

Ciarlet M, Gilger MA, Barone C, McArthur M, Estes MK, Conner ME (1998) Rotavirus disease, but not infection and development of intestinal histopathological lesions, is age restricted in rabbits. *Virology* **251**: 343-360.

Coconnier MH, Lorrot M, Barbat A, Laboisse C, Servin AL (2000) Listeriolysin O-induced stimulation of mucin exocytosis in polarized intestinal mucin-secreting cells: evidence for toxin recognition of membrane-associated lipids and subsequent toxin internalization through caveola. *Cell. Microbiol.* **6**:487-504.

Coulson BS, Londrigan SL, Lee DJ (1997) Rotavirus contains integrin ligand sequences and a disintegrin-like domain that are implicated in virus entry into cells. *Proc. Natl. Acad. Sci. USA* **94**: 5389-5394.

Delorme C, Brüssow H, Sidoti J, Roche N, Karlsson KA, Neeser JR, Teneberg S. (2001) Glycosphingolipid binding specificities of rotavirus: identification of acid-binding epitope. *J. Virol.* **75**: 2276-2287.

Denker BM, Nigam SK (1998) Molecular structure and assembly of the tight junction. *Am. J. Physiol.* **274**:F1-F9.

Dickman KG, Mempson SJ, Anderson J, Lippe S, Zhao L, Burakoff R, Shaw RD (2000) Rotavirus alters paracellular permeability and energy metabolism in Caco-2 cells. *Am. J. Physiol.* **279**: G757-G766.

Dong Y, Zeng CQY, Ball JM, Estes MK, Morris AP (1997) The rotavirus enterotoxin NSP4 mobilizes intracellular calcium in human intestinal cells by stimulating phospholipase C-mediated inositol 1,4,5-triphosphate production. *Proc. Natl. Acad. Sci. USA* **94**: 3960-3965.

Dowling W, Denisova E, Lamonica R, Mackow ER (2000) Selective membrane permeabilization by the rotavirus VP5* protein is abrogated by mutations in an internal hydrphobic domain. *J. Virol.* **74**: 6368-6376.

Estes MK, Morris AP (1999) A viral enterotoxin. A new mechanism of virus-induced pathogenesis. In: *Mechanisms in the Pathogenesis of Enteric Diseases,* (Eds: Kluwer P and Francis D.) pp 73-82. Academic/Plenum Publisher. New York.

Finlay BB, Falkow S (1997) Common themes in microbial pathogenicity revisited. *Microb. Mol. Biol. Rev.* **61**:136-169.

Gilbert JM, Greenberg HB (1997) Virus-like particle-induced fusion from without in tissue culture cells: role of outer-layer proteins VP4 and VP7. *J. Virol.* **71**: 4555-4563.

Gilbert JM, Greenberg HB (1998) Cleavage of rhesus rotavirus VP4 after arginine 247 is essential for rotavirus-like particle-induced fusion from without. *J. Virol.* **72**: 5323-5327.

Guerrero CA, Méndez E, Zárate S, Isa P, López S, Arias CF (2000) Integrin alpha(v)beta(3) mediates rotavirus cell entry. *Proc. Natl. Acad. Sci. USA* **97**: 14644-14649.

Guerrero CA, Zárate S, Corkidi GG, López S, Arias CF (2000) Biochemical characterization of rotavirus receptors in MA104 cells. *J. Virol.* **74**: 9362-9371.

Guignot J., Bernet-Camard MF, Poüs C, Plançon L, Le Bouguenec C, Servin. AL (2001) Polarized entry into human epithelial cells of uropathogenic Afa/Dr diffusely adhering *Escherichia coli* strain IH11128: Evidence for $\alpha_5\beta_1$ integrin recognition, and subsequent internalization through a pathway involving caveolae and dynamic unstable microtubules. *Infect. Immun.* **69**:1856-1868.

Guo CT, Nakagomi O, Mochizuki M, Ishida H, Kiso M, Ohta Y, Suzuki T, Miyamoto D, Hidar KI, Suzuki Y (1999) Ganglioside GM(1a) on the cell surface is involved in the infection by human rotavirus KUN and MO strains. *J. Biochem.* **126**: 683-688.

Halaihel N, Liévin V, Alvarado F, Vasseur M (2000a) Rotavirus infection impairs intestinal brush border membrane Na⁺-solute cotransport activities in young rabbits. *Am. J. Physiol.* 279: G587-G596.

Halaihel N, Liévin V, Ball J, Estes M, Alvarado F, Vasseur M (2000b) Direct inhibitory effect of rotavirus NSP4 (114-135) peptide on the Na⁺-D-glucose symporter of rabbit intestinal brush border. Membranes. *J. Virol.* **74**: 9464-9470.

Hewish MJ, Takada Y, Coulson BS (1999) Integrins alpha2beta1 and alpha4beta1 can mediate SA11 rotavirus attachment and entry into the cells. *J. Virol.* **74**: 228-236.

Horie Y, Nakagomi O, Koshumira Y, Nakagomi T, Suzuki Y, Oka T, Sasaki S, Matsuda Y, Watanabe S (1999) Diarrhea induction by rotavirus NSP4 in the homologous mouse model system. *Virology* **262**: 398-407.

Huang H, Schroeder F, Zeng C, Estes MK, Schoer JK, Ball JM (2001) Membrane interactions of a novel viral enterotoxin: rotavirus nonstructural glycoprotein NSP4. *Biochemisty* **40**: 4169-4180.

Jourdan N, Maurice M, Delautier D, Quéro AM, Servin AL, Trugnan G (1997) Rotavirus is released from the apical surface of cultured human intestinal cells through non-conventional vesicular transport that bypasses the Golgi apparatus. *J. Virol.* **71**:8268-8278.

Jourdan N, Brunet JP, Sapin C, Blais A, Cotte-Laffitte J, Forestier F, Quero AM, Trugnan G, Servin AL (1998) Rotavirus infection reduces sucrase-isomaltase expression in human intestinal cells by perturbing protein targeting and organization of microvillar cytoskeleton. J. *Virol.* **72**:7228-7236.

Kirkwood CD, Bishop RF, Coulson BS (1998) Attachment and growth of human rotaviruses RV-3 and S12/85 in Caco-2 cells depend on VP5. *J. Virol.* **72**: 9348-9352.

Kraehenbuhl JP, Neutra MR (2000) Epithelial M cells: differentiation and function. *Annu. Rev. Cell Dev. Biol.* **16**: 301-332.

LaMonica R, Kocer SS, Nazarova J, Dowling W, Geimonen E, Shaw RD, Mackow ER (2001) VP4 differentially regulates TRAF2 signaling, disengaging JNK activation while directing NF-kappaB to effect rotavirus-specific cellular responses. *J. Biol. Chem.* **276**: 19889-19896.

Londrigan SL, Hewish MJ, Thompson MJ, Sanders GM, Mustafa H, Coulson BS (2000) Growth of rotaviruses in continuous human and monkey cell lines that vary in their expression of integrins. *J. Gen. Virol.* **81**: 2203-2213.

Louvard D, Kedinger M, Hauri HP (1992) The differentiating intestinal epithelial cell: establishment and maintenance of functions through interactions between cellular structures. *Annu. Rev. Cell Biol.* **8**: 157-195.

Lundgren O, Peregrin AAT, Persson K, Kordasti S, Uhnoo I, Svensson L (2000) Role of enteric nervous system in the fluid and electrolyte secretion of rotavirus diarrhea. *Science* **287**: 491-495.

Martin S, Lorrot M, Aloaoui El Azher M, Vasseur M (2002) Ionic strength- and temperature-induced K_{Ca} shifts in the uncoating reaction of rotavirus strains RF and SA11: correlation with membrane permeabilization. *J. Virol.* **76**: 552-559.

Morris AP, Scott JK, Ball JM, Zeng CQY, O'Neal WK, Estes MK (1999) NSP4 elicits age-dependent diarrhea and Ca^{2+}-mediated I⁻ influx into intestinal crypts of CF mice. *Am. J. Physiol.* **277**: G431-G444.

Morris AP, Estes MK (2001) Microbes and microbial toxins: Paradigms for microbial-mucosal interactions. VIII. Pathological consequences of rotavirus infection and its enterotoxin. *Am. J. Physiol.* **281**: G303-G310.

Nejmedine M, Trugnan G, Sapin C, Kohli E, Svensson L, López S, Cohen J (2000) Rotavirus spike protein VP4 is present at the plasma membrane and is associated with microtubules in infected cells. *J. Virol.* **74**: 3313-3320.

Neutra MR, Frey A, Kraehenbuhl (1996) Epithelial M cells: gateways for mucosal infection and immunization. *Cell* **86**: 345-348.

Obert G, Peiffer I, Servin AL (2000) Rotavirus-induced structural and functional alterations in tight junctions of polarized intestinal Caco-2 cell monolayers. *J. Virol.* **10**: 4645-4651.

Orlandi PA, Fishman PH (1998) Filipin-dependent inhibition of cholera toxin: evidence for toxin internalization and activation through caveola-like domains. *J. Cell. Biol.* **141**: 905-915.

Perez JF, Chemello ME, Liprandi F, Ruiz MC, Michelangeli F (1998) Oncosis in MA104 cells is induced by rotavirus infection through and increase in intracellular Ca^{2+} concentration. *Virology* **252**:17-27.

Rollo EE, Kumar KP, Reich NC, Cohen J, Angel J, Greenberg HB, Sheth R, Anderson J, Oh B, Hempson SJ, Mackow ER, Shaw RD (1999) The epithelial cell response to rotavirus infection. *J. Immunol.* **163**: 4442-4452.

Rolsma MD, Kuhlenschmidt TB, Gelberg HB, Kuhlenschmidt MS (1998) Structure and function of a ganglioside receptor for porcine rotavirus. *J. Virol.* **72**: 9079-9091.

Ruiz MC, Cohen J, Michelangeli F (2000) Role of Ca^{2+} in the replication and pathogenesis of rotavirus and other viral infections. *Cell Calcium* **28**: 137-149.

Schulte R, Kerneis S, Klinke S, Bartels H, Preger S, Kraehenbuhl JP, Pringault E, Autenrieth IB (2000) Translocation of *Yersinia entrocolitica* across reconstituted intestinal epithelial monolayers is triggered by *Yersinia* invasin binding to beta1 integrins apically expressed on M-like cells. *Cell. Microbiol.* **2**:173-85

Sears CL (2000) Molecular physiology and pathophysiology of tight junctions. V. Assault of the tight junction by enteric pathogens. *Am. J. Physiol.* **279:** G1123-G1134.

Sheth R, Anderson J, Sato T, Oh B, Hempson SJ, Rollo E, Mackow ER, Shaw D (1996) Rotavirus stimulates IL-8 secretion from epithelial cells. *Virology* **221**: 251-259.

Shin JS, Gao Z, Abraham SN (2000) Involvement of cellular caveolae in bacterial entry into mast cells. *Nature* **289**: 785-788.

Superti F, Ammendolia MG, Tinari A, Bucci B, Giammarioli AM, Rainaldi G, Rivabene R, Donelli G (1996) Induction of apoptosis in HT-29 cells infected with SA-11 rotavirus. *J. Med. Virol.* **50**: 325-334.

Svensson L, Finlay BB, Bass D, von Bonsdorff CH, Greenberg HB (1991) Symmetric infection of rotavirus on polarized human intestinal epithelial (Caco-2) cells. *J. Virol.* **65**: 4190-4197.

Tafazoli F, Zeng CQ, Estes MK, Magnusson KE, Svensson L (2001) NSP4 enterotoxin of rotavirus induces paracellular leakage in polarized epithelial cells. *J. Virol.* **75**: 1540-1546.

Vachon PH, Durand J, Beaulieu JF (1993) Basement membrane formation and redistribution of the ß1 integrins in a human intestinal co-culture system. *Anat. Rec.* **236**: 567-576.

Vallance BA, Finlay BB. (2000) Exploitation of host cells by enteropathogenic *Escherichia coli*. *Proc. Natl. Acad. Sci. USA* **97**: 8799-8806.

Werling D, Hope JC, Chaplin P, Collins RA, Taylor G, Howard CJ (1999) Involvement of caveolae in the uptake of respiratory syncytial virus antigen by dendritic cells. *J. Leukocyte Biol.* **66**: 50-58.

Xu A, Bellamy AR, Taylor JA (2000) Immobilization of the early secretory pathway by a virus glycoprotein that binds to microtubules. *EMBO J.* **19**: 6465-6474.

Zhang M, Zeng CQ, Morris AP, Estes MK (2000) A functional NSP4 enterotoxin peptide secreted from rotavirus-infected cells. *J. Virol.* **74**: 11663-11670.

Zárate S, Espinosa R, Romero P, Guerrero CA, Arias CF, López S (2000) Integrin alpha2beta1 mediates the cell attachment of the rotavirus neuraminidase-resistant variant nar3. *Virology* **5**: 50-54.

Zárate S, Espinosa R, Romero P, Mendez E, Arias CF, López S (2000) The VP5 domain of VP4 can mediate attachment of rotavirus to cells. *J. Virol.* **74**: 593-599.

Zweibaum A, Laburthe M, Grasset E, Louvard D (1991) Use of cultured cell lines in studies of intestinal cell differentiation and function. In: *Handbook of Physiology. The Gastrointestinal System, vol. IV. Intestinal absorption and secretion.* (Eds Schultz SJ, Field M, Frizell RA), pp. 223-255. American Physiological Society, Bethesda. MD.

II, 9.   Microarrays and host-virus interactions: A transcriptional analysis of Caco-2 cells following rotavirus infection

Dino A. Feigelstock[1], Mariela A. Cuadras[2] and Harry B. Greenberg[3, 2]

[1] *Laboratory of Hepatitis and Related Emerging Agents, DETTD, OBRR, CBER, FDA,*
  8800 Rockville Pike, Building 29, Room 231, NIH, Bethesda, MD 20892
[2] *Departments of Microbiology and Immunology and of Medicine, Division of Gastroenterology,*
  Stanford University School of Medicine, Stanford, California 94305
[3] *V. A. Palo Alto Health Care System, Palo Alto California 94304*

Introduction

Technical considerations of microarrays

The use of microarrays to analyze gene expression on a global level has recently been broadly exploited in basic and clinical research. Traditionally, the standard techniques used to detect specific sequences of DNA or RNA have depended on the specificity and affinity of sequence complementarity between nucleic acids. Microarrays are not an exception and follow the same basic principle (59). The power of this technique relies on the fact that it has miniaturized and automated the standard hybridization procedures, allowing the simultaneous detection of thousands of specific DNA or RNA sequences in a single experiment.

Although several different approaches to building and analyzing microarrays have been developed by academic groups and private industry, two have prevailed:

DNA microarrays (Fig. 1a; Nature Cell Biology. 2001. 3:E190-95)

These are constructed by printing thousands of DNA sequences as spots (usually denominated "probes") on a glass microscope slide using a robotic arrayer (10); additional solid supports available include silicon, nylon and nitrocellulose membranes. The material for these DNA "probes" consists of DNA fragments such as products of the polymerase chain reaction (PCR) or cDNA library clones. To compare the relative abundance of each gene in two samples (e.g., non-infected versus infected cells), mRNA from each population is purified and used to produce a labeled sample frequently called the "target". The target is synthesized by a reverse transcription reaction (RT) in the presence of dye-labeled nucleotides (e.g. Cy5-dUTP or Cy3-dUTP).

The differentially labeled populations of cDNAs are mixed and hybridized to a single microarray slide containing the " probes". In cases in which the starting RNA concentration is limited, target amplification is achieved by *in vitro* transcription in combination with cDNA synthesis. After hybridization, fluorescent measurements of each dye are separately made with a scanner, and the intensity of the same spot for both dyes or channels (Cy3 and Cy5) is compared giving as a result the ratio of transcript level between the two samples for each gene printed on the microarray.

Oligonucleotide microarrays (Fig. 1b; Nature Cell Biology, Vol 3, August 2001)

In this method the microarrays are constructed by synthesizing short 20-25mer single stranded oligonucleotides *in situ* by the use of photolithography (Affymetrix) or by ink-jet technology (Agilent Technologies). The advantages of oligonucleotide microarrays are the high density of probes that can be arranged in the array (300,000 oligonucleotides on a 1.28×1.28 cm surface) and that they are designed and synthesized based on sequence information alone (there is no need of physical intermediates such as clones, PCR products or cDNAs) (38). The target preparation usually requires cDNA synthesis in a reaction that is primed with oligo (dT) containing a transcriptional start site for T7 RNA polymerase at its 5' end. The resulting cDNA is then used as a template to make biotinylated cRNA by *in vitro* transcription using biotin-labeled nucleotides. The labeled cRNA is fragmented and hybridized to the microarray. In this case each target sample is hybridized to a separate array, and target binding is detected by staining with a fluorescent dye coupled to streptavidin. The use of multiple oligonucleotides corresponding to different sequences of the same RNA, each with a corresponding mismatch that serves as internal control for hybridization specificity, makes quantitative measurement of transcript levels possible using a single labeled target. The reproducibility of *in situ* synthesis of the oligunocleotide microarrays allows accurate comparison of signals generated by samples hybridized to separate arrays.

Fig. 1. Schematic overview of probe array and target preparation for spotted cDNA microarrays and high-density oligonucleotide microarrays. **a**, cDNA microarrays. Array preparation: inserts from cDNA collections or libraries (such as IMAGE libraries) are amplified using either vector-specific primers. PCR products are printed at specific sites on glass slides using high-precision arraying robots. Through the use of chemical linkers, selective covalent attachment of the coding strand to the glass surface can be achieved. Target preparation: RNA from two different tissues or cell populations is used to synthesize single-stranded cDNA in the presence of nucleotides labelled with two different fluorescent dyes (for example, Cy3 and Cy5). Both samples are mixed in a small volume of hybridization buffer and hybridized to the array surface, usually by stationary hybridization under a coverslip, resulting in competitive binding of differentially labelled cDNAs to the corresponding array elements. High-resolution confocal fluorescence scanning of the array with two different wavelengths corresponding to the dyes used provides relative signal intensities and ratios of mRNA abundance for the genes represented on the array. **b**, High-density oligonucleotide microarrays.

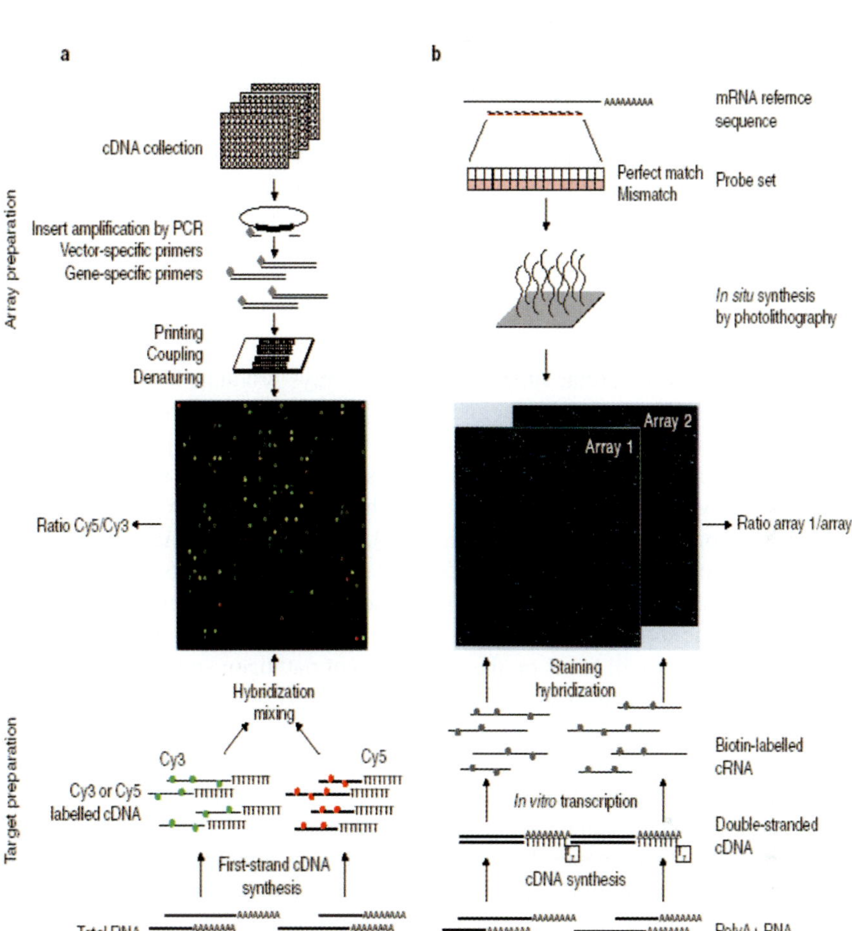

Array preparation: sequences of 16-20 short oligonucleotides (typically 25mers) are chosen from the mRNA reference sequence of each gene, often representing the most unique part of the transcript in the 5'-untranslated region. Light-directed, *in situ* oligonucleotides synthesis is used to generate high-density probe arrays containing over 300,000 individual elements. Target preparation: polyA⁺ RNA from different tissues or cell populations is used to generate double stranded cDNA carrying a transcriptional start site for T7 DNA polymerase. During *in vitro* transcription, biotin-labelled nucleotides are incorporated into the synthesized cRNA molecules. Each target sample is hybridized to a separate probe array, and target binding is detected by staining with a fluorescent dye coupled to streptavidin. Signal intensities of probe array element sets on different arrays are used to calculate relative mRNA abundance for the genes represented on the array. With premission from Nature Cell Biology.

The use of microarrays generates a massive amount of data and, as expected, false positive results can occur. The challenge is then to identify meaningful results in this massive amount of data and to create hypotheses from these results. Careful experimental design, multiple independent experiments and internal controls facilitate the interpretation of the results. Because of the complexity of the data generated, the use of appropriate data and image analysis computer tools is mandatory. Currently, several data-analysis software packages have been developed by academic (21, 57, 64) and commercial sources (e.g., GeneChip™ from Affymetrix and GeneSpring™ from Silicon genetics).

Applications

As the number of commercial suppliers of microarrays products has increased over the past few years, the cost of this technology has come down, and the use of arrays is becoming accessible for more research laboratories. Among the major applications of microarrays we mention the study of gene expression on a genomic scale (13, 14, 30), the determination of mutations and polymorphisms that underlie susceptibility or resistance to common diseases (genotyping) (23, 65), the measurement of DNA copy numbers of specific genes in tumor cells (51), the study of drug effects on cellular physiology (54), the genetic profiling of diverse tumor subtypes in different types of cancers (2, 48) and the study of the complex interaction between pathogens and their host (17, 29). This chapter will focus on the interaction of pathogen and host, but it should be noted that the other mentioned areas have been subject of intense research over the past few years, therefore the reader is directed to several excellent reviews (1, 17, 66).

The identification and analysis of cellular transcription pathways that are activated and/or repressed during viral infection could lead to a better understanding of the virus-host interaction. From this understanding we can identify host-defense mechanisms and virulence-associated viral genes that might help us to develop potential antiviral strategies. The first study, employing global gene expression methods applied to understand viral pathogenesis, was a time course of infection of primary human fibroblast cells with human cytomegalovirus (HCMV) (68). Analyzing more than 6,000 human sequences using oligonucleotide microarrays, the authors found that the virus-host interaction modulates the expression of 258 genes by a factor of 4 or more in comparison with uninfected cells. The authors discussed a possible role of some of the regulated genes emphasizing those genes related to induction of an immune response (68). Following this study, a similar experimental design was used to analyze the gene expression profiles during HIV (26), Marek's disease virus (43), and echovirus type 1 infection (49).

Similarly, the transcriptional response of fibroblasts to herpes simplex virus (44) and HeLa cells to influenza virus infection (25) has been studied using cDNA microarrays. In these studies the authors went a step further: by using replication incompetent viruses, they were able to distinguish changes in cellular transcripts that required viral replication from those that responded in a viral replication-independent manner.

Studies with animal models such as chimpanzees and mice have been done to analyze tissue-specific host-gene expression in response to hepatitis C (4) and coxsackievirus infection (63), respectively, at the transcriptional level. Recently, cDNA microarrays were used to examine the gene expression profiles in tissue lesions of patients with chronic hepatitis B and C. Clustering analysis of the gene expression profiles in all the patients, grouped them into two main categories determined principally by the type of infecting virus, HBV versus HCV. This suggested that the molecular mechanisms that control the pathogenesis of these two viruses may differ (28).

The most exciting aspect of these analyses has, in our opinion, been the ability to provide a global picture of expression patterns under a specified set of infection conditions. Although much of the interest in using microarrays has been focused on identifying individual genes that might play a key role in a given observation (perhaps not the most optimal use of this technique), it is important to emphasize that this is not the purpose of these analyses. The experiments described in these studies are exploratory and are designed to catalog the broad spectrum of transcriptional events that occur after an infection. For those genes with unknown functions the implications of a genome-wide study go even further. Since a gene is usually transcribed only in specific cells and under a specific set of conditions where its function is required, determining where and under which conditions a gene is expressed in conjunction with its grouping with other genes that share similarities in the pattern of expression allow us to make inferences concerning functional connections between gene products.

Microarray technology can also be used to obtain a better understanding of viral or bacterial physiology and the identification of virulence determinants. The design and construction of a DNA or oligonucleotides microarrays for any given pathogen is relatively easy. The first report of global gene analysis in a viral pathogen used oligonucleotide microarrays to monitor the expression of almost all known open reading frames (ORFs) of HCMV (68). Using drugs that block either synthesis of viral proteins or viral DNA replication, the authors drew a temporal map of expression profiles of the entire genome of HCMV classifying unknown ORFs into preexisting categories (immediate early, early, early late and late) (8).

Following the study of HCMV, nylon membrane DNA microarrays were used to characterize the expression of 88 ORFs of Kaposi's sarcoma-associated herpesvirus (KSHV) during latency and after induction of lytic replication. The authors found that the majority of the KSHV genes were downregulated during the latent state and enormously increased in transcriptional activity during lytic replication (31).

Viral microarrays have many other potential uses such as identification of virulence factors, analysis of viral mutations in the transcriptional response, identification of genomic deletions in mutant strains (genotyping), and studies of the effect of drugs on viral biology.

Obviously, gene expression analysis at the level of transcript abundance using microarrays is not restricted to the study of viral pathogens. Transcription of bacterial (e.g., *E. coli, Mycobacterium tuberculosis*) (29, 45, 52) and other prokaryoticc pathogens as *Listeria monocytogenes* (15) have been measured during infection using this technology.

Previous studies of cellular gene expression during rotavirus infection

During infection, rotavirus induces profound alterations in the morphology and bio-chemistry of the host cell; however, little is known about the requirement or involvement of host proteins in the diverse steps of the rotavirus replication cycle. The majority of our knowledge about the virus replication cycle comes from studies performed in tissue culture cells, to which rotaviruses have been adapted to grow. These studies showed that the entire viral replication cycle occurs in the cytoplasm of the infected cell (reviewed in 22). Knowledge is scare about the molecular mechanisms underling the cellular response to infection. A reduction in cellular protein synthesis has been report-ed in rotavirus infected cells (27). Changes in intracellular Ca^{2+} concentration (associ-ated with virus maturation and proposed to be related to cell death) (42), alterations in the organization of the cell cytoskeleton (6, 32), structural and functional alterations in the tight junctions (46) and secretion of virus particles prior to cellular lysis (33), have also been described following rotavirus infection. These phenotypic responses could be associated with alterations in mRNA expression of particular host cell genes.

Some limited attempts have been made to characterize the mRNA and protein expression levels of certain cellular genes in rotavirus infected cells. Previous work in MA104 cells demonstrated an upregulation, at the mRNA and protein level, of BiP (grp78) and endoplasmin (grp94, tumor rejection antigen 1), two proteins resident of the endoplasmic reticulum and members of a family of glucose-regulated chaperones (67).

Increases in cytokine gene expression in response to rotavirus infection had been reported (7, 53, 58). The transcriptional activity of genes encoding a variety of che-mokines in HT-29 cells was analyzed by RT-PCR. These authors reported the induc-tion of C-X-C chemokines including IL-8, IP-10 and GROγ and the induction of C-C chemokines such as RANTES and monocyte chemo-attractant protein 1 (MCP1). In addition, genes encoding Interferon alpha and granulocyte-macrophage colony-stimulating factor (GM-CSF) were found to be upregulated in this analysis. Conversely, the transcripts for TNF-α, IL-1α, IL-1β and IFN-β were not increased following rotavirus infection of HT-29 cells (53).

In vivo, animal models such as virus infected mice or fetal lambs have also been used to characterize the transcriptional response to rotavirus infection. Rollo and col-leagues found similarities between the murine chemokine response and the response observed in HT-29 cells (53). Recently, the identification of genes expressed during the induction of mucosal immune response was investigated by differential display (DD) using jejunal Peyer's patches (PP) of fetal lamb as a model system (62). RNA was purified from jejunal PP from mock- or rotavirus-infected fetal lambs, and mRNA DD was used to identify genes expressed following rotavirus infection. Ten cDNAs were identified by DD, and these cDNA were isolated, re-amplified and cloned for sequencing. The majority of the cDNA fragments did not have significant homology to reported sequences, only one of the cDNAs displayed homology to the gene encoding the sperm surface protein Sp17. The expression of Sp17 in a broad variety of mucosa-associated lymphoid tissues was confirmed by RT-PCR, and the possible relevance of this finding was discussed (62).

In order to better understand the molecular mechanisms involved in the events following rotavirus infection, we identified host cellular genes whose mRNA levels changed after infection. For this analysis, we used microarrays containing more than 38,000 human cDNAs to study the transcriptional response of the human intestinal cell line Caco-2 to rotavirus infection. The results presented in the following sections have been extensively described in (16) and provide a new picture of the global transcriptional regulation of the rotavirus infected cell.

Global transcriptional response to rotavirus infection

Infection of Caco-2 cells with RRV as a model to study the cellular response to rotavirus infection

We used human cDNA microarrays to study the global transcriptional response of Caco-2 cells infected with the RRV strain of rotavirus. We choose the human intestinal cell line Caco-2 (24) as a model that would mimic characteristics of rotavirus replication *in vivo*. Caco-2 cells originated from a human colon adenocarcinoma and display several characteristics of mature enterocytes, such as cellular polarization, development of an apical brush border membrane and expression of intestinal hydrolases on the apical domain (69). In addition, Caco-2 cells have been used to study the interaction of several enteropathogens with the host cell (40, 41, 60), and most rotavirus strains, including the laboratory strain RRV, grow efficiently in Caco-2 cells (34, 61). The rhesus monkey rotavirus strain was chosen because this is our laboratory strain and because RRV is very well characterized and adapted for growth in cell culture. In addition, RRV and monoreassortants thereof have been orally administered to hundreds of thousands of people as a live virus vaccine.

We choose one and sixteen hours as primary time points to evaluate the cellular transcriptional response to rotavirus infection. We then looked at the cellular transcripitonal response at intermediate and late time points during a time course infection. We choose one hour as an early time point after infection, and sixteen hours as our late point because Caco-2 cells are lysed at late times after infection (after 24-48 h.) (33, 46).

An important issue for this analysis was to determine the viability of the infected cells under the experimental conditions employed. The percentage of live and dead cells following infection was determined by FACS analysis using the Live/Dead Viability/ Cytotoxicity Kit (Molecular probes). Mock and RRV infected cells looked similar at 16 h.p.i.; 77% of the mock infected and 78% of the RRV infected cells were alive (not shown).

We also determined the percentage of cells infected under our experimental conditions. Caco-2 cells were mock or RRV infected at a multiplicity of infection (MOI) of 20. At 16 h.p.i. the cells were fixed and stained with a rabbit polyclonal antiserum against RRV. FACS analysis showed that more than 75% of the inoculated cells stained positively for RRV antigen (not shown). Hence mRNA from infected cells used for hybridization analysis was derived from a population of living cells in which more than 75% were alive and infected.

Materials and methods

Infection conditions and RNA extractions

For microarray experiments 17-26 days old Caco-2 cells (passages 27 to 36) were infected with trypsin-activated RRV (RRV was activated by treatment with 10 µg/ml of trypsin in the absence of FBS for 45 minutes at 37°C). After 1 hour the infected monolayers were washed, and the cells were harvested or incubated for an additional period of 15 hours. As a reference, mock-infected Caco-2 cells were treated under the same conditions as infected cells except that the "mock" inoculum was derived from a cleared lysate of uninfected MA104 cells. At the end of each incubation, total RNA from mock- or RRV- infected cells was extracted using RNAwiz™ (Ambion) according to the manufacturer's protocol. After RNA extraction, polyA mRNA was purified with the FastTrack 2.0 kit (Invitrogen). The concentration of mRNA was determined by absorbance at 260 nm.

Preparation of fluorescently labeled cDNA and hybridization

For analysis of rotavirus infection versus mock infection, labeled cDNA from rotavirus infected cells was generated by reverse transcription (RT) using the red-fluorescent dye Cy5 (Amersham Pharmacia); fluorescent cDNA from mock infected cells was prepared using the green fluorescent dye Cy3 (Amersham Pharmacia). For control comparisons (mock vs. mock and RRV vs. RRV), cDNA from mock- or RRV- infected cells was generated and fluorescently labeled during RT reaction using Cy5 or Cy3 (see Table 1, p. 268). For each RT reaction three micrograms of polyA mRNA were mixed with two micrograms of an achored oligo dT primer (MWG-Biotech), heat denatured and transferred to ice. Then, the reverse transcription reaction was done in the presence of labeled and unlabeled nucleotides (dCTP, dGTP, dATP and either Cy5-dUTP or Cy3-dUTP). At the end of the RT reaction, the residual RNA was degraded by alkaline hydrolysis, the two cDNAs (Cy5 and Cy3 labeled) were mixed and phenol-chloroform extracted. For final probe preparation, SSC was added at a 3x final concentration and SDS was added at a 0.25% final concentration; the mixture was heat denatured, incubated at room temperature and transferred to the microarray surface. The microarray and probe were covered and incubated overnight at 65 °C.

The cDNA microarrays were produced in the Microarray Production Facility of Stanford University (30, 55). These arrays contain 39,552 array elements of which 38,432 correspond to human sequence verified genes, 447 correspond to non-human genes and 673 correspond to unknown samples. Detailed protocols for the array production are available at (http://cmgm.stanford.edu/pbrown/).

Signal detection and data analysis

The fluorescent intensity for each dye was detected using a GenePix 4000b microarray scanner (Axon Instruments). Images were analyzed using the GenePix Pro 3.0 software provided with the scanner. First, each spot was defined automatically by a spot-indicator (grid). The software automatically discards (flags) a spot if: i) the intensity is not greater than the background threshold, ii) if the spot has an irregular size or iii) if the grid designed to that spot overlaps with an adjacent grid (See GenePix Pro 3.0 manual for more details). After the automatic gridding and flagging, a visual inspection was performed to eliminate from the analysis all spots with signals due to visually detectable array artifacts. GenePix Pro 3.0 displays the data in tables that can be exported in any standard spreadsheet program. The data files generated were entered into the Stanford Microarray Database (SMD), a custom database that maintains Web-accessible files for further analysis (57). After submission into SMD, the red signal was normalized by applying a single multiplicative factor to all intensities measured for the red dye Cy5. The normalization factor was computed so that the median Cy3/Cy5 fluorescent ratio of non-flagged spots on each array was 1.0.

For a detailed description of the data filtering criteria and the technical procedures described above, the reader is directed to (16). After filter criteria, genes whose spot intensities did not pass these filter criteria were eliminated from further analysis (between 20,093 and 24,926 of arrayed human genes).

To analyze levels of up- or downregulation we applied a hierarchical clustering algorithm implemented by the software "Cluster" as described in Eisen *et al.* (21). This software clusters genes according to their similarity in the pattern of gene expression and displays the data in a dendrogram that resembles a tree (TreeView). Each row represents genes, and each column represents a single experiment or microarray. In this TreeView image the computed red/green ratios for each spot are represented by color display, black cells represent red/green ratios of 1.0 (unchanged genes), red cells represent ratios with increasing intensities of red (upregulated genes) and green cells represent ratios with increasing intensities of green (downregulated genes). A detailed explanation of the Cluster and TreeView can be reviewed in Eisen *et al.* (21), and this software is available in http://rana.stanford.edu/software.

Experimental design

One of our concerns was to optimize our ability to identify genes regulated by rotavirus infection as opposed to other causes of transcriptional variation under our experimental conditions. Variability due to mRNA extraction and purification, cDNA synthesis, labeling and purification, and signal detection could generate changes in the amount of mRNA detected. Also, natural transcriptional variation is likely to occur in Caco-2 cells during culture. We utilized a variety of strategies to minimize background variability. We used duplicate or triplicate assays for each condition tested. For example, two sets of flasks were infected with RRV, harvested at 16 h.p.i., and mRNA from each set of flasks was extracted and tested separately and compared to three separate control flasks. Fluorescent dye-labeled cDNAs were synthesized using each separate extracted mRNA. In this way we obtained independent samples of labeled cDNAs from RRV-infected cells and mock-infected cells.

Results

Rotavirus infection induces changes in cellular gene expression of Caco-2 cells

Our approach for studying the cellular transcriptional response during rotavirus infection consisted of comparing the relative abundance of specific polyA mRNA in infected cells to the same specific polyA mRNA from mock infected cells using cDNA microarrays containing thousands of cellular genes (see Material and Methods). Preliminary experiments using Caco-2 cells from different passage levels, or different flasks at the same passage level under the same culture conditions, showed some variability in the pattern of mRNA expression. In order to specifically attribute transcriptional changes to virus infection (and not to background variability), we determined the background transcriptional variability of our cell culture system. To do this, we compared mock infected cells vs. mock infected cells at 1 h.p.i., mock infected cells vs. mock infected cells at 16 h.p.i. and RRV infected cells vs. RRV infected cells at 16 h.p.i. Each comparison was repeated one or two times (see Table 1, p 468). Representative plots of some of these comparisons are shown in Fig. 2.

We analyzed the number and percentage of gene transcripts that varied by more than 2-fold in this series of comparisons in order to identify the background rate of transcriptional variation in our system. As can be seen, when mock infected cells are compared to mock infected cells at 16 h.p.i. (Fig. 2a) or RRV infected cells compared to RRV infected cells at 16 h.p.i. (not shown), few transcripts varied by more than 2-fold: 0.9, 1.15, 1.68, 1.71 and 2.2% for the five control hybridizations in Table 1. This variability could be due to variability in the microarray technique (extraction and purification of the mRNA, cDNA synthesis and labeling, hybridization and/or signal detection) or to natural transcriptional variability that may occur in Caco-2 cells in culture. When we compared RRV infected cells vs. mock infected cells at 1 hour post infection (not shown), 1.0, 1.2 and 1.8% of the genes varied more than 2-fold (for the three RRV vs. mock hybridizations in Table 1), which was similar to the background rate. However, when we compared RRV infected cells vs. mock infected cells at 16 h.p.i., 8.6, 8.9 and 11% (for the three RRV vs. mock hybridizations in Table 1) of the gene transcripts varied more than 2-fold (Fig. 2b). These results indicate that rotavirus infection induces changes in cellular gene expression at 16 h.p.i..

Genes that respond to RRV infection

It was clear from our analysis (Fig. 2a) that there is an intrinsic variability in our cell culture system, since there were some changes in cellular gene expression which were detected when comparing RNA extractions from cells treated identically. In order to identify genes that specifically responded to RRV infection, we performed independent infections with corresponding independent controls (Table 1).

We undertook a series of steps to identify genes that responded specifically to RRV infection. First we selected genes that passed the filter criteria and did not change by more than 1.4-fold in control comparisons (mock vs. mock and infection vs. infection,

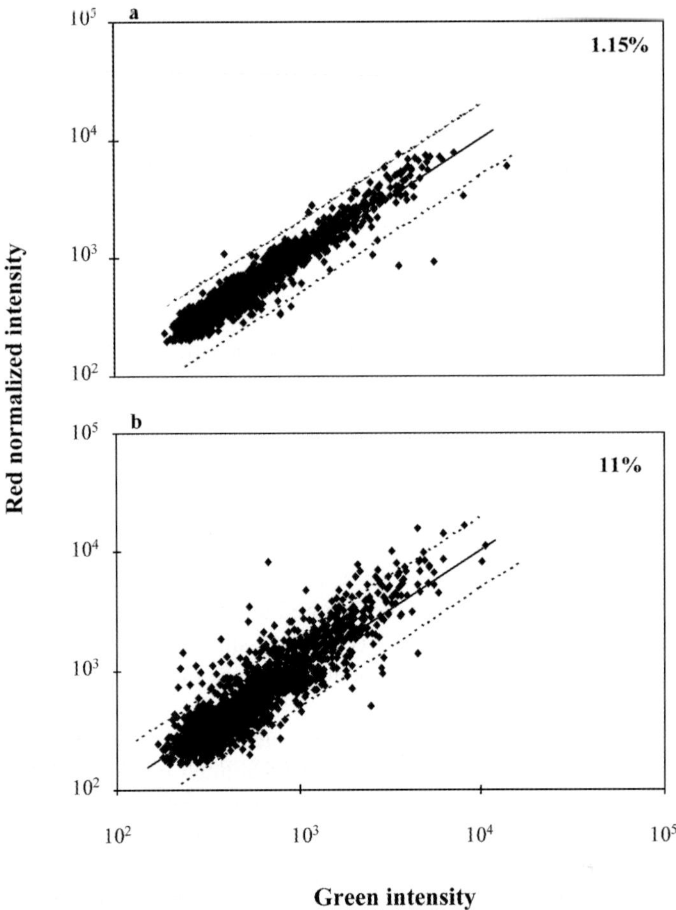

Fig. 2. Hybridization results from mock-infected cells vs mock-infected cells (a), and RRV-infected cells vs mock-infected cells (b) at 16 h.p.i. X and Y axes represent the intensity for the green fluorescent dye Cy3 and the normalized intensity (see Signal detection and data analysis) of the red fluorescent dye Cy5, respectively. Solid lines represent no change in the level of gene expression and dashed lines indicate a two fold change in the transcriptional response. The percentage of genes that changed more than twofold is indicated for each panel.

Table 1). We then determined how many of those genes passed the filter criteria in the experimental comparisons (RRV vs. mock) at 1 and 16 hours, and how many of these were up- or downregulated.

Between 35.1% and 47.7% of the human printed array elements (13,506 to 18,339 out of 38,432 human printed elements) were suitable for analysis after initial filtration. Following the elimination of genes that varied 1.4-fold or more in the control comparisons we identified a list of 8,528 genes for the 1 hour time point, and of 9,171 genes for the 16 hours time point that were suitable for further study. These genes were analyzed in the infectious comparisons (RRV vs. mock). Of the 8,528 genes, 7,263 gave a signal above background in the 1 hour infection comparisons (RRV vs. mock) and of the 9,171 genes from the 16 hours time point, 8,575 gave a signal above the background in the 16 hours infection comparisons (RRV vs. mock).

A similar analysis with the non-human genes (the arrays contain 448 non-human genes from yeast and bacteria, see Material and Methods) was also performed. Only one gene (0.22%) passed the filter criteria in more than 50% of the eleven arrays examined.

We used the following rationale to select the threshold for classifying genes as up- or downregulated. First, we wanted to ensure that genes identified were actually regulated by infection, and for this purpose, the higher the fold change selected as a threshold, the higher the likelihood of significance. However, we do not yet know what significance to assign to lower level changes in transcription. Therefore, we arbitrarily choose to classify those genes whose transcriptional level changed by more than 2-fold in at least two of the three experimental hybridizations (Table 1) as genes regulated by rotavirus infection. Two-fold or greater responses has been used as selection criteria by others as well to identify transcriptionally regulated genes by microarray analysis (9, 25, 30). As described above, none of the selected genes varied by more that 1.4 fold in any of the control comparisons. Five hundred and eight genes were up- or downregulated more than two fold at 16 h.p.i. (Fig. 3). A similar analysis at the 1 hour time point disclosed only one gene which changed by more than two fold (the potassium large conductance Ca^{2+} activated channel).

Several observations can be made from this analysis. It is clear that after one hour of rotavirus infection, Caco-2 cells showed a very limited transcriptional response to infection (only 1 gene was transcriptionaly regulated by more than 2-fold). In contrast, at 16 h.p.i., 579 array elements, representing 511 genes (some genes were printed more than once) were up- or downregulated. Of these, 375 (73.4%) were upregulated and 133 (26.02%) were downregulated. Of the 579 array elements identified only three (0.58%) gave inconsistent results, being upregulated in one hybridization, and down regulated in other hybridization. Of note, the fluorochrome signal used for labeling did not affect the results, since labeling the cDNAs in the opposite way (cDNA from mock infected cells labeled with Cy5 and cDNA from RRV infected cells with Cy3) had very little effect on the outcome (data not shown). The 579 array elements that passed the 2-fold change criteria represent 6.7% of the total analyzable population (8,575 genes).

An examination of the genes in Fig. 3 shows variation in the intensities of the green and red signals obtained. This reflects different levels of up- and downregulation among the various genes. We analyzed the level of change by identifying those

Fig. 3. Cluster analysis of genes that were differentially expressed after 16 hours of RRV infection. Each horizontal row represents a single cDNA and each vertical column represents a single microarray hybridization. The results are presented in color display; each square represents the ratio of hybridization signal of labeled cDNA prepared from mRNA of RRV-infected or mock-infected cells relative to mock-infected cells. Red squares denote upregulated genes, green squares downregulation, black squares denote no significant change in the level of gene expression and gray squares indicate missing data. The colored

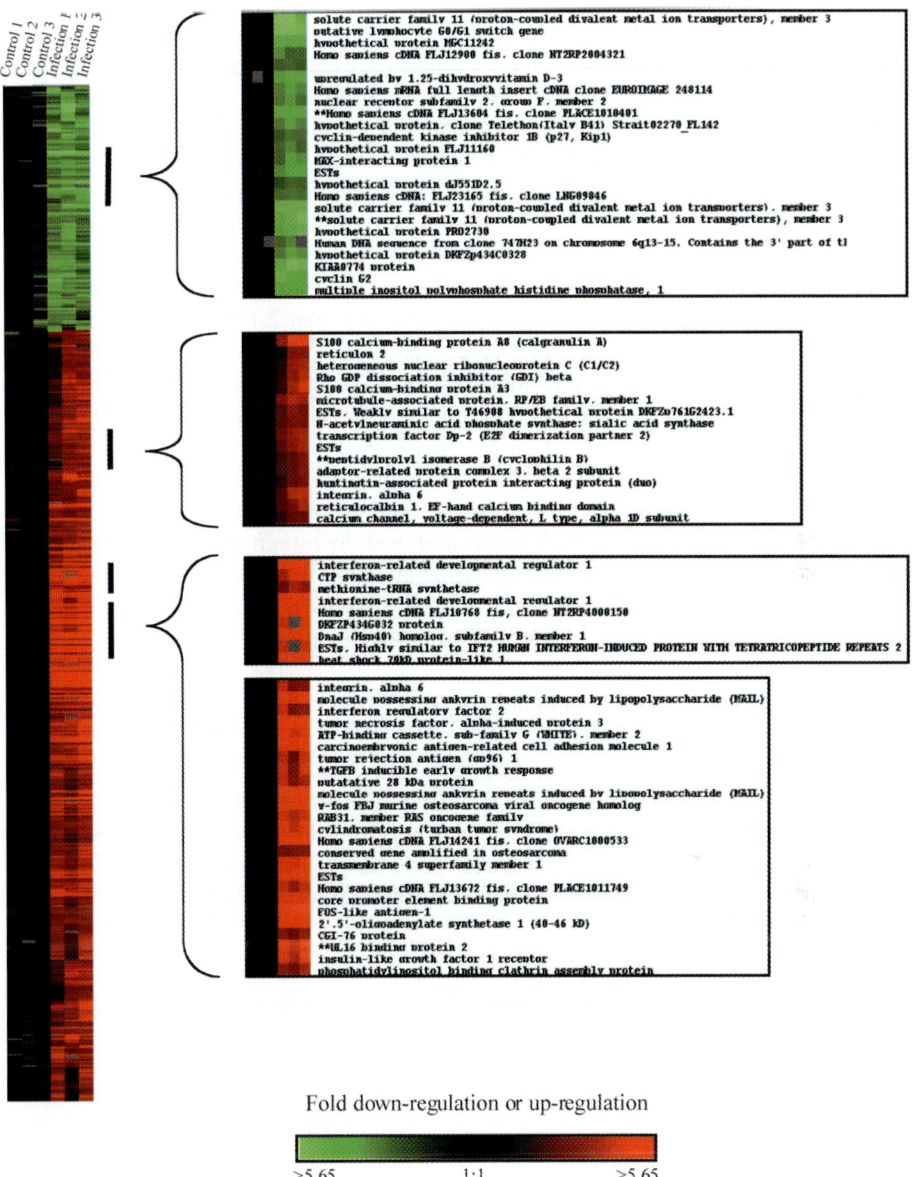

scale for the level of up- or downregulation is depicted at the bottom. The first three columns correspond to control hybridizations (mock infected vs mock infected or RRV infected vs RRV infected) and the last three columns correspond to experimental hybridizations (RRV infected vs mock infected). Genes shown are those whose transcript levels varied more than two fold in at least two of the three experimental comparisons. Some regions are amplified at the right to show the name and the expression profile of selected genes (with permission from Journals Department ASM[16]).

genes that changed their level of expression between two- and four-fold and those that changed more than four fold at 16 h.p.i. Changes between 2- and 4-fold are most frequent. 463 genes were regulated between 2- and 4-fold and only 45 genes were regulated more than 4-fold. Of those, 42 were upregulated, and three were downregulated. 5 genes were regulated more than 8-fold, and those were all upregulated.

The results observed were consistent across infection comparisons using several types of analysis. The regulated genes were initially identified by setting the program to select genes that were up or downregulated more than 2-fold in a least two of the three experimental hybridizations (Table 1). In total, 304 of the 576 array elements (52%) or 266 of the 508 genes were regulated more than 2-fold in all three experimental hybridizations (hybridization 4, 5 and 6, Table 1), and of the remaining array elements (272), 180 were regulated more than 1.6-fold in the third hybridization. This means that 84% of the genes showed the same transcriptional regulation in the three separate hybridizations (hybridization 4, 5 and 6, Table 1). Of the remaining 16% of genes, half were not analyzable in the third hybridization. Only 3 genes (less than 0.04% of the total analyzed population) showed inconsistent regulation, meaning upregulation in one experimental hybridization and downregulation in another. From this initial analysis we conclude that rotavirus infection induces changes in the levels of 6.7% of cellular analyzable mRNAs in Caco-2 cells at 16 h.p.i. and that the majority of these transcriptional responses correspond to upregulation of genes.

The majority of the changes in cellular gene expression occur at late time points after infection

The results of the microarray analysis at 1 and 16 h.p.i. showed a significant difference in the transcriptional response between the two time points, from one gene at 1 h.p.i. to 508 genes at 16 h.p.i. To determine if the genes responding at 16 hours were regulated at earlier times during infection, we carried out an additional time course experiment in which mock- or

Table 1

Experimental design to analyze the transcriptional response of Caco-2 cells to RRV infection using cDNA microarrays

Experiment description	mRNA from[§]	mRNA from[¶]	# of Hybridizations
Controls 1 and 2	Mock 16 h	Mock 16 h	2
Control 3	RRV 16 h	RRV 16 h	1
Infections 1, 2 and 3	RRV 16 h	Mock 16 h	3
Controls 4 and 5	Mock 1 h	Mock 1 h	2
Infections 4, 5 and 6	RRV 1 h	Mock 1 h	3
			Total 11

[§] corresponding cDNA was fluorescently labeled with red dye (Cy5) during the reverse transcription reaction.
[¶] corresponding cDNA was fluorescently labeled with green dye (Cy3) during the reverse transcription reaction.

RRV-infected Caco-2 cells were harvested at 1, 6, 12, and 24 h.p.i.. Labeled cDNAs synthesized from mRNA from RRV-infected cells at each time point were mixed with labeled cDNAs synthesized from mRNA from mock infected cells at the corresponding time points (RRV 1h. vs. mock 1h., RRV 6 h. vs. mock 6 h., etc.). We also carried out a control comparison, by mixing labeled cDNAs from mock infected cells with labeled cDNAs from the same batch of mock infected cells (mock vs. mock hybridization). The mixed cDNAs were hybridized to microarrays and analyzed as in the first series of hybridizations (Table 1), the difference being that in the time course experiment, we used only one infection sample and one mock-infected sample for each time point.

We focused our analysis on the genes that were regulated by more than 2-fold in the first experiment (Fig. 3 and Table 2) at 16 h.p.i. in order to find out whether the changes observed at 16 h.p.i. could be detected at earlier time points. We eliminated genes that changed their level of expression by 1.4-fold or more in the control comparison of the time course experiment. We selected 227 genes that passed the filtering criteria (described in Materials and Methods) for all time points (control, 1, 6, 12 and 24 hours). Cluster analysis was then performed with the five hybridizations of the time course experiments and one (RRV vs. control) hybridization from the 16 hours time point. Fig. 4 shows the pattern of expression for selected genes that were up- or downregulated more than 2-fold in at least two hybridizations in the time course analysis. A common pattern was found in the time course study; the number of up- and downregulated genes increased with the time from infection, starting with very few genes after one hour and reaching the maximum at the latest time points (16 and 24 h.p.i.). In the time course experiment, up- and downregulated genes were observed, and again upregulated genes were most frequent (Fig. 4). Three out of 227 genes examined were transcriptionally regulated (more than 2-fold change) at 6 h.p.i. and 33 genes were transcriptionally regulated at 12 h.p.i.. It is clear from these results that most of the responses observed at 16 h.p.i. in the first experiment (Fig. 3, Table 1) became detectable late, i.e. after 12 h.p.i..

Grouping of genes according to their biological function

We identified 375 upregulated and 133 downregulated genes following rotavirus infection in our primary analysis. A partial list of these genes is presented in Table 2. In an attempt to facilitate the analysis of our data we grouped the differentially regulated genes according to known biological functions. However, the classification of genes to specific cellular functions is difficult as many gene products participate in more than one biological process. We listed the regulated genes in only a single functional category for simplicity. Only two genes listed in Table 2 have been previously described as rotavirus-induced genes (endoplasmin and RANTES). The regulation of some genes involved in calcium homeostasis, cytoskeleton structure, interferon-regulation and stress response were anticipated since previous data had demonstrated the relationship between rotavirus infection and these biological processes (5, 6, 19, 47, 53, 67). Genes associated with other important cell functions including cell cycle and proliferation, protein degradation, viral receptors and membrane transporters are also regulated in response to rotavirus infection (Table 2).

Table 2

List of some of the transcriptionally regulated genes in Caco-2 cells after 16 hours of RRV infection. Genes are grouped according to a related biological function

Gene name	SwissProt Acc. No.[b]	FC[c]	Gene name	SwissProt Acc.No.	FC	Gene name	SwissProt Acc. No.	FC
Protein modification and degradation			ras homolog gene family member E	P52199	2.1	Heat shock 70 kD protein 1B	P08107	0.4
*Itchy E3 ubiquitin protein ligase	P46934	2.6	Rho GDP dissociation inhibitor (GDI) β	P52566	2.7			
Ubiquitin ligase SMURF2	P39940	2.2	Rap1b member of RAS oncogene family	P09526	2.1	***Interferon response***		
*Yeast homolog ubiquitin ligase Rsp5	P46935	3.5	Protein regulator of cytokinesis 1	O43663	2.2	***Interferon-related developmental regulator 1**	O00458	**6.3**
Proteasome 26S subunit ATPase 3	P17980	2	*Microtubule associated protein member 1	P40013	2.6	*Interferon regulatory factor 2	P14316	2.6
*Matrix metalloproteinase 14	P50281	3.3				***2'-5' oligo (A) synthetase 1 (OAS1)**	P00973	**5**
Disintegrin and metalloproteinase domain 9	Q61072	2.1	***Cell adhesion and intercellular junctions***			Interferon induced with tetraticopeptide repeats 1	P09914	3.8
*hnRNP methyltransferase 1-like 2	Q99873	2.4	*Cadherin 5 type 2, VE-cadherin	P33151	3.4	Interferon-γ inducible protein 16	Q16666	2.7
Cathepsin D (lysosomal aspartyl protease)	P07339	**8.8**	Protocadherin 18	Q9HCL0	2.6	***Guanylate binding protein 1, 67 kD**	P32455	**6.2**
*Lectin-like oxidized-LDL receptor 1	P78380	3.8	*CEA1	P13688	3.4			

Table 2

Continued

Gene name	SwissProt Acc. No.[b]	FC[c]	Gene name	SwissProt Acc.No.	FC	Gene name	SwissProt Acc. No.	FC
*Kallikrein 6	Q92876	2.3	*CEA5	P06731	2.2	*Immune response*		
*Ariadne (Drosophila) homolog	P36113	2.4	Catenin alpha-like 1	Q61301	2.4	Small inducible cytokine A5 (RANTES)	P13501	2.6
*Ubc6p homolog	P51965	2.5	Vinculin	P18206	2.5	*Interleukin 2 receptor beta	P14784	2.9
Transmembrane protease, serine 2	O15393	0.5	*Epithelial V-like antigen 1	O60487	3.4	*Interleukin 1 receptor accessory protein	Q02955	2.4
			Vascular cell adhesion molecule 1	P19320	2.2	*UL16 binding protein 2	Q9BZM5	3.5
Trasncriptional regulators			*Thrombospondin 1	P07996	3.9	Pregnancy specific β-1-glycoprotein 1	P11464	3
*Transcription factor Dp-2	Q14188	2.5	Cartilage linking protein 1	P10915	2.3	Chornichon-like	O95406	2.1
FOS	P01100	**6.6**	Occluding	Q16625	2	*Prostaglandin-endoperoxide synthase 2	P35354	3.5
Fos-like antigen 1	P15407	**15**	*Catenin delta 1	O60716	0.4	*Leukotriene A4 hydrolase	P09960	2.7
JUN	P05412	**17**				*Spermine synthase	P52788	3.4
*Activating trasncription factor 4	P18848	2.7	*Cell cycle*			**Spermidine/spermine N1-acetyltransferase**	P21673	**6.7**
NFkB1	P19838	2.4	Cyclin C	P24863	2	*S-adenosylmethionine decarboxylase 1	P17707	2.6
NFkBIE	O00221	**5.4**	*Cyclin D1	P24385	2.9	*Amiloride binding protein 1	P19801	0.4

Table 2

Continued

Gene name	SwissProt Acc. No.[b]	FC[c]	Gene name	SwissProt Acc.No.	FC	Gene name	SwissProt Acc. No.	FC
*E74-like factor 3	P78545	3.8	Cyclin E1	P24864	2.1			
Thyroid hormone receptor interactor 8	Q15652	2.6	**Cyclin L ania-6a**	Q9UK58	**5.1**	*Membrane transporters*		
Far upstream element binding protein 1	P57722	2.3	*Cyclin-dependent kinase 7	P50613	2.4	K$^+$ channel subfamily K member 1	O00180	2.2
Bromodomain adjacent to zinc finger domain 1A	Q9UIG1	2.4	Pescadillo homolog 1	P54258	2.1	*Zinc/iron regulated transporter-like	Q9BTV0	2.4
MAIL	Q9BYH8	**6.4**	*BTG family member 2	P78543	2.5	*ATP-binding cassette subfamily G member 2	Q9UNQ0	2.9
cAMP responsive element modulator	Q03060	**5.4**	*Cyclin G2	Q16589	0.3	Solute carrier family 2 member 3	P11169	2.5
Core promoter element binding	Q99612	**4.3**	Cyclin-dependent kinase inhibitor 1B	P46527	0.5	Solute carrier family 26 member 3	P40879	2.2
Zinc finger protein (LOC51042) (IMAGE:471200)		2.8	**Putative lymphocyte G0/G1 switch gene**	P27469	0.2	Kynurenine 3-monooxygenase	Q9H1Y5	2
*TGFB inducible early growth response	Q13118	2.9	Damage-specific DNA binding protein 1	Q16531	0.4	Transferrin receptor	P02786	2.3
*RNA polymerase II elongation factor ELL2	O00472	2.9	*Ataxia telangiectasia mutated	P42346	0.4	Ferredoxin 1	P10109	2.1

Table 2

Continued

Gene name	SwissProt Acc. No.[b]	FC[c]	Gene name	SwissProt Acc.No.	FC	Gene name	SwissProt Acc. No.	FC
La ribonucleoprotein	P05455	2.4				*Thioredoxin reductase 1	Q16881	2.5
Zinc finger protein 9	P20694	2.1	*Cellular development and differentiation*			*IMP dehydrogenase2	P12268	2.3
Zinc finger protein 216	O76080	2.3	**Retinoic acid induced 3**	O95357	**5.1**	**Steroid-5-aplha reductase, alpha polypeptide 1**	P18405	**6.4**
Zinc finger protein 313	ZNF313	2.6	*Fibroblast growth factor 7	P21781	2.7	*Methylene tetrahydrofolate dehydrogenase 2	P36776	2.8
*DNA damage-inducible transcript 3	P35638	2.5	*Enhancer filamentation 1	Q14511	2.3	*K$^+$voltage-gated channel subfamily H member3	Q12809	0.4
Nuclear receptor co-activator 2	P41738	0.4	Oncogene TC21	P10833	2.2	*Solute carrier family 11 member 3	Q9NRL0	0.3
*Transcriptional intermediary factor 1	Q64127	0.4	*N-myc downstream regulated	Q92597	3	*Duodenal cytochrome b	P49447	0.3
Zinc-fingers and homeoboxes 1	Q9UKY1	0.3	*Ras association (RalGDS/AF-6) domain family 1	O60539	2.2			
*p300/CBP-associated factor	P13941	0.3	*Tax interaction protein 1	O14907	2.8	*Signal transduction*		
*MAX-interacting protein 1	P50539	0.4	*Nucleolar protein p40	Q99848	3	*Integrin alpha 6	P23229	2.7
*Nuclear receptor subfamily 2, group F, member 2 (P24468)	P24468	0.35	**Epithelial membrane protein 1**	P54849	**4.2**	**Plasminogen activator urokinase receptor**	Q03405	**4.7**

Table 2

Continued

Gene name	SwissProt Acc. No.[b]	FC[c]	Gene name	SwissProt Acc.No.	FC	Gene name	SwissProt Acc. No.	FC
						Tetraspan NET-7	O95858	2.7
Apoptosis			*RNA processing and modification*			Zinc finger protein 259	O75312	2.3
Programmed cell death 5	P56812	2.8	*U3 snoRNP-associated 55-kDa protein	O43818	2.4	SBBI31 protein (IMAGE:429470)		2.3
Apoptosis antagonizing transcription factor	Q9UNX5	2.2	*Nucleolar protein 1 120 kDa	P46087	2.7	***TNF receptor superfamily member 11b**	P20333	**4.4**
*Cytochrome c	P00009	2.6	Splicing factor arginine/serine-rich 7	Q16629	2.2	*Oncogene ERBB3	P21860	0.4
***TNFα-induced protein 3**	P21580	**6.5**	Splicing factor arginine/serine-rich 3	P23152	2.3	Vav 3 oncogene	Q9UKW4	0.4
*Immediate early response 3	P46695	3.8	*Heterogeneous ribonucleoprotein A/B	Q99383	2.3	*Ral guanine nucleotide exchange factor	Q9NZ16	0.4
*Baculoviral IAP repeat-containing 3	Q13490	3.8	Heterogeneous ribonucleoprotein C	P07910	2.7	Epidermal growth factor receptor	P00533	0.5
*Insulin-like growth factor receptor	P08069	3.1	*phorbolin	Q9UH17	2.8			
SERPINB2	P05120	**5.7**				*Vesicle formation and intracellular transport*		

Table 2

Continued

Gene name	SwissProt Acc. No.[b]	FC[c]	Gene name	SwissProt Acc.No.	FC	Gene name	SwissProt Acc. No.	FC
*SH3GLB2	Q99961	0.5	**Kinases/phosphatases**			*Caveolin 1	Q03135	3
			NF-kappa-β-induced kinase	Q99558	2.4	*Caveolin 2	P51636	3.6
Translation machinery			PCTAIRE protein kinase 2	Q00537	2	*PI-clathrin assembly protein	Q13492	2.6
EIF1A	Q60872	2.4	*Serum inducible kinase	Q9NYY3	3.2	Adaptor-related protein complex 3, β2 subunit	P21851	2.2
EIF2S2	P20042	2.3	Serum/glucocorticoid regulated kinase	O00141	2.5	*ADP-ribosylation factor binding protein GGA3	P22892	2.2
EIF2B2	P49770	2.4	Protein tyrosine kinase 2β	Q14289	2.5	Myosin X	P19524	2
EIF3S1	O75822	2.5	Kinase insert domain receptor	P35968	2.7	*RAB31, RAS oncogene family	Q13636	2.7
EIF3S5	O00303	2.3	Uridine monophosphate kinase	P52623	2.3	Syntaxin 3A	Q13277	2.5
*EIF4A1	P04765	2.6	Hexokinase 2	P52789	2.6	*Myomegalin	P25386	0.4
SUI1	Q9UNQ9	2.4	Ethanolamine kinase (EKI1)	Q9HBU6	2.2			
Alanine-tRNA synthetase	P49588	2.2	Tyrosine 3-monooxygenase	P29312	2.5	**CNS-specific functions/ neuropeptides**		
Lysyl-tRNA synthetase	Q15046	2.1	*cAMP-dependent Protein kinase inhibitor beta	P27775	2.5	***Neuronal protein (NP25)**	Q9UI15	**6.3**
Phenylalanyl-tRNA synthetase β	Q9NSD9	2.3	Protein phophatase 2A, β-isoform	P11082	2	***SRY-box 11 (SOX11)**	P35716	**6.4**

Table 2

Continued

Gene name	SwissProt Acc. No.[b]	FC[c]	Gene name	SwissProt Acc.No.	FC	Gene name	SwissProt Acc. No.	FC
Tyrosyl-tRNA synthetase	P54577	2.1	*MKP-1 like protein tyrosine phosphatase	P28563	2.8	*Dunce-like phosphodiesterasa E4	Q07343	3.4
*RNA helicase	P40562	4.4	*Dual specificity phosphatase 6	Q16828	2.6	*Gamma-aminobutyric acid A receptor γ2	P18507	2.8
DEAD/H box polypeptide 3	O00571	2.3	Pyrophosphatase inorganic	Q15181	2.3	Huntingtin-associated protein interacting protein	O60229	2.2
*DEAD/H box polypeptide 16	O60231	2.4	*MINPP1 (IMAGE:592630)		0.4	*Cholecystokinin	P06307	4.4
DEAD/H box polypeptide 21	Q64060	2.3				*Glycoprotein hormones α polypeptide	P01215	14.7
DEAD/ H box polypeptide Y chromosome	O15523	2.2	Lipids biosynthesis and drug metabolism			*Gamma-aminobutyric acid A receptor ε	P78334	0.4
*hqp0256 protein	P38712	2.5	*Dual adaptor of phosphotyrosine	Q9UN19	5.8			
*Nucleophosmin 1	P06748	2.6	*Phosphoplipid scramblase 1	P47140	2.5	Enzymes		
Signal recognition particle 19kD	P09132	2.2	*Niemann-Pick disease type C1	O15118	2.4	N-acetylneuraminic acid phosphatase synthase	Q12999	2.3
			Stomatin-like 1	Q9UBI4	2.1	Iduronidase alpha L	P35475	2.1

Table 2

Continued

Gene name	SwissProt Acc. No.[b]	FC[c]	Gene name	SwissProt Acc.No.	FC	Gene name	SwissProt Acc. No.	FC
Cytoskeleton and cell structure			*Cytochrome b5 reductasa b5R.2	P20070	3.6	*Rab geranylgeranyltransferase, β subunit (RABGGTB)	P53611	2.7
*Actin alpha cardiac muscle	P04270	3.9	*Fatty acid CoA ligase long-chain 4	O60488	2.7	Phosphorylase, glycogen; liver	P06737	2.1
*Actin alpha 2 smooth muscle, aorta	P03996	3.3	*Alpha-methylacyl-CoA racemase	P70473	0.5	Enolase 1, alpha	P06733	2.2
Actinin alpha 4	P08640	2.3	*CTP synthase	P17812	3.3	*Glutathione S-transferase A2	P09210	2.6
Actin related protein 2/3 complex subunit 2	O15144	2.1	Cytochrome P450 subfamily IIB polypetide 6	P20813	2.3	*Glutathione S-transferase A3	Q16772	2.8
Calponin 3 acidic isoform	Q15417	2.6				Leucine aminopeptidase	P28838	2.4
Caldesmon 1	Q05682	2.9	**Stress response**			*Putative b,b-carotene-9',10'-dioxygenase	Q03393	2.9
Ectodermal-neural cortex	O14682	2.2	*HspC030	Q9Y6D0	2.4	*Solute carrier family 3, member 2	P40884	2.2
Thymosin beta 4 Y chromosome	O14604	3.1	**Hsp40**	Q9QYJ3	**7.2**	*Indolethylamine N-methyltransferase	O95050	3.1
*Lymphocyte cytosolic protein 1	P13796	3.3	**Hsp70 kD protein-like 1**	P34931	**6.7**	Aminopeptidase A	Q07075	2.2
*Vimentin	P08670	2.1	*Tumor rejection antigen 1	P14625	2.7	UDP-glucose ceramide glucosyltransferase	Q16739	2.3

Table 2

Continued

Gene name	SwissProt Acc. No.[b]	FC[c]	Gene name	SwissProt Acc.No.	FC	Gene name	SwissProt Acc. No.	FC
*Keratin 4	P19013	3	*GrpE-like protein cochaperone	P97576	2.5	Alpha 2,3-sialyltransferase	Q02734	0.5
*Keratin 5	P13647	3.6	*Sperm-associated antigen 1	P50503	2.6	*Mannosidase alpha class 1A, member 1	P33908	0.3
Keratin 15	P19012	2.4	**Crystalline alpha B**	O43416	**4.3**	Glycogen debranching enzyme	P35573	0.4
*Syndecan binding protein	O17583	2.9	**Small stress protein-like protein Hsp22**	Q9UJY1	**4.7**	PFKFB4	Q16877	0.5
Myosin light polypeptide 1	P05976	2.2	*Chaperonin containing TCP1 subunit 5ε	P48643	2.8	Sorbitol dehidrogenase	Q00796	0.4
Tropomyosin 4	P07226	2.6	Chaperonin containing TCP1 subunit 4δ	P50991	2.3			
Bystin-like	Q13895	2.5	*Chaperonin containing TCP1 subunit 6A (zeta 1)	P40227	2.2	*Viral receptors*		
Small proline-rich protein 1B	P22528	**7.1**	Peptidylprolyl isomerase B	P23284	2.2	*Decay accelerating factor for complement	P08174	2.7
*Testin	O35115	2.9	Oxidative-stress responsive 1	Q9Z1W9	2.3	Integrin alpha 2	P17301	3.1
Epithelial protein lost in neoplasma β	Q9UHB6	2.6	Clusterin	P10909	2.2	Integrin beta 1	P05556	2.8
Calcium homeostasis			Reticulon 2 (RTN2/NSP2/NSPL1)	O75298	2.4	Phorbolin like protein MDS019	Q9UH17	3.1
S100 calcium binding protein A8	P05109	**4.6**	*Cysteine and hystidine-rich domain (CHORDC1)	Q9NZ93	2.8	Hematological and neurological expressed 1	Q9UK76	2.3

Table 2

Continued

Gene name	SwissProt Acc. No.[b]	FC[c]	Gene name	SwissProt Acc.No.	FC	Gene name	SwissProt Acc. No.	FC
S100 calcium binding protein A3 (S100A3/S110E)	P33764	3.1	Semaphorin L	O75326	2.4	Reelin	P78509	3.5
Reticulocalbin 1, EF-hand calcium binding domain	Q15293	2.5	*Dedicator of cyto-kinesis 3 (IMAGE:39922)		2.3	*Golgin 67	Q9NZW0	0.5
Ca^{2+} channel voltage-dependent, L type subunit α1D	Q01668	2.1	*Schwannomin-interacting protein 1 (IMAGE:506143)		3.6	Synaptogyrin 1 (SYNGR1)	O43759	0.5
*Annexin A9	Q07936	0.4	Uncharacterized bone marrow protein BM040	Q9NZ82	2.2	**Semaphorin 4G**	Q9WUH7	**0.2**
*Stanniocalcin 1	P52823	0.4	*Enhancer of polycomb 1	Q9H7T7	2.6	*Disabled (Drosophila) homolog 1	O75553	0.4
			Human proteinase activated receptor-2 mRNA, 3'UTR	P39195	2.2	*H. sapiens LUCA-15 protein mRNA	P52756	0.3
Histones			*Arginine-rich, mutated in early stage tumors	Q9N3B0	3.1	*G protein-coupled receptor GPRC5B	Q9NZH0	0.3
H1 histone family member 2	P16403	2.2	*Epithelial stromal interaction 1 (IMAGE:1855351)		2.3	*Upregulated by 1,25-dihydroxyvitamin D-3	Q9H3M7	0.3
*H2A histone family member N	P02262	2.1	*PVT1 (murine) oncogene homolog (IMAGE:295410)		2.8	Wilm's tumor 1-associating protein	Q15007	0.4
			*GE36 gene GE36 gene	Q9H2B6	2.5	HP1-BP74	Q9UHY0	0.4

Table 2

Continued

Gene name	SwissProt Acc. No.[b]	FC[c]	Gene name	SwissProt Acc.No.	FC	Gene name	SwissProt Acc. No.	FC
Miscellaneous			*Desmoplakin	P15924	3	TACC2	O95359	0.4
WD repeat domain 1	O88342	2.6	*Laminin, gamma 2	Q13753	3.6	Mitochondrial ribosomal protein S14	O60783	0.4
*Transmembrane 4 superfamily member 1	P30408	3.5	*Cylindromatosis (turnban tumor syndrome)	Q9NQC7	3.1	Sorting nexin 17	Q15036	0.5
*Nucleolar phosphoprotein Nopp34	Q9BYG3	2.6	***Serine (or cysteine) proteinase inhibitor**	P05120	**5.7**	Lipin 2	Q92539	0.5
Sperm-associated antigen 9	P34609	2.5	***Brain specific protein**	Q9Y326	**4.6**	Amyotrophic lateral sclerosis 2	Q9C0K7	0.4
Albumin	P02768	3.8	Cornichon-like	O95406	2.1	*Pleckstrin homology domain interacting protein	Q9H261	0.3
Prion protein (p27-30)	P04156	2.4	Karyopherin alpha 2	P52292	2.2			

Only genes whose name and/or function are known are indicated. For a complete list of the 508 genes that changed their transcriptional level during rotavirus infection see Table 3 in (16). The functional groups of genes are ordered vertically (i.e. follow the left column from p 270 to p 280, then the middle and right columns in the same way.

[a] As registered in Stanford Microarray Database (57).

[b] For those genes that do not have a SwissProt accession number the I.M.A.G.E. number is indicated as resported in the human cDNA collection (37). The SWISS-PROT entry is copyright. It is produced through a collaboration between the Swiss Institute of Bioinformatics and the EMBL outstation -the European Bioinformatics Institute. There are no restrictions on its use by non-profit institutions as long as its content is in no way modified and this statement is not removed. Usage by and for commercial entities requires a license agreement (See http://www.isb-sib.ch/announce/).

[c] Fold change. Values greater than 2 and less than 0.5 correspond to up and downregulated genes respectively.

* Genes whose transcriptional level changed more than two-fold in all three experimental hybridizations (see Table 1). In bold are indicated those genes whose transcriptional level changed more than 4-fold.

Possible relationship between the host transcriptional response and the viral replication cycle

A clear understanding of the mechanistic relationship between transcriptional regulation changes and rotavirus replication will require additional studies. However, it is reasonable to begin to examine some of the data obtained here and to speculate on possible relationships between the host transcriptional response observed and the viral replication cycle.

Among the transcripts upregulated in infected Caco-2 cells, integrin α2, integrin β1 and HSPA1L (a homolog of HSP70) were identified. The mRNA transcript for sialic acid synthase, an enzyme involved in the biosynthetic pathway of sialic acids, was also upregulated. Interestingly, it has been shown that the products of those genes, and sialic acid, are involved in attachment and entry of rotavirus into cells (see Section II, Chapter 3 of this book). Although the protein expression of these putative viral receptor transcripts needs to be directly measured, upregulation of these receptor mRNAs raises several interesting possibilities. Rotavirus infection might upregulate expression of its own cellular receptor and/or enhance receptor upregulation in surrounding uninfected cells facilitating viral entry and the spread of the infection. On the other hand, overexpression of the rotavirus receptors in infected cells could mediate binding and "neutralization" of the newly formed and shed viral particles.

Several genes involved in protein synthesis were upregulated following rotavirus infection. These include 6 eukaryotic translation initiation factors (eIF1A, eIF2S2, eIF2B2, eIF3S1, eIF3S5, eIF4A1), four tRNA synthetases (alanyl, phenylalanyl, lysyl and tyrosyl tRNA synthases) and four DEAD box proteins (DEAD/H box polypeptide 3, 16, 21, Y chromosome) (putative RNA helicases, involved in translation initiation). No translation elongation factors were found to be modulated in this analysis. It has been shown that rotavirus infection mediates a reduction of cellular protein synthesis, favoring translation of viral proteins (27, 50). If the upregulation of the translation associated genes seen here is correlated with an upregulation of the corresponding host proteins, one might speculate that rotavirus induces an upregulation of the protein translation machinery of the cell and uses this machinery for the synthesis of its own proteins, while simultaneously blocking host cellular protein synthesis. In this manner, two mechanisms could be used by the virus for efficient translation of its own proteins. One would involve an upregulation of cellular translation factors for a more efficient translation of viral proteins; and the other would direct the cell translation machinery to specifically favor viral genes, due to specific interaction of NSP-3 and eIF4G1 (50).

Among the 508 genes differentially regulated after 16 hours of rotavirus infection, six are interferon inducible genes: 2'5'-oligoadenylate synthetase 1 (OAS1), interferon regulatory factor 2 (IRF2), interferon-related developmental regulator 1 (IFRD1), interferon-induced protein with tetratricopeptide repeats 1 (IFIT1), guanylate binding protein 1 (GBP1) and interferon gamma-inducible protein 16 (IFI16). The upregulation of the gene encoding OAS1 is potentially significant. This synthetase is stimulated by dsRNA to produce 2'-5' linked oligoadenylates. The principal function of these products is to activate the latent ribonuclease L which in turns degrades

Homo sapiens cDNA: FLJ21278 fis, clone COL01832
**Homo sapiens cDNA FLJ13604 fis, clone PLACE1010401
hypothetical protein DKFZp566G1424
cyclin-dependent kinase inhibitor 1B (p27, Kip1)
nuclear receptor subfamily 2, group F, member 2
upregulated by 1,25-dihydroxyvitamin D-3
hypothetical protein, clone Telethon(Italy B41) Strait02270_FL142
multiple inositol polyphosphate histidine phosphatase, 1
alpha-methylacyl-CoA racemase
solute carrier family 11 (proton-coupled divalent metal ion transporters), member 3
hypothetical protein FLJ23516
solute carrier family 11 (proton-coupled divalent metal ion transporters), member 3
solute carrier family 11 (proton-coupled divalent metal ion transporters), member 3
Homo sapiens, clone IMAGE:3450973, mRNA
hypothetical protein PRO2730
**solute carrier family 11 (proton-coupled divalent metal ion transporters), member 3
**Homo sapiens mRNA: cDNA DKFZp586B1722 (from clone DKFZp586B1722)
mannosidase, alpha, class 1A, member 1
hypothetical protein DKFZp434C0328
Homo sapiens, Similar to hypothetical protein FLJ10883, clone IMAGE:3855861, mRNA, partial cds
hypothetical protein FLJ13057 similar to germ cell-less
cyclin G2
ESTs
Homo sapiens cDNA FLJ14368 fis, clone HEMBA1001122
annexin A9
molecule possessing ankyrin repeats induced by lipopolysaccharide (MAIL), homolog of mouse
putative b,b-carotene-9',10'-dioxygenase
arginine-rich, mutated in early stage tumors
eukaryotic translation initiation factor 2B, subunit 2 (beta, 39kD)
solute carrier family 3 (activators of dibasic and neutral amino acid transport), member 2
chaperonin containing TCP1, subunit 5 (epsilon)
potassium channel, subfamily K, member 1 (TWIK-1)
heterogeneous nuclear ribonucleoprotein A/B
Tax interaction protein 1
nuclear FGF3 binding protein
cytochrome b5 reductase b5R.2
zinc finger protein
HKP-1 like protein tyrosine phosphatase
E74-like factor 3 (ets domain transcription factor, epithelial-specific)
R3H domain (binds single-stranded nucleic acids) containing
Rho GDP dissociation inhibitor (GDI) beta
transcription factor Dp-2 (E2F dimerization partner 2)
hypothetical protein MGC13007
caveolin 1, caveolae protein, 22kD
solute carrier family 2 (facilitated glucose transporter), member 3
hypothetical protein MGC14376
heat shock 70kD protein-like 1
steroid-5-alpha-reductase, alpha polypeptide 1 (3-oxo-5 alpha-steroid delta 4-dehydrogenase alpha 1)
TONDU
caveolin 2
serum-inducible kinase
ectodermal-neural cortex (with BTB-like domain)
v-jun avian sarcoma virus 17 oncogene homolog
Ubc6p homolog
oxidised low density lipoprotein (lectin-like) receptor 1
carcinoembryonic antigen-related cell adhesion molecule 1 (biliary glycoprotein)
calponin 3, acidic
vinculin
ESTs, Highly similar to KHHUD cathepsin D [H.sapiens]
lymphocyte cytosolic protein 1 (L-plastin)
laminin, gamma 2 (nicein (100kD), kalinin (105kD), BM600 (100kD), Herlitz junctional epidermolysis bullosa))
spermidine/spermine N1-acetyltransferase
spermine synthase

protein kinase H11; small stress protein-like protein HSP22
syntaxin 3A
ESTs
decay accelerating factor for complement (CD55, Cromer blood group system)
interferon-related developmental regulator 1
cyclin-dependent kinase 7 (homolog of Xenopus MO15 cdk-activating kinase)
transmembrane 4 superfamily member 1
ESTs, Weakly similar to I37356 epithelial microtubule-associated protein, 115K [H.sapiens]
integrin, alpha 6
dual specificity phosphatase 6
Homo sapiens mRNA: cDNA DKFZp564C2063 (from clone DKFZp564C2063)
nuclear factor of kappa light polypeptide gene enhancer in B-cells 1 (p105)
cartilage linking protein 1
splicing factor, arginine/serine-rich 3
myosin IB
KIAA0005 gene product
thioredoxin reductase 1
v-fos FBJ murine osteosarcoma viral oncogene homolog
KIAA0410 gene product

viral and cellular single stranded RNA (ssRNA). During its replication, rotavirus generates viral dsRNA as well as ssRNA with ds secondary structure. If the increase in mRNA of OAS1 detected in this analysis reflects an increase of this protein, we can suggest that the OAS1/RNAse L system is a mechanism of cellular defense against rotavirus infection. Several previous reports suggested a possible role for IFNs in host defense against rotavirus infectious (3, 11, 12, 18, 35, 36, 39, 53, 56).

Several studies indicate that calcium is a critical factor in rotavirus cytopathology. On the one hand this cation is required for rotavirus morphogenesis while on the other hand calcium accumulation appears to be responsible for the cytopathic effect and cell death observed at late stages of infection. A progressive increase in plasma membrane permeability to mono and divalent cations has been found a few hours after rotavirus infection (19, 47). The characterization of the Ca^{2+} entry pathway suggests that rotavirus infection activates an L-type Ca^{2+}channel of the plasma membrane in MA104 and HT-29 cells (47). Interestingly, in this study we reported the upregulation of the alpha 1D subunit of this L-type calcium channel supporting the involvement of this channel in the pathway of plasma membrane permeability observed during rotavirus infection.

The genes encoding for S100A3 and S100A8 were also upregulated. These proteins belong to a family of low molecular weight Ca^{2+}-binding proteins of the EF-hand type known as S100. Previous work has implicated S100 proteins in Ca^{2+}-dependent regulation of intracellular and extracellular functions such as protein phosphorylation, calcium homeostasis, inflammation and regulation of the dynamics of cytoskeleton components (20). The S100A8/S100A9 complex modulates Ca^{2+}-dependent interactions between vimentin, keratin intermediate filaments and membranes. S100A2 plays a role in the organization of the actin cytoskeleton by regulating F actin-tropomyosin interactions in epithelial cell lines in a Ca^{2+}-dependent manner. It is interesting to note that the rotavirus-induced increases in $[Ca^{2+}]_i$ has been shown to be directly responsible for the disassembly of microvillar F-actin and the microtubule network in differentiated Caco-2 cells at late time after rotavirus infection (5, 6). If the increase in mRNA of S100A3 and S100A8 reported in this analysis reflects an increase of these proteins, it seems possible that these proteins participate in the Ca^{2+}-dependent disorganization of cytoskeleton observed in RRV-infected cells.

Fig. 4. Cluster analysis of selected genes that were significantly up or downregulated at 16 h.p.i during a time course of rotavirus infection. An array from the 16 hours time point of the first experiment is included for comparison. Genes that passed filtering criteria in all the hybridizations and were not regulated more than 1.4-fold in the control hybridization were studied. Of the selected genes, genes that were regulated more than 2-fold in at least two arrays are shown. The color coding is the same as in Fig. 3 (with permission from Journals Department ASM[16]).

Conclusions

In order to better understand the molecular mechanisms involved in the events following rotavirus infection, we identified host cellular genes whose mRNA levels changed after infection. For this analysis, we used microarrays containing more than 38,000 human cDNAs to study the transcriptional response of the human intestinal cell line Caco-2 to rotavirus infection. We found that 508 genes were differentially regulated by more than two-fold at sixteen hours after rotavirus infection and only one gene was similarly regulated at one hour post infection. 73% of these transcriptional changes corresponded to upregulation of genes, with the majority of them occurring late, twelve or more hours post infection. Some of the regulated genes were classified according to known biological function and included genes encoding integral membrane proteins, interferon regulated genes, transcriptional and translational regulators, and calcium metabolism related genes. A new picture of global transcriptional regulation in the infected cell is presented, and families of genes which may be involved in viral pathogenesis are discussed. Although the relationships between cellular mRNA levels and the rotavirus replication cycle are not clear at present, further characterization of the response of individual genes can provide a better understanding of host/pathogen interaction.

Future directions

The potential applications of this new technology in basic and clinical research are vast. Future studies in rotavirus should focus on which steps of the viral replication cycle (e.g., binding, entry, transcription and translation, assembly, etc.) are responsible for the transcriptional changes observed. We have started a series of experiments with genetically inactivated virus, in an effort to differentiate between genes that respond solely to viral binding and penetration from those that respond to viral replication. The characterization of a wide variety of viral mutants will also help to identify which viral genes are responsible to mediate specific transcriptional changes in the host. The availability of mouse microarrays will allow us to examine whether the transcriptional response program identified in this cell culture model is representative of changes seen in the mouse intestine *in vivo*.

This work was supported by a V.A. merit review grant, by NIH grants AI21362 and DK38707 and by DDC grant DK56339. This work was published, in large part, in Journal of Virology, 2002; 76: 4467-82 (ref 16).

References

1. Aitman, T. J. 2001. DNA microarrays in medical practice. Brit Med J. 323:611-5.
2. Alizadeh, A. A., M. B. Eisen, R. E. Davis, C. Ma, I. S. Lossos, A. Rosenwald, J. C. Boldrick, H. Sabet, T. Tran, X. Yu, J. I. Powell, L. Yang, G. E. Marti, T.

Moore, J. Hudson, Jr., L. Lu, D. B. Lewis, R. Tibshirani, G. Sherlock, W. C. Chan, T. C. Greiner, D. D. Weisenburger, J. O. Armitage, R. Warnke, L. M. Staudt, *et al.* 2000. Distinct types of diffuse large B-cell lymphoma identified by gene expression profiling. Nature 403:503-11.

3. Bass, D. M. 1997. Interferon gamma and interleukin 1, but not interferon alfa, inhibit rotavirus entry into human intestinal cell lines. Gastroenterology 113:81-9.

4. Bigger, C. B., K. M. Brasky, and R. E. Lanford 2001. DNA microarray analysis of chimpanzee liver during acute resolving hepatitis C virus infection. J Virol. 75:7059-66.

5. Brunet, J. P., J. Cotte-Laffitte, C. Linxe, A. M. Quero, M. Geniteau-Legendre, and A. Servin 2000. Rotavirus infection induces an increase in intracellular calcium concentration in human intestinal epithelial cells: role in microvillar actin alteration. J Virol. 74:2323-32.

6. Brunet, J. P., N. Jourdan, J. Cotte-Laffitte, C. Linxe, M. Geniteau-Legendre, A. Servin, and A. M. Quero 2000. Rotavirus infection induces cytoskeleton disorganization in human intestinal epithelial cells: implication of an increase in intracellular calcium concentration. J Virol. 74:10801-6.

7. Casola, A., M. K. Estes, S. E. Crawford, P. L. Ogra, P. B. Ernst, R. P. Garofalo, and S. E. Crowe 1998. Rotavirus infection of cultured intestinal epithelial cells induces secretion of CXC and CC chemokines. Gastroenterology 114:947-55.

8. Chambers, J., A. Angulo, D. Amaratunga, H. Guo, Y. Jiang, J. S. Wan, A. Bittner, K. Frueh, M. R. Jackson, P. A. Peterson, M. G. Erlander, and P. Ghazal 1999. DNA microarrays of the complex human cytomegalovirus genome: profiling kinetic class with drug sensitivity of viral gene expression. J Virol. 73:5757-66.

9. Chang, Y. E., and L. A. Laimins 2000. Microarray analysis identifies interferon-inducible genes and Stat-1 as major transcriptional targets of human papillomavirus type 31. J Virol. 74:4174-82.

10. Cheung, V. G., M. Morley, F. Aguilar, A. Massimi, R. Kucherlapati, and G. Childs 1999. Making and reading microarrays. Nat Genet. 21:15-9.

11. Chieux, V., D. Hober, W. Chehadeh, J. Harvey, G. Alm, J. Cousin, H. Ducoulombier, and P. Wattre 1999. MxA protein in capillary blood of children with viral infections. J Med Virol. 59:547-51.

12. Chieux, V., D. Hober, J. Harvey, G. Lion, D. Lucidarme, G. Forzy, M. Duhamel, J. Cousin, H. Ducoulombier, and P. Wattre 1998. The MxA protein levels in whole blood lysates of patients with various viral infections. J Virol Methods 70: 183-91.

13. Cho, R. J., M. J. Campbell, E. A. Winzeler, L. Steinmetz, A. Conway, L. Wodicka, T. G. Wolfsberg, A. E. Gabrielian, D. Landsman, D. J. Lockhart, and R. W. Davis 1998. A genome-wide transcriptional analysis of the mitotic cell cycle. Mol Cell 2:65-73.

14. Chu, S., J. DeRisi, M. Eisen, J. Mulholland, D. Botstein, P. O. Brown, and I. Herskowitz 1998. The transcriptional program of sporulation in budding yeast. Science 282:699-705.

15. Cohen, P., M. Bouaboula, M. Bellis, V. Baron, O. Jbilo, C. Poinot-Chazel, S.

Galiegue, E. H. Hadibi, and P. Casellas 2000. Monitoring cellular responses to Listeria monocytogenes with oligonucleotide arrays. J Biol Chem. 275:11181-90.

16. Cuadras, M. A., D. A. Feigelstock, S. An, and H. B. Greenberg 2002. Gene expression pattern in Caco-2 cells following rotavirus infection. J Virol. 76:4467-82.

17. Cummings, C. A., and D. A. Relman 2000. Using DNA microarrays to study host-microbe interactions. Emerg Infect Dis. 6:513-25.

18. De Boissieu, D., P. Lebon, J. Badoual, Y. Bompard, and C. Dupont 1993. Rotavirus induces alpha-interferon release in children with gastroenteritis. J Pediatr Gastroenterol Nutr. 16:29-32.

19. del Castillo, J. R., J. E. Ludert, A. Sanchez, M. C. Ruiz, F. Michelangeli, and F. Liprandi 1991. Rotavirus infection alters Na^+ and K^+ homeostasis in MA-104 cells. J Gen Virol. 72:541-7.

20. Donato, R. 1999. Functional roles of S100 proteins, calcium-binding proteins of the EF- hand type. Biochim Biophys Acta 1450:191-231.

21. Eisen, M. B., P. T. Spellman, P. O. Brown, and D. Botstein 1998. Cluster analysis and display of genome-wide expression patterns. Proc Natl Acad Sci USA 95: 14863-8.

22. Estes, M. K. 2001. Rotaviruses and their replication. p. 1747-1786. In Fields Virology, 4th ed., Lippincott Williams and Wilkins, Philadelphia.

23. Fan, J. B., X. Chen, M. K. Halushka, A. Berno, X. Huang, T. Ryder, R. J. Lipshutz, D. J. Lockhart, and A. Chakravarti 2000. Parallel genotyping of human SNPs using generic high-density oligonucleotide tag arrays. Genome Res. 10: 853-60.

24. Fogh, J., J. M. Fogh, and T. Orfeo 1977. One hundred and twenty-seven cultured human tumor cell lines producing tumors in nude mice. J Natl Cancer Inst. 59: 221-6.

25. Geiss, G. K., M. C. An, R. E. Bumgarner, E. Hammersmark, D. Cunningham, and M. G. Katze 2001. Global impact of influenza virus on cellular pathways is mediated by both replication-dependent and -independent events. J Virol. 75: 4321-31.

26. Geiss, G. K., R. E. Bumgarner, M. C. An, M. B. Agy, A. B. van 't Wout, E. Hammersmark, V. S. Carter, D. Upchurch, J. I. Mullins, and M. G. Katze 2000. Large-scale monitoring of host cell gene expression during HIV-1 infection using cDNA microarrays. Virology 266:8-16.

27. Heath, R. L., and C. J. Birch 1988. Synthesis of human rotavirus polypeptides in cell culture. J Med Virol. 25:91-103.

28. Honda, M., S. Kaneko, H. Kawai, Y. Shirota, and K. Kobayashi 2001. Differential gene expression between chronic hepatitis B and C hepatic lesion. Gastroenterology 120:955-66.

29. Huang, Q., D. Liu, P. Majewski, L. C. Schulte, J. M. Korn, R. A. Young, E. S. Lander, and N. Hacohen 2001. The plasticity of dendritic cell responses to pathogens and their components. Science 294:870-5.

30. Iyer, V. R., M. B. Eisen, D. T. Ross, G. Schuler, T. Moore, J. C. Lee, J. M. Trent, L. M. Staudt, J. Hudson, Jr., M. S. Boguski, D. Lashkari, D. Shalon, D. Botstein,

and P. O. Brown 1999. The transcriptional program in the response of human fibroblasts to serum. Science 283:83-7.

31. Jenner, R. G., M. M. Alba, C. Boshoff, and P. Kellam 2001. Kaposi's sarcoma-associated herpesvirus latent and lytic gene expression as revealed by DNA arrays. J Virol. 75:891-902.

32. Jourdan, N., J. P. Brunet, C. Sapin, A. Blais, J. Cotte-Laffitte, F. Forestier, A. M. Quero, G. Trugnan, and A. L. Servin 1998. Rotavirus infection reduces sucrase-isomaltase expression in human intestinal epithelial cells by perturbing protein targeting and organization of microvillar cytoskeleton. J Virol. 72:7228-36.

33. Jourdan, N., M. Maurice, D. Delautier, A. M. Quero, A. L. Servin, and G. Trugnan 1997. Rotavirus is released from the apical surface of cultured human intestinal cells through nonconventional vesicular transport that bypasses the Golgi apparatus. J Virol. 71:8268-78.

34. Kitamoto, N., R. F. Ramig, D. O. Matson, and M. K. Estes 1991. Comparative growth of different rotavirus strains in differentiated cells (MA104, HepG2, and CaCo-2). Virology 184:729-37.

35. La Bonnardiere, C., J. Cohen, and M. Contrepois 1981. Interferon activity in rotavirus infected newborn calves. Ann Rech Vet. 12:85-91.

36. Lecce, J. G., J. M. Cummins, and A. B. Richards 1990. Treatment of rotavirus infection in neonate and weanling pigs using natural human interferon alpha. Mol Biother. 2:211-6.

37. Lennon, G., C. Auffray, M. Polymeropoulos, and M. B. Soares 1996. The I.M.A.G.E. Consortium: an integrated molecular analysis of genomes and their expression. Genomics 33:151-2.

38. Lipshutz, R. J., S. P. Fodor, T. R. Gingeras, and D. J. Lockhart 1999. High density synthetic oligonucleotide arrays. Nat Genet. 21:20-4.

39. Mangiarotti, P., F. Moulin, P. Palmer, S. Ravilly, J. Raymond, and D. Gendrel 1999. Interferon-alpha in viral and bacterial gastroenteritis: a comparison with C-reactive protein and interleukin-6. Acta Paediatr. 88:592-4.

40. Menard, R., C. Dehio, and P. J. Sansonetti 1996. Bacterial entry into epithelial cells: the paradigm of Shigella. Trends Microbiol. 4:220-6.

41. Mengaud, J., H. Ohayon, P. Gounon, R. M. Mege, and P. Cossart 1996. E-cadherin is the receptor for internalin, a surface protein required for entry of L. monocytogenes into epithelial cells. Cell 84:923-32.

42. Michelangeli, F., F. Liprandi, M. E. Chemello, M. Ciarlet, and M. C. Ruiz 1995. Selective depletion of stored calcium by thapsigargin blocks rotavirus maturation but not the cytopathic effect. J Virol. 69:3838-47.

43. Morgan, R. W., L. Sofer, A. S. Anderson, E. L. Bernberg, J. Cui, and J. Burnside 2001. Induction of host gene expression following infection of chicken embryo fibroblasts with oncogenic Marek's disease virus. J Virol. 75:533-9.

44. Mossman, K. L., P. F. Macgregor, J. J. Rozmus, A. B. Goryachev, A. M. Edwards, and J. R. Smiley 2001. Herpes simplex virus triggers and then disarms a host antiviral response. J Virol. 75:750-8.

45. Mysorekar, I. U., M. A. Mulvey, S. J. Hultgren, and J. I. Gordon 2002. Molecular

regulation of urothelial renewal and host defenses during infection with uropathogenic E. coli. J Biol Chem. 277: 7412-9.

46. Obert, G., I. Peiffer, and A. L. Servin 2000. Rotavirus-induced structural and functional alterations in tight junctions of polarized intestinal Caco-2 cell monolayers. J Virol. 74:4645-51.

47. Perez, J. F., M. C. Ruiz, M. E. Chemello, and F. Michelangeli 1999. Characterization of a membrane calcium pathway induced by rotavirus infection in cultured cells. J Virol. 73:2481-90.

48. Perou, C. M., T. Sorlie, M. B. Eisen, M. van de Rijn, S. S. Jeffrey, C. A. Rees, J. R. Pollack, D. T. Ross, H. Johnsen, L. A. Akslen, O. Fluge, A. Pergamenschikov, C. Williams, S. X. Zhu, P. E. Lonning, A. L. Borresen-Dale, P. O. Brown, and D. Botstein 2000. Molecular portraits of human breast tumours. Nature 406: 747-52.

49. Pietiainen, V., P. Huttunen, and T. Hyypia 2000. Effects of echovirus 1 infection on cellular gene expression. Virology 276:243-50.

50. Piron, M., P. Vende, J. Cohen, and D. Poncet 1998. Rotavirus RNA-binding protein NSP3 interacts with eIF4GI and evicts the poly(A) binding protein from eIF4F. EMBO J. 17:5811-21.

51. Pollack, J. R., C. M. Perou, A. A. Alizadeh, M. B. Eisen, A. Pergamenschikov, C. F. Williams, S. S. Jeffrey, D. Botstein, and P. O. Brown 1999. Genome-wide analysis of DNA copy-number changes using cDNA microarrays. Nat Genet. 23: 41-6.

52. Ragno, S., M. Romano, S. Howell, D. J. Pappin, P. J. Jenner, and M. J. Colston 2001. Changes in gene expression in macrophages infected with Mycobacterium tuberculosis: a combined transcriptomic and proteomic approach. Immunology 104:99-108.

53. Rollo, E. E., K. P. Kumar, N. C. Reich, J. Cohen, J. Angel, H. B. Greenberg, R. Sheth, J. Anderson, B. Oh, S. J. Hempson, E. R. Mackow, and R. D. Shaw 1999. The epithelial cell response to rotavirus infection. J Immunol. 163:4442-52.

54. Roses, A. D. 2000. Pharmacogenetics and the practice of medicine. Nature 405: 857-65.

55. Schena, M., D. Shalon, R. W. Davis, and P. O. Brown 1995. Quantitative monitoring of gene expression patterns with a complementary DNA microarray. Science 270:467-70.

56. Schwers, A., C. Vanden Broecke, M. Maenhoudt, J. M. Beduin, J. Werenne, and P. P. Pastoret 1985. Experimental rotavirus diarrhoea in colostrum-deprived newborn calves: assay of treatment by administration of bacterially produced human interferon (Hu-IFN alpha 2). Ann Rech Vet. 16:213-8.

57. Sherlock, G., T. Hernandez-Boussard, A. Kasarskis, G. Binkley, J. C. Matese, S. S. Dwight, M. Kaloper, S. Weng, H. Jin, C. A. Ball, M. B. Eisen, P. T. Spellman, P. O. Brown, D. Botstein, and J. M. Cherry 2001. The Stanford Microarray Database. Nucleic Acids Res. 29:152-5.

58. Sheth, R., J. Anderson, T. Sato, B. Oh, S. J. Hempson, E. Rollo, E. R. Mackow, and R. D. Shaw 1996. Rotavirus stimulates IL-8 secretion from cultured epithe-

lial cells. Virology 221:251-9.

59. Southern, E., K. Mir, and M. Shchepinov 1999. Molecular interactions on micro-arrays. Nat Genet. 21:5-9.

60. Stein, M. A., D. A. Mathers, H. Yan, K. G. Baimbridge, and B. B. Finlay 1996. Enteropathogenic Escherichia coli markedly decreases the resting membrane potential of Caco-2 and HeLa human epithelial cells. Infect Immun. 64:4820-5.

61. Svensson, L., B. B. Finlay, D. Bass, C. H. von Bonsdorff, and H. B. Greenberg 1991. Symmetric infection of rotavirus on polarized human intestinal epithelial (Caco-2) cells. J Virol. 65:4190-7.

62. Tatlow, D., R. Brownlie, L. A. Babiuk, and P. Griebel 2000. Differential display analysis of gene expression during the induction of mucosal immunity. Immunogenetics 52:73-80.

63. Taylor, L. A., C. M. Carthy, D. Yang, K. Saad, D. Wong, G. Schreiner, L. W. Stanton, and B. M. McManus 2000. Host gene regulation during coxsackievirus B3 infection in mice: assessment by microarrays. Circ Res. 87:328-34.

64. Tusher, V. G., R. Tibshirani, and G. Chu 2001. Significance analysis of micro-arrays applied to the ionizing radiation response. Proc Natl Acad Sci USA 98: 5116-21.

65. Wang, D. G., J. B. Fan, C. J. Siao, A. Berno, P. Young, R. Sapolsky, G. Ghandour, N. Perkins, E. Winchester, J. Spencer, L. Kruglyak, L. Stein, L. Hsie, T. Topaloglou, E. Hubbell, E. Robinson, M. Mittmann, M. S. Morris, N. Shen, D. Kilburn, J. Rioux, C. Nusbaum, S. Rozen, T. J. Hudson, E. S. Lander, *et al.*, 1998. Large-scale identification, mapping, and genotyping of single-nucleotide polymor-phisms in the human genome. Science 280:1077-82.

66. Wooster, R. 2000. Cancer classification with DNA microarrays is less more? Trends Genet. 16:327-9.

67. Xu, A., A. R. Bellamy, and J. A. Taylor 1998. BiP (GRP78) and endoplasmin (GRP94) are induced following rotavirus infection and bind transiently to an endoplasmic reticulum-localized virion component. J Virol. 72:9865-72.

68. Zhu, H., J. P. Cong, G. Mamtora, T. Gingeras, and T. Shenk 1998. Cellular gene expression altered by human cytomegalovirus: global monitoring with oligo-nucleotide arrays. Proc Natl Acad Sci USA 95:14470-5.

69. Zweibaum, A., N. Triadou, M. Kedinger, C. Augeron, S. Robine-Leon, M. Pinto, M. Rousset, and K. Haffen 1983. Sucrase-isomaltase: a marker of foetal and malignant epithelial cells of the human colon. Int J Cancer **32**:407-12.

II, 10. The rat model of rotavirus infection

Max Ciarlet[1], Margaret E. Conner[1,2] and Mary K. Estes[1]

[1] *Department of Molecular Virology and Microbiology, Baylor College of Medicine*
[2] *Veterans Affairs Medical Center, Houston, Texas 77030, U.S.A.*

Introduction

Group A rotavirus infection occurs in all age groups, but acute gastroenteritis is primarily restricted, but not limited, to young children (6 months to 2 years of age), and is also associated with diarrhea in the young of avian and many mammalian animal species (Conner and Ramig, 1996; Estes, 2001). The outcome of rotavirus infection in children varies in different countries. In developing countries, rotavirus infections alone are responsible for close to one million deaths annually, and death due to rotavirus infection is significant in Asia, Africa and South America. Although in the United States the impact of rotavirus infections accounts for 100 to 200 deaths annually, approximately 2% of infected children are hospitalized, and minimal annual economic losses due to rotavirus infections reach up to 1 billion dollars (Clark *et al.*, 1996; Glass *et al.*, 1996; Estes, 2001; Kapikian *et al.*, 2001; Clements-Mann *et al.*, 2001). These statistics support the global need for the development of a safe and effective rotavirus vaccine.

Development of a rotavirus vaccine is complicated by a number of factors related to the epidemiology of rotavirus infections. Currently, 15 G (defined by the glycoprotein VP7) and at least 13 P (defined by the spike protein VP4) serotypes are known, and an additional 7 P genotypes (designated within square brackets) have been recognized based on amino acid sequence homology of the spike protein VP4 (Kapikian *et al.*, 2001; Estes, 2001). In humans, 10 G (G1-G6, G8-G10, and G12) and 9 P (P1A[8], P1B[4], P2A[6], P2C[6], P3[9], P4[10], P5A[3], P8[11], and P11[14]) serotypes have been identified, but the majority of the infections in developed countries is caused by rotaviruses of serotypes G1 to G4 in association with P1A[8] or P1B[4] serotypes. In the developing world, in addition to the common strains, unusual rotavirus strains of G5, G9, or G10 serotype in combination with P2A[6] or P8[11] serotypes circulate frequently and are becoming prevalent (Kapikian *et al.*, 2001; Estes, 2001; Griffin *et al.*, 2000; Iturriza-Gómara *et al.*, 2000; Cunliffe *et al.*, 2001). Multiple rotavirus serotypes or strains often circulate in a given community at the same time or within the same year making predictions of prevalent rotavirus serotypes almost impossible. Since some studies have shown limited cross-protection between viruses of different serotypes, multivalent vaccines have been and are being developed (Kapikian *et al.*, 2001; Estes, 2001).

How can rotavirus infection, pathogenesis and immunity be studied?

Rotaviruses are ubiquitous, and virtually 100% of the adult population has antibody to rotavirus. Understanding rotavirus infection and immunity in humans is important, but studies in humans are difficult due to the inability to control for many experimental variables, such as prior exposure to virus. As an alternative, animal model systems of rotavirus infection have been developed and provided key insights in our understanding of the infection, pathogenesis and immunity of both human and animal rotaviruses. Seven models in four large (cow, pig, horse, and sheep), and three small laboratory (rabbit, mouse, and rat) animal species have been used to define parameters of rotavirus infection, pathology, disease, immune response or test vaccine efficacy (reviewed in Saif *et al.*, 1994; Conner and Ramig, 1996; Ciarlet and Conner, 2000; Guérin-Danan *et al.*, 2001; Ciarlet *et al.*, 2002).

Large animal models

Much of the early work on rotavirus pathogenesis utilized large animals, generally either colostrum-deprived or gnotobiotic models (reviewed in Conner and Ramig, 1996; Ciarlet and Conner, 2000; Greenberg *et al.*, 1994; Saif *et al.*, 1994). Most studies in calves used rotaviruses of bovine origin, but some heterologous (non-bovine) rotaviruses can also infect and induce mild disease in calves (Conner and Ramig, 1996; Greenberg *et al.*, 1994; Mebus *et al.*, 1971; Pearson *et al.*, 1978; Reynolds *et al.*, 1985; Saif *et al.*, 1994). Piglets are monogastric animals with intestinal physiology that resembles that of humans and are susceptible to infection and severe disease by some rotavirus isolates from other species, including humans (Conner and Ramig, 1996; Greenberg *et al.*, 1994; Crouch and Woode, 1978; Pearson and McNulty, 1977; Saif *et al.*, 1994; Theil *et al.*, 1978). Infection of horses with homologous (equine) rotaviruses results in induction of disease as well as intestinal lesions (Conner and Darlington, 1980; Browning *et al.*, 1991; Saif *et al.*, 1994). Compared with infections in piglets, foals, and calves, rotavirus infection in lambs results in less severe histopathologic changes and mild clinical disease (Conner and Ramig, 1996; Greenberg *et al.*, 1994; Saif *et al.*, 1994; Snodgrass *et al.*, 1979). Large animal models have limited use due to high costs, the restricted availability of isolation or germ-free facilities for large animals, and the need for specialized equipment and staff – all factors that preclude their use in large-scale studies.

Small animal models

Small animal models have several advantages over large animal models, including: cost effectiveness, the ability to incorporate large numbers of animals in studies, the feasibility of isolation of large numbers of infected animals, short gestations and multiparous births, and the availability of rotavirus-naive animals. The majority of rotavirus studies in small animal models have been predominantly performed in mice and rabbits, and since these models have been described widely, they will be only discussed briefly. The salient features of the rabbit, mouse, and rat models of rotavirus infection are summarized in Table 1 and compared to human rotavirus infection in Table 2. Compared to rabbits and mice, rats exhibit high permissivity for infection with several heterologous (non-rat) rotaviruses, including human strains, and no apparent sterilizing immunity (Table 1). Similar to rotavirus infection in children, rats can be infected with multiple serotypes, and immunity requires more than one infection (Table 2). Next, we will discuss the relative advantages and disadvantages of the different small animal models, giving emphasis to the rat model, to study human rotavirus infections.

The rabbit and mouse models

The rabbit model was the first small animal model developed to examine active humoral immunity and protection (Ciarlet et al., 1998a; 1998b; 1998c; 2000; Ciarlet and Conner, 2000; Conner et al., 1988; 1991; 1993), followed by the development of the adult mouse model (Burns et al., 1995; Ciarlet and Conner, 2000; Feng et al., 1994; Ward et al., 1990; 1992). Limitations of the rabbit and adult mouse models of rotavirus infection are that i) human rotavirus strains do not efficiently replicate in either animal, ii) clinical disease is only observed in animals ≤ 2 weeks of age, iii) only homologous virus strains (isolated from the same species) replicate efficiently and spread horizontally to uninoculated control animals, whereas heterologous virus strains (isolated from a different species) do not, and iv) the small size of the intestinal tract of neonatal mice does not allow pathophysiological studies, and while the rabbit's intestinal tract is an optimal size, studies in naive rabbits are expensive (Ciarlet et al., 1998a; 1998b; 1998c; 2000; Ciarlet and Conner, 2000; Conner et al., 1988; 1991; 1993; Burns et al., 1995; Feng et al., 1994; Ward et al., 1990; 1992). Early and recent studies demonstrated that rotavirus disease, but not infection, is age-restricted in mice and rabbits following inoculation of murine and lapine rotavirus strains, respectively (Ciarlet et al., 1998a; 1998b; 1998c; 2000; Ciarlet and Conner, 2000; Conner et al., 1988; 1991; 1993; Burns et al., 1995; Feng et al., 1994; Ward et al., 1990; 1992). We recently reported that among 27 different heterologous (non-lapine) or reassortant rotavirus strains, only simian rhesus rotavirus (RRV) strain and human rotavirus strains belonging to the P11[14] serotype replicated or were capable of horizontal transmission with efficiency in adult rabbits (Ciarlet et al., 1998a; 2000). In addition, only limited or abortive virus replication occurs in mice infected with heterologous (non-murine) rotaviruses as evidenced by lack of virus excretion above input titers and no transmission of infection occurs

Table 1

Comparison of the Rabbit, Mouse and Rat Animal Models of Rotavirus Infection

Parameter	Rabbit	Mouse	Rat
Cost effectiveness	Moderate	High	High
Availability (rotavirus antibody-negative)	Moderate	High	High
Utility for studies of mucosal immunity	High	High	High
Permissivity for human rotaviruses	Moderate	Low	High
Utility to study physiology	High	Moderate	High
Genetically altered animal strains for immunological studies	Low	High	Moderate
Feasibility of long-term immunity and protection studies	High	High	High
Sterilizing immunity	Yes	Yes	No
Diarrhea induced following rotavirus challenge in $\leq$15 day-old animals	Yes	Yes	Yes
Diarrhea induced following rotavirus challenge in >15 day-old animals	No	No	No
Histopathologic lesions in small intestine after rotavirus infection	Yes	Limited	Limited
Transplacental transfer of antibody	Yes	Yes	Yes
Vaccine studies			
(a) Homologous immunity	Yes	Yes	ND
(b) Heterologous immunity	Yes	Yes	Yes
Vaccines tested			
(a) Inactivated virus	Yes	Yes	ND
(b) Vaccinia-, adenovirus- or baculovirus-expressed proteins	ND	Yes	ND
(c) VLPs	Yes	Yes	ND
(d) DNA	ND	Yes	ND
(e) Live attenuated	Yes	Yes	Yes

ND, Not done

Table 2

Comparison of small animal (mouse, rabbit, and rat) models to study rotavirus infection in children

	Mice and rabbits	Rats	Children
Infection	Single serotype: G3	Multiple serotypes: G1, G3, G5, G6, G8, and G9.	Multiple serotypes: G1-G6, G8-G10, and G12
	Homologous rotavirus	Homologous rotavirus?	Heterologous rotavirus?
Immunity	Requires one infection	Requires more than one infection	Requires multiple infections
Physiological studies	Mice - extremely difficult Rabbits - not cost effective	Feasible and cost effective	N/A

N/A, not applicable.

from inoculated to control animals (Burns *et al.*, 1995; Conner and Ramig, 1996; Feng *et al.*, 1994; Greenberg *et al.*, 1994; Ramig, 1988). Therefore, the limitations of both the rabbit and mouse models highlight the need for other small animal models to study homologous and heterologous group A rotavirus infection, pathogenesis and immunity. Since new rotavirus studies are being performed in rats (Guérin-Danan *et al.*, 1998; 2001; Ciarlet *et al.*, 2002), the rest of this review focuses on the rat model of rotavirus infection.

The rat model

The rat model of rotavirus infection was only fully developed recently in spite of the ideal size of the rat intestine for physiological studies, including Ussing chamber analysis, the well established physiology of the gastrointestinal tract of rats, and the cost effectiveness of the use of rats to perform rotavirus experiments (Ciarlet *et al.*, 2002). The rat was not originally pursued as a model to study rotavirus infection because no homologous (rat) group A rotavirus had been identified, data on experimental infection of rats with group A rotaviruses were limited, and the epidemiology of natural group A rotavirus infections in rats indicated that rats might not be susceptible to group A rotavirus infection (Yolken *et al.*, 1988; Awang and Yap, 1990). One study failed to detect antibodies to group A rotavirus in unspecified albino laboratory rats, wild rats (*Rattus rattus*) and shrews (*Suncus murinus*) resident in an animal facility (Awang and Yap, 1990). However, another study detected a high incidence (76%) of group A rotavirus antibodies in serum specimens collected from wild rats in the Shizouka prefecture in Japan (Takahashi *et al.*, 1979), suggesting that group A rotavirus infections in rats occur. In addition, initial attempts to experimentally infect rats with group A rotavirus infection demonstrated that i) induction of diarrhea in 3- to 5-day old rats was limited to infection with the simian RRV strain, and did not occur with simian SA11, bovine NCDV or human Wa strains (Yolken *et al.*, 1988), ii) replication of simian RRV and Wa rotavirus strains in cell culture was inhibited by Wistar rat mucins (Yolken *et al.*, 1993), iii) Wistar rats lack the cellular group A rotavirus receptor (Bass *et al.*, 1991), and iv) Fisher 344 germfree neonatal rats can be infected with the simian SA11 strain (Guérin-Danan *et al.*, 1998; 2001). Therefore, the lack of a documented animal model may have obscured the fact that group A rotaviruses are capable of infecting rats and discouraged additional studies to search for rat rotaviruses. We recently examined in detail (summarized in Fig. 1) if heterologous (non-rat) group A rotaviruses replicate, spread, and induce disease, histopathology, and immune response in Lewis (LEW/SsNHsd) rats (Ciarlet *et al.*, 2002; Ciarlet *et al.*, manuscript in preparation). The newly developed rat model of rotavirus infection is unique among the small animal models because it allows human rotavirus infections to be analyzed.

Group A rotavirus infection in rats

No group A rat rotaviruses have been isolated to date. However, rats of 5 days to 9 months of age have been experimentally infected with several heterologous (non-rat) rotavirus strains of simian (RRV, SA11), lapine (ALA), bovine (WC3), porcine (OSU) and human (Wa, WI61, HAL1166) origin as described in Fig. 1 (Guérin-Danan *et al.*, 1998; Ciarlet *et al.*, 2002). Similar to infections of rabbits and mice, rotavirus-induced disease is restricted to rats of ≤ 2 weeks of age (Ciarlet *et al.*, 2002), and infection results in induction of a rotavirus-specific immune response (Ciarlet *et al.*, in preparation). In contrast to rabbits and mice, the majority of the heterologous rotavirus strains replicate in neonatal rats, and these strains are readily transmitted to uninoculated control dams or littermates (Ciarlet *et al.*, 2002). Although heterologous rotavirus replication is less restricted in rats than in mice and rabbits, the capacity of the rat intestine to support replication of a group A rotavirus may be limited as compared to the replication capability of the group B rotavirus IDIR (Vonderfecht *et al.*, 1984; 1988; Ciarlet *et al.*, 2002).

Table 3 summarizes the experiments carried out along an algorithm as described in Fig. 1. Although heterologous (non-rat) rotavirus strains infect rats, disease severity and transmission efficiency differ depending on the virus strain used (Table 3). Based on virus antigen or infectious virus shedding and severity of disease, among the group A rotavirus strains tested, simian RRV is the most efficient at replication and disease induction (Ciarlet *et al.*, 2002). The level of infectious virus excretion of neonatal rats inoculated with HAL1166, and ALA was slightly less than RRV, while levels of excretion of simian SA11, bovine WC3, porcine OSU, or human Wa or WI61 were decreased compared to RRV (Ciarlet *et al.*, 2002). Porcine OSU, shown to be attenuated in neonatal mice (Zhang *et al.*, 1998), induces mild disease in neonatal rats, suggesting that this strain may also be attenuated in rats (Ciarlet *et al.*, 2002).

Infection of neonatal rats with the different rotavirus strains results in different patterns of transmission to dams and control (PBS) mock-inoculated littermates (Table 3).

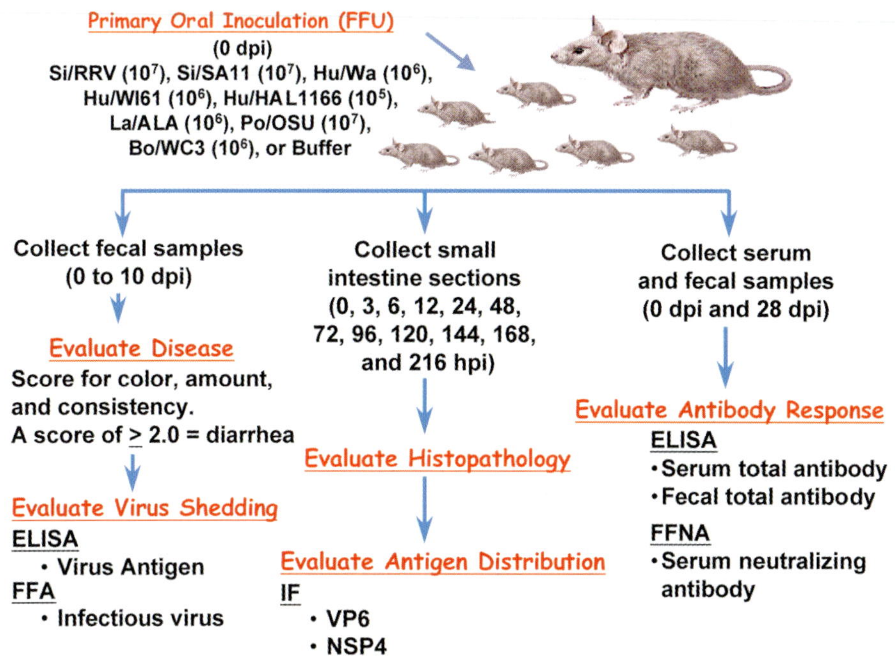

Fig. 1. Outline of experimental design to evaluate group A rotavirus infection, induction of disease, histopathology, and immune response in Lewis (LEW/SsNHsd) rats. Neonatal (5-day old) or young (21-day old) rats were orally inoculated with the indicated rotavirus strains or PBS. Individual fecal samples were collected 0 to 10 dpi. Virus antigen and infectious virus shedding was monitored by ELISA and FFA, respectively. Disease was scored for color, amount, and consistency of feces. A score of ≥2.0 was considered diarrhea. Serum and fecal samples of each dam or young rats were collected at 0 dpi. At 28 dpi, serum and fecal samples from the dams, pups and young rats were collected. Antibody responses were evaluated by ELISA for both serum and intestinal total antibodies. Also, these samples were screened for neutralizing antibodies by FFNA. In addition, rat pups inoculated with RRV strain were sacrificed at the indicated time points, and the small intestine sections were collected to evaluate histopathology and distribution of rotavirus antigen. Abbreviations: *FFU*, fluorescent focus units; *dpi*, days post inoculation; *ELISA*, enzyme-linked immunosorbent assay; *FFA*, fluorescent focus assay; *IF*, immunofluorescence; *VP6*, rotavirus protein 6; *NSP4*, rotavirus non-structural protein 4; *FFNA*, fluorescent focus neutralization assay; *RRV*, rhesus rotavirus.

Based on virus antigen or infectious virus shedding, simian RRV and lapine ALA rotavirus strains are readily transmitted to PBS-inoculated littermates, while bovine WC3, porcine OSU, and human Wa, WI61, and HAL1166 rotavirus strains do not spread to control littermates. However, all viruses, with the exception of OSU, are transmitted to dams. Therefore, horizontal transmission of rotavirus from infected pups to their dams is more efficient than that to PBS-inoculated littermates probably due to the altricious nature of rats (Henning *et al.*, 1994).

Table 3

Comparison of group A rotavirus infection by different rotavirus strains in rats

Rat Age	Inoculum	Virus Shedding[a] (days)	Diarrhea	Disease Severity[b]	Horizontal Transmission	Immune Response[c]	
						Serologic	Mucosal
Neonatal	PBS	No	No	_	No	No	No
	Si/RRV	9-10	Yes	++++	Yes [@]	Yes	Yes
	La/ALA	9-10	Yes	+++	Yes [@]	Yes	Yes
	Hu/HAL1166	9-10	Yes	+++	Yes [#]	Yes	Yes
	Bo/WC3	8-9	Yes	++	Yes [#]	Yes	No
	Si/SA11	8-9	Yes	++	Yes [#]	Yes	No
	Hu/Wa	7-8	Yes	++	Yes [#]	Yes	No
	Hu/WI61	5-6	Yes	+	Yes [#]	Yes	No
	Po/OSU	5-6	Yes	+	No	Yes	No
Adult	PBS	No	No	N/A	ND	No	No
	Si/RRV	1-2	No	N/A	ND	Yes	Yes
	La/ALA	1-2	No	N/A	ND	Yes	Yes
	Hu/HAL1166	1-2	No	N/A	ND	Yes	Yes
	Bo/WC3	1-2	No	N/A	ND	Yes	Yes
	Hu/Wa	1-2	No	N/A	ND	Yes	Yes

[a] Measured by ELISA (virus antigen) or FFA (infectious virus) (Ciarlet *et al.*, 2002).
[b] Mean disease severity was noted and scored from 1 to 4 based on color, consistency, and amount of stool.
(Ciarlet *et al.*, manuscript in preparation). Disease severity was noted according to the following scale: _, 0 to 1.5; +, 2 to 2.5; ++, 2.5 to 3; +++, 3 to 3.5; ++++, 3.5 to 4.
[c] Measured by ELISA at 28 dpi (Ciarlet *et al.*, manuscript in preparation).
[@] Spread to dams and control (PBS)-inoculated littermates.
[#] Spread to dams only.
Neonatal, 5-day old; *Adult*, ≥ 21-day old; *N/A*, not applicable; *ND*, Not determined.

Analysis of intestinal gut homogenates to measure infectious virus in neonatal rats infected with RRV demonstrates that at least one complete replication cycle of RRV occurs (Ciarlet *et al.*, 2002). Shortly after infection and for up to 3 hours post infection (hpi), only small amounts of RRV can be detected (eclipse phase). Rotavirus release quickly follows up to 12 hpi, which is the time at which virus titers reach a maximum. Thereafter, rotavirus titers in the small intestine of neonatal rats decline gradually, and likely represent the decay of viral progeny. Calculation of total intestinal yields of infectious RRV demonstrates that output virus is almost 10-fold more than that of input virus (Ciarlet *et al.*, 2002).

VP6 protein is detected by immunofluorescence in the epithelial cells located on the upper half of the intestinal villi in the jejunum and the ileum, but not in the duodenum of RRV- or SA11-infected neonatal rats (Guérin-Danan *et al.*, 1998; Ciarlet *et al.*, 2002). As with other animal models (Conner and Ramig, 1996; Saif *et al.*, 1994; Greenberg *et al.*, 1994), there is no evidence of rotavirus antigen present in the lower half of the intestinal villi or in the crypts in any region of the small intestine of virus-infected neonatal rats (Guérin-Danan *et al.*, 1998; Ciarlet *et al.*, 2002). In addition, the rotavirus non-structural protein 4 (NSP4) is detected by immunofluorescence in the epithelial cells located on the upper half of the intestinal villi in the jejunum and the ileum, but not in the duodenum of neonatal rats inoculated with RRV (Fig. 2). NSP4 mRNA must be synthesized and expressed in epithelial cells in the small intestine of RRV-inoculated neonatal rats to allow detection of NSP4, which provides further evidence that RRV replicates in neonatal rats (Ciarlet *et al.*, 2002).

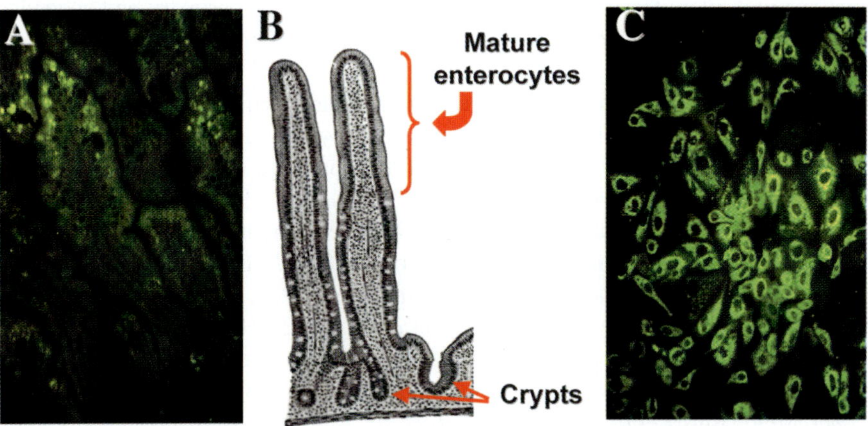

Fig. 2. Photomicrographs of the distribution of the rotavirus specific antigen NSP4 in the ileum of a 5-day old RRV-infected neonatal rat at 24 hpi (A) and in embryonic rhesus monkey kidney (MA104) cells infected with RRV at 18 hpi (C). Rotavirus NSP4 antigen, detected by immunofluorescence, is distributed in the epithelial cells of the upper half of the intestinal villi (mature enterocytes) of the ileum (A), and in the cytoplasm of cultured cells (C). The schematic in the middle panel (B) illustrates the mature enterocytes, which are the site of rotavirus replication *in vivo*, and the crypt cells, which are spared from rotavirus infection. Abbreviations: *hpi,* hours post inoculation; *RRV,* rhesus rotavirus; *NSP4,* rotavirus non-structural protein 4.

It is important to keep in mind that the effect of host factors, such as different rat strains, on rotavirus replication are not well defined. Different strains of rats may differ in their susceptibility to group A rotavirus infection. Other studies of group A rotavirus infection in rats demonstrate that germ-free Fisher 344 rats are susceptible to simian SA11 rotavirus infection (Guérin-Danan *et al.*, 1998; 2001). Wistar rats are susceptible to simian RRV, but not simian SA11 or human Wa infection (Yolken *et al.*, 1988). Wistar rat mucins inhibit replication of both RRV and Wa strains in cell culture (Yolken *et al.*, 1993). Additional work is required to determine the susceptibility of different rat strains to a broader repertoire of different group A heterologous rotavirus strains.

Rats up to at least 9 months of age are susceptible to infection with RRV (Ciarlet *et al.*, 2002), and develop a rotavirus-specific immune response (Ciarlet *et al.*, manuscript in preparation). The intensity of the rotavirus-specific immune response differs depending on the virus strain used for inoculation (Table 3). However, the immune response is not fully protective against a subsequent rotavirus infection. Preliminary data indicate that unlike the rabbit and mouse models, complete protection from a homologous or heterologous challenge is not achieved in the rat model. Like children, rats remain susceptible to infection by a second rotavirus serotype (Ciarlet *et al.*, manuscript in preparation). The long period of susceptibility of rats to multiple rotavirus infections will allow the use of the rat model to examine the primary and secondary active immune responses to the same or different group A rotaviruses (Ciarlet *et al.*, manuscript in preparation).

As observed in other small (mice, rabbits) and large (cows, pigs, horses, lambs) animal models, histopathological changes caused by rotavirus infection in rats are primarily restricted to the villus epithelium of the small intestine (Conner and Ramig, 1996; Greenberg *et al.*, 1994; Saif *et al.*, 1994; Mebus *et al.*, 1971; Pearson *et al.*, 1977; Reynolds *et al.*, 1985; Crouch and Woode, 1978; Pearson and McNulty, 1977; Theil *et al.*, 1978; Conner and Darlington, 1980; Snodgrass *et al.*, 1979; Ciarlet *et al.*, 1998; Burns *et al.*, 1995; Coelho *et al.*, 1981; Osborne *et al.*, 1988; Ramig, 1988; Starkey *et al.*, 1986; Guérin-Danan *et al.*, 1998; Ciarlet *et al.*, 2002). However, histopathologic changes in rats are limited and most similar to those observed in mice. Neonatal (≤ 2 weeks of age) or adult mice develop no or minimal histologic damage (Ramig and Conner, 1996; Burns *et al.*, 1995; Osborne *et al.*, 1988; Ramig, 1988; Starkey *et al.*, 1986). Infection of neonatal germ-free Fisher 344 or Lewis rats with SA11 or RRV, respectively, results in extensive vacuolization in the ileum without additional histopathology in the small intestine (Fig. 3) (Guérin-Danan *et al.*, 1998; Ciarlet *et al.*, 2002). Vacuolated enterocytes of infected neonatal rats are characterized by nuclei localized at their base and by a large supranuclear area occupying almost the whole apical cytoplasm (Ciarlet *et al.*, 2002). Enterocytes exhibiting the marked vacuolization in the ileum do not contain rotavirus antigen as detected by immunofluorescence or EM (Ciarlet *et al.*, 2002). Similar results are observed following oral inoculation of mice with heterologous (non-murine) rotaviruses (Ramig, 1988; Mori *et al.*, 2001). The origin and cause of the vacuolization in intestinal epithelial cells in either rats or mice are not known. Since the vacuoles do not contain rotavirus antigen, the vacuoles might not be the site of rotavirus replication, but are possibly indirect changes resulting from rotavirus infection or replication. The vacuoles in rat enterocytes do not contain lipid, as revealed by oil red O staining of neutral fats; the vacuoles may represent dilated vesicles resulting from virus infection (Ciarlet *et al.*, 2002).

Since vacuoles are considerably more numerous in SA11- or RRV-inoculated neonatal rats than in mock-inoculated neonatal rats, the vacuolization caused by rotavirus infection likely results in decreased intestinal absorption (Guérin-Danan *et al.*, 1998; Ciarlet *et al.*, 2002). In support of this hypothesis, the mean weight gain of RRV-inoculated neonatal rats is significantly lower ($P<0.001$, t-test for Equality of Means) than the mean weight gain of PBS-inoculated neonatal rats at 48 hpi (Ciarlet *et al.*, 2002).

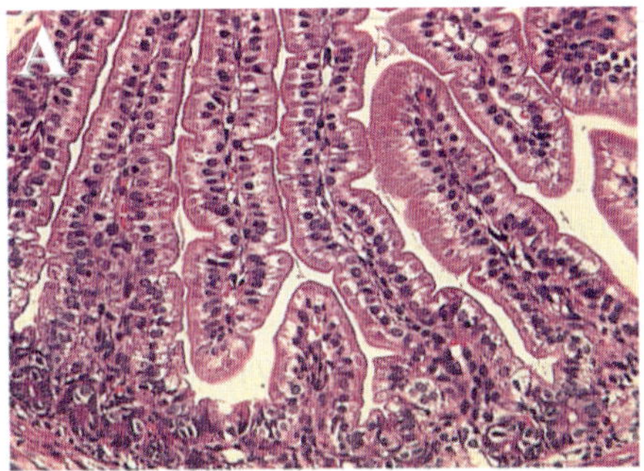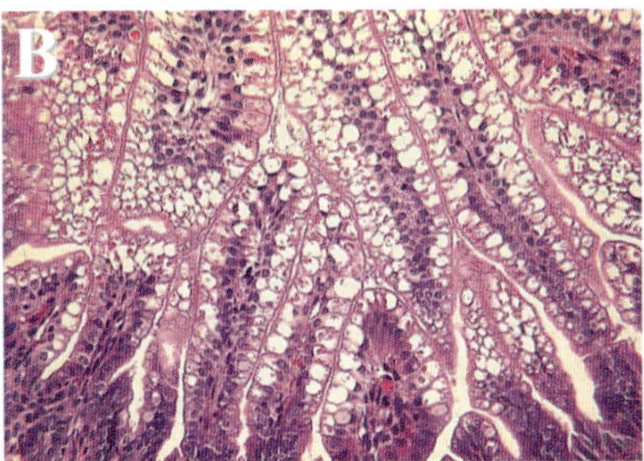

Fig. 3. Photomicrograph of small intestine of ileal mucosa of (A) 5-day old PBS mock-infected control neonatal rat, or (B) 5-day old simian RRV-infected neonatal rat at 72 hpi. In RRV-infected rat pups, histopathologic lesions are limited to large vacuoles in the enterocytes lining most of the surface of the villi in the ileum. Hematoxylin-eosin staining; original magnification X40. Abbreviations: *hpi,* hours post inoculation; *RRV,* rhesus rotavirus.

Group A RRV infection in neonatal rats is non-fatal and does not cause any long-term effects on growth and development, and the loss of weight, which occurs at the onset of disease and virus shedding, is overcome rather quickly (5-6 dpi) (Ciarlet *et al.,* 2002). Similar to human rotavirus infections in children, both group A and B rotavirus infections in rats cause a self limiting acute diarrheal disease and associated transient growth retardation (Mata, 1982; Salim *et al.,* 1995; Ciarlet *et al.,* 2002).

The rat model is currently limited to infection with heterologous group A rotaviruses. The utility of the model might be enhanced if rat group A rotaviruses became available. It is interesting to speculate on why a group A rotavirus has not been isolated from rats. The most obvious reason is the limited number of research groups that have thoroughly evaluated and properly analyzed this possibility or the prevalence of group A rotavirus infection in rats. The finding that group A rotaviruses are indeed capable of replicating in rats will stimulate research efforts directed toward the identification of group A rat rotaviruses. However, isolation of a group A rat rotavirus may be difficult because animals facilities obey strict husbandry regulations, and group A rotavirus infections may no longer be a veterinary problem in such facilities. Seroepidemiological studies also should reveal high rates of group A infection in rats. In time, the epidemiology, the interaction with the host, and the significance in nature of group A rat rotaviruses can be understood.

The newly established rat model of group A rotavirus infection possesses unique characteristics amenable to study rotavirus gastrointestinal pathophysiology and pathogenesis, and the molecular regulation of intestinal development and nutritional control of age-dependent disease. The rat provides the most practical animal model for physiological studies of rotavirus pathogenesis, such as understanding the mechanisms of rotavirus-induced diarrhea due to the convenience of the size of the intestinal tract

302

of neonatal rats over that of neonatal mice. This model should be useful to identify the cellular receptor for rotavirus since there are a number of cDNA libraries for rat intestinal cell lines, and monoclonal antibodies to rat intestinal cell markers (Kraml *et al.*, 1994; Takanaga *et al.*, 1995; Dodson *et al.*, 1996; Simister *et al.*, 1997; Murgia *et al.*, 1999). In addition, the sequence of the rat genome is projected to be available by 2005 (Summers *et al.*, 2001; Jacob and Kwitek, 2002). Finally, one major advantage that distinguishes the rat model from the rabbit and mouse models is the relative lack of host range restriction of the rat; rats are susceptible to infection by rotavirus strains isolated from many different species, including humans.

Acknowledgments

Work in the authors' laboratories has been supported by grants from the National Institutes of Health (DK30144, DK56338, and AI24998) and a grant from the Veterans Affairs Merit Review.

References

Awang, A., Yap, K. L. (1990). Group A rotavirus infection in animals from an animal house and in wild-caught monkeys. J. Diarrh. Dis. Res. **8:** 82-86.

Bass, D., Mackow, E., Greenberg, H. B. (1991). Identification and partial characterization of a rhesus rotavirus binding glycoprotein on murine enterocytes. Virology **183:** 602-610.

Burns, J.W., Krishnaney, A., Vo, P., Rouse, R., Anderson, L., Greenberg, H. B. (1995). Analyses of homologous rotavirus infection in the mouse model. Virology **207:** 143-153.

Browning, G. F., Chalmers, R., Snodgrass, D. R., Batt, R., Hart, C., Ormarod S., Leadon, D., Stoneham, S., Rossdale, P. D. (1991). The prevalence of enteric pathogens in diarrhoeic thoroughbred foals in Britain and Ireland. Equine Vet. J. **23:** 405-409.

Ciarlet, M., Estes, M. K., Barone, C., Ramig, R. F., Conner, M. E. (1998a). Analysis of host range restriction determinants in the rabbit model: Comparison of homologous and heterologous rotavirus infections. J. Virol. **72:** 2341-2351.

Ciarlet, M., Crawford, S. E., Barone, C., Bertolotti-Ciarlet, A., Ramig, R. F., Estes, M. K., Conner, M. E. (1998b). Subunit rotavirus vaccine administered parenterally to rabbits induces active protective immunity. J. Virol. **72:** 9233-9246.

Ciarlet, M., Gilger, M. A., Barone, C., McArthur, M., Estes, M. K., Conner, M. E. (1998c). Rotavirus disease, but not infection and development of intestinal histophatological lesions, is age-restricted in rabbits. Virology **251:** 343-360.

Ciarlet, M., Estes, M. K., Conner, M. E. (2000). Simian rhesus rotavirus (RRV) is a unique heterologous (non-lapine) rotavirus strain capable of productive replication and horizontal transmission in rabbits. J. Gen. Virol. **81:** 1237-1249.

Ciarlet, M., Conner, M. E. (2000). Evaluation of rotavirus vaccines in small animal models. *In:* Methods in Molecular Medicine, Vol. 34: Rotaviruses: Methods and Protocols, (Gray, J., and U. Desselberger, Eds.), pp. 147-187 Humana Press, Inc., Totowa, New Jersey, USA.

Ciarlet, M., Conner, M. E., Finegold, M. F., Estes, M. K. (2002). Group A rotavirus infection and age-dependent diarrheal disease in rats: a new animal model to study the pathophysiology of rotavirus infection. J. Virol. **76:** 41-57.

Ciarlet, M., Conner, M. E., Estes, M. K. Immunity to group A rotavirus infection in rats. Manuscript in preparation.

Clark, H. F., Offit, P. A., Ellis, R., Krah, D., Shaw, A., Eiden, J., Pichichero, M., Treanor, J. J. (1996). WC3 reassortant vaccines in children. Arch. Virol (Suppl.) **12:** 187-198.

Clements-Mann, M. L., Dudas, R., Hoshino, Y., Nehring, P., Sperber, E., Wagner, M., Stephens, I., Karron, R., Deforest, A., Kapikian, A. Z. (2001). Safety and immunogenicity of live attenuated quadrivalent human-bovine (UK) reassortant rotavirus vaccine administered with childhood vaccines to infants. Vaccine **19:** 4676-4684.

Coelho, K., Bryan, A., Hall, C., Flewett, T. (1981). Pathology of rotavirus infection in suckling mice: a study by conventional histology, immunofluorescence, ultra-thin sections, and scanning electron microscopy. Ultrastruct. Pathol. **2:** 59-69.

Conner, M. E., Darlington, R.W. (1980). Rotavirus infection in foals. Am. J. Vet. Res. 41: 1699-1703.

Conner M. E., Estes, M. K., Graham, D. Y. (1988). Rabbit model of rotavirus infection. J. Virol. **62:** 1625-1633.

Conner M.E., Gilger, M. A., Estes, M. K., Graham, D. Y. (1991). Serologic and mucosal immune response to rotavirus infection in the rabbit model. Journal of Virology **65:** 2562-2571.

Conner M. E., Crawford, S. E., Barone, C., Estes, M.K. 1993. Rotavirus vaccine administered parenterally induces protective immunity. J. Virol. **67:** 6633-6641.

Conner M. E., Ramig, R. F. (1996). Enteric Diseases. *In*: Viral Pathogenesis (Nathanson, N., Ahmed, R., González-Scarano, F., Griffin, D. E., Homes, K.V., Murphy, F. A., and Robinson, H. L., Eds.), pp. 713-743. Lippincott-Raven Publishers, Philadelphia.

Crouch, C. F., Woode, G. N. (1978). Serial studies of virus multiplication and intestinal damage in gnotobiotic piglets infected with rotavirus. J. Med. Microbiol. **11:** 325-334.

Cunliffe, N., Dove, W., Bunn, J., Ramadam, M., Nyangao, J., Riveron, R., Cuevas, L., Hart C. (2001). Expanding global distribution of rotavirus serotype G9: detection in Libya, Kenya, and Cuba. Emerg. Infect. Dis. **7:** 890-892.

Dodson, B., Wang, J., Swietlicki, E., Rubin, D. C., Levin, M. (1996). Analysis of cloned cDNAs differentially expressed in adapting remnant small intestine after partial resection. Am. J. Physiol. **271:** G347-G356.

Estes, M. K. 2001. Rotaviruses and their replication. *In*: Fields Virology (D. M. Knipe and P. M. Howley, Eds.), 4th edition, pp. 1747-1785. Lippincott Williams &

304

Wilkins Publishers, Philadelphia.

Feng, N., Burns, J., Bracey, L., Greenberg, H. B. (1994). Comparisons of the mucosal and systemic humoral immune responses and subsequent protection in mice orally inoculated with homologous or heterologous rotaviruses. J. Virol. **68:** 7766-7773.

Jacob, H., Kwitek, A. (2002). Rat genetics: attaching physiology and pharmacology to the genome. Nat. Rev. Genet. **3:** 33-42.

Glass, R., Kilgore, P., Holman, R., Jin, S., Smith, J., Woods, P., Clarke, M., Ho, M.-S., Gentsch, J. R. (1996). The epidemiology of rotavirus diarrhea in the United States: surveillance and estimates of disease burden. J. Infect. Dis. **174:** S5-S11.

Greenberg, H. B., Clark, H. F., Offit, P. A. (1994). Rotavirus pathology and pathophysiology. *In:* Rotaviruses (R. F. Ramig, Ed.), pp. 255-283. Springer-Verlag, Berlin.

Griffin, D., Kirkwood, C. D., Parashar, U. Woods, P., Bresee, J., Glass, R. I., Gentsch, J. R. (2000). Surveillance of rotavirus strains in the United States: identification of unusual strains. The National Rotavirus Strain Surveillance System collaborating laboratories. J Clin. Microbiol. **38:** 2784-2787.

Guérin-Danan, C., Meslin, J.-C., Lambre, F., Charpilienne, A., Serezat, M., Bouley, C., Cohen, J., Andrieux, C. (1998). Development of a heterologous model in germfree suckling rats for studies of rotavirus diarrhea. J. Virol. **72:** 9298-9302.

Guérin-Danan, C., Meslin, J-C., Chambard, A., Charpilienne, A., Relano, P., Bouley, C., Cohen, J., Andrieux, C. (2001). Food supplementation with milk fermented by *Lactobacillus casei* DN-114 001 protects suckling rats from rotavirus-associated diarrhea. J. Nutr. **131:** 111-117.

Henning, S. J., Rubin, D. C., Shulman, R. J. (1994). Ontogeny of the intestinal mucosa. *In*: Physiology of the Gastrointestinal Tract (L. R. Johnson, Ed.), pp. 571-610, 3rd edition. Raven Press, New York.

Iturriza-Gómara, M., Isherwood, B., Desselberger. U., Gray, J. (2001). Reassortment *in vivo*: driving force for diversity of human rotavirus strains isolated in the United Kingdom between 1995 and 1999. J. Virol. **75:** 3696-3705.

Kapikian, A. Z., Hoshino, Y., Chanock, R. M. (2001). Rotaviruses. *In*: Fields Virology (D. M. Knipe and P. M. Howley, Eds.), 4th edition, pp. 1787-1833. Lippincott Williams & Wilkins Publishers, Philadelphia.

Kraml, J., Kadlecova, L., Kolinska, J. (1994). Sialic acid – marker of development and differentiation. Sb. Lek. **95:** 271-275.

Mata, L. (1982). Diarrheal disease as a cause of malnutrition. Am. J. Trop. Med. Hyg. **47:** 16-27.

Mebus, C. A., Stair, L., Underdahl, N., Twiehaus, M. (1971). Pathology of neonatal calf diarrhea induced by a reo-like virus. Vet. Path. **8:** 490-505.

Mori, Y., Sugiyama, M., Takayama, M., Atoji, Y., Masegi, T., Minamoto, N. (2001). Avian-to-mammal transmission of an avian rotavirus: Analysis of its pathogenicity in a heterologous mouse model. Virology **288:** 63-70.

Murgia, C., Vespignani, I., Cerase, J., Nobili, F., Perozzi, G. (1999). Cloning, expression, and vesicular localization of zinc transporter Dri 27/ZnT4 in intestinal tissue and cells. Am. J. Physiol. **277**: G1231-G1239.

Osborne, M., Haddon, S., Spencer, A., Collins, J., Starsky, W., Wallis, T., Clarke, G., Worton, K., Candy, D., Stephen, J. (1988). An electron microscopic investigation of time-related changes in the intestine of neonatal mice infected with murine rotavirus. J. Pediatr. Gastroenterol. Nutr. **7**: 236-248.

Pearson, G., McNulty, M. S. (1977). Pathological changes in the small intestine of neonatal pigs infected with a pig reovirus-like agent (rotavirus). J. Comp. Path. **87**: 363-375.

Pearson, G., McNulty, M. S., Logan, E. (1978). Pathological changes in the small intestine of neonatal calves naturally infected with reo-like virus (rotavirus). Vet. Rec. **102**: 464-458.

Ramig, R.F. (1988). The effects of host age, virus dose, and virus strain on heterologous rotavirus infection of suckling mice. Microb. Pathogen. **4**: 189-202.

Reynolds, D., Hall, G., Debney, T., Parsons, K. (1985). Pathology of natural rotavirus infection in clinically normal calves. Res. Vet. Sci. **38**: 264-269.

Saif, L. J., Rosen, B., Parwani, A. (1994). Animal rotaviruses. *In*: Viral Infections of the Gastrointestinal Tract (A. Z. Kapikian, Ed.), 2nd edition, pp. 279-367. Marcel Dekker, Inc., New York.

Salim, A., Phillips, A., Walker-Smith, J., Farthing, M. (1995). Sequential changes in small intestinal structure and function during rotavirus infection in neonatal rats. Gut **36**: 231-238.

Simister, N., Jacobowitz Israel, E., Ahouse, J., Story, C. (1997). New functions of the MHC class I-related Fc receptor, FcRn. Biochem. Soc. Trans. **25**: 481-486.

Snodgrass, D. R., Ferguson, A., Allan, F., Angus, K., Mitchell, B. (1979). Small intestinal morphology and epithelial cell kinetics in lamb rotavirus infections. Gastroenterology **76**: 477-481.

Starkey, W., Collins, J., Wallis, T., Clarke, G., Spencer, A., Haddon, S., Osborne, M., Candy, D., Stephen, J. (1986). Kinetics, tissue specificity and pathological changes in murine rotavirus infection of mice. J. Gen. Virol. **67**: 2625-2634.

Summers, T., Thomas, J., Lee-Lin, S., Maduro, V., Idol, J., Green, E. (2001). Comparative physical mapping of the targeted regions of the rat genome. Mamm. Genome **12**: 508-512.

Takahashi, E., Inaba, Y., Sato, K., Kurogi, H., Akashi, H., Satoda, K., Omori, T. (1979). Antibody to rotavirus in various animal species. Nat. Inst. Anim. Hlth Quart. **19**: 72-73.

Takanaga, H., Tamai, I., Inaba, S., Sai, Y., Higashida, H., Yamamoto, H., Tsuji, A. (1995). cDNA cloning and functional characterization of rat intestinal monocarboxylate transporter. Biochem. Biophys. Res. Commun. **217**: 370-377.

Theil, K., Bohl, E., Cross, R., Köhler, E., Agnes, A. (1978). Pathogenesis of porcine rotaviral infection in experimentally inoculated gnotobiotic pigs. Am. J. Vet. Res. **39**: 213-220.

Vonderfecht, S., Huber, A., Eiden, J., Mader, L., Yolken, R. (1984). Infectious diarrhea of infant rats produced by a rotavirus-like agent. J. Virol. **52**: 94-98.

Vonderfecht, S., Eiden, J., Miskuff, R., Yolken, R. (1988). Kinetics of intestinal replication of group B rotavirus and relevance to diagnostics methods. J. Clin. Microbiol. **26:** 216-221.

Ward, R. L., McNeal, M. M., Sheridan, J. (1990). Development of an adult mouse model for studies on protection against rotavirus. J. Virol. **64:** 5070-5075.

Ward, R. L., McNeal, M. M., Sheridan, J. (1992). Evidence that active protection following oral immunization of mice with live rotavirus is not dependent on neutralizing antibody. Virology **188:** 57-66.

Yolken, R., Arango-Jaramillo, S., Eiden, J., Vonderfecht, S (1988). Lack of genomic reassortment following infection of infant rats with group A and group B rotaviruses. J. Infect. Dis. **158:** 1120-1123.

Yolken, R., Ojeh, C., Khatri, I., Sajjan, U., Forstner, J. (1993). Intestinal mucins inhibit rotavirus replication in an oligosaccharide-dependent manner. J. Infect. Dis. **169:** 1002-1006.

Zhang M., Zeng, C. Q.-Y., Dong, Y., Ball, J. M., Saif, L. J., Morris, A. P., Estes, M. K. (1998). Mutations in rotavirus nonstructural glycoprotein NSP4 are associated with altered virulence. J. Virol. **72:** 3666-3672.

Viral Gastroenteritis
U. Desselberger and J. Gray (editors)
© 2003 Elsevier Science B.V. All rights reserved

II, 11. Human adaptive immunity to rotaviruses: A model of intestinal mucosal adaptive immunity

Ana María Gonzalez[1,3], Maria C. Jaimes[2,1], Olga L. Rojas[1], Juana Angel[1], Harry B. Greenberg[2] and Manuel A. Franco[1]

[1] *Instituto de Genetica Humana Pontificia Universidad Javeriana, Bogotá Colombia*
[2] *Departments of Medicine, Microbiology and Immunology, Stanford University School of Medicine, 300 Pasteur Dr, CCSR Bldg Rm 3115, Stanford, CA 94305-5187*
[3] *Present address: Food Animal Health Research Program (FAHRP), Department of Veterinary Preventive Medicine,Ohio Agricultural Research and Development Center (OARDC), The Ohio State University,Wooster, Ohio, 44691-4096,USA*

Introduction

Rotaviruses (RV) are the single most important cause of severe diarrhea in young children, and there is an urgent need to develop effective vaccines against them (Franco and Greenberg, 2001). Because RV replication is highly restricted to enterocytes *in vivo*, the immune response against them originates in and exhibits its effector function directly at the intestinal mucosa. For this reason, RV infections are an excellent model to study intestinal mucosal immunity. Here we will review the human adaptative immune response to RV, first placing the immune response to RV in the context of the immune response to other mucosal viruses, and then highlighting studies of both RV-specific T and B cells.

Since children with T and/or B immunodeficiencies can develop chronic RV infection, prolonged symptoms and extraintestinal infection (Saulsbury *et al.*, 1980; Wood *et al.*, 1988; Gilger *et al.*, 1992) it is clear that both T and B cells are important for immunity to RV. Nonetheless, the T and B cells induced by natural RV infection are not sufficient in many cases to prevent symptomatic or asymptomatic reinfections in children (Velazquez *et al.*, 1996) or adults (Rodriguez *et al.*, 1987; Nakajima *et al.*, 2001). The severity and number of RV infections diminish with age, and life threatening infections are generally limited to the primary infection, suggesting that adaptive protective immune responses to RV develop gradually. Reinfection with many other viruses that replicate only at mucosal surfaces is also frequent (Murphy, 1999). This is in contrast to viruses that spread and replicate systemically for which the development of long lasting protective immunity is the rule (Murphy, 1999). The various reasons proposed to explain absence of complete immunity to mucosal viruses such as RV, following primary infection, include:

1) A short incubation period after viral exposure. Since the ultimate cellular target for replication of viruses like influenza viruses, rhinoviruses or RV is the same cell as

used for entry, these viruses do not require a period of recirculation through the organism. For this reason after secondary viral exposure, the immune system does not have time to amplify the protective specific responses against these viruses. Sustained high levels of antiviral effector mechanisms at the mucosal surface are thus necessary to achieve protection against these viruses (Roitt, 1997).

2) Difficulty in maintaining a high level of protective antibody at respiratory and gastrointestinal mucosal surfaces. This problem is due to the high volume, rapid transit time and hostile environment (for example proteases that can inactivate the antibodies) of these surfaces (Murphy, 1999).

3) A short lived protective humoral mucosal immune response. Although the serum antibody response to systemic viruses (e.g. mumps virus, rubella virus, or HBV) lasts for over a decade, the mucosal antibody responses to influenza virus, respiratory syncytial virus, coronavirus and RV last from 3 to 30 months (Doherty and Ahmed, 1997; Murphy, 1999). For a given virus (RV included) to which both systemic and mucosal antibody responses develop, the former seem to be more long lasting than the latter (Coulson *et al.*, 1990; Nishio *et al.*, 1990).

4) Viral mutations that permit the virus to escape the effect of protective neutralizing antibodies. Although this factor is certainly not exclusive to mucosal viruses, it probably plays an important role in conjunction with factors 2 and 3 mentioned above. Protection from reinfections with rhinoviruses seems to be highly dependent on the presence of serotype-specific antibodies but the existence of over 100 viral serotypes makes subsequent infections with viruses of different serotypes highly likely (Doherty and Ahmed, 1997). Susceptibility to severe influenza disease is also serotype dependent: the appearance of a completely new viral serotype (antigenic shift) causes pandemics. More frequently, minor changes that do not change the serotype of the hemagglutinin (antigenic drift) of a circulating virus seem to be responsible for the susceptibility of individuals to reinfections with influenza virus in each winter epidemic (Doherty and Ahmed, 1997). For this reason the design of annual influenza vaccines takes into considerations drift mutations in circulating viruses for a given period. RV-specific immunity seems to be, to some extent, serotype-specific (Kapikian *et al.*, 2001), and reinfections in children appear to be preferentially (but not exclusively) due to viruses of a different serotype (Velazquez *et al.*, 1996). However, serotype-specific immunity for RVs seems to be less important than for some other mucosal viruses such as rhinoviruses or influenza viruses. Unlike the multitude of serotypes found for rhinoviruses, RV of only five major serotype combinations circulate in most parts of the world (Desselberger *et al.*, 2001) suggesting that variation in viral serotype alone can not account for viral reinfections. In addition, as opposed to influenza virus, the appearance of a new RV serotype (e.g. emerging by reassortment with an animal RV) followed by wide spread disease in a population immune to RV of a different serotype seems to be a very rare event if it exists at all. The appearance of limited mutations in the neutralizing antigens that do not change the viral serotype (like influenza antigenic drift) have been proposed as a factor that could be favoring RV reinfection, but this hypothesis remains unproven at present (Jin *et al.*, 1996; Coulson, 1998).

Since serotype-specific immunity cannot completely explain protection against RV, other factors probably play a role in protection against this pathogen. For example, studies in animal models have suggested that non-neutralizing antibodies against other viral proteins (VP6 and or NSP4 (Ball *et al.*, 1996; Burns *et al.*, 1996)) or other mechanisms (like T cells) can also play a role in immunity (Kapikian *et al.*, 2001).

The above discussion underscores the importance of studying RV-specific memory T and B lymphocytes, and of investigating the ways these memory cells localize to the intestinal mucosal surfaces were they are required to mediate their protective effect. The understanding of RV-specific CD4+ T cell responses is a key issue because most of RV-specific antibody responses in animal models seem to be dependent on CD4 help (Franco and Greenberg, 1997). Studies of the T cell response to RV have used lympho-proliferation as a readout after *in vitro* restimulation with viral antigen (Table 1). It is generally assumed that results obtained using this method reflect the presence of CD4+ antigen specific T cells. These studies have shown that most, but not all, healthy adults and children and children in the convalescent phase of RV-induced diarrhea have RV-specific CD4+ T cells. The percentage of healthy elderly persons and healthy children below the age of six months that have a lymphoproliferative responses to RV antigen is lower than the percentage of healthy adults and children older than six months of age that have a lymphoproliferative responses to RV antigen (Totterdell *et al.*, 1988a; Offit *et al.*, 1992). The T cells that proliferate in response to RV most probably recognize cross-reactive epitopes since restimulation *in vitro* using simian and bovine RVs as antigens seems to be as efficient as human RV (Table 1). The fine specificity of this T cell response, either at a protein or epitope level, has not been thoroughly investigated. In fact, only one study has formally demonstrated that the proliferative responses were due, at least in part, to CD4 cells (Rott *et al.*, 1997).

Table 1

Studies of *in vitro* proliferative responses to RV antigen

Primary study population	Antigen used for *in vitro* restimulation of cells	Reference
Healthy adults	Tissue culture supernatant of simian RV	Totterdell *et al.*, 1988b
Elderly adults with and without RV diarrhea	Tissue culture supernatant of simian RV	Totterdell *et al.*, 1988a
Healthy adults	Infectious and inactivated human RV and infectious bovine RV	Yasukawa *et al.*, 1990
Healthy children and adults	Purified human RV and simian RRV	Offit *et al.*, 1992
Children with RV induced diarrhea	Purified human RV	Offit *et al.*, 1993
Healthy adults	Purified inactivated simian RV*	Rott *et al.*, 1997

* This study used purified subsets of CD4+T lymphocytes and is thus the only one for which the lymphoproliferative activity can directly be attributed to CD4 T cells.

As mentioned above, an important issue to address with RV-specific lymphocytes is the mechanism that permits these lymphocytes to localize and accumulate in the intestinal mucosa were they are required to perform their antiviral effect. Mature naive lymphocytes recirculate between various secondary lymphoid organs until they encounter the antigen that stimulates their specific receptor. After antigen-specific activation, the lymphocyte migration pattern changes, redirecting them preferentially to the tissue where the activating antigen was originally encountered (Butcher *et al.*, 1999). This tissue-specific migration provides lymphocytes with the best chance of re-encountering the antigen that originally stimulated them. The homing of lymphocytes stimulated by antigens first encountered in intestinal Peyer's patches back to the same or other Peyer's patches or the intestinal lamina propria is mediated by interactions between the integrin $\alpha 4\beta 7$, expressed on B and T lymphocytes and the cell adhesion molecule MadCAM 1, expressed on the vascular endothelium of the postcapillary venules in the intestine (Butcher *et al.*, 1999). To date, only one study has addressed the issue of the potential capacity of RV-specific CD4+ T cells from humans to localize to the intestine based on the presence or absence of the intestinal homing receptor (the integrin $\alpha 4\beta 7+$) on these lymphocytes: Rott *et al.* studied the ability of $\alpha 4\beta 7+$ and $\alpha 4\beta 7-$ CD4+ T cell subsets of memory phenotype (CD45RA-) (purified by flow cytometry) to proliferate in response to RV. $\alpha 4\beta 7+$ CD45RA- cells proliferated more than 2.3 fold over $\alpha 4\beta 7-$ CD45RA- CD4+ T cells to RV antigen (Rott *et al.*, 1997). In contrast, $\alpha 4\beta 7-$ memory cells were the predominant population responsive to mumps antigen after intramuscular vaccination. These findings suggest that RV-specific T cells, but not T cells specific for a viral antigen administered parenterally, have the capacity to migrate to the intestinal mucosa (Rott *et al.*, 1997).

The studies that have used lymphoproliferation to study the T cell response to RV have the limitations that they do not directly quantify or establish the phenotype (with the exception of the study by Rott *et al.*, 1997) of the cells responding to the RV antigen. We have recently adapted a flow cytometry based assay (Waldrop *et al.*, 1997) for the study of RV-specific T cells that detects the intracellular accumulation of cytokines after short term *in vitro* antigen stimulation (Jaimes *et al.*, 2002). Because this assay allows to analyse individual cells it permits the simultaneous quantification and characterization of the phenotype of the RV-specific T cells. Using this assay we have been able to characterize, for the first time, human RV-specific CD8+ T cells, and directly determined the frequencies of RV-specific CD4+T cells that secrete IFN-γ and IL-13 and CD8+ T cells that secrete IFN-γ in children and adults with RV-induced diarrhea and in healthy adults. A representative plot of a flow cytometric analysis of an adult in the convalescent phase of RV induced diarrhea is shown in Fig. 1. Table 2 shows the frequencies of IFN-γ secreting CD4 and CD8+ RV-specific T cells in healthy adults, and adults and children with RV induced diarrhea: while adults with RV induced diarrhea have greater numbers of RV-specific T cells than healthy adults, children with RV induced diarrhea have very low levels of RV-specific T cells. Using this assay we expect to be able to identify, quantitate and evaluate differences between mucosally primed RV T cells and T cells specific for viruses that replicate systemically. Some future studies will investigate: 1) the pattern and quantity of cytokines secreted by RV-specific lymphocytes; 2) the expression of various molecules involved in lymphocyte migration (Rott *et al.*, 1997) and 3) the kinetics of appearance and circulation in blood of RV-specific lymphocytes after RV infection.

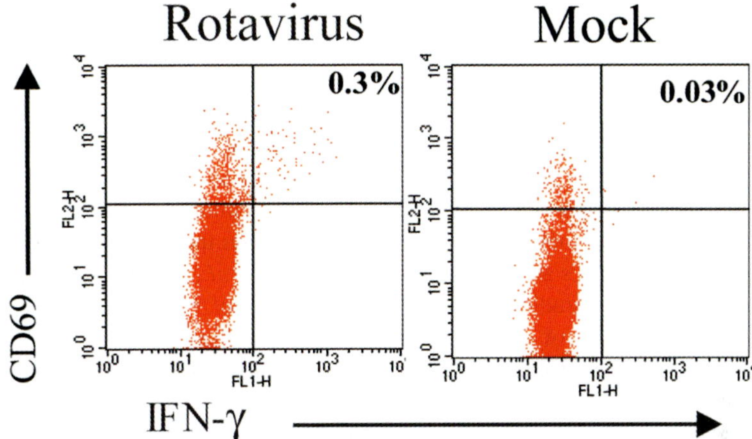

Fig. 1. Frequencies of CD8+ cells in an adult with RV induced diarrhea. Cell were restimulated *in vitro* with cesium chloride density gradient purified RF RV (right) or the supernatant of mock infected cells (left) in the presence of brefeldin A to permit the accumulation of intracellular cytokines. Cells were then fixed, permeabilized and stained with fluorochrome labeled monoclonal antibodies against CD8 and CD69 (an early T cell activation marker) and interferon gamma. The events in the graphs are gated on CD8+ lymphocytes. The percentage of RV-specific T cell (events that express both CD69 and interferon gamma) are shown in the upper right quadrant.

Table 2

Frequencies of IFN-γ secreting RV-specific CD4 and CD8 T cells in healthy adults and children and adults with RV diarrhea

Study population	Mean % RV specific CD4 T cells (SEM)*	Mean % RV specific CD8 T cells (SEM)*
Adults with RV diarrhea (n= 7)	0.18 (0.10)	0.28 (0.11)
Healthy adults (n= 8)	0.02 (0.006)	0.03 (0.01)
Children with RV diarrhea (n= 12)	0.01 (0.009)	0.02 (0.006)

* The percentages of RV-specific CD4 and CD8 T cells were calculated as indicated in the legend to Fig. 1.

Studies of the human antibody response to RV have highlighted the importance of serum IgA (Velazquez *et al.*, 2000) and particularly the intestinal IgA response in protection against RV (Coulson *et al.*, 1992). As for T cells, the detailed characterization of the B cells that produced these IgA antibodies may shed more light on the quantitative and qualitative aspect of intestinal mucosal B cells. Published studies that have characterized these B cells have been limited to the study of effector antibody secreting cells (ASC) identified by ELISPOT. Blood circulating ASC detected in children after acute natural infection or vaccination have been shown to secrete predominantly the IgM isotype (Kaila *et al.*, 1992; Isolauri *et al.*, 1995). Our current model to explain the detection of these RV-specific ASC in human peripheral blood is that these B cells, originally stimulated by RV antigen in Peyer's patches, are detected in blood during their transit from Peyer's patches, through the circulation and back to the intestine. A recent study supported this model by showing in older children without documented recent RV infection that there exists a correlation between the presence of blood circulating RV-specific ASCs and intestinal ASCs (Brown *et al.*, 2000). Furthermore, we have recently performed studies similar to those of Rott *et al.* (1997) but using purified α4β7+ and α4β7- subsets of peripheral blood B cells and then testing in an ELISPOT assay. As expected, RV-specific ASC were predominantly present in the α4β7+ subpopulation (Gonzalez *et al.*, 2002).

We have also recently developed a flow cytometry assay in the mouse model that identifies antigen activated (IgD-) B cells (CD19+) expressing surface RV-specific immunoglobulin (Kuklin *et al.*, 2001; Youngman *et al.*, 2002). With this assay we are able to detect both RV-specific ASCs and memory murine B cells. The principal of this assay consists of the specific binding of a fluorescent RV antigen to a B cell that expresses RV-specific Ig. As an antigen we chose recombinant virus-like particles (VLP) made from RV VP6 protein and a fusion protein consisting of green fluorescent protein (GFP) fused to the N terminus of RV VP2 deleted of its first 92 aminoacids (Charpilienne *et al.*, 2001). In addition to the RV/GPF-VLPs, the B cells are also stained with fluorochrome labeled monoclonal antibodies directed against a B cell marker (CD19), IgD (to distinguish antigen activated from naive B cells), and the intestinal homing receptor α4β7 and then analyzed in a flow cytometer. We have recently applied this assay to detect RV-specific B cells in both children and adults with an acute RV infection (Gonzalez *et al.*, 2002). An example of one of these experiments is shown in Fig. 2. Like for RV-specific ASCs, these studies have shown that B cells that express RV-specific surface immunoglobulin from both children and adults predominantly express α4β7 on their surface. These results open the possiblity of characterizing in detail the phenotype of mucosal effector and memory B cells and in particular the expression of other molecules in addition to the integrin α4β7 that have been implicated in migration of lymphocytes to the intestine (Cook *et al.*, 2000).

Acknowledgments

This work was supported by funds from the Pontificia Universidad Javeriana, ECOS/ICFES/COLCIENCIAS/ICETEX, Fundación para la Promoción de la Investigación y la Tecnología, Banco de la República grant 1103, Colciencias grants 1203-04-151-98

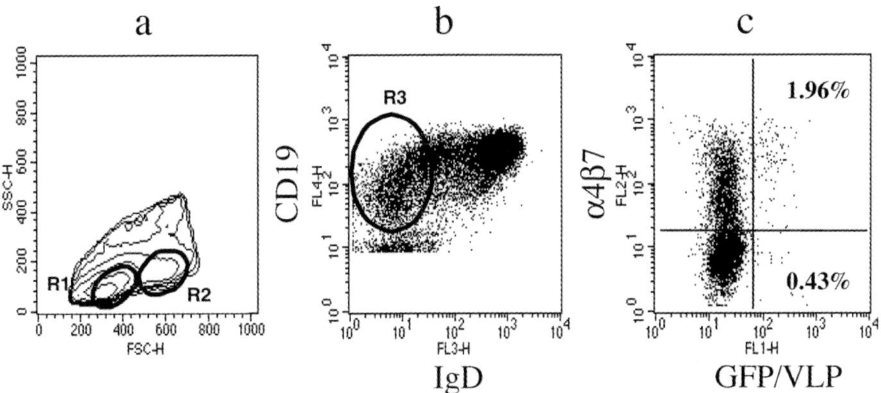

Fig. 2. Frequencies of CD19+ IgD-, α4β7 +/-, GPF-VLPS+ cells in an adult with RV induced diarrhea. PBMC were stained with fluorochrome labeled monoclonal antibodies against CD19, IgD, α4β7 and with GPF-VLPS and acquired in a flow cytometer. a) forward and light scatter plot showing two populations of cells (R1 small cells R2 large cells). b) Dot plot of Large cells (R2 in a) showing staining of CD19 and IgD and the creation of a gate (R3) of the IgD- CD19+ cells. Notice that CD19 low cells are not acquired for the experiment. c) Dot plot gated in R3 showing expression of the intestinal homing receptor α4β7 and binding of GPF-VLPS. The percentage of RV-specific B cells are shown in the left quadrants.

and 1203-04-966-98 and by ICGEB grant COL99-01(a1); by a grant from the National Institutes of Health R37AI21632 and by VA grants to H.B.G.. We would like to thank the subjects that participated in our study for their generosity and the personnel of the pediatrics department of the Hospital San Ignacio for their help in identifying children with diarrhea.

References

Ball, J. M., P. Tian, C. Q.-Y. Zeng, A. P. Morris and M. K. Estes (1996). Age-dependent diarrhea induced by a rotaviral nonstructural glycoprotein. Science 272: 101-104.

Brown, K. A., J. A. Kriss, C. A. Moser, W. J. Wenner and P. A. Offit (2000). Circulating rotavirus-specific antibody-secreting cells (ASCs) predict the presence of rotavirus-specific ASCs in the human small intestinal lamina propria. J Infect Dis 182: 1039-1043.

Burns, J. W., P. M. Siadat, A. A. Krishnaney and H. B. Greenberg (1996). Protective effect of rotavirus VP6-specific IgA monoclonal antibodies that lack neutralizing activity. Science 272: 104-107.

Butcher, E. C., M. Williams, K. Youngman, L. Rott and M. Briskin (1999). Lymphocyte trafficking and regional immunity. Adv Immunol 72: 209-253.

314

Charpilienne, A., M. Nejmeddine, M. Berois, N. Parez, E. Neumann, E. Hewat, G. Trugnan and J. Cohen (2001). Individual rotavirus-like particles containing 120 molecules of fluorescent protein are visible in living cells. J Biol Chem 276: 29361-29367.

Cook, D. N., D. M. Prosser, R. Forster, J. Zhang, N. A. Kuklin, S. J. Abbondanzo, X. D. Niu, S. C. Chen, D. J. Manfra, M. T. Wiekowski, L. M. Sullivan, S. R. Smith, H. B. Greenberg, S. K. Narula, M. Lipp and S. A. Lira (2000). CCR6 mediates dendritic cell localization, lymphocyte homeostasis, and immune responses in mucosal tissue. Immunity 12: 495-503.

Coulson, B. S. (1998). Longitudinal studies of neutralizing antibody responses to rotavirus in stools and sera of children following severe rotavirus gastroenteritis. Clin Diagn Lab Immunol 5: 897-901.

Coulson, B. S., K. Grimwood, I. L. Hudson, G. L. Barnes and R. F. Bishop (1992). Role of coproantibody in clinical protection of children during reinfection with rotavirus. J Clin Microbiol 30: 1678-1684.

Coulson, B. S., K. Grimwood, P. J. Masendycz, J. S. Lund, N. Mermelstein, R. F. Bishop and G. L. Barnes (1990). Comparison of rotavirus immunoglobulin A copro-conversion with other indices of rotavirus infection in a longitudinal study in childhood. J Clin Microbiol 28: 1367-1374.

Desselberger, U., M. Iturriza-Gómara and J. J. Gray (2001). Rotavirus epidemiology and surveillance. Novartis Found. Symp. 238: 125-147.

Doherty, P. C. and R. Ahmed (1997). Immune response to viral infection. In: Viral pathogenesis. N. Nathanson (Ed.), pp. 143-161. Philadelphia, Lippincott-Raven.

Franco, M. A. and H. B. Greenberg (1997). Immunity to rotavirus in T cell deficient mice. Virology 338: 169-179.

Franco, M. A. and H. B. Greenberg (2001). Challenges for rotavirus vaccines. Virology 281: 153-155.

Gilger, M. A., D. O. Matson, M. E. Conner, H. M. Rosenblatt, M. J. Finegold and M. K. Estes (1992). Extraintestinal rotavirus infections in children with immuno-deficiency. J Pediatr 120: 912-917.

Gonzalez, A. M., M. C. Jaimes, I. Cajiao, O. L. Rojas, J. Cohen, P. Pothier, E. Kholi, E. C. Butcher, H. B. Greenberg, J. Angel and M. A. Franco (2003). Rotavirus specific B cells induced by recent infection in adults and children predominantly express the intestinal homing receptor $\alpha4\beta7$. Virology 305: 93-105.

Isolauri, E., J. Joensuu, H. Suomalainen, M. Luomala and T. Vesikari (1995). Improved immunogenicity of oral D x RRV reassortant rotavirus vaccine by lactobacillus casei gg. Vaccine 13: 310-312.

Jaimes, M. C., O. L. Rojas, A. M. Gonzalez, I. Cajiao, A. Charpilienne, P. Pothier, E. Kholi, H. B. Greenberg, M. A. Franco and J. Angel (2002). Frequencies of virus specific CD4+ and CD8+ T lymphocytes secreting interferon gamma after acute natural rotavirus infection in children and adults. J Virol 76, 4741-4749.

Jin, Q., R. L. Ward, D. R. Knowlton, Y. B. Gabbay, A. C. Linhares, R. Rappaport, P. A. Woods, R. I. Glass and J. R. Gentsch (1996). Divergence of VP7 genes of G1 rotaviruses isolated from infants vaccinated with reassortant rhesus rotaviruses. Arch Virol 141: 2057-2076.

Kaila, M., E. Isolauri, E. Virtanen and H. Arvilommi (1992). Preponderance of IgM from blood lymphocytes in response to infantile rotavirus gastroenteritis. Gut 33: 639-642.

Kapikian, A. Z., Y. Hoshino and R. M. Chanock (2001). Rotaviruses. In: Fields Virology, 4th ed., D. M. Knipe, P. M. Howley *et al.* (Eds.), pp. 1787-1833. Lippincott Williams & Wilkins, Philadephia.

Kuklin, N. A., L. Rott, N. Feng, M. E. Conner, N. Wagner and H. B. Greenberg (2001). Protective intestinal anti-rotavirus B cell immunity is dependent on α4β7 integrin expression but does not require IgA antibody production. J Immunol 166: 1894-1902.

Murphy, B. R. (1999). Mucosal immunity to viruses. In: Mucosal immunology. P. L. Ogra, J. Mestecky, M. E. Lamm *et al.* (Eds.), pp. 695-707. Academic Press, San Diego CA.

Nakajima, H., T. Nakagomi, T. Kamisawa, N. Sakaki, K. Muramoto, T. Mikami, H. Nara and O. Nakagomi (2001). Winter seasonality and rotavirus diarrhoea in adults. Lancet 357: 1950.

Nishio, O., J. Sumi, K. Sakae, Y. Ishihara, S. Isomura and S. Inouye (1990). Fecal IgA antibody responses after oral poliovirus vaccination in infants and elder children. Microbiol Immunol 34: 683-689.

Offit, P. A., E. J. Hoffenberg, E. S. Pia, P. A. Panackal and N. L. Hill (1992). Rotavirus-specific helper T cell responses in newborns, infants, children, and adults. J. Infect. Dis. 165: 1107 - 1111.

Rodriguez, W. J., H. W. Kim, C. D. Brandt, R. H. Schwartz, M. K. Gardner, B. Jeffries, R. H. Parrott, R. A. Kaslow, J. I. Smith and A. Z. Kapikian (1987). Longitudinal study of rotavirus infection and gastroenteritis in families served by a pediatric medical practice: Clinical and epidemiologic observations. Pediatr Infect Dis J 6: 170-176.

Roitt, I. (1997). Immunity to infections. In: Essential immunology. I. Roitt (Ed.), pp. 282-283. Blackwell Science, Oxford.

Rott, L. S., J. R. Rosé, D. Bass, M. B. Williams, H. B. Greenberg and E. C. Butcher (1997). Expression of mucosal homing receptor alpha4beta7 by circulating CD4+ cells with memory for intestinal rotavirus. J Clin Invest 100: 1204-1208.

Saulsbury, F. T., J. A. Winkelstein and R. H. Yolken (1980). Chronic rotavirus infection in immunodeficiency. J Pediatr 97: 61-65.

Totterdell, B. M., J. E. Banatvala, I. L. Chrystie, G. Ball and W. D. Cubitt (1988a). Systemic lymphoproliferative responses to rotavirus. J Med Virol 25: 37-44.

Totterdell, B. M., S. Patel, J. E. Banatvala and I. L. Chrystie (1988b). Development of a lymphocyte transformation assay for rotavirus in whole blood and breast milk. J Med Virol 25: 27-36.

Velazquez, F. R., D. O. Matson, M. L. Guerrero, J. Shults, J. J. Calva, A. L. Morrow, R. I. Glass, L. K. Pickering and G. M. Ruiz-Palacios (2000). Serum antibody as a marker of protection against natural rotavirus infection and disease. J Infect Dis 182: 1602-1609.

Velazquez, R. F., D. O. Matson, J. J. Calva, L. Guerrero, A. L. Morrow, S. Carter-Campell, R. I. Glass, M. K. Estes, L. K. Pickering and G. M. Ruiz-Palacios (1996). Rotavirus infection in infants as protection against subsequent infections. N Engl J Med 335: 1022-1028.

Waldrop, S. L., C. J. Pitcher, D. M. Peterson, V. C. Maino and L. J. Picker (1997). Determination of antigen-specific memory/effector cd4+ t cell frequencies by flow cytometry: Evidence for a novel, antigen-specific homeostatic mechanism in HIV-associated immunodeficiency. J Clin Invest 99: 1739-1750.

Wood, D. J., T. J. David, I. L. Chrystie and B. Totterdell (1988). Chronic enteric virus infection in two T-cell immunodeficient children. J Med Virol 24: 435-444.

Yasukawa, M., O. Nakagomi and Y. Kobayashi (1990). Rotavirus induces proliferative response and augments non-specific cytotoxic activity of lymphocytes in humans. Clin Exp Immunol 80: 49-55.

Youngman, K. R., M. A. Franco, N. A. Kuklin, L. S. Rott, E. C. Butcher and H. B. Greenberg (2002). Correlation of tissue distribution, developmental phenotype, and intestinal homing receptor expression of antigen-specific B cells during the murine anti-rotavirus immune response. J Immunol 168: 2173-2181.

Viral Gastroenteritis
U. Desselberger and J. Gray (editors)
© 2003 Elsevier Science B.V. All rights reserved

II, 12. Molecular epidemiology of rotaviruses: Genetic mechanisms associated with diversity

Miren Iturriza Gómara, Ulrich Dessselberger and Jim Gray

Clinical Microbiology and Public Health Laboratory, Addenbrooke's Hospital, Hills Road, Cambridge CB2 2QW, United Kingdom

Introduction

Rotaviruses are the main cause of acute gastroenteritis in infants and young children worldwide and in the young of many animals. They are classified into at least 5 groups (A to E), and there are possibly 2 more groups (F and G) according to epitopes on the VP6 protein [Estes, 2001]. Group A rotaviruses are most commonly associated with infections in humans [Kapikian *et al.*, 2001]. Within group A, 4 subgroups (SGs) have been differentiated according to the epitopes on the VP6 middle layer protein. The two outer layer proteins, VP7 and VP4, form the basis of the current dual classification system of group A rotaviruses into G and P types [Estes, 2001]. At least 14 different G-serotypes, G1 to G14, have been identified among human and animal rotaviruses, based on differences in the neutralisation epitopes of the VP7 protein. Of these, at least 10 are associated with infection in humans. G serotypes correlate fully with G genotypes as determined by sequence analysis of the VP7 gene. Twenty different P types have been differentiated by sequence analysis of their VP4 genes [Estes, 2001], of which at least 10 have been found in humanisolates. To date, only 12 out of 20 P genotypes have been correlated with P serotypes [Estes, 2001]. This chapter will describe the variability of group A rotaviruses with regard to subgroups and G and P types, and review the mechanisms of genomic change underlying the evolution of these viruses.

Rotavirus classification and diversity

Subgroups

Rotavirus SGs I, II, I+II and nonI/nonII have been defined according to the presence or absence of two distinct epitopes reactive with one, both or neither of two monoclonal antibodies (Mabs), 255/60 (SGI-specific) and 631/9 (SGII-specific)[Greenberg *et al.*, 1983]. Subgrouping ELISAs have been used extensively in epidemiological studies and shown that in human infections the most commonly found rotaviruses are of SGII.

A total of 20 subgroup-specific amino acids (aa) have been identified, and studies using site-directed mutagenesis [Tang *et al.*, 1997] or recombinant VP6 proteins [López *et al.*, 1994] have shown that substitutions at aa positions 296-299, 305, 306, 308 and 315 were capable of changing reactivities with SGI- and SGII-specific Mabs. It has been proposed that an Ala residue at position 305 and the region between aa 296-299 contribute to the reactivity with Mab 255/60 (SGI), whereas Glu at aa position 315 contributes to reactivity with Mab 631/9 (SGII) [Tang *et al.*, 1997].

Analysis of a 379 bp region of the VP6 gene encompassing the previously defined subgroup-specific epitope coding region, has allowed subgrouping of rotaviruses according to the diversity within the VP6 gene. Sequence analysis of VP6 amplicons of rotavirus strains characterised serologically as SGI, II, I+II and nonI/nonII revealed the existence of two genetic clusters or genogroups (Fig. 1; Iturriza-Gómara *et al.*, 2002). Genogroup I correlated with strains serologically defined as SGI, and two distinct genetic lineages were identified among the strains in this genogroup. Genogroup II encompassed rotavirus strains serologically identified as SGII, I+II and nonI/nonII. Three genetic lineages were identified within genogroup II, but no correlation was found between these and the serologically defined SG-specificities.

It is still unclear whether the SG specificity is determined by linear epitopes or by conformational epitopes present only in the trimeric form of VP6. The VP6 region chosen for amplification (nt 747 -1126, coding for aa 241-367) encompassed the aa positions previously implicated in the recognition by SG-specific Mabs (see above), as well as those associated with trimerisation, aa 246-315 [Clapp and Patton, 1991; Affranchino and Gonzalez, 1997], and most of those necessary for the formation of double layered particles, aa 281-397 [Tosser *et al.*, 1994]. The deduced aa sequences of all the strains serologically determined as SGI had the Ala residue at position 305 characteristic of SGI. However, the deduced aa sequences of the amplicons obtained from strains serologically identified as SGII, SGI+II or SGnonI/nonII were indistinguishable from each other. None of the strains serologically determined as SGI+II or SGnonI/nonII differed from SGII strains in any of the residues identified as SGII determinants. None of the strains serologically determined as SGnonI/nonII had the insertion Pro-Glu at positions 299-300 or an aa substitution at position 308 which were previously shown to abolish reactivity with both 255/60 and 631/9 Mabs [Tang *et al.*, 1997].

The genetic similarity between strains subgrouped as SGII, SGI+II or SGnonI/nonII using serological methods would suggest either some serological cross-reactivity between SGI and SGII or the loss of a SG-determining epitope. The cross-reactivity would appear to be unidirectional, as strains of SGII may cross-react with the SGI-specific Mab, but strains of SGI do not appear to cross-react with anti-SGII. It is also possible that aa positions or regions outside the one analysed may contribute to the epitope recognised by the SGI-specific Mab. Also, the unsatisfactory definition of a subgroup by a lack of reactivity with either SG-specific Mab does not rule out the possibility that SG nonI/nonII are strains misclassified as a result of insufficient antigen being present in the ELISA.

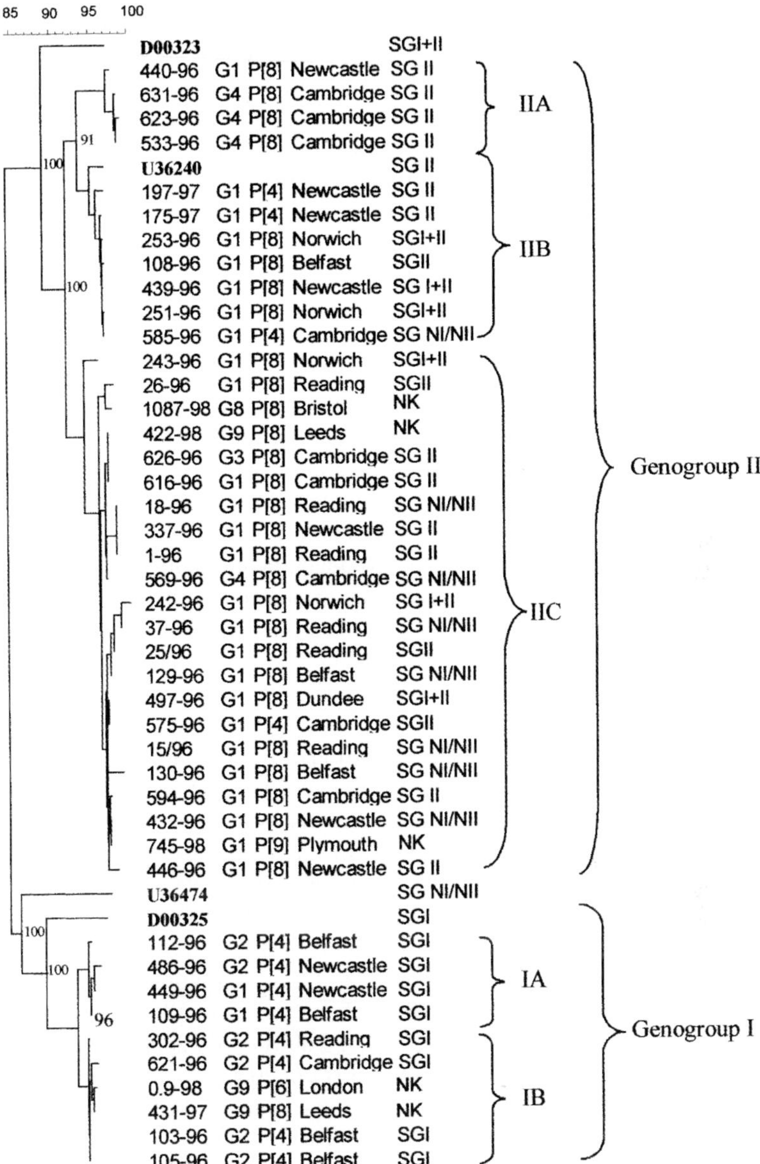

Fig. 1: Phylogenetic tree constructed using Clustal and Neighbour Joining methods from a 379bp fragment (nt 747-1126, aa 241-367) of the VP6 gene of rotavirus strains for which subgroup specificities had been determined using serological methods. Significant bootstrap values (>90%) are shown at the branching points of each cluster. Accession number of sequences of prototype strains of SGI, SGII, SGI+II and SGnonI/nonII obtained from GenBank are indicated in bold. Colums denote (from left). strain designation, year of isolation, G and P type, geographical location of isolate, subgroup determined serologically, genetic lineage and genogroup. The calibration bar indicates percent homology. (From Iturriza Gómara et al., 2002; with permission from the publisher).

320

Only one VP6 cDNA sequence of a SGI+II and two of SGnonI/nonII strains are available in the nucleic acid databases, none of which are derived from human rotavirus strains [Gorziglia *et al.*, 1988; Tang *et al.*, 1997]. These sequences were distinguishable by phylogenetic analysis of the VP6 region between nt 747 and 1126, and none of the VP6 cDNA fragments obtained from human rotavirus strains serologically determined as SGI+II or SGnonI/nonII clustered with them (Fig. 1). No confirmed SGI+II or SG nonI/nonII could be found in a subset of human rotavirus strains on the basis of molecular sub-grouping methods. The incidence of infection in the human population with SGI+II or SG nonI/nonII may be rare.

G and P types

The incidence and distribution of group A rotavirus G and P sero/genotypes varies between geographical areas during a rotavirus season, and from one season to the next. Globally, G1 to G4 and P[4] and P[8] are the most common G and P types of rotaviruses causing disease in humans, and numerous studies have reported that G1P[8], G2P[4], G3P[8] and G4P[8] are the most common G and P combinations [reviewed in Desselberger *et al.*, 2001]. However, the introduction and more extensive application of molecular typing methods has revealed the existence of other G and P types causing infection in humans, often as the predominant types and particularly in developing countries [reviewed in Desselberger *et al.*, 2001]. The initial studies that used serological methods to determine rotavirus G types found significant numbers of non-typable rotavirus strains. These findings may in part have been due to a lack of sensitivity of the methods used, but it is likely that a proportion of those samples were rotaviruses of G types other than G1-G4, particularly in those geographical regions where molecular typing has revealed a greater diversity of rotavirus genotypes. In Bangladesh, India, Malawi and Guinea-Bissau G1P[8] strains, which represent the most commonly found G/P combination in Europe and North America, constitute only a very small proportion of the rotavirus G/P types found. In these countries, G4P[8], G9P[6], G9P[11], G8P[6] and G2P[6] were the most frequently found G/P combinations. G9 strains have been found frequently in India dating back to 1980's [Ramachandran *et al.*, 1996; Das *et al.*, 1993b; 1994]. More recently, G9 strains have spread to all continents since 1995 [Ramachandran *et al.*, 1998; Bon *et al.*, 2000; Cubitt *et al.*, 2000; Iturriza-Gómara *et al.*, 2000a; Palombo *et al.*, 2000; Widdowson *et al.*, 2000].

Human group A rotavirus strains that possess genes commonly found in animal rotaviruses have been associated with infections in children, both in developed and developing countries. Strains such as G3 (found commonly in humans and also in many animal species such as cats, dogs, monkeys, pigs, mice, rabbits and horses), G5 (found in pigs and horses), G6 and G8 (both found in cattle), G9 (found in pigs and lambs) and G10 (found in cattle) have been detected in the human population throughout the world. Unusual P types have been found in humans and include P[6] (common porcine type), P[9] (common feline type), P[11] (common bovine type), P[14] (found in porcine and lapine rotavirus strains) and P[19] (found in porcine rotavirus strains) [reviewed in Desselberger *et al.*, 2001].

Mechanisms of rotavirus evolution and strain diversity

The epidemiology of rotavirus infections in the human population and their diversity reflect the mechanisms and forces that drive the evolution of these viruses.

Like many RNA viruses, rotaviruses display great diversity. As well as showing different G and P types and a variety of combinations of those, there is also intratypic variation. Three mechanisms, singly or in combination, are important for the evolution and diversity of rotaviruses: Gene rearrangements, the accumulation of genomic point mutations (drift), and reassortment of genome segments (shift).

Gene rearrangements

The concatemerisation or truncation of genome segments and their open reading frames (ORFs) have the potential to contribute to the evolution of rotaviruses through the abolition of normal proteins or the production of new proteins with altered functions [Desselberger, 1996]. Extraction and separation of the genome of rotaviruses by polyacrylamide gel electrophoresis (PAGE) has revealed different patterns of migration. The resulting patterns are referred to as electropherotypes [Estes, 2001], and rotaviruses of different groups exhibit distinct electropherotypes. Within group A rotaviruses the genome segments usually migrate in four size groups: I (segments 1-4), II (segments 5-6), III (segments 7-9) and IV (segments 10-11), in descending size order [Estes, 2001]. Two typical group A rotavirus electropherotypes, "short" and "long", which differ in the relative migration of RNA segment 11 are found commonly. However, atypical variants of the more common group A electropherotypes that arise through molecular rearrangements of individual gene segments, have also been observed [Desselberger, 1996]. The most common rearrangement consists of a partial duplication which maintains the normal ORF with an extended 3'-untranslated region. Other possible rearrangements may result in an ORF extended by partial duplication, or a truncated ORF resulting from the introduction of point mutations and premature stop codons [Desselberger, 1996; Taniguchi *et al.*, 1996; Patton *et al.* 2001]. Although most rearrangements occur in genes encoding non-structural proteins, gene segments 5, 8, 10 and most frequently 11 [Estes, 2001; Patton *et al.* 2001], rearrangements of gene 6, encoding the middle layer protein VP6, have also been described [Shen *et al.*, 1994; Xu *et al.*, 1996]. The contribution of gene rearrangements to the diversity and molecular epidemiology of rotaviruses in humans is believed to be relatively small.

Accumulation of point mutations

The rate of mutation within rotavirus genes is relatively high because RNA replication is error-prone. The mutation rate has been calculated to be appr. 5×10^{-5} per nucleotide per replication which implies that on average a rotavirus progeny genome differs from its parental genome by at least one mutation [Blackhall *et al.*, 1996]. Point mutations can accumulate and give rise to intratypic variation characterised by the emergence of genetic lineages within individual genes; VP6 encoding gene [Iturriza-Gomara *et al.*, 2002), NSP4 encoding gene [Cunliffe *et al.*, 1997; Horie *et al.*, 1997; Kirkwood *et al.*, 1999; Ciarlet *et al.*, 2000], and genes encoding VP7 and VP4 of particular G and P types [Palombo *et al.*, 1993; Xin *et al.*, 1993; Wen *et al.*, 1995; Jin *et al.*, 1996; Wen *et al.*, 1997; Maunula and von Bonsdorff, 1998; Diwarkarla and Palombo, 1999; Gouvea and Santos, 1999; Iturriza-Gómara *et al.*, 2000c; Iturriza-Gómara *et al.*, 2001a].

Point mutations associated with failure to genotype

Nucleotide sequence data of the VP4 and VP7 genes from rotavirus strains isolated in different countries have revealed considerable sequence variability between these and the prototype strains for which oligonucleotide primers for genotyping RT-PCRs were designed [Palombo *et al.*, 1993; Xin *et al.*, 1993; Jin *et al.*, 1996; Maunula and von Bonsdorff, 1998]. A study in Nigeria reported mutations at the G8-specific primer binding site of the VP7 sequence of G8 strains, which prevented the annealing of the G8-specific primer used at the time and led to the mistyping by RT-PCR of G8 strains as G3 [Adah *et al.*, 1997].

In 25 of 648 (4%) rotavirus isolates collected from different geographical locations in the UK during the 1995/96 season for which the G-type was succesfully determined by RT-PCR, a P-type could not be assigned using oligonucleotide primers and methods previously described [Iturriza-Gómara *et al.*, 2000c], despite the fact that amplicons were obtained using the VP4 consensus primers, Con2 and Con3, in the 1st round PCR. Nucleic acid sequences derived from these strains clustered with P[8] sequences ($\geq$95% homology at the nucleotide [nt] level). The nt sequence alignment showed a series of point mutations at the binding site of the P[8]-specific primer 1T-1 (Fig. 2). Mismatches between the P[8]-specific primer and its complementary region of the VP4 cDNA of these P[8] isolates explained the failure of the typing RT-PCR. The oligonucleotide primer used for P[8] typing (1T-1) had been designed using the VP4 sequence of the strain KU (G1P[8]) as a template. The primer 1T-1 showed 100% complementarity only with the VP4 gene of this strain (KU). Sequences derived from

Fig. 2. Alignment of fragments of the VP4 gene (corresponding to the VP8* subunit), and the reverse complementary sequences of the original primer 1T-1 and the degenerate primer 1T-1D. I. Sequences from strains which were not typable by RT-PCR using the primer 1T-1. II. Sequences from strains typed as P[8] using 1T-1 primer. III. Sequences from strains typed as P[8] with the degenerate primer 1T-1D. IV. Sequences of rotavirus P[8] strains obtained from GenBank/EMBL. Strains from the same geographical area (Cambridge) are marked with "*", the rest are from diverse areas. "-96" indicates strains collected during 95/96, "-97" during 96/97 and "-98" during 97/98 seasons. Residues that match primer 1T-1 are denoted by dots. Substitutions at the P[8]-primer binding site that are conserved in all the sequences are in bold. The reverse complement sequences of primers 1T-1 and *1T-1D* are shown with the changes of primer 1T-1D in bold [from Iturriza-Gómara *et al.* 2000c; with permission from the publisher].

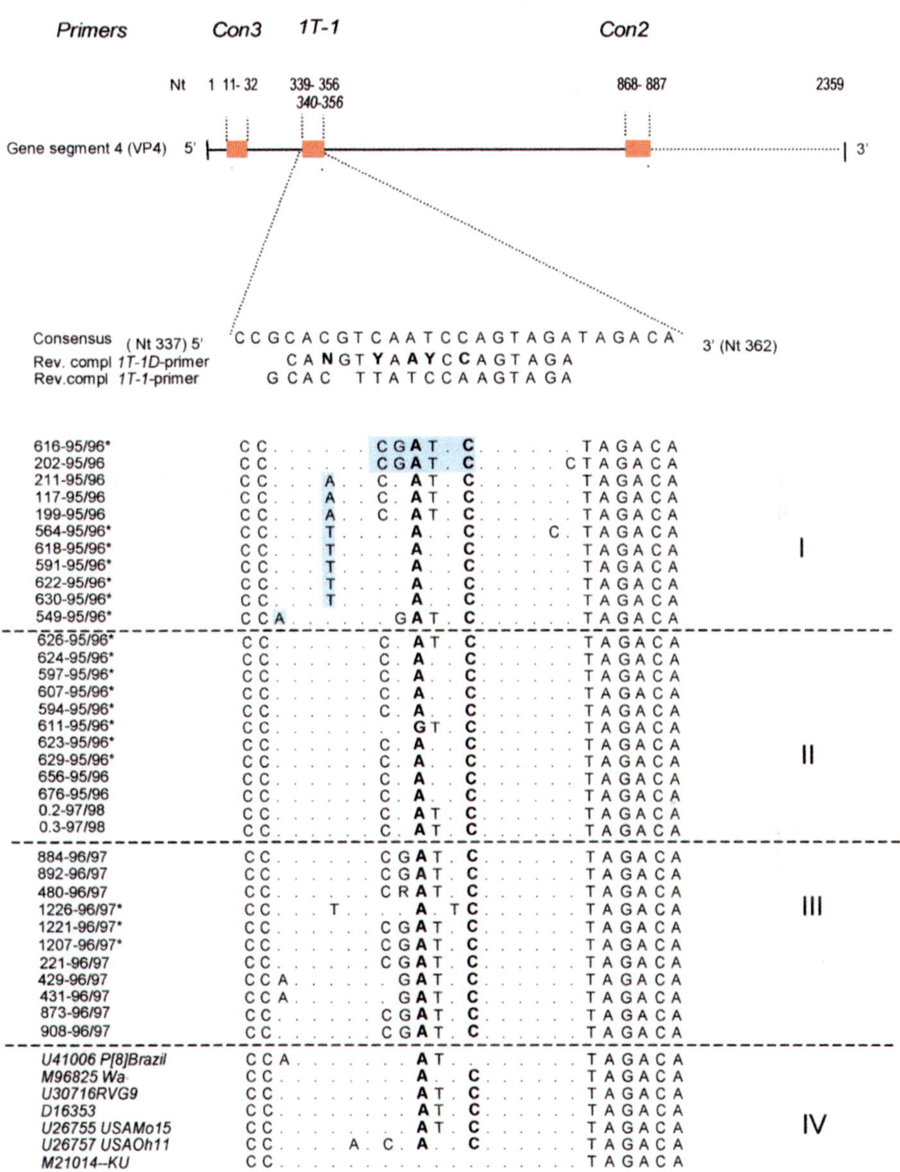

Abbreviations:
N: A or C or G or T
Y: C or T
R: A or G

all other strains, including those obtained from GenBank/EMBL, showed two consistent mismatches at nt positions 9 and 12 from the primer's 3' end, and the majority of strains also showed mismatches with primer 1T-1 at nucleotide positions 7 and 10 from the 3' end. Those strains that could not be typed by RT-PCR using primer 1T-1 showed a substitution at nt position 1 (C:A) or 4 (G:A or G:T) from the 3'end and/or the accumulation of 5 to 6 mismatches in total between primer and template cDNA (Fig. 2) [Iturriza-Gómara *et al.*, 2000c].

During the rotavirus season 1996/97 a significantly higher percentage of strains that gave dual P types in contrast with the other 3 seasons was observed in the UK (Table 1). These strains were predominantly G1 in combination with P[4] and P[8], and the majority (80.4%) were from a single geographical location. Partial sequencing of the VP4 genes (nt 11-887) of a subset of these strains revealed them to be P[8] strains. The sequence data showed an accumulation of point mutations at the P[8]-specific primer binding site, and also at the region between nt 474-494 which in P[4] strains is complementary to the P[4]-specific primer (2T-1) (Fig. 3) (Iturriza-Gómara *et al.*, unpublished data). The VP4 sequences of the 3 strains analysed showed a conserved point mutation at nt position 474: instead of the T which is typical of P[8] strains, they had G, typical of P[4] strains (Fig. 3). Two of the strains also showed a point mutation at nt position 482 (T as in P[4] strains, instead of the A of P[8]) and the third strain also possesed a point mutation at nt position 483 (C instead of T) (Fig. 3). Because these strains had been typed using the original P[8] specific primer (1T-1), a subset (15 strains in total) were tested again using the degenerate P[8] specific primer *1T-1D*, which compensates for the point mutations found at the P[8]-specific primer binding site. This allowed 13 out of 15 to be typed as P[8] and the remaining 2 as P[4].

The use of a degenerate primer in the multiplex typing-PCR allowed all of the previously untyped P[8] strains to be successfully typed and did not affect the specificity of the PCR or the ability to type strains previously characterised with the original P[8]-specific primer (1T-1). In addition, the incorporation of primer *1T-1D* allowed unequivocal typing of P[4] and P[8] strains that produced two amplicons using the original set of primers.

The nt sequence alignments of P[4], P[6] and P[8] strains showed that the variability was greater at the P type-specific binding site of P[8] strains than at the regions specific for the primers of the P[4] and P[6] strains analysed. All the strains that could not be typed and the majority of those characterised as mixed types using the original set of primers were P[8] strains.

Table 1

G and P genotype combinations of the rotavirus strains typed by RT-PCR. Samples were collected during the seasons 1995/96, 1996/97, 1997/98 and 1998/99 from up to 16 different catchment areas throughout the UK (adapted from Iturriza-Gómara *et al.* 2000b, 2001a)

G/P Combination	1995/96		1996/97		1997/98		1998/99		Total	
	No.	(%)	No.	(%)	No.	(%)	No.	(%)	No.	(%)
G1P[8]	407	55.4	919	82.2	830	72.9	632	91.7	**2789**	**75.8**
G2P[4]	128	17.4	20	1.8	180	15.8	27	3.9	**355**	**9.6**
G3P[8]	45	6.1	45	4.0	5	0.4	2	0.3	**97**	**2.6**
G4P[8]	83	11.3	45	4.0	53	4.7	13	1.9	**194**	**5.3**
Sub-Total	*663*	*90.2*	*1029*	*92.0*	*1068*	*93.8*	*674*	*97.8*	*3434*	*93.3*
G1P[4]	11	1.5	13	1.2	9	0.8	1	0.1	**34**	**0.9**
G1P[6]	2	0.3	0	0.0	0	0.0	0	0.0	**2**	**0.1**
G1P[9]	0	0.0	1	0.1	2	0.2	2	0.3	**5**	**0.1**
G2P[8]	6	0.8	0	0.0	3	0.3	0	0.0	**9**	**0.2**
G3P[6]	11	1.5	0	0.0	0	0.0	0	0.0	**11**	**0.3**
G3P[9]	1	0.1	0	0.0	0	0.0	0	0.0	**1**	**0.0**
G4P[4]	1	0.1	2	0.2	2	0.2	1	0.1	**6**	**0.2**
G4P[6]	1	0.1	0	0.0	0	0.0	0	0.0	**1**	**0.0**
G8P[8]	0	0.0	0	0.0	2	0.2	0	0.0	**2**	**0.1**
G9P[6]	18	2.4	0	0.0	8	0.7	0	0.0	**26**	**0.7**
G9P[8]	1	0.1	18	1.6	42	3.7	9	1.3	**69**	**1.9**
Sub-Total	*52*	*7.1*	*34*	*3.0*	*68*	*6.0*	*13*	*1.9*	*167*	*4.5*
G1+G2/P[4]	1	0.1	1	0.1	1	0.1	1	0.1	**4**	**0.1**
G1+G2/P[8]	1	0.1	0	0.0	0	0.0	0	0.0	**1**	**0.0**
G1+G3/P[8]	4	0.5	0	0.0	0	0.0	0	0.0	**4**	**0.1**
G1+G4/P[8]	1	0.1	0	0.0	0	0.0	1	0.1	**2**	**0.1**
G1+G9/P[8]	5	0.7	0	0.0	1	0.1	0	0.0	**6**	**0.2**
G1+G9/P[6]	2	0.3	0	0.0	0	0.0	0	0.0	**2**	**0.1**
G2+G4/P[4]	1	0.1	0	0.0	0	0.0	0	0.0	**1**	**0.0**
G3+G4/P[8]	1	0.1	1	0.1	0	0.0	0	0.0	**2**	**0.1**
G1/P[4]+P[8]	1	0.1	44	3.9	0	0.0	0	0.0	**45***	**1.2**
G2/P[4]+P[8]	0	0.0	1	0.1	0	0.0	0	0.0	**1**	**0.0**
G3/P[4]+P[8]	0	0.0	5	0.4	0	0.0	0	0.0	**5**	**0.1**
G4/P[4]+P[8]	1	0.1	3	0.3	0	0.0	0	0.0	**4**	**0.1**
G1+G2/P[4]+P[8]	2	0.3	0	0.0	0	0.0	0	0.0	**2**	**0.1**
Sub-Total	*20*	*2.7*	*55**	*4.9*	*2*	*0.2*	*2*	*0.3*	*79*	*2.1*
TOTAL	735		1118		1138		689		**3680**	

* The majority of these strains are predicted to be P[8] strains mistyped using the P[8]-specific primer 1T-1 due to the accumulation of point mutations at the primer binding sites (see page 322-324)

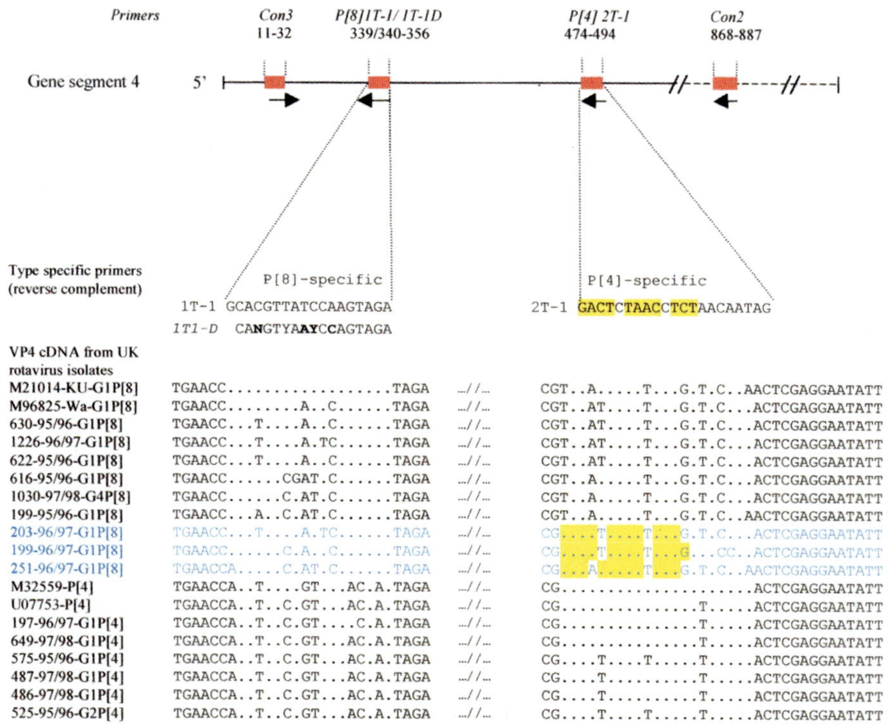

Fig. 3. Alignment of fragments of the VP4 encoding gene of P[8] and P[4] strains, and the reverse complementary sequences of the original P[8]-specific primer 1T-1, the degenerate primer *1T-1D* and the P[4] specific primer 2T-1. Nucleotides that differ from the reverse complemetary sequences of 1T-1 and 2T-1 are shown. Regions of complementarity between the P[8] strains originally typed as P[4]+P[8] (represented in blue) and the P[4]-specific primer *2T-2* are highlighted in yellow. A diagrammatic representation of the VP4 gene shows the sites complementary to primers Con2, Con3, 1T-1 / *1T-1D* and 2T-1.

No problems have so far been encountered in typing other P types by RT-PCR, suggesting that the type-specific primer binding sites of other P types are more conserved. The P[8] type is much more frequently found than any other P type, and only a relatively small percentage of P[8] strains, localised both geographically and temporally, failed to type or were mistyped as mixed P types. Therefore, greater numbers of other P types may need to be examined to assess whether or not other P types are as variable as P[8]. The greater diversity within P[8] strains may also explain the much higher incidence and continuous circulation of this genotype throughout the world, season after season, as speculated for G1 strains [Maunula and von Bonsdorff, 1998].

Point mutations associated with failure to serotype

Point mutations can also lead to antigenic changes, which may result in the emergence of escape mutants and render serological methods for subgrouping and serotyping of rotaviruses unreliable [Palombo *et al.*, 1997; Iturriza-Gómara *et al.*, 2001b].

Serotyping using G type-specific Mabs has been applied widely in rotavirus epidemiological studies [reviewed in Desselberger *et al.*, 2001]. However, the use of serotyping methods has increasingly been found to be limited by the existence of monotypes or antibody escape mutants within the different G types, which react with different degrees of affinity against different panels of G-specific Mabs [Coulson and Kirkwood, 1991; Raj *et al.*, 1992; Coulson, 1996; Coulson *et al.*, 1996]. The loss of reactivity with G2-specific Mabs of one G2 rotavirus strain isolated in Australia (strain 1076) has been correlated with aa substitutions in antigenic regions B and C of VP7, at aa positions 147 and 213 and 217, respectively [Coulson *et al.*, 1996].

Rotavirus strains collected in the UK during the 1995/96 season, and genotyped by RT-PCR as G2, failed to serotype in ELISAs using a panel of three G2-specific Mabs [Iturriza-Gómara *et al.*, 1999; Iturriza-Gómara *et al.*, 2001b]. Complementary DNAs of the VP7 genes of a subset of these strains were partially sequenced and compared to corresponding sequences of a subset of successfully serotyped G2 strains collected during 1990/91.

Alignment of the deduced aa sequences of the VP7 genes revealed aa substitutions at positions 87 (Ala→Thr) and 96 (Asp→Asn), both located within antigenic region A (aa 87-101) as the only consistent differences between the strains that were successfully serotyped and those that failed to serotype with all three G2-specific Mabs (Table 2) [Iturriza-Gómara *et al.*, 2001b]. The sequences of antigenic region A of the strains successfully serotyped were identical to those of the prototype G2 strains RV5, S2 and DS-1 (Table 2). The prototype serotype G2 strain HN126 [Green *et al.*, 1987] which was typable by G2-specific Mabs possessed a single substitution Ala→Thr at aa position 87 (Table 2), strongly suggesting that this aa substitution is not responsible for the antigenic change leading to a failure to react with G2-specific Mabs. The aa substitution at position 96 (Asp→Asn) was likely to be responsible for the failure to react with the G2-specific Mabs. This substitution induces a change in electric charge and may result in a conformational change of the epitopes recognised by neutralising antibodies. Although the epitopes recognised by rotavirus neutralising antibodies have not been completely defined, it has been proposed that antigenic regions A and C which are distant in the linear molecule interact closely together in the folded form of the VP7 molecule [Dyall-Smith *et al.*, 1986].

Table 2

Deduced aa sequences of the VP7 antigenic site A of rotavirus G2 strains. Prototype strains and sequences obtained from GenBank/EMBL are in italics, and strains found in the UK that were serotyped are in bold (from Iturriza Gomara *et al.*, 2001b; with permission from the publisher]

.Strain	Antigenic site A (aa 87-101)														
DS-1	A	E	A	K	N	E	I	S	D	D	E	W	E	N	T
S2	.	.	.	.	.	.	.	.	.	.	.	.	.	.	.
RV5	.	.	.	.	.	.	.	.	.	.	.	.	.	.	.
HN 126	T	.	.	.	.	.	.	.	.	.	.	.	.	.	.
Hu/Aus/5/77	.	.	.	.	.	.	.	.	.	.	.	.	.	.	.
92A-Aus	.	.	.	.	.	.	.	.	.	.	.	.	.	.	.
1076-Aus	.	.	.	.	.	.	.	.	.	.	.	.	.	.	.
DC3/91	.	.	.	.	.	.	.	.	.	.	.	.	.	.	.
DC2/91	.	.	.	.	.	.	.	.	.	.	.	.	.	.	.
DC4/91	.	.	.	.	.	.	.	.	.	.	.	.	.	.	.
DC12/91	.	.	.	.	.	.	.	.	.	.	.	.	.	.	.
DC14/91	.	.	.	.	.	.	.	.	.	.	.	.	.	.	.
DC15/91	.	.	.	.	.	.	.	.	.	.	.	.	.	.	.
DC17/91	.	.	.	.	.	.	.	.	.	.	.	.	.	.	.
DC24/91	.	.	.	.	.	.	.	.	.	.	.	.	.	.	.
DC22/91	.	.	.	.	.	.	.	.	.	.	.	.	.	.	.
543/96	T	.	.	.	.	.	.	.	.	N	.	.	.	.	.
544/96	T	.	.	.	.	.	.	.	.	N	.	.	.	.	.
563/96	T	.	.	.	.	.	.	.	.	N	.	.	.	.	.
583/96	T	.	.	.	.	.	.	.	.	N	.	.	.	.	.
538/96	T	.	.	.	.	.	.	.	.	N	.	.	.	.	.
621/96	T	.	.	.	.	.	.	.	.	N	.	.	.	.	.
625/96	T	.	.	.	.	.	.	.	.	N	.	.	.	.	.
366/96	T	.	.	.	.	.	.	.	.	N	.	.	.	.	.
617/96	T	.	.	.	.	.	.	.	.	N	.	.	.	.	.
351/96	T	.	.	.	.	.	.	.	.	N	.	.	.	.	.
399/99	T	.	.	.	.	.	.	.	.	N	.	.	.	.	.
721/99	T	.	.	.	.	.	.	.	.	N	.	.	.	.	.

Thus, these VP7 mutants can be regarded as antibody escape mutants that appear to be widely dispersed geographically and maintained over time. G2 strains isolated in the UK in 1998/99 possessed the same aa substitutions on the antigenic region A as the strains isolated in 1995/96, and similar G2 mutants were also isolated in an epidemic caused by G2 rotavirus strains in Taiwan in 1993 and were found in subsequent years [Zao *et al.*, 1999]. The epidemic re-emergence of G2 strains in Taiwan may have been due to immune-evasion as a consequence of altered antigenicity conferred by the aa substitution at position 96 of antigenic region A. The lack of cross-protection conferred from previous infections with G2 strains may also explain the higher incidence of infection with rotavirus G2 strains found in the older population in the UK between 1995 and 1999 [Iturriza-Gómara *et al.*, 2000b].

Rotavirus reassortment

Shuffling of gene segments through reassortment can occur after dual infection of one cell. Studies of mixed infections and reassortment *in vitro* are useful for the identification of gene-protein assignments, of gene functions and gene product interactions [Ramig, 2000]. Reassortment does not seem to be a random phenomenon. *In vitro* there is a non-random association of certain genes which is influenced by the host cell [Graham *et al.*, 1987]. *In vivo*, the host affects the combinations of gene segments present in reassortants isolated following mixed infection [Ramig, 1997], and experiments with animal models have shown that the immune status of the host also affects the progeny of reassortment [Gombold and Ramig, 1989]. While *in vitro* and *in vivo* reassortment among group A rotaviruses occurs at high frequency, reassortment between rotaviruses of different groups has never been reported [Ramig, 1997].

There is also evidence that reassortment, through alterations in protein-protein interactions, can possibly lead to changes in conformational epitopes, and can contribute to the evolution of antigenic types [Chen *et al.*, 1992; Lazdins *et al.*, 1995]. Interspecies transmission and the subsequent reassortment of animal and human rotavirus genes has the potential to increase the diversity of cocirculating rotaviruses enormously. Also, the ability to combine the replicative advantages of human rotavirus genes in the human host with genes of animal rotaviruses encoding proteins to which the human population is immunologically naïve may allow their rapid spread into the human population [Iturriza-Gómara *et al.*, 2000b].

The role of reassortment in rotavirus epidemiology

In a molecular epidemiological study of rotavirus infection carried out in the UK, the G and P types were determined by RT-PCR in 3680/ 4021 (92%) rotavirus-positive stool specimens collected over a period of 4 rotavirus seasons (1995/96 – 1998/99) [Iturriza-Gómara *et al.*, 2000b; Iturriza-Gómara *et al.*, 2001a]. Although G1P[8], G2P[4], G3P[8] and G4P[8] together represented 93% of all typed strains, other less commonly

observed combinations were also found in the UK over this period (Table 1):
1. Unusual combinations of common G and P types: G1P[4], G2P[8] and G4P[4];
2. An uncommon human G/P type combination, G9P[6], which appeared in the UK first in 1995/96; and
3. Combinations of common G types with uncommon P types, G1P[6], G1P[9], G3P[6], G3P[9] and G4P[6], and, *vice versa*, of uncommon G types with common P types, G8P[8] and G9P[8].

Reassortment between strains of common G and/or P types

The genetic analysis of rotavirus strains with common and uncommon G/P combinations found in this population provided strong evidence of reassortment between cocirculating rotavirus strains. The VP7 and VP4 genes of the G1P[8] strains were further classified by nucleic acid sequencing in order to identify genetic lineages. Three lineages were found of each, the VP7 and VP4 encoding genes, homologous to lineages previously reported and isolated throughout the world.

The comparison of sequence data obtained from G1P[8] strains and of the possible reassortant strains of G1P[4] showed that the VP7 sequences of G1P[4] strains clustered with VP7 sequences of two different lineages obtained from G1P[8] strains. VP7 nt sequences of G1P[4] strains were more closely related to the VP7 sequences of G1P[8] strains within their corresponding lineages (>96% homology) than to those of G1P[4] strains of a different lineage (≤95% homology). Similarly, VP4 sequences of G2P[4] and of G1P[4] clustered together and were highly homologous (≥96%). The VP4 sequences of G2P[4] and G1P[4] strains isolated in the same location and during the same season shared ≥98% homology at nt level. This strongly suggests that G2P[8], G1P[4] and G4P[4] strains originated through reassortment between the cocirculating G1P[8], G2P[4] and G4P[8] rotavirus strains [Iturriza-Gómara *et al.*, 2001a]. Also, putative reassortant strains were found more commonly in seasons in which G1P[8], G2P[4], G3P[8] and G4P[8] cocirculated at relatively high frequencies, and were found rarely in seasons in which G1P[8] strains were overwhelmingly predominant (Table 1).

Further evidence for reassortment was obtained by comparing partial VP4 nucleic acid sequences obtained from P[8] strains in combination with G1, G2, G3, G4, G8 or G9 types (Fig. 4). Lineage P[8]-II was found to contain rotavirus strains of G types G1, G2, G3, G4, G8 and G9, and the VP4 sequences of all these strains were highly homologous [Iturriza-Gómara *et al.*, 2001a]. The clustering within this lineage was

Fig. 4. Phylogenetic tree constructed from partial nt sequences of the VP8* fragment of VP4 genes (nt 11-887) of rotavirus strains G1P[8], G3P[8], G4P[8], and possible reassortants G2P[8], G8P[8] and G9P[8] using the *Clustal* method and *Megalign*. Laboratory number, rotavirus season, G/P combination and geographical origin of the strains are indicated. Sequences of strains Wa and KU (GenBank accession numbers M96825

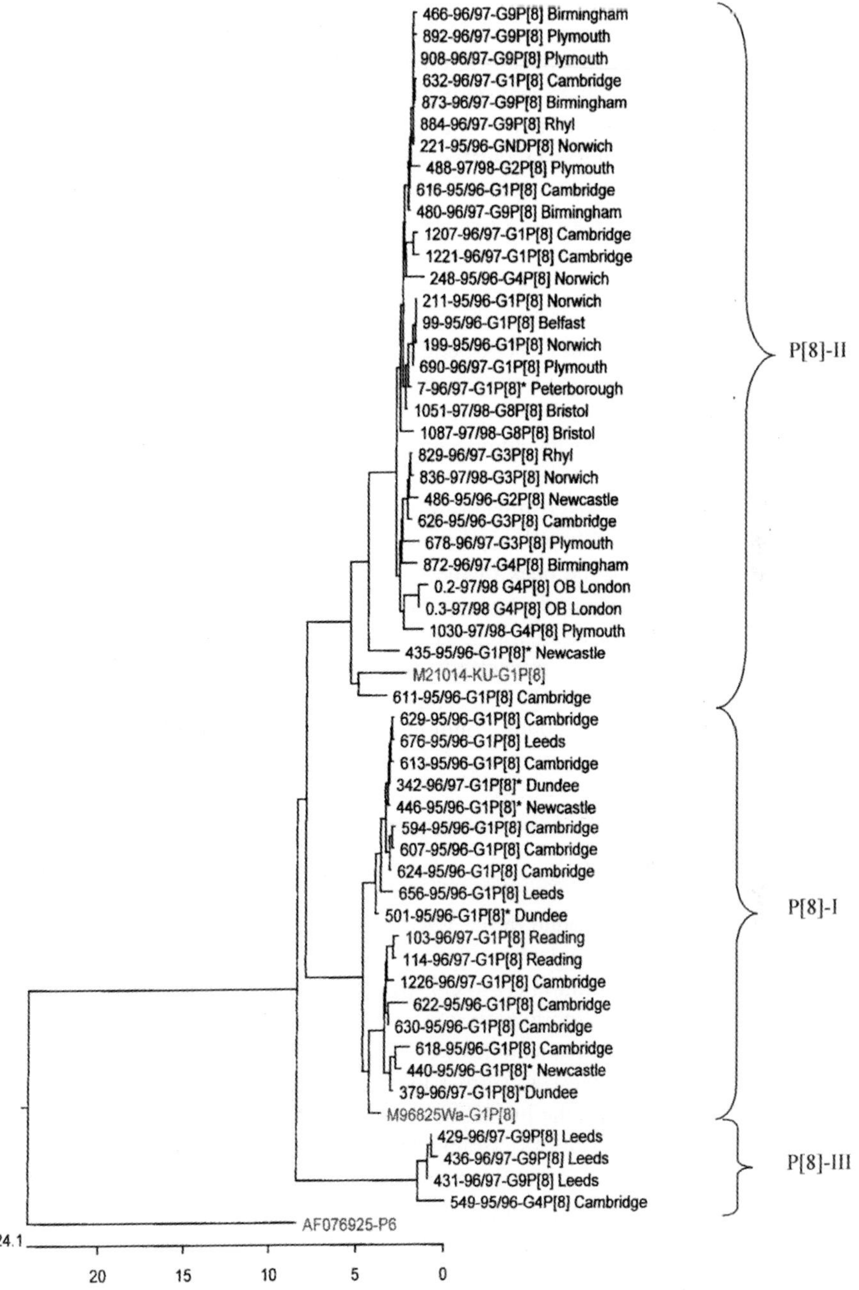

and M2114, respectively) and of a P[6] strain (accession number AF076925) were included for comparisons. Sequences obtained from GenBank are shaded. GND= G type not determined. The calibration bar indicates percent difference [from Iturriza Gomara *et al.*, 2001a; with permission from the publisher].

not associated with the G type of the strains. This may suggest that these strains have originated through reassortment during dual infection. By contrast, all the strains in lineage P[8]-I were G1, which may indicate some constraint to reassortment involving these strains. P[8] lineage-III was found at much lower frequency, and only G4 or G9 strains were identified in this cluster; the lack of G1 strains associated with this lineage may explain its lower incidence. This finding may also suggest that there are constraints against reassortment between P[8] strains of lineage-III and G1 strains [Iturriza-Gómara et al., 2001a].

Reassortment opportunities depend on the occurrence and frequency of double infections of cells/hosts. Those have been found at a frequency of 2% in the UK collection [Iturriza-Gómara et al., 2000b] and included G1+G2/P[4]+P[8], G1+G2/P[4], G1+G2/P[8], G2+G4/P[4], G1+G9/P[6] and G1+G9/P[8], which could have given rise to the reassortant strains found in the UK (G1P[4], G2P[8], G4P[4] and G9P[8]) (Table 1). It is possible that the four common G/P combinations are the genetically most fit rotavirus strains and that other combinations may be disadvantaged in evolutionary terms. However, mixed infections and putative reassortant strains have been found throughout the world, and at particularly high frequencies (up to 30%) in the developing world [e.g. at the peak of the monsoon season in Bangladesh; Ahmed et al., 1991]. There is evidence that putative reassortants can spread and become epidemiologically significant strains, e.g. G5P[8] in Brazil, G9P[11] in India; G9P[6] and G9P[8] in India and more recently in the USA and Europe [Desselberger et al., 2001; Jain et al., 2001], G1P[4] in Argentina [Argüelles et al., 2000] and G2P[8] in Bangladesh [Unicomb et al., 1999].

Emergence of G9P[6] strains in the UK and their subsequent reassortment with common rotavirus strains

Rotavirus G9 strains were found in the UK for the first time during the rotavirus season 1995/96. Initially the incidence of infection with G9 strains in the UK was low in comparison to the more common G1 to G4 strains. However, the incidence of infection with G9 strains increased significantly over time, from 1995/96 to 1997/98, in those geographical areas from which samples were collected in consecutive seasons. G9P[8] strains spread throughout the UK, from 1 location during 1995/96 to 8 different locations during 1997/98, whilst the incidence of G9P[6] strains decreased throughout this period. In some locations G9P[8] strains became the second most prevalent strain, accounting for up to 28% of the genotyped strains [Iturriza-Gómara et al., 2000b; Iturriza-Gómara et al., 2001a]. During the 1996/97 and 1997/98 seasons the G9P[8] strains showed a pattern of temporal distribution which was different to that of all other rotavirus strains, and this finding may be associated with their introduction into an immunologically naïve population [Iturriza-Gómara et al., 2000b].

G9 strains were first identified in the human population in the US in 1983 [Clark et al., 1987] but they were not reported again until more than a decade later [Ramachandran et al., 1998]. The first finding of G9 strains in the UK coincides with the emergence of

G9 strains in Bangladesh [Unicomb *et al.*, 1999] and the re-emergence and spread of G9 strains in the USA [Unicomb *et al.*, 1999; Griffin *et al.*, 2000].

Several findings provided evidence of a relatively recent introduction and spread of G9 rotaviruses into the human population:

- The relative lack of diversity among the VP7 nt sequences of G9P[6] and G9P[8], and among the VP4 sequences of G9P[6] strains in comparison to that observed in the VP7 and VP4 sequences of strains of G and P types more commonly found in the human population [Iturriza-Gómara *et al.*, 2000b; Iturriza-Gómara *et al.*, 2001a];
- The chronological clustering of the VP7 nucleic acid sequences of G9 strains into lineages (Fig. 5), and the increased diversity characterised by the coexistence of more than one lineage coinciding with the time at which the incidence and spread of G9P[8] strains reached their highest levels;
- The greater genetic variability of VP4 nt sequences of G9P[8] strains, clustering in the same global lineages identified for sequences of P[8] rotavirus strains occurring in combination with other G types (Fig. 4), and not correlating with geographical or temporal clustering [Maunula and von Bonsdorff, 1998; Gouvea *et al.*, 1999; Iturriza-Gómara *et al.*, 2001a];
- The observation that a significantly higher proportion of infections with G9P[6] and G9P[8] were associated with more severe disease in older children [Cubitt *et al.*, 2000], the association of G9P[6] rotaviruses with nursery outbreaks in Holland [Widdowson *et al.*, 2000], and the higher incidence of infections in urban populations, all supported the view that these strains had been recently introduced into an immunologically naïve population.

The G9P[8] strains probably emerged through reassortment in humans between the G9P[6] strains introduced recently and the more prevalent cocirculating G1P[8], G3P[8] or G4P[8] strains which contributed the VP4 gene [Iturriza-Gómara *et al.*, 2001a]. The incidence of infections with G9P[6] strains has remained consistently low and may suggest that they have a replicative disadvantage in humans, consistent with the growth behaviour in humans of strains of animal origin.

Dual infections were found in 1995/96, characterised by samples containing G1 and G9 in combination with P[6], and G1 and G9 in combination with P[8], which are a prerequisite for reassortment (Table 1). The replicative advantage or increased pathogenicity conferred on these reassortants, possibly by the VP4 gene of genotype P[8], could have made G9P[8] the predominant G9 strain in subsequent years. Previous studies have found that rotavirus RNA segments 3, 4, 5, 7, 9 and 10 can be determinants of pathogenicity [reviewed in Burke and Desselberger, 1996; Estes, Section II, Chapter 6 of this book], therefore a more detailed analysis of multiple genes of these isolates is required.

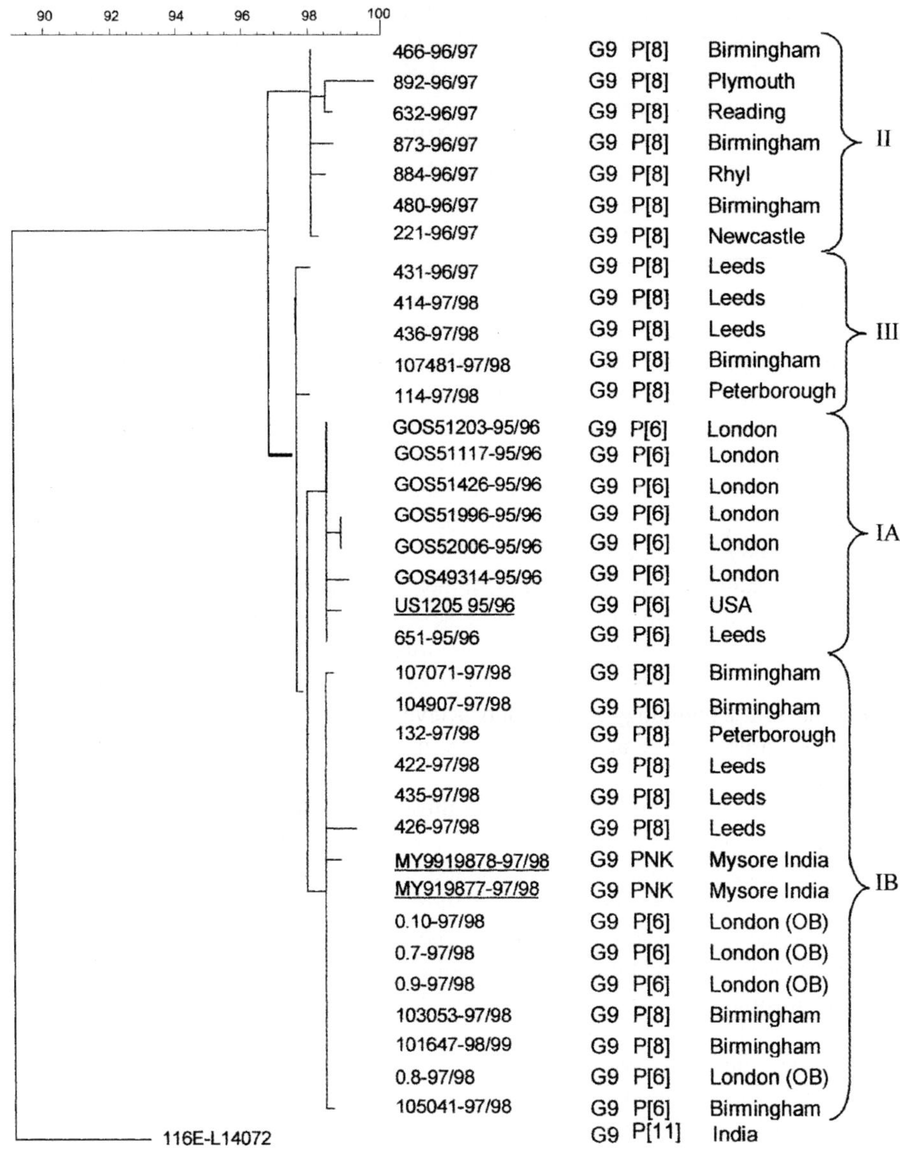

Fig. 5. Phylogenetic tree constructed from partial nt sequences of the VP7 gene (nt 166 to 782) of the G9 rotavirus strains using the *Maximum Likelihood* method. Nomenclature of strains indicates the laboratory number, season of isolation, the G and P types, and the geographical origin, OB indicates outbreak strains. The VP7 sequence of strain US1205 was provided by Dr. Kirkwood and that of strain 116E was obtained from GenBank (accession number L14072). Brackets indicate lineages I-A, I-B, II and III. The calibration bar indicates percent homology. Strains from the US and from India are underlined [from Iturriza Gomara *et al.*, 2001a; with permission from the publisher]. PNK= P type not known.

Human rotavirus strains with uncommon G or P types

Rotavirus strains with other G and/or P types characteristic of animal rotavirus strains have been isolated in different countries. G8 and G10 strains which are normaly associated with bovine strains have been found in the UK, India, Brazil and in several African countries [Das *et al.*, 1993a; Beards *et al.*, 1995; Santos *et al.*, 1998; Holmes *et al.*, 1999; Steele *et al.*, 1999; Cunliffe *et al.*, 2000; Iturriza-Gómara *et al.*, 2000b], and P[11] strains have been found in India [Gentsch *et al.*, 1993].

P[9] sequences in combination with G1 or G3 strains were also found in the UK and were closely related to those of rotaviruses of feline origin [Iturriza-Gómara *et al.*, 2001a], similar to recent findings in the USA [Griffin *et al.*, 2000].

Sequencing of the VP7 gene of a strain that failed to type (683/97) in the G-specific RT-PCR but from which a 1st round PCR amplicon was obtained revealed that it did not belong to any of the types for which primers were included in the typing PCR. However, it was more closely related to G4 strains than to any other known G type showing 84% and 89.2% homology at nt and aa levels, respectively. Homology of the VP7 cDNA between this strain and representative strains of G1, G2, G3, G8, G9 and G10 types was between 72% and 78.8% at the the nt level and between 75% and 82.4% at the aa level [Iturriza-Gómara *et al.*, unpublished data]. The best matches were an atypical strain isolated in Australia from a 2 month old baby with severe gastroenteritis [Palombo *et al.*, 1997], with 86.6% and 93.8% homology at the nt and aa levels, respectively, and a porcine strain, CB2185, isolated in Brazil [Racz *et al.*, 2000] with 85.7% and 93.2% homology at the nt and aa levels, respectively (Fig. 6). The UK strain was more closely related to the human strain isolated in Australia, and comparison of the deduced aa sequences showed 100% identity in the antigenic regions C and F, and a single aa difference in region B [Iturriza-Gómara *et al.*, unpublished data]. The three atypical strains showed greatest homology with G4 strains, and it was suggested that M3014 may have arisen from G4 strains by antigenic and genetic drift through the accumulation of point mutations [Palombo *et al.*, 1997]. However, strains of the same type share ≥91% homology at the aa level [Kapikian *et al.*, 2001], and as the homology between any of these 3 strains and G4 strains is <90% at the aa level, they may be considered as separate G types. The finding of these strains in three different continents and infecting different species would suggest that this G type evolved in the distant past. Northern hybridisation under stringent conditions using whole genome probes between the porcine IBC2185 and the human M3014 strains showed that the homology between these strains was limited to the VP7 gene [Racz *et al.*, 2000]. This suggests that interspecies rotavirus reassortment is the most likely mechanism for the introduction of this rotavirus type into the human population.

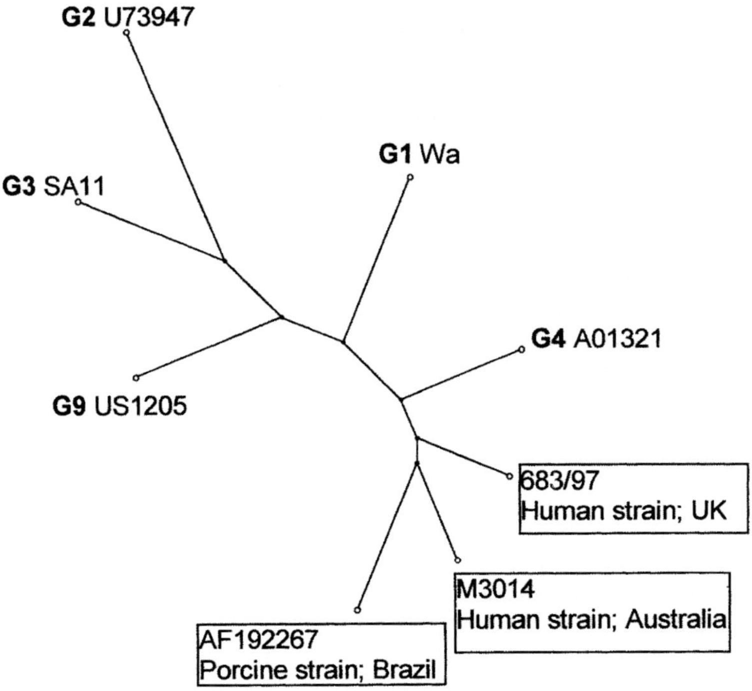

Fig. 6. Phylogenetic tree constructed from nt sequences of the VP7 genes of representative strains of different G types and a human strain of unknown type found in the UK (accession numbers K02033 [G1], U73947 [G2], K02028 [G3], A01321 [G4], AF192267 [a porcine strain found in Brazil], and X99126 [M3014; a human strain of unknown type found in Australia]) using the *Maximum Likelihood* method.

Uncommon subgroup, G and P type associations

Uncommon associations between G/P type and SG were found in 2% of strains analysed [Iturriza-Gómara *et al.*, 2001a]. Sequence analysis of VP7 and VP4 cDNA from rotaviruses of common G and P genotypes and of reassortant strains of uncommon G and/or P genotypes provided evidence for the independent segregation of the genes encoding VP7 and VP4 in reassortant rotavirus strains found in the UK [Iturriza-Gómara *et al.*, 2001a]. Rotavirus reassortant strains of genotype G1P[4] probably originated through reassortment between commonly circulating human rotavirus strains G1P[8] and G2P[4] [Iturriza-Gómara *et al.*, 2001a]. Usually G1P[8] strains are associated with a VP6 of genogroup (subgroup) II, whilst G2P[4] are associated with a VP6 of genogroup (subgroup) I [Iturriza-Gómara, M. *et al.*, 2001a; Iturriza-Gómara *et al.*, 2002]. The reassortant strains G1P[4] were found in association with VP6 of either genogroup I or II (Fig. 1), suggesting that the gene encoding VP6 segregated independently of the genes encoding VP7 and VP4 during reassortment. VP6 genes of SGI have been shown to be associated with G9P[6] strains [Ramachandran *et al.*, 1998; Jain *et al.*, 2001]. Analysis of the VP7 and VP4 genes of G9P[6] and G9P[8] strains provided evidence for the emergence of G9P[8] strains through reassortment [Iturriza-Gómara *et al.*, 2001a]. The detection of G9P[8] rotavirus strains with VP6 genes of either genogroups I or II provided further evidence of their reassortant origin, and of the independent segregation of the genes encoding VP4, VP6 and VP7. Sequence analysis of the VP4, VP6 and VP7 encoding genes provided molecular evidence for the involvement of VP6 in reassortment [Iturriza-Gómara *et al.*, 2002], strengthening previous reports that had described possible VP6 reassortants on the basis of serologically-defined subgroups [Krishnan *et al.*, 1996; Iturriza-Gómara *et al.*, 2001a]. Other strains found in the UK, G1P[9] and G8P[8], may also represent reassortants derived from human and animal strains [Iturriza-Gómara *et al.*, 2001a]. Both these reassortants were found in association with VP6 molecules of genogroup II which was possibly derived from the human parental strains carrying the gene segment 4 encoding P[8]. The presence of VP6 reassortants suggests that this mechanism of creating diversity is not confined to the VP4 and VP7 genes of rotaviruses [Iturriza-Gómara *et al.*, 2001a]. The reassortment potential of other rotavirus genes, possible gene associations and/or constrains during reassortment of human rotaviruses is still unknown, but this knowledge will be necessary in order to understand the epidemiology of rotaviruses infecting the human population.

Concluding remarks

The application of molecular methods (genotyping RT-PCRs, sequencing and phylogenetic analysis) the characterisation of rotavirus strains infecting humans is revealing an increasingly complex and diverse picture. Genetic and antigenic drift have been identified within the most frequently found rotavirus types, and are one of the main mechanisms for the diversification of rotaviruses within different G and P types. Reassortment which may result in antigenic shift plays a major role in generating the enormous diversity of rotavirus strains cocirculating in humans and animals. This is illustrated by the recent emergence of G9 strains into the human population in the UK and worldwide. The spread and increased incidence of infection with G9P[8] strains during 1997/98 have most likely been due to the acquisition of a replicative advantage through reassortment with human rotavirus strains. The introduction of an animal rotavirus into the human population by crossing the species barrier is not likely to have major impact *per se* as it is likely to replicate poorly in a human host. In a naturally occurring dual infection with human and bovine rotavirus strains, the bovine strain was recovered at significantly lower concentration from the faeces of the infected infant, which suggested a poorer replication of this strain in comparison with the coinfecting human strain [Nakagomi *et al.*, 1994]. As rotaviruses of multiple G/P combinations infect many different animal species, and as it appears that dual infections and reassortment in humans are relatively frequent events, there is an enormous reservoir of diverse rotavirus genotypes that could give origin to novel strains capable of infecting the human population. The introduction of novel types may have major implications for the successful implementation of a rotavirus vaccination strategy as it may lead to the selective increase of reassortant strains for which cross-protection may not be achieved.

By analogy to the influenza surveillance where the analysis of animal isolates is recognised as increasingly important for the assessment of human epidemiology [Schäfer *et al.*, 1993; Horimoto and Kawaoka, 2001], surveillance of rotavirus strains from animals in close contact with humans will have to be a vital complement to the surveillance of human rotavirus infections. When reviewing human infections emerging over the past 20-25 years, it became obvious that many human isolates are of proven, likely or possible zoonotic origin [Desselberger, 2002].

Sequence analysis of VP4, VP6 and VP7 cDNA from strains with unusual SG/G/P combinations provided evidence for independent segregation of the VP4, VP6 and VP7 encoding genes. However, it also appears that reassortment does not occur at random between the VP7 genes of different types and the VP4 genes of the different lineages of P[8] types. Studies of reassortment *in vitro* using these human strains should provide a better understanding of the possible gene associations and/or restrictions that may occur during reassortment.

The complex relationships among co-circulating rotavirus strains, the relative infectivities of different genotypes, and the correlates of immunological protection still require elucidation before an accurate model of the epidemiology of rotaviruses and their transmission dynamics can be established.

References

Adah, M. I., Rohwedder, A., Olaleyle, O. D., Werchau, H. 1997. Nigerian rotavirus serotype G8 could not be typed by PCR due to nucleotide mutation at the 3' end of the primer binding site. *Arch Virol* **142**: 1881-1887.

Affranchino, J. L., Gonzalez, S. A. 1997. Deletion mapping of functional domains in the rotavirus capsid protein VP6. *J Gen Virol* **78**: 1949-1955.

Ahmed, M. U., Urasawa, S., Taniguchi, K., Urasawa, T., Kobayashi, N., Wakasugi, F., Islam, A. I., Sahikh, H. A. 1991. Analysis of human rotavirus strains prevailing in Bangladesh in relation to nationwide floods brought by the 1988 monsoon. *J Clin Microbiol* **29**: 2273-2279.

Argüelles, M. H., Villegas, G. A., Castello, A., Abrami, A., Ghiringhelli, P. D., Semorile, L., Glikmann, G. 2000. VP7 and VP4 genotyping of human group A rotavirus in Buenos Aires, Argentina. *J Clin Microbiol* **38**: 252-259.

Beards, G., Graham, C. 1995. Temoporal distribution of rotavirus G-serotypes in the West Midlands region of the United Kingdom, 1983-1994. *J Diarrh Dis Res.* **13**: 235-237.

Blackhall, J., Fuentes, A., Magnusson, G. 1996. Genetic stability of a porcine rotavirus RNA segment during repeated plaque isolation. *Virology* **225**: 181-190.

Bon, F., Fromantin, C., Aho, S., Pothier, P., Kohli, E. 2000. G and P genotyping of rotavirus strains circulating in France over a three-year period: detection of G9 and P[6] strains at low frequencies. *J Clin Microbiol* **38**: 1681-1683.

Burke, B., Desselberger, U. 1996. Rotavirus pathogenicity. *Virology* **218**: 299-305.

Chen, D. Y., Estes, M. K., Ramig, R. F. 1992. Specific interactions between rotavirus outer capsid proteins VP4 and VP7 determine expression of a cross-reactive, neutralizing VP4-specific epitope. *J Virol* **66**: 432-439.

Ciarlet, M., Liprandi, F., Conner, M. E., Estes, M. K. 2000. Species specificity and interspecies relatedness of NSP4 genetic groups by comparative NSP4 sequence analyses of animal rotaviruses. *Arch Virol* **145**: 371-383.

Clapp, L. L., Patton, J. T. 1991. Rotavirus morphogenesis: domains in the major inner capsid protein essential for binding to single-shelled particles and for trimerization. *Virology* **180**: 697-708.

Clark, H. F., Hoshino, Y., Bell, L. M., Groff, J., Hess, P., Bachman, P., Offit, P. A. 1987. Rotavirus isolate WI61 representing a presumptive new human serotype. *J Clin Microbiol* **25**: 1757-1762.

Coulson, B. S. 1996. VP4 and VP7 typing using monoclonal antibodies. *Arch Virol Suppl* **12**: 113-118.

Coulson, B. S., Kirkwood, C. 1991. Relation of VP7 amino acid sequence to monoclonal antibody neutralization of rotavirus and rotavirus monotype. *J Virol* **65**: 5968-5974.

Coulson, B. S., Kirkwood, C. D., Masendycz, P. J., Bishop, R. F., Gerna, G. 1996. Amino acids involved in distinguishing between monotypes of rotavirus G serotypes 2 and 4. *J Gen Virol* **77**: 239-245.

Cubitt, W. D., Steele, A. D., Iturriza, M. 2000. Characterisation of rotaviruses from children treated at a London hospital during 1996: emergence of strains G9P2A[6] and G3P2A[6]. *J Med Virol* **61**: 150-154.

Cunliffe, N..A., Woods, P..A., Leite, J. P., Das, B. K., Ramachandran, M., Bhan, M. K., Hart, C. A., Glass, R. I., Gentsch, J. R. 1997. Sequence analysis of NSP4 gene of human rotavirus allows classification into two main genetic groups. *J Med Virol* **53**: 41-50.

Cunliffe, N..A, Gentsch, J. R., Kirkwood, C.D., Gondwe, J.S., Dove, W., Nakagomi, O., Nakagomi, T., Hoshino, Y., Bresee, J.S., Glass, R.I., Molyneux, M.E., Hart, C.A. 2000. Molecular and serologic characterization of novel serotype G8 human rotavirus strains detected in Blantyre, Malawi. *Virology* **274**: 309-320.

Das, M., Dunn, S. J., Woode, G. N., Greenberg, H. B., Rao, C. D. 1993a. Both surface proteins (VP4 and VP7) of an asymptomatic neonatal rotavirus strain (I321) have high levels of sequence identity with the homologous proteins of a serotype 10 bovine rotavirus. *Virology* **194**: 374-379.

Das, B. K., Gentsch, J. R., Cicirello, H. G., Woods, P. A., Gupta, A., Ramachandran, M., Kumar, R., Bhan, M. K., Glass, R. I. 1994. Characterization of rotavirus strains from newborns in New Delhi, India. *J Clin Microbiol* **32**: 1820-1822.

Das, B. K., Gentsch, J. R., Hoshino, Y., Ishida, S., Nakagomi, O., Bhan, M. K., Kumar, R., Glass, R. I. 1993b. Characterization of the G serotype and genogroup of New Delhi newborn rotavirus strain 116E. *Virology* **197**: 99-107.

Desselberger, U. 1996. Genome rearrangements of rotaviruses. *Adv. Virus Res.* **46**: 69-95.

Desselberger, U. 2003. Emerging and re-emerging viral pathogens. In: *Viral Infections and their Treatment*. Rübsamen-Waigmann H. *et al.* (Eds.). Marcel Dekker Inc, New York, in press.

Desselberger, U., Iturriza-Gómara, M., Gray, J. 2001. Rotavirus epidemiology and surveillance. *Novartis Found Symp* **238**: 125-147.

Diwarkarla, C., Palombo, E. 1999. Genetic and antigenic variation of capsid protein VP7 of serotype G1 human rotavirus isolates. *J Gen Virol* **80**: 341-344.

Dyall-Smith, M. L., Lazdins, I., Tregear, G. W., Holmes, I. H. 1986. Location of the major antigenic sites involved in rotavirus serotype-specific neutralization. *Proc Natl Acad Sci USA* **83**: 3465-3468.

Estes, M. 2001. Rotaviruses and their replication. *In: Fields Virology*, 4th edition, D. Knipe *et al.* (Eds), pp. 1747-1785. Lippincott Williams and Wilkins, Philadelphia.

Gentsch, J.R., Das, B.K., Jiang, B., Bhan, M.K., Glass, R.I. 1993. Similarity of the VP4 protein of human rotavirus strain 116E to that of the bovine B223 strain. *Virology* **194**:424-430.

Gombold, J. L., Ramig, R. F. 1989. Passive immunity modulates genetic reassortment between rotaviruses in mixedly infected mice. *J Virol* **63**: 4525-4532.

Gorziglia, M., Hoshino, Y., Nishikawa, K., Maloy, W. L., Jones, R. W., Kapikian, A. Z., Chanock, R. M. 1988. Comparative sequence analysis of the genomic segment 6 of four rotaviruses each with a different subgroup specificity. *J Gen Virol* **69**: 1659-1669.

Gouvea, V., Lima, R. C., Linhares, R. E., Clark, H. F., Nosawa, C. M., Santos, N. 1999. Identification of two lineages (WA-like and F45-like) within the major rotavirus genotype P[8]. *Virus Res* **59**: 141-147.

Gouvea, V., Santos, N. 1999. Rotavirus serotype G5: an emerging cause of epidemic childhood diarrhea. *Vaccine* **17**: 1291-1292.

Graham, A., Kudesia, G., Allen, A., Desselberger, U. 1987. Reassortment of human rotavirus possesing genome rearrangements with bovine rotavirus: evidence for host cell selection. *J Gen Virol* **68**: 115-122.

Green, K., Midthun, K., Gorziglia, M., Hoshino, Y., Kapikian, A., Chanock, R., Flores, J. 1987. Comparison of the amino acid sequences of the major neutralisation protein of four human rotavirus serotypes. *Virology* **161**: 153-159.

Greenberg, H. B., Valdesuso, J., van Wyke, K., Midthun, K., Walsh, M., McAuliffe, V., Wyatt, R. G., Kalica, A. R., Flores, J., Hoshino, Y. 1983. Production and preliminary characterization of monoclonal antibodies directed at two surface proteins of rhesus rotavirus. *J Virol* **47**: 267-275.

Griffin, D. D., Kirkwood, C. D., Parashar, U. D., Woods, P. A., Bresee, J. S., Glass, R. I., Gentsch, J. R. 2000. Surveillance of rotavirus strains in the United States: identification of unusual strains. *J Clin Microbiol* **38**: 2784-2787.

Holmes, J., L., Kirkwood, C.D., Gerna, G., Clemens, J. D., Rao, M.R., Naficy, A.B.,

Abu-Elyazeed, R., Savarino, S.J., Glass, R.I., Gentsch, J. R. 1999. Characterization of unusual G8 rotavirus strains isolated from Egyptian children. *Arch Virol* **144**:1381-1396.

Horie, Y., Masamune, O., Nakagomi, O. 1997. Three major alleles of rotavirus NSP4 proteins identified by sequence analysis. *J Gen Virol* **78**: 2341-2346.

Horimoto, T., Kawaoka, Y. 2001. Pandemic threat posed by avian influenza A viruses. *Clin Microbiol Rev* **14**: 129-149.

Iturriza-Gómara, M., Cubitt, D., Desselberger, U., Gray, J. 2001b. Amino acid substitution within the VP7 protein of G2 rotavirus strains associated with failure to serotype. *J Clin Microbiol* **39**: 3796-3798.

Iturriza-Gómara, M., Cubitt, D., Steele, D., Green, J., Brown, D., Kang, G., Desselberger, U., Gray, J. 2000a. Characterisation of rotavirus G9 strains isolated in the UK between 1995 and 1998. *J Med Virol* **61**: 510-517.

Iturriza-Gómara, M., Green, J., Brown, D., Ramsay, M., Desselberger, U., Gray, J. 2000b. Molecular epidemiology of human group A rotavirus infections in the UK between 1995 and 1998. *J Clin Microbiol* **38**: 4394-4401.

Iturriza-Gómara, M., Green, J., Brown, D. W., Desselberger, U., Gray, J. J. 1999. Comparison of specific and random priming in the reverse transcriptase polymerase chain reaction for genotyping group A rotaviruses. *J Virol* Methods **78**: 93-103.

Iturriza-Gómara, M., Green, J., Brown, D. W., Desselberger, U., Gray, J. J. 2000c. Diversity within the VP4 gene of rotavirus P[8] strains: implications for reverse transcription-PCR genotyping. *J Clin Microbiol* **38**: 898-901.

Iturriza-Gómara, M., Isherwood, B., Desselberger, U., Gray, J. 2001a. Reassortment *in vivo*: driving force for diversity of human rotavirus strains isolated in the United Kingdom between 1995 and 1999. *J Virol* **75**: 3696-3705.

Iturriza-Gómara, M., Wong, C., Blome, S., Desselberger, U., Gray, J. 2002. Molecular characterisation of VP6 genes of human rotavirus isolates: Correlation of genogroups with subgroups and evidence for independent segregation. *J Virol* **76**: 6596-6601.

Jain, V., Das, B. K., Bhan, M. K., Glass, R. I., Gentsch, J. R. 2001. Great diversity of group A rotavirus strains and high prevalence of mixed rotavirus infections in India. *J Clin Microbiol* **39**: 3524-3529.

Jin, Q., Ward, R. L., Knowlton, D. R., Gabbay, Y. B., Linhares, A. C., Rappaport, R., Woods, P. A., Glass, R. I., Gentsch, J. R. 1996. Divergence of VP7 genes of G1 rotaviruses isolated from infants vaccinated with reassortant rhesus rotaviruses. *Arch Virol* **141**: 2057-2076.

Kapikian, A., Hoshino, Y., Chanock, R. 2001. Rotaviruses. *In: Fields Virology*, 4th edition, D. Knipe *et al.* (Eds), pp. 1787-1834. Lippincott Williams and Willkins, Philadelphia.

Kirkwood, C., Gentsch, J., Glass, R. 1999. Sequence analysis of the NSP4 gene from human rotavirus strains isolated in the United States. *Virus Genes* **19**: 113-122.

Krishnan, T., Naik, T. N., Desselberger, U. 1996. Molecular epidemiology of human rotaviruses: reassortment *in vivo* as a mechanism for strain diversity? *J Infect*

342

32: 169-170.

Lazdins, I., Coulson, B. S., Kirkwood, C., Dyall-Smith, M., Masendycz, P. J., Sonza, S., Holmes, I. H. 1995. Rotavirus antigenicity is affected by the genetic context and glycosylation of VP7. *Virology* **209**: 80-89.

López, S., Espinosa, R., Greenberg, H. B., Arias, C. F. 1994. Mapping the subgroup epitopes of rotavirus protein VP6. *Virology* **204**: 153-162.

Maunula, L., von Bonsdorff, C. H. 1998. Short sequences define genetic lineages: phylogenetic analysis of group A rotaviruses based on partial sequences of genome segments 4 and 9. *J Gen Virol* **79**: 321-332.

Nakagomi, O., Isegawa, Y., Ward, R. L., Knowlton, D. R., Kaga, E., Nakagomi, T., Ueda, S. 1994. Naturally occurring dual infection with human and bovine rotaviruses as suggested by the recovery of G1P8 and G1P5 rotaviruses from a single patient. *Arch Virol* **137**: 381-388.

Palombo, E. A., Bishop, R. F., Cotton, R. G. 1993. Intra- and inter-season genetic variability in the VP7 gene of serotype 1 (monotype 1a) rotavirus clinical isolates. *Arch Virol* **130**: 57-69.

Palombo, E. A., Bugg, H. C., Masendycz, P. J., Bishop, R. F. 1997. Sequence of the VP7 gene of an atypical human rotavirus: evidence for genetic and antigenic drift. DNA Sequences **7**: 307-311.

Palombo, E. A., Masendycz, P. J., Bugg, H. C., Bogdanovic-Sakran, N., Barnes, G. L., Bishop, R. F. 2000. Emergence of serotype G9 human rotaviruses in Australia. *J Clin Microbiol* **38**: 1305-1306.

Patton, J.T., Taraporelawa, Z., Chen, D., Chizhikov, V., Jones, M., Elhelu, A., Collins, M., Kearney, K., Wagner, M., Hoshino, Y., Gouvea, V. 2001. Effect of intragenic rearrangement and changes in the 3' consensus sequence on NSP1 expression and rotavirus replication. *J Virol* **75**: 2076-2086.

Racz, M. L., Kroeff, S. S., Munford, V., Caruzo, T. A., Durigon, E. L., Hayashi, Y., Gouvea, V., Palombo, E. A. 2000. Molecular characterization of porcine rotaviruses from the southern region of Brazil: Characterization of an atypical genotype G[9] strain. *J Clin Microbiol* **38**: 2443-2446.

Raj, P., Matson, D. O., Coulson, B. S., Bishop, R. F., Taniguchi, K., Urasawa, S., Greenberg, H. B., Estes, M. K. 1992. Comparisons of rotavirus VP7-typing monoclonal antibodies by competition binding assay. *J Clin Microbiol* **30**: 704-711.

Ramachandran, M., Das, B. K., Vij, A., Kumar, R., Bhambal, S. S., Kesari, N., Rawat, H., Bahl, L., Thakur, S., Woods, P. A., Glass, R. I., Bhan, M. K., Gentsch, J. R. 1996. Unusual diversity of human rotavirus G and P genotypes in India. *J Clin Microbiol* **34**: 436-439.

Ramachandran, M., Gentsch, J. R., Parashar, U. D., Jin, S., Woods, P. A., Holmes, J. L., Kirkwood, C. D., Bishop, R. F., Greenberg, H. B., Urasawa, S., Gerna, G., Coulson, B. S., Taniguchi, K., Bresee, J. S., Glass, R. I. 1998. Detection and characterization of novel rotavirus strains in the United States. *J Clin Microbiol* **36**: 3223-3229.

Ramig, R. 2000. Mixed infections with rotaviruses: protocols for reasortment, complementation and other assays. *In: Rotaviruses: Methods and Protocols*, J. Gray and

U. Desselberger (Eds), pp. 79-100. Humana Press, Totowa, NJ.

Ramig, R. F. 1997. Genetics of the rotaviruses. Annu Rev Microbiol **51**: 225-255.

Santos, N., Lima, R.C., Pereira, C.F., Gouvea, V. 1998. Detection of rotavirus types G8 and G10 among Brazilian children with diarrhea. *J Clin Microbiol.* **36**: 2727-2729.

Schäfer, J., Kawaoka, Y., Bean, W., Suss, J., Senne, D., Webster, R. 1993. Origin of the pandemic 1957 H2 influenza A virus and the persistence of its possible progenitors in the avian reservoir. *Virology* **194**: 781-788.

Shen, S., Burke, B., Desselberger, U. 1994. Rearrangement of the VP6 gene of a group A rotavirus in combination with a point mutation affecting trimer stability. *J Virol* **68**: 1682-1688.

Steele, A. D, Parker, S. P., Peenze, I., Pager, C. T., Taylor, M. B., Cubitt, W. D. 1999. Comparative studies of human rotavirus serotype G8 strains recovered in South Africa and the United Kingdom. *J Gen Virol* **80**:3029-3304.

Tang, B., Gilbert, J. M., Matsui, S. M., Greenberg, H. B. 1997. Comparison of the rotavirus gene 6 from different species by sequence analysis and localization of subgroup-specific epitopes using site-directed mutagenesis. *Virology* **237**: 89-96.

Taniguchi, K., Kojima, K., Urasawa, S. 1996. Nondefective rotavirus mutants with an NSP1 gene which has a deletion of 500 nucleotides, including a cysteine-rich zinc finger motif-encoding region (nucleotides 156 to 248), or which has a nonsense codon at nucleotides 153-155. *J Virol* **70**: 4125-4130.

Tosser, G., Delaunay, T., Kohli, E., Grosclaude, J., Pothier, P., Cohen, J. 1994. Topology of bovine rotavirus (RF strain) VP6 epitopes by real-time biospecific interaction analysis. *Virology* **204**: 8-16.

Unicomb, L. E., Podder, G., Gentsch, J. R., Woods, P. A., Hasan, K. Z., Faruque, A. S., Albert, M. J., Glass, R. I. 1999. Evidence of high-frequency genomic reassortment of group A rotavirus strains in Bangladesh: emergence of type G9 in 1995. *J Clin Microbiol* **37**: 1885-1891.

Wen, L., Nakayama, M., Yamanishi, Y., Nishio, O., Fang, Z. Y., Nakagomi, O., Araki, K., Nishimura, S., Hasegawa, A., Muller, W. E., Ushijima, H. 1997. Genetic variation in the VP7 gene of human rotavirus serotype 3 (G3 type) isolated in China and Japan. *Arch Virol* **142**: 1481-1489.

Wen, L., Ushijima, H., Kakizawa, J., Fang, Z. Y., Nishio, O., Morikawa, S., Motohiro, T. 1995. Genetic variation in VP7 gene of human rotavirus serotype 2 (G2 type) isolated in Japan, China, and Pakistan. Microbiol Immunol **39**: 911-915.

Widdowson, M., van Doornum, G., van der Poel, W., de Boer, A., Mahdi, U., Koopmans, M. 2000. Emerging group-A rotavirus and a nosocomial outbreak of diarrhoea. *Lancet* **356**: 1161-1162.

Xin, K. Q., Morikawa, S., Fang, Z. Y., Mukoyama, A., Okuda, K., Ushijima, H. 1993. Genetic variation in VP7 gene of human rotavirus serotype 1 (G1 type) isolated in Japan and China. *Virology* **197**: 813-816.

Xu, Z., Tuo, W., Clark, K. I., Woode, G. N. 1996. A major rearrangement of the VP6 gene of a strain of rotavirus provides replication. Vet Microbiol **52**: 235-247.

[Erratum Vet Microbiol 1998; **60**:293].

Zao, C. L., Yu, W. N., Kao, C. L., Taniguchi, K., Lee, C. Y., Lee, C. N. 1999. Sequence analysis of VP1 and VP7 genes suggests occurrence of a reassortant of G2 rotavirus responsible for an epidemic of gastroenteritis. *J Gen Virol* **80**: 1407-1415.

Viral Gastroenteritis
U. Desselberger and J. Gray (editors)

345

II, 13. Current state of development of human rotavirus vaccines

Paul A. Offit[1,2], H Fred Clark[1] and Richard L. Ward[3]

[1] *Section of Infectious Diseases, The Children's Hospital of Philadelphia, Philadelphia, PA*
[2] *The Wistar Institute of Anatomy and Biology, Philadelphia, PA*
[3] *The Children's Hospital Medical Center, Cincinnati, OH*

Introduction

On August 31st, 1998 the first rotavirus vaccine for use in infants was licensed in the United States. The vaccine (Rotashield) was given by mouth to infants at 2, 4, and 6 months of age and contained 4 different viruses (CDC, 1999a). One of those viruses was simian rotavirus strain RRV (similar to human serotype G3). The other three viruses were reassortants containing 10 genes from RRV and one gene (gene segment 8 or 9) from human rotaviruses of serotypes G1, G2, or G4. The choice of simian and simian-human reassortant rotaviruses for use as a vaccine for infants was based on inclusion of the attenuated virulence characteristics of the nonhuman strain RRV, and inclusion of human rotavirus genes that encode a surface protein (VP7) responsible for evoking virus-neutralizing, protective antibodies. Rotashield replicated less efficiently in the human intestine than natural human rotaviruses, evoked virus-specific neutralizing antibodies, and had been shown to protect against challenge in prospective, controlled studies (Bernstein *et al.*, 1995; Rennels *et al.*, 1996; Joensuu *et al.*, 1997; Pérez-Schael *et al.*, 1997; Santosham *et al.*, 1997).

About 1 year after licensure, after Rotashield vaccine had been administered to approximately 1 million infants in the United States, 15 cases of intussusception following its use were reported to the Vaccine Adverse Events Reporting System (VAERS) (CDC, 1999b). In July of 1999, the Centers for Disease Control and Prevention (CDC) temporarily suspended the use of the vaccine pending the results of a case-control analysis. Subsequent studies found that Rotashield was a rare cause of intussusception in children (about 1 case of intussusception for every 10,000 children vaccinated) (Murphy *et al.*, 2001; Kramarz *et al.*, 2001), and the CDC withdrew its recommendation for use of the vaccine in October of 1999 (CDC, 1999c).

Two different vaccine strategies are currently under intensive study: bovine-human reassortant rotaviruses and an attenuated human rotavirus. Experience with Rotashield vaccine, possible etiologies for intussusception following its use, and the likelihood that differences in current vaccine strategies will result in differences in the capacity to cause intussusception will be discussed.

Experience with simian (RRV) rotavirus-based vaccines

Efficacy

Prior to licensure, large multicenter trials of Rotashield vaccine were performed in Venezuela, Finland, and the United States (Bernstein *et al.*, 1995; Rennels *et al.*, 1996; Joensuu *et al.*, 1997; Perez-Schael *et al.*, 1997; Santosham *et al.*, 1997; Vesikari, 1992). Protective efficacy against all rotavirus disease was 48% to 68%; protection against severe rotavirus disease was 70% to 100%.

Studies of safety performed before licensure

Placebo-controlled trials performed prior to licensure of Rotashield vaccine showed that the major side effects occurred after the first dose (Bernstein *et al.*, 1995; Rennels *et al.*, 1996; Joensuu *et al.*, 1997; Perez-Schael *et al.*, 1997; Santosham *et al.*, 1997). Side effects included an increase in fever greater than 38°C, an increase in fever greater than 39°C, irritability, decreased appetite, and decreased activity (CDC, 1999).

Intussusception was also observed in children inoculated with Rotashield before licensure (Rennels *et al.*, 1998). Five cases of intussusception were observed in approximately 11,000 children who received Rotashield vaccine and 1 case in approximately 4,500 children who received placebo. These differences were not found to be statistically significant. All cases of intussusception in children who received Rotashield in pre-licensure studies occurred after the second or third dose, and most cases occurred within 15 days of receipt of the vaccine. Studies examining the relationship between receipt of Rotashield vaccine and intussusception were required by the Food and Drug Administration (FDA) after licensure.

Studies of safety performed after licensure

A case-control analysis performed by the CDC found that 1) relative risks (RR) for intussusception 3-7 days after the first and second doses of Rotashield vaccine were 27.9 and 5.0, respectively, 2) the RR 8-14 days after the first and second doses were 3.9 and 1.5, respectively, and 3) the RR 15-21 days after the first and second doses were 0.7 and 0.9, respectively (Murphy *et al.*, 2001). Estimates of the attributable risk were 1 case of intussusception for approximately every 4,700 to 9,500 children vaccinated (Murphy *et al.*, 2001). A subsequent study found an attributable risk of 1 case of intussusception for every 11,000 children vaccinated (Kramarz *et al.*, 2001). Because Rotashield was found to be a cause of intussusception, recommendations for its use in infants were withdrawn by both the CDC (CDC, 1999b) and American

Academy of Pediatrics (AAP) (Committee on Infectious Diseases, 1999) in October of 1999.

Recent ecological studies failed to show an appreciable increase in the incidence of intussusception after introduction of Rotashield vaccine (Chang *et al.*, 2001; Simonsen *et al.*, 2001). These studies suggest that the attributable risk of intussusception may be far less than initial estimates. However, ecological studies are confounded by the use of hospital discharge diagnosis (rather than studies of medical records), and estimates of vaccine use that were based on doses distributed rather than doses administered.

Currently no plan is in place to reintroduce Rotashield vaccine into either developed or developing countries.

Experience with bovine (WC3) rotavirus-based vaccines.

Vaccine strategy and biologic distinction from RRV-based vaccine

RIT 4237

Rotavirus was first discovered as a cause of diarrhea in cattle. Shortly thereafter, a high-cell-culture passage, highly attenuated bovine rotavirus was developed in Nebraska for the protection of calves. This vaccine (termed RIT 4237) was the first rotavirus vaccine administered to infants. Given at a high dose (ca.10^8 TCID$_{50}$) its administration was totally free of adverse events in infants, possibly because of the greater phylogenetic distance of the cow from humans compared with primate strains. RIT 4237 was highly immunogenic and induced serum-neutralizing antibodies that were specific for this bovine (serotype G6) rotavirus. Although RIT 4237 protected against (predominately serotype G1) rotavirus disease in clinical trials in Finland, it was less effective in several other trials at other testing sites (Hanlon *et al.*, 1987; DeMol *et al.*, 1986; Santosham *et al.*, 1991).

WC3 rotavirus

Subsequently, another bovine rotavirus, designated WC3 (also of serotype G6), was isolated in Pennsylvania. It was evaluated as a vaccine candidate at the twelfth cell-culture-passage level, on the presumption that a low-cell-culture-passage virus might be less attenuated and might, therefore, elicit a more vigorous immune response protective for infants. In several trials (Clark *et al.*, 1986, Garbarg-Chenon *et al.*, 1989), WC3 rotavirus induced a serum-neutralizing antibody response at an incidence of between 70% and 100%. This antibody response was almost exclusively specific to bovine G6 virus.

Like RIT 4237, WC3 was protective in certain clinical trials but not in others (Clark *et al.*, 1988, Bernstein *et al.*, 1990, Georges-Courbot *et al.*, 1991). Nevertheless it was established that WC3, commonly given at a dose of ca.$10^{7.0}$ pfu, did not induce adverse events in infants (unlike RRV). WC3 also differed from RRV in that replication in the infant gut apparently occurred at a very low level. Although given at a high dose (100- to 1000-fold higher than that used for RRV), WC3 rotavirus was recovered in low incidence ($\leq 21\%$) from infant feces, and only by use of the very sensitive plaque assay (Clark *et al.*, 1986). In contrast, RRV was commonly (50% of infants after dose 1, 2, or 3) shed in infant feces at concentrations detected by the much less sensitive ELISA test, suggesting a much higher level of replication than bovine rotavirus (Ward et. al., 1998). All titers of WC3 recovered from infants were less than $10^{4.0}$ pfu per gm of feces.

Studies of safety and efficacy of WC3-human rotavirus reassortant vaccines

Because of the erratic efficacy results found with WC3, reassortants of this virus containing genes encoding neutralization proteins of human rotaviruses were also developed. Earliest efforts concentrated on evaluation of a serotype G1 reassortant of WC3 because of the worldwide high prevalence of G1 human rotaviruses. Reassortant strain WI79-9 was constructed bearing the gene of the serotype G1 VP7 surface protein of human rotavirus strain WI79 on a WC3 genome background.

In Phase I clinical trials in infants aged 2 to 11 months, consisting of the WC3 G1 reassortant given at two doses of $10^{7.5}$ pfu or the G1 reassortant followed by a WC3 dose, the G1 reassortant was shown to retain the non-reactogenic character of WC3. Fecal shedding of G1 reassortant vaccine virus was detected in only 8% of these infants. As in the case of WC3, when shedding of vaccine was detected, titers were less than $10^{4.0}$ pfu/gm of feces (Clark *et al.*, 1990b).

As was also found in infants administered WC3, the G1 reassortant was highly immunogenic. Although the most efficient serum neutralizing antibody response continued to be of WC3 specificity, an enhanced immune response to G1 was also induced (Clark *et al.*, 1990a).

Based on demonstrated safety and immunogenicity, a small double-blind, placebo-controlled efficacy trial with the G1 reassortant was performed using two doses of $10^{7.3}$ pfu (Clark *et al.*, 1990a). No adverse events were associated with vaccine administration but the immune response to G1 was a disappointing 22%. Nevertheless the vaccine was completely protective against severe infection.

The protective efficacy of G1 reassortant WI79-9 was confirmed in a larger trial in Philadelphia and Rochester. Infants were vaccinated with 3 doses of $10^{7.3}$ pfu at 2-month intervals. No adverse events were associated with the vaccine. Protective rates associated with the vaccine, in a primarily G1 serotype season, were 64% against all rotavirus disease episodes and 87% against clinically significant rotavirus disease (Treanor *et al.*, 1995).

Although the G1 rotavirus vaccine had a demonstrated protective efficacy, it was disappointing that immune response rates in infants always occurred at higher incidence against VP4 (serotype P7, genotype [5]) of WC3 (VP4 being less relevant than VP7-based human G serotypes) than immune response rates to WI79 (represented by G1 VP7 in the reassortant). Therefore a "reverse reassortant" was created with a human rotavirus VP4 (serotype P1A, genotype [8]) and a bovine rotavirus VP7 (serotype G6).

Surprisingly, infants given two doses of this P1A vaccine also developed a serum antibody response predominantly specific for WC3 (Clark *et al.*, 1996a).

A third approach was then to develop a vaccine carrying *both* P1A and G1 human surface proteins. A WC3 reassortant containing genes for both P1A and G1 serotypes (strain WI79[4+9]) was constructed. Two doses of this virus were administered to infants and compared with a cohort of infants given a mixture of the single gene reassortants P1A (WI79-4) and G1 (WI79-9). The double reassortant WI79(4+9) was poorly immunogenic whereas two doses of P1A reassortant mixed with G1 reassortant elicited the most effective serum antibody response yet observed (Clark *et al.*, 1996b).

Later, it was found that three doses of the G1 reassortant were nearly as potent as two doses of mixed P1A and G1 reassortants. It was determined that the third dose of G1 was necessary to obtain this level of response because antibodies specific for WC3 (VP4) appeared primarily after the first dose while antibodies specific for WI79 (VP7) appeared primarily after the second and third doses (Clark *et al.*, unpublished data).

Although the mixture of P1A and G1 reassortants of WC3 appeared to be ideal in terms of safety and immunogenicity, it was considered desirable to add antigens to this vaccine of G serotypes other than G1, which have predominated in other regions of the world and within certain seasons and locales in the United States. Reassortants were developed that contained the VP7 genes for G2- and a G3-specificity, each upon the background of the WC3 rotavirus genome.

The G1, G2, G3, and P1A reassortants, each at a concentration of approximately $10^{7.0}$ pfu, were combined into a quadrivalent vaccine that was administered in three doses in a multi-site, placebo-controlled efficacy trial (Clark *et al.*, 1995, 1996b). The vaccine was well tolerated. Only about 5% of vaccinees shed infectious vaccine virus in their feces. The rate of vaccine-induced protection against all rotavirus disease was 67% (p<0.001). Severe cases of rotavirus diarrhea were seen in the placebo cohort but not in vaccinated infants (protection 100%).

To complete a WC3-human rotavirus reassortant vaccine with antigens specific for all of the traditionally recognized major serotypes, a WC3 reassortant bearing a gene for a VP7 of G4 specificity has been generated and separately tested for safety, permitting the formulation of a quintavalent vaccine (serotypes G1, G2, G3, G4, and P1A). This vaccine is currently under study in a large, multicentered, placebo-controlled trial.

Experience with attenuated human (strain 89-12) rotavirus vaccine

Rationale for human rotaviruses as vaccine candidates

All rotavirus vaccine candidates evaluated in clinical trials have been live virus strains that are delivered orally. The rationale for this approach is that natural rotavirus infection has been reported to provide good and sometimes outstanding (i.e., 100%) protection against gastrointestinal disease following subsequent exposure to rotavirus (Bernstein *et al.*, 1991; Velazquez *et al.*, 1996; Ward *et al.*, 1992a). The goal for these vaccine candidates is to mimic the protection provided by natural infections. Because natural rotavirus infections in humans are due almost solely to strains characterized as human rotaviruses, it follows that human rotaviruses are almost solely responsible for the protection observed after natural rotavirus infection. Development of human rotaviruses into vaccine candidates is, therefore, a logical outgrowth of these observations.

Concerns regarding the use of human rotaviruses as live, orally deliverable vaccine candidates

The traditional primary aims for the use of live rotaviruses as vaccine candidates is that they are sufficiently attenuated, yet still provide excellent protection, and that this protection extend to illnesses elicited by infection with multiple serotypes of rotavirus. There are at least 14 G and 21 P types of group A rotaviruses (Gentsch *et al.*, 1996; Rao *et al.*, 2000). Although human rotavirus strains belong predominantly to any one of 5 of these G types and 3 of these P types, other G or P types have been reported to dominate in certain locales (Gentsch *et al.*, 1996; Kapikian and Chanock, 1996). Furthermore, it is anticipated that the predominance of specific serotypes may shift with time as has occurred in recent years with the emergence of G9 and P2 strains (Cubitt *et al.*; 2000; Gentsch *et al.*, 1996). If neutralizing antibodies are felt to be the primary if not sole effectors of protection, then the candidate vaccine must contain multiple rotavirus strains in order to provide broad protection. In contrast, if it is believed that effectors other than neutralizing antibodies, such as cross-reactive T cells, play important if not dominant roles in protection after live rotavirus infection, it is possible that a single rotavirus strain may be sufficient as a broadly protective vaccine.

One method to attenuate rotaviruses is by genetic modification of virus strains. Because gene rescue (reverse genetics) is not yet possible for rotaviruses, this cannot be done through systematic nucleotide manipulation but must rely on selective processes that occur under special growth conditions such as during passage in cell culture. This was the method used to attenuate a rotavirus vaccine candidate (strain 89-12) developed from a virulent strain of human rotavirus.

Justification for use of attenuated human rotaviruses as vaccines

Protection against rotavirus disease by Rotashield vaccine was compared with that provided by natural infection in a study performed over two years (Bernstein *et al.*, 1995; Ward and Bernstein, 1994). In that study, Rotashield provided 57% protection against rotavirus gastroenteritis while natural rotavirus infections in the placebo recipients were 93% protective. These results suggest that natural rotavirus infections may be more protective than immunization with rotaviruses derived from heterologous hosts.

Development of a virulent human rotavirus into a vaccine candidate

During the evaluation of the bovine rotavirus WC3 as a vaccine candidate in Cincinnati during the 1988-89 rotavirus season, administration of a single dose of the candidate vaccine to children aged 2-12 months was found to have no significant impact on development of subsequent rotavirus illnesses (Bernstein *et al.*, 1990, 1991). In contrast, children who had symptomatic or asymptomatic rotavirus infections during that season were fully protected against rotavirus illness during the following year. Furthermore, only 2 of 60 subjects infected during the first year had a detectable asymptomatic reinfection during the second year, while among the 82 subjects not infected during the first year, 9 had symptomatic and 20 had asymptomatic infections in the second year (Bernstein *et al.*, 1991). Exposure to a variety of rotavirus strains occurred during this second year, as determined by the electropherotypes of strains causing disease, but all analyzed strains belonged to serotype G1.

It was noted that during the first year of the study, all rotavirus isolates from ill subjects were also characterized as serotype G1. Surprisingly, however, almost all rotaviruses obtained from these subjects, as well as from >50 patients hospitalized at Children's Hospital Medical Center with acute gastroenteritis, were indistinguishable based on RNA electropherotype. Therefore, the protection elicited in study subjects against multiple G1 strains during year 2 appeared to be induced by infections with the same rotavirus strain in year 1. From these observations, it followed that if attenuation of this G1P[8] strain did not ablate its ability to elicit protection against rotavirus disease, it should make an excellent vaccine candidate, at least against G1 strains. Furthermore, if protection can be stimulated by vaccination with this attenuated strain through mechanisms other than neutralizing antibody, as has been suggested from a variety of studies (Vesikari, 1993; Ward *et al.*, 1992a, 1992b, 1997), it may be broadly protective as a single strain rotavirus vaccine. Based on this, rotavirus from the stool of one ill subject from the 1989 WC3 vaccine trial was developed as a vaccine candidate and designated 89-12 for the year of study and the subject from whom it was isolated. Attenuation of the virus was performed by 33 serial passages in cultured monkey kidney cells.

Evaluation of a candidate human rotavirus vaccine

The 89-12 rotavirus vaccine candidate was first evaluated for safety and immunogenicity in adults, in children with evidence of a previous rotavirus infection, and finally in infants (Bernstein *et al.*, 1998). Once found to be generally safe and highly immunogenic in infants, it was tested for efficacy in healthy infants aged 10-16 weeks at four centers in the USA (Bernstein *et al.*, 1999). Two doses of either 10^5 pfu or placebo were administered to 108 or 107 subjects, respectively, separated by a 6-10 week interval prior to the 1998 rotavirus season. Low-grade fever after the first dose was the only side effect more common in the vaccine than in the placebo group (21 vs. 5 subjects; $p < 0.001$). An immune response to rotavirus was detected in 94% of vaccinees and in only 4% of the placebo recipients. During the first rotavirus season, rotavirus disease was detected in 18 of the placebo recipients and 2 vaccinees (yielding a vaccine efficacy of 89%). Ten infants in the placebo group but none of the vaccinees presented for medical care associated with rotavirus disease. In the second year, efficacy decreased to 59%, but only one case of severe rotavirus gastroenteritis was found in vaccinees while 10 severe cases were found in placebo recipients. The overall efficacy for the

2 years was 76% against any rotavirus gastroenteritis, 84% against severe disease, and 100% against very severe rotavirus G1 disease. Because serotype G1 rotaviruses predominated during both years, the efficacy against heterotypic rotaviruses was not determinable in the field. The 89-12 candidate vaccine is being further evaluated in larger trials and in countries where non-G1 community strains are typically isolated.

Relationship between pathogenesis of intussusception following Rotashield vaccine and success of alternative vaccine strategies

The pathogenesis of intussusception following infection with Rotashield remains unclear. Epidemiologic studies, however, reveal clues to the possible etiologies of intussusception following administration of the Rotashield vaccine. Each will be discussed below.

The "unique strain" hypothesis

Whereas natural rotavirus infection occurs almost solely during winter months in temperate climates, intussusception observed prior to the use of the Rotashield vaccine, occurred year-round; no increase in rates of intussusception had been observed during winter months (Rennels *et al.*, 1998). Therefore, natural rotavirus infection does not appear to be an important cause of intussusception in the United States. If natural rotavirus infection does not cause intussusception, then infection with vaccine viruses must differ critically from infection with wild-type viruses.

Simian strain RRV does have several biological features that are unique among rotaviruses. First, RRV has the capacity to replicate in the small intestines and cause diarrhea in several species other than monkeys (Ciarlet *et al.*, 2000). Second, RRV causes hepatitis in certain strains of inbred and immunodeficient mice (Uhnoo *et al.*, 1990). Third, RRV has the capacity to replicate *in vitro* in the absence of trypsin (HF Clark, unpublished data). The degree to which these biological differences are important in the pathogenesis of intussusception in human infants is unclear. If RRV causes intussusception, then vaccine strategies that do not include RRV (including the possibility of RRV-human reassortants) are more likely to avoid the consequence of intussusception.

The "bolus dose" hypothesis

Another difference between Rotashield immunization and natural infection is the quantity of virus to which the infant is initially exposed. During natural infection a child is usually infected with small to very small quantities of infectious virus that are amplified in multiple cycles of replication. Replication probably occurs initially in the proximal small intestine and then extends to the distal small intestine (Starkey *et al.*, 1986). During immunization with Rotashield children are exposed to a larger quantity of infectious virus in one dose. A "bolus" of rotavirus may be taken up at a site or presented to gut-associated lymphoid tissues in a manner different from that which occurs after natural infection.

The "viral replication" hypothesis

The greatest risk for intussusception occurs 3-7 days after administration of Rotashield vaccine (Murphy *et al.*, 2001). This time interval corresponds with the time at which rotavirus vaccine replicates in the small intestinal mucosal surface (Ward *et al.*, 1998). If rotavirus replication at the intestinal surface alone is capable of causing intussusception, then infection with wild-type rotavirus strains (which are better adapted to growth in the intestinal mucosal surface than vaccine viruses) would also likely cause intussusception. Although no data clearly support the hypothesis that natural rotavirus infection is a cause of intussusception, it is possible that wild-type rotaviruses are a rare cause of this disease. If rotavirus replication is an important factor in the etiology of intussusception, bovine-human reassortant rotaviruses (which are the least well adapted to growth at the intestinal surface when compared to attenuated human rotavirus or simian-human reassortant vaccines) are most likely to avoid the risk of intussusception following vaccination.

Other possible candidates that may avoid the risk of intussusception associated with oral inoculation with live viruses include virus-like particles (VLPs), DNA-based vaccines, or individual rotavirus proteins expressed in non-pathogenic bacterial vectors. None of these approaches have been evaluated in humans.

Conclusions

Current rotavirus vaccine candidates include the use of bovine-human reassortant rotaviruses and attenuated human rotavirus. Both bovine rotaviruses and attenuated human rotaviruses are biologically distinct from simian rotaviruses. Ongoing trials with these second-generation rotavirus vaccines are likely to shed light on the pathogenesis of intussusception following the use of simian and simian-human reassortant rotaviruses.

References

Bernstein DI, Smith V, Sander D, *et al.* (1990). Evaluation of WC3 rotavirus vaccine and correlates of protection in healthy infants. J Infect Dis 162, 1055-1062.

Bernstein DI, Sander DS, Smith V, *et al.* (1991). Protection from rotavirus reinfection: Two-year prospective study. J Infect Dis 164, 277-283.

Bernstein DI, Glass R, Rodgers G, *et al.* (1995). Evaluation of rhesus rotavirus monovalent and tetravalent reassortant vaccines in U.S. children. J Am Med Assoc 273, 1191-1196.

Bernstein DI, Smith VE, Sherwood JR, *et al.* (1998). Safety and immunogenicity of

354

live, attenuated human rotavirus vaccine 89-12. Vaccine 16, 381-387.

Bernstein DI, Sack DA, Rothstein E. *et al.* (1999). Efficacy of live, attenuated, human rotavirus vaccine 89-12 in infants: a randomised placebo-controlled trial. Lancet 354, 287-290.

Centers for Disease Control and Prevention (1999a). Rotavirus vaccine for the prevention of rotavirus gastroenteritis among children - recommendations of the Advisory Committee on Immunization Practices. MMWR 48:1-23.

Centers for Disease Control and Prevention (1999b). Intussusception among recipients of rotavirus vaccine - United States, 1998-1999. MMWR 48:577-581.

Centers for Disease Control and Prevention (1999c). Withdrawal of rotavirus vaccine recommendation. MMWR 48:1007.

Ciarlet M, Estes MK, Connor ME. (2000). Simian rhesus rotavirus is a unique heterologous (non-lapine) rotavirus strain capable of productive replication and horizontal transmission in rabbits. J Gen Virol 81, 1237-1249.

Chang HH, Smith PF, Ackelsberg, J, *et al.* (2001). Intussusception, rotavirus diarrhea, and rotavirus vaccine use among children in New York State. Pediatrics 108: 54-60.

Clark HF, Furukawa T, Bell LM, *et al.* (1986). Immune response of infants and children to low passage bovine rotavirus (strain WC3). Am J Dis Child 140, 350-356.

Clark HF, Borian FE, Bell LM, *et al.* (1988). Protective effect of WC3 vaccine against rotavirus diarrhea in infants during a predominantly serotype 1 rotavirus season. J Infect Dis 158, 570-587.

Clark HF, Borian FE, Modesto K, Plotkin SA. (1990a). Serotype 1 reassortant of bovine rotavirus WC3, strain WI79-9, induces a polytypic antibody response in infants. Vaccine 8, 327-332.

Clark HF, Borian FE, Plotkin SA. (1990b). Immune protection of infants against rotavirus gastroenteritis by a serotype 1 reassortant of bovine rotavirus WC3. J Infect Dis 161, 1099-1104.

Clark HF, White CJ, Offit PA, *et al.* (1995). Preliminary evaluation of safety and efficacy of quadrivalent human-bovine reassortant rotavirus vaccine. Ped Res 37, 172A.

Clark HF, Offit PA, Ellis RW, *et al.* (1996a). The development of multivalent bovine rotavirus (strain WC3) reassortant vaccine for infants. J Infect Dis 174 [suppl. 1], S73-S80.

Clark HF, Offit PA, Ellis RW, *et al.* (1996b). WC3 reassortant vaccines in children. Arch Virol [suppl.] 12, 187-198.

Committee on Infectious Diseases. American Academy of Pediatrics (1999). Possible association of intussusception with rotavirus vaccination. Pediatrics 104:575.

Cubitt WD, Steele AD, Iturriza M. (2000). Characterisation of rotaviruses from children treated at a London hospital during 1996: Emergence of strains G9P2A[6] and G3P2A[6]. J Med Virol 61, 150-154.

DeMol P, Zissis G, Butzler JP, *et al.* (1986). Failure of live, attenuated oral rotavirus vaccine. Lancet 2:108.

Garbarg-Chenon A, Fontaine JL, Lasfargues G, *et al.* (1989). Reactogenicity and immunogenicity of rotavirus WC3 vaccine in 5-12 month old infants. Ann Inst Pasteur: Res Virol 140, 207-217.

Gentsch JR, Woods PA, Ramachandran M., *et al.* (1996). Review of G and P typing results from a global collection of rotavirus strains: Implications for vaccine development. J Infect Dis 174, S30-S36.

Georges-Courbot MC, Monges J, Siopathis MR, *et al.* (1991). Evaluation of the efficacy of a low passage bovine rotavirus vaccine (strain WC3) in children in Central Africa. Ann Inst Pasteur: Res Virol 142, 405-411.

Hanlon P, Hanlon K, Marsh V, *et al.* (1987). Trial of an attenuated bovine rotavirus vaccine (RIT 4237) in Gambian infants. Lancet 1:1342-1345.

Joensuu J, Koskenniemi E, Pang XL, *et al.* (1997). Randomized placebo-controlled trial of rhesus human reassortant rotavirus vaccine for prevention of severe rotavirus gastroenteritis. Lancet 35, 1205-1209.

Kapikian AZ, Chanock RM. (1996). Rotaviruses. *In* Fields Virology, BN Fields, DM Knipe, PH Howley, *et al.* (eds.). 3rd ed., pp.1657-1708, Philadelphia, Lippincott-Raven Press.

Kramarz P, France EK, Destefano F, *et al.* (2001). Population-based study of rotavirus vaccination and intussusception. Pediatr Infect Dis J 20, 410-416.

Murphy TV, Gargiullo PM, Massoudi MS, *et al.* (2001). Intussusception among infants given an oral rotavirus vaccine. N Engl J Med 344, 564-572.

Pérez-Schael I, Guntiñas MJ, Pérez M, *et al.* (1997). Efficacy of the rhesus rotavirus-based quadrivalent vaccine in infants and young children in Venezuela. N Engl J Med 337, 1181-1187.

Rao CD, Gowda K, Reddy BSY. (2000). Sequence analysis of VP4 and VP7 genes of nontypeable strains identifies a new pair of outer capsid proteins representing novel P and G genotypes in bovine rotaviruses. Virology 276, 104-113.

Rennels MB, Glass RI, Dennehy PH, *et al.* (1996). Safety and efficacy of high-dose rhesus-human reassortant rotavirus vaccines – report of the National Multicenter Trial. Pediatrics 97, 7-13.

Rennels MB, Parasjar U, Holman R, *et al.* (1998). Lack of an apparent association between intussusception and wild or vaccine rotavirus infection. Pediatr Infect Dis J 17. 924-925.

Santosham M, Letson GW, Wolff M, *et al.* (1991). A field study of the safety and efficacy of two candidate rotavirus vaccines in a native American population. J Infect Dis 163:483-487.

Santosham M, Moulton LH, Reid R, *et al.* (1997). Efficacy and safety of high-dose rhesus-human reassortant rotavirus vaccine in Native American populations. J Pediatr 131, 632-638.

Simonsen L, Morens DM, Elixhauser A, *et al.* (2001). Incidence trends in infant hospitalization for intussusception: impact of the 1998-1999 rotavirus vaccination program in 10 U.S. states. Lancet 358:1224-1229.

Starkey W, Collins J, Wallis T, *et al.* (1986). Kinetics, tissue specificity, and pathological changes in murine rotavirus infection of mice. J Gen Virol 67, 2625-2634.

Treanor J, Clark HF, Pichichero M, *et al.* (1995). Evaluation of the protective efficacy of a serotype 1 bovine-human rotavirus reassortant vaccine in infants. Ped Infect Dis J 14, 301-307.

Uhnoo I, Riepenhoff-Talty M, Dharakul T, *et al.* (1990). Extramucosal spread and development of hepatitis in immunodeficient and normal mice infected with rhesus rotavirus. J Virol 64, 361-368.

Velazquez FR, Matson DO, Calva JJ, *et al.* (1996). Rotavirus infection in infants as protection against subsequent infections. N Engl J Med 335, 1022-1028.

Vesikari T, Ruuska T, Green K, *et al.* (1992). Protective efficacy against serotype 1 rotavirus diarrhea by live oral rhesus-human reassortant rotavirus vaccines with human rotavirus VP7 serotype 1 or 2 specificity. Pediatr Infect Dis J 11, 535-542.

Vesikari T. (1993). Clinical trials of live oral rotavirus vaccines: The Finnish experience. Vaccine 11, 255-261.

Ward RL, Clemens JD, Knowlton DR, *et al.* (1992a). Evidence that protection against rotavirus diarrhea after natural infection is not dependent on serotype-specific neutralizing antibody. J Infect Dis 166, 1251-1257.

Ward R L, McNeal MM, Sheridan J. (1992b). Evidence that active protection following oral immunization of mice with live rotavirus is not dependent on neutralizing antibody. Virology 188, 57-66.

Ward RL, Bernstein DI. (1994). Protection against rotavirus disease following natural rotavirus infection. J Infect Dis 169, 900-904.

Ward RL, Knowlton DR, Zito ET, *et al.* (1997). Serologic correlates of immunity in a tetravalent reassortant rotavirus vaccine trial. J Infect Dis 176, 570-577.

Ward RL, Dinsmore AM, Goldberg G, *et al.* (1998). Shedding of rotavirus after administration of the tetravalent rhesus rotavirus vaccine. Ped Infect Dis J 17, 386-390.

II, 14. Rotavirus-like particle vaccines evaluated in a pig model of human rotavirus diarrhea and in cattle

Lijuan Yuan[1] and Linda J. Saif[2]

[1] *Epidemiology Section, Laboratory of Infectious Diseases,*
National Institutes of Allergy and Infectious Diseases,
National Institutes of Health, Bethesda MD 20892, USA
[2] *Food Animal Health Research Program,*
Ohio Agricultural Research and Development Center,
Veterinary Preventive Medicine Department, The Ohio State University,
Wooster, OH 44691, USA

Introduction

Virus-like particles (VLPs) are commonly generated using baculovirus expression vectors for coexpression of multiple viral protein genes in insect cells (e.g. Crawford *et al.*, 1994; Conner *et al.*, 1996; Roy *et al.*, 1992; Roy, 1996). VLPs are non-infectious because they lack genetic material (nucleic acid), but they are morphologically and antigenically similar to the native virus (Conner *et al.*, 1996; Roy, 1996). Bluetongue virus VLPs administered to sheep with various adjuvants were the first VLP vaccines developed in the *Reoviridae* family and were evaluated in animals for the induction of protective immunity against infectious wild type virus challenge (Roy *et al.*, 1992). The success in producing different formulations of rotavirus VLPs by coexpression of different combinations of rotaviral structural proteins (Crawford *et al.*, 1994) has facilitated studies of safer and potentially more cost-effective rotavirus subunit vaccines (Conner *et al.*, 1996). The major advantages of VLP vaccines for humans and animals include: (*i*) increased safety by avoiding side effects seen after oral immunization with some live rotavirus vaccines in infants (fever, diarrhea, and possibly intussusception) (Kramarz *et al.*, 2001); (*ii*) lack of revertance to virulence as may occur with attenuated live virus vaccines; (*iii*) the lack of adventitious agents potentially present during cultivation of live vaccines; and (*iv*) the consistent expression of various serotype VP4s and VP7s containing neutralizing epitopes on VLPs in a stable and immunogenic form to confer protection against multiple rotavirus serotypes (Crawford *et al.*, 1999; Jiang *et al.*, 1999). This latter aspect was an advantage over conventional inactivated vaccines for bovine rotavirus, because cell culture propagation of bovine rotavirus lead to the inconsistent production of triple-layered rotavirus containing VP4 and VP7 (Fernandez *et al.*, 1996).

Studies in adult mice and rabbits have shown that various formulations of VLPs (VP2/6, VP2/6/7, VP2/4/6/7), administered via intramuscular (i.m.) or intranasal (i.n.) routes with or without adjuvants, induced complete or significant partial protection against rotavirus infection (Ciarlet *et al.*, 1998a; Crawford *et al.*, 1999; Jiang *et al.*, 1999; O'Neal *et al.*, 1997, 1998; Siadat-Pajouh and Cai, 2001). However, only protection against infection, but not against diarrhea, could be assessed in the adult mouse or rabbit models as these species are only susceptible to rotavirus-induced diarrhea during the first two weeks of life (Ciarlet *et al.*, 1998b; Ward *et al.*, 1990). In comparison, neonatal gnotobiotic pigs present a number of important advantages for investigating immune responses to human rotavirus (HRV) and for evaluating vaccine efficacy: they closely resemble humans in gastrointestinal physiology and in the development of mucosal immunity; they are susceptible up to at least 8 weeks of age to infection *and* disease with several HRV strains (Saif *et al.*, 1996, 1997; Schaller *et al.*, 1992; Ward *et al.*, 1996a; Yuan *et al.*, 1998); they develop pronounced histopathologic lesions (e.g., villous atrophy) in the small intestine following HRV infection (Ward *et al.*, 1996a), similar to that seen after rotavirus infection in infants (Davidson and Barnes, 1979); they are born devoid of maternal antibodies but are immunocompetent, allowing assessment of true primary immune responses (Kim, 1980; Mehrazar and Kim, 1988); and their gnotobiotic status assures that exposure to extraneous rotaviruses or other enteric pathogens can be eliminated as potential confounding variables.

The neonatal gnotobiotic pig model of rotavirus infection and diarrhea (Saif *et al.*, 1996, 1997) has been used in our laboratory for studies of pathogenesis and immunity to rotaviruses and evaluation of attenuated and inactivated rotavirus vaccines (Chen *et al.*, 1995; Hoshino *et al.*, 1995; Ward *et al.*, 1996a, 1996b; Yuan *et al.*, 1996, 1998, 2001a). We previously examined antibody-secreting cell (ASC) and memory B cell responses in the intestinal and systemic lymphoid tissues and the correlates of protection against rotavirus diarrhea in gnotobiotic pigs inoculated with virulent HRV (mimic natural infection), live attenuated HRV (mimic oral attenuated vaccines), or inactivated HRV. Our data indicated that protection rates against rotavirus diarrhea upon challenge positively correlated with the magnitude of IgA ASC and memory B cell responses in the intestinal lymphoid tissues, but not with the IgG ASC or memory B cell responses in the systemic lymphoid tissues (Yuan *et al.*, 1996, 1998, 2001a). Studies of natural rotavirus infections in children showed a similar positive correlation of IgA antibodies and protective immunity against rotavirus infection and diarrhea (Matson *et al.*, 1993; Velazquez *et al.*, 2000). Thus, the capacity to induce sufficient levels of intestinal IgA ASCs or antibodies and memory B cell responses appears to be critical for the efficacy of rotavirus vaccines in pigs and humans. This is in contrast to the findings from evaluations of rotavirus vaccines using the adult mouse model of rotavirus infection that have indicated that none of the known effectors of acquired immune responses (antibodies, CD4+, CD8+, IFN-γ, etc) were essential for protective immunity against rotavirus infection in mice (Franco and Greenberg, 1999, 2000; O'Neal *et al.*, 2000). Because of the similarities in pathogenesis and protective immunity to rotavirus diarrhea in neonatal pigs and human infants, the evaluation of VLP vaccines in pigs offers information that may be more relevant for the development of rotavirus vaccines for humans.

Although rotavirus diarrhea is a major disease problem in 1- to 3-week-old calves, bovine rotavirus (BRV) vaccines are of variable efficacy in the field (Saif and Fernandez, 1996). Currently licensed BRV vaccines include an attenuated live vaccine given orally to calves shortly after birth. However, this vaccine was of limited efficacy due to neutralization of the vaccine virus by BRV-specific maternal antibodies transferred passively to calves in colostrum (de Leeuw *et al.*, 1980; Saif and Fernandez, 1996). Inactivated BRV vaccines administered to cows twice during pregnancy also provided variable efficacy to suckling calves or colostrum-supplemented dairy calves in the field (Castrucci *et al.*, 1984, 1993; Saif *et al.*, 1983). Alternative vaccines and vaccination strategies are desirable to improve BRV vaccine efficacy. Thus, we also prepared 2/6- and 2/4/6/7-VLP vaccines for use in pregnant cows and evaluated the ability of colostrum from the vaccinated cows to passively protect calves from rotavirus infection and diarrhea (Fernandez *et al.*, 1996, 1998; Kim *et al.*, 2002).

VLP vaccines evaluated in neonatal gnotobiotic pigs

The immunogenicity and protective efficacy of rotavirus 2/6-VLPs and combined 2/6-VLPs and live attenuated rotavirus vaccine regimens were evaluated in gnotobiotic pigs (Yuan *et al.*, 2000, 2001b). A mutant *Escherichia coli* heat labile enterotoxin R192G (mLT) was used as a mucosal adjuvant for i.n. immunization of gnotobiotic pigs with VLPs. The mLT and other LT or *Cholera toxin* (CT) derivatives have been shown to retain the adjuvant effects of the native LT or CT to enhance immune responses to unrelated antigens when coadministered orally or i.n. through induction of both Th1- and Th2-type immune responses (LT) or by promoting Th2-type immune responses (CT) (Freytag and Clements, 1999).

Double-layered SA11 2/6-VLPs or Wa 2/6-VLPs were evaluated using the i.n. immunization route (Yuan *et al.*, 2000). The rotavirus-specific ASC and memory B cell responses and the protection rates induced by the SA11 2/6-VLP vaccines are summarized in Table 1, along with historical data from pigs orally inoculated with virulent or 3 doses of attenuated Wa HRV (Yuan *et al.*, 2000, 2001a). In this VLP vaccine study, four groups of gnotobiotic pigs received 250 ug per dose of SA11 2/6-VLPs plus 5 ug mLT in 2 or 3 i.n. dose regimens (SA11 2/6-VLP+mLT IN 2x and 3x), 2 doses of SA11 2/6-VLPs alone (SA11 2/6-VLPs IN 2x), or mLT alone as controls, respectively. Intranasal inoculation of pigs with SA 11 2/6-VLPs plus mLT induced higher numbers of intestinal IgA and IgG ASC and memory B cell responses than three oral doses of live attenuated Wa HRV, but lower IgA and IgG ASC and memory B cells in the spleen (Table 1). The mLT adjuvant enhanced intestinal ASC and memory B cell responses to the VLP vaccines, as evident by the greatly increased numbers of intestinal IgA and IgG ASC and memory B cells in pigs vaccinated with SA11 2/6-VLPs+mLT compared to the SA11 2/6-VLPs alone at PID21 (PCD 0). However, SA11 2/6-VLPs administered i.n. in two or three dose regimens with or without mLT did not

360

confer protection to pigs against diarrhea upon challenge (Table 1). Similarly, 2 or 3 doses of Wa 2/6-VLPs with (Table 1) or without (data not shown) mLT also failed to induce protection. The failure of 2/6-VLP vaccines suggests that protective immunity to rotavirus diarrhea in neonatal pigs requires the presence of intestinal IgA neutralizing antibodies to the outer capsid rotavirus proteins, VP4 and VP7. This concept is supported by comparing studies of adult mice evaluating the protective efficacy of VLP vaccines for the induction of active protective immunity against rotavirus infection (O'Neal et al., 1998) and passive protection against diarrhea in the offspring of VLP-vaccinated dams (Coste et al., 2000). An i.n. SA11 2/6-VLP vaccine conferred complete protection against infection (shedding) in adult mice (O'Neal et al., 1998); however, protection against diarrhea in neonatal mice offspring was only conferred by i.n. immunization of dams with BRV RF strain (G6) 2/6/7-VLPs, but not RF 2/6-VLPs (Coste et al., 2000). The latter study also demonstrated that protection against diarrhea in neonatal mice was mediated by IgG and IgA antibodies present in the gut lumen (acquired via milk intake), but not IgG antibodies in the circulation (acquired transplacentally), and the protection was only against challenge with the homotypic rotavirus strain (same G type) (Coste et al., 2000).

Combined VLP vaccination regimens

Gut- and nasal-associated lymphoid tissues (GALT and NALT, respectively) are two major inductive sites for mucosal immune responses. The use of combined oral and i.n. vaccination routes stimulates multiple mucosal inductive sites, i.e., both GALT and NALT in contrast to the individual oral or i.n. vaccination route alone. Combined vaccine regimens exploiting the advantages of multiple mucosal immunization routes (oral and i.n.) and vaccine types (replicating virus and non-replicating VLPs) have the potential to optimally stimulate the mucosal immune system and confer maximal protective efficacy (Yuan et al., 2001b). Two combined vaccine regimens were evaluated in gnotobiotic pigs. In the first combined vaccine regimen group, gnotobiotic pigs were inoculated with one oral dose of attenuated Wa HRV at 3-5 days of age to prime the GALT inductive sites by live replicating virus. This was then followed by two i.n. boosting doses of Wa 2/6-VLP plus mLT (AttHRV/VLP2x) at 10-day intervals. In the second combined vaccine regimen group, pigs were inoculated with two i.n. doses of Wa 2/6-VLP plus mLT followed by one oral dose of live attenuated Wa HRV (VLP2x/AttHRV1x). A single oral dose of attenuated Wa HRV (AttHRV1x) and three i.n. doses of Wa 2/6-VLP plus mLT (Wa 2/6-VLP+mLT 3x) groups were included as controls. The ASC responses and protective efficacies of the combined regimens and the controls are summarized in Table 1. The AttHRV/VLP2x regimen induced similar mean numbers of intestinal IgA ASC as induced after virulent Wa HRV infection (135 versus 171 ASC per 5×10^5 MNC), and the IgA ASC numbers were significantly higher than those induced by a single oral dose of attenuated Wa HRV or three i.n. doses of Wa 2/6-VLP plus mLT at PID 28 (PCD 0) (Table 1).

Table 1

Rotavirus-specific intestinal and systemic IgA and IgG antibody-secreting cells (ASC) and memory B cell responses at postinoculation day 21-28/postchallenge day 0 and protection in gnotobiotic pigs[a]

Vaccine group	Mean No. ASC/5 x10^5 MNC[b]				Mean No. Memory B cells/5 x10^5 MNC				Protection rate[c] against challenge[d]	
	Intestinal lamina propria		Spleen		Intestinal lamina propria		Spleen		Diarrhea	Shedding
	IgA	IgG	IgA	IgG	IgA	IgG	IgA	IgG	(%)	(%)
Historical data (Yuan *et al.* 2001a):										
Attenuated Wa HRV 3x	8	3	2	10	8	55	135	3700	62	67
Virulent Wa HRV	171	114	4	13	364	3637	58	2288	87	100
VLPs (Yuan *et al.* 2000):										
SA11 2/6-VLPs IN 2x	6	4	0	1	9	3	1	3	0	0
SA11 2/6-VLP+mLT IN 2x and 3x	29	16	0	1	75	149	4	26	0	10
Combined regimens (Yuan *et al.* 2001b):										
AttHRV/VLP2x	135	67	6	4					44	58
VLP2X/AttHRV	16	11	4	15					25	17
Controls:										
Wa 2/6-VLP+mLT IN 3x	2	52	0	0					0	0
AttHRV1x oral	3	2	1	6					33	0

[a] Data summarized from Yuan *et al.* (2000, 2001a, 2001b). The ASC and memory B cell responses were measured with ELISPOT assay using semi-purified attenuated Wa HRV as antigen.
[b] Mononuclear cells
[c] Calculation of protection rate (Yuan *et al.*, 1996)
[d] The challenge virus for all the vaccine groups were virulent Wa HRV at a dose of 1 x10^6 ID$_{50}$.

The AttHRV/VLP2x regimen induced the highest mean numbers of intestinal IgA ASC at challenge among all rotavirus vaccines tested to date in gnotobiotic pigs including one to three oral doses of live attenuated Wa HRV, two or three i.m. doses of inactivated Wa HRV with incomplete Freund's adjuvant, and two or three i.n. doses of SA11 or Wa 2/6-VLPs with or without mLT (Yuan *et al.,* 1996, 1998, 2000, 2001a, 2001b). These results suggest that the combined immunization routes (oral priming followed by intranasal boosting), and vaccine types (live attenuated Wa HRV for priming followed by 2/6-VLPs plus mLT for boosting), were more effective in stimulating intestinal IgA ASC responses than similar doses of each individual vaccine. The protective efficacy of the AttHRV/VLP2x regimen was slightly lower than that of 3 doses of attenuated Wa HRV (Table 1) with a 58% protection rate against virus shedding and a 44% protection rate against diarrhea upon challenge of pigs with virulent Wa HRV. Although the AttHRV/VLP2x regimen did not confer a higher rate of protection compared to the AttHRV3x vaccine, the potential advantage of the combined AttHRV/VLP2x vaccine was indicated by the significantly higher ASC responses induced in the intestinal lymphoid tissues and the potential greater safety of using a non-replicating vaccine as boosters to reduce the risk of intussusception associated with a live oral reassortant HRV vaccine in infants (Kramarz *et al.,* 2001), as well to reduce other risks as mentioned earlier associated with live vaccines. The absence of the outer capsid VP4 and VP7 rotavirus neutralizing antigens from the 2/6 VLPs used for i.n. boosting in the AttHRV/VLP2x regimen presumably was again a determinant in the failure of this regimen to confer higher protection rates and in the complete failure of i.n. 2/6-VLP vaccines tested in pigs to confer any protection. If a triple-layered 2/4/6/7-VLP vaccine had been used for boosting in the combined regimen, a much higher protection rate might have been achieved, based on the high numbers of IgA ASC in the intestinal lymphoid tissues induced by this vaccine approach and based on the positive correlation between the numbers of IgA ASC in the intestinal lymphoid tissues induced by intact rotavirus and the degree of protection (Yuan *et al.,* 1996, 1998, 2001a). The VLP2x/AttHRV regimen was less efficacious than the AttHRV/VLP2x regimen in inducing intestinal ASC responses and protection against virus shedding or diarrhea, but more efficacious than the i.n. Wa 2/6-VLP+mLT alone (Table 1). The ineffectiveness in inducing intestinal ASC responses by the VLP2x/AttHRV vaccine regimen was possibly partially due to the functional compartmentalization of the primary mucosal inductive site (NALT) involved in this vaccination approach, suggesting that oral priming (at the portal of infection and the site of virus replication) with a live replicating vaccine is more effective for inducing intestinal IgA responses. Thus, the use of a replicating vaccine to prime lymphocytes in the major inductive site (GALT) followed by boosting with triple-layered VLPs (2/6/7 or 2/4/6/7) at a second mucosal inductive site (NALT) may be a highly effective approach to stimulate the mucosal immune system and induce protective immunity against rotavirus infection and disease.

VLP vaccines to induce passive immunity in cattle

Passive protection conferred by VLP vaccines against rotavirus diarrhea in newborn mice was reported in studies in which dams were immunized i.m. with BRV C-486 strain (P6[1]G6) 4/6-VLPs or 6/7-VLPs (Redmond *et al.*, 1993), or i.n. with BRV RF strain (P7[5]G6) 2/6/7-VLPs (Coste *et al.*, 2000). In our studies, BRV seropositive (due to almost universal exposure to field rotaviruses) pregnant cows were immunized with the two most prevalent BRV serotypes, IND (P[5]G6) and 2292B (P[11]G10) 2/4/6/7-VLPs, simian SA11 (P[2]G3) 2/4/6/7-VLPs or SA11 2/6-VLPs or inactivated SA11 or IND BRV to determine if VLPs could boost the titers of rotavirus antibodies in the colostrum and milk (Fernandez *et al.*, 1996; Saif and Fernandez, 1996; Kim *et al.*, 2002) and confer passive protection to newborn calves against diarrhea after challenge by inoculation with virulent BRV at two days of age (Fernandez *et al.*, 1998). Data from the SA11 and BRV VLP cow vaccination and SA11 VLP passive protection studies are summarized in Table 2. Cows were immunized twice using sequential i.m. and intramammary gland routes with the VLP vaccines in incomplete Freund's adjuvant. The BRV antibody titers, especially IgG1 in serum, colostrum and milk were significantly enhanced by immunization with the SA11 2/6-VLPs and SA11 and BRV 2/4/6/7-VLPs; however, virus neutralizing antibody titers were only boosted by the 2/4/6/7-VLPs. In addition, all the 2/4/6/7 VLP vaccines induced significantly higher antibody responses in cows than the inactivated virus vaccines. The effectiveness of SA11 2/4/6/7 VLPs for boosting the antibody responses to heterotypic BRV strains in rotavirus seropositive cows probably relates to two important factors. First, the existence of G3 BRV strains in cattle as reported previously (Hussein *et al.*, 1993) could account for the high antibody titers induced to the serotypically-related G3 (VP7) of the SA11 2/4/6/7 VLP vaccine. Second, previous observations from our laboratory and by others (Snodgrass *et al.*, 1984) have confirmed that even heterotypic strains of rotavirus such as simian SA11 can enhance both the homotypic (SA11 strain) and heterotypic (BRV strains) virus neutralizing antibody responses in rotavirus seropositive cows. Newborn calves fed pooled colostrum from cows vaccinated with the SA11 2/4/6/7-VLPs were also fully protected against diarrhea after inoculation with the virulent heterotypic BRV IND strain (Table 2, Fernandez *et al.*, 1998). In comparison, calves fed colostrum from cows vaccinated with SA11 2/6-VLPs or inactivated IND BRV were only partially protected against virus shedding and diarrhea. Evaluation of the passive protection conferred to calves by the BRV 2/4/6/7-VLPs including IND-VLP, 2292B-VLP or the combined IND and 2292B-VLP vaccines are in progress.

Table 2

Virus neutralization (VN) and ELISA IgG1 antibody titers to IND bovine rotavirus (BRV) in colostrum pools from cows vaccinated with SA11 VLPs or inactivated SA11 rotavirus vaccines and passive protection in calves fed each colostrum type and challenged with virulent IND BRV

Vaccine	Dose	BRV Antibody Titer		Percent Protection [a, b] against	
		VN	IgG1	Diarrhea	Virus Shedding
		Pooled colostrum (n=3-6)[a]			
SA11 2/6-VLP	250 ug	1,400	1,048,576	80%	0
SA11 2/4/6/7-VLP	250 ug	98,000	1,048,576	100%	60%
Inactivated SA11	5x10[7]PFU (pre-inactivation)	25,000	262,144	40%	20%
Control	-	1,800	16,384	0	0

[a] Data summarized from Fernandez et al. (1998) for 3-6 cows per vaccine group.

[b] Calves (n = 5 per group) were fed 1% supplemental pooled colostrum from birth to 7 days of age and challenged at 20-30 hr of age with virulent IND BRV.

In rotavirus seropositive cows, the virus neutralizing antibody titers in the colostrum were boosted by immunization with SA11 2/4/6/7-VLPs, but not 2/6-VLPs (Table 2, Fernandez et al., 1996). These findings concur with observations of pigs immunized with the combined attenuated Wa HRV and RF VP2/Wa VP6-VLP (2/6-VLP) vaccine regimen, that 2/6-VLP vaccines are effective to boost antibody responses in animals previously orally primed with live rotavirus (either through live oral vaccines in pigs or natural infection of cows; Yuan et al., 2001b; Fernandez et al., 1996). Our results further confirm that VLP vaccines are highly immunogenic when administered with adjuvants; however, 2/6-VLPs alone do not induce or boost neutralizing antibody responses (mice, pigs, calves) and do not confer substantial active or passive protection against rotavirus diarrhea in pigs or mice, respectively. To induce maximal protective immunity against rotavirus diarrhea, apparently antigens that elicit neutralizing antibodies (VP4 and VP7) are required for VLP vaccines or 2/6 VLP vaccines need to be used as booster vaccines in animals orally primed with live rotavirus.

Accumulating evidence from studies of neonatal pigs, cattle and mice have shown that the immunologic mechanisms differ for protective immunity induced against rotavirus infection in adults versus induction of protective immunity against rotavirus diarrhea in neonates. These findings have important implications for the design, formulation and evaluation of candidate rotavirus vaccines for the prevention of rotavirus diarrhea in human infants and young farm animals (See also Offit et al., Section II, Chapter 13 of this book).

References

Castrucci G, Frigeri F, Ferrari M, Cilli V, Caleffi F, Aldrovandi V, Nigrelli A (1984). The efficacy of colostrum from cows vaccinated with rotavirus in protecting calves to experimentally induced rotavirus infection. *Comp. Immunol. Microbiol. Infect. Dis.* **7:** 11-18.

Castrucci G, Ferrari M, Angelillo V, Rigonat F, Capodicasa L (1993). Field evaluation of the efficacy of Romovac 50, a new inactivated, adjuvanted bovine rotavirus vaccine. *Comp. Immunol. Microbiol. Infect. Dis.* **16:** 235-239.

Chen WK, Campbell T, VanCott J, Saif LJ (1995). Enumeration of isotype-specific antibody-secreting cells derived from gnotobiotic piglets inoculated with porcine rotaviruses. *Vet. Immunol. Immunopathol.* **45:** 265-284.

Ciarlet M, Crawford SE, Barone C, A. Bertolotti-Ciarlet A, Ramig RF, Estes MK, Conner ME (1998a). Subunit rotavirus vaccine administered parenterally to rabbits induces active protective immunity. *J. Virol.* **72:** 9233-9246.

Ciarlet M, Gilger MA, Barone C, McArthur M, Estes MK, Conner ME (1998b). Rotavirus disease, but not infection and development of intestinal histopathological lesions, is age restricted in rabbits. *Virology* **251:** 343-360.

Conner ME, Zarley CD, Hu B, Parsons S, Drabinski D, Greiner S, Smith R, Jiang B, Corsaro B, Barniak V, Madore HP, Crawford SE, Estes MK (1996). Virus-like particles as a rotavirus subunit vaccine. *J. Infect. Dis.* **174 (Suppl 1):** S88-S92.

Coste A, Sirard JC, Johansen K, Cohen J, Kraehenbuhl JP (2000). Nasal immunization of mice with virus-like particles protects offspring against rotavirus diarrhea. *J. Virol.* **74:** 8966-8971.

Crawford SE, Estes MK, Ciarlet M, Barone C, O'Neal CM, Cohen J, Conner ME (1999). Heterotypic protection and induction of a broad heterotypic neutralization response by rotavirus-like particles. *J. Virol.* **73:** 4813-4822.

Crawford SE, Labbé M, Cohen J, Burroughs MH, Zhou YJ, Estes MK (1994). Characterization of virus-like particles produced by the expression of rotavirus capsid proteins in insect cells. *J. Virol.* **68:** 5945-5952.

Davidson GP, Barnes GI (1979). Structural and functional abnormalities of the small intestine in infants and young children with rotaviral enteritis. *Acta Paediatr. Scand.* **68:** 181-186.

de Leeuw PW, Ellens DJ, Talmon FP, Zimmer GN, Kommerij R (1980). Rotavirus infections in calves: efficacy of oral vaccination in endemically infected herds. *Res. Vet. Sci.* **29:** 142-147.

Fernandez FM, Conner ME, Parwani AV, Todhunter D, Smith KL, Crawford SE, Estes MK, Saif LJ (1996). Isotype-specific antibody responses to rotavirus and virus proteins in cows inoculated with subunit vaccines composed of recombinant SA11 rotavirus core-like particles (CLP) or virus-like particles (VLP). *Vaccine* **14:** 1303-1312.

Fernandez F, Conner ME, Hodgins DC, Parwani A, Nielsen PR, Crawford SE, Estes MK, Saif LJ (1998). Passive immunity to bovine rotavirus in newborn calves fed colostrum supplements from cows immunized with recombinant SA11 rotavirus core-like particle (CLP) or virus-like particle (VLP) vaccines. *Vaccine* **16:** 507-516.

Franco MA, Greenberg HB (1999). Immunity to rotavirus infection in mice. *J. Infect. Dis.* **179:** S466-S469.

Franco MA, Greenberg HB (2000). Immunity to homologous rotavirus infection in adult mice. *Trends Microbiol.* **8:** 50-52.

Freytag LC, Clements JD (1999). Bacterial toxin as mucosal adjuvants. *Curr. Top. Microbiol. Immunol.* **236:** 215-236.

Hoshino Y, Saif LJ, Kang SY, Sereno MM, Chen WK, Kapikian AZ (1995). Identification of group A rotavirus genes associated with virulence of a porcine rotavirus and host range restriction of a human rotavirus in the gnotobiotic piglet model. *Virology* **209:** 274-280.

Hussein H, Parwani A, Rosen B, Lucchelli, Saif LJ (1993). Detection of rotavirus serotypes G1, G2, G3, and G11 in feces of diarrheic calves by using PCR-derived cDNA probes. *J. Clin. Microbiol.* **31:** 2491-2496.

Jiang B, Estes MK, Barone C, Barniak V, O'Neal CM, Ottaiano A, Madore HP, Conner ME (1999). Heterotypic protection from rotavirus infection in mice vaccinated with virus-like particles. *Vaccine* **17:** 1005-1013.

Kim YM (1980). Development of immunity in the piglet. *Birth Defects* **11:** 549-557.

Kim Y, Nielsen PR, Hodgins D, Chang KO, Saif LJ (2002). Lactogenic antibody responses in cows vaccinated with recombinant bovine rotavirus-like-particles (VLPs) of two serotypes or inactivated bovine rotavirus vaccines. *Vaccine* **20:** 1248-1258.

Kramarz P, France EK, Destefano F, Black SB, Shinefield H, Ward JI, Chang EJ, Chen RT, Shatin D, Hill J, Lieu T, Ogren JM (2001). Population-based study of rotavirus vaccination and intussusception. *Pediatr. Infect. Dis. J.* **20:** 410-416.

Matson DO, O'Ryan ML, Herrera I, Pickering LK, Estes MK (1993). Fecal antibody responses to symptomatic and asymptomatic rotavirus infections. *J. Infect. Dis.* **167:** 577-583.

Mehrazar K, Kim YB (1988). Total parenteral nutrition in germfree colostrum-deprived neonatal miniature piglets: A unique model to study the ontogeny of the immune response. *J. Parenter. Enter. Nutr.* **12:** 563-568.

O'Neal CM, Clements JD, Estes MK, Conner ME (1998). Rotavirus 2/6 virus-like particles administered intranasally with cholera toxin, *Escherichia coli* heat-labile toxin (LT), and LT-R192G induce protection from rotavirus challenge. *J. Virol.* **72:** 3390-3393.

O'Neal CM, Crawford SE, Estes MK, Conner ME (1997). Rotavirus virus-like particles administered mucosally induce protective immunity. *J. Virol.* **71:** 8707-8717.

O'Neal CM, Harriman GR, Conner ME (2000). Protection of the villus epithelial cells of the small intestine from rotavirus infection does not require immunoglobulin A. *J. Virol.* **74:** 4102-4109.

Redmond MJ, Ijaz MK, Parker MD, Sabara MI, Dent D, Gibbons E, Babiuk LA (1993). Assembly of recombinant rotavirus proteins into virus-like particles and assessment of vaccine potential. *Vaccine* **11:** 273-281.

Roy P (1996). Genetically engineered particulate virus-like structures and their use as vaccine delivery systems. *Intervirology* **39**: 62-71.

Roy P, French T, Erasmus BJ (1992). Protective efficacy of virus-like particles for bluetongue disease. *Vaccine* **10**: 28-32.

Saif LJ, Fernandez FM (1996). Group A rotavirus veterinary vaccines. *J. Infect. Dis.* 174 (Suppl) **1**: S98-106.

Saif LJ, Redman DR, Smith KL, Theil KW (1983). Passive immunity to bovine rotavirus in newborn calves fed colostrum supplements from immunized or nonimmunized cows. *Infect. Immun.* **41**: 1118-1131.

Saif LJ, Ward LA, Yuan L, Rosen BI, To TL (1996). The gnotobiotic pig as a model for studies of disease pathogenesis and immunity to rotavirus. *Arch. Virol. (Suppl)* **12**: 153-161.

Saif LJ, Yuan L, Ward LA, To TL (1997). Comparative studies of the pathogenesis, antibody immune responses, and homologous protection to porcine and human rotaviruses in gnotobiotic piglets. In: *Mechanisms in the Pathogenesis of Enteric Diseases* (Paul PS, Francis DH, Benfield DA, eds), pp 397-403, Plenum Press, New York.

Schaller JP, Saif LJ, Cordle CT, Candler E Jr, Winship TR, Smith KL (1992). Prevention of human rotavirus-induced diarrheal in gnotobiotic piglets using bovine antibody. *J. Infect. Dis.* **165**: 623-630.

Siadat-Pajouh M, Cai L (2001). Protective efficacy of rotavirus 2/6-virus-like particles combined with CT-E29H, a detoxified cholera toxin adjuvant. *Viral Immunol.* **14**: 31-47.

Snodgrass DR, Ojeh OK, Campbell I, Herring AJ (1984). Bovine rotavirus serotypes and their significance for immunization. *J. Clin. Microbiol.* **20**: 342-346.

Velazquez FR, Matson DO, Guerrero ML, Shults J, Calva JJ, Morrow AL, Glass RI, Pickering LK, Ruiz-Palacios GM (2000). Serum antibody as a marker of protection against natural rotavirus infection and disease. *J. Infect. Dis.* **182**: 1602-1609.

Ward RL, McNeal MM, Sheridan JF (1990). Development of an adult mouse model for studies on protection against rotavirus. *J. Virol.* **64**: 5070-5075.

Ward LA, Rosen BI, Yuan L, Saif LJ (1996a). Pathogenesis of an attenuated and a virulent strain of group A human rotavirus in neonatal gnotobiotic pigs. *J. Gen. Virol.* **77**: 1431-1441.

Ward LA, Yuan L, Rosen BI, To TL, Saif LJ (1996b). Development of mucosal and systemic lymphoproliferative responses and protective immunity to human group A rotaviruses in a gnotobiotic pig model. *Clin. Diagn. Lab. Immunol.* **3**: 342-350.

Yuan L, Ward LA, Rosen BI, To TL, Saif LJ (1996). Systemic and intestinal antibody-secreting cell responses and correlates of protective immunity to human rotavirus in a gnotobiotic pig model of disease. *J. Virol.* **70**: 3075-3083.

Yuan L, Kang SY, Ward LA, To TL, Saif LJ (1998). Antibody-secreting cell responses and protective immunity assessed in gnotobiotic pigs inoculated orally or intramuscularly with inactivated human rotavirus. *J. Virol.* **72**: 330-338.

Yuan L, Geyer A, Hodgins DC, Fan Z, Qian Y, Chang KO, Crawford SE, Parreño V, Ward LA, Estes MK, Conner ME, Saif LJ (2000). Intranasal administration of

2/6-rotavirus-like particles with mutant *Escherichia coli* heat-labile toxin (LT-R192G) induces antibody-secreting cell responses but not protective immunity in gnotobiotic pigs. *J. Virol.* **74:** 8843-8853.

Yuan L, Geyer A, Saif LJ (2001a). Short-term immunoglobulin A B cell memory resides in intestinal lymphoid tissues but not in bone marrow of gnotobiotic pigs inoculated with Wa human rotavirus. *Immunology* **103:** 188-198.

Yuan L, Iosef C, Azevedo MMS, Kim Y, Qian Y, Geyer A, Chang KO, Saif LJ (2001b). Protective immunity and antibody-secreting cell responses elicited by combined oral attenuated Wa human rotavirus and intranasal Wa 2/6-virus-like-particles with mutant *Escherichia coli* heat-labile toxin (LT-R192G) adjuvant in gnotobiotic pigs. *J. Virol.* **75:** 9229-9238.

Viral Gastroenteritis
U. Desselberger and J. Gray (editors)
© 2003 Elsevier Science B.V. All rights reserved

II, 15. DNA-based rotavirus vaccines

John E. Herrmann

*Division of Infectious Diseases, University of Massachusetts Medical School,
Worcester, MA 01605-2324*

Introduction

Group A rotaviruses are the most important cause of diarrhea in infants and young children worldwide. The exact number of deaths due to rotavirus infections is not known, but some estimates are in the hundreds of thousands per year in developing countries (Parashar *et al.*, 1998). Deaths resulting from rotavirus infections are uncommon in developed countries but the morbidity is similar to that of developing countries. In the United States they cause approximately 50,000 hospitalizations per year, with an estimated cost of $274 million in medical care (Parashar *et al.*, 1998). Because the morbidity in developed countries is similar to that in developing countries, it is thought that diarrhea in developing countries is not likely to be reduced only by improved sanitation and water supplies; thus, control measures will require effective vaccines (Bishop, 1993). The principal approach taken in the past as well as in more recent years has been active immunization with live, attenuated rotavirus.

Because the vaccines produced to date do not provide complete protection and may have adverse effects associated with them (see Offit *et al.*, Section II, Chapter 13 of this book), new approaches to immunization will continue to be needed. New approaches that have shown promise in animal models of rotavirus infection include the use of virus-like particles (VLPs) prepared from recombinant rotavirus proteins (Conner *et al.*, 1996; Fernandez *et al.*, 1998; Crawford *et al.*, 1999; Yuan and Saif, Section II, Chapter 14 of this book), chimeric rotavirus proteins (Choi *et al.*, 1999; Choi *et al.*, 2001), and plasmid DNA vaccines (Herrmann *et al.*, 1996a, b; Chen *et al.*, 1997, 1998; Herrmann *et al.*, 1999; Yang *et al.*, 2001).

Properties of DNA vaccines

DNA vaccines are plasmid DNAs (pDNAs) encoding specific proteins that can be expressed in cells of an inoculated host. The plasmids used are eukaryotic expression vectors, which contain the necessary elements for expression in eukaryotic cells.

The development of DNA vaccines was based on the finding that plasmid DNAs encoding beta-galactosidase, luciferase, or acetylcholine transferase expressed these enzymes in muscle cells after injection of the plasmid into the quadriceps of mice (Wolff *et al.*, 1990). Enzyme production in muscle tissue was detected up to two months after inoculation. Tang *et al.* (1992) described the production of antibodies to human growth hormone by gene gun immunization with DNA-coated gold microparticles, demonstrating that plasmid DNAs could also be administered by this route and elicit antibody production. The first DNA vaccines that demonstrated protective immunity in animal models were described in 1993 for plasmids encoding influenza virus hemagglutinin (Robinson *et al.*, 1993; Fynan *et al.*, 1993) and influenza virus nucleoprotein (Ulmer *et al.*, 1993).

DNA vaccines have an inherent advantage over traditional inactivated whole microbial vaccines and subunit vaccines in that these traditional vaccines do not provide endogenously synthesized proteins and generally do not elicit, or only weakly elicit, cytotoxic T lymphocyte (CTL) responses that are important in controlling many infections. The use of DNA vaccines encoding specific microbial proteins is a subunit vaccine approach that allows for the expression of immunizing proteins by host cells that take up the inoculated DNA. Expression of the immunizing proteins in host cells results in the presentation of normally processed proteins to the immune system which is important for raising immune responses against the native forms of proteins (Webster *et al.*, 1994). Expression of the immunogen in host cells also results in the immunogen having access to class I major histocompatibility complex presentation, which is necessary for eliciting CD8$^+$ CTL responses.

Most studies to date have administered DNA vaccines by intramuscular, intravenous, or intradermal injection, or by gene-gun delivery of DNA-coated particles into the epidermis. Of these methods, gene-gun delivery is the most efficient one, requiring less DNA to obtain protective immune responses (Fynan *et al.*, 1993; Robinson *et al.*, 1997). We later showed that oral administration of DNA vaccines encapsulated in poly (lactide-co-glycolide) (PLG) microparticles could also elicit protective immune responses (Chen *et al.*, 1998; Herrmann *et al.*, 1999). The concept of DNA vaccines for control of infectious diseases is still a relatively new one, but inoculation of DNA has been shown to generate protective immunity against several viral, bacterial, and parasitic agents in animal models (Robinson and Torres, 1997).

Immune responses to DNA vaccines

In animal models, DNA immunization has been shown to be an effective means of eliciting both humoral and cellular immunity against infectious agents (reviewed in Robinson and Torres, 1997). Gene gun bombardment of the surface of skin with DNA-coated gold beads and intramuscular (i.m.) injection of DNA vaccines have been the methods most often used for immunization. Although both methods have been found to generate both CTL and antibody responses, the two methods give different types of T helper (Th) cell responses. Th1-type responses have been generally obtained with i.m. immunizations, indicated by low levels of serum IgG1 antibodies, high levels of serum IgG2a antibodies, and Th1-related cytokine profiles. Th2-type responses have been obtained after gene gun immunization, indicated by a high ratio of serum IgG1 to IgG2a antibodies and Th2-related cytokine profiles (Pertmer *et al.*, 1996; Raz *et al.*, 1996; Feltquate *et al.*, 1997). Later studies have also shown that different forms of antigen expressed by DNA vaccines can modulate the type of Th response generated by i.m. immunization (Cardoso *et al.*, 1998; Torres *et al.*, 1999).

Rotavirus DNA vaccines

Rotavirus virions have a three-layered protein capsid. The protein-coated, RNA-containing core is coated by VP6, a protein that is antigenically relatively conserved among group A rotaviruses but does not elicit antibodies that neutralize rotavirus *in vitro*. The two outer capsid surface proteins, VP4 and VP7, elicit neutralizing antibodies. Because DNA vaccines can be made specific for individual rotavirus proteins, it permits the study of the immune responses and protection that might be generated by a particular rotavirus protein.

All of the reported studies to date specific to rotavirus DNA vaccines have been in the mouse model. In our first studies, we found that rotavirus DNA vaccines encoding either VP4, VP7, or VP6 gave partial protection after virus challenge when the vaccines were administered to mice by gene gun delivery of the DNA to the epidermis (Herrmann *et al.*, 1996a,b; Chen *et al.*, 1997). Both rotavirus-specific antibody responses and CTL responses were obtained with the DNA vaccines given by gene gun. We have since found that partial protection was obtained with VP6 DNA vaccines given i.m. (Yang *et al.*, 2001). Other reports also found that DNA vaccines encoding either VP4, VP7, or VP6 gave antibody responses in mice but did not give protective immunity (Choi *et al.*, 1997, 1998). A bovine rotavirus VP4 DNA vaccine given i.m. did not lead to detectable bovine rotavirus-specific antibody responses in mice, but did result in a Th1-type response as measured by IFN-gamma production (Suradhat *et al.*, 1997).

In a later study using gene gun delivery we found that less DNA was required to give protective immunity when given to the anorectal epithelium rather than to the abdominal epidermis, the site generally targeted (Chen *et al.*, 1999). Vaccines administered by either route elicited rotavirus-specific ELISA antibodies, and analysis of the IgG subtypes indicated that Th2-type responses were generated by both routes of administration, in contrast to Th1-type responses generated by live rotavirus infection. The protection obtained by immunization of the anorectal epithelium was greater than that by epidermal immunization at the same vaccine dose. These results suggested that mucosal immunization of DNA vaccines could be an effective means to generate protective immunity against mucosal pathogens. Both inoculation routes resulted in enhanced intestinal IgA responses after rotavirus challenge, but neither induced detectable intestinal IgA prior to challenge. Protective immune responses against rotavirus infections have been correlated with production of rotavirus-specific fecal IgA *in vivo* in human and porcine studies as well as in the murine model (Coulson *et al.*, 1992; Feng *et al.*, 1994; Matson *et al.*, 1993; Ward, 1996; Yuan *et al.*, 1996). Thus, induction of intestinal IgA is likely to be an important correlate in the development of rotavirus vaccines.

Oral delivery of rotavirus DNA vaccines

Targeting of rotaviruses to the gut-associated lymphoid tissue (GALT) by oral administration of an aqueous-based system of microencapsulated noninfectious rotaviruses generated serum IgG and intestinal IgA antibody responses (Khoury *et al.*, 1995). This suggested that mucosal targeting of DNAs expressing rotavirus proteins might also generate similar immune responses. A method for encapsulation of plasmid DNA encoding insect luciferase had been developed which permitted the DNA to be orally administered. Plasmid DNA was encapsulated in poly (lactide-co-glycolide) (PLG) microparticles, and oral administration of these PLG microparticles stimulated serum IgG, IgM, and IgA antibodies to luciferase (Jones *et al.*, 1997). Luciferase-specific IgA was also detected in stool samples, indicating a mucosal response. We adapted this method for encapsulating rotavirus protein-specific DNA vaccines and tested them in adult mice for induction of serum and mucosal antibody responses and for their ability to provide protection against rotavirus challenge.

We first demonstrated that a rotavirus VP6 DNA vaccine encapsulated in PLG microparticles would elicit rotavirus-specific serum antibody responses after oral administration of the microparticles. We further found that IgA was produced locally in the intestinal tract prior to virus challenge as determined by detection of antibody in feces (Chen *et al.*, 1998). This did not occur with DNA vaccines we gave i.m. (Yang *et al.*, 2001) or by gene gun (Herrmann *et al.*, 1996a; Chen *et al.*, 1999). Protection against virus challenge was demonstrated as well, indicating that the oral route was a promising approach for administration of DNA vaccines generally and rotavirus DNA vaccines specifically. In a second study, we showed that PLG-encapsulated rotavirus VP4 and VP7 DNA vaccines also provided protective immunity and generated similar types of antibody responses to those obtained with the PLG-encapsulated VP6 DNA vaccine (Herrmann *et al.*, 1999).

CTL responses

We did not assay for CTL responses in our work reported on PLG-encapsulated rotavirus DNA vaccines (Chen *et al.*, 1998; Herrmann *et al.*, 1999), but have since demonstrated in preliminary studies that CTL responses are generated (Herrmann, unpublished data). Cellular immunity was determined by measuring CTL activity in BALB/c mice that were vaccinated with PLG-encapsulated VP4 + VP7 DNA vaccine or were given live EDIM rotavirus, by methods we had used previously for determining CTL responses in BALB/c mice immunized with DNA vaccines given by gene gun (Herrmann *et al.*, 1996a). Co-encapsulation of VP4 +VP7 DNA vaccines in PLG was done as described (Herrmann *et al.*, 1999). Memory CTL activity was measured after *in vitro* stimulation. Splenic lymphocytes from DNA vaccine-immunized or EDIM rotavirus-infected adult BALB/c mice were stimulated *in vitro* with EDIM rotavirus, and the activity of these effector lymphocytes was measured in a chromium release assay using EDIM rotavirus-infected P815 (H-2^d) cells as target cells (Herrmann *et al.*, 1996a). The results are shown in Fig. 1.

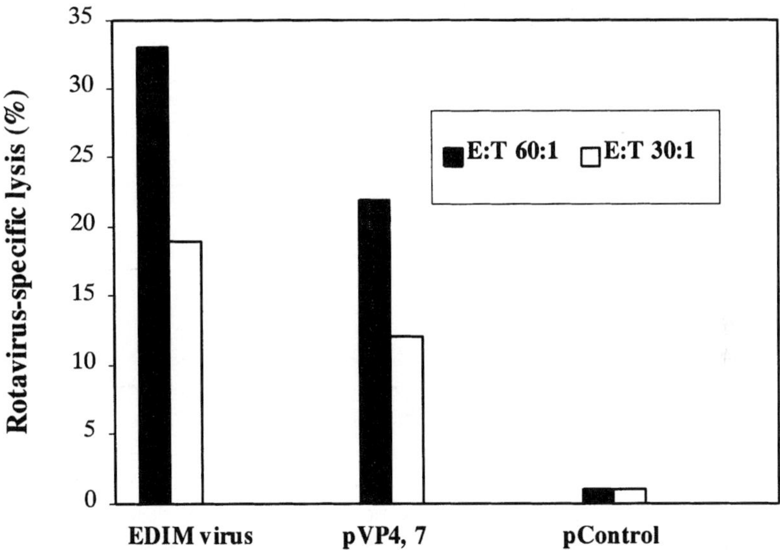

Fig.1. Rotavirus-specific CTL responses after oral immunization with a PLG-encapsulated rotavirus DNA vaccine. Adult BALB/c mice were immunized orally (by gavage) with PLG-encapsulated murine rotavirus VP4 plus VP7 DNA vaccine (pVP4, 7), PLG-encapsulated control plasmid DNA (pControl), or live EDIM rotavirus (EDIM virus). The effector cell to target cell (E:T) ratios used were 60:1 and 30:1.

As shown in the figure, rotavirus-specific CTL activity was generated by mice inoculated with PLG-encapsulated pVP4+pVP7 at effector cell to target cell (E:T) ratios of both 60:1 and 30:1. The magnitude of the response was not as high at E:T ratios of 60:1 as that obtained by infection with EDIM virus, but there were effective CTL responses. There was minimal lysis of uninfected target cells by the effectors (not shown). Thus, protection induced by the PLG-encapsulated rotavirus DNA vaccines could be due to cellular as well as antibody responses.

Cross protective rotavirus VP6 DNA vaccines

Because VP6 is conserved among all group A rotaviruses, it would be highly advantageous if a VP6 DNA vaccine prepared from one rotavirus strain would protect against other rotavirus strains, including those from other species. We tested this by use of a rotavirus VP6 DNA vaccine prepared from a bovine rotavirus gene (Indiana strain), looking for protection against challenge with a murine rotavirus (Herrmann, unpublished data). The preparation of the bovine VP6 DNA vaccine was previously described (Yang *et al.*, 2001) and in that report we found partial protection with this vaccine given i.m. For the PLG-encapsulated DNA vaccine studies, orally immunized adult BALB/c mice were challenged with EDIM rotavirus at 12 weeks post immunization. The results of these experiments, expressed as ELISA values for detection of rotavirus antigen shed in stools, are shown in Fig. 2. Significant reductions in virus antigen shed were noted on days 2-4. This demonstrated cross protection obtained by a VP6-based vaccine, and holds promise for a vaccine that is protective against various rotavirus serotypes.

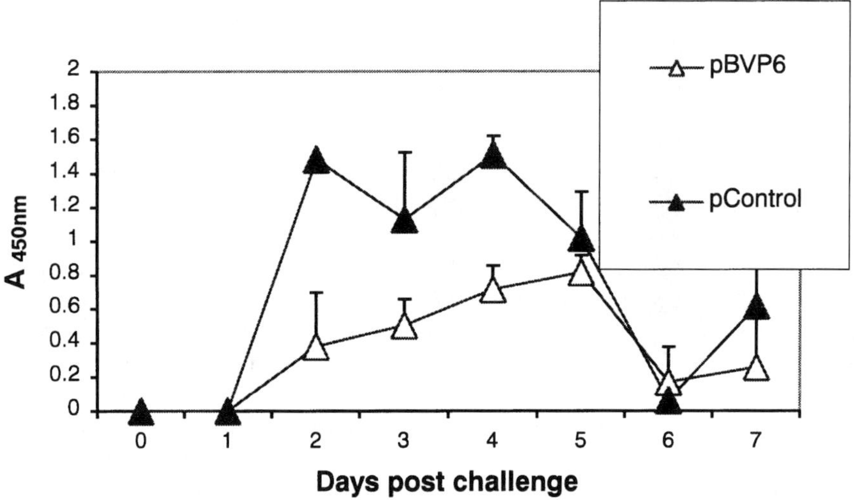

Fig. 2. Cross protection against rotavirus challenge obtained with a bovine rotavirus VP6 DNA vaccine. Adult BALB/c mice were challenged with 100 ID_{50} of EDIM rotavirus per mouse 12 weeks after receiving PLG-encapsulated bovine VP6 DNA vaccine (pBVP6) or PLG-encapsulated control plasmid DNA (pControl) by oral gavage. Virus shedding in feces, determined by an ELISA for detecting rotavirus antigen, is given as A_{450} values $\pm$ Standard Deviation. A positive test is one in which the A_{450} value is ≥ 0.1. There were significant differences (p<0.01) in viral shedding between the mice receiving the plasmid encoding bovine VP6 (N=5) and the control plasmid (N=4) on days 2-4.

In this and previous studies we have not determined the target cell for oral DNA vaccines, but based on work done with PLG-encapsulated peptide and other vaccines, it is likely to be the intestinal M cell (discussed in detail in Chen *et al.*, 1998 and Herrmann *et al.*, 1999). The mechanisms of protection seen with either the murine or bovine VP6 DNA vaccine are not known. Among potential mechanisms are cell-mediated immunity and IgA-mediated intracellular neutralization of virus that is undergoing assembly. In studies with rotavirus VP2 plus VP6 VLPs (O'Neal *et al.,* 1997), protective immunity was obtained by co-administration of cholera toxin, which is known to enhance both mucosal antibody responses and CTL responses. Thus, it is possible that either or both types of immune responses are involved. IgA-mediated intracellular virus neutralization has been shown for Sendai virus and influenza virus (Mazanec *et al.,* 1992; Mazanec *et al.,* 1995), and studies with IgA monoclonal antibodies to VP6 suggest that IgA mediated intracellular neutralization may also occur with rotaviruses (Burns *et al.,* 1996). Based on these findings and our demonstration of enhanced IgA responses in mice vaccinated orally with PLG-encapsulated DNA vaccines both before and after virus challenge (Chen *et al.,* 1998; Herrmann *et al.,* 1999), IgA-mediated intracellular neutralization in the intestinal mucosa may indeed be a factor in the protective immunity we have obtained. However, a direct proof of intracellular neutralization of rotavirus with VP6-specific IgA antibodies is still lacking.

DNA vaccines against other rotavirus proteins

VP2 is an internal protein on the innermost capsid. It accounts for approximately 15% of the virion protein mass and appears to act as a scaffold for the proper assembly of the viral core (Lawton *et al.,* 1997). When VP2 is coexpressed with VP6 in insect cells with baculovirus vectors, particles similar to native single-shelled particles are spontaneously formed (Lawton *et al.,* 1997). Virus-like particles (VLPs) have been formed with VP4, VP6, and VP7 as well, when VP2 is present. Subsequently it has been found that VLPs containing VP2, VP6 or VP2, VP6, VP7 were immunogenic in mice and, when cholera toxin or other adjuvants were given, could protect mice against virus challenge (O'Neal CM *et al.,* 1997, 1998; Jiang *et al.,* 1999). We have not required the use of cholera toxin or other adjuvants to get protective immune responses to our DNA vaccines, likely to be due to the manner of protein expression discussed above. To test VP2 as a protective antigen, a VP2 DNA vaccine was prepared from a bovine rotavirus strain (RF) (obtained from Dr. Jean Cohen, INRA, Jouy-en-Josas, France). If VLPs are formed by coinoculation of DNA vaccines encoding VP2 and DNA vaccines encoding other rotavirus proteins, it might result in enhanced immunity. In initial tests we also looked to see if PLG-encapsulated pDNA encoding VP2 might generate protection when given alone.

DNA vaccines were prepared using the same vector and methods previously described (Herrmann *et al.,* 1996a,b; Yang *et al.,* 2001). We encapsulated pVP2 alone or co-encapsulated pVP2 and pVP6 in PLG microparticles, and tested the vaccines for induction of protective immunity (Herrmann, unpublished data). Plasmid DNA encoding the rotavirus proteins was encapsulated in PLG microparticles by the solvent extraction technique as previously described (Chen *et al.,* 1998; Herrmann *et al.,* 1999). For immunization, BALB/c mice were inoculated orally (by gavage) with PLG-encapsulated DNA vaccines. The PLG microparticles were suspended in a solution of 0.1 M sodium bicarbonate in distilled water, pH 8.5, and given at 0.5 ml/mouse. The DNA dose administered was approximately 90 micrograms DNA per mouse.

376

Mice were challenged with 100 ID_{50} of EDIM rotavirus at 12 weeks post immunization to determine if the immunizations had provided protection. The challenge virus used was given by oral gavage. Protection was assessed by testing for the reduction of rotavirus antigen shedding in stools in an ELISA as we have done previously (Herrmann *et al.,* 1999). The results are shown in Fig. 3. The PLG-encapsulated pVP2+pVP6 DNA vaccine gave protection equivalent to that found with the PLG-encapsulated murine VP6 DNA vaccine given alone, but it was difficult to determine if there was any contribution of pVP2 to the protection seen in this first attempt. Also, there was no protection by the PLG-encapsulated VP2 DNA vaccine when given alone.

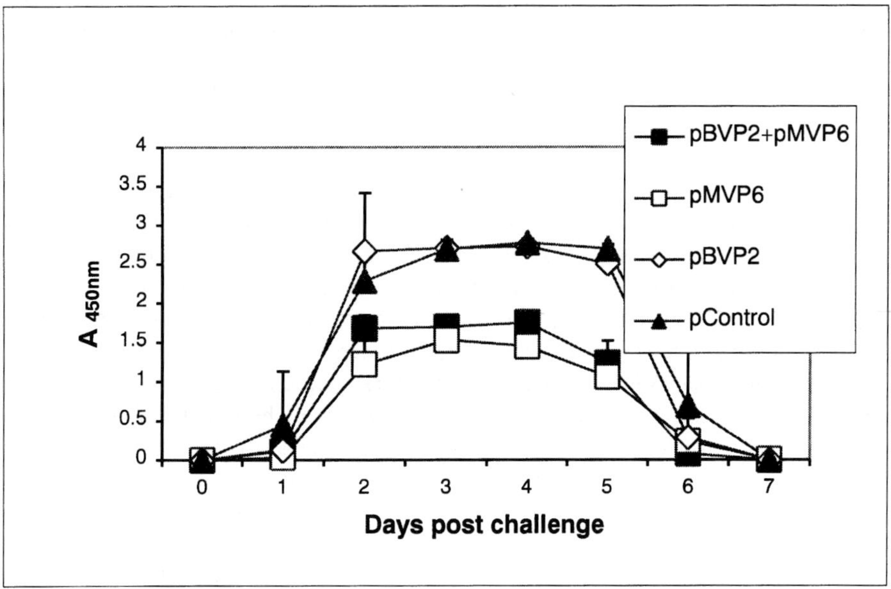

Fig. 3. Protection against rotavirus challenge in adult BALB/c mice immunized with a VP2 DNA vaccine alone and in combination with a VP6 DNA vaccine. Mice were challenged with 100 ID_{50} of EDIM rotavirus per mouse 12 weeks after receiving (by oral gavage) PLG-encapsulated murine VP6 DNA vaccine (pMVP6), bovine VP2 DNA vaccine (pBVP2), co-encapsulated pBVP2 + pMVP6, or control plasmid DNA (pControl). Virus shedding in feces, determined by an ELISA for detecting rotavirus antigen, is given as A_{450} values $\pm$ Standard Deviation. A positive test is one in which the A_{450} value is ≥ 0.1. There were significant differences in viral shedding between the mice receiving pMVP6 or pBVP2 + pMVP6 and the control plasmid or pBVP2 on days 2-5. N=5 mice for pBVP2+pMVP6 and pBVP2; N=6 mice for pMVP6 and pControl.

Application of DNA vaccines

Although the list of DNA vaccines tested in animal models is extensive and continues to increase, this has not translated into DNA vaccines available for routine use in humans. One concern has been the safety of DNA vaccines, and the other has been effectiveness. Safety concerns of DNA vaccines include two major areas, possible integration of vaccine DNA into host cell genomes and the generation of antibodies to vaccine DNA. Studies to date have not shown development of anti-vaccine DNA antibodies (Mor *et al.*, 1997). Integration of plasmid DNA targeted to B cells (Xiong *et al.*, 1997) has been shown, but the consequences of this are unknown. Martin *et al.* (1999) did not detect integration of vaccine DNA after i.m. injection of mice, and based on the amount of pDNA that was associated with purified genomic DNA, it was not considered a potential safety threat. Others have also concluded that the risk of integration of DNA into the host cell genome is low (Ledwith *et al.*, 2000). There has been another concern about possible induction of tolerance. It was reported (Mor *et al.*, 1996) that a DNA vaccine against the circumsporozoite protein of the malaria parasite that induced protection in adult mice induced tolerance when given to 2-6 day old BALB/c mice. However, tolerance in viral infections has not been demonstrated. Immunization of neonatal mice with plasmid DNA encoding rabies virus glycoprotein did not induce tolerance (Wang *et al.*, 1997), and immunization of newborn mice with plasmid DNA encoding influenza virus nucleoprotein developed protective immunity as measured by reduction of virus lung titer after challenge (Bot *et al.*, 1996).

The clinical effectiveness of DNA vaccines remains to be determined, and there have been so far only a few reports of clinical trials from the standpoint of protection induced by DNA vaccines or their therapeutic use. One encouraging report has been the finding that a hepatitis B virus DNA vaccine inoculated into volunteers by gene gun resulted in raising of serum antibodies to hepatitis B virus surface antigen at antibody concentrations considered to be protective (Roy *et al.*, 2000). How well DNA vaccines will be accepted and used in the future remains to be determined. For purposes of investigations into mechanisms of immunity to rotaviruses, the use of plasmid DNAs encoding various rotavirus proteins provides an approach to examine immune responses and protection induced by specific rotavirus proteins. It may also provide an approach to new rotavirus vaccines. As to the best route for administration of rotavirus DNA vaccines, we have shown that protective responses can be obtained using various routes. Although we have not made direct comparisons, the results we have found to date with rotavirus DNA vaccines encapsulated in PLG microparticles suggest that this route is the most effective one, and it combines the potential advantages of DNA-based vaccination with the ease of administration by the oral route.

Acknowledgements

The studies reported here were supported by grant R01 AI47393 from the National Institutes of Health.

References

Bishop, R.F. (1993). Development of candidate rotavirus vaccines. *Vaccine* **11**, 247-254.

Bot, A., Bot, S., García-Sastre, A., Bona, C. (1996). DNA immunization of newborn mice with a plasmid-expressing nucleoprotein of influenza virus. *Viral Immunol.* **9**, 207-210.

Burns, J. W., Pajouh, M.S., Krishnaney, A.A., Greenberg, H.B. (1996). Protective effect of rotavirus VP6-specific IgA monoclonal antibodies that lack neutralizing activity. *Science* **272**, 104-107.

Cardoso, A.I., Sixt, N., Vallier, A., Fayolle, J., Buckland, R., Wild, T.F. (1998). Measles virus DNA vaccination: antibody isotype is determined by the method of immunization and by the nature of both antigen and the coimmunized antigen. *J. Virol.* **72**, 2516-2518.

Chen, S.C., Fynan, E.F., Robinson, H.L., Lu, S., Greenberg, H.B., Santoro, J.C., Herrmann, J.E. (1997). Protective immunity induced by rotavirus DNA vaccines. *Vaccine* **15**, 899-902.

Chen, S.C., Jones, D.H., Fynan, E.F., Farrar, G.H., Clegg, J.C., Greenberg, H.B., Herrmann, J.E. (1998). Protective immunity induced by oral immunization with a rotavirus DNA vaccine encapsulated in microparticles. *J. Virol.* **72**, 5757-5761.

Chen, S.C., Fynan E.F., Greenberg H.B., Herrmann J.E. (1999). Immunity obtained by gene-gun inoculation of a rotavirus DNA vaccine to the abdominal epidermis or anorectal epithelium. *Vaccine* **17**, 3171-3176.

Choi, A.H., Knowlton, D.R., McNeal, M.M., Ward, R.L. (1997). Particle bombardment-mediated DNA vaccination with rotavirus VP6 induces high levels of serum rotavirus IgG but fails to protect mice against challenge. *Virology* **232**, 129-138.

Choi, A.H., Basu, M., Rae, M.N., McNeal, M.M., Ward, R.L. (1998). Particle-bombardment-mediated DNA vaccination with rotavirus VP4 or VP7 induces high levels of serum rotavirus IgG but fails to protect mice against challenge. *Virology* **250**, 230-240.

Choi, A.H., Basu, M., McNeal, M.M., Clements, J.D., Ward, R.L. (1999). Antibody-independent protection against rotavirus infection of mice stimulated by intranasal immunization with chimeric VP4 or VP6 protein. *J. Virol.* **73**, 7574-7581.

Choi, A.H., Basu, M., McNeal, M.M., Flint, J, VanCott, J.L., Clements, J.D., Ward, R.L. (2000). Functional mapping of protective domains and epitopes in the rotavirus VP6 protein. *J. Virol.* **74**, 11574-11580.

Conner, M.E., Crawford, S.E., Barone, C., O'Neal, C., Zhou, Y.J., Fernandez, F.M., Parwani, A.V., Saif, L.J., Cohen, J., Estes, M.K. (1996). Rotavirus subunit vaccines. *Arch. Virol. Suppl.* **12**, 199-206.

Coulson, B. S., Grimwood, K., Hudson, I.L., Barnes, G.L., Bishop, R.F. (1992). Role of coproantibody in clinical protection of children during reinfection with rotavirus. *J. Clin. Microbiol.* **30**, 1678-1684

Crawford, S.E., Estes, M.K., Ciarlet, M., Barone, C., O'Neal, C.M., Cohen, J., Conner, M.E. (1999). Heterotypic protection and induction of a broad heterotypic neutralization response by rotavirus-like particles. *J. Virol.* **73**, 4813-5822.

Feng, N., J. W. Burns, J.W., L. Bracy, L., Greenberg, H.B. (1994). Comparison of mucosal and systemic humoral immune responses and subsequent protection in mice orally inoculated with a homologous or a heterologous rotavirus. *J. Virol.* **68**, 7766-7773.

Feltquate, D.M., Heaney, S., Webster, R.G., Robinson, H.L. (1997). Different T helper cell types and antibody isotypes generated by saline and gene gun DNA immunization. *J. Immunol.* **158**, 2278-2284

Fernandez, F.M., Conner, M.E., Hodgins, D.C., Parwani, A.V., Nielsen, P.R., Crawford, S.E., Estes, M.K., Saif, L.J. (1998). Passive immunity to bovine rotavirus in newborn calves fed colostrum supplements from cows immunized with recombinant SA11 rotavirus core-like particle (CLP) or virus-like particle (VLP) vaccines. *Vaccine* **16**, 507-516.

Fynan, E.F., Webster, R.G., Fuller, D.H., Haynes, J.R., Santoro, J.C., Robinson, H.L. (1993). DNA vaccines: Protective immunizations by parenteral, mucosal, and gene-gun inoculations. *Immunology* **90**, 11478-11482.

Herrmann, J.E., Chen, S.C., Fynan, E.F., Santoro, J.C., Greenberg, H.B., Robinson, H.L. (1996a) DNA vaccines against rotavirus infections. *Arch. Virol. Suppl.* **12**, 207-215.

Herrmann, J.E., Chen, S.C., Fynan, E.F., Santoro, J.C., Greenberg, H.B., Wang, S., Robinson, H.L. (1996b). Protection against rotavirus infections by DNA vaccination. *J. Infect. Dis.* **174** (Suppl. 1), S93-S97.

Herrmann, J.E., Chen, S.C., Jones, D.H., Tinsley-Bown, A., Fynan, E.F., Greenberg, H.B., Farrar, G.H. (1999). Immune responses and protection obtained by oral immunization with rotavirus VP4 and VP7 DNA vaccines encapsulated in microparticles. *Virology* **259**, 148-153.

Jiang, B., Estes, M.K., Barone, C., Barniak, V., O'Neal, C.M., Ottaiano, A., Madore, H.P., Conner, M.E. (1999). Heterotypic protection from rotavirus infection in mice vaccinated with virus-like particles. *Vaccine* **17**, 1005-1113.

Jones, D. H., Corris, S., McDonald, S., Clegg, J.C.S., Farrar, G.H. (1997). Poly (DL-lactide co-glycolide)-encapsulated plasmid DNA elicits systemic and mucosal antibody responses to encoded protein after oral administration. *Vaccine* **15**, 814-817.

Khoury, C. A., Moser, C.A., Speaker, T.J., Offit, P.A. (1995). Oral inoculation of mice with low doses of microencapsulated, noninfectious rotavirus induces virus-specific antibodies in gut-associated lymphoid tissue. *J. Infect. Dis.* **172**, 870-874.

Lawton, J.A., Zeng, C.Q.Y., Mukhurjee, S.K., Cohen, J., Estes, M.K., Prasad, B.V.V. (1997). Three-dimensional structural analysis of recombinant rotavirus-like particles with intact and amino-terminal-deleted VP2: implications for the architecture of the VP2 capsid layer. *J. Virol.* **71**, 7353-7360.

Ledwith, B.J., Manam, S., Troilo, P.J., Barnum, A.B., Pauley, C.J., Griffiths, T.G. II, Harper, L.B., Beare, C.M., Bagdon, W.J., Nichols, W.W. (2000). Plasmid DNA vaccines: investigation of integration into host cellular DNA following intramuscular injection in mice. *Intervirology* **43**, 258-72.

Martin, T., Parker, S.E., Hedstrom, R., Le, T., Hoffman, S.L., Norman, J., Hobart, P., Lew, D. (1999). Plasmid DNA malaria vaccine: the potential for genomic integration after intramuscular injection. *Hum. Gene Ther.* **10**, 759-768.

Matson, D. O., O'Ryan, M.L., Herrera, I., Pickering, L.K., Estes, M.K. (1993). Fecal antibody responses to symptomatic and asymptomatic rotavirus infections. *J. Infect. Dis.* **167**, 577-583

Mazanec, M. B., Kaetzel, C.S., Lamm, M.E., Fletcher, D.R., Nedrud, J.G. (1992). Intracellular neutralization of virus by immunoglobulin A antibodies. *Proc. Natl. Acad. Sci. U.S.A.* **89**, 6901-6905.

Mazanec, M. B., C. L. Coudret, C.L., Fletcher, D.R. (1995). Intracellular neutralization of influenza virus by immunoglobulin A anti-hemagglutinin monoclonal antibodies. *J. Virol.* **69**, 1339-1343.

Mor, G., Yamshchikov, G., Sedegah, M., Takeno, M., Wang, R., Houghten, R.A., Hoffman, S., Klinman, D.M. (1996). Induction of neonatal tolerance by plasmid DNA vaccination of mice. *J. Clin. Invest.* **98**, 2700-2705.

Mor, G., Singla, M., Steinberg, A.D., Hoffman, S.L., Okuda, K., Klinman, D.M. (1997). Do DNA vaccines induce autoimmune disease? *Hum. Gene Ther.* **8**, 293-300.

O'Neal, C.M., Crawford, S.E., Estes, M.K., Conner, M.E. (1997). Rotavirus virus-like particles administered mucosally induce protective immunity. *J. Virol.* **71**, 8707-8717.

O'Neal, C.M., Clements, J.D., Estes, M.K., Conner, M.E. (1998). Rotavirus 2/6 virus-like particles administered intranasally with cholera toxin, Escherichia coli heat-labile toxin (LT), and LT-R192G induce protection from rotavirus challenge. *J. Virol.* **72**, 3390-3393.

Parashar, U.D., Bresee, J.S., Gentsch, J.R., Glass, R.I. (1998). Rotavirus. *Emerging Infect. Dis.* **4**, 561-570.

Pertmer, T.M., Roberts, T.R., Haynes, J.R. (1996). Influenza virus nucleoprotein-specific immunoglobulin G subclass and cytokine responses elicited by DNA vaccination are dependent on the route of vector DNA delivery. *J. Virol.* **70**, 6119-6125.

Raz, E., Tighe, H., Sato, Y., Corr, M., Dudler, J.A., Roman, M., Swain, S.L., Spiegelberg, H.L., Carson, D.A. (1996). Preferential induction of a Th1 immune response and inhibition of specific IgE antibody formation by plasmid DNA immunization. *Proc. Natl. Acad. Sci. U.S.A.* **93**, 5141-5145.

Robinson, H.L., Hunt, L.A., Webster, R.G. (1993). Protection against a lethal influenza virus challenge by immunization with a haemagglutinin-expressing plasmid DNA. *Vaccine* **11**, 957-960.

Robinson, H.L., Torres, C.A. (1997). DNA vaccines. Semin. Immunol. **5**, 271-283.

Robinson, H.L., Boyle, C.A., Feltquate, D.M., Morin, M.J., Santoro, J.C., Webster, R.G. (1997). DNA immunization for influenza virus: studies using haemagglutinin and nucleoprotein-expressing DNAs. *J. Infect. Dis.* 176 (Suppl. 1), S50-S55.

Roy, M.J., Wu, M.S., Barr, L.J., Fuller, J.T., Tussey, L.G., Speller, S, Culp, J., Burkholder, J.K., Swain, W.F., Dixon, R.M., Widera, G., Vessey, R., King, A.,

Ogg, G., Gallimore, A., Haynes, J.R., Heydenburg Fuller, D. (2000). Induction of antigen-specific CD8+ T cells, T helper cells, and protective levels of antibody in humans by particle-mediated administration of a hepatitis B virus DNA vaccine. *Vaccine* **19**, 764-78.

Suradhat, S., Yoo, D., Babiuk, L.A., Griebel, P., Baca-Estrada, M.E. (1997). DNA immunization with a bovine rotavirus VP4 gene induces a Th1-like immune response in mice. *Viral Immunol.* **10**, 117-127.

Tang, D.C., DeVit, M., Johnston, S.A. (1992). Genetic immunization is a simple method for eliciting an immune response. Nature **356**, 152-154.

Torres, C.A., Yang, K., Mustafa, F., Robinson, H.L. (1999). DNA immunization: Effect of secretion of DNA–expressed hemagglutinins on antibody responses. *Vaccine* **18**, 805-814.

Ulmer, J.B., Donnelly, J.J., Parker, S.E., Rhodes, G.H., Felgner, P.L., Dwarki, V.J., Gromkowski, S.H., Deck, R.R., DeWitt, C.M., Friedman, A., Hawe, L.A., Leander, K.R., Martinez, D., Perry, H.C., Shiver, J.W., Montogmery, D.L., Liu, M.A. (1993). Heterologous protection against influenza by injection of DNA encoding a viral protein. *Science* **259**, 1745-1749.

Wang, J., Xiang, Z.., Pasquini, S., Ertl, H.C. (1997). Immune response to neonatal genetic immunization. *Virology* **228**, 278-284.

Ward, R. L. (1996). Mechanism of protection against rotavirus in humans and mice. *J. Infect. Dis.* **174** (Suppl. 1), S51-S58.

Webster, R.G., Fynan, E.F., Santoro, J.C., Robinson, H.L. (1994). Protection of ferrets against influenza challenge with a DNA vaccine to the haemagglutinin. *Vaccine* **12**, 1495-1498.

Wolff, J.A., Malone, R.W., Williams, P., Chong, W., Acsadi, G., Jani, A., Felgner, P.L. (1990). Direct gene transfer into mouse muscle *in vivo*. *Science* **247**, 1465-1468.

Xiong, S., Gerloni, M., Zanetti, M. (1997). *In vivo* role of B lymphocytes in somatic transgene immunization. Proc.Natl. Acad. Sci. U.S.A. **94**, 6352-6357.

Yang, K.J., Wang, S., Chang, K-O, Lu, S., Saif, L.J., Greenberg, H.B., Brinker, J.P., Herrmann, J.E. (2001). Immune responses and protection obtained with rotavirus VP6 DNA vaccines given by intramuscular injection. *Vaccine* **19**, 3285-3291.

Yuan, L., Ward, L.A., Rosen, B.I., To, T.L., Saif, L.J. (1996). Systematic and intestinal antibody-secreting cell responses and correlates of protective immunity to human rotavirus in a gnotobiotic pig model of disease. *J. Virol.* **70**, 3075-3083.

Viral Gastroenteritis
U. Desselberger and J. Gray (editors)

SECTION III

Enteric adenoviruses

Introduction

Human adenoviruses are members of the *Adenoviridae* family and occur in 51 distinct serotypes, classified into six different subgroups (A-F; recently termed species, van Regenmortel *et al.*, 2000) according to immunological, biochemical and biological differences. Within those, Adenoviruses of subgroups F, serotypes 40 and 41, have been found to be regularly associated with gastroenteritis in infants and young children (for general review see Ruuskanen *et al.*, 1997; Russell, 1998; Wadell, 2000).

Adenoviruses are non-enveloped icosahedral particles of a diameter of 80 nm. The viral capsid is formed by 252 capsomeres termed hexons (as most of them are surrounded by six other capsomers); each hexon is formed by a trimer of polypeptide 2. From 12 capsomers which are located at the vertices of the virus particles, antennae-like projections extend (fibre). These morphological units are called pentons (as they are surrounded by 5 capsomers and exhibit 5 fold symmetry). Each penton capsomer is formed of 5 copies of polypeptide 3. The fibres are glycoproteins consisting of trimers of polypeptide 4. The genome consists of a linear double-stranded DNA molecule of approximately 35,000 base pairs (bp) encoding eight different transcription units which produce early proteins (E1A, E1B, E2A, E3, E4), intermediate proteins (P9 and 4A2) and late proteins (L1-L5).

Adenovirus particles are stable at low pH, and unaffected by bile and gut enzymes, allowing them to replicate to high titers in the gastrointestinal tract.

Viral replication starts with binding of virions to receptors (members of the immunoglobulin superfamily) on the surface of host cells, mediated through the head region of the fibre protein. Proteins of the integrin family have been implicated as co-receptors (Wickham *et al.*, 1994). After uptake by receptor-mediated endocytosis the pentons are degraded in the acidic environment of endosomes. The viral core containing DNA enters the nucleus in which various mRNA are produced and released in a strictly controlled fashion. Early protein E1A acts as a transcriptional activator, and E1B binds to p53, thus inhibiting p53-mediated tumour suppression and inducing the host cell to enter the S phase of the cell cycle. E3 protein inhibits TNF-mediated apoptosis and binds to the heavy chain of MHC class 1 molecules, thus impairing its transport to the cell surface and helping immune evasion of the virus-infected cell. It was in the adenovirus replication system that the fundamental mechanism of mRNA splicing was discovered by JE Darnell Jr, P. Sharp, TR Broker and their colleagues, (Bachenheimer and Darnell, 1975; Berget *et al.*, 1977; Chow *et al.*, 1977, 1979). Viral

DNA synthesis starts at 7 hours post-infection, and synthesis of viral structural proteins (capsid components) 2 hours after that. At the same time inhibition of host cell protein synthesis occurs. Newly formed viral structural polypeptides are transported into the nucleus where the formation of particles starts, identifiable as paracrystalline arrays in intranuclear inclusion bodies. Viral morphogenesis begins with the formation of capsomers from monomeric polypeptide subunits. Hexon base and fibre are combined to form the penton unit. A *cis*-acting packaging element has been localized in the left end of the viral DNA, immediately upstream of the E1A transcription unit which promotes preferential encapsidation of viral DNA into viral capsid. Newly synthesized capsids mature through a number of proteolytic cleavage steps involving five viral polypeptide precursors. Virus particles are released from infected cells, aided by a very late protein of the E3 transcription unit which promotes cell lysis. Up to 10^5 particles are assembled per cell within 30 hours after infection. Enteric adenoviruses have a very restricted host cell range *in vitro* compared to adenoviruses of other subgroups. F Stevenson and V Mautner (Section III, Chapter 1) have explored this by investigating the EIA genome region and found that the EIA promoter is weak in most cell lines compared to EIA promotors of adenovirus type 5 (subgroup C).

Adenoviruses replicate in the epithelia of the human respiratory and gastrointestinal tracts as well as in the conjunctiva and in lymphocytes. They are most easily recovered from nasopharyngeal aspirates, throat swabs, conjunctival swabs, and faeces. Various viral factors have been implicated in pathogenesis. Whilst the pentons are directly cytotoxic, several early proteins play a role in the ability of adenoviruses to persist in lymphoid cells. These proteins counteract TNF, downregulate the expression of MHC Class I molecules and counteract apoptosis. The pathogenesis of persistent (possibly life-long) and often symptomless adenovirus infections is poorly understood (Horwitz, 2001).

Whilst there are good animal models for respiratory adenovirus infections (mouse, Ginsberg *et al.*, 1991; cotton rat, Ginsberg and Prince, 1994), such a model is not available for the enteric adenoviruses. The immune response is both humoral and cell-mediated (Nishio *et al.*, 1992). The correlates of protection have not been definitely identified, but humoral neutralizing antibodies clearly play a significant role to prevent reinfection with the same type.

Clinically adenoviruses, mainly those of subgroups B and C, cause respiratory tract infections (tonsillitis, laryngitis, bronchitis and pneumonia) but also otitis media and keratoconjunctivitis. However, there are many subclinical infections. In the past enteric adenovirus (subgroup F) infections were thought to be the second most common cause of infantile diarrhoea (after rotaviruses; Krajden *et al.*, 1990; Ruuska and Vesikari, 1991). More recently Norwalk- and Sapporo-like viruses have been recognised as more frequent (see Section IV). Of the adenoviruses detected in faeces, 30-80% are represented by types 40 and 41 (Uhnoo *et al.*, 1984; Krajden *et al.*, 1990; Lew *et al.*, 1991). In large surveys, adenovirus infections were associated with diarrhoea in chidren in approximately 3% of all cases tested (Bon *et al.*, 1999; Waters *et al.*, 2000). Adenovirus infections with gastroenteritis are a frequent event after bone marrow transplantation (Hale *et al.*, 1999; Chakrabarti *et al.*, 2000)

Adenovirus-caused diarrhoeas are mainly found in children below the age of 2 years. Diarrhoea occurs after a short incubation period of about 2 days, more often than not accompanied by fever and vomiting. The diarrhoeas last on average longer (8 days, range 3-11 days) than diarrhoeas caused by rotaviruses. Usually, adenovirus gastroenteritis is mild but severe colitis has been found in immunocompromised patients (Krajden *et al.*, 1990; Janoff *et al.*, 1991). Many enteric adenovirus infections are asymptomatic (Van *et al.*, 1992).

The diagnosis by electron-microscopy or ELISA is not difficult, as large numbers of particles are shed during the acute stage of diarrhoea. Enteric adenoviruses grow best in Graham 293 cells (human embryonic kidney cell line transformed by adenovirus type 5 DNA). The sensitivity of enzyme immunoassays for enteric adenoviruses is 85-100% compared to that of electron microscopy. Nucleic acid amplification techniques can also be applied to the diagnosis of adenovirus infection.

Adenovirus infections occur worldwide as epidemic, endemic or sporadic infections. Most respiratory adenoviruses are both endemic and epidemic, causing outbreaks in closed communities (boarding schools, daycare centres, military camps etc.). Enteric adenoviruses mainly occur endemically, but outbreaks in hospitals and boarding schools have been reported. Whilst most infections are in infants and young children between 6 months and 5 years of age, they continue throughout life. Serologically repeated infections with adenoviruses of different types can be demonstrated. The incidence of enteric adenovirus infections is between 4 and 7 per 100 person years in small children (Mistchenko *et al.*, 1992). Whilst respiratory adenovirus infections can show seasonal preference (winter and spring) in temperate climates, adenovirus gastroenteritis does not exhibit a seasonal preference. Within serotypes adenovirus 40 and 41, different genomic subtypes have been distinguished (de Jong *et al.*, 1993). J de Jong has produced a large review (Section III, Chapter 2) on the epidemiology of enteric and non-enteric adenoviruses and particularly investigated adenovirus infections in immunodeficient patients (e.g. recipients of bone marrow or solid organ transplants, patients with HIV infection).

Vaccines have only been produced specific for adenovirus types 4 and 7 for application in US military personnel. Vaccines against adenoviruses causing gastroenteritis have so far not been developed due to the mostly mild course of the disease. Careful hand washing, both before and after contact with patients, is the most effective preventive measure in blocking transmission of such infections. Hospital infections have been controlled by cohort nursing of patients, use of gloves, gowns and goggles, and exclusion of symptomatic staff from wards. Disinfectants to clean environmental surfaces are sodium hypochlorite and chloramine T. There is no specific antiviral treatment, although ribavirin (Kapelushnik *et al.*, 1995) has been used in a few cases and cidofovir is active against adenovirus *in vitro* (DeClercq, 1996).

References

Bachenheimer S, Darnell JA (1975). Adenovirus 2 mRNA is transcribed as part of a high molecular weight precursor RNA. *Proc. Natl. Acad. Sci. USA* **72:** 4445-4449.

Berget SM, Moore C, Sharp PA (1977). Spliced segments at the 5' terminus of adenovirus 2 late mRNA. *Proc. Natl. Acad. Sci. USA* **74:** 3171-3175.

Bon F, Fascia P, Dauvergne M *et al.* (1999). Prevalence of group A rotavirus, human calicivirus, astrovirus and adenovirus type 40 and 41 infections among children with acute gastroenteritis in Dijon, France. *J. Clin. Microbiol.* **37:** 3055-3058.

Chakrabarti S, Collingham KE, Stevens RH, Pillay D, Fegan CD, Milligan DW (2000). Isolation of viruses from stools in stem cell transplant recipients: a prospective surveillance study. *Bone Marrow Transplant.* **25:** 277-282.

Chow LT, Broker TR, Lewis JB (1979). Complex splicing patterns of RNAs from the early regions of adenovirus 2. *J. Mol. Biol.* **134:** 265-303.

Chow LT, Roberts JM, Lewis JB, Broker TR (1977). A map of cytoplasmic RNA transcripts from lytic adenovirus type 2, determined by electron microscopy of RNA:DNA hybrids. *Cell* **11:** 819-836.

DeClercq E (1996). Therapeutic potential of Cidofovir (HPMPC, Vistide) for the treatment of DNA virus (i.e. herpes-, papova-, pox- and adenovirus) infections. *Verh. K. Acad. Geneeskd. Belg.* **58:** 19-47.

DeJong JC, Bijlsma K, Wermenbol AG *et al.* (1993). Detection, typing and subtyping of enteric adenoviruses 40 and 41 from fecal samples and observation of changing incidences of infection with these types or subtypes. *J. Clin. Microbiol.* **31:** 1562-1569.

Ginsberg HS, Prince GA. (1994). The molecular basis of adenovirus pathogenesis. *Infect. Agent. Dis.* **3:** 1-8.

Ginsberg HS, Moldawer LL, Sehgal PB *et al.* (1991). A mouse model for investigating the molecular pathogenesis of adenovirus pneumonia. *Proc. Natl. Acad. Sci. USA* **88:** 1651-1655.

Hale GA, Heslop HE, Krance RA, Brenner MA *et al.* (1999). Adenovirus infection after pediatric bone marrow transplantation. *Bone Marrow Transplant.* **23:** 277-282.

Horwitz MS (2001). Adenoviruses. In: *Fields Virology*, 4th edition (DM Knipe, PM Howley *et al.*, eds), pp.2301-2326. Lippincott Williams and Wilkins, Philadelphia.

Janoff EN, Orenstein JM, Manischewitz JF, Smith PD (1991). Adenovirus colitis in the acquired immunodeficiency syndrome, *Gastroenterology* **100:** 976-979.

Kapelushnik OR, Delukina M, Nagler A, Livni N, Engelhard D (1995). Intravenous ribavirin therapy for adenovirus gastroenteritis after bone marrow transplantation. *J. Pediatr. Gastroenterol. Nutr.* **21:** 110-112.

Krajden M, Brown M, Petrasek A, Middleton PJ (1990). Clinical features of adenovirus enteritis: a review of 127 cases. *Pediatr. Infect. Dis. J.* **9:** 636-641.

Lew JF, Moe CL, Monroe SS *et al.* (1991). Astrovirus and adenovirus associated with diarrhea in children in day care settings. *J. Infect. Dis.* **164:** 673-678.

Mistchenko AS, Huberman KH, Gomez JA, Grinstein S (1992). Epidemiology of enteric adenovirus infection in prospectively monitored Argentine families. *Epidemiol. Infect.* **109**: 539-546.

Nishio O, Sakae K, Ishihara Y *et al.* (1992). Adenovirus infection and specific secretory IgA responses in the intestine of infants. *Microbiol. Immunol.* **36**: 623-631.

Russell WC (1998). Adenoviruses. In: *Topley and Wilson's Microbiology and Microbial Infections.* Vol. 1: *Virology,* Ninth Edition, (Mahy BWJ and Collier L, eds), pp 281-307. E Arnold, London.

Ruuska T, Vesikari T (1991). A prospective study of acute diarrhoea in Finnish children from birth to 2½ years of age. *Acta Paediatr. Scand.* **80**: 500-507.

Ruuskanen O, Meurman O, Akusjärvi G (1997). Adenoviruses. In: *Clinical Virology* (Richman DD, Whitley RJ, Hayden FG, eds), pp 525-547. Churchill Livingstone, New York etc.

Uhnoo I, Wadell G, Svensson L, Johansson ME (1984). Importance of enteric adenoviruses 40 and 41 in acute gastroenteritis in infants and young children. *J. Clin. Microbiol.* **20**: 365-372.

Van RR, Wun C-C, O'Ryan ML *et al.* (1992): Outbreaks of human enteric adenovirus types 40 and 41 in Houston day care centers. *J. Pediatr.* **120**: 516-521.

Van Regenmortel M, Fauquet CM, Bishop DHL (Eds) (2000). *Virus Taxonomy. Classification and Nomenclature of Viruses.* pp 227-238, Academic Press, San Diego.

Wadell G (2000). Adenoviruses. In: *Clinical Virology,* Fourth Edition (Zuckerman AJ, Banatvala J, Pattison J, eds), pp 307-327. J. Wiley and Sons, Chichester etc.

Waters V, Ford-Jones EL, Petric M *et al.* (2000). Etiology of community-acquired pediatric viral diarrhea: a prospective longitudinal study in hospitals, emergency departments, pediatric practices and child care centres during the winter rotavirus outbreak, 1997 to 1998. *Pediatr. Infect. Dis. J.* **19**: 843-848.

Wickham TJ, Filardo EJ, Cheresh DA *et al.* (1994). Integrin $\alpha v\beta 5$ selectively promotes adenovirus mediated cell membrane permeabilization. *J. Cell Biol.* **127**: 257-264.

III, 1. Aspects of the molecular biology of enteric adenoviruses

Fiona Stevenson and Vivien Mautner

Cancer Research UK Institute for Cancer Studies, The University of Birmingham, Edgbaston, Birmingham B15 2TT

The *Adenoviridae* family

The *Adenoviridae* family is divided into two genera, *Mastadenovirus* and *Aviadenovirus*. The *Mastadenovirus* genus includes human, simian, bovine, equine, porcine, ovine and canine viruses. The first human adenoviruses (Ad) were isolated from explants of adenoid tissue (Rowe *et al.*, 1953), and to date there have been 51 serotypes of human adenoviruses identified, which are classified into six subgroups, A-F, based upon their immunological properties, oncogenicity in rodents, DNA homologies and morphological properties (Hierholzer *et al.*, 1988; van Regenmortel *et al.*, 2000). The human adenoviruses are associated with a number of infectious diseases, which affect the respiratory, urinary and gastrointestinal tracts and the eyes (Horwitz, 2001).

The enteric adenoviruses

The enteric adenoviruses were first identified from stool samples of infants with acute gastroenteritis (Flewett *et al.*, 1973). The subgroup F adenoviruses were established as causal agents of gastroenteritis in children through epidemiological and clinical studies (Uhnoo *et al.*, 1984; Kidd *et al.*, 1986; Kotloff *et al.*, 1989; Tiemessen *et al.*, 1989; Kim *et al.*, 1990; Cruz *et al.*, 1990; Lew *et al.*, 1991) and are important besides rotaviruses and caliciviruses as a causal agent (Estes *et al.*, 1983; Uhnoo *et al.*, 1984; Brandt *et al.*, 1985).

The enteric adenoviruses are distinct from other adenovirus subgroups, by serology and DNA restriction patterns (Jacobson *et al.*, 1979; Johansson *et al.*, 1980; Kidd and Madeley, 1981), and studies with large numbers of clinical isolates have revealed the presence of two enteric adenovirus serotypes, 40 and 41 (de Jong *et al.*, 1983; Uhnoo *et al.*, 1983) which possess different DNA restriction patterns (Uhnoo *et al.*, 1983; Takiff *et al.*, 1984; Adrian *et al.*, 1986). Within each serotype a number of variants were observed through restriction enzyme analysis, including eleven variants of Ad40 and twenty-eight variants of Ad41 (Kidd, 1984; Kidd *et al.*, 1984; van der Avoort *et al.*, 1989) for which a classification scheme has been presented (van der Avoort *et al.*, 1989; Kidd *et al.*, 1993).

Epidemiology

Epidemiological studies detected enteric adenoviruses in stool samples collected from infants and young children with acute gastroenteritis, in the developed and developing world. These are associated with approximately 5 to 20 percent of cases of paediatric diarrhoea (reviewed in Uhnoo *et al.*, 1990; Wadell *et al.*, 1994). The enteric adenoviruses are prevalent throughout the year with little seasonal variation, suggesting that they are endemic (de Jong *et al.*, 1983, 1993; Uhnoo *et al.*, 1984; Brandt *et al.*, 1985; Johansson *et al.*, 1985; Tiemessen *et al.*, 1989; Kotloff *et al.*, 1989; de Jong, Section III, Chapter 2 of this book).

Pathogenesis and clinical course

The enteric adenoviruses cause gastrointestinal infection in children (generally under the age of three years: Uhnoo *et al.*, 1983), in immunocompromised patients (Krajden *et al.*, 1990; Janoff *et al.*, 1991) or in post bone marrow transplantation patients (Hale *et al.*, 1999; Chakrabarti *et al.*, 2001). The incubation period for the disease is approximately 7-8 days (Richmond *et al.*, 1979); and typically the infection lasts 5-12 days (Uhnoo *et al.*, 1984; Kotloff *et al.*, 1989) longer than other types of viral gastroenteritis, but produces a milder infection with reduced frequency of vomiting. The most prominent feature of the disease is watery diarrhoea which is followed by 1-2 days of vomiting (Uhnoo *et al.*, 1984). The mean duration of the diarrhoea in Ad41 (12.2 days) is longer than in Ad40 infection (8.6 days), however both viruses elicit similar symptoms. Other symptoms include low grade fever (lasting 2-3 days), dehydration and respiratory tract symptoms (occuring in less than 20% of cases) (Yolken *et al.*, 1982; Uhnoo *et al.*, 1984). Enteric adenoviruses have rarely been associated with fatal disease (Whitelaw *et al.*, 1977; Johansson *et al.*, 1985), but Ad41 virus particles were isolated from cells in the small intestine of a fatal case of gastroenteritis. On propagation in tissue culture this strain did not appear to be unusually virulent (Johansson *et al.*, 1985). Nosocomial outbreaks do occur (Rodriguez *et al.*, 1985; Kotloff *et al.*, 1989), but spread to adults is not common (Chiba *et al.*, 1983).

Diagnosis

Enteric adenovirus infection is diagnosed by electron microscopy (EM), enzyme linked immunosorbent assay (ELISA), polymerase chain reaction (PCR) or immunofluorescence (see Wadell *et al.*, 1999; Allard *et al.*, 2001).

Treatment

After diagnosis, treatment involves maintenance of fluid balance and prevention of transmission between patients by the use of gloves, gowns and goggles; exclusion of symptomatic staff from the ward; and use of disinfectants, such as sodium hypochlorite and chloramine T, to clean environmental surfaces. There are currently no specific antiviral or immunotherapy treatments available for the enteric adenoviruses, however the antiviral drug ribavirin (Kapelushnik *et al.,* 1995) has been used to treat enteric adenoviruses in immunocompromised and post-transplantation patients (Maslo *et al.,* 1997, Jurado *et al.,* 1998). Cidofovir has been shown to be active against adenovirus *in vitro* (Afouna *et al.,* 1996) (See also Bass, Section I, Chapter 5 of this book).

Growth properties of enteric adenoviruses in tissue culture

The enteric adenoviruses are shed in large numbers from the gut, but they fail to propagate in cells normally used to propagate other human adenoviruses, as discussed in detail previously (Mautner *et al.,* 1995). Overall the Ad41 growth restriction in tissue culture appears to be less severe than that of Ad40, as a number of cell lines will support the growth of Ad41 but not of Ad40 (de Jong *et al.,* 1983; Uhnoo *et al.,* 1983, 1984; van Loon *et al.,* 1985). The Ad41 block in replication has been shown to occur within the early phase of the infectious cycle (Tiemessen *et al.,* 1996).

Ad40 can be propagated in 293 cells (Takiff *et al.,* 1981), a human embryo kidney cell line transformed with the Ad5 Early (E) region 1 (Graham *et al.,* 1977; Kidd and Madeley, 1981), albeit at reduced levels compared to other serotypes. This suggests that E1 functions are poorly expressed, and therefore implicated in the growth restriction of the enteric adenoviruses. A review of published and unpublished data follows to explain our understanding of why E1 functions are poorly expressed, and how this contributes to the growth restriction of the enteric adenoviruses in tissue culture.

The Ad40 genome

The Ad40 genome of 34,214 base pairs has been sequenced (Davison *et al.,* 1993) and described in some detail (Mautner *et al.,* 1995). The main features that distinguish Ad40 from the other human adenovirus serotypes are the presence of two distinct fibre genes, a single VA gene (involved in late translation) and a highly divergent E3 region (Fig. 1a).

The E1 region

Our understanding of the function of the E1 region in lytic virus infection has mostly been obtained from studies of Ad2 and Ad5 and the oncogenic serotype Ad12, and it has largely been assumed that the enteric adenovirus E1 proteins possess identical functions. Given

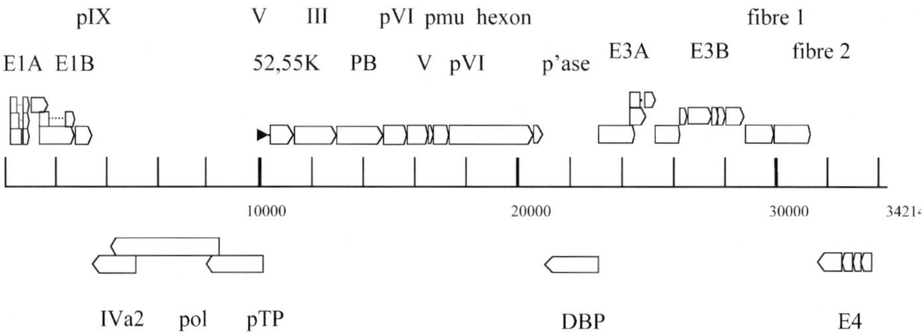

Fig. 1a. The Ad40 genome (Genbank L19443). For a full description and annotated sequence see Davison *et al.*, 1993; Mautner *et al.*, 1995. Open arrows indicate open reading frames. For abbreviations of gene products see Davison *et al.*, 1993.

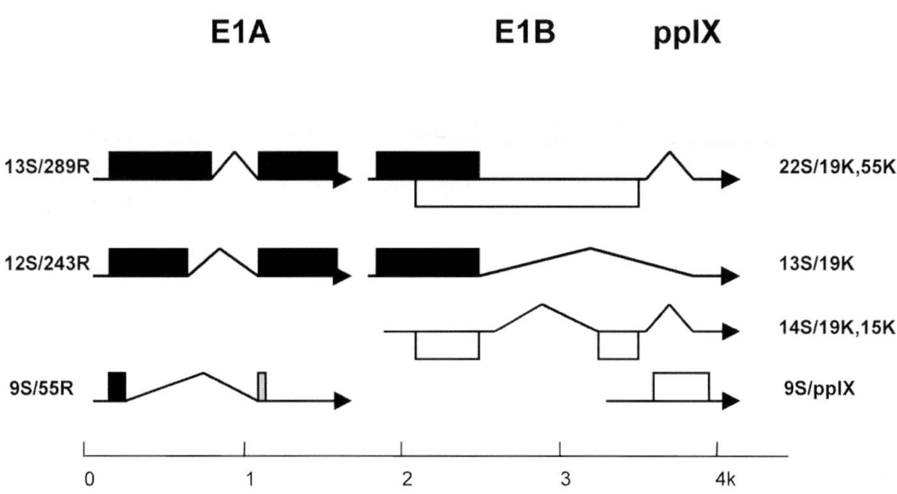

Fig. 1b. E1 transcription map (based on Ad2). The sedimentation values (S) for mRNAs and molecular weights (K) or residue number (R) for each protein are shown. E1A functions: Binds p300, blocking histone acetyltransferase activity and stimulating cell DNA synthesis. Binds pRB family, releasing E2F transcription factor which stimulates cell DNA synthesis. Loss of pRB function increases p14 ARF expression, which prevents MDM2 mediated degradation of p53, resulting in apoptosis. 289R transactivates its own and other Ad promoters. 289R and 243R act as transcriptional repressors. E1B functions: E1B 55K binds p53, and prevents its tumour suppressor functions. E1B 55K in concert with E4 34K modulates viral mRNA transport, protein synthesis and host cell shut-off. E1B 19K is a homologue of the anti-apoptotic protein bcl-2. ppIX: Structural protein with transcriptional regulatory activity

that the organisation of theAd40 E1 region is similar to that of other human adenoviruses (Figs 1b and 1c) this seems a reasonable assumption. However, although the E1 sequences of both Ad40 strain Dugan and Ad41 strain Tak are 85% identical to each other, they are only 52% homologous to the Ad5 E1 region (van Loon *et al.*, 1987a).

For Ad5, the E1A proteins, which are the first viral proteins to be expressed during infection, are important regulators of viral and host cell transcription and act to stimulate cells into S phase, in preparation for viral DNA replication. There are two major gene products, which differ only in their use of alternate splice donor sites, resulting in an internal 43 amino acid domain unique to the larger protein. The resultant 289 residue (R) product of the 13S mRNA is a a *trans*-activator whilst the 243R product of the 12S mRNA functions acts as a transcriptional repressor and a *trans*-activator. These functions are executed via interactions with a number of cell proteins (reviewed by Gallimore and Turnell, 2001). Ad40 encodes equivalent proteins 249R and 221R.

The E1B region encodes two unrelated proteins designated E1B 19 kilo Dalton (kD) and E1B 55K, generated from the same 22S mRNA by the use of different initiation codons. Another mRNA generated by alternative splicing (see Figs 1b and 1c) also encodes the smaller protein. Both proteins have anti-apoptotic properties, while the 55K protein has additional functions, mostly involved in the transition from the early phase (pre-DNA replication) to the late phase of the virus lytic cycle (see White, 2001).

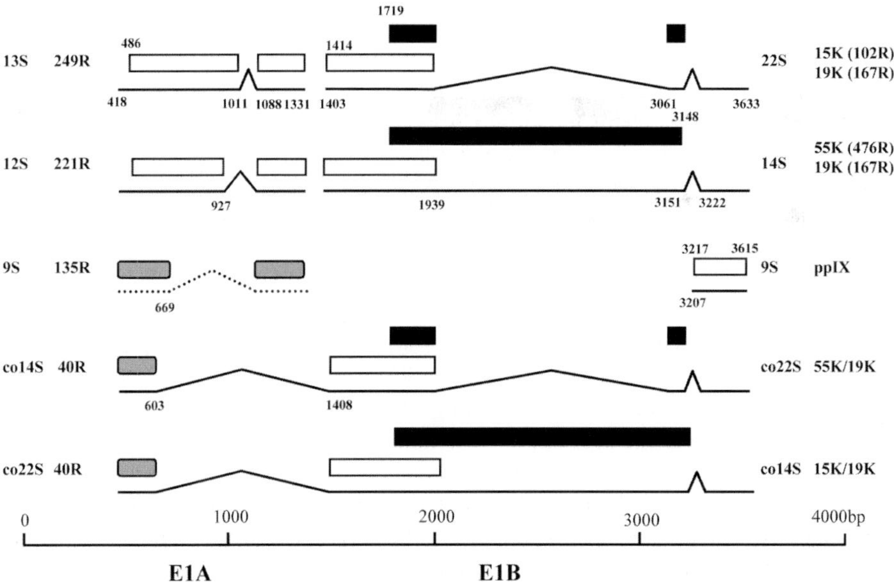

Fig. 1c. Map of transcription from the Ad40 (Dugan) E1 region. Solid lines denote RNA with introns indicated by carets; rectangles are open reading frames; shading indicates reading frame; rectangles with rounded corners are proteins which have not been confirmed experimentally (adapted from Steinthorsdottir, 1991). Designations at left and right of Fig. are names for E1A and E1B mRNA transcripts (S), co-transcripts (co..S) and proteins (R, K, pp).

394

Early region 1B: functions

The poor expression of Ad40 E1 has mainly been attributed to the E1B region: there is indirect evidence such as the failure of Ad40 and Ad41 to completely transform BRK cells (van Loon *et al.*, 1985, Cousin *et al.*, 1991), and more direct evidence of restricted propagation of Ad40 in stably transformed KB cell lines which constitutively expressed Ad2 E1A, E1B, both E1A *and* E1B, or neither (Babiss *et al.*, 1983; Mautner *et al.*, 1989); Ad40 was able to grow in cells expressing E1B, but not in cells expressing E1A alone. This is supported by coinfection assays on HeLa cells, where there is reciprocal complementation between Ad40 and the Ad5 E1A mutant *dl*312, but not with Ad5 or Ad12 mutants which lack the E1B region or are defective for E1B 55K protein expression (Steinthorsdottir and Mautner, 1991; Gomes *et al.*, 1992).

To further characterise expression of the Ad40 E1B region, Ad5 recombinants were constructed containing the Ad40 E1B coding region in place of the Ad5 E1B coding region, under the control of either the Ad5 E1B promoter (sub5P) or the Ad40 E1B promoter (sub40P) (Bailey *et al.*, 1994). Both recombinants were able to plaque on 293 cells but not on HeLa cells, and the virus yields and growth rates were reduced in HeLa cells compared with 293 cells. This suggested that even under the control of the Ad5 E1B promoter, Ad40 E1B gene expression and/or protein function is impaired. E1B transcription from sub5P and sub40P is cell-type specific, in that Ad40 E1B promoter activity is lower than the Ad5 E1B promoter in all cell types, and the pattern of E1B RNA splicing and temporal regulation is characteristic of cell type, but the same for each recombinant. Thus, host cell type and E1B promoter activity are important in determining the growth characteristics of Ad40 (Bailey *et al.*, 1994).

The Ad40 E1B promoter has minimal basal activity and, compared to the Ad5 E1B promoter, is poorly *trans*-activated by the Ad40 and Ad5 E1A proteins and by the promiscuous varicella zoster virus (VZV) *trans*-activator p140 (Steinthorsdottir, 1991; Mautner *et al.*, 1999). The Ad5 E1B core promoter consists of a TATA box and a binding site for the Sp1 transcription factor (Wu *et al.*, 1987), although upstream sequences are known to modulate its efficient functioning (Parks *et al.*, 1988; Parks and Spector, 1990; Spector *et al.*, 1993). The Ad40 E1B promoter like the Ad5 E1B promoter consists of a TATA box and an Sp1 binding site. The Sp1 site differs from that found in Ad5 but is identical to the Ad12 Sp1 binding site. It also has a reduced affinity for the Sp1 protein, relative to the Ad5 site, which along with differences in Sp1 binding site core sequence influences Sp1 recognition. Sequences upstream of the Sp1 binding site mediate downregulation of the Ad40 E1B promoter; these sequences contain two protein binding sites that have been implicated in this effect (Mautner *et al.*, 1999).

In a lytic infection, Ad40 and Ad41 produce transcripts equivalent to the Ad5 14S and 22S mRNA, but no 13S mRNA is found, although it is the major transcript from Ad5 E1B (Steinthorsdottir and Mautner, 1991; Allard and Wadell, 1992). Ad12 E1B also produces the E1B 14S mRNA as a major transcript, with no detectable 13S mRNA (Virtanen *et al.*, 1982; Virtanen and Pettersson, 1985). In Ad41 an additional small exon is detected in the 14S mRNA which is not observed in Ad40 (Steinthorsdottir and Mautner, 1991; Allard and Wadell, 1992). E1A-E1B co-transcript counterparts of the

14S and 22S mRNAs are made by Ad40, containing the first 40 codons of E1A spliced to a site 4-5 nucleotides (nt) downstream of the E1B cap site (see Fig. 1c). The splice junction is unusual in not conforming to splice consensus sequences (Steinthorsdottir and Mautner, 1991; Ishida *et al.*, 1994).

In a productive Ad5 infection, E1B 22S mRNA is made early and E1B 13S mRNA after the onset of DNA replication (Montell *et al.*, 1984; Glenn and Ricciardi, 1988). In non-permissive HeLa cells infected with Ad40, no E1B mRNA is detected early, and only low levels of the 22S and 14S mRNAs are made at late times (Mautner *et al.*, 1990; Bailey *et al.*, 1994). However in 293 cells, KB16 and INT407 cells (which all support a productive Ad40 infection), 22S mRNA is detected before DNA replication, and 14S mRNA at the onset of DNA replication (Bailey *et al.*, 1994).

Consistent with this, the E1B 19K and 55K proteins can be detected in 293 and KB16 cells (Bailey *et al.*, 1993, 1994), whereas 55K is not made in HeLa cells (Mautner *et al.*, 1990).

Early region 1A: functions

Although the poor expression of the Ad40 E1 function has mainly been attributed to the E1B region, the Ad40 E1A region may also be subject to aberrant expression as Ad40 E1A mRNA is first detected at approximately 36h pi in Ad40 infected cells (Ullah, 1997), compared with detection at approximately 3h pi in Ad5 infected cells (Glenn and Ricciardi, 1988). It has also been shown that the Ad40 E1A promoter has a weaker *cis*-acting potential than the Ad5 E1A promoter (Ishino *et al.*, 1988) and that the Ad40 E1A proteins are weaker *trans*-activators of viral promoters (Ad2 E4, Ad5 E2 and E3) than the equivalent Ad5 and Ad12 E1A proteins (van Loon *et al.*, 1987b; Ishino *et al.*, 1988). Taken together these data suggest that the Ad40 E1A region is either subject to aberrant expression or of intrinsically lower activity. To investigate the involvement of the E1A region in the restricted growth of Ad40, we undertook a characterisation of the Ad40 E1A promoter, with the objectives of defining the *cis*-acting sequences important for basal transcription and evaluating the *trans*-activating effect of the Ad40 E1A proteins on the Ad40 E1A promoter (see Fig. 2).

The Ad5 E1A core promoter consists of a TATA element, an initiator sequence (cap site) and a downstream promoter element, which binds transcription factor TFIID to initiate the assembly of the preinitiation complex containing RNA polymerase II and its associated basal transcription factors (Hoey *et al.*, 1990; Verrijzer *et al.*, 1995; Zawel and Reinberg, 1995; Burke and Kadonaga, 1997). The upstream enhancer elements are bound to a variety of cellular transcription factors, which are bridged by coactivators to the basal transcription factors bound at the core promoter, supporting enhancer-dependent transcription (reviewed by Hampsey, 1998; Fig. 2). To identify possible transcription factor binding sites within the Ad40 E1A promoter, sequences from the start of the Ad40 genome to the core promoter (nt 1-452), were matched to the transcription factor database held at EMBL (using a single mismatch algorithm; Ghosh, 1990). This revealed a number of possible transcription factor binding sites

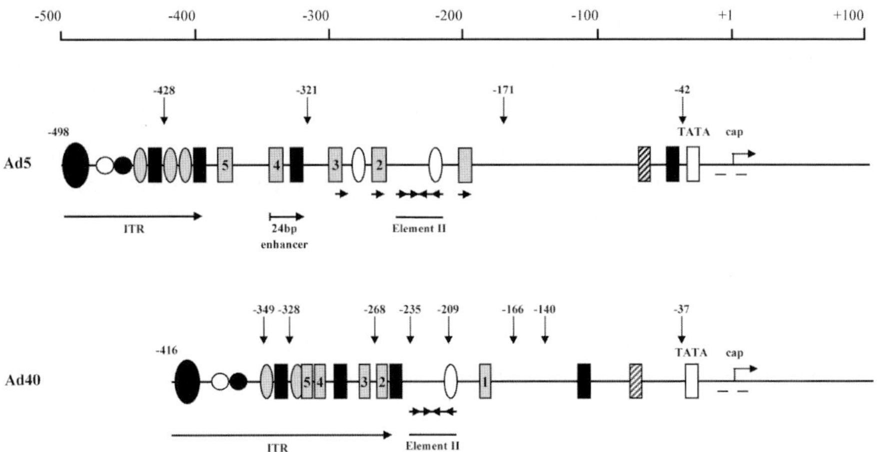

Fig. 2. Comparison of the intact and deleted Ad5 and Ad40 E1A promoters. Transcription factor binding sites are indicated by: ↓ denotes deletion point; █ denotes ORP-A; ○ denotes NF I; ● denotes NF III; ◖ denotes SP1; ▐ denotes ATF; ▨ denotes EF-1A; ▨ denotes E4F1; ◯ denotes E2F; ▢ denotes TATA; ➤►◄◄ denotes Element II in the Ad5 sequence and an Element II-like sequence in the Ad40 sequence.

with striking similarity to the sites in the Ad5 E1A promoter (Fig. 2) and provided a basis for the design of a series of deletion mutants which could be used to define *cis*-acting sequences important in basal transcription from the Ad40 E1A promoter.

Basal transcription of the Ad40 E1A promoter was compared to the Ad5 E1A promoter by cloning PCR-generated DNA fragments which encompassed the Ad5 E1A (nt 1-533) and Ad40 E1A (nt 1-452) promoter sequences, upstream of the firefly luciferase gene into a pGL3 vector (Promega). Luciferase gene expression from the Ad40 and the Ad5 E1A promoters was analysed in transient transfection assays of HeLa cells, in parallel with a positive control containing the SV40 promoter upstream of the firefly luciferase gene and a negative control which was promoterless. The inclusion of a plasmid, pRL-SV40, which contained the SV40 promoter upstream of the *Renilla* luciferase gene, in each transfection provided a single tube dual luciferase reporter system, allowing normalisation of results within a single experiment. The results are expressed as the ratio of the firefly luciferase luminescence (FLL) to *Renilla* luciferase luminescence (RLL). This revealed that basal transcription from the Ad40 E1A promoter was approximately 6 fold lower than from the Ad5 E1A promoter (Fig. 3).

In order to delineate the sequences important for basal transcription, *Bal* 31 deletion mutants were constructed for the Ad40 and Ad5 E1A promoters. Transient transfection assays revealed an almost identical pattern of expression (Fig. 4), albeit at reduced levels for Ad40. Deletion of sequences -498 to -428 in Ad5 (Ad5.428) and -416 to -349 in Ad40 (Ad40.349) resulted in a 1.6 fold increase in expression from both promoters. This was unexpected, as the Ad5 ITR had shown intrinsic promoter and enhancer

FLL:RLL

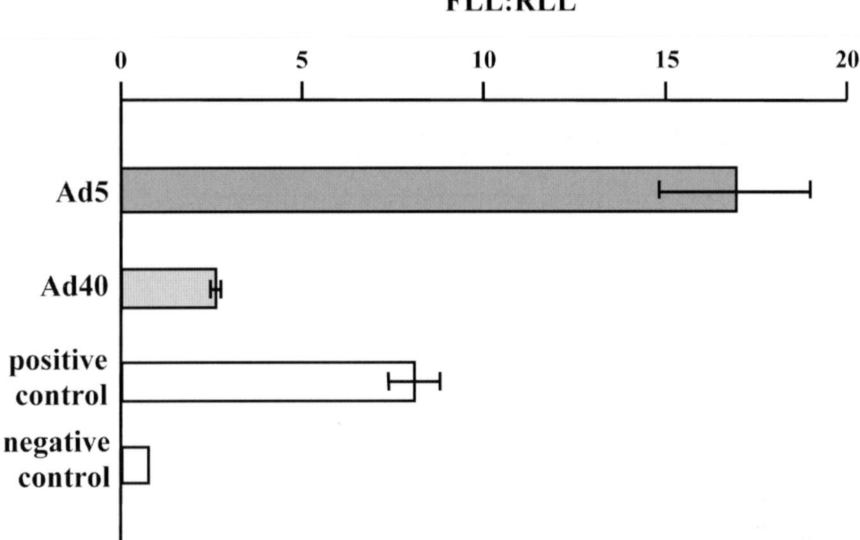

Fig. 3. Analysis of Ad40 and Ad5 E1A basal transcription. Two independent experiments were set up with triplicate 60mm plates of WS HeLa cells (seeded at 1x10⁶ cells/plate), which were transfected with independently derived calcium phosphate precipitates containing 10µg of experimental DNA (Ad40 or Ad5 promoter controlling the firefly luciferase gene) plus 200ng of an internal control (pRL-SV40: SV40 promoter/*Renilla* luciferase), in parallel with a positive control (SV40 promoter/firefly luciferase) and negative control (promoterless firefly luciferase). At 48h post transfection the cells were lysed and luciferase activity measured by luminometry. The ratio of firefly luciferase luminescence (FLL) to *Renilla* luciferase luminescence (RLL) is shown.

activities when placed adjacent to the E1A TATA box (Hatfield and Hearing, 1991). The simplest model to account for the increase in transcription from both promoters is that the deleted sequences contain a binding site for a transcriptional repressor. However, no such activity has been demonstrated within the Ad5 E1A promoter. Further deletion of sequences between -428 and -171 in Ad5 (Ad5.321 and Ad5.171) and between -349 and -140 in Ad40 (Ad40.268 and Ad40.140) resulted in a reduction in transcription almost to background levels (Fig. 4), showing that these sequences are important for basal transcription. Finally, neither the Ad5 nor the Ad40 TATA box alone (Ad5.42 and Ad40.37) are able to support basal transcription.

To further define the region important for basal transcription within the Ad40 E1A promoter, a series of specific deletion mutants was constructed by restriction digest, which deleted predicted transcription factor binding sites known to be important for basal transcription from the Ad5 E1A promoter (Fig. 2, Fig. 4). Deletion of -416 to -328 of the Ad40 E1A promoter (Ad40.328) mediated a 2.5 fold increase in gene expression compared to the wt promoter, whereas deletion of sequences between -328

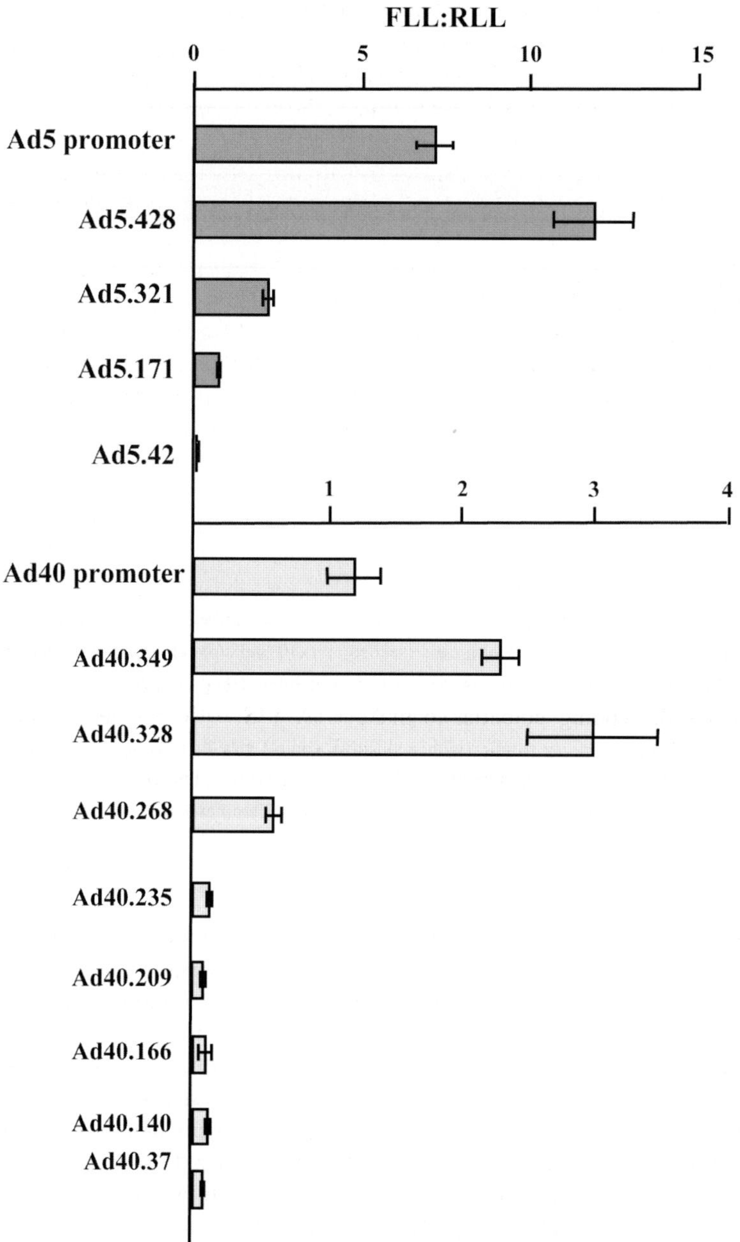

Fig. 4. Basal transcription from deleted Ad5 or Ad40 E1A promoters. The experiment was set up as described in Fig. 2, using plasmid constructs with truncated promoter regions (as in Fig. 2). The ratio of firefly luciferase luminescence (FLL) to *Renilla* luciferase luminescence (RLL) is shown for the intact and deleted Ad5 and Ad40 E1A promoters. For description of deletion mutants see text.

to -268 and -268 to -235 (Ad40.268 and Ad40.235) decreased expression by a factor of 2.3 and 12, respectively, resulting in a promoter unable to support basal transcription in HeLa cells. We can therefore conclude that the Ad40 E1A enhancer region maps to sequences within -328 to -235 relative to the Ad40 E1A cap site.

The region between -328 to -268 is predicted to contain three EF-1A-related sequences which would bind alternative members of the *ets* gene oncogene family known as Elk 1, Elk 1 and GABP (sites 5, 4 and 3 respectively, Fig. 2). The *ets* gene family of transcription factors comprises seven subfamilies, which show tissue specificity (reviewed in Wasylyk *et al.*, 1993), suggesting a role for tissue specificity within the Ad40 E1A promoter.

The region 268 to 235 is predicted to contain an ATF binding site, one copy of element I and an element II-related sequence (Fig. 2). The element II-related sequence shows 75% homology to the Ad5 element II, deletion of which mediates a decrease in Ad5 E1A mRNA expression (Hearing and Shenk, 1986). A similar decrease is seen for the equivalent Ad40 deletion. We can therefore conclude that the Ad40 E1A enhancer region maps to sequences within -328 to -235 relative to the Ad40 E1A cap site.

Alternative mechanisms for basal transcription from the Ad40 E1A promoter *in vivo* cannot be ruled out. A sequence comparison of the Ad40 E1A promoter with TRANSFAC (Quandt *et al.*, 1995) reveals that the promoter contains motifs related to binding sites of gut-specific transcription factors from the HNF and HFH families within the region -209 to 37. It is possible that Ad40 has evolved to utilise gut-specific transcription factors, and it would therefore be interesting to analyse basal transcription from the intact and deleted promoter constructs in the context of gut cells.

Ad40 has the coding potential to make E1A 13S, 12S and 9S transcripts (van Loon *et al.*, 1987b; Ishino *et al.*, 1988; Allard and Wadell, 1988), however transcription maps have only been reported for E1 plasmid-transformed cells, where all three mRNAs were detected (van Loon *et al.*, 1987b). The Ad40 E1A 13S, 12S and 9S transcripts encode proteins of 249R, 221R and 135R that correspond to the Ad5 289R, 243R and 55R proteins (Fig. 1). Conserved regions CR1, CR2 and CR3 identified within adenovirus E1A proteins (Kimelman *et al.*, 1985) are also present in Ad40 (van Loon *et al.*, 1987a; Allard and Wadell, 1988; Ishino *et al.*, 1988), and all have a role in *trans*-activation, by interaction with a range of cellular proteins.

As the Ad40 E1A proteins are weaker *trans*-activators of other early adenovirus promoters they are also likely to be weaker *trans*-activators of their own promoter, which would contribute to the aberrant expression or the intrinsically lower activity of the E1 region. To investigate this, Ad40 E1A cDNA sequences were generated and cloned into an expression plasmid under the control of the cytomegalovirus immediate early promoter (White and Cipriani, 1990). A library of Ad40 E1A cDNAs was made by reverse transcription-polymerase chain reaction of cytoplasmic RNA harvested at 42h pi from 293 cells infected with Ad40 and blocked in the early phase by cytosine arabinoside treatment. The library was cloned into pCRII (Invitrogen TA cloning kit), and the Ad40 E1A 13S cDNA identified by sequence analysis and shown to encode the predicted 249R protein. The 12S protein coding sequences were not isolated; however the cDNA cloning was from cells blocked in early phase, and it is possible that the 12S mRNA is made only at late times in infection. A number of other Ad40 E1A specific

400

cDNAs were obtained from the Ad40 infected 293 cells which could encode truncated proteins. The abundance of these RNAs in the cytoplasm suggests that they are likely to be functional rather than by-products, but their role remains to be established. It is possible that Ad40 has evolved to use components of the RNA processing machinery that are unique to intestinal cells.

To investigate whether the Ad40 E1A 249R protein was a weaker *trans*-activator of the Ad40 E1A promoter, luciferase plasmid constructs containing the Ad5 and the Ad40 E1A promoters were co-transfected with expression vectors encoding either the Ad5 E1A 289R protein or the Ad40 E1A 249R protein. Fig. 5 shows that the Ad5 E1A 289R protein increases transcription 10 fold over the basal level, whereas the Ad40 E1A 249R protein only gives a marginal increase. This is consistent with *trans*-activation levels previously reported for plasmids expressing the whole Ad40 E1A region, i.e. plasmids expressing all the Ad40 E1A proteins (van Loon *et al.*, 1987b; Ishino *et al.*, 1988). Interestingly, the Ad5 E1A 289R protein *trans*-activates the Ad40 and Ad5 E1A promoters to a comparable extent. This indicates that the Ad40 E1A promoter does indeed contain transcription factor binding sites sufficient for *trans*-activation by the Ad5 E1A 289R protein.

FLL:RLL

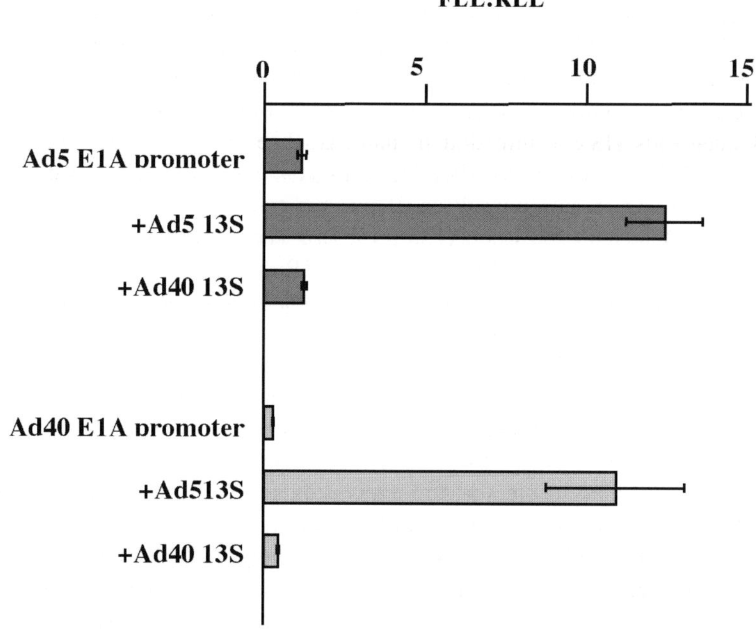

Fig. 5 Analysis of *trans*-activation by the Ad5 E1A 289R protein and the Ad40 E1A 249R protein on the Ad5 and the Ad40 E1A promoters. The experiment was set up with triplicate 60mm plates of WS HeLa cells (seeded at 1x10⁶ cells/plate), which were co-transfected with independently derived calcium phosphate precipitates containing 5μg Ad5 or Ad40 E1A promoters with either 5μg Ad5 E1A 13S cDNA or Ad40 E1A 13S equivalent cDNA, plus an internal control (SV40 promoter/renilla luciferase). The ratio of firefly luciferase luminescence (FLL) to *Renilla* luciferase luminescence (FLL) is shown for each transfection.

Summary

The enteric adenoviruses display a very restricted host range phenotype in tissue culture and, as previously reviewed (Mautner *et al.,* 1995), this is a complex and poorly understood phenomenon. In this chapter we have described some properties of the Ad40 E1A region, which suggest that the basal activity of the E1A promoter, and the activity of the E1A 249R protein are limited in cell lines normally permissive for adenovirus propagation. A fuller understanding of the potential of Ad40 will require the development of gut cell cultures, which more closely mirror the conditions found in the intestine.

The work described in this chapter was carried out while the authors were in the Medical Research Council Virology Unit, Institute of Virology, Glasgow G11 5JR, SCOTLAND.

References

Adrian, T., Wadell, G., Hierholzer, J.C. and Wigand, R. (1986). DNA restriction analysis of adenovirus prototypes 1 to 41. *Arch. Virol.* **91**, 277-290.

Afouna, M.J., Mehta, S.C., Ghanem, A.H., Higuchi, W.I., Kern, E.R., DeClerq, E. and El-Shattaway, H.H. (1999). Influence of the treatment protocol upon *in vivo* efficacy of cidofovir (HPMPC) and of acyclovir (ACV) formulations in topical treatment of cutaneous HSV-1 infection in hairless mice. *J. Pharm. Sci.* **88**, 530-534.

Allard, A. and Wadell, G. (1988). Physical organisation of the enteric adenovirus type 41 early region 1A. *Virology* **164**, 220-229.

Allard, A and Wadell, G. (1992). The E1B transcription map of the enteric adenovirus type 41. *Virology* **183**, 319-330.

Allard, A., Albinsson, B. and Wadell, G. (2001). Rapid typing of human adenoviruses by a general PCR combined with restriction endonuclease analysis. *J Clin. Microbiol.* **39**, 498-505.

Babiss, L.E., Young, C.S.H., Fisher, P.B. amd Ginsberg, H.S. (1983). Expression of adenovirus E1A and E1B gene products and the *Escherichia coli* XGPRT gene in KB cells. *J. Virol.* **46**, 454-465.

Bailey, A, Mackay, N. and Mautner, V. (1993). Enteric adenovirus type 40: Expression of EIB proteins *in vivo. Virology* **193**, 631-641.

Bailey, A., Ullah, R. and Mautner, V. (1994). Cell type specific regulation of expression from the Ad40 E1B promoter in recombinant Ad5/Ad40 viruses. *Virology* **202**, 695-706.

Brandt, C. D., Kim, H. W., Rodriguez, W. J., Arobio, J. O., Jeffries, B. C., Stallings, E. P., Lewis, C., Miles, A. J., Gardner, M. K. and Parrott, R. H. (1985). Adenoviruses and paediatric gastroenteritis. *J. Infect. Dis.* **151**, 437-443.

Burke, T.W. and Kadonaga, J.T. (1997). The downstream core promoter element, DPE, is conserved from *Drosophila* to humans and is recognized by TAFII60 of *Drosophila. Genes Dev.* **11**, 3020-3031.

Chakrabarti, S., Mackinnon, S., Kottaridis, P., Hale, G., Osman, H., Peggs, K., Fegan, C., Linch, D. and Milligan, D. (2001). Community respiratory virus and adenovirus infections in nonmyeloablative transplant recipients conditioned with Campath-1H: High incidence but low mortality. *Blood* **98**, 1649.

Chiba, S., Nakata, S., Nakamura, I., Taniguchi, K., Urasawa, S., Fujinaga, K. and Nakao, T. (1983). Outbreak of infantile gastroenteritis due to type 40 adenovirus. *Lancet* **2**, 954-957.

Cousin, C., Winter, N., Gomes, S.A. and D'Halluin, J.C. (1991). Cellular transformation by E1 genes of enteric adenoviruses. *Virology* **181**, 277-287.

Cruz, J. R., Caceres, P., Cano, F., Flores, J., Bartlett, A. and Torun, B. (1990). Adenovirus types 40 and 41 and rotaviruses associated with diarrhea in children from Guatemala. *J. Clin. Microbiol.* **28**, 1780-1784.

Davison, A. J., Telford, E. A. R., Watson, M. S., McBride, K. and Mautner, V. (1993). The DNA sequence of adenovirus type 40. *J. Mol. Biol.* **234**, 1308-1316.

de Jong, J. C., Wigand, R., Kidd, A. H., Wadell, G., Kapsenberg, J. G., Muzerie, C. J., Wermenbol, A. G. and Firtzlaff, R.-G. (1983). Candidate adenoviruses 40 and 41: fastidious adenoviruses from infant stool. *J. Med. Virol.* **11**, 215-231.

de Jong, J. C., Bijlsma, K., Wermenbohl, A. G., Verweij-Uijterwaal, M. W., van der Avoort, H. G. A. M., Wood, D. J., Bailey, A. S. and Osterhaus, A. D. M. E. (1993). Detection, typing and subtyping of enteric adenoviruses 40 and 41 from faecal samples and observations of changing incidence of infections with these types and subtypes. *J. Clin. Microbiol.* **31**, 1562-1569.

Estes, M. K., Palmer, E. L. and Obijeski, J. F. (1983). Rotaviruses: a review. *Curr. Top. Microbiol. Immunol.* **105**, 123-184.

Flewett, T.H., Bryden, A.S., Davies, H.A. and Morris, C.A. (1973). Epidemic viral enteritis in a long-stay childrens ward. *Lancet* **1**, 4-5.

Gallimore, P.H. and Turnell, A.S. (2001). Adenovirus E1A: remodelling the host cell, a life or death experience. *Oncogene* **20**, 7824-7835.

Ghosh, D. (1990). A relational database of transcription factors. *Nucleic Acids Res.* **18**, 1749-1756.

Glenn, G. M. and Ricciardi, R. P. (1988). Detailed kinetics of adenovirus type 5 steady state transcripts during early infection. *Virus Res.* **9**, 73-91.

Gomes, S. A., Neil, C. and D'Halluin, J. C. (1992). Growth of fastidious adenovirus serotype 40 in HRT 18 cells: interactions with E1A and E1B deletion mutants of subgenus C adenoviruses. *Arch. Virol.* **124**, 45-56.

Graham, F. L., Smiley, J. S., Russell, W. C. and Nairn, R. (1977). Characterisation of a human cell line transformed by DNA from human adenovirus type 5. *J. Gen. Virol.* **36**, 59-72.

Hale, G. A., Heslop, H. E., Krance, R. A., Brenner, M. A., Jayawardene, D., Srivastava, D. K. and Patrick, C. C. (1999). Adenovirus infection after paediatric bone marrow transplantation. *Bone Marrow Transplantation* **23**, 277-282.

Hampsey, M. (1998). Molecular genetics of the RNA polymerase II general transcriptional machinery. *Microbiol. Mol. Biol. Rev.* **62**, 465-503.

Hatfield, L. and Hearing, P. (1991). Redundant elements in the adenovirus type 5 inverted terminal repeat promote bi-directional transcription *in vitro* and are important for virus growth *in vivo*. *Virology* **184**, 265-276.

Hearing, P. and Shenk, T. (1986). The adenovirus type 5 E1A enhancer contains two functionally distinct domains: One is specific for E1A and the other modulates all early units in *cis*. *Cell* **45**, 229-236.

Hierholzer, J. C., Wigand, R. and de Jong, J. C. (1988). Evaluation of human adenoviruses 38, 39, 40 and 41 as new serotypes. *Intervirology* **29**, 1-10.

Hoey, T., Dynlacht, B.D., Peterson, M.G., Pugh, B.F. and Tijan, R. (1990). Isolation and characterization of the *Drosophila* gene encoding the TATA box binding protein, TFIID. *Cell* **61**, 1179-1186.

Horwitz, M.S. (2001) Adenoviruses. In: *Fields' Virology*, 4th edn. (Eds DM Knipe, PM Howley *et al*), pp 2301-2326. Lippincott Williams and Wilkins, Philadelphia.

Ishida, S., Fujinaga, Y., Fujinaga, K., Sakamoto, N. and Hashimoto, S. (1994). Unusual splice sites in the E1A-E1B cotranscripts synthesised in adenovirus type-40 infected A549 cells. *Arch. Virol.* **139**, 389-402.

Ishino, M., Ohashi, Y., Emoto, T., Sawada, Y. and Fujinaga, K. (1988). Characterisation of adenovirus type 40 E1 region. *Virology* **165**, 95-102.

Jacobson, P. A., Johansson, M. E. and Wadell, G. (1979). Identification of an enteric adenovirus by immuno-electroosmophoresis (IEOP) technique. *J. Med. Virol.* **3**, 307-312.

Janoff, E. N., Orenstein, J. M., Manischewitz, J. F. and Smith, P. D. (1991). Adenovirus colitis in the acquired immunodeficiency syndrome. *Gastroenterology* **100**, 976-979.

Johansson, M. E., Uhnoo, I., Kidd, A. H., Madeley, C. R. and Wadell, G. (1980). Direct identification of adenovirus, a candidate new serotype associated with infantile gastroenteritis. *J. Clin. Microbiol.* **12**, 95-100.

Johansson, M. E., Uhnoo, I., Svensson, L., Pettersson, C. A. and Wadell, G. (1985). Enzyme linked immunosorbent assay for detection of enteric adenovirus 41. *J. Med. Virol.* **17**, 19-27.

Jurado, C.M., Hernandez, M.F., Navarro, MJM., Ferrer, C.C., Escobar, V.JL, de Pablos, G.J.M. (1998). Adenovirus pneumonitis successfully treated with intravenous ribavirin. *Haematologica* **83**,1128-9.

Kapelushnik, J., Delukina, M., Nagler, A., Livni, N. and Engelhard, D. (1995). Intravenous ribavirin therapy for adenovirus gastroenteritis after bone-marrow transplantation. *J. Ped. Gastroenterol. Nutr.* **21**, 110-112.

Kidd, A. H. (1984). Genome variants of adenovirus 41 (subgroup G) from children with diarrhoea in South Africa. *J. Med. Virol.* **14**, 49-59.

Kidd, A. H. and Madeley, C. R. (1981). *In vitro* growth of some fastidious adenoviruses from stool specimens. *J. Clin. Path.* **34**, 213-216.

Kidd, A. H., Berkowitz, F. E., Blaskovic, P. J. and Schoub, B. D. (1984). Genome variants of human adenovirus 40 (subgroup F). *J. Exp. Med.* **14**, 235-246.

Kidd, A. H., Rosenblatt, A., Besselaar, T. G., Erasmus, M. J., Tiemessen, C. T., Berkowitz, F. E. and Schoub, B. D. (1986). Characterisation of rotaviruses and subgroup F adenoviruses from acute summer gastroenteritis in South Africa.

J. Med. Virol. 18, 159-168.

Kidd, A. H., Chroboczek, J., Cusack, S. and Ruigrok, R. W. H. (1993). Adenovirus type 40 contains two distinct fibers. *Virology* **192**, 73-84.

Kim, K. H., Yang, J. M., Joo, S. I., Cho, Y. G., Glass, R. I. And Cho, Y. J. (1990). Importance of rotavirus and adenovirus types 40 and 41 in acute gastroenteritis in Korean children. *J. Clin. Microbiol.* **28**, 2279-2284.

Kimelman, D., Miller, J. S., Porter, D. and Roberts, B. E. (1985). E1A regions of the human adenoviruses and of the highly oncogenic simian adenovirus 7 are closely related. *J. Virol.* **53**, 399-409.

Kotloff, K. L., Losonsky, G. A., Morris, J. G. Jr., Wasserman, S. S., Singh-Naz, N. and Levine, M. M. (1989). Enteric adenovirus infection and childhood diarrhea: an epidemiologic study in three clinical settings. *Pediatrics* **84**, 219-225.

Krajden, M., Brown, M., Petrasek, A. and Middelton, P. J. (1990). Clinical features of adenovirus enteritis – a review of 127 cases. *Ped. Infect. Disease J.* **9**, 636-641.

Lew, J. F., Moe, C. L., Monroe, S. S., Allen, J. R., Harrison, B. M., Forrester, B. D., Stine, S. E., Woods, P. A., Hierholzer, J. C., Hermann, J. E., Blacklow, N. R., Bartlett, A. V. and Glass, R. I. (1991). Astrovirus and adenovirus associated with diarrhoea in children in day care settings. *J. Infect. Dis.* **164**, 673-678.

Maslo C, Girard PM, Urban T *et al.* (1997) Ribavirin therapy for adenovirus pneumonia in an AIDS patient. *Am. J. Respir. Crit. Care Med.* **156**, 1263-1264.

Mautner, V., Mackay, N. and Steinthorsdottir, V. (1989). Complementation of enteric adenovirus type 40 for lytic growth in tissue culture by E1B 55K function of adenovirus types 5 and 12. *Virology* **171**, 619-622.

Mautner, V., Mackay, N. and Morris, K. (1990). Enteric adenovirus type 40: Expression of E1B mRNA and proteins in permissive and nonpermissive cells. *Virology* **179**, 129-138.

Mautner, V., Steinthorsdottir, V. and Bailey, A. (1995). Enteric adenoviruses. *Curr. Top. Microbiol. Immunol.* **199**/III, 229-282.

Mautner, V., Bailey, A., Steinthorsdottir, V., Ullah, R. and Rinaldi, A. (1999). Properties of the adenovirus type 40 E1B promoter that contribute to its low transcriptional activity. *Virology* **265**, 10-19.

Montell, C., Fisher, E. F., Caruthers, M. H. and Berk, A. J. (1984). Control of adenovirus E1B messenger-RNA synthesis by a shift in the activities of RNA splice sites. *Mol. Cell. Biol.* **4**, 966-972.

Parks, C. L. and Spector, D. J. (1990). Cis-dominant defect in activation of adenovirus type 5 E1B early RNA synthesis. *J. Virol.* **64**, 2780-2787.

Parks, C. L., Banerjee, S. and Spector, D. J. (1988). Organisation of the transcriptional control region of the E1B gene of adenovirus type 5. *J. Virol.* **62**, 54-67.

Quandt, K., Frech, K., Karas, H., Wingender, E. and Werner, T. (1995). MatInd and MatInspector - New fast and versatile tools for detection of consensus matches in nucleotide sequence data. *Nucleic Acids Res.* **23**, 4878-4884.

Richmond, S. J., Caul, E. O., Dunn, S. M., Ashley, C. R., Clarke, S. R. and Seymour, N. R. (1979). An outbreak of gastroenteritis in young children caused by adenoviruses. *Lancet* **1**, 1178-1180.

Rodriguez, W. J., Kim, J. W. and Brandt, C. D. (1985). Faecal adenoviruses from a longitudinal study of families in metropolitan Washington DC: laboratory, clinical and epidemiology observations. *J. Pediatr.* **107**, 514-520.

Rowe, W. P., Huebner, A. J., Gilmore, L. K., Parrott, R. N. and Wand, T. G. (1953). Isolation of a cytopathic agent from human adenoids undergoing spontaneous degeneration in tissue culture. *Proc. Soc. Exp. Biol. Med.* **84**, 570-573.

Spector, D. J., Parks, C. L. and Knittle, R. A. (1993). A multicomponent cis-activator of transcritpion of the E1B gene of adenovirus type 5. *Virology* **194**, 128-136.

Steinthorsdottir, V. (1991). Adenovirus type 40 host range in tissue culture: A study of the E1B region. *PhD Thesis*, University of Glasgow.

Steinthorsdottir, V. and Mautner, V. (1991). Enteric adenovirus type 40: E1B transcription map and novel E1A-E1B co-transcripts in lytically infected cells. *Virology* **181**, 139-149.

Takiff, H. E., Straus, S. E. and Garon, C. F. (1981). Propagation and *in vitro* studies of previously non-cultivable enteral adenovirus in 293 cells. *Lancet* **2**, 832-834.

Takiff, H. E., Reinhold, W., Garon, C. F. and Straus, S. E. (1984). Cloning and physical mapping of enteric adenoviruses (candidate types 40 and 41). *J. Virol.* **51**, 131-136.

Tiemessen, C. T., Wegerhoff, F. O., Erasmus, M. J. and Kidd, A. H. (1989). Infection by enteric adenoviruses, rotaviruses, and other agents in a rural African environment. *J. Med. Virol.* **28**, 176-182.

Tiemessen, C. T., Nel, M. J. and Kidd A. H. (1996). Adenovirus 41 replication: Cell-related differences in viral gene transcription. *Mol. Cell. Probes* **10**, 279-287.

Uhnoo, I., Wadell, G., Svensson, L. and Johansson, M. (1983). Two new serotypes of enteric adenoviruses causing infantile diarrhoea. *Dev. Biol. Standard.* **53**, 311-318.

Uhnoo, L., Wadell, G., Svensson, L. and Johansson, M. E. (1984). Importance of enteric adenoviruses 40 and 41 in acute gastroenteritis in infants and young children. *J. Clin. Microbiol.* **20**, 365-372.

Uhnoo, L., Svensson, L. and Wadell, G. (1990). Enteric adenoviruses. *Bailliere's Clin. Gastroenterol.* **4**, 627-642.

Ullah, R. R. (1997). Adenovirus type 40 host range in tissue culture: replication and gene expression in INT407 cells. *PhD Thesis,* University of Glasgow.

van der Avoort, H. G. A. M., Wermenbol, A. G., Zomerdijk, T. P. L., Kleijne, J. A. F. W., van Asten, J. A. A. M., Jensma, P., Osterhaus, A. D. M. E., Kidd, A. H. and de Jong, J. C. (1989). Characterisation of fastidious adenovirus types 40 and 41 by DNA restriction enzyme analysis and by neutralising monoclonal antibodies. *Virus Res.* **12**, 139-157.

van Loon, A. E., Mass, R., Vaessen, R. T. M. J., Reemst, A. M. C. B., Sussenbach, J. M. and Rozijn, T. H. (1985). Cell transformation of the left terminal regions of the adenovirus 40 and 41 genomes. *Virology* **147**, 227-230.

van Loon, A. E., Ligtenberg, M., Reemst, A. M. C. B., Sussenbach, J. S. and Rozijn, T. H. (1987a). Structure and organisation of the left-terminal DNA regions of fastidious adenovirus types 40 and 41. *Gene* **58**, 109-126.

van Loon, A. E., Gilardi, P., Perricaudet, M., Rozijn, Th. H. and Sussenbach, J. S.

(1987b). Transcriptional activation by the E1A regions of adenovirus types 40 and 41. *Virology* **160**, 305-307.

van Regenmortel, M.H.V., Fauquet, C.M., Bishop, D.H.L., Carstens, E.B., Estes, M.K., Lemon, S.M., Maniloff, J., Mayo, M.A., McGeoch, D.J., Pringle, C.R., and Wickner, R.B. (2000) *Virus Taxonomy. Classification and Nomenclature of Viruses*. Seventh Report of the International Committee on Taxonomy of Viruses pp. 227-238. Academic Press, San Diego, USA.

Verrijzer, C.P., Chen, J.L., Yokomori, K. and Tijan, R. (1995). Binding of TAFs to core elements directs promoter selectivity by RNA polymerase II. *Cell* **81**, 1115-1125.

Virtanen, A. and Pettersson, U. (1985). Organisation of early region 1B of human adenovirus type 2: Identification of four differentially spliced mRNAs. *J. Virol.* **54**, 383-391.

Virtanen, A., Pettersson, U., Le Moullec, J. M., Tiollais, P. and Perricaudet, M. (1982). Different mRNAs from the transforming region of highly oncogenic and nononcogenic human adenoviruses. *Nature* **295**, 705-707.

Wadell, G., Allard, A., Johansson, M., Svensson, L. and Uhnoo, I. (1994). Enteric adenoviruses. In Kapikian A. Z. (Ed.). Viral Infections of the Gastrointestinal Tract, 2nd. ed., p. 519-547. Marcel Dekker, New York.

Wadell, G., Allard, A. and Hierholzer, J.C. (1999).Adenoviruses In: Murray, P.R. (ed.) Manual of Microbiology, 7th ed. p 970-982. American Society for Microbiology, Washington, D.C.

Wasylyk, B., Hahn, S.L. and Giovane, A. (1993). The Ets family of transcription factors. *Eur. J. Biochem.* **211**, 7-18.

White, E. (2001). Regulation of the cell cycle and apoptosis by the oncogenes of adenovirus. *Oncogene* **20**, 7836-7846.

White, E. and Cipriani, R. (1990). Role of adenovirus E1B proteins in transformation: altered organization of intermediate filaments in transformed cells that express the 19-Kilodalton protein. *Mol. Cell. Biol.* **10**, 120-130.

Whitelaw, A., Davies, H. and Parry, J. (1977). Electron microscopy of fatal adenovirus gastroenteritis. *Lancet* **1**, 361.

Wu, L., Rosser, D. S. E., Schmidt, M. C. and Berk, A. (1987). A TATA box implicated in E1A transcriptional activation of a simple adenovirus 2 promoter. *Nature* **326**, 512-515.

Yolken, R. H., Lawrence, F., Leister, F., Takiff, H. E. and Straus, S. E. (1982). Gastroenteritis associated with enteric type adenovirus in hospitalised infants. *J. Paediatr.* **101**, 21-26.

Zawel, L. and Reinberg, D. (1995). Advances in RNA polymerase II transcription. *Curr. Opin. Cell Biol.* **4**, 488-495.

III, 2. Epidemiology of enteric adenoviruses 40 and 41 and other adenoviruses in immunocompetent and immunodeficient individuals

J.C. de Jong

*Department of Virology, Erasmus University Rotterdam, Dr Molewaterplein 50,
3015 GE, Rotterdam, the Netherlands*

Summary

Adenovirus infections are frequent events but they do not usually cluster in distinct epidemics and are therefore somewhat inconspicuous. In the immunocompetent host they are rarely associated with severe illness or mortality. Syndromes commonly associated with adenovirus infection include respiratory illness, ranging from the common cold to pneumonia, (kerato)conjunctivitis and diarrhoea. On the basis of 21 publications, the frequency of occurrence of enteric adenoviruses 40 (Ad40) and Ad41 in stool specimens from immunocompetent children with diarrhoea and controls was evaluated in the perspective of other enteric pathogens. The median observed (all studies) and mean control-corrected (controlled studies) frequencies were 4.0% and 2.9% for enteric adenovirus, 3.4% and 0% for non-enteric adenovirus, 21% and 19% for rotavirus, 15% and 11% for calicivirus, 6.7% and 7.9% for astrovirus and 14% and 17% for nonviral microorganisms, respectively. An interesting epidemiological observation was the gradual change in the predominance among enteric adenoviruses from Ad40 in the early 1980s to Ad41 in the late 1980s, which occurred first in Australia and the Netherlands, later in Canada and lastly in Japan. More serious illness is caused by adenovirus infections in immunocompromized individuals. In 19 studies involving bone marrow transplant recipients with intestinal disease, on average 4.4% of the stool samples contained an adenovirus. A median percentage of 12% of the recipients contracted an adenovirus infection of any organ, 17% of whom died as a possible or probable result of this infection. There are indications that these rates are rising.

Introduction

Adenovirus infections are common, especially in children. The most frequent manifestations are respiratory illnesses (ranging from the common cold to pneumonia), diar-

408

rhoea and (kerato)conjunctivitis. These conditions are relatively mild and self-limited, and do not cause permanent damage. They may occur in epidemics, especially among military recruits. Effective vaccination is possible but this has only been implemented for military recruits in the USA. Adenovirus is also isolated frequently from faecal specimens but its role in diarrhoea and other gastrointestinal disorders is still controversial. Only adenovirus types 40 (Ad40) and Ad41 have a proven causal relationship with infantile diarrhoea. Together, these types are often called *"enteric"* or *"fastidious"* adenoviruses. Sometimes the term "enteric adenovirus" is used to include other types which are isolated exclusively from stool samples, such as Ad31. In this review, however, the use of the term is restricted to Ad40 and Ad41.

General information about (enteric) adenoviruses can be found in text books and reviews (Christensen, 1989; Foy, 1989; Hierholzer, 1992; Mautner *et al.*, 1995; Lukashok and Horwitz 1998; Goodgame, 1999 and 2001; Wadell *et al.*, 1999; Horwitz, 2001; Shenk, 2001). In this review, enteric adenoviruses are discussed in the perspective of other adenoviruses and other enteric viruses.

General properties of adenoviruses

Structure and classification of adenoviruses

Adenoviruses are nonenveloped icosahedral viruses with a diameter of 70-90 nm (Figs 1 and 2; Shenk, 2001). The outer capsid consists of 240 hexons and 12 pentons which form the 12 vertices. The penton consists of a base and a thin fibre bearing a terminal knob. Inside the capsid there is a genome of contiguous, linear double-stranded DNA. The structure of *enteric adenovirus* is similar, except that it has two fibres of different lengths. Members of the family *Adenoviridae* occur throughout the animal kingdom. Two genera are recognized, *Mastadenovirus* and *Aviadenovirus,* which infect mammals and birds, respectively. At present, the mastadenoviruses are grouped into 20 species (Benkö *et al.*, 2000). The six human species (A-F) comprize 51 adenovirus types (Table 1).

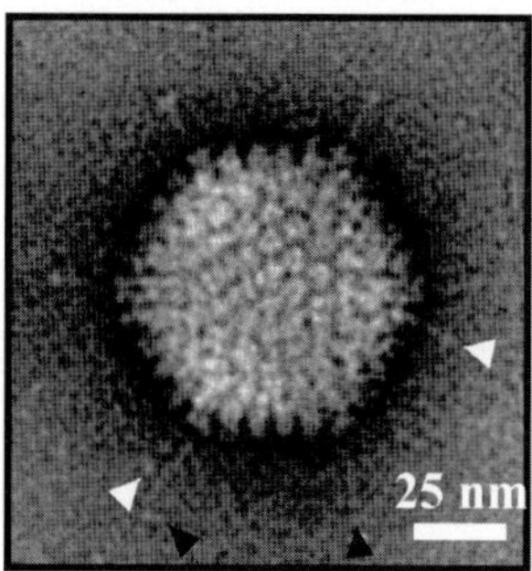

Fig. 1. Adenovirus 41 particle, showing long (black arrows) and short (white arrows) fibres. Image kindly provided by Dr G. Schoehn, CNRS/EMBL Grenoble Outstation, Grenoble, France.

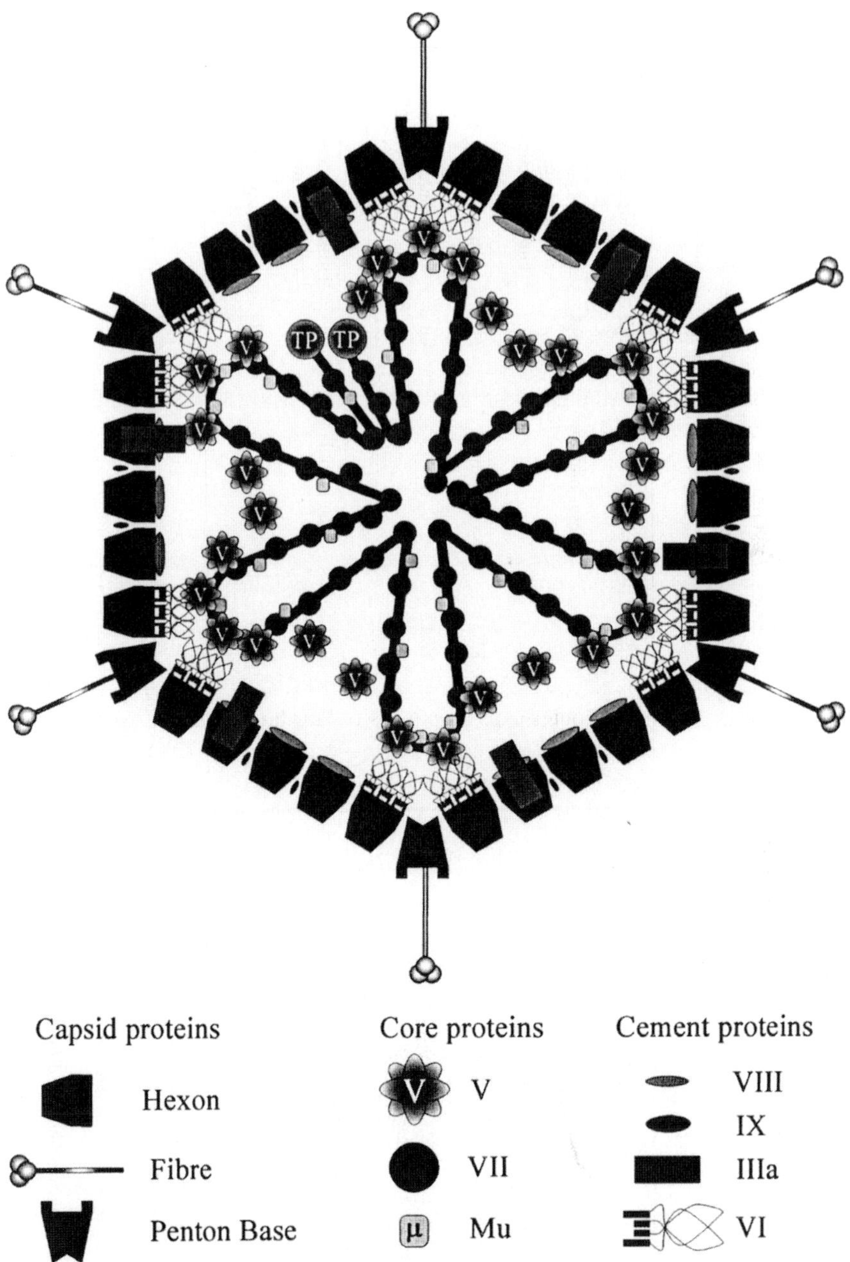

Fig. 2. Schematic diagram of an adenovirus particle, showing the icosahedral capsid, six of the fibres and the major virus proteins. Reproduced from W.C. Russell (2000), with permission.

Table 1

Properties of human adenoviruses species A to F [1]

Species	Types	DNA		G + C (%)	Length of fibres	Haemagglutination pattern	Oncogeni-city [2]	Most common infection/disease	Mode of spread
		Homology (%)							
		Intraspecies	Interspecies						
A	12, 18, 31	48-69	8-20	48	28-31	rat [3]	high	infantile diarrhoea?	endemic
B1 [4]	3, 7, 16, 21, 50	89-94	9-20	51	9-11	rhesus monkey	weak	respiratory disease, conjunctivitis	epidemic
B2 [4]	14, 11, 34, 35							kidney infection	epidemic
C	1, 2, 5, 6	99-100	10-16	58	23-31	rat, partial	nil	respiratory disease	endemic
D	8-10, 13, 15, 17, 19, 20, 22-30, 32, 33, 36-39, 42-49, 51	94-99	4-17	58	12-13	rat	nil	keratoconjunctivitis, diarrhoea in AIDS?	epidemic
E	4		4-23	58	17	rat, partial	nil	respiratory disease, conjunctivitis	epidemic
F	40, 41	62-69	15-22	52	28-33 [5]	rat [3]	nil	infantile diarrhoea	endemic

[1] Adapted with permission from Wadell *et al.* (1999).
[2] Oncogenicity for the pouch of newborn hamsters. High/weak: tumours in most animals in four months/only after 4-18 months.
[3] Agglutination of rat erythrocytes by these species discernible only after addition of heterotypic antiserum.
[4] Members of species B are divided into two clusters (subspecies) based on pronounced differences in DNA restriction sites (Wadell *et al.*, 1999). The predominant illnesses differ.
[5] Species F particles have two fibres of different length, only one of which is attached to each penton base.

Serotypes of adenoviruses

Adenovirus infection results in an immune response that probably renders lasting protection against reinfection with the same virus type and reactivation of persistent infection. Adenoviruses have partly evaded this acquired immunity through the evolution of a large number of serotypes, also called types for short, that induce no or only minor cross-protection. According to the seventh report of the International Committee on Taxonomy of Viruses, in *neutralization* assays a serotype has either no cross-reaction with other types or shows a homologous-to-heterologous titre ratio of >16 in both directions. If this ratio is 8 or 16 in either or both directions, distinctiveness of serotype is still established if: (i) the haemagglutinins are unrelated, as shown by lack of cross-reaction in haemagglutination inhibition assays, or (ii) substantial biophysical/biochemical differences in the DNAs exist (Francki *et al.*, 1991). To date, 51 adenovirus types have been recognized (de Jong *et al.*, 1999).

In principle, neutralization and haemagglutination inhibition assays could support two independent systems for the classification of adenoviruses. This is apparent from the existence of many "intermediate" types such as Ad15/9, which is Ad15 in neutralization tests and is, therefore, formally Ad15 but behaves similar to Ad9 in haemagglutination inhibition tests (Wigand and Adrian, 1989). These hybrids are only known to exist between two types of the same species, in agreement with the relatively low homology between the DNAs of different species.

Tissue tropism and clinical features of adenovirus species

Adenoviruses have mainly been isolated from the respiratory and the intestinal tract, and the eye. Less frequently, they infect the urinary bladder and liver and, even more rarely, the pancreas, myocardium and central nervous system (Horwitz, 2001). The various (sub)species are associated with different illnesses (Schmitz *et al.*, 1983; Wadell *et al.*, 1999; Table 1). All species can infect the gut, whereas species B, C, D and E can also multiply in the airways and the eye. Most members of subspecies B2 have a predilection for infecting the kidney, usually without symptoms but the infection may become persistent. On average, as many as 50% of adenovirus infections have been reported to be subclinical (Brandt *et al.*, 1969; Fox *et al.*, 1969; Van *et al.*, 1992). The only primary illness caused by *enteric adenovirus* is diarrhoea, which generally presents with less pronounced vomiting, fever and diarrhoea than rotavirus diarrhoea (Uhnoo *et al.*, 1986).

Persistence and oncogenicity of adenovirus

A characteristic feature of adenovirus is its propensity to establish (perhaps life-long) persistent symptomless infections (Horwitz, 2001). Adenoviruses have been re-isolated for two years after initial infection (Fox *et al.*, 1969). In most cases, species C and sometimes species B viruses, but not *enteric adenoviruses*, are involved. The faculty of persistence might be of epidemiological importance, and reactivation may lead to severe disease in immunodeficient subjects (See below).

Adenovirus can be detected in tonsils, appendices and mesenteric lymph nodes from normal, healthy individuals, a phenomenon which led to the discovery of the adenovirus family in 1953 and which has been confirmed in many studies (Neumann *et al.*, 1987; Horwitz, 2001). Usually, infectious virus has not been recovered directly from adenoid or tonsillar tissue but has been isolated after cultivation of such tissue in more than 50% of the tissue samples examined. One study estimated that only one in 10^7 to 10^9 cells harbours the virus or the virus genome (Strohl and Schlesinger, 1965), which presumably is kept in check by immune mechanisms. Some authors have suggested that fibroblasts are the long-term reservoir of the adenovirus or its genome (Strohl and Schlesinger, 1965), while others favoured a role for the lymphocyte, especially the B cell (Abken *et al.*, 1987).

The molecular nature of the persistent state is not known. *True latent infection*, like the persistent state of herpes simplex virus in ganglion cells, is defined as persistent infection in which the virus is not detectable, although the complete virus genome is present, allowing reactivation and virus shedding under certain conditions; disease is absent except for periods of reactivation (White and Fenner, 1986). True latency of adenovirus has not been firmly established, although suggestive evidence has been obtained. In peripheral blood lymphocytes and tonsillar tissues, adenovirus nucleic acid sequences can be demonstrated in the absence of virus (Abken *et al.*, 1987; Neumann e*t al.*, 1987). However, whether these sequences cover the whole virus genome remains to be determined. The virus isolated from lymphoid tissues after culture might originate from a low-level, conventional lytic virus replication.

Persistence of adenovirus or adenovirus DNA sequences may play a role in several, mostly chronic, diseases. Adenovirus DNA has been detected in lung tissue from patients with asthma, bronchiectasis and chronic obstructive lung disease, in heart tissue from patients with acute myocarditis and in brain tissue from senile patients (Horwitz, 2001). However, as before, it is not known whether the complete adenovirus genome is present in these cells, and the role of adenovirus DNA in the illnesses mentioned remains uncertain at this time.

Adenoviruses have an *oncogenic potential*. All six species, including enteric adenovirus (species F), can transform rat cells and species A and B are able to induce tumours in the pouch of the newborn hamster. Despite extensive searches, however, no convincing evidence has been obtained that adenovirus of any species can cause cancer in humans (Horwitz, 2001).

Diagnostic methods for adenovirus infections

Traditionally, adenoviruses have been detected and typed by virus culture and neutralization assays. This method is still the gold standard, but long periods of time are required (one to three weeks). Rapid serological methods, including immunofluorescence, enzyme immunoassay (EIA) and latex agglutination, reduce the time span of detection to hours (Clarke, 1992; Hierholzer, 1995). Generally, these methods do not distinguish between species or serotypes.

During the last few years, a multitude of PCR-based techniques have been devised. These tests take one or two days to complete and generally reduce the detection limit to about 10-100 virus particles (Allard *et al.*, 1990; Echavarria *et al.*, 1998; Xu *et al.*, 2000; Avellón *et al.*, 2001). To be optimally useful for the physician, the PCR should be able to distinguish between the seven (sub)species of adenovirus. Many PCR protocols meeting that demand have been published (Kidd *et al.*, 1996; Echavarria *et al.*, 1998; Takeuchi *et al.*, 1999; Xu *et al.*, 2000; Allard *et al.* 2001). Unfortunately, up to now none of them is free from some shortcomings in the set-up or evaluation.

Most PCR systems were not evaluated for *specificity and sensitivity* using all known prototypes and field isolates or clinical samples including at least adenoviruses of all frequently circulating types and representing all seven (sub)species of adenovirus. Failure to amplify five prototype strains of species B and seven out of 62 field isolates of this species by a published multiplex PCR system, presumably due to mismatches in the species B primers, probably resulted from this type of oversight (Xu *et al.*, 2000). Sometimes the evaluation of the PCR system did not include comparing the (sub)species identifications with the results of typing by *neutralization*. One study found that 9% of the results of a PCR species identification assay differed from the results of serotyping by neutralization (Elnifro *et al.*, 2000). Most PCR assays were not tested for possible reactivity to *other viruses* that can be expected to be present in clinical specimens and do not include a check for the absence of *inhibitors* of PCR amplification in clinical specimens in order to avoid possible false-negative results. Most PCR systems have not been evaluated in *controlled studies* including clinical specimens from healthy controls, to establish the predictive value of a positive result. This is especially important because a highly sensitive PCR tends to score positive more frequently than classical assays in cases of illness due to other causes, when simultaneous infections with adenovirus occur that are subclinical or in the incubation or recovery phase. Thus, PCR may in fact more often yield misleading results.

Identification, type designation and diagnosis of enteric adenoviruses

In the mid 1970s, electron microscopy demonstrated that in many cases of infantile diarrhoea the stool contained large numbers of adenovirus particles that induced typical adenovirus cytopathic effects in cell culture but could not be passaged or typed (Flewett *et al.*, 1975; Schoub *et al.*, 1975; Christensen, 1989). The numbers of particles can amount to up to 10^{11} virus particles per gram of faeces (Bryden *et al.*, 1975). Strangely, the chance of growing these viruses seemed inversely related to the amount of virus seen in the electron microscope (Madeley *et al.*, 1977). In a four-year study in the USA, these *"uncultivable"* adenoviruses were seen in 3.9% of 804 faecal samples from young children with diarrhoea, significantly more frequently than in those from young children without that condition, 0.6% of 812, strongly suggesting a *causal relationship* between uncultivable adenoviruses and disease (Brandt *et al.*, 1979). In contrast, in the same study, cultivable adenoviruses were found in roughly equal percentages in faecal samples from children with or without diarrhoea. Subsequently, with a newly developed type-specific EIA it was shown that the uncultivable adenoviruses were *antigenically distinct* from the 35 types known at the time, but closely related to each other (Johansson *et al.*, 1980).

Eventually, the 293 cell line, created by transformation of primary human embryonic kidney cells by Ad5 was found to permit satisfactory growth of the uncultivable adenoviruses (Takiff *et al.*, 1981). With the help of *neutralization assays* in 293 and tertiary cynomolgus monkey kidney (tMK) cells, the enteric adenoviruses were shown to differ from the 39 types known at the time and to meet the criteria for a separate serotype (de Jong *et al.*, 1983). The same study compared the growth-sustaining qualities of a number of cell culture systems, and two variants were distinguished. For the prototype strains Dugan and Tak representing these variants, the differences between the homologous and heterologous neutralization titres for the two antisera were 64 and 256, respectively, in tMK cells, but only 4 and 8, respectively, in 293 cells. The variants did not, therefore, differ consistently by a factor of >16, and according to the type definition used at the time the variants belonged to the same serotype. In *haemagglutination inhibition* assays, the two variants were different from all other types but could not be distinguished from each other (de Jong *et al.*, 1983).

DNA restriction enzyme fragment analysis, however, revealed a large difference between the DNAs of the two variants (Uhnoo *et al.*, 1984), and many virologists considered it a short-coming that great genetic differences were not taken into account in the type definition. An amendment to the definition was introduced which allowed separation of the two enteric adenovirus variants into the present two serotypes, *Ad40 and Ad41* (de Jong *et al.*, 1983). The degree of homology of the DNAs of Ad40 and Ad41 proved to be 62-69%, more than among the adenovirus types of species A and justifying classification as a single *species F* (Table 1; van Loon *et al.*, 1985). From a medical point of view, Ad40 and Ad41 behave like a single serotype, and usually EIAs aim to detect enteric adenovirus without type differentiation.

For a while, *diagnosis* of the novel types was cumbersome as growth and neutralization tests remained difficult. Most laboratories, therefore, used DNA restriction enzyme fragment analysis for typing directly from stool specimens, which is reliable but laborious. Later, monoclonal antibodies were developed that detected and distinguished the two enteric adenoviruses from each other and from the non-enteric adenoviruses, allowing the construction of convenient and sensitive EIAs (Herrmann *et al.*, 1987; Wood *et al.*, 1989; de Jong *et al.*, 1993). Still later, PCR systems have been designed for this purpose (see above).

Epidemiology of adenoviruses in immunocompetent individuals

Occurrence of adenovirus types in general

Adenoviruses are estimated to account for about 3% of the infections in the general population and for 7% of febrile infectious illnesses (Fox *et al.*, 1969). For young children, these figures are 5% and 10%, respectively. Most of the 51 types known at present are rarely isolated and other types are found with widely variable frequency, as appears from the data for 3972 strains isolated in the general diagnostic laboratory of the National Institute of Public Health (RIVM) in the Netherlands during the period 1961-1992 (Table 2). More than 95% of these adenovirus strains belonged to 11 types, namely, in order of decreasing numbers of isolates, 2, 41, 1, 7, 3, 5, 40, 4, 31, 21 and 8. These data are roughly in accordance with the seroprevalence of antibodies to the various adenovirus types as assessed in Italy (D'Ambrosio *et al.*, 1982). From an epidemiological point of view, the various species can be divided into *"endemic"* (A, C and F) and *"epidemic"* (B, D and E) species that display a constant and variable incidence, respectively, when considered over many years (Table 2).

Adenovirus types have proven to be antigenically relatively stable, at least since their discovery. Sometimes, however, new types or variants emerge and spread worldwide, e.g., Ad7, Ad19, and Ad37 (Desmyter *et al.*, 1974; Wadell and de Jong, 1980; de Jong *et al.*, 1981; Wadell *et al.*, 1981a and 1981b; Kajon *et al.*, 1996). Some evidence suggests the possible occurrence of a sudden increase in pathogenicity, as in the case of a cluster of fatal cases in young children in Malaysia associated with an adenovirus species B virus (Cardosa *et al.*, 1999).

Frequency of detection of enteric adenovirus in diarrhoea (Table 3a)

Ad40 and Ad41 are common pathogens in infants and young children. Table 3a summarizes the data from 21 papers describing the results from 27 studies, mostly conducted among 0 to 5-year-old children in 16 countries distributed over all five continents. In these surveys, the median observed proportion of enteric adenovirus-positive diarrhoeic stool samples was 4.0%. The control-corrected proportion was 2.9%. Factors that influence the frequency of enteric adenovirus infections are hard to identify in this collection of data. Possibly the *climate* plays a role: the three highest frequencies (13-31%) were reported from tropical regions. However, the median value for the frequency observed in the studies of Table 3a was not higher for investigations in tropical than for those in temperate areas, being 3.8% and 4.0%, respectively. The frequency in *hospitalized patients* ("inpatients") was found to be equal to that in non-hospitalized patients ("outpatients") in the three studies in which these two groups were compared: 5.2% and 5.3%, respectively. Similarly, comparison between the five surveys carried out in *rural* tropical areas with the four conducted in *urbanized* tropical populations does not give any indication that urbanization would affect the frequency of enteric adenovirus infections: the median values were 3.8% and 4.6%, respectively.

The median observed frequency value of 4.0% was mainly derived from publications dealing with sporadic hospital-related cases of diarrhoea. Only one study compared the frequency in *families* with that among hospital-related patients and did not find a significant difference (Mistchenko *et al.*, 1992). Two year-round surveillance studies of *day care centres/homes* yielded frequencies of 3.9% in 127 and 8.0% in 524 patients with sporadic diarrhoea, respectively (Paerregaard *et al.*, 1990; Lew *et al.*, 1991), which is again not significantly different from the overall mean value. As expected, the proportion of enteric adenovirus infections can be much higher during outbreaks.

Table 2

Adenoviruses isolated at RIVM, Bilthoven, the Netherlands, by serotype and year of isolation

Year	Serotype																					OT [1]	NT [2]	Total
	1	2	3	4	5	6	7	8	9	11	12	15	17	19	21	26	31	37	40	41			
1961-1962	33	38	14		30	4	13		1		2	6			32	1	nd [3]	nd	nd	nd	2	5	181
1963-1964	43	60	59	3	38	3	2		2		3		1		11	1	nd	nd	nd	nd	1	14	241
1965-1966	51	54	21	15	22	2	4				9	1	2				3	nd	nd	nd		12	196
1967-1968	58	55	67	8	35	2	15		1	2	1	1	2				5	nd	nd	nd	3	11	267
1969-1970	45	69	17	3	30		129		2		8	5	7				8	nd	nd	nd	5	3	331
1971-1972	72	61	19	10	36	2	99		1		4	7	1			1	3	nd	nd	nd	1	5	322
1973-1974	45	72	19	8	24	1	123	4			6	1	1	1	1		7	nd	nd	nd	1	4	318
1975-1976	59	55	27	6	23		24			1	2	3	2	1			4	1	nd	nd	1	2	211
1977-1978	49	65	28	18	24		46				2			1			1	2	nd	nd	1	13	250
1979-1980	58	59	26		26	4	79	39	1		2	1		1	1		5	1	19	2		4	328
1981-1982	32	44	26		17		18	11			2				3		1	8	4	46	26	1	15	254
1983-1984	22	41	10	4	19	1	12			2								2	2	14	38	2	14	183
1985-1986	36	50	17	4	22		10	3			1				8	1	8	1	4	35	3	19	222
1987-1988	23	32	24	1	15	2	8			1	2	1					5		4	44	3	21	186

Table 2

Continued

Year	Serotype																						
	1	2	3	4	5	6	7	8	9	11	12	15	17	19	21	26	31	37	40	41	OT [1]	NT [2]	Total
1989-1990	29	47	22	9	14	1	11		1		5					1	12		3	60	2	2	219
1991-1992	30	56	31	4	20	2	14		1	8	1		1		6	3	6		3	67	5	5	263
Total	685	858	427	93	395	24	607	57	10	14 [4]	50	26 [5]	17 [6]	7	59	10	77	11	93	272	31	149	3972
Total '81-'92 [7]	172	270	130	22	107	6	73	14	2	11 [4]	11	1		3	14	6	41	7	74	270	17	76	1327

[1] OT: Other types.

[2] NT: Not typed.

[3] nd: type not recognized because it was not yet discovered.

[4] Including 5 isolates of intermediate type Ad11+35 (in neutralization tests reactive as Ad11 as well as Ad35).

[5] Including 18 isolates of intermediate type Ad15/H9 (in neutralization tests reactive as Ad15 and in haemagglutination inhibition tests as Ad9).

[6] Including 10 isolates of intermediate type Ad17/H29 (in neutralization tests reactive as Ad17 and in haemagglutination inhibition tests as Ad29).

[7] Totals for 1981-1992 are given to allow comparison of frequencies of types including Ad40 and Ad41, which were regularly recognized only after 1980.

418

Table 3a

Frequency of detection of enteric adenovirus Ad40 and Ad41 in faecal samples from children with sporadic diarrhoea and controls ^

Country	Reference	Study population	Detection of (enteric) adenovirus [1]	Typing of Ad40 and Ad41 or Ad40/41	Enteric adenovirus Ad40/41	
					With diarrhoea	No diarrhoea
Argentina	Mistchenko *et al.*, 1992	newborns in urban families	EIA (PA)	dot-blot hybridization	3.3% of 180 pat.**	0.8% of 766 sub.**
idem	idem	urban outpatient children	idem	idem	5.4% of 129 pat.	
Australia	Barnes *et al.*, 1998	urban hospitalized children	EM, EIA	EIA (MoAb, de Jong)	6.0% of 4637 pat.	
Bangladesh	Jarecki-Khan *et al.*, 1993	rural hospitalized children [2]	EIA (PA)	EIA (MoAb, de Jong)	3.7% of 4027 pat.	
Denmark	Paerregaard *et al.*, 1990	urban day care children	EM	DREA [3]	3.9% of 127 pat.*	0.0% of 51 sub.*
France	Bon *et al.* 1999	urban outpatient children	EIA (MoAb, Adenoclone)	EIA (MoAb, Adenoclone)	3.1% of 414 pat.*	0% of 50 sub.*
Guatemala	Cruz *et al.*, 1990	rural non-hospitalized children	EIA (MoAb, Adenoclone)	EIA (MoAb, Adenoclone)	14% of 385 pat.**	4.7% of 191 sub.**
idem	idem	urban hospitalized children	idem	idem	31% of 48 pat.	
India	Bhan *et al.*, 1988	children at rural health centre	EIA (PA)	EIA (PA)	3.8% of 340 pat.*	2.5% of 315 sub.*
idem	idem	urban hospitalized children	idem	idem	0.9% of 330 pat.*	2.5% of 319 sub.*
Iran	Saderi *et al.*, 2002	urban children	EIA (MoAb, de Jong)	EIA (MoAb, de Jong)	6.7% of 872 pat.	
Italy	Caprioli *et al.*, 1996	urban children	latex agglutination	EIA (MoAb, de Jong)	2.1% of 618 pat.**	0.0% of 135 sub.**

Table 3a

Continued

Country	Reference	Study population	Detection of (enteric) adenovirus 1)	Typing of Ad40 and Ad41 or Ad40/41	Enteric adenovirus Ad40/41	
					With diarrhoea	No diarrhoea
Italy	Vizzi *et al.*, 1996	urban hospitalized children	EIA (MoAb, de Jong)	EIA (MoAb, de Jong)	2.6% of 273 pat.*	0.0% of 137 sub.*
Japan	Shinozaki *et al.*, 1991	urban children	various [4]	EIA (MoAb, Adenoclone)	2.6% of 2223 pat.	
Mexico	Maldonado *et al.*, 1998	rural non-hospitalized children	EIA (MoAb, Adenoclone)	EIA (MoAb, Adenoclone)	2.0% of 305 pat.*	1.4% of 1949 sub.*
Netherlands	de Jong *et al.*, 1993	hospitalized patients	EIA (MoAb, de Jong)	EIA (MoAb, de Jong)	4.0% of 6666 pat.	
South Africa	Kidd *et al.*, 1986	urban hospitalized children	EIA (PA)	DREA or dot-blot hybridization	6.5% of 616 pat.	
South Africa	Tiemessen *et al.*, 1989	rural children	dot-blot hybridization	dot-blot hybridization	13% of 310 pat.**	2.5% of 122 sub.**
South Africa	Steele *et al.*, 1998	out- and inpatient children	EIA (MoAb, Adenoclone)	EIA (MoAb, Adenoclone)	2.7% of 225 pat.*	1.8% of 56 sub.*
Sweden	Uhnoo *et al.*, 1986	outpatient children	EIA (PA)	EIA (PA) and DREA	8.5% of 272 pat.	
idem	idem	urban hospitalized children	idem	idem	6.9% of 144 pat.**	0.0% of 200 sub.**
Thailand	Herrmann *et al.*, 1991	urban outpatient children	EIA (MoAb, Herrmann)	EIA (MoAb, Herrmann)	2.6% of 1691 pat.**	0.5% of 1459 sub.**
USA	Brandt *et al.*, 1979	urban outpatient children	EM, IEM	no growth in cell culture		
idem	idem	urban inpatient children	idem	idem		

Table 3a

Continued

Country	Reference	Study population	Detection of (enteric) adenovirus 1)	Typing of Ad40 and Ad41 or Ad40/41	Enteric adenovirus Ad40/41	
					With diarrhoea	No diarrhoea
USA	Kotloff *et al.*, 1989	urban outpatient children	EIA (PA)	EIA (MoAb, Singh-Naz)	4.1% of 246 pat.**	0.0% of 155 sub.**
idem	idem	urban inpatient children	idem	idem	6.2% of 292 pat.**	2.3% of 217 sub.**
idem	idem	pooled out- and inpatients	idem	idem	5.2% of 538 pat.**	1.3% of 372 sub.**
USA	Lew *et al.*, 1991	day care + household children	EIA (MoAb anti-hexon)	EIA (MoAb, de Jong)	2.8% of 565 pat.*	1.6% of 129 sub.*
Median value					**4.0%**	**1.1%**

^ Abbreviations: "pat.": patients, "sub.": subjects.

* Difference between patients and controls not statistically significant by a chi-square test. If necessary, Fisher's exact test was applied.

** Difference between patients and controls statistically significant, P<0.05.

1) MoAb-based EIAs: "Adenoclone": Adenoclone Type 40/41 from Cambridge Bioscience, Worcester, MA, USA, or Meridian Diagnostics Inc., Cincinnaty, Ohio, USA. "de Jong": MoAbs specific for Ad-genus, Ad40, and Ad41, respectively, developed in the Netherlands: de Jong, *et al.*, 1993;

2) The positivity rate for *Vibrio cholerae* (382/4409 samples = 8.7%) in the Jarecki-Khan study was not included in the calculations for this table.

3) "DREA" means DNA restriction enzyme analysis.

4) "Various" methods include EM, latex agglutination and monoclonal-based EIA (Adenoclone).

The *burden* of enteric adenovirus diarrhoea was calculated in five papers (Rodriguez *et al.*, 1985; Kotloff *et al.*, 1989; Mistchenko *et al.*, 1992; Van *et al.*, 1992; Maldonado *et al.*, 1998). Under-two-year-olds had a median value of 6.4 episodes of enteric adenovirus diarrhoea per 100 person-years, range: 2.5 to 8.9. At higher ages, the incidence is decreased. In the family study referred to above, the rate in 0-2 year-olds was 7.4 and in 0-5 year-olds 3.9 per 100 person-years (Mistchenko *et al.*, 1992).

Quality of diagnostic methods for detection of enteric adenoviruses

Some caution is necessary when comparing individual aetiological studies. Apart from the composition of the study population and the method of sample collection, the *threshold* of detection of enteric adenovirus may also differ (Grandien *et al.*, 1987; Clarke, 1992; Noel *et al.*, 1994). However, high sensitivity is relatively unimportant for the diagnosis of enteric adenovirus infection, given the usually large amounts of virus shed in stools of enteric adenovirus-infected symptomatic children. As a consequence, virus culture, electron microscopy, EIA, latex agglutination and PCR have roughly the same sensitivity for diagnosing enteric adenovirus infections (Grandien *et al.*, 1987; Wood *et al.*, 1989). Still, an interesting instance of greatly insufficient sensitivity has been reported. In 1990, it was published that the dominant variant of Ad41 was not detected by Adenoclone, a commercial monoclonal antibody (MoAb)-based EIA specific for Ad40 and Ad41 and marketed by Cambridge Bioscience (Scott-Taylor *et al.*, 1990). This failure – which was rectified shortly after its detection - was the consequence of the replacement of the epidemiologically prevailing subtype of Ad41/M1 by another subtype, Ad41/M3, during the 1980s. In fact, such a risk is inherent in all MoAb-based virus assays, because all viruses are continuously subject to mutations which may eliminate the epitope of the virus to which the MoAb is targeted. This example illustrates that even adequately validated MoAb-based EIAs can fail and need post-marketing surveillance for assured maintenance of efficacy.

Latex agglutination tests for (enteric) adenovirus may lack sensitivity as well as specificity. In an Italian study, only 13 of 24 latex-positive stool samples reacted in an EIA specific for Ad40/41 (Caprioli *et al.*, 1996). In Belgium, a number of stool samples from young babies without diarrhoea in a neonatal intensive care unit were latex-positive whereas the results of EIAs and electron microscopy on these specimens proved negative (Ieven *et al.*, 1994).

Although in the cases described above the manufacturers concerned have rectified the failures of their products, such events may have affected the reliability of individual frequency data in Table 3a. Because, however, the median observed frequency for enteric adenovirus given in this Table is based on 26 studies using different methods of detection and different sources of EIA, this value is probably reliable.

Causal role of enteric adenovirus in disease and duration of virus shedding

In 9 of the 12 controlled studies on enteric adenovirus infections listed in Table 3a the virus was significantly more often detected in stool samples from patients with diar-

422

rhoea than in stool samples from control subjects. The studies mentioned therefore firmly established the ability of enteric adenovirus infection to cause diarrhoea. On the other hand they have also revealed the occurrence of subclinical infections.

The relatively low rate of enteric adenovirus-positives among control subjects is explainable by the short-lived nature of *virus shedding*, being 8-13 days in a Swedish hospital study (Uhnoo *et al.*, 1986) and 1-14 days (mean: 3.9 days) in an American day care survey (Van *et al.*, 1992). This short excretion period reduces the chance that the virus will be found during the incubation period or after the recovery from the illness, or in subclinical infections. This aspect of enteric adenovirus infection contrasts with that of some non-enteric adenoviruses (see section 2.7).

Outbreaks of enteric adenovirus diarrhoea have provided additional evidence for a causal potential for this virus. Such events have been reported rarely in hospitals (Kotloff *et al.*, 1989) but more often in day care centres. During such outbreaks, attack rates of infection can be as high as 38%. Of the 249 infected children involved in ten outbreaks of diarrhoea, 54% had symptoms (Van *et al.*, 1992).

Spread of enteric adenovirus from symptomatic patients

Facilitated by the large numbers of enteric adenovirus shed from the gut of infected children with diarrhoea, often extending from several days before until several days after the disease (Van *et al.*, 1992), outbreaks of enteric adenovirus in day care centres are common events. Yet, family studies indicate a low infectiousness of the virus in this setting. In two studies (Rodriguez *et al.*, 1985; Mistchenko *et al.*, 1992), none of the 9 and 29 family contacts, respectively, of such patients proved to have enteric adenovirus in their stool. In contrast, rotavirus spread readily to exposed children (56% of 62) as well as adults (25% of 65) when a child in a similar family had a rotavirus infection (Rodriguez *et al.*, 1985).

Changing proportions of Ad40 and Ad41 infections

As described above, Ad40 and Ad41 are antigenically related. Interestingly, evidence for the expected immunological competition between the two types is provided by the occurrence of variation in the ratio of the frequencies of infections with Ad40 and Ad41. In the Netherlands, the proportion of Ad41 infections compared to the total number of Ad40/Ad41 infections rose from 12% in 1980 to 90-100% in the period 1987 to 1995 (de Jong *et al.*,1993; Fig. 3). During the same period, there was no significant change in the frequency of the two types taken together. Concomitantly with the described change in the proportion of Ad41 infections, the proportion of subtype Ad41/M3 within Ad41 in the Netherlands increased from about 40% in 1981 to 80% in 1984 (Fig. 4). Possibly, this more successful subtype not only superseded the other two subtypes of Ad41, but also Ad40.

The described change in the ratios of Ad40 and Ad41 infections seems to have been a general occurrence (Fig. 3). Similar transitions were observed in Australia in about the same period as in the Netherlands (Grimwood *et al.*, 1995), in Canada one or two

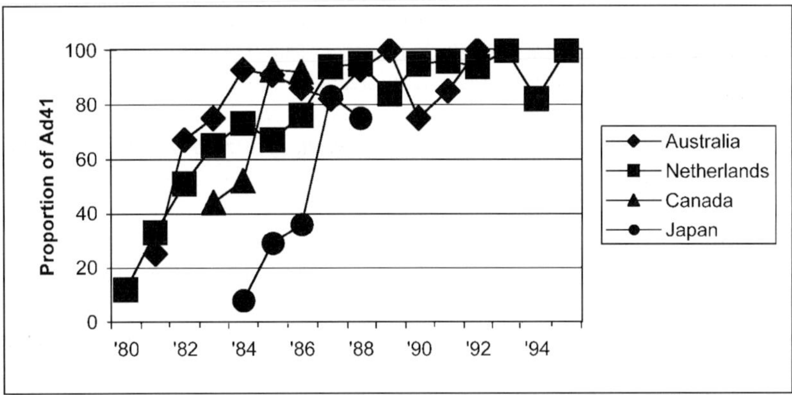

Fig. 3. Proportion of Ad41 infections relative to the total number of enteric adenovirus Ad40 and Ad41 infections reported in separate studies in Australia, the Netherlands, Canada and Japan for the period 1980 up to and including 1995 (See Grimwood *et al.*, 1995; de Jong *et al.*, 1993; Brown, 1990; Shinozaki *et al.*, 1991, respectively).

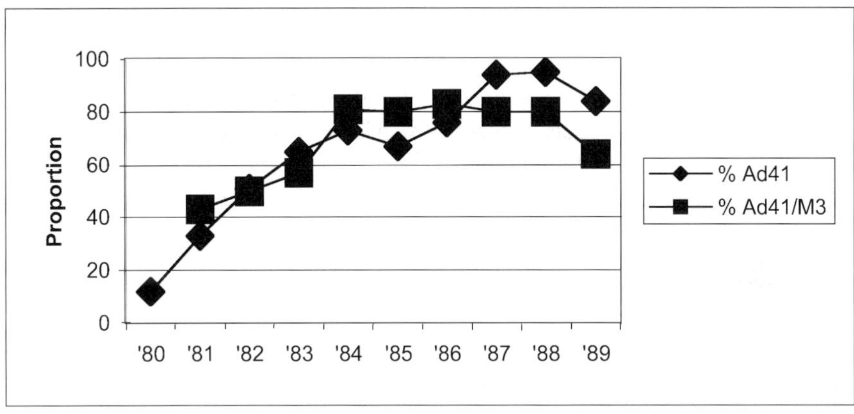

Fig. 4. Proportion of Ad41 infections relative to the total number of enteric adenovirus Ad40 and Ad41 infections, and proportion of Ad41/M3 infections relative to the total number of enteric adenovirus Ad41 infections in the Netherlands for the period 1980 up to and including 1989.

years later (Brown, 1990) and in Japan four to five years later (Shinozaki *et al.*, 1991). Still, the change may not have taken place in all countries or may have been reversed later. Of seven enteric adenoviruses detected in children with diarrhoea in Italy in

1993 and 1994, five were Ad40 and two Ad41 (Vizzi et al., 1996), and of 59 enteric adenoviruses detected in children with diarrhoea in Iran in 1999, 29 were Ad40 and 30 were Ad41 (Saderi et al., 2002).

Frequency of enteric adenovirus infections in comparison to that of other enteric pathogens observed in controlled studies

Seventeen of the 27 studies on enteric adenovirus infection listed in Table 3a included a group of healthy control subjects, and some of these 17 studies also provided frequencies for non-enteric adenovirus, rotavirus, calicivirus and astrovirus (entered in Table 3b, except calicivirus) and pathogenic microorganisms and the presence of any pathogen (entered in Table 3c) in the stool samples examined. By subtracting the frequency found in the controls from that found in the patients, the frequency of cases in which the diarrhoea was actually caused by the virus concerned — the mean value of the *"control-corrected frequency"* — was estimated for each of the mentioned viruses. Table 4, seventh column, gives both the observed and the control-corrected frequencies.

Table 3b

Frequency of detection of nonenteric adenovirus, rotavirus and astrovirus in faecal samples from children with sporadic diarrhoea and controls ^

Country	Reference	Nonenteric adenovirus		Rotavirus		Astrovirus	
		With diarrhoea	No diarrhoea	With diarrhoea	No diarrhoea	With diarrhoea	No diarrhoea
Argentina	Mistchenko *et al.*, 1992	14% of 180 pat.*	13% of 766 sub.*				
idem	idem	11% of 129 pat.					
Australia	Barnes *et al.*, 1998	5.1% of 827 pat.		40% of 4637 pat.			
Bangladesh	Jarecki-Khan *et al.*, 1993	0.5% of 4027 pat.		26% of 4027 pat.			
Denmark	Paerregaard *et al.*, 1990	0.0% of 127 pat.*	0% of 51 sub.*	19% of 127 pat.**	2.0% of 51 sub.**		
France	Bon *et al.* 1999			61% of 414 pat.**	10% of 50 sub.**	6.3% of 414 pat.*	2.0% of 50 sub.*

Table 3b
Continued

Country	Reference	Nonenteric adenovirus		Rotavirus		Astrovirus	
		With diarrhoea	No diarrhoea	With diarrhoea	No diarrhoea	With diarrhoea	No diarrhoea
Guatemala	Cruz *et al.*, 1990			4.7% of 385 pat.**	0.5% of 191 sub.**		
idem	idem			51% of 57 pat.			
India	Bhan *et al.*, 1988	8.8% of 340 pat.**	1.0% of 315 sub.**	16% of 340 pat.**	2.9% of 315 sub.**		
idem	idem	1.5% of 330 pat.*	0.9% of 319 sub.*	15% of 330 pat.**	1.9% of 319 sub.**		
Iran	Saderi *et al.*, 2002	2.0% of 872 pat.					
Italy	Caprioli *et al.*, 1996	3.9% of 618 pat.**	0.0% of 135 sub.**	24% of 618 pat.**	2.2% of 135 sub.**		
Italy	Vizzi *et al.*, 1996	3.2% of 273 pat.*	6.5% of 137 sub.*				
Japan	Shinozaki *et al.*, 1991	1.1% of 2223 pat.					
Mexico	Maldonado *et al.*, 1998			0.7% of 305 pat.**	0.1% of 1949 sub.**	26% of 305 pat.**	19% of 1949 sub.**
Netherlands	de Jong *et al.*, 1993			22% of 6666 pat.			
South Africa	Kidd *et al.*, 1986	3.4% of 616 pat.		14% of 616 pat.			
South Africa	Tiemessen *et al.*, 1989			13% of 310 pat.**	2.5% of 122 sub.**		

Table 3b
Continued

Country	Reference	Nonenteric adenovirus		Rotavirus		Astrovirus	
		With diarrhoea	No diarrhoea	With diarrhoea	No diarrhoea	With diarrhoea	No diarrhoea
South Africa	Steele *et al.*, 1998			20% of 225 pat.**	3.6% of 56 sub.**	7.1% of 225 pat.*	1.8% of 56 sub.*
Sweden	Uhnoo *et al.*, 1986	4.4% of 272 pt.		41% of 272 pat.			
idem	idem	7.6% of 144 pat.**	1.5% of 200 sub.**	53% of 144 pat.**	2.0% of 200 sub.**		
Thailand	Herrmann *et al.*, 1991	1.5% of 1691 pat.*	1.1% of 1459 sub.*	19% of 1691 pat.**	1.0% of 1459 sub.**		
USA	Brandt *et al.*, 1979	0% of 200 pat.*	0.3% of 290 sub.*	22% of 200 pat.**	3.4% of 290 sub.**		
idem	idem	0.8% of 604 pat.*	1.0% of 522 sub.*	39% of 604 pat.**	4.0% of 522 sub.**		
USA	Kotloff *et al.*, 1989	3.3% of 246 pat.**	30% of 155 sub.**				
idem	idem	7.2% of 292 pat.*	6.9% of 217 sub.*				
USA	Kotloff *et al.*, 1989	5.4% of 538 pat.**	16% of 372 sub.**	21% of 538 pat.**	7.3% of 372 sub.**		
idem	idem	4.7% of 565 pat.*	6.0% of 129 sub.*	3.0% of 496 pat.*	1.8% of 55 sub.*	4.0% of 524 pat.**	0.7% of 138 sub.**
Median value		**3.4%**	**1.1%**	**21%**	**2.2%**	**6.7%**	**1.9%**

^   Abbreviations: "pat.": patients, "sub.": subjects.

*   Difference between patients and controls not statistically significant by a chi-square test. If necessary, Fisher's exact test was applied.

**   Difference between patients and controls statistically significant, P<0.05.

***   "Pathogenic microorganisms" include *Salmonella* spp., *Shigella* spp., *Campylobacter jejuni*, *Escherichia coli*, *Yersinia enterocolitica*, *Aeromonas*, *Cryptosporidium* spp., *Giardia* spp. and *Entamoeba* spp.

Table 3c

Frequency of detection of pathogens in faecal samples from children with sporadic diarrhoea and controls ^

Country	Reference	Pathogenic microorganisms ***		Positive for any pathogen	
		With diarrhoea	No diarrhoea	With diarrhoea	No diarrhoea
Argentina	Mistchenko *et al.*, 1992				
idem	idem				
Australia	Barnes *et al.*, 1998	10% of 3785 pat.		56% of 4637 pat.	
Bangladesh	Jarecki-Khan *et al.*, 1993			55% of 4409 pat.	
Denmark	Paerregaard *et al.*, 1990				
France	Bon *et al.* 1999				
Guatemala	Cruz *et al.*, 1990				
idem	idem	65% of 57 pat.		90% of 57 pat.	
India	Bhan *et al.*, 1988				
idem	idem				
Iran	Saderi *et al.*, 2002				
Italy	Caprioli *et al.*, 1996	39% of 618 pat.**	8.9% of 135 sub.**	59% of 618 pat.**	10% of 135 sub.**
Italy	Vizzi *et al.*, 1996				
Japan	Shinozaki *et al.*, 1991				
Mexico	Maldonado *et al.*, 1998	1.7% of 305 pat.*	2.9% of 1949 sub.*		

Table 3c

Continued

Country	Reference	Pathogenic microorganisms ***		Positive for any pathogen	
Netherlands	de Jong *et al.*, 1993				
South Africa	Kidd *et al.*, 1986				
South Africa	Tiemessen *et al.*, 1989				
South Africa	Steele *et al.*, 1998				
Sweden	Uhnoo *et al.*, 1986	14% of 272 pat. **	8% of 200 sub.**	63% of 272 pat.**	8.0% of 200 sub.**
idem	idem	19% of 144 pat.**	8% of 200 sub.**	77% of 144 pat.**	8.0% of 200 sub.**
Thailand	Herrmann *et al.*, 1991				
USA	Brandt *et al.*, 1979			25% of 200 pat.**	5% of 290 sub.**
idem	idem			46% of 604 pat.**	7% of 522 sub.**
USA	Kotloff *et al.*, 1989				
idem	idem				
idem	idem	10% of 538 pat.**	1.1% of 372 sub.**	33% of 538 pat.**	9.4% of 372 sub.**
USA	Lew *et al.*, 1991				
Median value		14%	4.0%	56%	8.0%

^ Abbreviations: "pat.": patients, "sub.": subjects.

* Difference between patients and controls not statistically significant by a chi-square test. If necessary, Fisher's exact test was applied.

** Difference between patients and controls statistically significant, P<0.05.

*** "Pathogenic microorganisms" include *Salmonella* spp., *Shigella* spp., *Campylobacter jejuni*, *Escherichia coli*, *Yersinia enterocolitica*, *Aeromonas*, *Cryptosporidium* spp., *Giardia* spp. and *Entamoeba* spp.

A causal relationship with diarrhoea was indicated in at least two-thirds of the controlled studies for rotavirus, calicivirus, astrovirus, and enteric adenovirus, but only in a quarter for non-enteric adenovirus (Table 4, last column). The role of these *non-enteric adenoviruses* in diarrhoea is unclear. The calculated control-corrected frequency of 0% certainly does not mean that these viruses cannot cause this condition. Non-enteric adenoviruses are often present in stool for many months after infection. This phenomenon may obscure a possible pathogenic role for the virus. Indeed, serological evidence indicates that non-enteric adenoviruses can also induce diarrhoea (Uhnoo *et al.*, 1984), especially adenoviruses of species A in young children (Adrian *et al.*, 1987).

In three studies presented in Tables 3a and 3b using a broad series of diagnostic assays (Uhnoo *et al.*, 1984 and 1986; Kotloff *et al.*, 1989; Caprioli *et al.*, 1996, all three conducted in industrialized countries), on average, a pathogen was identified in only 59% of the patients. The large "diagnostic deficit" of 41% is partly due to the lack of assays for the detection of astrovirus and calicivirus in the three studies mentioned, but means also that all presented frequencies should be considered as minimum values.

Age distribution of adenovirus infections

Similar to infections with rotavirus, astrovirus and calicivirus, those with adenovirus species C and F occur early in life. Species B1 virus infections can occur in the same age group but have a higher frequency of occurrence in older children and adults, especially recently assembled military recruits (Horwitz, 2001). These conclusions are corroborated and extended by seroprevalence studies. In one study, antibodies to the "endemic" adenovirus types 1, 2 and 5 proved to be most common and were present in about 40% of five-year-old children (D'Ambrosio *et al.*, 1982). Antibodies to the endemic types 18 and 31 reached prevalence levels of 20% at this age. Antibodies to the epidemic types 3, 4 and 7 were absent or present at low titre in this age group and developed during the next five years of life.

Examination of sera from widely different geographical areas showed that among two- and five-year-old children, 14% and 50%, respectively, possess neutralizing antibodies to *enteric adenoviruses* (Kidd *et al.*, 1983). In Japan, EIA-reactive antibodies were found in 20% of children 1-6 months old and in 50% of those 37-48 months old (Shinozaki *et al.*, 1987). In most studies, the highest frequency of shedding of enteric adenovirus (as well as rotavirus, calicivirus and astrovirus) has been shown with children less than 12 months of age. Combining eight reports, the age of the patients varied from 7-17 months, median: 10 months. Rarely do children over five years of age contract enteric adenovirus diarrhoea. Only one of the publications consulted for this review mentioned infection of an adult (Mistchenko *et al.*, 1992).

Seasonal distribution of adenovirus infections

Adenoviruses of all types are circulating year-round. Sometimes, however, marked seasonal prevalence has been reported. For instance, the seasons for highest rates of respiratory adenovirus infection, usually with the epidemic types 3, 4 or 7, among mili-

Table 4

Frequencies of detection in controlled studies of enteric pathogens in faecal specimens from children with sporadic diarrhoea and from controls

Pathogen	Number of studies	Patients (diarrhoea)		Controls (no diarrhoea)		Control-corrected mean frequency [2]	Significant studies [3]
		Mean frequency [1]	Range	Mean frequency [1]	Range		
Rotavirus	15	21% (1438/6727)	0.7-61%	2.0% (119/6086)	0.1-10%	19%	14/14
Calicivirus [4]	2	15% (80/529)	14-19%	4.3% (5/116)	0.0-7.6%	11%	2/2
Astrovirus	4	9.7% (142/1468)	4.0-26%	1.8% (40/2193)	0.7-2.0%	7.9%	2/2
Enteric adenovirus	16	4.1% (264/6445)	0.9-14%	1.2% (76/6251)	0.0-4.7%	2.9%	9/12
Non-enteric adenovirus	13	3.4% (190/5610)	0.0-14%	4.6% (217/4695)	0.0-30%	0%	2/8
Microorganisms [5]	5	20% (367/1877)	1.6-39%	3.0% (80/2656)	1.1-8.9%	17%	not applicable

[1] Total numbers of patients (controls) examined with the indicated pathogen / the total numbers of patients (controls) examined. The two numbers are given in parentheses.

[2] To obtain the "control-corrected mean frequency", the mean frequency in the controls was subtracted from that in the patients.

[3] "Significant studies": number of studies that indicated a significantly higher percentage of pathogen-positive faecal samples from patients with diarrhoea compared with control subjects/number of controlled studies large enough to achieve that all numbers of patients and controls statistically expected in a chi-square test were greater than five.

[4] The data for calicivirus were derived from Bon *et al.*, 1999 and Farkas *et al.*, 2000.

[5] "Microorganisms" included *Salmonella* spp., *Shigella* spp., *Campylobacter jejuni*, *Escherichia coli*, *Yersinia enterocolitica*, *Aeromonas*, *Cryptosporidium* spp., *Giardia* spp. and *Entamoeba* spp.

tary recruits in the USA in the 1960s were winter and spring (Foy, 1989). In contrast, in a general diagnostic setting in the Netherlands during the period 1981-1989, a sharp peak of isolations of the epidemic adenovirus types 3 and 7 was observed in August (de Jong *et al.*, 1993). The endemic adenovirus types 1 and 2 were detected throughout the year. In the same setting and period, Ad40 and Ad41 were isolated year-round, with some enhanced frequency during autumn and winter. No consistent seasonal pattern of enteric adenovirus infection was observed in Australia in the period 1980-1993 (Barnes *et al.*, 1998).

Adenovirus infections in immunodeficient patients

Introduction

During the last decades, advances in transplantation technology and treatment options for cancer as well as the current AIDS epidemic have created novel opportunities for a number of pathogens, including adenovirus, to invade humans and set up persistent and generalized infections in the immunocompromized host (Hierholzer, 1992; Carrigan, 1997; Lukashok and Horwitz, 1998; Goodgame, 2001). Generally, such infections are difficult to treat, tend to be long-term, add to the burden of debilitation in the patient and sometimes rapidly overwhelm the patient and result in death.

Adenovirus types in immunodeficient patients

All seven adenovirus (sub)species A, B1, B2, C, D, E and F have been isolated from immunodeficient patients. Of 29 typed isolates from the adenovirus-associated *fatal cases* in bone marrow transplant (BMT) recipients, one isolate was of species A (Ad12/31), two of B1 (2×Ad7), 10 of B2 (8×Ad11, 2×Ad35), 14 of C (5×Ad1, 4×Ad2, 5×Ad5), two of D (Ad29, Ad32) and none of E or F (Table 5). Generally, the serotypes involved reflect the (past) exposure of these patients to the common childhood viruses in the environment (Hierholzer, 1992). Most adenoviruses from immunocompetent individuals belong to either species C or subspecies B1 (Table 2). In comparison, in patients with any kind of immunodeficiency, infections with *subspecies B2* viruses tend to be more prevalent, especially when the kidney is involved (Hierholzer, 1992; Schirm *et al.*, 1991). In fact, a new member of this subspecies, namely Ad35, was discovered in a renal transplant recipient.

Strikingly, most of the adenoviruses infecting the gut of *HIV-infected patients* represent serotypes of species D which are rarely encountered in the stools of immunocompetent patients or in patients suffering from other kinds of immunodeficiencies (Hierholzer, 1992; Khoo *et al.*, 1995; de Jong *et al.*, 1999). Typical AIDS-associated adenoviruses include Ad types 9, 17, 20, 22, 23, 26, 27, and 42-51. The cause of this predilection is not known. It has been suggested that the long-term infection characteristic for AIDS patients may have provided the opportunity for mutations to occur within a strain or for recombinational events between coinfecting serotypes to

Table 5

Frequencies of detection of adenovirus infections in bone marrow transplant recipients

Reference	Study period	All adenovirus infections	Adenovirus in diarrhoea [1]	DAD (disseminated adenovirus disease)	Fatal cases from adenovirus [2]	Fatality rate [3]
Yolken et al., 1982	1980/81		15% (12/78)		6.4% (5/78)	42% (5/12)
Shields et al., 1985	1976-1982	4.9% (51/1051)	2.0% (21/1051)	1.0% (10/1051)	0.5% (5/1051)	9.8% (5/51)
Ambinder et al., 1986	1977-1984	3.8% (19/502)		0.4% (2/502)	0.4% (2/502)	11% (2/19)
Wasserman et al., 1988	1979-1986	18% (17/96)				
Flomenberg et al., 1994	1987-1990	21% (42/201)	4.0% (8/201)	3.0% (6/201)	3.5% (7/201)	17% (7/42)
idem, 0-18-year-olds	idem	31% (26/83)	6.0% (5/83)	2.4% (2/83)	4.8% (4/83)	15% (4/26)
idem, >18-year-olds	idem	14% (16/118)	2.5% (3/118)	3.4% (4/118)	2.5% (3/118)	19% (3/16)
Cox et al., 1994	1989/90		1.7% (5/296)			
Troussard et al., 1993	1984-1991		11% (10/94)	1.1% (1/94)	1.1% (1/94)	10% (1/10)
Blanke et al., 1995	1990-1992	14% (10/74)	6.8% (5/74)	9.5% (7/74)	8.1% (6/74)	60%(6/10)
Bruno et al., 1997	1976-1994	9.1% (434/4768)	4.9% (234/4768)	0.4% (19/4768)	1.1% (52/4768)	12% (52/434)
Hale et al., 1999	1990-1994	6.3% (13/206)	1.9% (4/206)	2.4% (5/206)	2.9% (6/206)	46% (6/13)
Howard et al., 1999	1986-1997	12% (64/532)			2.1% (11/532)	17% (11/64)
idem, 0-18-year-olds	idem	23% (29/127)				
idem, 18-58-year-olds	idem	8.6% (35/405)				

Table 5

Continued

Reference	Study period	All adenovirus infections	Adenovirus in diarrhoea [1]	DAD (disseminated adenovirus disease)	Fatal cases from adenovirus [2]	Fatality rate [3]
Baldwin *et al.*, 2000	1987-1998	17% (100/572)	13% (74/572)	1.9% (11/572)	1.0% (6/572)	6% (6/100)
idem, 0-19-year-olds	idem	20% (85/436)				
idem, >19-year-olds	idem	11% (15/136)				
Chakrabarti *et al.*, 2000	1997-1998	5.3% (4/75)	1.3% (1/75)	1.3% (1/75)	1.3% (1/75)	25% (1/4)
Echavarria *et al.*, 2001	1985-1999	12% (38/328)				
Median values		**12%**	**4.4%**	**1.3%**	**1.3%**	**17%**

[1] Number of patients with diarrhoea, colitis or other gastrointestinal symptoms and adenovirus in the stool/total number of transplant recipients in the study.
[2] Number of cases in which adenovirus had possibly or probably contributed significantly to death.
[3] Number of cases in which adenovirus had possibly or probably contributed significantly to death/number of adenovirus-infected patients.

take place (Hierholzer, 1992). However, a number of the "typically" AIDS-associated adenoviruses are not strictly confined to this group of patients. The mentioned proto-types up to Ad42 were first isolated from immunocompetent humans long before AIDS was discovered. Neither has it been established that they circulate more frequently in HIV-infected patients than in immunocompetent individuals. In this regard, we should realize that data relating to adenovirus serotype distribution in the gut of immuno-competent adults are scanty. For example, at the National Institute of Public Health (RIVM) in the Netherlands, only seven of the 1478 strains isolated during the period 1981-1992 from faecal specimens of non-HIV-infected patients were from individuals aged 25-59 years. Perhaps, therefore, the "new" adenoviruses have just come to light as stool samples from that particular age group are now examined more often with AIDS patients frequently suffering from intestinal disorders of varied etiology. Possibly phylogenetic analysis can produce evidence for or against these various hypotheses.

Enteric adenoviruses – which are widespread among young children in the com-munity - have rarely been isolated from immunocompromized individuals (Hierholzer, 1992; Khoo *et al.*, 1995; Hale *et al.*, 1999; de Jong *et al.*, 1999). Hierholzer (1992) only mentioned the detection of enteric adenovirus in four of such individuals. This paucity of reports, however, may be due to a lack of appropriate diagnosis of these "unculti-vable" viruses. For instance, 11 of 105 cases of adenovirus infection in BMT patients were diagnosed by electron microscopy of stool specimens and not by virus culture (Baldwin *et al.*, 2000). A considerable proportion of these uncultivable adenoviruses was probably Ad40 or Ad41. Two unpublished cases of gastrointestinal Ad41 infection in young BMT patients came to the knowledge of the author by personal communica-tion (C. de Bel, A.C.M. Kroes and M.J.D. van Tol, Leiden, the Netherlands). A true indication that enteric adenovirus rarely infects HIV-infected patients comes from an Ad40/41-specific EIA study of 82 of such individuals with diarrhoea, among whom only one enteric adenovirus infection was detected (Giordano *et al.*, 1999).

Incidence of adenovirus infections in bone marrow transplant (BMT) patients

Perhaps because of the more aggressive immunosuppressive treatments of BMT recipi-ents, adenovirus infections are especially frequent and sometimes fatal in these patients. In 14 papers, adenovirus infections were reported to occur in 3.8-21% (median value 12%) of BMT recipients (Table 5). BMT children (0-18 years-olds) had a higher chance of becoming infected with adenovirus compared to adults, with a relative risk of 1.8-2.7 (Flomenberg *et al.*, 1994; Howard *et al.*, 1999; Baldwin *et al.*, 2000). Adenovirus was possibly or probably responsible for the death of 1.3% of all BMT recipients and of 17% of the adenovirus-infected BMT patients. From 0.4 to 9.5% (median value 1.3%) of the BMT recipients developed disseminated adenovirus disease (DAD).

In ten studies, adenoviruses were isolated from 1.3-15% (median value 4.4%) of stool samples from patients with various gastrointestinal symptoms (Table 5). A comparative MoAb-based EIA study involving 296 BMT patients, 150 of whom had diarrhoea, showed astrovirus in seven samples from these individuals, adenovirus in five, rotavirus in two and cytomegalovirus in two, whereas six samples contained *Clostridium difficile* and its cytotoxin, and two contained *Aeromonas* species (Cox *et al.*, 1994).

In recent years, there has been a trend toward applying larger proportions of HLA-mismatched and unrelated donors for BMT. As a consequence, *more powerful immunosuppressive conditioning* — including T cell depletion and total body irradiation — is used. These treatments have been shown to promote adenovirus and other virus infections. Of 92 recipients of an autologous BMT, one was infected by an adenovirus and recovered, whereas of 114 recipients of an allogenic BMT, 11 contracted an adenovirus infection, seven of whom died ($p = 0.01$; Hale *et al.*, 1999). In the same study, total body irradiation increased the chance of adenovirus infection by a factor of 14. As adenoviruses are more often searched for and more sensitive methods are applied, and also the proportion of higher-risk transplants and the intensiveness of immunosuppression are on the increase, the incidence of adenovirus infections in BMT patients could have been expected to *increase* with time. Still, for the data in Table 5 covering the period 1976-1999, no rising trend can be discerned for the incidence of adenovirus infections in general or for those in the gastrointestinal tract, nor for DAD or for fatal cases where adenovirus infection was presumed to have played a major role. Probably, however, the large variation in the incidences shown in Table 5 would obscure any moderate positive or negative trend. Analysis of the incidence of adenovirus infections in the same hospital over longer periods would avoid this problem. The two studies following this approach which were available for this review did show a rising trend. Bruno *et al.* (1997) divided the incidence of adenovirus infections in their 4768 BMT patients in Seattle, USA, into six periods of three years. While maintaining a rather homogeneous transplant population, they demonstrated a gradual increase in incidence of adenovirus infections from 3.5% during 1976-1978 to 12.4% during 1991-1993. In this study, 50% more adenovirus infections were observed in recipients of transplants from HLA-mismatched related or unrelated donors than in recipients of transplants from HLA-matched related donors. Similarly, in Leiden, the Netherlands, the incidence of adenovirus infections in recipients of BMTs from HLA-mismatched related or unrelated donors was 9.1% of 66 recipients in the 1985-1994 period and 41% of 51 recipients in the 1995-1998 period (chi-square test: $p < 0.01$) (A.C.M. Kroes and M.J.D. van Tol, personal communication).

Incidence of adenovirus infections in recipients of solid organ transplants

Adenovirus enteritis has been reported in 4% (3 of 70) of small bowel transplant patients (Parizhskaya *et al.*, 2001). Of 484 paediatric liver transplant recipients, 49 (10%) had adenovirus infection, most often of species C (types 1, 2 and 5). Invasive adenovirus infection occurred in 20 children, leading to death in nine (Michaels *et al.*, 1992). Of 191 adult liver transplant recipients, 11 (5.8%) had adenovirus infections; four patients developed disseminated disease and three of these died (McGrath *et al.*, 1998). Generally, however, recipients of solid organ transplants have a lower risk of DAD than BMT patients (Carrigan, 1997).

Incidence of adenovirus infection in HIV-infected patients

The *frequency* of detection of adenovirus in HIV patients ranged from 7-29% (Khoo *et al.*, 1995; Giordano *et al.*, 1999; Sabin *et al.*, 1999). The actuarial risk of adenovirus infection was reported to be 28% in one year, and the most frequent site was the gastrointestinal tract (17/18 patients) (Khoo *et al.*, 1995). This observation is the more significant because diarrhoea is an important cause of morbidity and mortality in untreated HIV infection, occurring in 30-60% of European and North American patients and in nearly 90% of patients in developing countries (Grohmann *et al.*, 1993). Prolonged faecal excretion was associated with low CD4 cell counts (<150/mm^3; Khoo *et al.*, 1995) and stool carriage of Ad40 in one AIDS patient was found to increase with increasing immunosuppression (Dionisio *et al.*, 1997). Many adenovirus infections are probably missed because they are not tested for. For example, adenovirus was found retrospectively in 16 of 135 cases of suspected cytomegalovirus colitis in patients with AIDS (Yan *et al.*, 1998).

Recently, the widespread introduction of *highly active antiretroviral therapy* (HAART) has dramatically reduced the incidence of and mortality from opportunistic infections among HIV-infected patients living in industrialized countries (Pollok, 2001). The development of viral resistance and difficulties with compliance, however, may lead to breakthrough HIV viraemia and the re-emergence of opportunistic gastrointestinal infections. Moreover, implementation of HAART on a large scale is costly and requires the existence of a good public health system and this treatment is therefore unavailable to the vast majority of HIV patients worldwide.

Clinical significance of adenoviruses in immunodeficient patients

For *HIV-infected individuals*, some investigators claim a causal relationship between adenovirus infections and diarrhoea, whereas others deny adenoviruses such a role for the majority of cases (Pollok, 2001). To the author's knowledge, with *transplant patients* no attempt has been made to address the causality problem in a scientific way. Goodgame (2001) stated in his review: "There is convincing evidence that adenovirus causes enteritis in bone marrow transplant patients (Hale *et al.*, 1999; Shields *et al.*, 1985; Yolken *et al.*, 1982)". It is difficult to understand how he reached that opinion. The authors he referred to did not produce nor comment on any evidence regarding this question, except for Shields *et al.*, 1985, who wrote: "It was unclear whether there was a specific relation between these symptoms [including diarrhoea] and adenovirus infection."

In fact, any causal role of adenovirus infections in diarrhoea in immunodeficient subjects would be hard to prove against a background of the predominant contribution of acute intestinal graft-versus-host disease to this syndrome, reportedly occurring in a high proportion of diarrhoeal episodes (Cox *et al.*, 1994; Baldwin *et al.*, 2000). The many enteric co-infections with other viruses and microorganisms are additional confounding factors.

Shorter *survival* has been reported for HIV-positive patients with diarrhoea who excreted adenovirus from the gut (1.0 year of survival) compared to patients with diarrhoea in whom adenovirus was not detected (2.4 years of survival) (Sabin *et al.*, 1999). The difference remained significant (p = 0.008) after controlling for differences in CD4 counts between the groups. Especially, DAD has a poor prognosis, even for individuals with a normally functioning immune system. In a small study in the USA, mortality rates of 60% were reported for immunocompetent and 80% for immunocompromized children with DAD (Munoz *et al.*, 1998). In the studies in Table 5, DAD was associated with an average death rate of 75%. Suggestive as such data may be, even in fatal cases of DAD it is often difficult to say more than that the adenovirus infection probably contributed to the death of the patient.

Epidemiology and control of adenoviruses in transplant recipients

Possible modes of adenovirus infection of transplant recipients include transmission from donor organ to recipient, reactivation of persistent virus and exogenous respiratory or faecal-oral transmission, possibly occurring outside or in the hospital. Molecular epidemiological studies should be able to clarify the relative importance of the various sources and routes. The proportion of exogenous infections is indicative of the success of *hygienic measures*. Indeed, some authors reported nosocomial infections with common pathogens including adenovirus as a likely cause of diarrhoea in their BMT patients (Cox *et al.*, 1994), and some studies suggested that certain hygienic precautions lower the incidence of graft-versus-host disease and prevent infection. However, most investigators have not observed clustering of adenovirus-infected BMT patients or a protective effect of laminar air flow treatment (Shields *et al.*, 1985; Wasserman *et al.*, 1988; Troussard *et al.*, 1993; Flomenberg *et al.*, 1994), suggesting that reactivation of persistent rather than exogenous infection was usually involved.

At any rate, *virological monitoring* of transplant patients is important for their treatment. "Isolation of an enteric virus, particularly an adenovirus, may be a surrogate marker for a poor prognosis in the presence of continued immunosuppression" (Chakrabarty *et al.*, 2000). Mitigation of the immunosuppressive regimen should also be considered when adenovirus DNA is detected in serum of the transplant patient (Echavarria *et al.*, 2001). Up until now, adenoviruses have been refractory to *antiviral treatment* and have therefore gained relative importance in such patients. Although there are anecdotal reports of successful therapy of such infections in BMT patients using cidofovir, ribavirin and ganciclovir, as yet none of these drugs has reproducibly proven clinical value (Horwitz, 2001; Baldwin *et al.*, 2000).

Acknowledgements

The virological advice and linguistic corrections of Dr A.H. Kidd and the provision of Fig. 1 by Dr R.W.H. Ruigrok are gratefully acknowledged.

438

References

Abken, H., Butzler, C., Willecke, K. (1987). Adenovirus type 5 persisting in human lymphocytes is unlikely to be involved in immortalization of lymphoid cells by fusion with cytoplasts or by transfection with DNA of mouse L cells. *Anticancer Res.* **7**, 553-558.

Adrian, T., Wigand, R., Richter, J. (1987). Gastroenteritis in infants, associated with a genome type of adenovirus 31 and with combined rotavirus and adenovirus 31 infection. *Eur. J. Pedriatr.* **146**, 38-40.

Allard, A., Girones, R., Juto, P, Wadell, G. (1990). Polymerase chain reaction for detection of adenoviruses in stool samples. J. Clin. Microbiol. 28, 2659-2667.

Allard, A., Albinsson, B., Wadell, G. (2001). Rapid typing of human adenoviruses by a general PCR combined with restriction endonuclease analysis. *J. Clin. Microbiol.* **39**, 498-505.

Ambinder, R.F., Burns, W., Forman, M., Charache, P., Arthur, R., Beschorner, W., Santos, G., Saral, R. (1986). Hemorrhagic cystitis associated with adenovirus infection in bone marrow transplantation. *Arch. Intern. Med.* **146**, 1400-1401.

Avellón, A., Pérez, P., Aguilar, J.C., Ortiz de Lejarazu, R., Echevarría, J.E. (2001). Rapid and sensitive diagnosis of human adenovirus infections by a generic polymerase chain reaction. *J. Virol. Methods* **92**, 113-120.

Baldwin, A., Kingman, H., Darville, M., Foot, A.B.M., Grier, D., Cornish, J.M., Goulden, N., Oakhill, A., Pamphilon, D.H., Steward, C.G., Marks, D.I. (2000). Outcome and clinical course of 100 patients with adenovirus infection following bone marrow transplantation. *Bone Marrow Transplant.* **26**, 1333-1338.

Barnes, G.L., Uren, E. Stevens, K.B., Bishop, R.F. (1998). Etiology of acute gastroenteritis in hospitalized children in Melbourne, Australia, from April 1980 to March 1993. *J. Clin. Microbiol.* **36**, 133-138.

Benkö, M., Harrach, B., Russell, W.C. (2000). Family Adenoviridae. In: Van Regenmortel, M.H.V., Fauquet, C.M., Bishop, D.H.L. *et al.* (eds): Virus Taxonomy. Seventh Report of the International Committee on Taxonomy of Viruses. Academic Press, New York, San Diego, pp. 227-228.

Bhan, M.K., Raj, P., Bhandari, N., Svensson, L., Stintzing, G., Prasad, A.K., Jayashree, S., Srivastava, R. (1988). Role of enteric adenoviruses and rotaviruses in mild and severe acute enteritis. *Pediatr. Infect. Dis. J.* **7**, 320-323.

Blanke, C., Clark, C., Broun, R., Tricot, G., Cunningham, I., Cornetta, K., Hedderman, A., Hromas, R. (1995). Evolving pathogens in allogeneic bone marrow transplantation: Increased fatal adenoviral infections. *Am. J. Med.* **99**, 326-328.

Bon, F., Fascia, P., Dauvergne M., Tenenbaum, D., Planson, H., Petion, A.M., Pothier, P., Kohli, E. (1999). Prevalence of group A rotavirus, human calicivirus, astrovirus, and adenovirus type 40 and 41 infections among children with acute gastroenteritis in Dijon, France. *J. Clin. Microbiol.* **37**, 3055-3058.

Brandt, C.D., Kim, H.W., Vargosko, A.J., Jeffries, B.C., Arrobio, J.O., Rindge, B., Parrott, R.H., Chanock, R.M. (1969). Infections in 18,000 infants and children in a controlled study of respiratory tract disease. I. Adenovirus pathogenicity in relation to serologic type and illness syndrome. Am. J. Epidemiol. 90, 484-500.

Brandt, C.D., Kim, H.W., Yolken, R.H., Kapikian, A.Z., Arrobio, J.O., Rodriguez, W.J., Wyatt, R.G., Chanock, R.M., Parrott, R.H. (1979). Comparative epidemiology of two rotavirus serotypes and other viral agents associated with pediatric gastroenteritis. *Am. J. Epidemiol.* **110**, 243-254.

Brown, M. (1990). Laboratory identification of adenoviruses associated with gastroentritis in Canada from 1983 to 1986. *J. Clin. Microbiol.* **28**, 1525-1529.

Bruno, B., Boeckh, M., Davis, C., Gooley, T., Hackman, R.C. (1997). Adenovirus infections in patients undergoing bone marrow transplantation. Blood 90, Suppl. Abstract 1867.

Bryden, A.S., Davies, A.H., Hadley, R.E., Flewett, T.H. (1975). Rotavirus enteritis in the West Midlands during 1974. *Lancet* **ii**, 241-243.

Caprioli, A., Pezzella, C., Morelli, R., Giammanco, A., Arista, S., Crotti, D., Facchini, M., Guglielmetti, P., Piersimoni, C., Luzzi, I. (1996). Enteropathogens associated with childhood diarrhea in Italy. *Infect. Pediatr. Dis. J.* **15**, 876-883.

Carrigan, D.R. (1997). Adenovirus infections in immunocompromised patients. *Am. J. Med.* **102**(3A), 71-74.

Cardosa, M.J., Krishnan, S., Tio, P.H., Perera, D., Wong, S.C. (1999). Isolation of subgenus B adenovirus during a fatal outbreak of enterovirus 71-associated hand, foot, and mouth disease in Sibu, Sarawak. *Lancet* **354**, 987-991.

Chakrabarti, S., Collingham, K.E., Stevens R.H., Pillay, D., Fegan, C.D., Milligan, D.W. (2000). Isolation of viruses from stools in stem cell transplant recipients: a prospective surveillance study. Bone Marrow Transplant. 25, 277-282.

Christensen, M.L. (1989). Human viral gastroenteritis. *Clin. Microbiol. Rev.* **2**, 51-89.

Clarke, LM., ed. (1992 and updates). Viruses, Rickettsiae, Chlamydiae, and Mycoplasmas. In: Clinical Microbiology Procedures Handbook. Ed. in Chief: Isenberg H.D. Volume 2, section 8, subsections 8.1-8.26. American Society for Microbiology, Washington, D.C.

Cox, G.J., Matsui, S.M., Lo, R.S., Hinds, M., Bowden, R.A., Hackman, R.C., Meyer, W.G., Mori, M., Tarr, P.I., Oshiro, L.S., Ludert, J.E., Meyers, J.D., McDonald, G.B. (1994). Etiology and outcome of diarrhea after marrow transplantation: A prospective study. *Gastroenterology* **107**, 1398-1407.

Cruz, J.R., Caceres, P., Cano, F., Flores, J., Bartlett, A., Torun, B. (1990). Adenovirus types 40 and 41 and rotaviruses associated with diarrhea in children from Guatemala. *J. Clin. Microbiol.* **28**, 1780-1784.

D'Ambrosio, E., Del Grosso, N., Chicca, A., Midulla, M. (1982). Neutralizing antibodies against 33 human adenoviruses in normal children in Rome. *J. Hyg., Camb.,* **89**, 155-161.

De Jong, J.C., Bijlsma, K., Wermenbol, A.G., Verweij-Uijterwaal, M.W., van der Avoort, H.G.A.M., Wood, D.J., Bailey, A.S., Osterhaus, A.D.M.E. (1993). Detection, typing, and subtyping of enteric adenoviruses 40 and 41 from fecal samples and observation of changing incidences of infections with these types and subtypes. *J. Clin. Microbiol.* **31**, 1562-1569.

De Jong, J.C., Wermenbol, A.G., Verweij-Uijterwaal, M.W., Slaterus, K.W.,

Wertheim-van Dillen, P., van Doornum, G.J.J., Khoo, S.H., Hierholzer, J.C. (1999). Adenoviruses from human immunodeficiency virus-infected individuals, including two strains that represent new candidate serotypes Ad50 and Ad51 of species B1 and D, respectively. *J. Clin. Microbiol.* **37**, 3940-3945.

De Jong, J.C., Wigand, R., Kidd, A.H., Wadell, G., Kapsenberg, J.G., Muzerie, C.J., Wermenbol, A.G., Firtzlaff, R.G. (1983). Candidate adenoviruses 40 and 41: fastidious adenoviruses from human infant stool. *J. Med. Virol.* **11**, 215-31.

De Jong, J.C, Wigand, R., Wadell, G., Keller, D., Muzerie, C.J., Wermenbol, A.G., Schaap, G.J.P. (1981). Adenovirus 37: identification and characterization of a medically important new adenovirus type of subgroup D. *J. Med. Virol.* **7**, 105-118.

Desmyter, J., de Jong, J.C., Slaterus, K.W., Verlaeckt, H. (1974). Keratoconjunctivitis caused by adenovirus type 19. *Brit. Med. J.* **4**, 406.

Dionisio, D., Arista, S., Vizzi, E., Manneschi, L.I., Di Lollo, S., Trotta, M., Sterrantino, G., Mininni, S., Leoncini, F. (1997). Chronic intestinal infection due to subgenus F type 40 adenovirus in a patient with AIDS. Scand. *J. Infect. Dis.* **29**, 305-307.

Echavarria, M., Forman, M., Ticehurst, J., Dumler, J.S., Charache, P. (1998). PCR method for detection of adenovirus in urine of healthy and human immunodeficiency virus-infected individuals. *J. Clin. Microbiol.* **36**, 3323-3326.

Echavarria, M., Forman, M., van Tol, M., Vossen, J.M., Charache, P., Kroes, A.C.M. (2001). Prediction of severe disseminated adenovirus infection by serum PCR. *Lancet* **358**, 384-385.

Farkas, T., Jiang, X., Guerrero, M.L., Zhong, W., Wilton, N., Berke, T., Matson, D.O., Pickering, L.K., Ruiz-Palacios, G. (2000). Prevalence and genetic diversity of human caliciviruses (HuCVs) in Mexican children. *J. Med. Virol.* **62**, 217-223.

Flewett, T.H., Bryden, A.S., Davies, H., Morris, C.A. (1975). Epidemic viral enteritis in a long-stay children's ward. *Lancet* **i**, 4-5.

Flomenberg, P., Babbitt, J., Drobyski, W.R., Ash, R.C., Carrigan, D.R., Sedmak, G.V., McAuliffe, T., Camitta, B., Horowitz, M.M., Bunin, N., Casper, J.T. (1994). Increased incidence of adenovirus disease in bone marrow transplant recipients. *J. Infect. Dis.* **169**, 775-781.

Fox, J.P., Brandt, C.D., Wassermann, F.E., Hall, C.E., Spigland, I., Kogon, A., Elveback, L.R. (1969). The Virus Watch Program: A continuing surveillance of viral infections in metropolitan New York families. VI. Observations of adenovirus infections: virus excretion patterns, antibody response, efficiency of surveillance, patterns of infection and relation to illness. *Am. J. Epidemiol.* **89**, 25-50.

Foy, H.M. (1989). Adenoviruses. In: Evans, A.S. (ed): Viral Infections of Humans. Epidemiology and Control. Plenum Medical Book Company, New York, pp 77-94.

Francki, R.I.B., Fauquet, C.M., Knudson, D.L., Brown, F. (1991). Classification and nomenclature of viruses. Fifth report of the International Committee on Taxonomy of Viruses. *Arch. Virol.* **Suppl 2**, 140-144.

Giordano, M.O., Martinez, L.C., Rinaldi, D., Espul, C., Martinez, N., Isa, M.B., Depetris, A.R., Medeot, S.I., Nates, S.V. (1999). Diarrhea and enteric emerging viruses in HIV-infected patients. AIDS Res. Hum. Retroviruses 15, 1427-1432.

Goodgame, R.W. (1999). Viral infections of the gastrointestinal tract. *Curr. Gastroenterol. Reports* **1**, 292-300.

Goodgame, R.W. (2001). Viral causes of diarrhea. *Gastroenterol. Clinics North America* **30**, 779-795.

Grandien, M., Pettersson, C.-A., Svensson, L., Uhnoo, I. (1987). Latex agglutination test for adenovirus diagnosis in diarrheal disease. *J. Med. Virol.* **23**, 311-316.

Grimwood, K., Carzino, R., Barnes, G.L., Bishop, R.F. (1995). Patients with enteric adenovirus gastroenteritis admitted to an Australian pediatric teaching hospital from 1981 to 1992. *J. Clin. Microbiol.* **33**, 131-136.

Grohmann, G.S., Glass, R.I., Pereira, H.G., Monroe, S.S., Hightower, A.W., Weber, R., Bryan, R.T. (1993). Enteric viruses and diarrhea in HIV-infected patients. *New Engl. J. Med.* **329**, 14-20.

Hale, G.A., Heslop, H.E., Krance, R.A., Brenner, M.A. Jayawardene, D., Srivastava, D.K., Patrick, C.C. (1999). Adenovirus infection after pediatric bone marrow transplantation. Bone Marrow *Transplant* **23**, 277-282.

Herrmann, J.E., Perron-Henry, D.M., Blacklow, N.R. (1987). Antigen detection with monoclonal antibodies for the diagnosis of adenovirus gastroenteritis. *J. Infect. Dis.* **155**, 1167-1171.

Hierholzer, J.C. (1992). Adenoviruses in the immunocompromised host. *Clin. Microbiol. Rev.* **5**, 262-274.

Hierholzer, J.C. (1995). Adenoviruses, p.169-188. *In* E.H. Lennette, D.A. Lennette, E.T. Lennette (eds.), Diagnostic Procedures for Viral, Rickettsial, and Chlamydial Infections, 7th ed. American Public Health Assn, Washington, DC.

Horwitz, M.S. (2001). Adenoviruses. In: Knipe, D.M., Howley, P.M. *et al.* (eds). Fields Virology, 4th ed., Lippincott, Williams and Wilkins, Philadelphia, pp 2301-2326.

Howard, D.S., Phillips II, G.L., Reece, D.E., Munn, R.K., Henslee-Downey, J., Pittard, M., Barker, M., Pomeroy, C. (1999). Adenovirus infections in hematopoietic stem cell transplant recipients. *Clin. Infect. Dis.* **29**, 1494-1501.

Ieven, M., van Reempts, P., van Overmeire, B., Provinciael, D., Vael, K, Pattyn, S. (1994). A pseudoepidemic of adenoviruses in a neonatal care unit. *Diagn. Microbiol. Infect. Dis.* **18**, 157-159.

Jarecki-Khan, K., Tzipori, S.R., Unicomb, L.E. (1993). Enteric adenovirus infection among infants with diarrhea in rural Bangladesh. *J. Clin. Microbiol.* **31**, 484-489.

Johansson, M.E., Uhnoo, I., Kidd, A.H., Madeley, C.R., Wadell, G. (1980). Direct identification of enteric adenovirus, a candidate new serotype, associated with infantile gastroenteritis. *J. Clin. Microbiol.* **12**, 95-100.

Kajon, A.E., Mistchenko, A.S., Videla, C., Hortal, M., Wadell, G., Avendano, L.F. (1996). Molecular epidemiology of adenovirus acute lower respiratory infections of children in the south cone of South America (1991-1994). *J. Med. Virol.* **48**, 151-156.

Khoo, S.H., Bailey, A.S., de Jong, J.C., Mandal, B.K. (1995). Adenovirus infections in human immunodeficiency virus-positive patients: Clinical features and molecular epidemiology. *J. Infect. Dis.* **172**, 629-637.

Kidd, A.H., Banatvala, J.E. de Jong, J.C. (1983). Antibodies to fastidious faecal adenoviruses (species 40 and 41) in sera from children. *J. Med. Virol.* **11**, 333-341.

Kidd, A.H., Cosgrove, B.P., Brown, R.A., Madeley, C.R. (1982). Faecal adenoviruses from Glasgow babies. Studies on culture and identity. *J. Hyg.* **88**, 463-474.

Kidd, A.H., Jönsson, M., Garwicz, D., Kajon, A.E., Wermenbol, A.G., Verweij, M.W., de Jong, J.C. (1996). Rapid subgenus identification of human adenovirus isolates by a general PCR. *J. Clin. Microbiol.* **34**, 622-627.

Kidd, A.H., Rosenblatt, A., Besselaar, T.G., Erasmus, M.J., Tiemessen, C.T., Berkowitz, F.E., Schoub, B.D. (1986). Characterization of rotaviruses and subgroup F adenoviruses from acute summer gastroenteritis in South Africa. *J. Med. Virol.* **18**, 159-168.

Kotloff, K.L., Losonsky, G.A., Glenn Moris, J., Wasserman, S.S., Singh-Naz, N., Levine, M.M. (1989). Enteric adenovirus infection and childhood diarrhea: An epidemiologic study in three clinical settings. *Pediatrics* **84**, 219-225.

Lew, J.F., Moe, C.L., Monroe, S.S., Allen, J.R., Harrison, B.M., Forrester, B.D., Stine, S.E., Woods, P.A., Hierholzer, J.C., Herrmann, J.E., Blacklow, N.R., Bartlett, A.V., Glass, R.I. (1991). Astrovirus and adenovirus associated with diarrhea in children in day care settings. *J. Infect. Dis.* **164**, 673-678.

Lukashok, S.A., Horwitz, M.S. (1998). New perspectives in adenoviruses. *Curr. Clin. Top. Infect. Dis.* **18**, 286-305.

Madeley, C.R., Cosgrove, B.P., Bell, E.J., Fallon, R.J. (1977). Stool viruses in babies in Glasgow. 1. Hospital admissions with diarrhoea. *J. Hyg.* **78**, 261-273.

Maldonado, Y., Cantwell, M., Old, M., Hill, D., Sanchez, M., Logan, L., Millan-Velasco, F., Valdespino, J.L., Sepulveda, J., Matsui, S. (1998). Population-based prevalence of symptomatic and asymptomatic astrovirus infection in rural Mayan infants. *J. Infect. Dis.* **178**, 334-339.

Mautner, V., Steinthorsdottir, V., Bailey, A. (1995). Enteric adenoviruses. *Curr. Top. Microbiol. Immunol.* **199 III**, 229-282.

McGrath, D., Falagas, M.E., Freeman, R., Rohrer, R., Fairchild, R., Colbach, C., Snydman, D.R. (1998). Adenovirus infection in adult orthotopic liver transplant recipients: Incidence and clinical significance. *J. Infect. Dis.* **177**, 459-462.

Michaels, M.G., Green, M., Wald, E.R., Starzl, T.E. (1992). Adenovirus infection in pediatric liver transplant recipients. *J. Infect. Dis.* **165**, 170-174.

Mistchenko, A.S., Huberman, K.H., Gomez, J.A., Grinstein, S. (1992). Epidemiology of enteric adenovirus infection in prospectively monitored Argentine families. *Epidemiol. Infect.* **109**, 539-546.

Munoz, F.M., Piedra, P.A., Demmler, G.J. (1998). Disseminated adenovirus disease in immunocompromised and immunocompetent children. *Clin. Infect. Dis.* **27**, 1194-1200.

Neumann, R., Genersch, E., Eggers, H.J. (1987). Detection of adenovirus nucleic acid sequences in human tonsils in the absence of infectious virus. *Virus Res.* **7**, 93-97.

Noel, J., Mansoor, A., Thaker, U., Herrmann, J., Perron-Henry, D., Cubitt, W.D. (1994). Identification of adenoviruses in faeces from patients with diarrhoea at the Hospitals for Sick Children, London, 1989-1992. *J. Med. Virol.* **43**, 84-90.

Paerregaard, A., Hjelt, K., Genner, J., Moslet, U., Krasilnikoff, P.A. (1990). Role of enteric adenoviruses in acute gastroenteritis in children attending day-care centres. *Acta Paediatr. Scand.* **79**, 370-371.

Parizhskaya, M., Walpusk, J., Mazariegos, G., Jaffe, R. (2001). Enteric adenovirus infection in pediatric small bowel transplant recipients. *Pediatr. Developm. Pathol.* **4**, 122-128.

Pollok, R.C.G. (2001). Viruses causing diarrhoea in AIDS. *Novartis Found. Symp.* **238**, 276-288.

Rodriguez, W.J., Kim, H.W., Brandt, C.D., Schwartz, R.H., Gardner, M.K., Jeffries, B., Parrott, R.H., Kaslow, R.A., Smith, J.I., Takiff, H. (1985). Fecal adenoviruses from a longitudinal study of families in metropolitan Washington, D.C.: laboratory, clinical, and epidemiologic observations. *J. Pediatr.* **107**, 514-520.

Russell, W.C. (2000). Update on adenovirus and its vectors. *J. Gen. Virol.* **81**, 2573-2604.

Sabin, C.A., Clewley, C.S., Deayton, J.R., Mocroft, A., Johnson, M.A., Lee, C.A., McLaughlin, J.E., Griffiths, P.D. (1999). Shorter survival in HIV-positive patients with diarrhoea who excreted adenovirus from the GI tract. *J. Med. Virol.* **58**, 280-285.

Saderi, H., Roustai, M.H., Sabahi, F., Sadeghizadeh, M., Owlia, P., de Jong, J.C. (2002). Incidence of enteric adenovirus gastroenteritis in Iranian children. *J. Clin. Virol.* **24**, 1-5.

Schirm, J., Manson, W.L., Schröder, F.P., Tegzess, A.M., van der Avoort, H.G.A.M., de Jong, J.C. (1991). [An adenovirus epidemic in patients with kidney transplantation]. [Dutch with English summary] *Ned. Tijdschr. Geneeskd.* **135**, 1310-1314.

Schmitz, H., Wigand, R, Heinrich, W. (1983). Worldwide epidemiology of human adenovirus infections. *Am. J. Epidemiol.* **117**, 455-466.

Schoub, B.D., Koornhof, H.J., Lecatsas, G., Prozesky, O.W., Freiman, I., Hartman, E., Kassel, H. (1975). Viruses in acute summer gastroenteritis in black infants. *Lancet* **i**, 1093-1094.

Scott-Taylor, T., Ahluwalia, G., Klisko, B., Hammond, G.W. (1990). Prevalent adenovirus variant not detected by commercial monoclonal antibody enzyme immunoassay. *J.Clin. Microbiol.* **28**, 2797-2801.

Shenk, T.E. (2001). Adenoviridae: The viruses and their replication. In: Knipe, D.M., Howley, P.M. *et al.* (eds). Fields Virology, 4th ed., Lippincott, Williams and Wilkins, Philadelphia, pp 2265-2300.

Shields, A.F., Hackman, R.C., Fife, K.H., Corey, L., Meyers, J.D. (1985). Adenovirus infections in patients undergoing bone-marrow transplantation. *New Engl. J. Med.* **312**, 529-533.

Shinozaki, T., Araki, K., Fujita, Y., Kobayashi, M., Tajima, T., Abe, T. (1991). Epidemiology of enteric adenoviruses 40 and 41 in acute gastroenteritis in infants and young children in the Tokyo area. *Scand. J. Infect. Dis.* **23**, 543-547.

Shinozaki, T., Araki, K., Ushijima, H., Fujii, R. (1987). Antibody response to enteric adenovirus types 40 and 41 in sera from people in various age groups. *J. Clin. Microbiol.* **25**, 1679-1682.

Steele, A.D., Basetse H.R., Blacklow, N.R., Herrmann, J.E. (1998). Astrovirus infection in South Africa: a pilot study. *Ann. Trop. Paediatr.* **18**, 315-319.

Strohl, W.A., Schlesinger, R.W. (1965). Quantitative studies of natural and experimental adenovirus infections of human cells. II. Primary cultures and the possible role of asynchronous viral multiplication in the maintenance of infection. *Virology* **26**, 208-220.

Takiff, H.E., Straus, S.E., Garon, C.F. (1981). Propagation and *in vitro* studies of previously non-cultivable enteral adenoviruses in 293 cells. *Lancet* **ii**, 832-834.

Tiemessen, C.T., Wegerhoff, F.O., Erasmus, M.J., Kidd, A.H. (1989). Infection by enteric adenoviruses, rotaviruses, and other agents in a rural African environment. *J. Med. Virol.* **28**, 176-182.

Troussard, X., Bauduer, F., Gallet, E., Freymuth, F., Boutard, P., Ballet, J.J., Reman, O., Leporrier, M. (1993). Virus recovery from stools of patients undergoing bone marrow transplantation. *Bone Marrow Transplant.* **12**, 573-576.

Uhnoo, I., Wadell, G., Svensson, L., Johansson, M. (1984). Importance of enteric adenoviruses Ad40 and Ad41 in acute gastroenteritis in infants and young children. *J. Clin. Microbiol.* **20**, 365-372.

Uhnoo, I., Wadell, G., Svensson, L., Olding-Stenkvist, E., Ekwall, E., Mölby, R. (1986). Aetiology and epidemiology of acute gastroenteritis in Swedish children. *J. Infect.* **13**, 73-89.

Van, R., Wun, C.-C., O'Ryan, M.L., Matson, D.O., Jackson, L., Pickering, L.K. (1992). Outbreaks of human enteric adenovirus types 40 and 41 in Houston day care centers. *J. Pediatr.* **120**, 516-521.

Van Loon, A.E., Rozijn, T.H., de Jong, J.C., Sussenbach, J.S. (1985). Physicochemical properties of the DNAs of the fastidious adenoviruses species 40 and 41. *Virology*, **140**, 197- 200.

Vizzi, E., Ferraro, D., Cascio, A., Di Stefano, R., Arista, S. (1996). Detection of enteric adenoviruses 40 and 41 in stool specimens by monoclonal antibody-based enzyme immunoassays. *Res. Virol.* **147**, 333-339.

Wadell, G., Allard, A., Hierholzer, J.C. (1999). Adenoviruses. In: Murray, P.R. (ed): Manual of Clinical Microbiology. American Society of Microbiology Press, Washington, DC, pp 970-982.

Wadell, G., de Jong, J.C. (1980). Restriction endonucleases in identification of a genome type of adenovirus 19 associated with keratoconjunctivitis. *Infect. Immun.* **27**, 292-6.

Wadell, G., de Jong, J.C., Wolontis, S. (1981a). Molecular epidemiology of adenoviruses: alternating appearance of two different genome types of adenovirus 7 during epidemic outbreaks in Europe from 1958 to 1980. *Infect. Immun.* **34**, 368-372.

Wadell, G., Sundell, G., de Jong, J.C. (1981b). Characterization of candidate adenovirus 37 by SDS-polyacrylamide gel electrophoresis of virion polypeptides and DNA restriction site mapping. *J. Med. Virol.* **7**, 119-125.

Wasserman, R., August, C.S., Plotkin, S.A. (1988). Viral infections in pediatric bone marrow transplant patients. *Pediatr. Infect. Dis. J.* **7**, 109-115.

White, D.O., Fenner, F. (1986). Persistent infections. In: Medical Virology. Academic Press, London, pp 193-215.

Wigand, R., Adrian, Th. (1989). Intermediate adenovirus strains occur in extensive variety. *Med. Microbiol. Immunol.* **178**, 37-44.

Wood, D.J., Bijlsma, K., de Jong, J.C., Tonkin, C. (1989). Evaluation of a commercial monoclonal antibody-based enzyme immunoassay for detection of adenovirus types 40 and 41 in stool specimens. *J. Clin. Microbiol.* **27**, 1155-1158.

Xu, W., McDonough, M.C., Erdman, D.D. (2000). Species-specific identification of human adenoviruses by a multiplex PCR assay. *J. Clin. Microbiol.* **38**, 4114-4120.

Yan, Z., Nguyen, S., Poles, M., Melamed, J., Scholes, J.V. (1998). Adenovirus colitis in human immunodeficiency virus infection. *Am. J. Surgical Pathol.* **22**, 1101-1106.

Yolken, R.H., Lawrence, F., Leister, F., Takiff, H.E., Strauss, S.E. (1982). Gastroenteritis associated with enteric type adenovirus in hospitalized infants. *J. Pediatr.* **101**, 21-26.

SECTION IV

Norwalk- and sapporo-like viruses (human caliciviruses)

Introduction

Besides rotaviruses and enteric adenoviruses, a number of smaller viruses have increasingly been recognized as important causes of viral gastroenteritis. With the application of molecular techniques it has become apparent that both the so-called small round structured viruses (SRSVs), also known as Norwalk virus (NV) and Norwalk-like viruses (NLVs), and the "classic" caliciviruses (Sapporo-like viruses, SLVs) which represent 2 of the 4 genera of the *Caliciviridae* family (Clarke and Lambden, 1997, 2001; Green *et al.*, 2000) are major causes of outbreaks of diarrhoea and vomiting in various populations (Fankhauser *et al.*, 1998). The genera NLVs and SLVs have recently been renamed as *Norovirus* and *Sapovirus*, respectively (Mayo, 2002).

The three-dimensional structure of NV virus-like particles (VLPs, see below) has been studied by cryoelectron microscopy and image processing which demonstrated that the VLPs have a T=3 symmetry and are composed of 90 dimers of capsid protein. Further analysis of this protein by x-ray crystallography has shown that it consists of a shell domain (S), joined by a flexible hinge to a protruding (P) domain which is further subdivided into subunits P1 and P2 (Prasad *et al.*, 1994, 1999). An update of the structural features by A Bertolotti-Ciarlet *et al.* (BVV Prasad's group) is found in Section IV, Chapter 1.

Human caliciviruses possess a single stranded RNA genome of positive polarity and approximately 7,400 – 7,800 nucleotides in length. The genome encodes 3 open reading frames (ORFs) (Fig. 3 in Section IV, Chapter 2): ORF1 at the 5' end encodes a large nonstructural polyprotein, followed by ORF2 encoding a single viral capsid protein, and ORF3 at the 3' terminus encodes a basic protein of as yet unknown biological function but shown to be a minor structural component of virions (Glass *et al.*, 2000). There is a major difference between the genomes of NLVs and SLVs in that for SLVs the structural protein (ORF2) is in the same reading frame as ORFI and contiguous with the RNA dependent RNA polymerase (RdRP). For NLVs, ORFs 1 and 2 overlap by a few nucleotides with a -2 frameshift for ORF2 (Liu *et al.*, 1995; Clarke and Lambden, 2001). The overall sequences of NLVs and SLVs are sufficiently different to justify their classification as two different genera of the *Caliciviridae* (Clarke and Lambden, 2001). The 5' end non-structural polyprotein is processed to smaller proteins with helicase, protease and RdRP activities, containing sequence motifs similar to those of the 2C helicase, 3C protease and 3D polymerase of picornaviruses,

448

respectively (Lambden and Clarke, 1995). The complete sequences of several NLVs have been obtained (Jiang *et al.*, 1993; Lambden *et al.*, 1993; Dingle *et al.*, 1995; Seah *et al.*, 1999) as well as SLVs (Liu *et al.*, 1995).

There is no *in vitro* cell culture system and no suitable animal model for infection with human caliciviruses. During natural infection only small numbers of virus particles are shed in faeces which has made their detection difficult. Only with the advent of molecular techniques (reverse transcription-polymerase chain reaction, RT-PCR) and their use in diagnostic laboratories has the full extent become apparent to which these viruses infect humans and cause outbreaks. Sequence comparison has allowed to divide the NLVs into two genogroups (genogroup 1:NV and Southampton virus, SV; genogroup 2: Lorsdale virus, LV; Camberwell virus, CV, and others). The expression of NV ORF2 cDNA by baculovirus recombinants in insect cells was shown to lead to the export and self-assembly of viral capsid protein into the cell culture supernatant where virus-like particles (VLPs) are found (Jiang *et al.*, 1995). VLPs have been expressed of both genogroup 1 and genogroup 2 viruses and are also formed upon expression of ORF2 in transgenic potatoes and tobacco (Mason *et al.*, 1996; Arntzen, 1997); these products have become the prototype of "edible vaccines" (Arntzen, 1997).

Little is known about the replication of human caliciviruses as a productive *in vitro* propagation system is still lacking. *In vitro* assays have been used and interaction of cellular proteins with the 5' end of viral RNA been found; the cellular proteins are probably involved in replication and translation (Guiterrez-Escolano *et al.*, 2000). However, feline calicivirus (FCV) has been found to grow in CRFK cells, and an infectious cDNA clone of FCV has been constructed (Sosnovtsev and Green, 1995), allowing the investigation of the replication cycle and of structure/function relationships in much more detail (Neill *et al.*, 2000). SV Sosnovtsev and K Green have produced an update of their work in Section IV, Chapter 2.

Whilst FCV represents a very useful model for calicivirus replication, it belongs to a different genus (Vesivirus) of the *Caliciviridae* and is associated with respiratory tract infections; thus, the relevance of conclusions drawn from the analysis of this virus to the study of human enteric caliciviruses should not be overstated. Other caliciviruses detected in mammals (e.g. the bovine Jena virus, JV; porcine enteric calicivirus, PEC) are more similar to human caliciviruses. The bovine JV is more closely related to genogroup 1 NLVs (Liu *et al.*, 1999), whilst the PEC is related to SLVs, but in a separate genogroup (Vinjé *et al.*, 2000a).

Knowledge of the pathogenesis of human caliciviruses is very limited due to the reasons outlined above. MZ Guo and L Saif have summarised their recent work on the pathogenesis of calicivirus infections of pigs and cattle (Section IV, Chapter 3) and have come up with some quite novel results.

Clinically, human calicivirus infections start after a short incubation period of 1-2 days with diarrhoea and vomiting, projectile vomiting being a characteristic feature. The clinical symptoms are observed in general over a shorter period than that of rotavirus or enteric adenovirus infections. Besides oral or intravenous rehydration therapy there is no specific treatment (see Section I, Chapter 5).

The epidemiology of human calicivirus infections is complex. Infections and outbreaks are observed throughout the year (Green, 1997). The virus is normally transmitted via the faeco-oral route but also readily by food and water. Outbreaks in hospitals, nursing homes, schools and restaurants due to infection with these viruses are frequently recorded (e.g. Hedlund *et al.*, 2000; Nakata *et al.*, 2000; Inouye *et al.*, 2000; Koopmans *et al.*, 2000). Genomically the viruses are quite diverse (Lew *et al.*, 1994; Wang *et al.*, 1994; Ando *et al.*, 1994; Jiang *et al.*, 1997; Maguire *et al.*, 1999; Green *et al.*, 2000a, b). However, development of a consensus RT-PCR with primers directed against highly conserved regions of the polymerase gene have allowed to recognize more than 90% of outbreaks in which caliciviruses have been diagnosed by electron microscopy (Vinjé *et al.*, 2000a). Moreover, as RT-PCR is very sensitive, allowing recognition of less than 100 RNA-containing particles, it has helped to identify NLVs as the cause of infection in many outbreaks which were negative by electron microscopy (e.g. Huang *et al.*, 2000). Thus, RT-PCR has now become the 'gold standard' of diagnostic assessment. Using this tool, human calicivirus infections have now taken second place (after rotaviruses) as a major etiologic agent of childhood diarrhoea (Bon *et al.*, 1999; Pang *et al.*, 2000; Nakata *et al.*, 2000; de Wit *et al.*, 2001; Koopmans *et al.*, Section IV, Chapter 5 of this book). Age-related seroprevalence studies have shown that NLV infections occur from infancy onwards and are likely to be in part asymptomatic (Gray *et al.*, 1993; Jing *et al.*, 2000). ELISAs to detect human caliciviruses are under development (Jiang *et al.*, 2000; Vipond *et al.*, 2000). X Jiang has reviewed the development of serological and molecular tests for the diagnosis of calicivirus infections (Section IV, Chapter 4).

Extensive sequencing of RT-PCR amplicons and comparison of those sequences in phylogenetic trees have allowed to distinguish at least 15 genotypes within genogroup 1 (NLVs) and at least 4 within genogroup 2 (SLVs) (Jiang *et al.*, 1997; Noël *et al.*, 1997b, Vinjé and Koopmans 2000; Vinjé *et al.*, 2000b, Koopmans *et al.*, 2000; Ando *et al.*, 2000; Berke and Matson, 2000). Genotyping according to capsid sequencing data yields a classification of strains which is consistent with that derived from serotyping (Vinjé *et al.*, 2000b). The central region of the capsid was found to be hypervariable, suggesting that it might be exposed on the surface of particles (see Section IV, Chapter 1). Within genotypes, there seems to be considerable genomic drift which may lead to lack of recognition by established PCR primers after some time (Koopmans *et al.*, 2001). It has been firmly established by several groups that different lineages of NLVs and SLVs may co-circulate at any one time in the community (Gray *et al.*, 1997; Koopmans *et al.*, 2000). M Koopmans *et al.* have reviewed the present knowledge on the epidemiology of human calicivirus infections (Seection IV, Chapter 5) and presented evidence on the usefulness of molecular data for the tracking of transmission pathways.

It is often not clear to what extent outbreaks are driven by spread from a common food or water sources (Lodder *et al.*, 1999) or by human-to-human transmission. The possibility of an animal reservoir for infections of humans will need to be explored (Sugieda *et al.*, 1998; Guo *et al.*, 1999; Van der Poel *et al.*, 2000), requiring surveillance of animals, particularly domestic animal populations, as much as human surveillance.

In some cases strain clustering was different depending on the genomic region compared, suggesting that such strains may have arisen by recombination which is possibly more common than previously thought (Jiang *et al.*, 1999; Vinjé *et al.*, 2000b; Vinjé and Koopmans 2000; Green *et al.*, 2000). There is evidence for coinfection of individuals with different caliciviruses as a prerequisite of viral recombination (Gray *et al.*, 1997). D Matson has reviewed the data on evidence for RNA recombination in caliciviruses and also produced some provocative thoughts (Section IV, Chapter 6).

The development of (a) vaccine(s) is in its infancy, due to difficulties with establishing the true correlate of protection and the genomic diversity (Estes *et al.*, 2000; Matsui and Greenberg, 2000). However, Norwalk virus-like particles expressed in transgenic plants have pioneered the interesting approach of "edible vaccines" (Mason *et al.*, 1996; Arntzen (1997; Tacket *et al.*, 2000).

References

Ando T, Mulders MN, Lewis DC, Estes MK, Monroe SS, Glass RI (1994). Comparison of the polymerase region of small round structured virus strains previously classified in three antigenic types by solid-phase immune electron microscopy. *Arch. Virol.* **135:** 217-226.

Ando T, Nöel JS, Fankhauser RL (2000). Genetic classification of Norwalk-like viruses. *J. Infect. Dis.* **181 (Suppl. 2):** 5336-5348.

Arntzen CJ (1997). Edible Vaccines. *Public Health Rep.* **112:** 190-197.

Berke T, Matson DO (2000). Reclassification of the *Caliciviridae* into distinct genera and exclusion of hepatitis E virus from the family or the basis of comparative phylogenetic analysis. *Arch. Virol.* **145:** 1421-1436.

Bon F, Fascia P, Dauvergne M *et al.* (1999). Prevalence of group A rotaviruses, human caliciviruses, astrovirus and adenoviruses type 40 and 41 among children with acute gastroenteritis in Dijon, France. *J. Clin. Microbiol.* **37:** 3055-3058.

Clarke IN, Lambden PR (1997). The molecular biology of caliciviruses. *J. Gen. Virol.* **78:** 291-301.

Clarke IN, Lambden PR (2001). The molecular biology of human caliciviruses. *Novartis Found. Sympos.* **238:** 180-191 (Discussion 191-196).

De Wit MA, Koopmans MP, Kortbeck LM, Wannet WJ, Vinjé J, van Leusden F, Bartelds AI, van Duynhoven YT (2001). Sensor, a population-based cohort study on gastroenteritis in the Netherlands: incidence and etiology. *Am. J. Epidermiol.* **154:** 666-674.

Dingle KE, Lambden PR, Caul EO, Clarke IN (1995). Human enteric *Caliciviridae:* the complete genome sequence and expression of virus-like particles from a genetic group II small round structured virus. *J. Gen. Virol.* **76:** 2349-2355.

Estes MK, Ball KM, Guerrero RA, Opekun AR, Gilger MA, Pasheco SS, Graham DY (2000). Norwalk virus vaccines: challenges and progress. *J. Infect. Dis.* **181 (Suppl 2):** S367-S373.

Fankhauser RL, Noël JS, Monroe SS, Ando T, Glass RI (1998). Molecular epidemiology of 'Norwalk-like viruses' in outbreaks of gastroenteritis in the United States, *J. Infect. Dis.* **178:** 1571-1578.

Glass PJ, White LJ, Ball KM, Leparc-Goffart I, Hardy ME, Estes MK (2000). Norwalk virus open reading frame 3 encodes a minor structural protein. *J. Virol.* **74:** 6581-6591.

Gray JJ, Green J, Cunliffe C, Gallimore C, Lee JV, Neal K, Brown DWG (1997). Mixed genogroup SRSV infections among a party of canoeists exposed to contaminated recreational water. *J. Med. Virol.* **52:** 425-429.

Gray JJ, Jiang X, Morgan-Capner P, Desselberger U, Estes MK (1993). Prevalence of antibodies to Norwalk virus in England: Detection by enzyme-linked immunosorbent assay using baculovirus-expressed recombinant Norwalk virus capsid antigen. *J. Clin. Microbiol.* **31:** 1022-1025.

Green J, Vinjé J, Gallimore C, Koopmans M, Hale A, Brown D (2000b). Capsid protein diversity among small round structured viruses. *Virus Genes* **20:** 227-236.

Green KY (1997). The role of human caliciviruses in epidemic gastroenteritis. *Arch. Virol (Suppl)* **13:** 153-165.

Green KY, Ando T, Balayan MS *et al.* (2000a). Taxonomy of the caliciviruses. *J. Infect. Dis.* **181 (Suppl. 2):** S322-S330.

Guiterrez-Escolano AL, Brito ZU, del Angel RM, Jiang X (2000). Interaction of cellular proteins with the 5' end of Norwalk virus genomic RNA. *J. Virol.* **74:** 8558-8562.

Guo M, Chang KO, Hardy ME, Zhang O, Parwani AV, Saif LK (1999) Molecular characterization of a porcine enteric calicivirus genetically related to Sapporo-like human caliciviruses. *J. Virol.* **73:** 9625-9631.

Hardy ME, Prasad BVV, Dokland T, Bella J, Rossmann MG, Estes MK (1999). X-ray crystallographic structure of the Norwalk virus capsid. *Science* **286:** 287-290.

Hedlund KO, Rubilar-Abren E, Svensson L (2000). Epidemiology of caliciviruses in Sweden, 1994-1998. *J. Infect. Dis.* **181 (Suppl. 2):** S275-S280.

Huang PW, Laborde D, Land IR, Matson DO, Smith AW and Xiang X (2000). Concentration and detection of caliciviruses in water samples by reverse transcription - PCR. *Appl. Environm. Microbiol.* **66:** 4383-4388.

Inouye S, Yamashita S, Yamadera S, Yoshikawa M *et al.* (2000). Surveillance of viral gastroenteritis in Japan: pediatric cases and outbreak incidents. *J. Infect. Dis.* **181 (Suppl. 2):** S270-S274.

Jiang X, Cubitt WD, Berke T *et al.* (1997). Sapporo-like human caliciviruses are genetically and antigenically diverse. *Arch. Virol* **142:** 1813-1827.

Jiang X, Espul C, Zhong WM, Cuello H, Matson DO (1999a). Characterization of a novel human calicivirus that may be a naturally occurring recombinant. *Arch. Virol.* **144:** 2377-2387.

Jiang X, Huang PW, Zhong WM, Farkas T, Cubitt DW, Matson DO (1999b). Design and evaluation of a primer pair that detects both Norwalk- and Sapporo-like caliciviruses by RT-PCR. *J. Virol. Methods* **83:** 145-154.

Jiang X, Matson DO, Ruiz-Palacios GM, Hu J, Treanor J, Pickering LK (1995). Expression, self-assembly, and antigenicity of a Snow Mountain Agent-like

452

calicivirus capsid protein. *J. Clin. Microbiol.* **33:** 1452-1455.

Jiang X, Wang M, Wang K, Estes MK (1993). Sequence and genomic organization of Norwalk virus. *Virology* **195:** 51-61.

Jiang X, Wilton N, Zhong WM *et al.* (2000). Diagnosis of human calicivirus by use of enzyme immunoassays. *J. Infect. Dis.* **181 (Suppl 2):** S349-S359.

Jiang X, Zhong WM, Kaplan M, Pickering LK, Matson DO (1999c) Expression and characterization of Sapporo-like human calicivirus capsid proteins in baculovirus. *J. Virol. Methods* **78:** 81-91.

Jing Y, Qian Y, Huo Y, Wang LP, Jiang X (2000). Seroprevalence against Norwalk-like human caliciviruses in Beijing, China. *J. Med. Virol.* **60:** 97-101.

Koopmans M (2001). Molecular epidemiology of human enteric caliciviruses in the Netherlands. *Novartis Found.Symp.* **238:** 197-214 (Discussion 214-218).

Koopmans M, Vinjé J, de Wit M, Leenen I, van der Poel W, van Duynhoven Y (2000). Molecular epidemiology of caliciviruses in The Netherlands. *J. Infect. Dis.* **181 (Suppl. 2):** S262-S269.

Lambden PR, Caul EO, Ashley CR, Clarke IN (1993). Sequence and genome organization of a human small round structured (Norwalk-like) virus. *Science* **259:** 516-519.

Lambden PR, Clarke IN (1995). Genome organization in the caliciviridae. *Trends Microbiol.* **3:** 261-265.

Lew JF, Kapikian AZ, Valdesuso J, Green KY (1994). Molecular characterization of Hawaii virus and other Norwalk-like viruses: evidence for genetic polymorphism among human caliciviruses. *J. Infect. Dis.* **170:** 535-542.

Liu BL, Clarke IN, Caul EO, Lambden PR (1995). Human enteric caliciviruses have a unique genome structure and are distinct from the Norwalk-like viruses. *Arch. Virol.* **140:** 1345-1356.

Liu BL, Lambden PR, Günther H, Otto P, Elschner M, Clarke IN (1999). Molecular characterization of a bovine enteric calicivirus: relationship to the Norwalk-like viruses. *J. Virol.* **73:** 819-825.

Lodder WJ, Vinjé J, Van der Heide R, de Roda Husman AM, Leenen EJ, Koopmans MP (1999). Molecular detection of Norwalk-like caliciviruses in sewage. *Appl. Environm. Microbiol.* **65:** 5624-5627.

Maguire AJ, Green J, Brown DW, Desselberger U, Gray JJ (1999). Molecular epidemiology of outbreaks of gastroenteritis associated with small round-structured viruses in East Anglia, United Kingdom, during the 1996-1997 season, *J. Clin. Microbiol.* **37:** 81-89.

Mason HS, Ball JM, Shi J-J, Jiang X, Estes MK, Arntzen CJ (1996). Expression of Norwalk virus capsid protein in transgenic tobacco and potato and its oral immunogenicity in mice. *Proc. Natl. Acad. Sci. USA* **93:** 5335-5340.

Matsui SM, Greenberg HB (2000). Immunity to calicivirus infection. *J. Infect. Dis.* **181 (Suppl 2):** S331-S335.

Mayo MA (2002). Virus Taxonomy - Houston 2002. *Arch. Virol.* **147:** 1071-1076.

Nakata S, Honma S, Numata KK, *et al.* (2000). Members of the family *Caliciviridae* (Norwalk virus and Sapporo virus) are the most prevalent cause of gastroenteritis outbreaks among infants in Japan. *J. Infect. Dis.* **181:** 2029-2032.

Neill JD, Sosnovtsev SV, Green KY (2000). Recovery and altered neutralization specificities of chimeric viruses containing capsid protein domain exchanges from antigenically distinct strains of feline calicivirus. *J. Virol.* **74:** 1079-1084.

Noël JS, Liu BL, Humphrey CD *et al.* (1997). Parkville virus: a novel genetic variant of human calicivirus in the Sapporo virus clade, associated with an outbreak of gastroenteritis in adults. *J. Med. Virol.* **52:** 173-178.

Pang XL, Honma S, Nakata S, Vesikari T (2000). Human calicivirus in acute gastroeneritis of young children in the community. *J. Infect. Dis.* **181 (Suppl 2):** S288-S294.

Prasad BV, Hardy ME, Dokland T, Bella J, Rossmann MG, Estes MK (1999). X-ray crystallographic structure of the Norwalk virus capsid. *Science* **286:** 287-290.

Prasad BVV, Rothnagel R, Jiang X, Estes MK (1994). Three-dimensional structure of baculovirus-expressed Norwalk virus capsids. *J. Virol.* **68:** 5117-5125.

Seah EL, Marshall JA, Wright PJ (1999) Open reading frame 1 of the Norwalk-like virus Camberwell: completion of sequence and expression in mammalian cells. *J. Virol.* **73:** 10531-10535.

Sosnovtsev S, Green KY (1995). RNA transcripts derived from a cloned full-length copy of the feline calicivirus genome do not require VpG for infectivity. *Virology* **210:** 383-390.

Sugieda M, Nagaoka H, Kakishima Y, Ohshita T, Nakamura S, Makajima S (1998). Detection of Norwalk-like virus genes in the caecum contents of pigs. *Arch. Virol.* **143:** 1215-1221.

Tacket CO, Mason HS, Losonsky G, Estes MK, Levine MM, Arntzen CJ (2000). Human immune response to a novel Norwalk virus vaccine delivered in transgenic potatoes. *J. Infect. Dis.* **182:** 302-305.

Van der Poel W, Vinjé J, van der Heide R, Herrera M, Vivo A, Koopmans M (2000). Norwalk-like calicivirus genes in farm animals. *Emerg.Infect. Dis.* **6:** 36-41.

Vinjé J, Deijl H, van der Heide R *et al.* (2000a) Molecular detection and epidemiology of Sapporo-like viruses. *J. Clin. Microbiol.* **38:** 530-536.

Vinjé J, Green J, Lewis D, Gallimore CI, Brown D, Koopmans M (2000b). Genetic polymorphism across the three open reading frames of Norwalk-like caliciviruses. *Arch. Virol.* **145:** 223-241.

Vinjé J, Koopmans M (2000) Simultaneous detection and genotyping of Norwalk-like viruses by oligonucleotide array in a reverse-line blot hybridisation format. *J. Clin. Microbiol.* **38:** 2595-2601.

Vipond IB, Pelosi E, Williams J, Ashley CR, Lambden PR, Clarke IN, Caul EO (2000). A diagnostic EIA for detection of the prevalent SRSV strains in United Kingdom outbreaks of gastroenteritis. *J. Med. Virol.* **61:** 132-137.

Wang J, Jiang X, Madore HP *et al.* (1994). Sequence diversity of small round structured viruses in the Norwalk virus group. *J. Virol.* **68:** 5982-5990.

IV, 1. Structure of Norwalk virus: the prototype human calicivirus

Andrea Bertolotti-Ciarlet [1,2,3], Rong Chen [2],

Mary K. Estes [3] and B. V. Venkataram Prasad [2]

[1] *Department of Microbiology, University of Pennsylvania, Philadelphia, PA 19104, USA*
[2] *Verna and Marrs Mclean Department of Biochemistry and Molecular Biology,*
  Baylor College of Medicine, Houston, TX 77030, USA
[3] *Department of Molecular Virology & Microbiology*
  Baylor College of Medicine, Houston, TX77030, USA

Introduction

Norwalk virus (NV) is the prototype human virus in the *Caliciviridae*, a viral family of non-enveloped positive-sense single-stranded RNA viruses (Green *et al.*, 2001) and is classified in the genus called "Norwalk-like viruses" (NLVs). The recent accessibility of molecular diagnostic assays has demonstrated that NLVs are the leading cause of outbreaks of non-bacterial gastroenteritis worldwide (Glass *et al.*, 2000b). The calicivirus RNA genome is single stranded, of positive polarity and about 7-8 kb in size. It is enclosed in a capsid which is composed of a single major structural protein, and a few molecules of a second structural protein (Green *et al.*, 2001; Glass *et al.*, 2000a; Prasad *et al.*, 1994b). When examined by negative stain electron microscopy, caliciviruses exhibit characteristic cup-shaped depressions on the surface (Schaffer *et al.*, 1980; Carter *et al.*, 1991). The name for the virus family is derived from the Latin word *calyx*, meaning cup, which makes reference to these surface depressions.

Although tissue culture systems have been developed for some animal caliciviruses, the human caliciviruses have not been propagated in cell culture yet (Green *et al.*, 2001; Sosnostsev and Green, Section IV, Chapter 2 of this book). In addition, NV virions are shed into the feces of patients with gastroenteritis in low concentrations (Jiang *et al.*, 1992a). Because of these difficulties, alternative approaches have been used to study these human viruses. Expression of the capsid protein of several caliciviruses in insect cells using the baculovirus expression system resulted in the self-assembly of the capsid protein into virus-like particles (VLPs) that are antigenically, and structurally similar to native virions (Green *et al.*, 1997; Hale *et al.*, 1999; Jiang *et al.*, 1992b; Kobayashi *et al.*, 2000; Nagesha *et al.*, 1995; Nagesha *et al.*, 1999). These VLPs have been useful to gain structural, antigenic and biological information about these non-

456

cultivable pathogens. The simple architecture of the NV capsid made from a single major gene product and the ability of the expressed protein to spontaneously assemble into a T=3 capsid make NV an excellent model system to further our understanding of the structural requirements for the assembly of a T=3 icosahedral capsid. Here, we review the structure of NV VLPs and new insights about NV capsid assembly and composition.

Norwalk Virus Structure

Overall NV structure

Norwalk virus is a prototypical T=3 icosahedral virus, with 180 copies of the capsid protein organized into 90 dimers. The capsid has a contiguous protein shell of radii between 100 Å and 145 Å, which encloses the genome, and prominent protrusions that emanate from this shell, at all the local and strict twofold axes, to an external radius of ~190 Å. Large hollows are present at all the five- and three-fold axes (Prasad *et al.*, 1994b; Prasad *et al.*, 1999; Prasad *et al.*, 2000) (Fig. 1A).

Structure of the capsid protein subunit

To form a T=3 icosahedral structure, the capsid protein has to adapt to three quasi-equivalent positions. The subunits located at these positions are referred to as A, B and C. The structures of all these three subunits are similar, having two principle domains, the S domain that forms the icosahedral shell and the P domain that forms the protrusions (Fig.1B). A flexible hinge links the S and P domains. The N-terminal 225 amino acid residues constitute the S domain, and amino acid residues 226 to 530 form the P domain. The amino acid residues 226 to 278 and 406 to 530 of this domain lie juxtaposed, connected by a globular region composed of residues from 278 to 406. Thus the P domain can be considered as having two subdomains, P1 formed by amino acid residues 226 to 278 and 406-530, and P2 consisting of residues 406 to 530, which is a large insertion between residues 278 and 406 (Fig. 1B). In addition, based on sequence alignments of the caliciviruses capsid proteins, the P2 subdomain is the most variable region of the capsid. The structure of the NV capsid protein exhibits both classical and novel features. Although the S domain has a canonical 8-stranded β-barrel structure (ESAB), the P2 subdomain in the P domain has a fold similar to that seen in domain 2 of the elongation factor Tu (EF-Tu). This is the first time that this structure has been observed in a viral capsid protein (Prasad *et al.*, 1999). The fold of the P1 subdomain is unlike any other polypeptide fold observed so far. The S domain is exclusively involved in the icosahedral contacts, whereas the P domains are involved in the dimeric contacts (Fig. 1C). The P domains of the A and B subunits interact across the quasi two-fold axes, and those of the C subunits interact across the strict two-fold axes to form the dimeric protrusions. The A/B dimers surround the five-fold axis, and C/C and A/B dimers alternate around the three-fold axis.

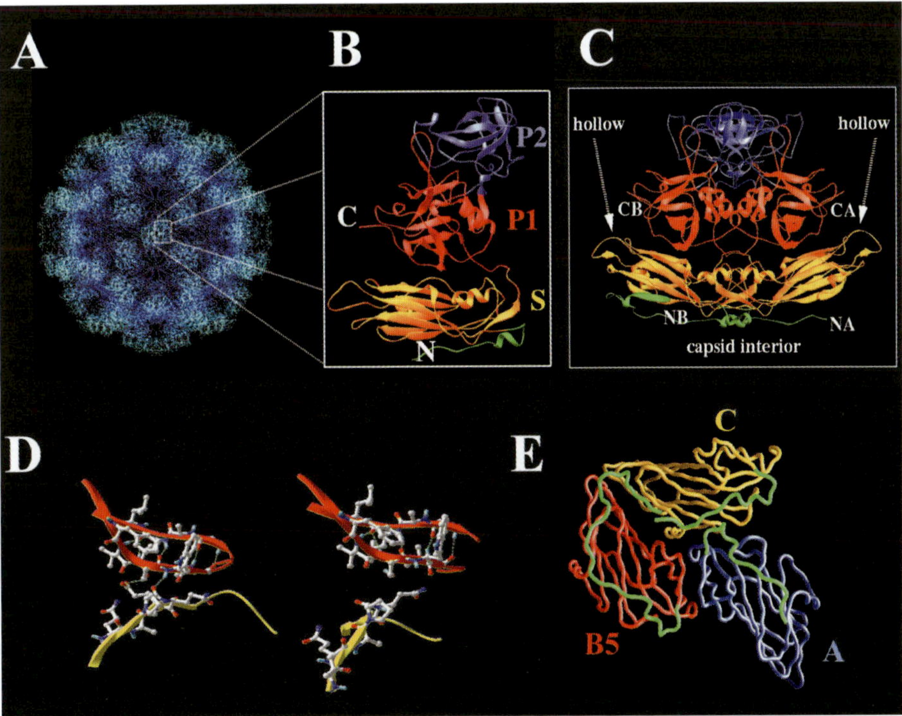

Fig. 1. rNV structure. (A) X-ray crystallographic structure of rNV capsid at 3.4Å resolution, as viewed along the icosahedral 2-fold axis. Only the backbone atoms of the 180 subunits are depicted. The structure is depth-cued, with deeper blue at lower radii and lighter blue at higher radii. (B) Ribbon representation of the structure of the rNV capsid unit (subunit C). The N-terminal arm, S domain, P1 and P2 subdomains are colored in green, yellow, red, and blue respectively. The N- and C-termini of the capsid protein are indicated. (C) A dimer of two C-subunits. The C-terminus faces the hollow, whereas the N-terminal arm faces the interior of the capsid and the P2 subdomain faces the exterior of the capsid. (D) Interactions between the P (red) and S (yellow) domains. Left. In the A or B subunits residues 504 and 506 of the P domain make hydrogen bond interactions (shown in dotted lines) with residues 64 and 65 of the S domain. Right. Such interactions are absent in the C subunits. (E) The N-terminal residues (10-14) of B subunit (red) interact with the ESAB of the neighboring C subunit (yellow). The N-terminal residues of the subunits are shown in green. Equivalent interactions between C and A or A and B are not present as the first 29 residues of A and C subunits are disordered.

Subunit association

The packing of the S domains in the A/B dimer has a bent conformation, whereas in the C/C dimer it has a flat conformation. Switching between the bent and the flat conformations facilitates the appropriate curvature for the formation of a closed shell.

The bent conformation of the A/B dimer appears to be stabilized by the hydrogen bond interactions between residues 505 and 506 in the P1 subdomain and residues 64 and 65 in the S domain of the respective A and B subunits (Fig. 1D). These interactions are absent in the C subunits as the S domain is rotated away from the P domain. It is possible that the flat conformation of the C/C dimer is stabilized by the interactions with the ordered N-terminal residues of the neighboring B subunits. The N-terminal 29 residues in A and C subunits are disordered, while only nine N-terminal residues are disordered in B subunits. Residues 10-14 in a B subunit interact with a neighboring C subunit (Fig. 1E) (Prasad *et al.*, 1999).

NV capsid assembly requirements

What is the role of the NV capsid protein N-terminal arm in assembly?

A switching region that is postulated to control the variations in the conformation of the coat protein of the T=3 viruses involves either the N-terminal arm of the capsid protein or the genomic RNA (Rossmann & Johnson, 1989). Elimination of this "molecular switch" is expected to result in a loss of precision in viral assembly. Indeed, in southern bean mosaic virus (SMBV) deletion of the N-terminal arm results in the formation of a T=1 capsid. In tombusviruses and sobemoviruses, which are plant viruses, the N-terminal arm of the S domain in the C subunit is ordered, whereas it is disordered in the A and B subunits (Harrison *et al.*, 1978; Rossmann *et al.*, 1983; Harrison, 1989). The ordered N-terminal arms of the C/C dimer form a wedge preventing the C/C dimer to adopt a bent conformation. A different mechanism is observed in the case of nodaviruses. In these viruses, ordered RNA interacting with C/C dimers plays an important role in modulating the capsid curvature by constraining the C/C dimers to a flat conformation (Fisher *et al.*, 1993).

NV particles are different from the T=3 capsids of sobemoviruses, tombusviruses and nodaviruses and represent a unique T=3 assembly model because neither the interactions involving the ordered N-terminal residues of the C subunit nor the interactions with RNA are observed. The rNV capsid protein readily forms T=3 capsids without RNA. In the native NV VLPs, the N-terminal residues (amino acids 10 to 14) of the B subunit interact with the F-strand of the ESAB of the neighboring C subunit in the T=3 lattice (Fig. 1E) (Prasad *et al.*, 1999). Similar (quasi-equivalent) interactions between the neighboring A subunits around the 5-fold axis or the C subunits around the 3-fold axis are lacking because the first 29 residues of the A and C subunits are disordered. The effect of deleting N-terminal residues has been studied in a number of other T=3 viruses. The N-terminal 60 residues in tomato bushy stunt virus (TBSV) (Harrison 1980), 52 residues in turnip crinkle virus (TCV) (Sorger *et al.*, 1986; Hogle *et al.*, 1986), and 25 residues in cowpea chlorotic mottle virus (CCMV) (Zhao *et al.*, 1995), which are not part of the ESAB domain are oriented towards the interior of the particle and have been shown to be required for the encapsidation of RNA but not for the assembly of empty particles (Zhao *et al.*, 1995; Hogle *et al.*, 1986).

Proteolytic digestion of the N-terminal portions of the capsid protein of SBMV leads to the assembly of T=1 particles instead of T=3 particles (Savithri & Erickson 1983; Erickson *et al.*, 1985). Likewise, deletions of 31 amino acid residues from the N-terminus of flock house virus (FHV), a T=3 virus member of the family *Nodaviridae*, results in the formation of variously shaped particles (Dong *et al.*, 1998). Recent biochemical and structural studies have shown that deletion of the N-terminal 20 residues of the NV capsid (NT20) does not prevent assembly of the particles (Fig. 2A), suggesting that the determinants for the T=3 capsid assembly for NV may lie outside of the N-terminus and that the interaction between the B and the C subunits is not obligatory for the formation of the T=3 capsid (Bertolotti-Ciarlet *et al.*, 2002). In the NT20 mutant, it is quite likely that a compensatory conformational change such as further ordering of the C subunit that interacts with the B subunit may retain the requirements for icosahedral assembly.

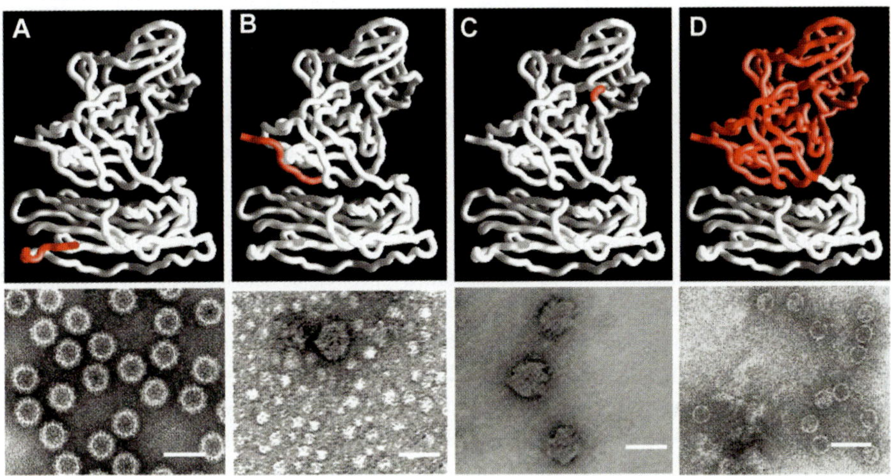

Fig. 2. EM analysis of NV capsid mutants. For each mutant the top panel shows the region mutated (red) on a rope representation (white) of the FL NV capsid protein crystal structure (Prasad *et al.*, 1999), and the lower panel shows an electron micrograph of the different mutant particles purified from undiluted supernatant material of the Sf9 cell cultures infected with the respective baculovirus recombinants. The EM pictures show a representative area of grids prepared with gradient fractions. Ammonium molybdate (1%) was used for staining. (A) NT20 forms VLPs that resemble full length NV capsid protein VLPs. (B) CT20 forms VLPs that are 45 nm in diameter. (C) ID285 forms VLPs that are 45nm in diameter. (D) CT303 forms VLPs that lack the characteristic arches formed by the protruding domains and resemble "smooth particles". The bar represents 50 nm. Modified from Bertolotti-Ciarlet *et al.*, 2002.

460

Does the P domain play a role in the NV capsid assembly?

Of the T=3 viruses, the closest structural relatives of NV are the tombusviruses. The capsid structure in these viruses also has a modular architecture with distinct S and P domains. In contrast, the capsid structure of other T=3 viruses, like sobemoviruses and nodaviruses, such as SBMV and FHV, respectively, have only one domain which is structurally homologous to the S domain (Harrison, 2001). Therefore, what is the role of the protruding domain in the NV capsid assembly? Does the S domain by itself possess all the requirements of a T=3 assembly, similar to SBMV or FHV?

Based on the rNV structure, deletion mutants in the P domain were designed to examine the role of P domain in the stability and assembly of rNV particles (Bertolotti-Ciarlet *et al.*, 2002). These studies included deletions in the regions that are involved in the P and S interactions, and in the regions that are involved in the dimeric contacts. Small deletions in the protruding domain, near the C-terminus, that affect the interactions between the P1 subdomains and S domains, and also those that affect the dimeric interactions, resulted in the production of particles that are 45 nm in diameter [20 amino acid deletion at the C-terminus (CT20) shown in Fig. 2B, and 15 amino acids deletion between amino acids 285 and 300 (ID285) shown in Fig. 2C]. Notice that all these mutants still retain a significant portion of the P domain. It is possible that interfering with either the P-S interactions (Fig. 1E) or the dimeric contacts forces the P domain to adopt alternate conformations, which directly influence the packing of the S domains and result in increased size and lower stability. Then, what is the role of the P-S interactions in the assembly processes and can the S domain by itself assemble into particles? When the entire P domain is deleted (mutant CT303), eliminating the interactions between the P and S domains, the truncated capsid protein produces well-formed "smooth particles" that lack the characteristic arches formed by the P domain dimers at the surface of the particle (Fig. 2D). Cryo-EM analysis of the "smooth particles" shows that the particles formed are indeed icosahedral but with a noticeably smaller diameter (27 nm) than the diameter expected from the rNV shell structure [30nm (Bertolotti-Ciarlet *et al.*, 2002)]. Thus, in all the particles produced from deletions at the P domain, an absence of interactions between the P and the S domains alters the size of the particles. It is plausible that the hydrogen bond interactions between the C-terminal residues of the P domain with the S domain residues, as observed in the crystal structure of the rNV capsid (Fig. 1E), control the size of the icosahedral capsid, and the lack of such interactions may be responsible for the observed alterations in size.

Relating structure to function for the NV capsid protein

Our extended knowledge of the NV capsid structure may help us to understand the structure of calicivirus capsids in general. In addition, some predictions about the location of functional domains in the NV capsid protein can be made by relating information from the cultivatable animal caliciviruses and the NV capsid structure. A multiple sequence alignment of caliciviruses shows that the sequences of other caliciviruses corresponding to the S domain of NV are highly conserved (Hardy *et al.*, 1997). The shell in the structure of primate calicivirus has a similar diameter and thickness to that of rNV (Prasad *et al.*, 1994b). Therefore, as for NV, other caliciviruses may have an eight-stranded β-barrel forming their S domain. Then, could the S domain, a highly conserved region of the calicivirus capsid that is sufficient for particle assembly form "smooth particles" during the life cycle of caliciviruses and could these particles have any biological relevance? Indeed, the ability of the S domain alone to form an icosahedral structure may have some biological implications. In the liver of infected rabbits, rabbit haemorrhagic disease virus (RHDV) calicivirus forms archless particles (25 to 27 nm in diameter), which are believed to be composed of the N-terminus of the capsid protein (Granzow *et al.*, 1996) and which resemble CT303 "smooth particles". Furthermore, chymotrypsin digestion of calicivirus-like viruses that infect worms has been reported to transform the particles into smooth particles (Hillman *et al.*, 1982). Although smooth particles have this far not been described for NV, NV may be similar to RHDV and the virus described by Hillman *et al.* For example, the 32K soluble protein present in high concentration in stool samples of NV-infected patients represents the C-terminus of the capsid protein that forms the protruding domains. The trypsin cleavage site in the protein is at amino acid 227 and is highly conserved in NV-like caliciviruses (Hardy *et al.*, 1995). Although, trypsin cleavage of the capsid *in vitro* is not possible when the particles are intact, there may be factors during natural infection that modify the conformation of the capsid and expose the trypsin cleavage site and result in the production of smooth particles. Could these smooth particles be the actual infectious form of the virus? Further biochemical analysis of the viral particles in stool using antibodies directed to the shell domain and development of a tissue culture system for human caliciviruses may help to answer this question.

Finally, what is the role of the P domain in the viral cycle? A multiple sequence alignment of different calicivirus capsid proteins has shown that the most highly variable regions are located in the P domain, the most exposed region of the capsid protein (Hardy *et al.*, 1997). Numerous biochemical data suggest that the P domain may be the one responsible for calicivirus binding to the cellular receptor. Indeed, a monoclonal antibody that recognizes a region between the residues 300 and 384 in the P2 subdomain on the NV capsid protein inhibits the binding of rNV capsids to cells (White *et al.*, 1996). In addition, a three-dimensional structure of RHDV VLPs binding a neutralizing antibody was obtained, showing that the antibody binds to the top of the arches that correspond to the P2 subdomain (Thouvenin *et al.*, 1997). Moreover, neutralizing epitopes have been mapped for feline caliciviruses (FCV) using monoclonal antibody neutralization-resistant variants of FCV. Amino acids substitutions were located in the hypervariable regions of the FCV capsid protein that correspond to the P2 subdomain in the NV capsid (Neill 1992; Milton *et al.*, 1992; Tohya *et al.*, 1997). Furthermore, a tissue culture adapted form of the porcine enteric calicivirus (PEC), the only cultivatable enteric calicivirus, has been characterized. This attenuated virus, that exhibits limited propagation in inoculated pigs, possesses one distant and three clustered amino acids substitutions in the region of the PEC capsid protein corresponding

462

to the P2 subdomain (Guo *et al.*, 2001). Together, these observations suggest that the hypervariable region on the P2 subdomain may determine tissue tropism and binding of the virus to the cell.

Other calicivirus structures

As for NV, three-dimensional structural studies of other members of the *Caliciviridae* family have been undertaken by using electron cryomicroscopy and computer image-processing techniques to understand structure-function relationship in these viruses (Prasad *et al.*, 1994a; Prasad *et al.*, 1999). The first three-dimensional structure of an infectious calicivirus virion was determined for a primate calicivirus (Prasad *et al.*, 1994a), which can be propagated in Vero cells. The virion structure shows that in addition to the shell (at a radius of 115 to 150 Å), and protruding domain (that extend from a radius of 150 to 200 Å) as seen in rNV capsids, there is another inner shell (at the radius of 85 to 110 Å) and additional density enclosed by the inner shell that is assumed to be the RNA genome. The primate calicivirus capsid is composed of 180 copies of capsid protein organized into 90 arch-like dimers arranged on a T=3 icosahedral lattice. Prominent hollows are present at all the five-fold and three-fold axes. Compared to the structure of the rNV, the top of the arch (P2) has a different shape. In addition, the connection between the S domain and the P1 subdomain shows noticeable differences among the A, B and C subunits, which means that this connection - the hinge region – provide the necessary flexibility for the coat protein to adapt to quasi-equivalent environments. Besides rNV and a primate calicivirus, the structure of the RHDV VLPs has been determined by cryo-EM, and viruses have also a T=3 icosahedral symmetry (Thouvenin *et al.*, 1997). Like the other caliciviruses, the three-dimensional reconstructed RHDV VLP is composed of a spherical shell with 180 copies of the VP60 capsid protein and 90 protruding arches. Moreover, the RHDV capsid protein consists, as does the NV capsid protein, of two domains the S and the P domains (Thouvenin *et al.*, 1997).

Conclusions

Biochemical and structural studies of NV VLPs have helped to understand the structure-function relationships of the calicivirus capsid protein. Mutational studies designed on the basis of the structure of NV have revealed the role of the domains of the capsid in the assembly of the NV capsid showing that the S domain by itself has all the attributes to assemble a stable icosahedron. However, the interactions between the entire P domain and the S domain enhance the stability of the assembled particles and may help in conferring the appropriate size to the particle. In addition, the sequence similarities between the calicivirus capsid proteins suggest that the atomic resolution structure of the NV capsid will be similar in other caliciviruses.

Future studies may lead to the resolution of the x-ray structure for a calicivirus virion containing RNA. This would reveal RNA-protein interactions that could not have been obtained from low-resolution electron cryomicroscopy studies or from x-ray structures of empty VLPs. In addition, further biochemical characterization of calicivirus, such as the development of a tissue culture system, analysis of viral interactions with the cell and study of viral protein functions, in combination with the resolution of new calicivirus structures may lead to the identification of the receptor, localization of epitopes that define the different antigenic groups, and further understanding of calicivirus biology. Finally, a more comprehensive understanding of the calicivirus capsid structure may help the development of antiviral strategies and diagnostic systems to permit a broad range of caliciviruses to be detected.

Acknowledgements

Work in the authors' laboratories has been supported by the R. Welch Foundation and Public Health Service grants AI38036 and AI46581.

References

Bertolotti-Ciarlet, A., White, L. J., Chen, R., Prasad, B. V., & Estes, M. K. (2002) Structural requirements for the assembly of Norwalk virus-like particles. *J.Virol.* **76:** 4044-4055

Carter, M.J., Milton, I. D., & Madeley, C. R. (1991) Caliciviruses. *Rev.Med.Virol.* **1:** 177-186

Dong, X.F., Natarajan, P., Tihova, M., Johnson J.E., & Schneemann, A. (1998) Particle polymorphism caused by deletion of a peptide molecular switch in a quasiequivalent icosahedral virus. *J. Virol.* **72:** 6024-6033

Erickson, J.W., Silva, A. M., Murthy, M. R., Fita, I., & Rossmann, M. G. (1985) The structure of a T = 1 icosahedral empty particle from southern bean mosaic virus. *Science* **229:** 625-629

Fisher, A.J., McKinney, B. R., Schneemann, A., Rueckert, R. R., & Johnson, J. E. (1993) Crystallization of viruslike particles assembled from flock house virus coat protein expressed in a baculovirus system. *J.Virol.* **67:** 2950-2953

Glass, P.J., White, L. J., Ball, J. M., Leparc-Goffart, I., Hardy, M. E., & Estes, M. K. (2000a) The Norwalk virus ORF 3 encodes a minor structural protein. *J.Virol.* **74:** 6581-6591

Glass, R.I., Noel, J., Ando, T., Fankhauser, R., Belliot, G., Mounts, A., Parashar, U. D., Bresee, J. S., & Monroe, S. S. (2000b) The epidemiology of enteric caliciviruses from humans: a reassessment using new diagnostics. *J.Infect.Dis.* **181 Suppl 2:** S254-S261

Granzow, H., Weiland, F., Strebelow, H.-G., Liu, C. M., & Schirrmeier, H. (1996) Rabbit hemorrhagic disease virus (RHDV): ultrastructure and biochemical studies of typical and core-like particles present in liver homogenates. *Virus Res.* **41:** 163-172

Green, K.Y., Chanock, R. M., & Kapikian, A. Z. (2001) Human Caliciviruses. In: *Fields Virology,* 4th Ed. (eds Knipe, D.M., Howley, P.M. *et al.*): 841-874. Lippincott Williams & Wilkins, Philadelphia

Green, K.Y., Kapikian, A. Z., Valdesuso, J., Sosnovtsev, S., Treanor, J. J., & Lew, J. F. (1997) Expression and self-assembly of recombinant capsid protein from the antigenically distinct Hawaii human calicivirus. *J.Clin.Microbiol.* **35:** 1909-1914

Guo, M., Hayes, J., Cho, K. O., Parwani, A. V., Lucas, L. M., & Saif, L. J. (2001) Comparative pathogenesis of tissue culture-adapted and wild-type Cowden porcine enteric calicivirus (PEC) in gnotobiotic pigs and induction of diarrhea by intravenous inoculation of wild-type PEC. *J.Virol.* **75:** 9239-9251

Hale, A.D., Crawford, S. E., Ciarlet, M., Green, J., Gallimore, C., Brown, D. W., Jiang, X., & Estes, M. K. (1999) Expression and self-assembly of Grimsby virus: antigenic distinction from Norwalk and Mexico viruses. *Clin.Diagn.Lab.Immunol.* **6:** 142-145

Hardy, M.E., Kramer, S. F., Treanor, J. J., & Estes, M. K. (1997) Human calicivirus genogroup II capsid sequence diversity revealed by analyses of the prototype Snow Mountain agent. *Arch.Virol.* **142:** 1469-1479

Hardy, M.E., White, L. J., Ball, J. M., & Estes, M. K. (1995) Specific proteolytic cleavage of recombinant Norwalk virus capsid protein. *J.Virol.* **69:** 1693-1698

Harrison, S.C. (1980) Protein interfaces and intersubunit bonding. The case of tomato bushy stunt virus. *Biophys. J.* **32:** 139-153

Harrison, S.C. (1989) Common features in the design of small RNA viruses. *Concepts in Viral Pathogenesis* (eds Oldstone, M.B.A. & Notkins, A.): 3-19. Springer-Verlag, New York

Harrison, S.C. (2001) Principles of Virus Structure. In: *Fields Virology,* 4th Ed. (eds Knipe, D.M., Howley, P.M. *et al.*): 53-85. Lippincott Williams & Wilkins, Philadelphia

Harrison, S.C., Olson, A., Schutt, C. E., Winkler, F. K., & Bricogne, G. (1978) Tomato bushy stunt virus at 2.9 angstrom resolution. *Nature* **276:** 368-373

Hillman, B., Morris, T. J., Kellen, W. R., Hoffman, D., & Schlegl, D. E. (1982) An invertebrate calici-like virus. Evidence for partial virion disintegration in host excreta. *J.Gen.Virol.* **60:** 115-123

Hogle, J.M., Maeda, A., & Harrison, S. C. (1986) Structure and assembly of turnip crinkle virus. I. X-ray crystallographic structure analysis at 3.2 angstrom resolution. *J.Mol.Biol.* **191:** 625-638

Jiang, X., Wang, J., Graham, D. Y., & Estes, M. K. (1992a) Detection of Norwalk virus in stool by polymerase chain reaction. *J.Clin.Microbiol.* **30:** 2529-2534

Jiang, X., Wang, M., Graham, D. Y., & Estes, M. K. (1992b) Expression, self-assembly, and antigenicity of the Norwalk virus capsid protein. *J.Virol.* **66:** 6527-6532

Kobayashi, S., Sakae, K., Suzuki, Y., Ishiko, H., Kamata, K., Suzuki, K., Natori, K., Miyamura, T., & Takeda, N. (2000) Expression of recombinant capsid proteins of Chitta virus, a genogroup II Norwalk virus, and development of an ELISA to detect the viral antigen. *Microbiol.Immunol.* **44**: 687-693

Milton, I.D., Turner, J., Teelan, A., Gaskell, R., Turner, P. C., & Carter, M. J. (1992) Location of monoclonal antibody binding sites in the capsid protein of feline calicivirus. *J.Gen.Virol.* **73**: 2435-2439

Nagesha, H.S., Wang, L. F., & Hyatt, A. D. (1999) Virus-like particles of calicivirus as epitope carriers. *Arch. Virol.* **144**: 2429-2439

Nagesha, H.S., Wang, L. F., Hyatt, A. D., Morrissy, C. J., Lenghaus, C., & Westbury, H. A. (1995) Self-assembly, antigenicity, and immunogenicity of the rabbit haemorrhagic disease virus (Czechoslovakian strain V-351) capsid protein expressed in baculovirus. *Arch.Virol.* **140**: 1095-1108

Neill, J.D. (1992) Nucleotide sequence of the capsid protein gene of two serotypes of San Miguel sea lion virus: identification of conserved and non-conserved amino acid sequences among calicivirus capsid proteins. *Virus Res.* **24**: 211-222

Prasad, B.V.V., Hardy, M. E., Dokland, T., Bella, J., Rossmann, M. G., & Estes, M. K. (1999) X-ray crystallographic structure of the Norwalk virus capsid. *Science* **286**: 287-290

Prasad, B.V.V., Hardy, M. E., & Estes, M. K. (2000) Structural studies of recombinant Norwalk capsids. *J. Infect. Dis.* **181(Suppl. 2)**: S317-S321

Prasad, B.V.V., Matson, D. O., & Smith, A. J. (1994a) Three-dimensional structure of calicivirus. *J.Mol.Biol.* **240**: 256-264

Prasad, B.V.V., Rothnagel, R., Jiang, X., & Estes, M. K. (1994b) Three-dimensional structure of baculovirus-expressed Norwalk virus capsids. *J.Virol.* **68**: 5117-5125

Rossmann, M.G., Abad-Zapatero, C., Hermodson, M. A., & Erickson, J. W. (1983) Subunit interactions in southern bean mosaic virus. *J.Mol.Biol.* **166**: 37-73

Rossmann, M.G. & Johnson, J. E. (1989) Icosahedral RNA virus structure. *Annu. Rev. Biochem.* **58**: 533-573

Savithri, H.S. & Erickson, J. W. (1983) The self-assembly of the cowpea strain of southern bean mosaic virus: formation of $T = 1$ and $T = 3$ nucleoprotein particles. *Virology* **126**: 328-335

Schaffer, F.L., Bachrach, H. L., Brown, F., Gillespie, J. H., Burroughs, J. N., Madin, S. H., Madeley, C. R., Povey, R. C., Scott, F., Smith, A. W., & Studdert, M. J. (1980) Caliciviridae. *Intervirology* **14**: 1-6

Sorger, P.K., Stockley, P. G., & Harrison, S. C. (1986) Structure and assembly of turnip crinkle virus. II. Mechanism of reassembly *in vitro*. *J.Mol.Biol.* **191**: 639-658

Thouvenin, E., Laurent, S., Madelaine, M. F., Rasschaert, D., Vautherot, J. F., & Hewat, E. A. (1997) Bivalent binding of a neutralising antibody to a calicivirus involves the torsional flexibility of the antibody hinge. *J.Mol.Biol.* **270**: 238-246

Tohya, Y., Yokoyama, N., Maeda, K., Kawaguchi, Y., & Mikami, T. (1997) Mapping of antigenic sites involved in neutralization on the capsid protein of feline calicivirus. *J.Gen.Virol.* **78**: 303-305

White, L.J., Ball, J. M., Hardy, M. E., Tanaka, T. N., Kitamoto, N., & Estes, M. K. (1996) Attachment and entry of recombinant Norwalk virus capsids to cultured human and animal cell lines. *J.Virol.* **70**: 6589-6597

Zhao, X., Fox, J. M., Olson, N. H., Baker, T. S., & Young, M. J. (1995) *In vitro* assembly of cowpea chlorotic mottle virus from coat protein expressed in *Escherichia coli* and *in vitro*-transcribed viral cDNA. *Virology* **207**: 486-494

IV, 2. Feline calicivirus as a model for the study of calicivirus replication

Stanislav V. Sosnovtsev and Kim Y. Green

Laboratory of Infectious Diseases, National Institute of Allergy and Infectious Diseases, National Institutes of Health, Bethesda, MD 20892-8026

Introduction

Caliciviruses are small, nonenveloped, positive strand RNA viruses that infect a broad range of hosts. The "Norwalk-like" caliciviruses are the major etiologic agents of nonbacterial epidemic gastroenteritis in humans (Cubitt *et al.*, 1979; Kapikian *et al.*, 1972). In animals, caliciviruses have been associated with an upper respiratory tract infection in cats caused by feline calicivirus (FCV) (Gaskell, 1985), a vesicular exanthema in swine caused by vesicular exanthema of swine virus (VESV) (Wawrzkiewicz *et al.*, 1968), a severe liver disease in rabbits caused by rabbit hemorrhagic disease virus (RHDV) (Ohlinger *et al.*, 1990), and diarrhea in swine caused by porcine enteric calicivirus (Saif *et al.*, 1980). Caliciviruses have also been associated with various conditions in canine, simian, mink, bovine, pinniped, reptilian, amphibian, and avian species (Green *et al.*, 2000).

The family *Caliciviridae* is comprised of four genera: *Vesivirus*, *Lagovirus*, "Norwalk-like viruses" (NLVs), and "Sapporo-like viruses" (SLVs) (Green *et al.*, 2000)*. The positive strand RNA genome of caliciviruses ranges in size from 7.3 to 8.3 kb and is covalently linked to a VPg protein at the 5'-end (Burroughs and Brown, 1978; Dunham *et al.*, 1998; Herbert *et al.*, 1997; Meyers *et al.*, 1991b) and polyadenylated at the 3'-end (Ehresmann and Schaffer, 1977). The viral capsid (27-40 nm in diameter) contains 180 copies of a major capsid protein (designated VP1), and the surface structure exhibits characteristic cup-like surface depressions. The capsid surrounds the VPg-linked RNA genome and 1-2 copies of a minor capsid protein, VP2 (Glass *et al.*, 2000; Sosnovtsev and Green, 2000; Wirblich *et al.*, 1996). The packaging of a VPg-linked subgenomic RNA has been demonstrated for RHDV (Meyers *et al.*, 1991b).

The genomes of caliciviruses are organized into either two or three major open reading frames (ORFs). In the genera *Lagovirus* and the SLVs, the nonstructural polyprotein and VP1 are encoded in the same ORF. In the genera *Vesivirus* and the NLVs, the VP1 is encoded in a separate ORF. All caliciviruses possess a small ORF near

---

* The NLVs and SLVs have been renamed *Noroviruses* and *Sapoviruses*, respectively (Mayo, 2002).

the 3'-end of the genome that encodes the minor structural protein, VP2. The 5'-end nontranslated regions (NTRs) of caliciviruses are relatively short, and the 3'-end NTRs vary in length from 42 to 178 nucleotides (Green *et al.*, 2001).

Elucidation of the replication strategy of certain caliciviruses, such as the NLVs and RHDV, has been hampered by the lack of an efficient cell culture system. Even so, expression of recombinant proteins from cDNA clones has allowed the generation of proteolytic processing maps for the nonstructural polyproteins of Southampton virus (SHV, in the genus NLV) and RHDV (Liu *et al.*, 1996; Wirblich *et al.*, 1996). Analysis of individual recombinant proteins from these noncultivatable viruses identified NTPase and "3C-like" cysteine proteinase activities for RHDV and the NLVs (Boniotti *et al.*, 1994; Liu *et al.*, 1996; Marin *et al.*, 2000; Pfister and Wimmer, 2001) and a "3D-like" RNA dependent RNA polymerase for RHDV (Lopez Vazquez *et al.*, 1998). The first detailed structural information for the caliciviruses was obtained from structural studies of Norwalk virus-like particles (VLPs) derived by expression of the capsid protein in the baculovirus system (Prasad *et al.*, 1994b, 1999; Bertolotti-Ciarlet *et al.*, Section IV, Chapter 1 of this book).

Although significant progress can be made in the study of the noncultivatable caliciviruses using *in vitro* expression systems, we recognized the need several years ago for the development of a cultivatable calicivirus model for the study of replication (Sosnovtsev and Green, 1995). We selected FCV for our studies because it grew efficiently in cultured feline cells and was widely available. Furthermore, the virus exhibited a strong growth restriction for feline cells, and study of the mechanism responsible for this restriction might facilitate the development of a cell culture system for the NLVs. This is a review of the basic features of FCV and its contribution thus far as a model for the study of calicivirus replication.

Feline calicivirus disease

Feline calicivirus is a common cause of upper respiratory tract disease in cats (Gaskell, 1985). Symptoms of the virus-induced disease include ulceration of the mucosa of the oral cavity and the oropharyngeal area. The disease can also be associated with rhinitis and conjunctivitis. The infected animal often exhibits signs of depression, loss of appetite, and pyrexia. Certain FCV strains have been associated with pulmonary lesions and a febrile lameness syndrome (Dawson *et al.*, 1994; Hoover and Kahn, 1973; Pedersen *et al.*, 1983). There have also been reports of "highly virulent" FCV strains associated with increased mortality rates in disease outbreaks (Pedersen *et al.*, 2000; Turnquist and Ostlund, 1997). In addition to acute clinical manifestations, several studies have linked FCV to a persistent infection. The virus was isolated from up to 92% of cats with chronic stomatitis (Knowles *et al.*, 1989). Feline calicivirus can also be isolated from healthy, asymptomatic animals. It was demonstrated that up to 18.5% of the cat population shows signs of a carrier-state infection (Harbour *et al.*, 1991). The virus exists in a large variety of antigenically different strains; however, cross-neutralization studies conducted in the 1970s suggested the presence of only one serotype (Kalunda *et al.*, 1975; Povey, 1974). Later studies reported evidence for serotypic differences using a panel of hyperimmune sera prepared against purified viruses (Neill *et al.*, 2000). The FCV F9 strain was recognized as inducing a broadly-reactive antibody response (Bittle and Rubic, 1976; Kalunda *et al.*, 1975) and has been used for vaccination of cats against calicivirus disease for the last two decades.

Feline calicivirus growth in cell culture

Feline calicivirus was first described in 1957 when a viral agent with a strong cytopathic effect (CPE) was isolated from a cat with panleukopenia (Fastier, 1957). The virus particles were shown to be temperature-sensitive (with inactivation of infectivity at 50°C) and relatively ether-resistant (Fastier, 1957; Lee and Gillespie, 1973). FCV was also found to be sensitive to low pH; however, strains of the virus demonstrated varying sensitivities to acidification (Lee and Gillespie, 1973; Studdert *et al.*, 1970). The virus displayed a narrow range of cell tropism, primarily infecting cells of feline origin (Crandell and Madin, 1960). Kreutz *et al.* (1994) reported evidence for the presence of receptors involved in the specific binding of virus particles, and proposed that replication of FCV in nonpermissive cells might be restricted at the level of virus interaction with the cell surface. The number of putative attachment sites was estimated to be in the range of 1.5 to 3×10^3 per cell. Virus binding to permissive cells was reduced by 30% when cells were treated with O-glycanase, consistent with the possible involvement of a glycoprotein receptor (Kreutz *et al.*, 1994). Little is known about the steps of the virus replication cycle that follow the virus attachment to cell receptors such as uptake and uncoating. It was suggested that FCV could enter cells by endocytosis (Kreutz and Seal, 1995). In addition, low pH was proposed to play an important role in the process of uncoating of virus particles. Kreutz and Seal (1995) demonstrated that a small concentration of chloroquine, a reagent that raises pH in endocytic vesicles, prevented virus replication when added to cells at an early stage of virus infection.

FCV growth in tissue culture is accompanied by a rapid CPE that includes the appearance of multiple foci of round-shaped cells that eventually detach and lyse. The formation of plaques can be observed within 18-60 hours post-infection (Fig. 1A). FCV replication occurs in the cytoplasm of cells, and cytoplasmic membranes of the infected cells undergo significant rearrangements resulting in the production of numerous vesicles (Fig. 1B). In addition, the nuclear chromatin condenses into a dense round body (Crandell and Madin, 1960; Studdert and O'Shea, 1975). Assembled FCV virions are distributed throughout the cytoplasm and may appear as non-regular accumulations of individual particles. The virus particles can also be seen as paracrystalline arrays associated with smooth membrane vesicles or as linear arrays associated with microfibrils (Love and Sabine, 1975; Peterson and Studdert, 1970). Progeny virus can be detected in the cell culture medium within three to four hours after infection (Fastier, 1957). The mechanism of virus release is not known; however, it was suggested that virus particles could be released from infected cells as they lysed (Peterson and Studdert, 1970).

470

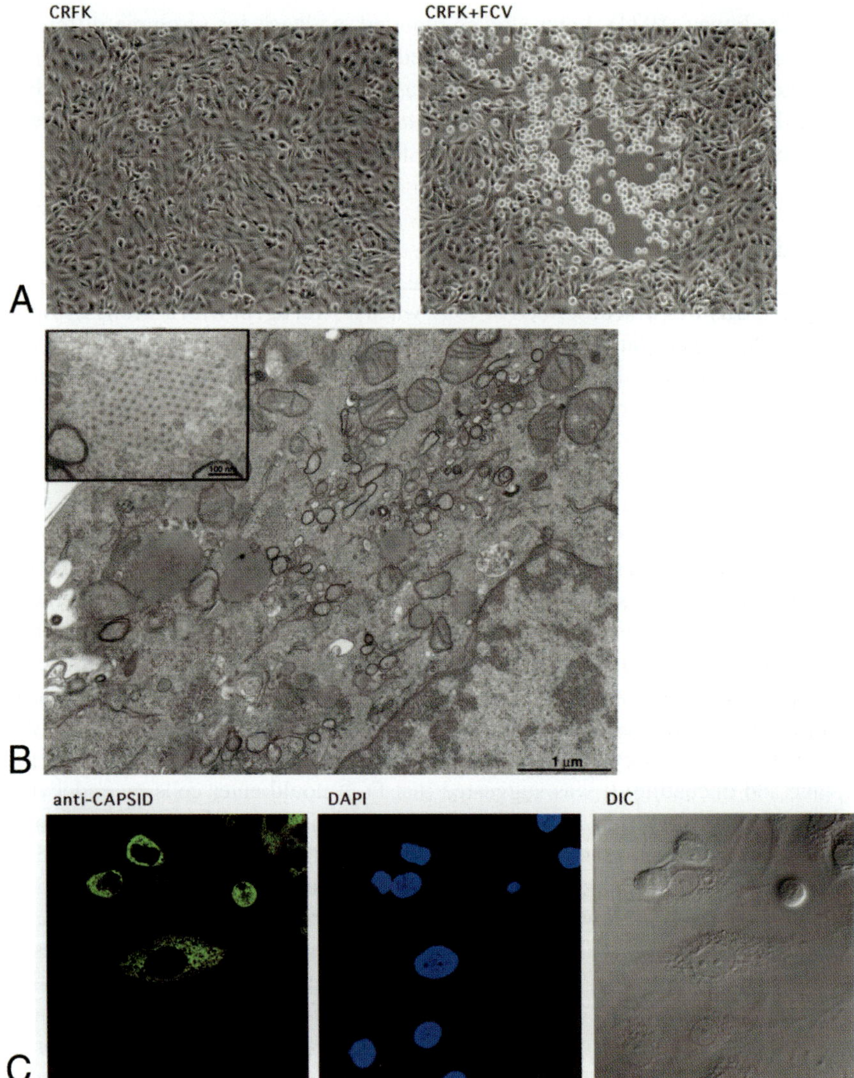

Fig. 1. Replication of FCV in CRFK cells. A) FCV infection of cultured Crandell-Rees feline kidney cells (CRFK) is accompanied by a rapid cytopathic effect that includes rounding of cells and plaque formation. Plaque formation can be observed within 24 hours after infection. B) Transmission electron microscopy of an FCV-infected cell, 6 hours post infection. Virus replication is accompanied by rearrangement of membranous structures in the cytoplasm of infected cells. Virus particles can be observed as paracrystalline arrays (inset) (courtesy of A. Weisberg). C) Immunofluorescent staining of FCV-infected cells (6 hours post infection) shows distribution of capsid protein throughout the cytoplasm. The capsid protein was detected by staining with hyperimmune serum prepared against purified FCV particles (anti-CAPSID) followed by staining with secondary FITC-conjugated antibodies. Nuclei localized by staining with 4'-6-diamidino-2-phenylindole (DAPI). Cells were visualized using differential interference contrast (DIC) (courtesy of O. Schwartz).

Feline calicivirus virions

FCV virions are small (~35 nm), icosahedral particles that have a "classical" calicivirus appearance. Distinct cup-like depressions can be detected by electron microscopy on the surface of the negatively stained FCV virions (Fig. 2). The virion shell is comprised of 32 subunits containing 180 copies of a single capsid protein (Peterson and Studdert, 1970; Prasad *et al.*, 1994a). When Geissler *et al.* (1999) expressed mature-sized FCV capsid protein in a transient vaccinia virus encoded T7 RNA polymerase-driven system, the capsid protein alone could self-assemble into antigenically-authentic empty capsids. Analysis of native FCV virions in infected cells suggested that there are two stages in the process of morphogenesis. The first stage consists of the formation of several different types of aggregates of capsid protein sedimenting at 5S, 9S, 11S, and 13S. These aggregates then assemble into stable 15S subunits. The second step includes the association of 15S subunits with the virus RNA and the assembly of infectious virions (Komolafe and Jarrett, 1986). The intracellular sites of capsid assembly are not known. Considering the ubiquitous cytoplasmic distribution of the synthesized *de novo* FCV capsid protein (Fig. 1C) and its ability to self-assemble, it is possible that the assembly process is concentration-dependent and can occur throughout the entire cytoplasm. The presence of "empty" particles suggests that some assembly does occur without packaging of the viral genome (Zhou *et al.*, 1994). It is possible that RNA or other viral proteins (such as VPg or VP2) could play a catalytic role in the assembly of the virions. The latter assumption is supported by observations that extensive accumulations of virus particles such as paracrystalline arrays are sometimes closely surrounded by smooth membrane vesicles, which are the putative sites of virus RNA replication (Love and Sabine, 1975; Studdert and O'Shea, 1975).

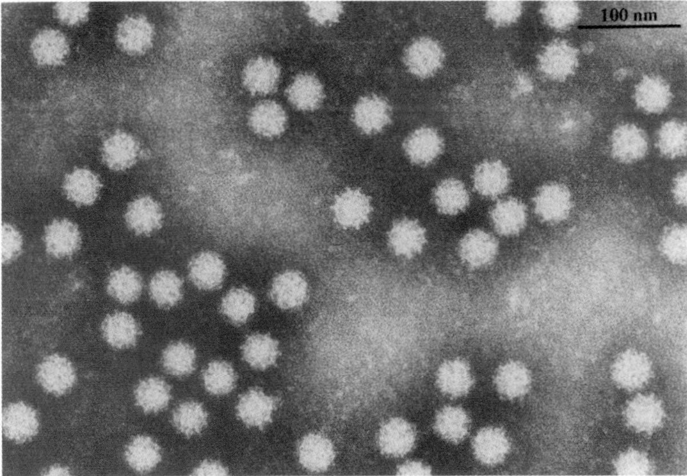

Fig. 2. Feline calicivirus virions. Purified FCV virions were stained with 2% phosphotungstic acid and visualized by electron microscopy. The characteristic cup-like structures of caliciviruses can be seen on some particles (courtesy of A. Kapikian).

The feline calicivirus genome

The first complete nucleotide sequence of FCV genome was published in 1992 (Carter *et al.*, 1992a), and several genomes have been sequenced since then (Glenn *et al.*, 1999; Oshikamo *et al.*, 1994; Sosnovtsev and Green, 1995). The FCV genome is organized into three major ORFs: ORF1 (nucleotides (nt) 20-5311), ORF2 (nt 5314-7320), and ORF3 (nt 7317-7637) (numbering according to Urbana strain, GenBank#L40021) (Fig. 3). The FCV ORF1 contains regions that share sequence similarity with the 2C NTPase, 3C cysteine proteinase, and 3D RNA-dependent RNA polymerase regions of the picornaviruses (Neill, 1990). The FCV ORF2 is separated from ORF1 by two nucleotides and encodes a precursor of the capsid protein with a molecular mass of approximately 73 kDa. The production of a capsid protein precursor protein appears to be unique to members of the genus *Vesivirus*. The FCV capsid protein precursor (preVP1) undergoes rapid proteolytic cleavage during the maturation process into the mature capsid protein (VP1) and the leader sequence (designated LC, for leader of capsid) (Carter, 1989; Carter *et al.*, 1992b; Sosnovtsev *et al.*, 1998). The ORF3 located at the 3'-end of the subgenomic RNA has a four-nucleotide overlap with the ORF2 and encodes a small protein (VP2) of 106 amino acids. The VP2 has been recently identified as a minor structural protein of FCV virions (Sosnovtsev and Green, 2000). The coding regions of the FCV genome are flanked by a short, highly conserved 19-nucleotide sequence at the 5'-end and by a less conserved (73.9-87.2% identity among FCV strains) 40-46-nucleotide sequence at the 3'-end. The sequence of the FCV 5'-end noncoding region contains a stretch of nucleotides that is repeated in the sequence corresponding to the beginning of the virus subgenomic RNA (Carter, 1990). This is a feature common among calicivirus genomes, and it had been suggested that related mechanisms might be involved either in the synthesis of the genomic and subgenomic RNAs or in the initiation of translation from the positive strand templates (Clarke and Lambden, 1997; Green *et al.*, 2001). A predicted stem-loop structure identified in the 3'-end noncoding region of the FCV genome could be involved in the formation of a recognition signal for the initiation of negative strand synthesis (Seal *et al.*, 1994).

Replication of FCV includes the synthesis of two major types of polyadenylated RNA molecules (Herbert *et al.*, 1996), a plus-sense genomic RNA of approximately 7.7 kb, and a subgenomic RNA of 2.4 kb. The genomic RNA serves as a template for synthesis of the nonstructural polyprotein encoded by the ORF1. The FCV subgenomic RNA begins 17 nucleotides upstream of ORF2, extends through the 3'-end poly (A) tract and is a template for translation of the structural proteins (Carter *et al.*, 1992b; Herbert *et al.*, 1996; Neill *et al.*, 1991; Sosnovtsev and Green, 2000). In addition to the 2.4 kb subgenomic RNA, at least six RNA transcripts ranging in size from full-length to 0.5 kb with 3'-co-terminal ends have been detected in FCV-infected cells (Carter, 1990; Neill and Mengeling, 1988).

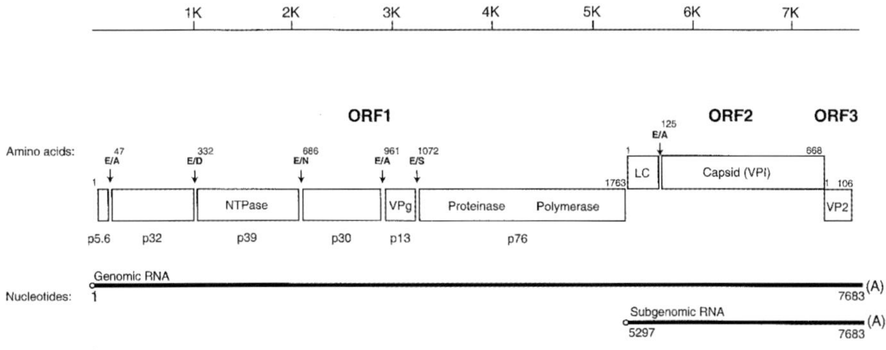

Fig. 3. Genome organization and cleavage map of FCV. The RNA genome of the Urbana strain of FCV is 7,683 nucleotides in length and organized into ORF1 (nt 20-5311), ORF2 (nt 5314-7320), and ORF3 (nt 7317-7637). The nucleotide numbering is given according to Urbana strain, GenBank accession number L40021 (Sosnovtsev and Green, 1995). The proteolytic cleavage sites recognized by the virus-encoded proteinase that define the borders of the nonstructural proteins and the mature capsid protein are indicated. The amino acid numbering is shown separately for each ORF. Protein coding sequences are drawn as boxes and designations of the nonstructural proteins are given below boxes. Abbreviations: NTPase, nucleoside triphosphatase; VPg, virion protein-genome linked; LC, leader sequence of capsid protein precursor; VP1, capsid protein; VP2, virion protein encoded by ORF3. Two major species of polyadenylated RNAs have been observed in infected cells that correspond to the full-length genome and the 3'-terminal end of the genome. A circle at the 5'-end of the full-length and subgenomic RNAs represents the covalently-linked VPg protein.

Development of a reverse genetics system for feline calicivirus

An advance in the study of FCV replication was achieved with the development of a reverse genetics system in 1995 (Sosnovtsev and Green, 1995). Like many positive strand RNA viruses, calicivirus RNA extracted from virus particles is infectious when transfected into permissive cells (Adldinger et al., 1969; Bachrach and Hess, 1973). However, removal of the calicivirus VPg by proteinase K digestion results in loss of infectivity (Burroughs and Brown, 1978; Dunham et al., 1998). An explanation for this effect was proposed when Herbert et al. (1997) showed that proteinase K removal of VPg bound to the 5'-end of the FCV RNA genome drastically reduced the efficiency of its translation in a rabbit reticulocyte lysate. The possible involvement of VPg at an early stage of the virus life cycle, such as initiation of translation, was perceived as a potential obstacle in the development of an infectious-RNA-based reverse genetics system for the caliciviruses. The initial presence of VPg, however, was not required for the infectivity of RNA transcripts derived from a cDNA clone of the FCV strain Urbana (Sosnovtsev and Green, 1995). The cDNA clone was engineered so that the FCV genome sequence was placed immediately downstream of the T7 bacteriophage RNA polymerase promoter and transcription with T7 RNA polymerase would begin from the precise 5'-end of the FCV genome. Termination of transcription was achieved by using a DNA template linearized with a restriction enzyme. It was noteworthy that FCV RNA transcripts were infectious only when they were synthesized in the presence of the 5'-end cap analog, $m^7G(5')ppp(5')G$. This suggested that the 5'-end cap structure could substitute for the presumed function of the VPg in the early stages of replication. Because the eukaryotic mRNA cap structure plays a crucial role in formation of the translation initiation complex (Gingras *et al.*, 1999), the synthetic cap structure of the FCV RNA synthesized *in vitro* might simply allow an adequate level of the initial translation of the viral nonstructural proteins needed to begin replication. This conclusion was supported by observations that *in vitro* translatability of FCV RNA is significantly reduced by removal of VPg (as noted above) and that the uncapped synthetic genomic RNA is not an efficient template for translation *in vitro* (Sosnovtsev and Green, unpublished).

Transfection of the capped RNA transcribed from the full-length FCV cDNA clone was shown to provide a relatively low level of infectivity. Even under optimized conditions, the infectivity of the capped RNA transcripts was close to 300 PFU/μg when approximately 10^6 cells were transfected. In comparison, the infectivity of synthetic RNA for other groups of positive-sense single-strand RNA viruses was observed to be at much higher levels (Table 1). To simplify the procedure for rescue of mutant viruses and to perhaps improve its efficiency, we employed vaccinia virus (MVA/T7) that carries the gene for the T7 RNA polymerase as a helper virus (Wyatt *et al.*, 1995). This recombinant vaccinia virus has lost its ability to achieve a productive infection in many mammalian cell lines, presumably due to the interrupted assembly of virus

Table 1.

Infectivity of full-length genome RNA transcripts reported for selected positive strand RNA viruses.

Virus	Infectivity of RNA (PFU/μg)	Reference
FCV *(Caliciviridae)*	3×10^2	Sosnovtsev & Green, 1995
Poliovirus *(Picornaviridae)*	10^5	van der Werf *et al.*, 1986
Sindbis virus *(Togaviridae)*	4×10^4	Rice *et al.*, 1987
Rubella virus *(Togaviridae)*	10^4	Pugachev *et al.*, 1997
Bovine viral diarrhea virus *(Flaviviridae)*	4×10^6	Frolov *et al.*, 1998

particles at an immature and noninfectious stage. However, MVA/T7 provides efficient expression of T7 RNA polymerase and its own capping enzymes inside the infected host cell. Transfection of a plasmid containing full-length FCV cDNA under the control of the T7 RNA polymerase promoter into MVA/T7-infected CRFK cells resulted in intracellular transcription of the genome and subsequent production of infectious FCV (Sosnovtsev *et al.*, 1996). This system bypassed time-consuming steps such as preparation of a linear DNA template for run-off transcription, RNA synthesis, and RNA purification. In addition, virus recovery could be obtained with "mini-prep" quality purified plasmids that were not linearized. The availability of a straightforward recovery system opened prospects for rapid screening of mutants for viability and biological characterization. Moreover, it allowed analysis of virus replication in non-permissive cells. When the plasmid-based system was used, the infectious FCV virus particles could be recovered from cells of animal and human origin that normally do not support FCV infection (Sosnovtsev *et al.*, 1996).

The proteolytic cleavage map of the feline calicivirus nonstructural polyprotein

When the FCV reverse genetics system was developed in 1995, knowledge of the basic molecular biology and replication strategy of this virus was very limited. In order to fully exploit the potential of a reverse genetics system, it was important to precisely define the coding regions of the genome and their products. The expression of the FCV VP1 capsid protein (encoded in ORF2, Fig. 3) from a subgenomic RNA was well established (Carter *et al.*, 1992b; Neill *et al.*, 1991), but the synthesis and processing of the nonstructural proteins encoded in ORF1 were poorly understood. Early studies of the proteins synthesized in feline kidney cells following infection with FCV had shown the presence of nonstructural proteins with molecular weights of approximately 96, 75, 39, 36 and 27 kDa (Carter, 1989). It was proposed that these proteins represented potential cleavage products (intermediate-sized and mature forms) from the ORF1 polyprotein.

Initial *in vitro* translation studies of a cDNA clone of the FCV ORF1 found evidence for the production of an active protease (Sosnovtsev *et al.*, 1998). The similarities between the products of this autocatalytic cleavage *in vitro* and the proteins in FCV-infected cells suggested that the protease sequences encoded in ORF1 were sufficient to mediate cleavage of the ORF1 polyprotein in the absence of cellular factors. In addition, the majority of proteins synthesized *in vitro* were identical in size to mature forms of the viral nonstructural proteins in infected cells (Sosnovtsev *et al.*, 1998). The latter finding was confirmed in recent experiments (Sosnovtsev *et al.*, 2002) when virus proteins generated *in vitro* and *in vivo* were compared in an immunoprecipitation analysis using a panel of region-specific antisera (Fig. 4). The cysteine 3C-like proteinase responsible for the processing of the FCV nonstructural proteins has been mapped to the C-terminal part of the ORF1 polyprotein (Sosnovtsev *et al.*, 1998; Sosnovtseva *et al.*, 1999). Substitution of Cys[1193] (amino acid numbering system used here and subsequently is that of intact FCV ORF1 polyprotein) in the active site of the

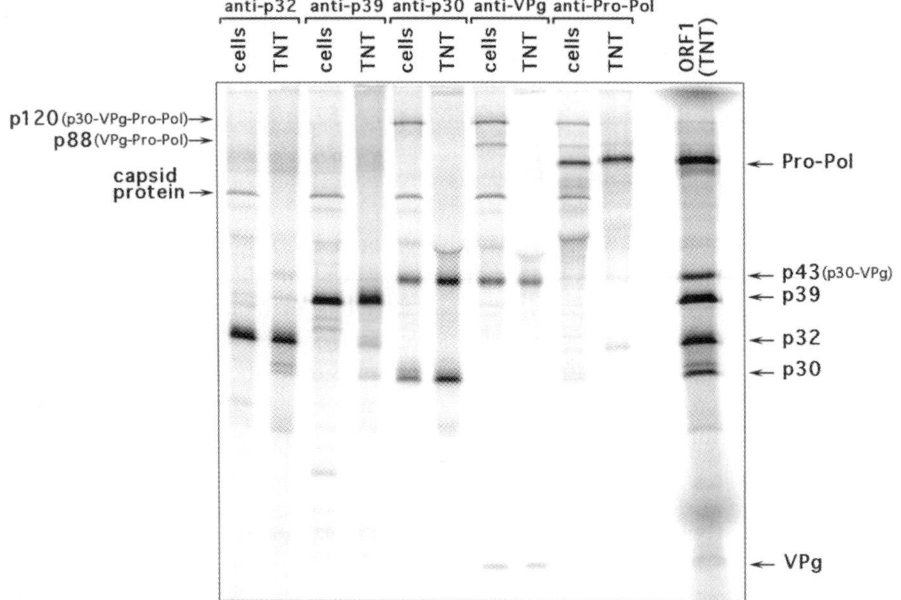

Fig. 4. Comparative immunoprecipitation analysis of virus-specific proteins derived from an ORF1 cDNA clone with those synthesized in FCV-infected cells. SDS-PAGE analysis of the radiolabeled FCV ORF1 polyprotein cleavage products shows similarity in profiles for proteins generated *in vitro* and in infected cells. Nonstructural proteins were immunoprecipitated using ORF1 region-specific antisera from FCV-infected cells (cells) or from an ORF1 *in vitro* translation mixture (TNT). Immunoprecipitation of radiolabeled nonstructural proteins from FCV-infected cells often leads to co-precipitation of the abundant 60 kDa capsid protein. Lane ORF1(TNT) represents radiolabeled TNT products derived from a clone encoding the entire FCV ORF1.

FCV 3C-like proteinase led to synthesis of the full-length polyprotein *in vitro* indicating that this was the only virus-encoded proteinase in ORF1 (Sosnovtsev and Green, unpublished).

The proteolytic cleavage map of the FCV nonstructural polyprotein was elucidated in an approach that employed direct N-terminal sequence analysis of ORF1 cleavage products expressed in bacteria, immunoprecipitation of cleavage products with region-specific antisera (Fig. 4), and microsequence analysis of radiolabeled cleavage products. The importance of each cleavage site in virus growth was then investigated by mutagenesis of the infectious cDNA clone. The map contains a total of five cleavage sites and shares similarities with other caliciviruses. The gene order is as follows: NH_2-p5.6-p32-p39-p30-VPg-Pro-Pol-COOH (Fig. 3) (Sosnovtsev *et al.*, 2002). The first protein encoded by the very beginning of the FCV ORF1 has a predicted molecular size of 5.6 kDa and strong basic features (pI=11.75) (Table 2). Release of this protein from the N-terminal part of the nonstructural polyprotein was demonstrat-

Table 2

Nonstructural proteins encoded by the FCV ORF1

	Nucleotides (position)	Amino acids (position)	M.W. (kDa)	pI	Functions
p5.6	138 (20-157)	46 (1-46)	5.6	11.75	?
p32	855 (158-1012)	285 (47-331)	31.99	6.19	?
p39	1062 (1013-2074)	354 (332-685)	38.88	9.08	putative NTPase
p30	825 (2075-2899)	275 (686-960)	30.09	9.25	?
p13	333 (2900-3232)	111 (961-1071)	12.65	7.24	VPg
p76	2076 (3233-5308)	692 (1072-1763)	75.74	7.6	3C-like proteinase 3D-like polymerase

ed in *in vitro* translation studies of ORF1 (Sosnovtsev *et al.*, 2002). However, attempts to identify this protein in infected cells have been unsuccessful so far. The amino acid sequence of p5.6 does not contain any known sequence motifs that could help assign its function, and the significance of this protein remains unclear. Of interest, RHDV contains a mapped cleavage site that releases a 16 kDa protein from the N-terminus of the RHDV ORF1 polyprotein. A corresponding cleavage site has not been found in the N-terminal 45 kDa protein of the SHV nonstructural polyprotein in *in vitro* translation studies. This uncleaved 45 kDa protein is likely to be comparable to the FCV p5.6 plus p32 (the next protein in linear arrangement of the FCV ORF1 polyprotein map) and the RHDV 16 kDa plus 23 kDa proteins. It was suggested that further cleavage of the NLV N-terminal protein might take place in infected cells (Liu *et al.*, 1996; Liu *et al.*, 1999; Seah *et al.*, 1999). The amino acid similarity among the SHV 45 kDa protein and corresponding proteins of RHDV and FCV ranges from 25-31%, making it one of the most divergent regions of the calicivirus nonstructural polyprotein. Of interest, significant sequence divergence for this region of the polyprotein is observed not only among the members of different genera of the *Caliciviridae*, but among viruses representing different genetic groups within a genus (Dingle *et al.*, 1995; Pletneva *et al.*, 2001; Rinehart-Kim *et al.*, 1999). The function of the FCV p32 in virus replication, as well as the function of the analogous proteins from other caliciviruses, are not known. However, it was suggested that the 23 kDa RHDV protein might have a functional similarity with the 2B protein of picornaviruses (Wirblich *et al.*, 1996).

The picornavirus 2B protein alone or as a part of its precursor, 2BC protein, is involved in membrane modifications in the virus-infected cells that include rearrangements and vesicle formation, changes in membrane permeability, and disassembly of the Golgi complex (Racaniello, 2001).

The p39 is next in the gene order of the FCV ORF1 (Fig. 3). The corresponding proteins from the genomes of RHDV and SHV are similar in size (37-41 kDa), and their amino acid sequences are relatively conserved among caliciviruses. All of them contain the predicted nucleoside triphosphate (NTP)-binding domain conserved in picornavirus 2C proteins (Gorbalenya *et al.*, 1990; Neill, 1990; Meyers *et al.*, 1991a; Jiang *et al.*, 1993; Lambden *et al.*, 1993). The picornavirus 2C is a multifunctional protein and has been implicated in the rearrangement of intracellular membranes and formation of membrane-associated RNA-replication complexes. This protein has also been characterized as an RNA-binding protein with NTPase activity (Racaniello, 2001). Recently, evidence of NTP-hydrolysis activity of this protein has been found for RHDV and SHV (Marin *et al.*, 2000; Pfister and Wimmer, 2001). In addition to the similarity of its NTPase sequence motifs, the FCV p39 shares an interesting structural feature with the corresponding RHDV and SHV proteins – the presence of an N-terminal cluster of hydrophobic amino acids. Transmembrane helices that might anchor the p39 in the intracellular membranes are predicted in the protein sequence for amino acids 335 to 353 and for amino acids 366 to 388 (Sonnhammer *et al.*, 1998). Based on their localization in the protein sequence, one might suggest that the p39 is anchored to intracellular membranes via its N-terminus.

The C-terminal part of the FCV ORF1 polyprotein includes p30, p13 (VPg), and the p76 (Pro-Pol) proteins (Fig. 4). A protein with a molecular mass of approximately 120 kDa (p120) corresponds to the potential precursor of these proteins and can be detected in virus-infected cells as early as 2-3 hrs post infection (Sosnovtseva *et al.*, 1999). It is likely that this large polyprotein undergoes further processing in at least two different ways. Immunoprecipitation analysis suggested that p120 can be cleaved into p30 and p88 (VPg-Pro-Pol) or into p43 (p30-VPg) and p76 (Pro-Pol) (Fig. 4) (Sosnovtsev *et al.*, 2002). The function of p30 is not known; it maps to the second most variable region among calicivirus ORF1 polyproteins. Computer analysis predicts an amphipathic helix in the C-terminal part of the protein (amino acids 920 – 939) (Rost and Sander, 1993) that might be responsible for the association of the protein with intracellular membranes. In agreement with this prediction, p30 can be detected in its mature form as well as in the form of the p43 precursor in membrane-associated proteins isolated from FCV-infected CRFK cells (Green *et al.*, unpublished). Following the picornavirus nomenclature, this protein corresponds to the 3A protein and, similar to the picornaviruses, as part of the p43 precursor (p30-VPg) it may serve as membrane anchor for the VPg protein. The picornavirus VPg is thought to be involved in the initiation of viral RNA replication and serves as a primer for the viral RNA polymerase (Nomoto *et al.*, 1977; Paul *et al.*, 1998). The ability of a recombinant polymerase to uridylylate the VPg protein was recently demonstrated for RHDV (Machin *et al.*, 2001). The release of the VPg from the p43 precursor might be dependent on several factors including RNA synthesis initiation as was suggested for picornaviruses (Kuhn and Wimmer, 1987). Mutagenesis experiments demonstrated that cleavage between these two proteins was required for productive virus replication (Sosnovtsev *et al.*, 2002). However, there are some indications that release of the VPg protein can occur independently of RNA synthesis because free nonmodified VPg is found in abundance in FCV-infected cells (Sosnovtsev and Green, 2000). Additional functions suggested for calicivirus VPg include packaging of RNA molecules into the virus particles and a "cap-like" function in the initiation of translation of the virus RNAs (Dunham *et al.*, 1998; Herbert *et al.*, 1997; Sosnovtsev and Green, 1995). Recently, the existence of sequence similarity between the calicivirus VPg and eukaryotic initiation factor (eIF) of translation, eIF1A, was found, suggesting that the "cap-like" function of VPg protein might be related to a direct interaction of VPg with ribosomal proteins of the host cell (Sosnovtseva *et al.*, 1999).

The final protein encoded in ORF1 is Pro-Pol, which is a stable precursor of the FCV proteinase and polymerase (Sosnovtseva *et al.*, 1999; Wei *et al.*, 2001). The proteinase region of the Pro-Pol protein is consistent with a chymotrypsin-like cysteine proteinase with the conserved His^{1110}, Glu^{1131}, and Cys^{1193} residues characteristic of the active site of picornavirus 3C proteinases (Gorbalenya *et al.*, 1989; Neill, 1990). Substitution of Cys^{1193} with Gly abolished the processing of the polyprotein, thus indicating the catalytic role of this residue (Sosnovtseva *et al.*, 1999). In addition, the identity of the FCV enzyme as a cysteine proteinase was confirmed by its inhibition with cysteine proteinase inhibitors (Sosnovtsev *et al.*, 1998). The FCV 3C-like proteinase was shown to be responsible for the processing of both nonstructural and structural proteins of the virus. The enzyme mediates removal of the leader sequence (LC) from the precursor of the virus capsid protein (Sosnovtsev *et al.*, 1998).

A cleavage event within the FCV Pro-Pol protein sequence that releases a "mature" form of the virus proteinase and polymerase has not been identified. The Pro-Pol protein appears to be a stable precursor that accumulates in infected cells (Sosnovtseva *et al.*, 1999). Cleavage sites identified in FCV Pro-Pol over-expression studies are likely aberrant because processing of the Pro-Pol protein at these sites would produce a truncated form of polymerase that is lacking some of the conserved functional motifs (Sosnovtseva *et al.*, 1999; Wei *et al.*, 2001). It is possible that virus replication does not require processing of the FCV Pro-Pol protein. Wei *et al.* (2001) demonstrated that the FCV Pro-Pol had properties expected of a replicative polymerase. Their findings did not rule out the possibility that shorter than full-length Pro-Pol active forms of polymerase may exist in infected cells (Wei *et al.*, 2001). In addition, they suggested that different forms of the polymerase might have different functions in RNA replication (Wei *et al.*, 2001). Enzymatically active replication complexes have recently been isolated from FCV-infected cells (Green *et al.*, 2002).

Mutagenesis of the FCV genome and its applications

The ability to introduce specific mutations into the FCV genome at the cDNA level demonstrated that the majority of proteolytic cleavage sites in the ORF1 nonstructural

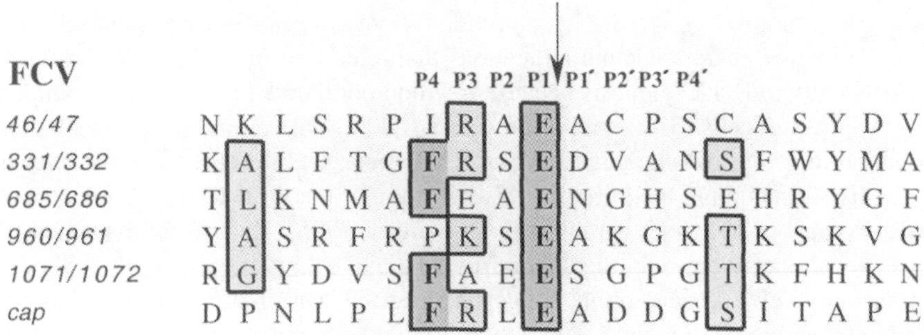

Fig. 5. Comparison of cleavage sites identified in the polyproteins encoded by the ORF1 of FCV and RHDV. A) Alignment of the amino acid sequences flanking the cleavage sites of the FCV ORF1 polyprotein and the sequence of the FCV capsid precursor cleavage site is shown. Identical and similar amino acids are shadowed in dark and light tones, respectively. B) Alignment of the amino acid sequences flanking cleavage sites of the RHDV ORF1 polyprotein is shown. The source for the RHDV sequence and cleavage site data was Meyers *et al.* (1991a) and Meyers *et al.* (2000), respectively. Mapped cleavage sites are indicated with an arrow. The four amino acids upstream of the cleavage site are labeled as P1 through P4. The four amino acids downstream of the cleavage site are labeled as P1′ through P4′.

polyprotein mapped *in vitro* are essential for a productive virus infection (Sosnovtsev *et al.*, 2002). Mutagenesis studies showed also that proteolytic processing of the capsid precursor protein by the virus-encoded proteinase is essential for a productive virus infection (Sosnovtsev *et al.*, 1998). The coordination of proteolytic cleavages and release of proteins with correct borders is clearly an important aspect of calicivirus replication. The mechanism of cleavage site recognition is not known for caliciviruses. Analysis of the cleavage sites processed by the FCV proteinase showed that the enzyme shares certain features with other calicivirus proteinases. For example, it was found that FCV proteinase exhibits a strict specificity for pairs of amino acids with glutamic acid residue in the P1 position (Sosnovtsev *et al.*, 1998, 2002). Similar substrate specificity is shared by RHDV and SHV proteinases that process cleavage sites with glutamic acid or glutamine residues in the P1 position (Liu *et al.*, 1999; Wirblich *et al.*, 1995). Moreover, comparison of the sequences adjacent to several of the essential FCV cleavage sites indicated the presence of an additional structural determinant. Most of the cleavage site sequences contained a hydrophobic phenylalanine residue at the P4 position (Fig. 5). It was suggested that RHDV proteinase may also have a minor preference for a large, hydrophobic amino acid residue just upstream of the proteinase cleavage site (Wirblich *et al.*, 1995). Comparison of sequences adjacent to the cleavage sites described for the RHDV ORF1 polyprotein shows the presence of phenylalanine or tyrosine residues next to the cleavage dipeptide (Fig. 5). In contrast to FCV, aromatic residues are found conserved in the P2 position of RHDV cleavage sites.

The presence of additional determinants, e.g. an amino acid in the P4 position, might reflect the existence of a special mechanism for recognition by the virus proteinase of important cleavage sites. Of interest, a similar pattern of recognition was described for poliovirus 3C proteinase. The enzyme recognizes glutamine-glycine pairs as cleavage sites, however, it cleaves only 9 out of 13 glutamine-glycine pairs present in the poliovirus polyprotein (Pallansch *et al.*, 1984). Most of the recognized cleavage sites have an alanine residue in the P4 position (Kräusslich *et al.*, 1988). The P4-alanine preference of poliovirus 3C proteinase is explained by the existence of a small hydrophobic P4-binding pocket (Mosimann *et al.*, 1997). Identification of the optimal recognition sequence for the calicivirus proteinase will require additional studies. Knowledge of the structural features of the cleavage sites as well as the structural characteristics of the proteinase itself should lead to a better understanding of calicivirus replication.

Another useful application of the FCV infectious clone relates to vaccine development strategies for the caliciviruses. In collaboration with Dr. John Neill, antigenic domain swaps were engineered between FCV strains with varying neutralization specificities (Neill *et al.*, 2000). The resulting chimeric viruses were analyzed by neutralization. Exchange of the "E" region, which is a variable region among capsid proteins (Neill, 1992), resulted in major changes in the antigenic phenotype of the virus, and viruses could be engineered that induced broadly-reactive neutralizing antibodies (Neill *et al.*, 2000). The ability to engineer such strains may be useful in FCV vaccine development, since recent reports have questioned the efficacy of the current vaccine (Baulch-Brown *et al.*, 1997). New prospects for improvement of today's FCV vaccine were opened when it was demonstrated that an entire capsid protein gene could be exchanged between two distinct FCV strains using the reverse genetics system (Sosnovtsev and Green, unpublished). The human caliciviruses have several antigenic types (Green *et al.*, 2001), and the ability to engineer chimeric FCV particles expressing human calicivirus capsid antigens might allow studies of neutralization *in vitro*. Furthermore, if a human calicivirus replication system were developed, FCV studies might serve as a model for the construction of engineered human calicivirus capsid proteins that contain multiple antigenic specificities.

482

The development of a FCV reverse genetics system indicates that such systems are on the horizon for other caliciviruses, including the NLVs associated with epidemic gastroenteritis. Although intense efforts to grow the NLVs in cell culture for nearly three decades have failed thus far, the ability to measure NLV infectivity and study virus replication remains an important research goal for public health. In the absence of a cell culture system, it may be possible to develop an NLV cDNA-based replicon system that could yield important new information about the basic features of RNA replication. Continued studies with the FCV reverse genetic system may assist these efforts. Finally, the availability of a full-length FCV cDNA clone encoding sequences with known authenticity is invaluable in mapping essential regions of the viral proteins and RNA involved in replication, and this work will add to our basic understanding of caliciviruses. Such studies are in progress in several laboratories, and the next few years should be a fruitful and exciting time for calicivirus research.

References

Adldinger, H. K., Lee, K. M. and Gillespie, J. H. (1969). Extraction of infectious ribonucleic acid from a feline picornavirus. *Arch Ges Virusforsch* **28**, 245-7.

Bachrach, H. L. and Hess, W. R. (1973). Animal picornaviruses with a single major species of capsid protein. *Biochem Biophys Res Commun* **55**, 141-9.

Baulch-Brown, C., Love, D. N. and Meanger, J. (1997). Feline calicivirus: a need for vaccine modification? *Aust Vet J* **75**, 209-13.

Bittle, J. L. and Rubic, W. J. (1976). Immunization against feline calicivirus infection. *Am J Vet Res* **37**, 275-8.

Boniotti, B., Wirblich, C., Sibilia, M., Meyers, G., Thiel, H. J. and Rossi, C. (1994). Identification and characterization of a 3C-like protease from rabbit hemorrhagic disease virus, a calicivirus. *J Virol* **68**, 6487-95.

Burroughs, J. N. and Brown, F. (1978). Presence of a covalently linked protein on calicivirus RNA. *J Gen Virol* **41**, 443-6.

Carter, M. J. (1989). Feline calicivirus protein synthesis investigated by western blotting. *Arch Virol* **108**, 69-79.

Carter, M. J. (1990). Transcription of feline calicivirus RNA. *Arch Virol* **114**, 143-52.

Carter, M. J., Milton, I. D., Meanger, J., Bennett, M., Gaskell, R. M. and Turner, P. C. (1992a). The complete nucleotide sequence of a feline calicivirus. *Virology* **190**, 443-8.

Carter, M. J., Milton, I. D., Turner, P. C., Meanger, J., Bennett, M. and Gaskell, R. M. (1992b). Identification and sequence determination of the capsid protein gene of feline calicivirus. *Arch Virol* **122**, 223-35.

Clarke, I. N. and Lambden, P. R. (1997). The molecular biology of caliciviruses. *J Gen Virol* **78**, 291-301.

Crandell, R. A. and Madin, S. H. (1960). Experimental studies on a new feline virus. *Am J Vet Res* **1**, 551-6.

Cubitt, W. D., McSwiggan, D. A. and Moore, W. (1979). Winter vomiting disease caused by calicivirus. *J Clin Pathol* **32**, 786-93.

Dawson, S., Bennett, D., Carter, S. D., Bennett, M., Meanger, J., Turner, P. C., Carter, M. J., Milton, I. and Gaskell, R. M. (1994). Acute arthritis of cats associated with feline calicivirus infection. *Res Vet Sci* **56**, 133-43.

Dingle, K. E., Lambden, P. R., Caul, E. O. and Clarke, I. N. (1995). Human enteric Caliciviridae: the complete genome sequence and expression of virus-like particles from a genetic group II small round structured virus. *J Gen Virol* **76**, 2349-55.

Dunham, D. M., Jiang, X., Berke, T., Smith, A. W. and Matson, D. O. (1998). Genomic mapping of a calicivirus VPg. *Arch Virol* **143**, 2421-30.

Ehresmann, D. W. and Schaffer, F. L. (1977). RNA synthesized in calicivirus-infected cells is atypical of picornaviruses. *J Virol* **22**, 572-6.

Fastier, L. B. (1957). A new feline virus isolated in tissue culture. *Am J Vet Res* **18**, 382-9.

Frolov, I., McBride, M. S. and Rice, C. M. (1998). *Cis*-acting RNA elements required for replication of bovine viral diarrhea virus-hepatitis C virus 5' nontranslated region chimeras. *RNA* **4**, 1418-35.

Gaskell, R. M. (1985). Viral-induced upper respiratory tract diseases. In *Feline Medicine and Therapeutics*, edited by E. A. Chandler, C. J. Gaskell and A. D. R. Hilbery, pp. 257-70. Oxford: Blackwell Scientific Publications.

Geissler, K., Schneider, K., Fleuchaus, A., Parrish, C. R., Sutter, G. and Truyen, U. (1999). Feline calicivirus capsid protein expression and capsid assembly in cultured feline cells. *J Virol* **73**, 834-8.

Gingras, A. C., Raught, B. and Sonenberg, N. (1999). eIF4 initiation factors: effectors of mRNA recruitment to ribosomes and regulators of translation. *Annu Rev Biochem* **68**, 913-63.

Glass, P. J., White, L. J., Ball, J. M., Leparc-Goffart, I., Hardy, M. E. and Estes, M. K. (2000). Norwalk virus open reading frame 3 encodes a minor structural protein. *J Virol* **74**, 6581-91.

Glenn, M., Radford, A. D., Turner, P. C., Carter, M., Lowery, D., DeSilver, D. A., Meanger, J., Baulch-Brown, C., Bennett, M. and Gaskell, R. M. (1999). Nucleotide sequence of UK and Australian isolates of feline calicivirus (FCV) and phylogenetic analysis of FCVs. *Vet Microbiol* **67**, 175-93.

Gorbalenya, A. E., Donchenko, A. P., Blinov, V. M. and Koonin, E. V. (1989). Cysteine proteases of positive strand RNA viruses and chymotrypsin-like serine proteases. A distinct protein superfamily with a common structural fold. *FEBS Lett* **243**, 103-14.

Gorbalenya, A. E., Koonin, E. V. and Wolf, Y. I. (1990). A new superfamily of putative NTP-binding domains encoded by genomes of small DNA and RNA viruses. *FEBS Lett* **262**, 145-8.

Green, K. Y., Ando, T., Balayan, M. S., Clarke, I. N., Estes, M. K., Matson, D. O., Nakata, S., Neill, J. D., Studdert, M. J. and Thiel, H.-J. (2000). *Caliciviridae*. In: *Virus Taxonomy: the Classification and Nomenclature of Viruses. The seventh report of the International Committee on Taxonomy of Viruses*, edited by M. H.

V. Regenmortel, C. M. Fauquet, D. H. L. Bishop, E. B. Carsten, M. K. Estes, S. M. Lemon, J. Maniloff, M. A. Mayo, D. J. McGeoch, C. R. Pringle and R. B. Wickner, pp. 725-35. San Diego, Academic Press.

Green, K. Y., Chanock, R. M. and Kapikian, A. Z. (2001). Human caliciviruses. In: *Fields Virology*, 4th edition, edited by D. M. Knipe, P. M. Howley *et al.*, pp. 841-874. Philadelphia, PA: Lippincott Williams and Wilkins.

Green, K. Y., Mory, A., Fogg, M. H., Weisberg, A., Belliot, G., Wagner, M., Mitra, T., Ehrenfeld, E., Cameron, C. E., and Sosnovtsev, S. V. (2002). Isolation of enzymatically active replication complexes from feline calicivirus-infected cells. *J Virol* **76**, 8582-95.

Harbour, D. A., Howard, P. E. and Gaskell, R. M. (1991). Isolation of feline calicivirus and feline herpesvirus from domestic cats 1980 to 1989. *Vet Rec* **128**, 77-80.

Herbert, T. P., Brierley, I. and Brown, T. D. (1996). Detection of the ORF3 polypeptide of feline calicivirus in infected cells and evidence for its expression from a single, functionally bicistronic, subgenomic mRNA. *J Gen Virol* **77**, 123-7.

Herbert, T. P., Brierley, I. and Brown, T. D. (1997). Identification of a protein linked to the genomic and subgenomic mRNAs of feline calicivirus and its role in translation. *J Gen Virol* **78**, 1033-40.

Hoover, E. A. and Kahn, D. E. (1973). Lesions produced by feline picornaviruses of different virulence in pathogen-free cats. *Vet Pathol* **10**, 307-22.

Jiang, X., Wang, M., Wang, K. and Estes, M. K. (1993). Sequence and genomic organization of Norwalk virus. *Virology* **195**, 51-61.

Kalunda, M., Lee, K. M., Holmes, D. F. and Gillespie, J. H. (1975). Serologic classification of feline caliciviruses by plaque-reduction neutralization and immunodiffusion. *Am J Vet Res* **36**, 353-6.

Kapikian, A. Z., Wyatt, R. G., Dolin, R., Thornhill, T. S., Kalica, A. R. and Chanock, R. M. (1972). Visualization by immune electron microscopy of a 27-nm particle associated with acute infectious nonbacterial gastroenteritis. *J Virol* **10**, 1075-81.

Knowles, J. O., Gaskell, R. M., Gaskell, C. J., Harvey, C. E. and Lutz, H. (1989). Prevalence of feline calicivirus, feline leukaemia virus and antibodies to FIV in cats with chronic stomatitis. *Vet Rec* **124**, 336-8.

Komolafe, O. O. and Jarrett, O. (1986). A possible maturation pathway of calicivirus particles. *Microbios* **46**, 103-11.

Kräusslich, H. G., Nicklin, M. J., Lee, C. K. and Wimmer, E. (1988). Polyprotein processing in picornavirus replication. *Biochimie* **70**, 119-30.

Kreutz, L. C. and Seal, B. S. (1995). The pathway of feline calicivirus entry. *Virus Res* **35**, 63-70.

Kreutz, L. C., Seal, B. S. and Mengeling, W. L. (1994). Early interaction of feline calicivirus with cells in culture. *Arch Virol* **136**, 19-34.

Kuhn, R. J. and Wimmer, E. (1987). The replication of picornaviruses. In: *The Molecular Biology of the Positive Strand RNA Viruses*, edited by D. J. Rowlands, M. A. Mayo and B. W. J. Mahy, pp. 17-51. London: Academic Press.

Lambden, P. R., Caul, E. O., Ashley, C. R. and Clarke, I. N. (1993). Sequence and genome organization of a human small round-structured (Norwalk-like) virus. *Science* **259**, 516-9.

Lee, K. M. and Gillespie, J. H. (1973). Thermal and pH stability of feline calicivirus. *Infect Immun* **7**, 678-9.

Liu, B., Clarke, I. N. and Lambden, P. R. (1996). Polyprotein processing in Southampton virus: identification of 3C-like protease cleavage sites by *in vitro* mutagenesis. *J Virol* **70**, 2605-10.

Liu, B. L., Viljoen, G. J., Clarke, I. N. and Lambden, P. R. (1999). Identification of further proteolytic cleavage sites in the Southampton calicivirus polyprotein by expression of the viral protease in *E. coli*. *J Gen Virol* **80**, 291-6.

Lopez Vazquez, A., Martin Alonso, J. M., Casais, R., Boga, J. A. and Parra, F. (1998). Expression of enzymatically active rabbit hemorrhagic disease virus RNA-dependent RNA polymerase in Escherichia coli. *J Virol* **72**, 2999-3004.

Love, D. N. and Sabine, M. (1975). Electron microscopic observation of feline kidney cells infected with a feline calicivirus. *Arch Virol* **48**, 213-28.

Machin, A., Martin Alonso, J. M. and Parra, F. (2001). Identification of the amino acid residue involved in rabbit hemorrhagic disease virus VPg uridylylation. *J Biol Chem* **276**, 27787-92.

Marin, M. S., Casais, R., Alonso Martin, J. M. and Parra, F. (2000). ATP binding and ATPase activities associated with recombinant rabbit hemorrhagic disease virus 2C-like polypeptide. *J Virol* **74**, 10846-51.

Mayo, M. A. (2002). Virus Taxonomy - Houston 2002. *Arch Virol* **147**, 1071-6.

Meyers, G., Wirblich, C. and Thiel, H.-J. (1991a). Rabbit hemorrhagic disease virus - molecular cloning and nucleotide sequencing of a calicivirus genome. *Virology* **184**, 664-76.

Meyers, G., Wirblich, C. and Thiel, H. J. (1991b). Genomic and subgenomic RNAs of rabbit hemorrhagic disease virus are both protein-linked and packaged into particles. *Virology* **184**, 677-86.

Meyers, G., Wirblich, C., Thiel, H. J. and Thumfart, J. O. (2000). Rabbit hemorrhagic disease virus: genome organization and polyprotein processing of a calicivirus studied after transient expression of cDNA constructs. *Virology* **276**, 349-63.

Mosimann, S. C., Cherney, M. M., Sia, S., Plotch, S. and James, M. N. (1997). Refined X-ray crystallographic structure of the poliovirus 3C gene product. *J Mol Biol* **273**, 1032-47.

Neill, J. D. (1990). Nucleotide sequence of a region of the feline calicivirus genome which encodes picornavirus-like RNA-dependent RNA polymerase, cysteine protease and 2C polypeptides. *Virus Res* **17**, 145-60.

Neill, J. D. (1992). Nucleotide sequence of the capsid protein gene of two serotypes of San Miguel sea lion virus: identification of conserved and non-conserved amino acid sequences among calicivirus capsid proteins. *Virus Res* **24**, 211-22.

Neill, J. D. and Mengeling, W. L. (1988). Further characterization of the virus-specific RNAs in feline calicivirus infected cells. *Virus Res* **11**, 59-72.

Neill, J. D., Reardon, I. M. and Heinrikson, R. L. (1991). Nucleotide sequence and expression of the capsid protein gene of feline calicivirus. *J Virol* **65**, 5440-7.

Neill, J. D., Sosnovtsev, S. V. and Green, K. Y. (2000). Recovery and altered neutralization specificities of chimeric viruses containing capsid protein domain exchanges from antigenically distinct strains of feline calicivirus. *J Virol* **74**, 1079-84.

Nomoto, A., Detjen, B., Pozzatti, R. and Wimmer, E. (1977). The location of the polio genome protein in viral RNAs and its implication for RNA synthesis. *Nature* **268**, 208-13.

Ohlinger, V. F., Haas, B., Meyers, G., Weiland, F. and Thiel, H. J. (1990). Identification and characterization of the virus causing rabbit hemorrhagic disease. *J Virol* **64**, 3331-6.

Oshikamo, R., Tohya, Y., Kawaguchi, Y., Tomonaga, K., Maeda, K., Takeda, N., Utagawa, E., Kai, C. and Mikami, T. (1994). The molecular cloning and sequence of an open reading frame encoding for non-structural proteins of feline calicivirus F4 strain isolated in Japan. *J Vet Med Sci* **56**, 1093-9.

Pallansch, M. A., Kew, O. M., Semler, B. L., Omilianowski, D. R., Anderson, C. W., Wimmer, E. and Rueckert, R. R. (1984). Protein processing map of poliovirus. *J Virol* **49**, 873-80.

Paul, A. V., van Boom, J. H., Filippov, D. and Wimmer, E. (1998). Protein-primed RNA synthesis by purified poliovirus RNA polymerase. *Nature* **393**, 280-4.

Pedersen, N. C., Elliott, J. B., Glasgow, A., Poland, A. and Keel, K. (2000). An isolated epizootic of hemorrhagic-like fever in cats caused by a novel and highly virulent strain of feline calicivirus. *Vet Microbiol* **73**, 281-300.

Pedersen, N. C., Laliberte, L. and Ekman, S. (1983). A transient febrile limping syndrome of kittens caused by two different strains of feline calicivirus. *Feline Practice* **13**, 26-35.

Peterson, J. E. and Studdert, M. J. (1970). Feline picornavirus. Structure of the virus and electron microscopic observations on infected cell cultures. *Arch Ges Virusforsch* **32**, 249-60.

Pfister, T. and Wimmer, E. (2001). Polypeptide p41 of a Norwalk-like virus is a nucleic acid-independent nucleoside triphosphatase. *J Virol* **75**, 1611-9.

Pletneva, M. A., Sosnovtsev, S. V. and Green, K. Y. (2001). The genome of Hawaii virus and its relationship with other members of the Caliciviridae. *Virus Genes* **23**, 5-16.

Povey, R. C. (1974). Serological relationships among feline caliciviruses. *Infect Immun* **10**, 1307-14.

Prasad, B. V., Hardy, M. E., Dokland, T., Bella, J., Rossmann, M. G. and Estes, M. K. (1999). X-ray crystallographic structure of the Norwalk virus capsid. *Science* **286**, 287-90.

Prasad, B. V., Matson, D. O. and Smith, A. W. (1994a). Three-dimensional structure of calicivirus. *J Mol Biol* **240**, 256-64.

Prasad, B. V., Rothnagel, R., Jiang, X. and Estes, M. K. (1994b). Three-dimensional structure of baculovirus-expressed Norwalk virus capsids. *J Virol* **68**, 5117-25.

Pugachev, K. V., Abernathy, E. S. and Frey, T. K. (1997). Improvement of the specific infectivity of the rubella virus (RUB) infectious clone: determinants of cytopathogenicity induced by RUB map to the nonstructural proteins. *J Virol* **71**, 562-8.

Racaniello, V. R. (2001). Picornaviridae: the viruses and their replication. In: *Fields Virology*, 4th edition, edited by D. M. Knipe, P. M. Howley *et al.*, pp. 685-722. Philadelphia, PA: Lippincott Williams and Wilkins.

Rice, C. M., Levis, R., Strauss, J. H. and Huang, H. V. (1987). Production of infectious RNA transcripts from Sindbis virus cDNA clones: mapping of lethal mutations, rescue of a temperature-sensitive marker, and *in vitro* mutagenesis to generate defined mutants. *J Virol* **61**, 3809-19.

Rinehart-Kim, J. E., Zhong, W. M., Jiang, X., Smith, A. W. and Matson, D. O. (1999). Complete nucleotide sequence and genomic organization of a primate calicivirus, Pan-1. *Arch Virol* **144**, 199-208.

Rost, B. and Sander, C. (1993). Prediction of protein secondary structure at better than 70% accuracy. *J Mol Biol* **232**, 584-99.

Saif, L. J., Bohl, E. H., Theil, K. W., Cross, R. F. and House, J. A. (1980). Rotavirus-like, calicivirus-like, and 23-nm virus-like particles associated with diarrhea in young pigs. *J Clin Microbiol* **12**, 105-11.

Seah, E. L., Marshall, J. A. and Wright, P. J. (1999). Open reading frame 1 of the Norwalk-like virus Camberwell: Completion of sequence and expression in mammalian cells. *J Virol* **73**, 10531-5.

Seal, B. S., Neill, J. D. and Ridpath, J. F. (1994). Predicted stem-loop structures and variation in nucleotide sequence of 3' noncoding regions among animal calicivirus genomes. *Virus Genes* **8**, 243-7.

Sonnhammer, E. L., von Heijne, G. and Krogh, A. (1998). A hidden Markov model for predicting transmembrane helices in protein sequences. *Proc Int Conf Intell Syst Mol Biol* **6**, 175-82.

Sosnovtsev, S. V., Garfield, M. and Green, K. Y. (2002). Processing map and essential cleavage sites of the nonstructural polyprotein encoded by ORF1 of the feline calicivirus genome. *J Virol* **76**, 7060-72.

Sosnovtsev, S. and Green, K. Y. (1995). RNA transcripts derived from a cloned full-length copy of the feline calicivirus genome do not require VPg for infectivity. *Virology* **210**, 383-90.

Sosnovtsev, S. V. and Green, K. Y. (2000). Identification and genomic mapping of the ORF3 and VPg proteins in feline calicivirus virions. *Virology* **277**, 193-203.

Sosnovtsev, S., Sosnovtseva, S. and Green, K. Y. (1996). Recovery of feline calicivirus from plasmid DNA containing a full-length copy of the genome. In: *The 1st International Symposium on Caliciviruses*, edited by D. Chasey, R. M. Gaskell and I. N. Clarke, pp. 125-30. Reading, UK: European Society for Veterinary Virology and Central Veterinary Laboratory.

Sosnovtsev, S. V., Sosnovtseva, S. A. and Green, K. Y. (1998). Cleavage of the feline calicivirus capsid precursor is mediated by a virus-encoded proteinase. *J Virol* **72**, 3051-9.

Sosnovtseva, S. A., Sosnovtsev, S. V. and Green, K. Y. (1999). Mapping of the feline calicivirus proteinase responsible for autocatalytic processing of the nonstructural polyprotein and identification of a stable proteinase-polymerase precursor protein. *J Virol* **73**, 6626-33.

Studdert, M. J., Martin, M. C. and Peterson, J. E. (1970). Viral diseases of the respiratory tract of cats: isolation and properties of viruses tentatively classified as picornaviruses. *Am J Vet Res* **31**, 1723-32.

Studdert, M. J. and O'Shea, J. D. (1975). Ultrastructural studies of the development of feline calicivirus in a feline embryo cell line. *Arch Virol* **48**, 317-25.

Turnquist, S. E. and Ostlund, E. (1997). Calicivirus outbreak with high mortality in a Missouri feline colony. *J Vet Diagn Invest* **9**, 195-8.

van der Werf, S., Bradley, J., Wimmer, E., Studier, F. W. and Dunn, J. J. (1986). Synthesis of infectious poliovirus RNA by purified T7 RNA polymerase. *Proc Natl Acad Sci USA* **83**, 2330-4.

Wawrzkiewicz, J., Smale, C. J. and Brown, F. (1968). Biochemical and biophysical characteristics of vesicular exanthema virus and the viral ribonucleic acid. *Arch Ges Viruforsch* **25**, 337-51.

Wei, L., Huhn, J. S., Mory, A., Pathak, H. B., Sosnovtsev, S. V., Green, K. Y. and Cameron, C. E. (2001). Proteinase-polymerase precursor as the active form of feline calicivirus RNA-dependent RNA polymerase. *J Virol* **75**, 1211-9.

Wirblich, C., Sibilia, M., Boniotti, M. B., Rossi, C., Thiel, H. J. and Meyers, G. (1995). 3C-like protease of rabbit hemorrhagic disease virus: identification of cleavage sites in the ORF1 polyprotein and analysis of cleavage specificity. *J Virol* **69**, 7159-68.

Wirblich, C., Thiel, H. J. and Meyers, G. (1996). Genetic map of the calicivirus rabbit hemorrhagic disease virus as deduced from *in vitro* translation studies. *J Virol* **70**, 7974-83.

Wyatt, L. S., Moss, B. and Rozenblatt, S. (1995). Replication-deficient vaccinia virus encoding bacteriophage T7 RNA polymerase for transient gene expression in mammalian cells. *Virology* **210**, 202-5.

Zhou, L., Yu, Q. and Luo, M. (1994). Characterization of two density populations of feline calicivirus particles. *Virology* **205**, 530-3.

Viral Gastroenteritis
U. Desselberger and J. Gray (editors)

IV, 3. Pathogenesis of enteric calicivirus infections

Mingzhang Guo [1] and Linda J. Saif [2]

[1] *Departments of Surgery and Medical Physiology, Cardiovascular Research Institute,
Texas A and M University System Health Science Center, 702 Southwest HK Dodgen Loop,
Temple, TX 76504, USA*
[2] *Food Animal Health Research Program, Ohio Agricultural Research and Development Center,
Veterinary Preventive Medicine Department, The Ohio State University, Wooster, OH 44691, USA*

Introduction

Caliciviruses are members of the *Caliciviridae* family. They are small, nonenveloped viruses of 27-40 nm in diameter and possess a single-stranded, plus-sense RNA genome of 7-8 kb in length and a single capsid protein of 56 to 71 kDa (Kapikian *et al.*, 1996; Estes *et al.*, 1997). These genetically diverse viruses are suspected or confirmed causes of a wide spectrum of diseases in their respective hosts and are divided into four genera: 1) *Vesivirus*, 2) *Lagovirus*, 3) *"Norwalk-like viruses"* (*NLVs*), and 4) *"Sapporo-like viruses"* (*SLVs*) (Green *et al.*, 2000).* Vesiviruses include the classical animal caliciviruses such as vesicular exanthema of swine virus (VESV), San Miguel sea lion viruses (SMSV), feline caliciviruses (FCV), bovine caliciviruses (BCV), mink caliciviruses (MCV), skunk caliciviruses (SCV), primate caliciviruses (PCV) and canine caliciviruses (CaCV). The VESV and SMSV cause skin vesicular lesions and/or reproductive failure in swine and marine animals, respectively. The FCV and BCV are responsible for respiratory infections in cats and calves, respectively. These cultivable classical caliciviruses are not considered to be enteric pathogens, although they may be isolated from fecal samples. Lagoviruses include rabbit hemorrhagic disease virus (RHDV) and European brown hare syndrome virus (EBHSV). The RHDV causes fatal systemic hemorrhages and liver necroses in rabbits, reaching mortality rates of up to 100% (Ohlinger *et al.*, 1993). The RHDV only affects adult rabbits or rabbits over 2 months of age. Recently RHDV has been used as a biological agent to efficiently control wild rabbit populations in Australia and New Zealand (Drollette, 1997, Mutze *et al.*, 1998). The EBHSV causes a similar hemorrhagic disease in a different species, the European brown hare.

Human caliciviruses (HuCV) have been recognized as the leading cause of food- and waterborne acute, nonbacterial gastroenteritis in humans worldwide (Lewis *et al.*, 1997; Vinjé *et al.*, 1997; Fankhauser *et al.*, 1998; Wright *et al.*, 1998). These non-cultivable enteric caliciviruses belong to either the *"Norwalk-like virus"* (*NLV*)

---

* NLVs and SLVs have recently been renamed as *Noroviruses* and *Sapoviruses*, respectively (Mayo, 2002).

or *"Sapporo-like virus" (SLV)* genera (Green *et al.*, 2000). The NLVs are commonly identified as causative pathogens in outbreaks of acute gastroenteritis in humans of all ages (Kapikian *et al.*, 1996; Vinjé *et al.*, 1997; Fankhauser *et al.*, 1998). Based on genetic similarities and divergence, the NLVs are further divided into 2 genogroups represented by prototype Norwalk virus (NV) and Snow Mountain virus (SMV). The SLVs are mainly associated with sporadic, acute gastroenteritis in infants and young children (Kapikian *et al.*, 1996), but also cause outbreaks of acute gastroenteritis in other age groups (Noel *et al.*, 1997; Vinjé *et al.*, 2000). The genetically diverse HuCV strains are continuously identified in gastroenteritis outbreaks, and currently there are at least 15 different genotypes in the NLV genus (Ando *et al.*, 2000) and another 4 in the SLV genus (Vinjé *et al.*, 2000), excluding animal enteric caliciviruses.

Enteric caliciviruses of animals are emerging pathogens associated with diarrhea in pigs, calves, chickens, mink, dogs and cats (Saif *et al.*, 1980; Bridger *et al.*, 1984, 1990; Guo *et al.*, 2001a). The porcine enteric caliciviruses (PEC) and bovine enteric caliciviruses (BEC, Newbury agent-2 and Jena virus) are genetically related to human SLVs and NLVs, respectively (Dastjerdi *et al.*, 1999; Guo *et al.*, 1999; Liu *et al.*, 1999). Both PEC and BEC infect villous enterocytes in the proximal small intestine, induce villous atrophy and crypt hyperplasia, and produce diarrhea in their respective hosts (Hall *et al.*, 1984; Flynn *et al.*, 1988; Bridger, 1990; Kapikian *et al.*, 1996; Guo *et al.*, 2001b). The HuCVs remain refractory to cell culture propagation, and no susceptible animal models are available. These limitations have hampered our understanding of their replication strategies, pathogenesis, natural history, biological properties, and host immunity. Some knowledge about the replication and pathogenesis of HuCV infections is derived mainly from early volunteer studies. Intestinal villous atrophy and crypt hyperplasia were observed in jejunal biopsies from infected volunteers with illness (Kapikian *et al.*, 1996), but the extent of small intestinal involvement remains unknown. Because PEC induces illness and small intestinal lesions in gnotobiotic pigs similar to those of HuCV infections in volunteers (Flynn *et al.*, 1988; Bridger, 1990; Kapikian *et al.*, 1996), and also because pigs are monogastrics with their gastrointestinal physiology resembling that of humans, infections of gnotobiotic pigs with PEC may be a useful animal model for studies of comparative pathogenesis and replication strategies of HuCVs. In this chapter, we summarize recent developments and advances in the understanding of the pathogenesis of HuCVs and animal enteric caliciviruses, with an emphasis on the pathogenesis of PEC infections in gnotobiotic pigs.

Clinical manifestations

Clinical manifestations of NV infections during natural outbreaks were similar to those observed in volunteer studies. Sudden onset of vomiting and/or diarrhea are the main clinical features, but a wide spectrum of symptoms may be seen in affected individuals or volunteers. The clinical symptoms observed in 38 outbreaks associated with NV included nausea (79%), vomiting (69%), diarrhea (66%), fever (37%), chills (32%), abdominal cramps (30%), myalgias (26%), headache (22%) and sore throat (18%)

(Kaplan *et al.*, 1982). Vomiting occurred more frequently than diarrhea in children, but the reverse was seen in adults. The illness is generally mild and self-limited, and usually no hospitalization or rehydration is required for adults. The incubation period ranges from 4 to 77 h with a mean (or median) of 24 to 48 h, and the symptoms usually last 24 to 48 h, but they may last up to several days. In 6 outbreaks, about 15% of affected individuals had illness lasting more than 3 days (Kaplan *et al.*, 1982). The diarrheal stools are often liquid, but do not contain excess mucus, blood or leukocytes (Dolin *et al.*, 1976).

In one volunteer study, of 52 adult volunteers orally administered NV, 31 (59.6%) developed illness, and the major clinical symptoms were anorexia (90%), headache (81%), diarrhea (81%), abdominal discomfort (68%), vomiting (65%), myalgias (58%) and fever (45%) (Wyatt *et al.*, 1974). The incubation period was 10 to 51 h, and illness lasted 24-48 h. Virus shedding detected by IEM was maximal around the onset of illness and occurred only infrequently after 72h following onset. Illness caused by Hawaii virus (HV) and SMV in volunteers was clinically indistinguishable from that observed with the NV (Wyatt *et al.*, 1974; Dolin *et al.*, 1975; Kapikian *et al.*, 1996). Subclinical infections with NV or HV were observed in infected volunteers and during natural outbreaks. It has been estimated that about 50% of people exposed to NV develop illness, and secondary infections often occur. In another study, 50 volunteers were orally given NV, and 41 (82%) became infected, among whom 68% were symptomatic and 32% were asymptomatic (Graham *et al.*, 1994). The seroepidemiology of NLV infections also points towards the likelihood of frequent occurrence of inapparent infections (Gray *et al.*, 1993). The virus shedding in stools was detected from 15 h to 7 days postinoculation, with a peak between 25 and 72 h, and it could persist up to 2 weeks (Okhuysen *et al.*, 1995). Recently virus shedding in stools was detected by RT-PCR and Southern hybridization for up to 28 days in patients naturally infected with a genogroup II NLV during a long-term study of a hospital outbreak of gastroenteritis (Hazelton *et al.*, 2000). The prolonged virus shedding in stools and a high rate of asymptomatic infections have implications for the understanding of virus transmission, natural history and epidemiology, and for disease diagnosis and outbreak control strategies (Graham *et al.*, 1994; Estes *et al.*, 1997).

The SLVs are predominantly associated with acute gastroenteritis in infants and young children (Cubitt, 1994; Pang *et al.*, 1999; Chiba *et al.*, 2000), but their association with gastroenteritis outbreaks in adults and the elderly has also been documented (Noel *et al.*, 1997; Vinjé *et al.*, 2000). The HuCVs with typical calicivirus morphology were first identified as an etiologic agent for "winter vomiting disease" in school children in 1978 in London, England (Cubitt, 1994). The clinical symptoms in ill children included predominantly vomiting, and occasionally diarrhea, whereas during the Sapporo virus (SV)-associated outbreak in 1977 in Sapporo, Japan, the affected infants had predominantly diarrhea (95%), followed by vomiting (44%) and fever (18%) (Chiba *et al.*, 2000). A recent survey of gastroenteritis episodes indicated that vomiting was the predominant symptom in affected young children of 2 months to 2 years of age in Finland (Pang *et al.*, 1999). Thus, the clinical symptoms for SLV-associated acute gastroenteritis include vomiting, diarrhea, nausea, malaise, aching limbs and headache (Cubitt, 1994), which were indistinguishable from those recorded in NV-infected and ill patients. The predominance of the symptoms varied between different outbreaks, which may be related to the average infectious dose per patient and the age of the patients. In infected volunteers, the incubation period was 12-72 h and illness lasted 1-11 days. The duration of detectable viral shedding paralleled the appearance of symptoms (Levy *et al.*, 1976).

Like HuCVs, enteric caliciviruses of animals such as PEC and BEC cause diarrhea in their respective hosts (Saif *et al.*, 1980; Bridger, 1990). In orally inoculated gnotobiotic pigs, the PEC/Cowden strain caused diarrhea at postinoculation day (PID) 2 to 3, and illness persisted for 3 to 7 days (Flynn *et al.*, 1988; Guo *et al.*, 2001b). The diarrhea ranged in color and consistency from tan semiliquid to pale yellow liquid. Some pigs developed liquid diarrhea for about 2 days. Virus shedding was detected by IEM up to and including PID 7 in fecal samples or in intestinal contents collected from the inoculated pigs. By using RT-PCR and an antigen-ELISA, fecal virus shedding was detected up to PID 28 and 15, respectively (Guo *et al.*, 2001b). The long duration of fecal virus shedding may facilitate virus transmission, causing secondary infections or multipoint source outbreaks. Proper planning for intervention strategies and handling of outbreaks associated with HuCVs in hospitals, nursing homes and the food industry may help to control virus transmission and disease spread.

Histopathology and pathophysiology

A common characteristic of HuCVs and animal enteric caliciviruses is their ability to cause cytolytic infections in the villous enterocytes but not in the crypt enterocytes of the proximal small intestine, leading to destruction of the terminally differentiated, absorptive villous enterocytes and resulting in villous atrophy and a malabsorptive diarrhea. Limited knowledge about the hisopathologic lesions and mechanisms for diarrhea induction by HuCVs was mainly derived from early volunteer studies using Norwalk, Montgomery County and Hawaii viruses (Blacklow *et al.*, 1972; Agus *et al.*, 1973; Schreiber *et al.*, 1973, 1974; Dolin *et al.*, 1975; Cubitt, 1994; Kapikian *et al.*, 1996). The histopathological lesions in the proximal jejunal biopsies from ill volunteers included a broadening and blunting of the intestinal villi, crypt cell hyperplasia, cytoplasmic vacuolization, and infiltration of polymorphonuclear and mononuclear cells into the lamina propria. However, the mucosa itself remained intact. When examined by transmission EM, the epithelial cells were intact but the microvilli were profoundly shortened. It was suggested that virus replication might occur in the mucosal epithelium of the small intestine. However, virus particles were not observed in these cells. The histologic lesions correlated with the appearance of symptomatic illness. Some volunteers showed similar jejunal lesions after oral inoculation with NV or HV, but did not show clinical illness (Agus *et al.*, 1973; Dolin *et al.*, 1975; Schreiber *et al.*, 1974). Normal intestinal villi and epithelial cells were observed in the jejunal biopsies from convalescent volunteers (Widerlite *et al.*, 1975). Histologic lesions were not observed in the gastric fundus, antrum, or rectal mucosa of volunteers who developed illness after oral inoculation with NV (Widerlite *et al.*, 1975). Because only the proximal small intestine was examined in these studies, the extent of small intestinal involvement remains unknown, and the site of HuCV replication has not been precisely determined (Kapikian *et al.*, 1996; Estes *et al.*, 1997).

In clinical studies it had been observed that the activity of small intestinal brush-border enzymes (trehalase, alkaline phosphatase and sucrase) decreased significantly (Kapikian *et al.*, 1996; Estes *et al.*, 1997). A transient intestinal malabsorption of fat, D-xylose, and lactose was observed in infected volunteers with illness. The adenylate cyclase activity in the jejunum was not elevated (Agus *et al.*, 1973; Levy *et al.*, 1976), and gastric secretion of hydrochloric acid, pepsin, and intrinsic factor did not change following HV- or NV-induced illness (Meeroff *et al.*, 1980). However, marked delays in gastric emptying were observed in infected volunteers who developed illness after ingestion of NV and HV (Meeroff *et al.*, 1980). It has been suggested that the abnormal gastric motor function may be responsible for the nausea and vomiting associated with calicivirus-induced gastroenteritis. Interferon was not detected in serum, jejunal aspirates, and jejunal biopsies of NV- or HV-infected volunteers (Dolin *et al.*, 1975).

Like NV and HV, the PEC/Cowden induced similar intestinal lesions in orally inoculated gnotobiotic pigs (Flynn *et al.*, 1988; Guo *et al.*, 2001b). Moderate to severe villous atrophy and fusion, crypt cell hyperplasia and reduction of villus:crypt ratios, cytoplasmic vacuolization, and infiltration of polymorphonuclear and mononuclear cells into the lamina propria were observed in the duodenum and jejunum, but only mild or no villous atrophy was seen in the ileum. The pronounced shortening, blunting, or fusion of villi was also observed in the duodenum and jejunum by scanning electron microscopy. Similar proximal small intestinal lesions coincided with the appearance of clinical illness in infected human volunteers (Blacklow *et al.*, 1972; Agus *et al.*, 1973; Schreiber *et al.*, 1973, 1974; Dolin *et al.*, 1975). Decreased villous height indicates the loss or sloughing of absorptive cells, resulting in malabsorption and increased enterocyte mitosis. The diarrhea is aggravated by crypt cell hyperplasia causing an increase in the number of cells that actively secrete Cl^- ions. This is probably the major mechanism for diarrhea induction by enteric caliciviruses. PEC-infected enterocytes were primarily detected by immunofluorescent (IF) staining of mucosal impression smears of the duodenum and jejunum. These findings indicate that PEC mainly infects villous epithelial cells of the proximal small intestine and induces lesions in the duodenum and jejunum of infected gnotobiotic pigs, suggesting that these areas are the major sites for PEC replication in the small intestine. Colon and extraintestinal tissues or organs are unlikely to support PEC replication, because PEC was not detected by IF of impression smears of those tissues including lungs, liver, spleen and kidneys, and because no lesions were evident in these tissues. The PEC/Cowden is most closely related genetically to human SLVs, and both of them tend to infect younger hosts. Understanding of the pathogenesis of PEC infections may have implications or relevance for human SLV infections. To date there have been no pathogenesis studies of human SLVs in the gastrointestinal tract. However, adult volunteers who were orally/nasally administered human SLVs developed similar clinical (mild to moderately severe) symptoms such as vomiting, diarrhea, nausea, pyrexia and abdominal pains (Cubitt, 1994) to those of human NLV infections, although some had inapparent infections. It is probable that human SLVs may share similar replication strategies, tissue tropisms (villous enterocytes) and virus distribution patterns in the gut with PEC or human NLVs in their respective hosts. Because PEC induced similar lesions in infected gnotobiotic pigs to those seen in NV and HV-infected human volunteers, PEC infection of pigs may be a useful model for studies of the comparative pathogenesis of the HuCVs.

Induction of diarrhea by intravenous inoculation of PEC and detection of viremia in pigs

The HuCVs and animal enteric caliciviruses are transmitted primarily through the fecal-oral route, and most of the gastroenteritis outbreaks with HuCV are linked to ingestion of contaminated food or water. Airborne transmission was reported, but requires further confirmation (Kapikian *et al.*, 1996), and transmission via the respiratory route may be less likely for these enteric pathogens in natural settings. In comparison, some cultivable, non-enteric animal caliciviruses (VESV, SMSV, and FCV) are transmitted through direct contact, infected fomites or the respiratory route (Smith and Madin, 1986; Carter *et al.*, 1991). Inoculation of pigs with VESV via intradermal, subcutaneous, intramuscular or intravenous (IV) routes produced vesicular disease in swine (Smith and Madin, 1986). The RHDV can be transmitted via many inoculation routes (Ohlinger *et al.*, 1993), and viremia appears within 24 h after inoculation and is likely to play an important role in virus spread to the target organ or tissues.

Based on these findings for non-enteric caliciviruses, it was of interest to determine whether or not enteric caliciviruses induce viremia and whether inoculation of enteric caliciviruses via alternative routes such as IV could cause infection or illness. In a recent study, gnotobiotic pigs were inoculated by the IV route with wild type (WT) PEC: all inoculated pigs developed diarrhea, and characteristic histologic lesions, and PEC antigens were detected in the proximal small intestine (mainly the duodenum and jejunum), similar to observations in gnotobiotic pigs inoculated orally with WT PEC (Table 1, Guo *et al.*, 2001b). One apparent difference between oral and IV-inoculated pigs was the consistently more severe villous atrophy in the jejunum of IV-inoculated pigs. Other major differences were the longer incubation periods (1 to 2 days) for clinical diarrhea and the correspondingly later appearance of fecal PEC antigens (PID 3 instead of PID 1) for IV-inoculated pigs (Table 1). High titers of virus shed in feces and from intestinal contents and high numbers of PEC-infected enterocytes detected by IF in the duodenal and jejunal mucosal smears correlated with marked villous atrophy and fusion of villi in the proximal small intestine as a result of the damage, exfoliation and loss of villous enterocytes. No diarrhea and small intestinal histologic lesions were evident in mock- or formalin-inactivated WT PEC-IV-inoculated pigs. Thus, WT PEC administered by the IV route reached the small intestine (presumably via the bloodstream) where it localized, propagated efficiently and induced characteristic small intestinal lesions. How the WT PEC reaches the small intestine from the bloodstream and infects the villous enterocytes is unknown at present. In type 1 reovirus infection

Table 1

Onset of diarrhea and fecal virus shedding and histopathologic findings in the small intestine of gnotobiotic pigs after inoculation with TC PEC or WT PEC.

Exposure[a] and inoculation route	Gn Pig #	PID[b] at euthanasia	Onset of diarrhea at PID	Onset of fecal virus shedding at PID		Villous atrophy score[c]		
				ELISA	RT-PCR	Duodenum	Jejunum	Ileum
TC PEC Oral	4-7	2	N/A[d]	1	1	0	0	0
	4-6	4	N/A	1	1	1	0	0
	9-4	4	N/A	1	1	0	0	0
	4-8	7	N/A	2	1	1	1	1
WT PEC Oral	9-6	3	3	1	1	3	0	0
	4-4	4	4	1	1	4	4	1
	9-7	4	4	1	1	2	0	0
	13-9	4	4	2	1	2	0	0
	8-8	4	4	1	1	3	3	0
	8-9	4	2	1	1	1	4	1
	4-3	7	4	1	1	1	3	1
	13-2	9	4	2	1	1	1	0
	13-3	9	4	1	1	1	2	0
WT PEC IV	13-11	6	4	3	1	4	4	0
	8-1	6	5	3	1	2	4	0
	8-2	8	5	3	1	2	4	0
WT PEC in sera[e] Oral	14-4	28	5	3	1	ND[f]	ND	ND
WT PEC in sera IV	14-5	7	4	4	2	2	4	0

[a]All pigs were 4 to 6 days of age at inoculation. [b]PID, postinoculation days. [c]Score is designated based on villus/crypt ratios (an indication of the lesion severity) as follows for: 0 = normal (0 ≥ 6:1), 1 = mild (1 = 5.0 − 5.9:1), 2 = moderate (2 = 4.0 − 4.9:1), 3 = marked (3 = 3.0 − 3.9:1), 4 = severe (4 ≤ 3.0:1). [d]N/A, not applicable. [e]WT PEC in sera was derived from WT PEC-positive acute-phase sera from Gn pigs IV or orally inoculated with WT PEC/Cowden. [f]ND, not determined. Data modified from Guo *et al.*, 2001b.

496

of mice, reoviruses in the bloodstream after viremia or IV inoculation are transported to the ileum, where they infect crypt cells possibly via attachment to the basolateral membrane, but not via the luminal surface (Organ and Rubin, 1998). Other entero-pathogenic animal viruses, such as adenovirus, parvovirus, and bovine viral diarrhea virus (BVDV) may infect crypt enterocytes after viremia in a similar manner (Saif, 1990). The PEC may be unique in this regard since IV inoculation leads to infection of mainly villous and not crypt enterocytes as detected by IF staining of small intestinal impression smears.

PEC RNA (by RT-PCR) and low titers of virus antigen (by ELISA) were detected in sera from the WT PEC-IV-inoculated pigs and from 7 of 9 pigs inoculated with WT PEC orally. It is unlikely that the PEC RNA or antigen in serum was from the initial inoculum because the PEC RNA or antigens were also detected in sera from WT PEC-orally-inoculated pigs. Further inoculation of gnotobiotic pigs with PEC-positive acute sera from both the WT PEC-IV- and orally-inoculated pigs proved that these sera contained infectious PEC as the inoculated pigs also developed diarrhea, typical small intestinal lesions and fecal virus shedding (Table 1). Thus, viremia may occur follow-ing natural PEC infection, and acute sera may contain infectious virus. Because of the low virus titers detected by ELISA in PEC-positive acute sera which induced illness in inoculated gnotobiotic pigs, the WT PEC/Cowden strain transmitted in serum must be highly infectious for pigs. A low infectious dose is an important common characteristic of PEC, NV and other related HuCVs that are associated with food- and waterborne viral gastroenteritis (Kapikian *et al.*, 1996). It is of interest to know whether HuCV infection also induces viremia, whether, in so doing, HuCV interacts with erythrocytes in the blood (see next section), and how the enteric caliciviruses are transferred from the bloodstream to the gut and vice versa.

Virus-cell interactions of caliciviruses

For non-cultivable caliciviruses, little is known about the early events of virus rep-lication in infected cells, including the virus-receptor interactions which precede and initiate productive infection. The non-cultivable RHDV infects rabbits over 2 months of age (but not younger), causing acute liver necrosis, systemic hemorrhage and widespread disseminated intravascular coagulation. RHDV agglutinates human erythrocytes (Pu *et al.*, 1985; Yang *et al.*, 1991), but not erythrocytes of other mam-mals, leading to the characterization of a viral hemagglutinin receptor on human red blood cells and the development of diagnostic assays such as the hemagglutination (HA) and hemagglutination inhibition (HI) tests for RHDV antigens and antibodies, respectively. The RHDV hemagglutinin receptor on human erythrocytes belongs to developmental antigens (ABH antigens) which are mainly carried by polyglycosyl-ceramides (Ruvoen-Clouet *et al.*, 1995) but are not expressed on human fetal eryth-rocytes nor on red blood cells from other mammals. Hemagglutination by RHDV is dependent on the presence of ABH blood group antigens and was inhibited by saliva from secretor individuals but not from the nonsecretors whose saliva lack H antigen

(Ruvoen-Clouet *et al.*, 2000), suggesting that the hemagglutinin receptor is identical with or closely related to the H antigen. Furthermore, RHDV specifically bound to synthetic A and H type 2 blood group oligosaccharides, confirming the previous findings.

The ABH histo-blood group antigens are widely expressed on various tissues or cells including enterocytes. The native RHDV and recombinant VLPs could bind to adult rabbit epithelial cells of the upper respiratory and digestive tracts because of the presence of A and H type 2 antigens on their surface as determined by immunohisto-chemical staining (Ruvoen-Clouet *et al.*, 2000). In contrast, the same tissues in young rabbits showed little virus binding, and the H type 2 antigen was expressed at much lower levels on epithelial cells of young rabbits, whereas the A antigen expression was not detected until rabbits reached 8 weeks of age (Ruvoen-Clouet *et al.*, 2000). Native RHDVs or VLPs did not bind to sections from liver, spleen, kidney or heart of adult rabbits, correlating with lack of detectable ABH antigens on rabbit hepatocytes. Moreover, these organs are the major sites for RHDV infection and propagation. Thus it is most likely that RHDV does not use the ABH antigens as receptors for entry into target cells in these tissues. How RHDV enters and penetrates target cells remains unknown. Because the A and H type 2 antigens are present on epithelial cells of the upper respiratory and digestive tracts in susceptible adult rabbits, but not on the corresponding cells in young rabbits that are resistant to RHDV infection, it was suggested the ABH histo-blood group antigens may be involved in the initial virus infection at the primary entry sites in the upper respiratory tract or intestine (Ruvoen-Clouet *et al.*, 2000). Further studies are needed to address if the binding of RHDV to the ABH antigens leads to viral entry and productive infection in susceptible cells.

For HuCVs, recent data suggest that susceptibility to NV infection and disease may be associated with an individual's ABO phenotype; this association was recognized after NV VLPs showed different binding properties to red blood cells based on ABO type (A.M. Hutson and Dr. M.K. Estes, personal communication). These observations of binding of caliciviruses and NV VLPs to erythrocytes are interesting in light of the recent demonstration that PEC induces viremia in gnotobiotic pigs (Guo *et al.*, 2001b). Thus calicivirus-erythrocyte interactions might play a role in calicivirus pathogenesis.

Cell culture adaptation of PEC

A wide range of cell lines and human embryonic intestinal organ cultures have been used in attempts to grow the HuCVs, but no success has been reported (Kapikian *et al.*, 1996, Estes *et al.*, 1997). Recently, radiolabeled recombinant Norwalk virus-like particles (rNV VLPs) were found to bind specifically to 13 cultured human and animal cells lines (White *et al.*, 1996). The differentiated human intestinal cell line Caco-2 bound the rNV VLPs most efficiently. The rNV VLPs produced in baculoviruses resembled the native NV in morphology, antigenicity and possibly also in their inter-action with cellular receptor(s) that may determine the viral tissue tropism and host specificity. However, only a few prebound VLPs (1.4 to 6.8%) became internalized into cells, suggesting a possible explanation for the failure of NV to grow in cell culture. Experimental inoculation of the NV into the alimentary tract of mice, guinea pigs, rabbits, kittens, calves, chimpanzees, baboons, rhesus monkeys, marmosets, owl monkeys, patas monkeys or cebus monkeys produced no illness (Kapikian *et al.*, 1996). However, a 33 kDa soluble NV antigen was detected by radioimmunoassay (RIA) in feces of the inoculated chimpanzees (Greenberg and Kapikian, 1978).

To date the PEC/Cowden is the only cultivable enteric calicivirus (Flynn and Saif, 1988; Parwani *et al.*, 1991), but for growth it requires an intestinal contents preparation (ICP) from uninfected gnotobiotic pigs as a medium supplement. The PEC Cowden strain was initially propagated in primary porcine kidney cells and then in a continuous porcine kidney cell line (LLC-PK) in the presence of ICP in the medium. Different intestinal enzymes of porcine origin such as trypsin, pancreatin, alkaline phosphatase, enterokinase, elastase, protease and lipase were also incorporated into the cell culture medium, but none of them alone promoted PEC growth in cell culture (Parwani *et al.*, 1991). An ICP from uninfected chickens and gnotobiotic calves had no effect on PEC growth, and bile from uninfected gnotobiotic pigs had a reduced effect compared to ICP (Flynn and Saif, 1988). It was speculated that some enzymes or factors in the porcine ICP could activate the viral receptor or promote signaling of host cells or affect the virion surface structure to enhance virus binding and internalization, thereby promoting viral growth. It is also possible that ICP might be implicated in cleavage of the viral capsid for successful uncoating. Thus one of the priorities for studies of cell culture propagation of enteric caliciviruses is to determine the identity of the PEC growth promotion factors present in the swine ICP. This novel approach to propagation of PEC and delineation of the factors involved may also be useful for adaptation of HuCVs to propagation in cell culture.

Virulence attenuation and its genetic basis

For many viruses serial passage in cell culture often leads to virulence attenuation, tissue tropism changes and even production of attenuated vaccines. After serial passage 19 times each in primary porcine kidney cells and in a continuous porcine kidney cell line (LLC-PK), the tissue culture adapted PEC/Cowden (TC PEC) was used to orally inoculate gnotobiotic pigs to examine its virulence and potential as a candidate vaccine to prevent PEC infection (Guo *et al.*, 2001b). No diarrhea developed in the TC PEC-orally-inoculated gnotobiotic pigs, and only mild or no villous atrophy was observed in the small intestine. In contrast, all WT PEC-orally-inoculated gnotobiotic pigs developed diarrhea and all that were euthanized at PID 3-4 had moderate to severe villous atrophy and fusion in the proximal small intestine as described previously (Flynn and Saif, 1988; Guo *et al.*, 2001b). The ELISA absorbance values of fecal PEC antigen shed were consistently higher in the WT PEC-inoculated pigs than in the TC PEC-inoculated pigs, in spite of the consistent detection of fecal PEC RNA in pigs of both groups. These data indicate that the TC PEC apparently infected gnotobiotic pigs and induced limited proximal small intestinal lesions and fecal virus shedding (with much lower virus titers), without causing clinical illness, suggesting the less efficient replication and growth of the TC PEC in gnotobiotic pigs following infection. Thus the virulence of the TC PEC was at least partially attenuated after serial passage in cell culture. Genomic sequence analysis revealed that the TC PEC and WT PEC have 100% nucleotide sequence identities in the genomic 5' terminal region, the 2C helicase region, the entire ORF2, and the 3' nontranslated region (Guo *et al.*, 1999). However, the TC PEC has one distant (C178 to S) and three closely clustered amino acid (aa) substitutions (Y289 to H, N291 to D, K295 to R) in the hypervariable region 2 of the predicted capsid protein, and 2 aa changes (Y1252 to H and R1379 to K) in the RNA dependent RNA polymerase region, compared with the WT PEC. The 3 clustered aa substitutions in a short region of 7 aa (aa 289 to 295 for PEC numbering) lead to a localized higher hydrophilicity in the hypervariable capsid region. Interestingly, this hypervariable region with 3 aa changes is predicted to form the externally located P2 subdomain on the virus surface and corresponds to the binding region of NV capsid (rNV VLPs) to human and animal cells *in vitro* (White *et al.*, 1996; Prasad *et al.*, 1999). It is likely that this P2 subdomain determines host specificity and tissue tropism. For some viruses, a few aa changes at critical positions of the attachment protein(s) can dramatically alter viral tissue tropisms or virulence. An example is the infectious bursal disease virus (IBDV), a birnavirus. The 2 aa substitutions (D279 to N and A284 to T) in the variable domain of a major structural protein, VP2, resulted in the cell culture adaptation and virulence attenuation of the uncultivable, highly virulent IBDV (Lim *et al.*, 1999). The limited propagation of TC PEC in the small intestine of inoculated gnotobiotic pigs and its reduced virulence may be the result of a potential change in tissue tropism or binding that may be related to the aa substitutions in the hypervariable capsid region. Thus the aa changes in the capsid hypervariable region may be associated with both the cell culture adaptation and the virulence attenuation of the TC PEC in gnotobiotic pigs. However, we cannot exclude the possibility that the aa changes observed in the RNA dependent RNA polymerase may also influence those properties of the TC PEC. The TC PEC may be a useful candidate vaccine for future evaluation for prevention of PEC infections of swine. WT PEC and TC PEC apparently differ considerably in their *in vitro* and *in vivo* growth kinetics. Therefore, an infectious clone of TC PEC may be a starting point for site-directed mutagenesis towards elucidating the functions of those substituted amino acids in the capsid and the RNA polymerase regions of TC PEC (See Section IV, Chapter 2). Because rPEC VLPs have recently been produced in baculovirus (Guo *et al.*, 2001c), it would also be of interest to study the early events of PEC replication in cell culture by using the rPEC VLPs for competitive binding to cellular receptor(s) or interaction with growth promotion factors in the ICP.

References

Agus, S. G., R. Dolin, R. G. Wyatt, A. J. Tousimis, R. S. Northrup. 1973. Acute infectious nonbacterial gastroenteritis: intestinal histopathology. Ann. Intern. Med.

79:18-25.

Ando, T., J. S. Noel, and R. L. Fankhauser. 2000. Genetic classification of "Norwalk-like viruses". J. Infect. Dis. 181(Suppl 2):S336-S348.

Blacklow, N. R., R. Dolin, D. S. Fedson, H. Dupont, R. S. Northrup, R. B. Hornick, and R. M. Chanock. 1972. Acute infectious nonbacterial gastroenteritis:etiology and pathogenesis. Ann. Intern. Med. 76:993-1008.

Bridger, J. C. 1990. Small viruses associated with gastroenteritis in animals. *In* L. J. Saif, and K. W. Theil (ed.), *Viral diarrheas of man and animals*. p. 161-182. CRC Press, Boca Raton, Fla.

Bridger, J. C., G. A. Hall, and J. F. Brown. 1984. Characterization of a calici-like virus (Newbury agent) found in association with astrovirus in bovine diarrhea. Infect. Immun. 43:133-8.

Carter, M. J., I. D. Milton, and C. R. Madeley. 1991. Caliciviruses. Rev. Med. Virol. 1: 177-186.

Chiba, S., S. Nakata, K. Numata-Kinoshita, and S. Honma. 2000. Sapporo virus: history and recent findings. J. Infect. Dis. 181(Suppl 2):S303-S308.

Cubitt, W. D. 1994. Caliciviruses. *In* A. Z. Kapikian (ed.), *Viral infections of the gastrointestinal tract*. 2nd edition, p. 549-568. Marcel Dekker Inc., New York, NY.

Dastjerdi, A. M., J. Green, C. I. Gallimore, D. W. G. Brown, and J. C. Bridger.1999. The bovine Newbury agent-2 is genetically more closely related to human SRSVs than to animal caliciviruses. Virology 254:1-5.

Dolin, R., A. G. Levy, R. G. Wyatt, T. S. Thornhill, and J. D. Gardner. 1975. Viral gastroenteritis induced by the Hawaii agent. Jejunal histopathology and serologic response. Am. J. Med. 59:761-8.

Dolin, R., R. C. Reichman, and A. S. Fauci. 1976. Lymphocyte populations in acute viral gastroenteritis. Infect. Immun. 14:422-428.

Drollette, D. 1997. Wide use of rabbit virus is good news for native species. Science 275:154.

Estes, M. K., R. L. Atmar, and M. E. Hardy. 1997. Norwalk and related diarrhea viruses. *In* D. D. Richman, R. J. Whitley, and F. G. Hayden (ed.), *Clinical Virology*, 3rd edition, p. 1073-1095. Churchill Livingstone, New York, N.Y.

Fankhauser, R. L, J. S. Noel, T. Ando, S.S. Monroe, and R. I. Glass. 1998. Molecular epidemiology of "Norwalk-like viruses" in outbreaks of gastroenteritis in the United States. J. Infect. Dis. 178:1571-1578.

Flynn, W. T., and L. J. Saif. 1988. Serial propagation of porcine enteric calicivirus-like virus in porcine kidney cells. J. Clin. Microbiol. 26:206-212.

Flynn, W. T, L. J. Saif, and P. G. Moorhead. 1988. Pathogenesis of porcine enteric calicivirus in four-day-old gnotobiotic piglets. Am. J. Vet. Res. 49:819-825.

Graham, D.Y., X. Jiang, T. Tanaka, A. R. Opekun, H. P. Madore, and M. K. Estes. 1994. Norwalk virus infection of volunteers: new insights based on improved assays. J. Infect. Dis. 170:34-43.

Green, K. Y., T. Ando, M. S. Balayan, T. Berke, I. N. Clarke, M. K. Estes, D. O. Matson, S. Nakata, J. D. Neill, M. J. Studdert, and H-J. Thiel. 2000. Taxonomy of the caliciviruses. J. Infect. Dis. 181(Suppl 2):S322-S330.

Greenberg, H. B. and A. Z. Kapikian. 1978. Detection of Norwalk agent antibody and antigen by solid-phase radioimmunoassay and immune adherence hemagglutination assay. J. Am. Vet. Med. Assoc. 173:620-623.

Guo, M., K.-O. Chang, M. E. Hardy, Q. Zhang, A. V. Parwani, and L. J. Saif. 1999. Molecular characterization of a porcine enteric calicivirus genetically related to Sapporo-like human caliciviruses. J. Virol. 73:9625-9631.

Guo, M., J. F. Evermann, and L. J. Saif. 2001a. Detection and molecular characterization of cultivable caliciviruses from clinically normal mink and enteric caliciviruses associated with diarrhea in mink. Arch Virol. 146:479-493.

Guo, M., J. Hayes, K. O. Cho, A. V. Parwani, L. M. Lucas and L. J. Saif. 2001b. Comparative pathogenesis of tissue culture adapted and wild type Cowden porcine enteric calicivirus (PEC) in gnotobiotic pigs and induction of diarrhea by intravenous inoculation of wild type PEC. J. Virol. 75:9239-9251.

Guo, M., Y. Qian, K.-O. Chang, and L. J. Saif. 2001c. Expression and self-assembly in baculovirus of porcine enteric calicivirus capsids into virus-like particles and their use in ELISA for antibody detection in swine. J. Clin. Microbiol. 39:1487-1493.

Hall, G. A., J. C. Bridger, B. E. Brooker, K. R. Parsons, and E. Ormerod. 1984. Lesions of gnotobiotic calves experimentally infected with a calicivirus-like (Newbury) agent. Vet. Pathol. 21:208-215.

Hazelton, P. R., K. M. Coombs, T. B. Ball, L. Klass, and P. Plourde. 2000. Detection of calicivirus shedding and viral loads in patients during a long-term care hospital outbreak, abstr. #W14-4, p.80. *In* Abstracts of the 19th Annual Meeting of the American Society for Virology 2000. Fort Collins, CO.

Kapikian, A.Z., M. K. Estes, and R. M. Chanock. 1996. Norwalk Group of Viruses, *In* B. N. Fields, D. M. Knipe, P. M. Howley *et al.* (ed.), *Fields Virology*, 3rd edition, p. 783-810. Lippincott-Raven, Philadephia, PA.

Kaplan, J. E., R. Feldman, D. S. Campbell, C. Lookabaugh, and G. W. Gary. 1982. The frequency of a Norwalk-like pattern of illness in outbreaks of acute gastroenteritis. Am. J. Public Health 72:1329-1332.

Levy, A. G., L. Widerlite, C. J. Schwartz, R. Dolin, N. R. Blacklow, J. D. Gardner, D. V. Kimberg, and J. S. Trier. 1976. Jejunal adenylate cyclase activity in human subjects during viral gastroenteritis. Gastroenterology 70:321-325.

Lewis, D. C, A. Hale, X. Jiang, R. Eglin, D. W. Brown. 1997. Epidemiology of Mexico virus, a small round-structured virus in Yorkshire, United Kingdom, between January 1992 and March 1995. J. Infect. Dis. 175:951-954.

Lim, B. L., Y. Cao, T. Yu, and C. W. Mo. 1999. Adaptation of very virulent infectious bursal disease virus to chicken embryonic fibroblasts by site-directed mutagenesis of residue 279 and 284 of viral coat protein VP2. J. Virol. 73:2854-2862.

Mayo, M.A. 2002. Virus Taxonomy-Houston 2002. Arch. Virol. 147:1071-1076.

Liu, B. L, P. R. Lambden, H. Günther, P. Otto, M. Elschner and I. N. Clarke. 1999. Molecular characterization of a bovine enteric calicivirus: relationship to the Norwalk-like viruses. J. Virol. 73:819-825.

502

Meeroff, J. C., D. S. Schreiber, J. S. Trier, and N. R. Blacklow. 1980. Abnormal gastric motor function in viral gastroenteritis. Ann. Intern. Med. 92:370-373.

Mutze, G., B. Cooke, and P. Alexander. 1998. The initial impact of rabbit hemorrhagic disease on European rabbit populations in South Australia. J. Wildl. Dis. 34: 221-227.

Noel, J. S., B. L. Liu, C. D. Humphrey, E. M. Rodriguez, P. R. Lambden, I. N. Clarke, D. M. Dwyer, T. Ando, R. I. Glass and S. S. Monroe. 1997. Parkville virus: a novel genetic variant of human calicivirus in the Sapporo virus clade, associated with an outbreak of gastroenteritis in adults. J. Med. Virol. 52:173-178.

Ohlinger, V. F., B. Haas, and H. J. Thiel. 1993. Rabbit hemorrhagic disease (RHD): characterization of the causative calicivirus. Vet. Res. 24:103-16.

Okhuysen, P. C, X. Jiang, L. Ye, P. C. Johnson, and M. K. Estes. 1995. Viral shedding and fecal IgA response after Norwalk virus infection. J. Infect. Dis. 171: 566-569.

Organ, E. L., and D. H. Rubin. 1998. Pathogenesis of reovirus gastrointestinal and hepatobiliary disease. Curr. Top. Microbiol. Immunol. 233 (Vol. II):67-83.

Pang, X. L., J. Joensuu, and T. Vesikari. 1999. Human calicivirus-associated sporadic gastroenteritis in Finnish children less than two years of age followed prospectively during a rotavirus vaccine trial. Pediatr. Infect. Dis. J. 18:420-426.

Parwani, A. V., W. T. Flynn, K. L. Gadfield, and L. J. Saif. 1991. Serial propagation of porcine enteric calicivirus: effects of medium supplementation with intestinal contents or enzymes. Arch. Virol. 120:115-122.

Prasad, B. V. V, M. E. Hardy, T. Dokland, J. Bella, M. G. Rossmann, and M. K. Estes. 1999. X-ray crystallographic structure of the Norwalk virus capsid. Science 286: 287-290.

Pu, B. Q., N. J. Qian, and S. J. Cui. 1985. HA and HI tests for the detection of antibody titers to so called "hemorrhagic pneumonia" in rabbits. Chinese J. Vet. Med. 11: 16-17.

Ruvoën-Clouet, N., D. Blanchard, G. André-Fontaine, and J. P. Ganière. 1995. Partial characterization of the human erythrocyte receptor for rabbit hemorrhagic disease virus. Res. Virol. 146:33-41.

Ruvoën-Clouet, N., J. P. Ganière, G. André-Fontaine, D. Blanchard, and J. L. Pendu. 2000. Binding of rabbit hemorrhagic disease virus to antigens of the ABH histo-blood group family. J. Virol. 74:11950-11954.

Saif, L. J. 1990. Comparative aspects of enteric viral infections. *In* L. J. Saif, and K. W. Theil (ed.), *Viral Diarrheas of Man and Animals.* p.10-31. CRC Press, Boca Raton, Fla.

Saif, L. J., E. H. Bohl, K. W. Theil, R. F. Cross, and J. A. House. 1980. Rotavirus-like, calicivirus-like and 23 nm virus-like particles associated with diarrhea in young pigs. J. Clin. Microbiol. 12: 105-111.

Schreiber, D. S., N. R. Blacklow, and J. S. Trier. 1973. The mucosal lesion of the proximal small intestine in acute infectious nonbacterial gastroenteritis. N. Engl. J. Med. 288: 1318-1323.

Schreiber, D. S., N. R. Blacklow, and J. S. Trier. 1974. The small intestinal lesion induced by Hawaii agent acute infectious nonbacterial gastroenteritis. J. Infect. Dis. 124:705-708.

Smith, A. W., and S. H. Madin. 1986. Vesicular exanthema of swine. *In* A. D. Leman, B. Straw, R. D. Glock, W. L. Mengeling, R. H. C. Penny, and E. Scholl (eds.), *Diseases of Swine.* 6th edition. p. 358-368. Iowa State Press, Ames, IA.

Vinjé, J., S. A. Altena, and M. P. G. Koopmans. 1997. The incidence and genetic variability of small round-structured viruses in outbreaks of gastroenteritis in the Netherlands. J. Infect. Dis. 176:1374-1378.

Vinjé, J., H. Deijl, R. van der Heide, D. Lewis, K-O. Hedlund, L. Svensson, and M. P. G. Koopmans. 2000. Molecular detection and epidemiology of Sapporo-like viruses. J. Clin. Microbiol. 38:530-536.

White, L. J, J. M. Ball, M. E. Hardy, T. N. Tanaka, N. Kitamoto, and M. K. Estes. 1996. Attachment and entry of recombinant Norwalk virus capsids to cultured human and animal cell lines. J. Virol 70:6589-6597.

Widerlite, L., J. S. Trier, N. R. Blacklow, and D. S. Schreiber. 1975. Structure of the gastric mucosa in acute infectious bacterial gastroenteritis. Gastroenterology 68: 425-430.

Wright, P. J., I. C. Gunesekere, J. C. Doultree, and J. A. Marshall. 1998. Small round-structured (Norwalk-like) viruses and classical human caliciviruses in southeastern Australia, 1980-1996. J. Med. Virol. 55:312-20.

Wyatt, R. G., R. Dolin, N. R. Blacklow, H. L. DuPont, R. F. Buscho, T. S. Thornhill, A. Z. Kapikian, and R. M. Chanock. 1974. Comparison of three agents of acute infectious nonbacterial gastroenteritis by cross-challenge in volunteers. J. Infect. Dis. 129:709-714.

Yang, H., W. Y. Xu and N. Du. 1991. Hemagglutination characteristics of rabbit hemorrhagic disease virus, p. 67. Proceedings of the International Symposium on Rabbit Hemorrhagic Disease, Beijing, China. Chinese Association of Animal and Veterinary Sciences, Beijing, China.

Viral Gastroenteritis
U. Desselberger and J. Gray (editors)

IV, 4. Development of serological and molecular tests for the diagnosis of calicivirus infections

Xi Jiang

Cincinnati Children's Hospital Medical Center, Cincinnati, OH 45229-3039

Introduction

Methods for diagnosis of infection with human caliciviruses (HuCVs) have undergone two major stages of development: before and after molecular cloning of Norwalk virus (NV). Methods developed before the cloning of NV mainly relied on reagents derived from clinical sources or volunteer challenge studies, because there is no cell culture and animal model for propagating HuCVs. These methods included immune electron microscopy (IEM) (45), radioimmunoassays (RIAs) (20), immune adherence hemagglutination assays (IAHAs) (20), solid phase IEM (52), enzyme-linked immunosorbent assays (ELISAs) (27, 70) and Western blot analysis (25). These techniques greatly advanced our knowledge of HuCVs and associated illness. However, due to a limited supply of reagents from these sources, these techniques were not widely used. The direct electron microscopy method, which does not require special reagents, has been used more frequently. However, direct electron microscopy is expensive, demands well-trained personnel, and the technique is relatively less sensitive (requiring at least 10^6 viral particles/ml).

Following the cloning and sequencing of NV (32, 40) and the Southampton virus (47) in the early 1990s, two major types of assays for the diagnosis of HuCVs have been developed. One is to detect the viral antigens or antibodies against the antigens by recombinant enzyme immune assays (EIAs), and the other is to detect the viral RNA by reverse transcription-polymerase chain reaction (RT-PCR) (9, 38, 39). Both types of assays no longer rely on clinical sources of reagents. The RT-PCR is readily available and has been widely adapted by many laboratories in HuCV researches. The recombinant EIAs using molecular engineered reagents are promising to be used for simple and rapid clinical diagnoses and large-scale epidemiology studies. However, because HuCVs are antigenically diverse, additional research is needed to generate antigens for more types of HuCVs and to define the antigenic relationships among different strains before such assays become available.

506

Diagnosis of HuCVs by enzyme immune assays

The first recombinant EIA was developed for NV, which was based on recombinant NV capsid proteins expressed in the baculovirus-expression system (39). A high yield of the NV recombinant capsid proteins was produced using the baculovirus polyhedron promoter, and the expressed proteins self-assembled into virus-like particles (VLPs) that were morphologically and antigenically similar to the authentic viruses found in stool specimens. These VLPs are easily purified from cell cultures, and the purified VLPs are stable at 4°C and frozen conditions, providing excellent reagents for the development of immunological diagnostic assays. The structure of the NV VLP has been determined by cryo-electron microscopy and crystallography, and these studies showed that NV exhibits a $T = 3$ icosahedral symmetry with 180 molecules of the capsid proteins organized into 90 dimers (62, 63; Bertolotti-Ciarlet *et al.*, Section IV, Chapter 1 of this book).

Following the expression of NV, a number of recombinant NLV capsid proteins representing different genetic clusters of NLVs have been expressed in baculovirus, including Mexico virus (MxV), Grimsby virus (GrV), Hawaii virus (HV), Desert Shield virus (DSV), Southampton virus (SV), Bristol virus, Chiba virus, and many other strains (11, 19, 21, 34, 42, 43, 46, 50, 51). Most of these expressed proteins formed VLPs similar to that of NV, but a few strains did not form VLPs or formed VLPs inefficiently. The morphogenesis of NLV capsids in the insect cell culture remains to be described. Even those proteins that did not form VLPs were found to be antigenically competent, resembling soluble viral antigens found in stool specimens of NV-infected patients. Several laboratories are dedicated to the expression of recombinant capsid proteins for different genetic clusters of HuCVs, including searching for new strains of HuCVs. This work is important because currently available recombinant capsid antigens represent limited types of HuCVs, and new genetic and antigenic types continue to be found.

The baculovirus expression system has been significantly improved since the report of the cloning and expression of the NV capsid antigen. Several commercial kits that provide highly efficient cloning vectors and simple cloning procedures for recombinant baculoviruses are available. The cell culture system for maintaining the insect cell culture and for production of recombinant proteins has also been significantly improved. The commercially available serum-free insect culture medium is particularly useful for large-scale production of the recombinant proteins with a minimum contamination of unrelated proteins.

Recombinant NLV capsid antigens also have been generated from other systems, such as bacterial and eukaryotic expression systems (4, 74). Recombinant proteins produced in *E. coli* usually have a high yield but may not have the same conformation or modification as their native counterparts. One study showed that several NLV recombinant capsid proteins expressed in *E. coli* were antigenic when they were tested against serum samples obtained from NLV-infected patients (74). Hyperimmune antisera raised against these recombinant antigens also reacted with NLV antigens purified from patients' stools by Western blot analysis. The NV capsid protein also was expressed using Venezuelan equine encephalitis virus replicon particles (VRP-NV1) (4). This expression system resulted in large numbers of recombinant NV-like particles that were primarily cell-associated and were indistinguishable from NV particles produced from baculoviruses.

Antigen detection EIAs

The baculovirus-expressed recombinant NV capsid antigens were used to raise hyperimmune antibodies in laboratory animals following purification of the capsid antigens from the insect cell culture (15). To establish a sandwich type assay, antibodies from two species, rabbits and guinea pigs, were produced. High titers of antibodies against the antigens were obtained from both species. In the assay, the rabbit serum was used to coat the microtiter plates as a capture antibody, and the guinea pig serum was used as a detector antibody. To ensure specificity of the assay, duplicates of each stool specimen were tested in wells coated with pre- and post-immunized rabbit sera, respectively. This allows determination of a positive result based on a significant P/N ratio (e.g. $\geq$ 2) between the OD values in the two wells. This is important because some stool specimens give false-positive results when only hyperimmune serum is used to coat a single well.

The NV antigen EIA was highly sensitive. It has been used successfully in detection of NV in stool specimens of NV-infected volunteers (15). The rNV EIA was as sensitive as dot-blot hybridization and RT-PCR and more sensitive than RIA (15, 58). The high sensitivity of the assay allowed the prolonged detection of excretion of NV antigen in the stools of volunteers, up to two weeks following challenge (58). The specificity of the assay for NV has been shown by the lack of reactivity with other enteric viruses, including rotaviruses, adenoviruses, astroviruses, hepatitis A viruses, and enteroviruses (15).

When the rNV EIA was used in surveys of HuCVs in field samples, unexpectedly low detection rates were obtained. In a summary of surveillance of HuCVs by different laboratories, 6084 stool specimens from sporadic cases of diarrhea in children from different populations and countries were tested, and the rNV EIA only detected 16 (0.3%) positive samples (41). Samples from several countries did not reveal a single positive sample, and the highest detection rate was 2.1% among the countries where positive results were found. This detection rate was lower than rates observed in the 1970s using EM and was thought to be due to antigenic changes of circulating strains because the prototype NV was found three decades ago. However, similar low detection rates were obtained when the specimens from the same populations were tested with an EIA specific for a currently circulating strain, the MxV, a genogroup II NLV that was found in a child with diarrhea in Mexico in 1989. The rMxV EIA detected 43 (0.9%) positives from the 6084 samples with the highest detection rate of 4.3% among the positive countries (41).

Similar low detection rates using recombinant NLV antigen detection EIAs were also obtained in outbreaks of acute gastroenteritis in adult populations. In a three-year survey, 216 stool specimens from 25 outbreaks of acute gastroenteritis in Virginia in 1996-1998 were studied. Of those, 10 specimens from 3 outbreaks were positive by the rMxV antigen EIA; three specimens from two outbreaks were positive for the rNV antigen EIA; and none of the 216 specimens was positive by the rHV antigen EIA (41). When the stool specimens were tested with a recombinant antigen EIA specific for a highly prevalent strain, the GrV, 6 of the 25 outbreaks were positive. In total, 11 (44%) of the 25 outbreaks were positive using four EIAs, with one EIA (rHV) not detecting a single positive sample. However, when the outbreak stool specimens were tested for HuCVs by RT-PCR, 22 (88%) of the 25 outbreaks were positive using a combination of five primer pairs (41).

The above results suggested that HuCVs are antigenically diverse and that multiple antigenic types of HuCVs co-circulate in the community, a common phenomenon that has been observed in many geographical locations by different laboratories (12, 71). The antigen detection EIAs are highly specific, i.e. hyperimmune antibodies generated against individual strains are not useful for broad detection. NV was found to be antigenically distinct from HV and SMV by cross-challenge studies in volunteers before the cloning of NV (73). The rNV capsid was found to be antigenically distinct from the rMxV capsid immediately after the two antigens were expressed (34). The rGrV capsid also was shown to be antigenically distinct from rNV and rMxV after the establishment of the rGrV antigen EIA (21). On the other hand, the rMxV capsid was found to share minor antigenic epitopes with HV when stool specimens from volunteers infected with HV were tested with the rMxV antigen EIA (31, 37). However, little cross-reactivity between rMxV and rHV was found when a recombinant antigen EIA for HV was developed (Jiang, unpublished data). The rMxV antigen EIA detected a number of MxV-like viruses (31, 37), which is likely to include the Toronto virus.

The antigenic relationships among different HuCVs were recognized better when more recombinant NLV capsid proteins became available. One study compared recombinant capsid antigens from 8 strains of NLVs representing eight genetic clusters (43) (Table 1). The antigenic relationships of the strains were determined based on cross-reactivities of the 8 recombinant antigens with available hyperimmune antibodies against the antigens. The results showed significant high homologous reactivities for individual strains, with low levels of cross-reactivities among certain clusters (Table 1). In general, strains within genogroups are antigenically closer to each other than strains between genogroups. For example, NV, DSV and virus 115 belong to genogroup I; hyperimmune antibodies generated against these strains cross-reacted higher with genogroup I strains than genogroup II strains (MxV, HV, 387, 207 and MOH) and *vice versa*. The eight strains may be divided into 8 antigenic types based on the data of Table 1.

The antigenic relationships of NLVs have also been characterized using monoclonal antibodies. Monoclonal antibodies against the native SMV have been developed before recombinant EIAs were developed (69). Monoclonal antibodies against rNV, rMxV, rHV and r387 are now available (24, 26). Most of these monoclonal antibodies were type-specific, mainly reacting with homologous antigens used in immunization to generate the antibodies. One study used a cross-immunization approach with three recombinant capsid antigens to immunize mice to attempt to generate monoclonal antibodies against potential shared epitopes. The mice revealed high titers of antibodies against all three antigens following the cross-immunization, but most of the hybridoma clones obtained were type-specific (Jiang, unpublished).

Table 1

Antibody titers of hyperimmune sera from guinea pigs immunized with different NLV antigens

Antigen*	EIA titer of hyperimmune antisera from guinea pigs immunized with						
	rNV	r115	rMxV	rHV	r387	r207	rMOH
rNV	**4,096,000#**	256,000	2,000	<1,000	8,000	16,000	16,000
r115	256,000	**512,000**	<1,000	<1,000	16,000	8,000	8,000
rDSV	32,000	128,000	2,000	<1,000	8,000	<1,000	<1,000
rMxV	32,000	1,000	**512,000**	8,000	128,000	64,000	64,000
rHV	16,000	2,000	64,000	**1,024,000**	256,000	128,000	128,000
r387	4,000	4,000	8,000	8,000	**8,192,000**	64,000	64,000
r207	16,000	8,000	4,000	4,000	256,000	**2,048,000**	128,000
rMOH	8,000	32,000	16,000	16,000	256,000	128,000	**8,192,000**

* All antigens used in the assays were adjusted to contain 0.25 ug/ml for coating.
\# Bold numbers indicate homologous antibody titers.

In a separate study, 10 monoclonal antibodies were isolated against rNV (23). When the antibodies were tested against a genetically distinct strain of genogroup I virus, the Chiba virus, three monoclonal antibodies reacted with that strain. The Chiba virus shared 75% amino acid identity with NV. One of the three monoclonal antibodies was then tested with additional stool specimens containing HuCVs by RT-PCR. The antibody reacted with strains representing 4 of 5 clusters of genogroup I viruses. Strains in the five clusters shared 63 to 70% amino acid identities in the capsid gene with NV. The one genetic cluster that was negative in the assay had the lowest sequence identity with NV. Therefore, shared antigenic epitopes are likely to exist within genogroup I NLVs. Whether there are shared antigenic epitopes between genogroups or genera remains to be determined.

Because of the narrow spectrum of NLV hyperimmune antibodies, several approaches have been used to overcome this limitation in developing a broadly reactive assay for HuCVs. One approach was to use a pool of antibodies against different antigenic types of HuCVs. This approach succeeded when the rNV, rMxV and rHV hyperimmune antibodies were pooled for both capture and detecting antibodies in a sandwich EIA similar to those using single antibody pairs; recombinant antigens from

510

all the three strains were detected by the assay (41). The usefulness of the assay was also demonstrated by the detection of viral antigens in stool specimens, but the assay was limited to specimens containing only closely related strains. The sensitivity of the assay was also slightly decreased compared with those of single antigen tests for the respective strains, possibly due to a competition of antibodies for binding sites in the microtiter plates during the coating step.

An alternative approach was to generate hyperimmune antibodies against multiple strains of HuCVs by cross-immunization of rabbits and guinea pigs with different NLV antigens. This approach was based on the fact that high titers of antibodies were generated against the three strains of NLVs in mice following cross-immunization as described earlier. Recombinant antigens from nine NLVs were divided into three groups; each was used to immunize a group of rabbits and guinea pigs. After 3-4 intramuscular injections of a single or a mix of different antigens in a two-week interval between injections, the animals were bled. High titers of antibodies were observed in all animals against corresponding antigens used in the immunization (Table 2). The hyperimmune antisera were pooled following purification of immunoglobulin (Ig) by a protein A column. The rabbit Ig pool was used as the coating antibody and the guinea pig Ig pool as the detecting antibody. Preliminary results showed that this EIA detected all nine recombinant capsid antigens used in the immunization as well as viral antigens in stool specimens containing NLVs by RT-PCR.

Table 2

EIA titers of hyperimmune antisera from animals to nine strains of HuCVs that were used in the immunization.

Antigen groups	Antigens	Animal groups					
		Group 1		Group 2		Group 3	
		Rabbits	Guinea pigs	Rabbits	Guinea pigs	Rabbits	Guinea pigs
Group 1	r115	**256,000***	**1,024,000**	64,000	64,000	8,000	32,000
	rC59	**512,000**	**256,000**	256,000	256,000	64,000	128,000
	rNV	**512,000**	**2,048,000**	8,000	256,000	4,000	64,000
Group 2	r387	32,000	64,000	**2,048,000**	**2,048,000**	256,000	1,024,000
	rGrV	8,000	32,000	**2,048,000**	**2,048,000**	128,000	512,000
	rHA	8,000	16,000	**512,000**	**2,048,000**	128,000	512,000
	rMX	2,000	8,000	**512,000**	**512,000**	128,000	256,000
Group 3	r207	64,000	128,000	512,000	1,024,000	**1,024,000**	**4,096,000**
	rMOH	64,000	128,000	256,000	512,000	**512,000**	**4,096,000**

* Group-homologous antibody titers are indicated in bold.

Antibody detection EIAs

The baculovirus-expressed HuCV capsid antigens were first used to detect antibody in serum specimens by coating the antigens in wells of a microtiter plate to capture antibody (39). Because the VLPs used for the coating were highly purified, high assay sensitivity and specificity were obtained. This format was initially developed to measure total Ig in human serum specimens. The same format was later adapted for detection of Ig isotypes (A, G, and M) in the serum specimens and then for detection of secretory IgA in fecal and milk specimens (58).

The direct coating EIAs have been used most widely to study the seroprevalence of HuCV infection in different populations and countries (8, 10, 16, 17, 22, 35, 44, 59, 61, 68). When the rNV and rMxV antibody EIAs were first introduced, high seroprevalence against both antigens was found in both developing and developed countries (Fig. 1).

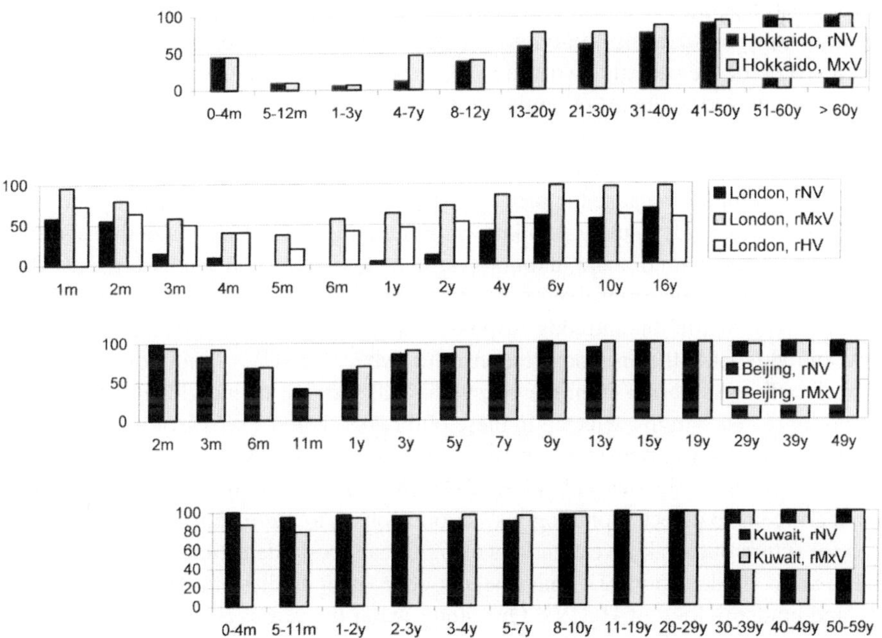

Fig. 1. Seroprevalence to HuCVs in four countries as determined by rNV, rMxV, and rHV antibody detection EIAs. The percentage of individuals who had antibodies against the three HuCV recombinant antigens in each age group was re-plotted based on published data obtained from Hokkaido, Japan (28, 57), London, England (7, 60, 61), Beijing, China (44), and Kuwait (10). Some age groups were deleted during the re-plotting for a better comparison among populations.

512

The age of acquisition of antibody to these antigens was older in developed than in developing countries, but all children were infected before reaching adulthood. The seroprevalence to rMxV was slightly higher than that to rNV, and seroprevalence to both antigens was higher in developing than in developed countries. One study in the UK tested serum samples against three (rNV, rMxV, rHV) antigens and found that seroprevalence to MxV was highest, to HV lower, and to NV lowest in that population (7) (Fig. 1).

In contrast to the antigen detection EIAs, the antibody EIAs are more broadly reactive as the sera tested contain antibodies against different viruses from repeated infections. Therefore, the antibody detected by an EIA may not represent infection with that particular strain, but with any of a group of strains with shared epitopes. Some recombinant antigens (e.g. rNV and rMxV) have been used in many serologic studies. Therefore, some study results are comparable; however, variation of results may be seen among studies performed in different laboratories from different countries. To avoid variation, standard assay protocols and control reagents are needed.

The antibody detection EIAs have been used to investigate outbreaks of acute gastroenteritis (8, 36, 54, 56). In general, the frequency and titer of antibody responses by patients correlates with the degree of genetic identity between the strains used in the assay and that causing the outbreak (54, 56). For example, in the Virginia outbreaks, 100% of the patients involved in three MxV-associated outbreaks had antibody responses detected in the rMxV antibody EIA.

The meaning of heterologous antibody responses is more difficult to interpret. Heterologous responses among genetically closely related strains are particularly difficult to differentiate because the past exposure history to different strains by individuals involved in a study usually is unknown. For example, six Virginia outbreaks, from which stools and paired acute and convalescent sera were collected were caused by LV-like strains. Stools from five of the six outbreaks were positive by the rGrV antigen EIA (41). While the antibody responses have not yet been determined using the rGrV antibody EIA, the antibody responses to rMxV and rHV of individuals involved in these outbreaks varied significantly, with 100% to 0% of individuals mounting a seroresponse. The strains detected in these outbreaks shared 71-76% of nucleotide (nt) identity of the RNA polymerase gene with MxV and 80-87% with HV, suggesting that they are types heterologous to MxV and HV (41). Therefore, development of recombinant antigens to detect "homologous" seroresponses to more antigenic types is a key need for future outbreak investigations using the antibody detection EIAs.

The antibody detection EIAs have been used in other studies of infection and immunity and epidemiology of HuCVs. One study using these assays found differences in risk factors for NV and MxV infection in populations with different socioeconomic and geographical locations (59). EIAs specific for detection of IgA and IgM have been used to describe infection and immunity in volunteers and in patients involved in outbreaks of acute gastroenteritis (16, 22, 60, 61). These assays were found to be more specific than the IgG detection EIAs and may be useful to define the type of a recent outbreak (16, 22, 53, 60, 61).

A capture-format of IgM EIAs was developed for rNV and rMxV (6, 16, 22, 61). In this format, an anti-human IgM antiserum was used to capture the IgM molecules from human sera. HuCV-specific IgM was then detected by addition of recombinant capsid antigens, followed by detection with monoclonal or hyperimmune antibodies against the recombinant capsid antigens. This capture IgM EIA was designed to increase test sensitivity by removing excess IgG in serum samples. The monoclonal antibody-based IgM EIAs have been reported to be more sensitive than the direct antigen coating IgM EIAs (6). The IgM detection EIAs may distinguish the "type" of infecting viruses better than other assays. The direct antigen-coating method is simple and less time consuming, making it more suitable for large-scale studies.

Diagnosis of HuCVs by RT-PCR

The first HuCV RT-PCR was developed for NV (9, 38). Soon it was found that RT-PCR with primers designed for NV also detects other strains of HuCVs that are genetically related. With the rapid progress of molecular technology in the early 1990s, this technique has been widely adapted by many laboratories in HuCV research. The wide involvement of many laboratories not only advanced our knowledge on the viruses, but also helped the improvement of the methodology of the assay. Due to the high degree of antigenic variation of HuCVs and due to the lack of a simple, broadly reactive antigenic assay for HuCVs, the RT-PCR technique will continue to be used in HuCV research. RT-PCR is also the method of choice for detection of HuCVs in environmental samples.

Techniques for extraction of viral RNA and removal of inhibitors

RT-PCR is an enzymatic method; the quality of viral RNA is critical for the test. Methods used for extraction of viral RNA from stool specimens have been paid special attention for high efficiency. This is because stool specimens may contain unknown substances that may inhibit the reverse transcriptase and *Taq* DNA polymerase activities. In addition, stool specimens may contain RNAses that can degrade the viral RNA. Finally, HuCVs usually have a low concentration in human stools. Therefore, a method for maximal recovery of the viral RNA and efficient removal of inhibitors has been a major goal.

The first step is to disrupt the virions to release the viral RNA. Methods used for this step include protease digestion, guanidinium isothiocyanate (GTC) treatment and heat denaturation. The proteinase K method is commonly used with high efficiency for removing the viral capsid. Proteinase K treatment also helps to remove RNAses from the samples. The GTC is a strong protein denaturation reagent, which disrupts the viral capsid and denatures any proteins in the samples including the viral capsid proteins and RNAses. This method has now been modified in several commercial kits that including a mixture of GTC, phenol and chloroform to extract the samples (RNAzol, Ultraspec and Trizol). Both the proteinase K and the GTC methods are highly effective, although the former needs an additional incubation step. The heat release method is simple and rapid but the efficiency may be variable, and it does not remove inhibitors. Heat release followed by a dilution of the samples to reduce inhibiting factors to RT-PCR could be used for large-scale rapid screening. Heat release also is used for immune capture RT-PCR, in which the viruses have been captured by antibodies.

Released viral RNA is then recovered by different procedures including phenol/ chloroform extraction, silica adsorption or exclusion chromatography. The major purpose of this step is to separate the viral RNA from other materials in the samples, including inhibitors. The phenol/chloroform method was one of the earliest used for HuCVs. In this method, a detergent cetyltrimethylammonium bromide (CTAB) was added during the phenol/chloroform extraction step (38). The addition of this detergent facilitates the recovery of viral RNA by a selective separation of the RNA from inhibitors in the samples during the extraction and the ethanol precipitation steps.

Specially prepared silica beads are used to adsorb viral RNA. Following washing of the beads, the adsorbed RNA is eluted, and a high quality of viral RNA with minimum contamination is usually obtained. Methods of exclusion chromatography include spin columns containing Sephadex G200 or chelation of multivalent cation impurities. Although these methods are highly sensitive for detection of HuCVs, variable results were obtained for removal of inhibitors from clinical specimens. A further method to selectively purify HuCVs is to use an antibody-based affinity adsorption. This method has been successful in the laboratory but is not available for field study because currently available antibodies have a narrow spectrum to limited antigenic types. Broadly reactive monoclonal antibodies maybe another choice for this method.

To prevent viral RNA degradation by RNAses, care is particularly important after the RNA is released from virion during the sample process. The samples should be processed as quickly as possible after the proteinase K or GTC treatment and during the extraction steps. The RNA extraction procedures can be stopped at the ethanol precipitation step. RNA stored in ethanol at −70°C is fairly stable.

Methods used for the detection of HuCVs in environmental samples require more steps because environmental samples are contaminated, and the virus concentration is usually low. Samples are usually concentrated for viruses before being processed for extraction of viral RNA (29). The general principle of extraction of viral RNA from environmental concentrates should be the same as that for clinical samples except that the volumes of environmental samples may be larger and repeated treatment of the samples may be necessary. Detection of HuCVs in food is also difficult because of unknown factors present. Methods for certain types of food, such as shellfish, cooked meat, salad, etc, have been described (3, 5, 49, 64-67).

RT-PCR is highly sensitive, and cross-contamination of testing samples is a common problem. The major source of contamination has been found to come from carry-over of products between samples, therefore, the procedure to perform pre- and post-RT-PCR in different laboratory spaces is critical. In addition, techniques to reduce contamination by reducing the time of opening the test samples, such as using thermostable RT and PCR enzymes or using wax layers to separate the two reactions have been used. Furthermore, techniques to improve the efficiency by simplifying and automating the assay, such as real-time PCR and detection of RT-PCR products by enzyme assay, have been developed.

For clinical and environmental samples, non-specific reactions could occur due to unrelated nucleic acids in the samples to be tested. Stringent conditions of RT-PCR including optimal reannealing temperature are necessary. In addition, confirmation of the RT-PCR products by hybridization (using an internal probe) and sequencing is necessary. Hybridization usually increases the detection sensitivity by 10- to 100-fold compared with the ethidium bromide staining method. Sequencing the RT-PCR products has been used frequently for confirmation; the determination of the genetic identity of the strains is particularly important for studies of the molecular epidemiology of HuCVs.

Designing primers for broad detection

HuCVs are genetically diverse. Designing proper primers for RT-PCR is critical. In the past decade, many laboratories were involved in this effort, and a large number of primers has been designed. The usefulness of most of the primers has not been fully assessed yet. Primers that were used more frequently have been mainly judged by the detection rates in surveillance of HuCVs in different populations (2, 18, 33, 71). In many studies, multiple primer sets were used to improve the detection rates, because there is no single primer pair suitable for detection of all HuCVs. In conclusion, continuous update of primer design for HuCVs remains important.

Most of primers designed in the past for HuCVs were targeted to the RNA polymerase regions because of its high conservation compared with other regions of the genome. The capsid region is more important for understanding the biological property of the viruses. However, due to limited regions of high conservation, fewer primers were designed in the capsid region. A highly conserved region was observed at the 5' end of the capsid gene of HuCVs (43). This region was found to be useful for amplification of the 3' region of the genome, including the entire capsid gene, by pairing with an oligo-dT primer. In many cases, strains that were detected by primers in the RNA polymerase region were further characterized in the capsid region. Primers in other regions of the genome, such as the 2C-like region (72) were also used, but much less commonly.

In addition to the selection of highly conserved regions, other parameters have also been considered for designing a primer pair. For example, an optimal primer should be of high GC content (30-50%) within the primer sequence, be of a length of at least 18 nucleotides, and have low sequence matches with other primers used in a single reaction. If mismatches of sequences between the primer and the target must be included, they should be at the 5' end; keeping a high degree of conservation at the 3' end of the primer is critical. Finally, the size of RT-PCR products needs to be considered. In general, a smaller product is easier to amplify and primer pairs for a <100 base product have been used. However, a slightly larger one (300 to 500 bases) will not significantly affect the detection sensitivity but will provide more useful sequence information.

Due to the difficulty for a single primer pair to detect all HuCVs, different approaches have been taken for broad detection of HuCVs. One study described an approach to use a single primer pair for the reverse transcrtiption reaction followed by multiple primers for the PCR step (2). Each of the primers in the PCR step was targeted to a subset of HuCVs. This approach has been successfully applied in surveillance of HuCVs in many populations by several laboratories (1, 2, 12-14, 30, 55). Similar multiplex approaches were used in other laboratories searching for new strains, in which a specimen usually was tested repeatedly by multiple primers until a positive result was obtained. A disadvantage of the multiplex approaches is that a limited number of primers can be used in a single reaction. In addtion, our current understanding of HuCV genetic diversity remains limited, and new strains with distinct genetic identities continue to be described; therefore, keeping a complete set of primers for all HuCVs is difficult.

An alternative approach is to design degenerate primers. This approach has been successfully used in clinical diagnosis of several pathogens. A degenerate primer set also was described for HuCVs (48). Like for the multiplex RT-PCR, if the degree of degeneration of a primer set is too high, interference among primers can occur which may decrease the detection sensitivity. To avoid this, a minimum degeneration of the primers should be considered.

Conclusions

In conclusion, the sandwich antigen detection EIAs using hyperimmune antibodies are highly sensitive. Previous low detection rates of HuCVs by the assays were due to the high specificity of the antibodies and limited numbers of assays used. A panel of hyperimmune antisera to multiple HuCV antigenic types is needed before the assays are useful in clinical laboratories. Cross-immunization of animals with multiple strains of NLVs representing different genetic clusters may be a simple approach for generating broadly reactive diagnostic reagents. Monoclonal antibodies against putative shared antigenic epitopes within genogroups of NLVs are promising for development of assays for broad detection.

The direct antigen-coating EIA for detection of antibody in different clinical specimens is simple, sensitive, and therefore useful for large-scale epidemiologic studies. The antibody detection assays are more broadly reactive. The IgM and IgA detection EIAs may be more specific because they appear to measure recent infection and may be formulated to distinguish infections by different virus types.

The RT-PCR method will continue to play an important role in HuCV research, including surveillance of HuCVs in different populations, outbreak investigation, molecular epidemiology, and monitoring environmental samples for public health safety. A higher efficiency of the assay is expected with the continued improvement of the methods for extraction of viral RNA and for improvement of broadly reactive primer sets.

References

1. Ando, T., Q. Jin, J. R. Gentsch, S. S. Monroe, J. S. Noel, S. F. Dowell, H. G. Cicirello, M. A. Kohn, and R. I. Glass 1995. Epidemiologic applications of novel molecular methods to detect and differentiate small round structured viruses (Norwalk-like viruses) *J Med Virol.* **47**:145-52.

2. Ando, T., S. S. Monroe, J. R. Gentsch, Q. Jin, D. C. Lewis, and R. I. Glass 1995. Detection and differentiation of antigenically distinct small round-structured viruses (Norwalk-like viruses) by reverse transcription-PCR and southern hybridization *J Clin Microbiol.* **33**:64-71.

3. Atmar, R. L., T. G. Metcalf, F. H. Neill, and M. K. Estes 1993. Detection of enteric viruses in oysters by using the polymerase chain reaction *Appl Environ Microbiol.* **59**:631-5.

4. Baric, R. S., B. Yount, L. Lindesmith, P. R. Harrington, S. R. Greene, F. C. Tseng, N. Davis, R. E. Johnston, D. G. Klapper, and C. L. Moe 2002. Expression and self-assembly of Norwalk virus capsid protein from Venezuelan equine encephalitis virus replicons *J Virol.* **76**:3023-30.

5. Beuret, C., D. Kohler, and T. Luthi 2000. Norwalk-like virus sequences detected by reverse transcription-polymerase chain reaction in mineral waters imported into or bottled in Switzerland *J Food Prot.* **63**:1576-82.

6. Brinker, J. P., N. R. Blacklow, M. K. Estes, C. L. Moe, K. J. Schwab, and J. E. Herrmann 1998. Detection of Norwalk virus and other genogroup 1 human caliciviruses by a monoclonal antibody, recombinant-antigen-based immunoglobulin M capture enzyme immunoassay *J Clin Microbiol.* **36**:1064-9.

7. Cubitt, W. D., K. Y. Green, and P. Payment 1998. Prevalence of antibodies to the Hawaii strain of human calicivirus as measured by a recombinant protein based immunoassay *J Med Virol.* **54**:135-9.

8. Cubitt, W. D., and X. Jiang 1996. Study on occurrence of human calicivirus (Mexico strain) as cause of sporadic cases and outbreaks of calicivirus-associated diarrhoea in the United Kingdom, 1983-1995 *J Med Virol.* **48**:273-7.

9. De Leon, R., S. M. Matsui, R. S. Baric, J. E. Herrmann, N. R. Blacklow, H. B. Greenberg, and M. D. Sobsey 1992. Detection of Norwalk virus in stool specimens by reverse transcriptase-polymerase chain reaction and nonradioactive oligoprobes *J Clin Microbiol.* **30**:3151-7.

10. Dimitrov, D. H., S. A. Dashti, J. M. Ball, E. Bishbishi, K. Alsaeid, X. Jiang, and M. K. Estes 1997. Prevalence of antibodies to human caliciviruses (HuCVs) in Kuwait established by ELISA using baculovirus-expressed capsid antigens representing two genogroups of HuCVs *J Med Virol.* **51**:115-8.

11. Dingle, K. E., P. R. Lambden, E. O. Caul, and I. N. Clarke 1995. Human enteric Caliciviridae: the complete genome sequence and expression of virus-like particles from a genetic group II small round structured virus *J Gen Virol.* **76**: 2349-55.

12. Fankhauser, R. L., J. S. Noel, S. S. Monroe, T. Ando, and R. I. Glass 1998. Molecular epidemiology of "Norwalk-like viruses" in outbreaks of gastroenteri-

tis in the United States *J Infect Dis.* **178**:1571-8.

13. Fleischer, J., C. Wagner Wiening, and P. Kimmig 2000. [Surveillance of group illnesses with Norwalk-like viruses (NLV) in Baden-Württemberg. Isolation and detection of Norwalk-like viral RNA from fecal samples with RT-PCR] *Gesundheitswesen* **62**:604-8.

14. Glass, R. I., J. Noel, T. Ando, R. Fankhauser, G. Belliot, A. Mounts, U. D. Parashar, J. S. Bresee, and S. S. Monroe 2000. The epidemiology of enteric caliciviruses from humans: a reassessment using new diagnostics *J Infect Dis.* **181 Suppl 2**:S254-61.

15. Graham, D. Y., X. Jiang, T. Tanaka, A. R. Opekun, H. P. Madore, and M. K. Estes 1994. Norwalk virus infection of volunteers: new insights based on improved assays *J Infect Dis.* **170**:34-43.

16. Gray, J. J., C. Cunliffe, J. Ball, D. Y. Graham, U. Desselberger, and M. K. Estes 1994. Detection of immunoglobulin M (IgM), IgA, and IgG Norwalk virus-specific antibodies by indirect enzyme-linked immunosorbent assay with baculovirus-expressed Norwalk virus capsid antigen in adult volunteers challenged with Norwalk virus *J Clin Microbiol.* **32**:3059-63.

17. Gray, J. J., X. Jiang, P. Morgan-Capner, U. Desselberger, and M. K. Estes 1993. Prevalence of antibodies to Norwalk virus in England: detection by enzyme-linked immunosorbent assay using baculovirus-expressed Norwalk virus capsid antigen *J Clin Microbiol.* **31**:1022-5.

18. Green, J., C. I. Gallimore, J. P. Norcott, D. Lewis, and D. W. Brown 1995. Broadly reactive reverse transcriptase polymerase chain reaction for the diagnosis of SRSV-associated gastroenteritis *J Med Virol.* **47**:392-8.

19. Green, K. Y., A. Z. Kapikian, J. Valdesuso, S. Sosnovtsev, J. J. Treanor, and J. F. Lew 1997. Expression and self-assembly of recombinant capsid protein from the antigenically distinct Hawaii human calicivirus *J Clin Microbiol.* **35**:1909-14.

20. Greenberg, H. B., R. G. Wyatt, J. Valdesuso, A. R. Kalica, W. T. London, R. M. Chanock, and A. Z. Kapikian 1978. Solid-phase microtiter radioimmunoassay for detection of the Norwalk strain of acute nonbacterial, epidemic gastroenteritis virus and its antibodies *J Med Virol.* **2**:97-108.

21. Hale, A. D., S. E. Crawford, M. Ciarlet, J. Green, C. Gallimore, D. W. Brown, X. Jiang, and M. K. Estes 1999. Expression and self-assembly of Grimsby virus: antigenic distinction from Norwalk and Mexico viruses *Clin Diagn Lab Immunol.* **6**:142-5.

22. Hale, A. D., D. C. Lewis, X. Jiang, and D. W. Brown 1998. Homotypic and heterotypic IgG and IgM antibody responses in adults infected with small round structured viruses *J Med Virol.* **54**:305-12.

23. Hale, A. D., T. N. Tanaka, N. Kitamoto, M. Ciarlet, X. Jiang, N. Takeda, D. W. Brown, and M. K. Estes 2000. Identification of an epitope common to genogroup 1 "Norwalk-like viruses" *J Clin Microbiol.* **38**:1656-60.

24. Hardy, M. E., T. N. Tanaka, N. Kitamoto, L. J. White, J. M. Ball, X. Jiang, and M. K. Estes 1996. Antigenic mapping of the recombinant Norwalk virus capsid protein using monoclonal antibodies *Virology* **217**:252-61.

25. Hayashi, Y., T. Ando, E. Utagawa, S. Sekine, S. Okada, K. Yabuuchi, T. Miki, and M. Ohashi 1989. Western blot (immunoblot) assay of small, round-structured virus associated with an acute gastroenteritis outbreak in Tokyo *J Clin Microbiol.* **27**:1728-33.

26. Herrmann, J. E., N. R. Blacklow, S. M. Matsui, T. L. Lewis, M. K. Estes, J. M. Ball, and J. P. Brinker 1995. Monoclonal antibodies for detection of Norwalk virus antigen in stools *J Clin Microbiol.* **33**:2511-3.

27. Herrmann, J. E., N. A. Nowak, and N. R. Blacklow 1985. Detection of Norwalk virus in stools by enzyme immunoassay *J Med Virol.* **17**:127-33.

28. Honma, S., S. Nakata, K. Numata, K. Kogawa, T. Yamashita, M. Oseto, X. Jiang, and S. Chiba 1998. Epidemiological study of prevalence of genogroup II human calicivirus (Mexico virus) infections in Japan and Southeast Asia as determined by enzyme-linked immunosorbent assays *J Clin Microbiol.* **36**:2481-4.

29. Huang, P. W., D. Laborde, V. R. Land, D. O. Matson, A. W. Smith, and X. Jiang 2000. Concentration and detection of caliciviruses in water samples by reverse transcription-PCR *Appl Environ Microbiol.* **66**:4383-8.

30. Iritani, N., Y. Seto, K. Haruki, M. Kimura, M. Ayata, and H. Ogura 2000. Major change in the predominant type of "Norwalk-like viruses" in outbreaks of acute nonbacterial gastroenteritis in Osaka City, Japan, between April 1996 and March 1999 *J Clin Microbiol.* **38**:2649-54.

31. Jiang, X., D. Cubitt, J. Hu, X. Dai, J. Treanor, D. O. Matson, and L. K. Pickering 1995. Development of an ELISA to detect MX virus, a human calicivirus in the Snow Mountain agent genogroup *J Gen Virol.* **76**:2739-47.

32. Jiang, X., D. Graham, K. Wang, and M. Estes 1990. Norwalk virus genome cloning and characterization *Science* **250**:1580-3.

33. Jiang, X., P. W. Huang, W. M. Zhong, T. Farkas, D. W. Cubitt, and D. O. Matson 1999. Design and evaluation of a primer pair that detects both Norwalk- and Sapporo-like caliciviruses by RT-PCR *J Virol Methods* **83**:145-54.

34. Jiang, X., D. O. Matson, G. M. Ruiz-Palacios, J. Hu, J. Treanor, and L. K. Pickering 1995. Expression, self-assembly, and antigenicity of a snow mountain agent- like calicivirus capsid protein *J Clin Microbiol.* **33**:1452-5.

35. Jiang, X., D. O. Matson, F. R. Velazquez, J. J. Calva, W. M. Zhong, J. Hu, G. M. Ruiz-Palacios, and L. K. Pickering 1995. Study of Norwalk-related viruses in Mexican children *J Med Virol.* **47**:309-16.

36. Jiang, X., E. Turf, J. Hu, E. Barrett, X. M. Dai, S. Monroe, C. Humphrey, L. K. Pickering, and D. O. Matson 1996. Outbreaks of gastroenteritis in elderly nursing homes and retirement facilities associated with human caliciviruses *J Med Virol.* **50**:335-41.

37. Jiang, X., J. Wang, and M. K. Estes 1995. Characterization of SRSVs using RT-PCR and a new antigen ELISA *Arch Virol.* **140**:363-74.

38. Jiang, X., J. Wang, D. Y. Graham, and M. K. Estes 1992a. Detection of Norwalk virus in stool by polymerase chain reaction *J Clin Microbiol.* **30**:2529-34.

39. Jiang, X., M. Wang, D. Y. Graham, and M. K. Estes 1992b. Expression, self-assembly, and antigenicity of the Norwalk virus capsid protein *J Virol.* **66**:6527-32.

520

40. Jiang, X., M. Wang, K. Wang, and M. K. Estes 1993. Sequence and genomic organization of Norwalk virus *Virology*. **195**:51-61.

41. Jiang, X., N. Wilton, W. M. Zhong, T. Farkas, P. W. Huang, E. Barrett, M. Guerrero, G. Ruiz-Palacios, K. Y. Green, J. Green, A. D. Hale, M. K. Estes, L. K. Pickering, and D. O. Matson 2000. Diagnosis of human caliciviruses by use of enzyme immunoassays *J Infect Dis*. **181 Suppl 2**:S349-59.

42. Jiang, X., W. Zhong, M. Kaplan, L. K. Pickering, and D. O. Matson 1999. Expression and characterization of Sapporo-like human calicivirus capsid proteins in baculovirus *J Virol Methods* **78**:81-91.

43. Jiang, X., W. M. Zhong, N. Wilton, T. Farkas, E. Barrett, D. Fulton, R. Morrow, and D. M. Matson 2002. Baculovirus expression and antigenic characterization of the capsid proteins of three Norwalk-like viruses. *Arch Virol*. **147**:119-30.

44. Jing, Y., Y. Qian, Y. Huo, L. P. Wang, and X. Jiang 2000. Seroprevalence against Norwalk-like human caliciviruses in Beijing, China *J Med Virol*. **60**:97-101.

45. Kapikian, A. Z., R. G. Wyatt, R. Dolin, T. S. Thornhill, A. R. Kalica, and R. M. Chanock 1972. Visualization by immune electron microscopy of a 27-nm particle associated with acute infectious nonbacterial gastroenteritis *J Virol*. **10**: 1075-81.

46. Kobayashi, S., K. Sakae, Y. Suzuki, K. Shinozaki, M. Okada, H. Ishiko, K. Kamata, K. Suzuki, K. Natori, T. Miyamura, and N. Takeda 2000. Molecular cloning, expression, and antigenicity of Seto virus belonging to genogroup I Norwalk-like viruses *J Clin Microbiol*. **38**:3492-4.

47. Lambden, P. R., E. O. Caul, C. R. Ashley, and I. N. Clarke 1993. Sequence and genome organization of a human small round-structured (Norwalk-like) virus *Science* **259**:516-9.

48. Le Guyader, F., M. K. Estes, M. E. Hardy, F. H. Neill, J. Green, D. W. Brown, and R. L. Atmar 1996a. Evaluation of a degenerate primer for the PCR detection of human caliciviruses *Arch Virol*. **141**:2225-35.

49. Le Guyader, F., F. H. Neill, M. K. Estes, S. S. Monroe, T. Ando, and R. L. Atmar 1996b. Detection and analysis of a small round-structured virus strain in oysters implicated in an outbreak of acute gastroenteritis *Appl Environ Microbiol*. **62**: 4268-72.

50. Leite, J. P., T. Ando, J. S. Noel, B. Jiang, C. D. Humphrey, J. F. Lew, K. Y. Green, R. I. Glass, and S. S. Monroe 1996. Characterization of Toronto virus capsid protein expressed in baculovirus *Arch Virol*. **141**:865-75.

51. Lew, J. F., A. Z. Kapikian, X. Jiang, M. K. Estes, and K. Y. Green 1994. Molecular characterization and expression of the capsid protein of a Norwalk-like virus recovered from a Desert Shield troop with gastroenteritis *Virology* **200**: 319-25.

52. Lewis, D., T. Ando, C. D. Humphrey, S. S. Monroe, and R. I. Glass 1995. Use of solid-phase immune electron microscopy for classification of Norwalk-like viruses into six antigenic groups from 10 outbreaks of gastroenteritis in the United States *J Clin Microbiol*. **33**:501-4.

53. Maguire, A. J., J. Green, D. W. Brown, U. Desselberger, and J. J. Gray 1999.

Molecular epidemiology of outbreaks of gastroenteritis associated with small round-structured viruses in East Anglia, United Kingdom, during the 1996-1997 season *J Clin Microbiol.* **37**:81-9.

54. Monroe, S. S., S. E. Stine, X. Jiang, M. K. Estes, and R. I. Glass 1993. Detection of antibody to recombinant Norwalk virus antigen in specimens from outbreaks of gastroenteritis *J Clin Microbiol.* **31**:2866-72.

55. Nakayama, M., Y. Ueda, H. Kawamoto, Y. Han-jun, K. Saito, O. Nishio, and H. Ushijima 1996. Detection and sequencing of Norwalk-like viruses from stool samples in Japan using reverse transcription-polymerase chain reaction amplification *Microbiol Immunol.* **40**:317-20.

56. Noel, J. S., T. Ando, J. P. Leite, K. Y. Green, K. E. Dingle, M. K. Estes, Y. Seto, S. S. Monroe, and R. I. Glass 1997. Correlation of patient immune responses with genetically characterized small round-structured viruses involved in outbreaks of nonbacterial acute gastroenteritis in the United States, 1990 to 1995 *J Med Virol.* **53**:372-83.

57. Numata, K., S. Nakata, X. Jiang, M. K. Estes, and S. Chiba 1994. Epidemiological study of Norwalk virus infections in Japan and Southeast Asia by enzyme-linked immunosorbent assays with Norwalk virus capsid protein produced by the baculovirus expression system *J Clin Microbiol.* **32**:121-6.

58. Okhuysen, P. C., X. Jiang, L. Ye, P. C. Johnson, and M. K. Estes 1995. Viral shedding and fecal IgA response after Norwalk virus infection *J Infect Dis.* **171**: 566-9.

59. O'Ryan, M. L., P. A. Vial, N. Mamani, X. Jiang, M. K. Estes, C. Ferrecio, H. Lakkis, and D. O. Matson 1998. Seroprevalence of Norwalk virus and Mexico virus in Chilean individuals: assessment of independent risk factors for antibody acquisition *Clin Infect Dis.* **27**:789-95.

60. Parker, S. P., W. D. Cubitt, and X. Jiang 1995. Enzyme immunoassay using baculovirus-expressed human calicivirus (Mexico) for the measurement of IgG responses and determining its seroprevalence in London, UK *J Med Virol.* **46**: 194-200.

61. Parker, S. P., W. D. Cubitt, X. J. Jiang, and M. K. Estes 1994. Seroprevalence studies using a recombinant Norwalk virus protein enzyme immunoassay *J Med Virol.* **42**:146-50.

62. Prasad, B. V., M. E. Hardy, T. Dokland, J. Bella, M. G. Rossmann, and M. K. Estes 1999. X-ray crystallographic structure of the Norwalk virus capsid *Science* **286**:287-90.

63. Prasad, B. V., R. Rothnagel, X. Jiang, and M. K. Estes 1994. Three-dimensional structure of baculovirus-expressed Norwalk virus capsids *J Virol.* **68**:5117-25.

64. Schwab, K. J., F. H. Neill, M. K. Estes, T. G. Metcalf, and R. L. Atmar 1998. Distribution of Norwalk virus within shellfish following bioaccumulation and subsequent depuration by detection using RT-PCR *J Food Prot.* **61**:1674-80.

65. Schwab, K. J., F. H. Neill, R. L. Fankhauser, N. A. Daniels, S. S. Monroe, D. A. Bergmire-Sweat, M. K. Estes, and R. L. Atmar 2000. Development of methods to detect "Norwalk-like viruses" (NLVs) and hepatitis A virus in delicatessen

522

foods: application to a food-borne NLV outbreak *Appl Environ Microbiol.* **66**: 213-8.

66. Schwab, K. J., F. H. Neill, F. Le Guyader, M. K. Estes, and R. L. Atmar 2001. Development of a reverse transcription-PCR-DNA enzyme immunoassay for detection of "Norwalk-like" viruses and hepatitis A virus in stool and shellfish *Appl Environ Microbiol.* **67**:742-9.

67. Shieh, Y., S. S. Monroe, R. L. Fankhauser, G. W. Langlois, W. Burkhardt, 3rd, and R. S. Baric 2000. Detection of Norwalk-like virus in shellfish implicated in illness *J Infect Dis.* **181 Suppl 2**:S360-6.

68. Smit, T. K., P. Bos, I. Peenze, X. Jiang, M. K. Estes, and A. D. Steele 1999. Seroepidemiological study of genogroup I and II calicivirus infections in South and southern Africa *J Med Virol.* **59**:227-31.

69. Treanor, J., R. Dolin, and H. P. Madore 1988. Production of a monoclonal antibody against the Snow Mountain agent of gastroenteritis by *in vitro* immunization of murine spleen cells *Proc Natl Acad Sci USA.* **85**:3613-7.

70. Treanor, J. J., H. P. Madore, and R. Dolin 1988. Development of an enzyme immunoassay for the Hawaii agent of viral gastroenteritis *J Virol Methods* **22**: 207-14.

71. Vinjé, J., and M. P. Koopmans 1996. Molecular detection and epidemiology of small round-structured viruses in outbreaks of gastroenteritis in the Netherlands *J Infect Dis.* **174**:610-5.

72. Wang, J., X. Jiang, H. P. Madore, J. Gray, U. Desselberger, T. Ando, Y. Seto, I. Oishi, J. F. Lew, K. Y. Green, and *et al.* 1994. Sequence diversity of small, round-structured viruses in the Norwalk virus group *J Virol.* **68**:5982-90.

73. Wyatt, R. G., R. Dolin, N. R. Blacklow, H. L. DuPont, R. F. Buscho, T. S. Thornhill, A. Z. Kapikian, and R. M. Chanock 1974. Comparison of three agents of acute infectious nonbacterial gastroenteritis by cross-challenge in volunteers *J Infect Dis.* **129**:709-14.

74. Yoda, T., Y. Terano, A. Shimada, Y. Suzuki, K. Yamazaki, N. Sakon, I. Oishi, E. T. Utagawa, Y. Okuno, and T. Shibata 2000. Expression of recombinant Norwalk-like virus capsid proteins using a bacterial system and the development of its immunologic detection *J Med Virol.* **60**:475-81.

Viral Gastroenteritis
U. Desselberger and J. Gray (editors)

IV, 5. Molecular epidemiology of human caliciviruses

Marion Koopmans, Elisabeth van Strien and Harry Vennema

Diagnostic Laboratory for Infectious Diseases, National Institute of Public Health and the Environment, Bilthoven, The Netherlands

Abstract

Caliciviruses are among the most common infections in humans, and are a major cause of illness in the community, and of outbreaks of gastroenteritis in institutions such as nursing homes and hospitals. While the course of illness is generally mild, the sheer numbers of infections and the high attack rate result in a substantial burden of illness. In recent years, molecular detection techniques have been developed and used initially to study the incidence of caliciviruses. With increasing refinement, strain typing is now being used to further our understanding of the epidemiology of calicivirus infections. By the powerful combination of epidemiological investigations and molecular strain typing, novel information has emerged which adds to our knowledge of the modes of transmission of these viruses and their evolutionary mechanisms.

Introduction

In recent years, several studies have addressed the importance of caliciviruses as causes of gastroenteritis in different selections of the population (Fig. 1). There are data from community surveys in the UK, Sweden, and The Netherlands, from physician-based studies in the UK, The Netherlands and France, from special surveys in young children in Finland, Canada and South-Africa, and from outbreak surveillance in many other countries (Tables 1-3 and references therein). The results of these studies paint the picture that caliciviruses, and especially the Norwalk-like group of caliciviruses (NLV), are among the most common causes of illness in the community, with the highest incidence in young children, but also a substantial burden of illness in adults. Surveys of outbreaks almost all show that the majority of outbreaks investigated is attributed to NLV (Table 1). For community-based studies, the positivity rates range more widely (1- 22%; Table 2). These differences are likely to be related to differences in methods used for virus detection and to differences in selection of the study population. The highest community incidence was reported from Finland, where a group of children was monitored during a rotavirus vaccine trial

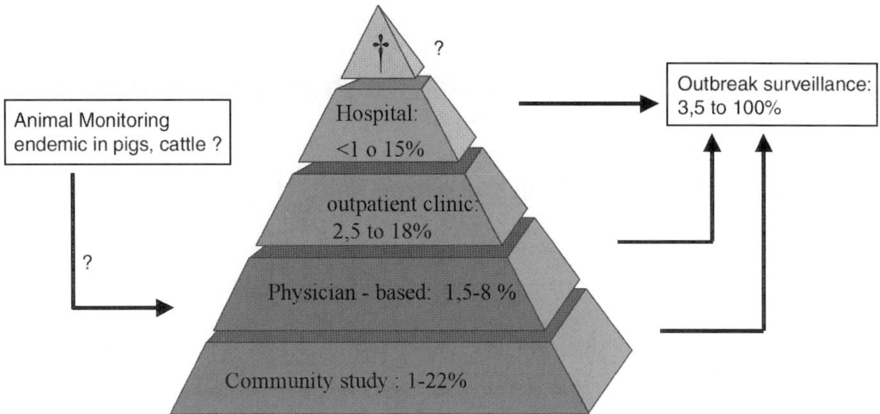

Fig. 1. The population pyramid indicating the selection of the population addressed by community studies, physician-based studies, and outbreak surveys. Percentages indicated are the reported proportions of cases of gastroenteritis in the study population in which caliciviruses were detected in the stool specimens, based on the studies listed in Table 1.

(Pang *et al.*, 1999). Although the incidence of NLV in the community survey in The Netherlands was significantly lower, the rates were much higher in young children (19% in the 1-4 year olds) (de Wit *et al.*, 2001a). Overall, illness due to caliciviruses is relatively mild, which is reflected by the lower incidence of caliciviruses in patients that are consulting their physician. However, the few hospital-based studies that have been done indicate that NLVs infection may lead to hospitalisation (Table 3). In special surveys of outbreaks, NLVs are seen as the most common cause of outbreaks of gastroenteritis in institutions such as nursing homes, hospitals, homes for the mentally handicapped, and as a prominent cause of food-related outbreaks. National differences in the selection of outbreaks that are reported and investigated make it difficult to compare rates between countries, and the proportion of outbreaks attributed to caliciviruses varies greatly (Lopman *et al.*, submitted for publication).

Molecular detection and typing assays have been developed for caliciviruses. They are increasingly used in epidemiological studies of caliciviruses, to provide insight in the diversity of caliciviruses, trace the source of outbreaks, and study virulence traits. In this paper, we will review the state of the art of calicivirus molecular epidemiology.

Taxonomy of the family *Caliciviridae*

The *Caliciviridae* are a family of positive-strand RNA viruses that infect a wide range of species including humans (Green *et al.*, 2000). There are now 4 recognised genera within the family, with most animal caliciviruses belonging to two genera named *Vesivirus* and *Lagovirus,* and the human pathogens along with some animal viruses grouped into the genera *Norwalk-like virus* (NLV) and *Sapporo-like virus* (SLV). The names for the last two genera are tentative, and will change in the near future (pending the adoption of a proposed nomenclature in March 2002 by the International Committee for Taxonomy of Viruses)*. Type strains for the 2 genera of calicivirus that infect humans are Norwalk virus (Hu/NLV/I/Norwalk/1968/US), and Sapporo virus (Hu/SLV/Sapporo/1982/JP). Well-known representatives of the animal caliciviruses are the vesicular exanthema of swine virus (type strain Po/VV/VESV-A48/1948/US), feline caliciviruses (both belonging to the genus *Vesivirus*), and the rabbit hemorrhagic disease viruses (genus *Lagovirus;* Ra/LV/RHDV/GH/1988/GE). In this chapter we will focus on the NLVs and SLVs.

Structure of the viruses

The calicivirus genome is approximately 7.5 kb long, with the polarity of messenger RNA (positive-strand RNA, polyadenylated 3'-end) (Jiang *et al.*, 1993; Lambden *et al.*, 1993). Across the NLV genome, 3 open reading frames (ORFs) are recognised, with the largest one (ORF1) located at the 5'-end of the genome, coding for the non-structural proteins (Fig. 2). The 2 ORFs (ORF2 and ORF3) located at the 3'-end encode the virion-associated proteins. The genome is packaged into an icosahedral particle, built by assembly of 90 dimers of the major capsid protein expressed from ORF2 (Jiang *et al.*, 1992b). The Mr of the major capsid protein is approximately 60 kDa. In addition, a second minor capsid protein of 10 kDa was identified, encoded by ORF3, which is probably involved in RNA binding and is located on the inner surface of the virus capsid. Additional coding assignments have been recognised for an RNA-dependent RNA polymerase (RdRp), a 2C-like nucleoside triphosphatase (NTPase), a VPg, and a proteinase, based on conserved amino-acid motifs located within ORF1 (Fig. 2). For NLV, the ORF2 overlaps by a few nucleotides with ORF1. For SLV, the genome organisation is similar to that of viruses in the genus *Lagovirus*, in that the capsid protein is in frame and contiguous with the non-structural polyprotein. The capsid protein has been expressed in baculovirus expression systems, yielding virus-like particles (VLPs) by self-assembly of the capsid proteins (Jiang *et al.*, 1992b). These VLPs have been used to study the structure of the Norwalk virus particle by cryo-eletron microscopy and computer imaging (Prasad *et al.*, 1994, 1996). Based on these observations, the capsid protein is thought to consist of a shell domain, involved in the basic structure of the icosahedral particle, and a protruding domain. A detailed overview of the genome organisation and structure of caliciviruses is given by Green *et al.* (2001) (See also Section IV, Chapter 1 and Section IV, Chapter 2 of this book).

---

* The NRVs and SRVs have been renamed *Noroviruses* and *Sapoviruses*, respectively (Mayo, 2002).

Table 1

Summary of studies of outbreaks of (viral) gastroenteritis, listing methods used for detection, and results of typing (N = number of outbreaks; n = number of individual cases; % pos = % of outbreaks/cases attributed to calicivirus; GG = genogroup; GGII.4 = viruses belonging to the genotype of Bristol-like viruses, nPCR = nested PCR (after reverse transcription), HMA = hybrid-mobility assay; EM = electron microscopy; EIA = enzyme immunoassay; nr = not reported); * = proportion of strains belonging to this category.

Study population	Country	Period	Design	N	% pos	Method	GGI*	GGII*	GGII.4*	Reference
National	Japan	89-98	Outbreaks	190	100	nPCR	14	87	<32	Kawamoto *et al.*, 2001
City	Japan	95-98	non-bacterial outbreaks	13	100	nPCR	0	62	21	Ohyama *et al.*, 1999
City	Japan	4/96-4/99	Outbreaks	64	73	PCR	11	79	7	Iritani *et al.*, 2000
City	Japan	4/99-3/00	Outbreaks	26	65	PCR	12	41	35	Iritani *et al.*, 2002
Infant home	Japan	76-95	infant home survey	36	42	PCR	nr	nr	nr	Nakata *et al.*, 2000
State	Australia	80-96	reference laboratory	6226	3,5	EM	5	95	68	Wright *et al.*, 1998
National	New Zealand	8/95-7/96	outbreaks	83	100	PCR	22	63	31	Greening *et al.*, 2001
State	US	88	Outbreaks	20	90	PCR	10	90	70	Green *et al.*, 2002
National	US	1/96-6/97	non-bacterial outbreaks	90	100	PCR	6	93	50	Fankhauser *et al.*, 1998
Regional	UK	7/92-6/98	Outbreaks	706	100	EM, EIA	nr	nr	nr	Hale *et al.*, 2000
Regional	UK	92-95	Outbreaks	9	100	EM, PCR	33	67	33	Green *et al.*, 1997
Regional	UK	97-98	Outbreaks	130	nr	PCR, HMA	nr	nr	57	Mattick *et al.*, 2000
Regional	UK	96-97	Outbreaks	94	76	EM, PCR	2	98	96	Maguire *et al.*, 1999
Regional	Ireland	93-98	Outbreaks	nk	nk	PCR	12	15	nr	Foley *et al.*, 2001

Table 1

Continued

Study population	Country	Period	Design	N	% pos	Method	GGI*	GGII*	GGII.4*	Reference
National	Norway	11/98-5/99	Outbreaks	52	40	nPCR	13	87	91	Vainio *et al.*, 2001
Regonal	Finland	97-98	Outbreaks	16	100	PCR	31	70	50	Maunula *et al.*, 1999
Regional	Germany	11/97-5/98	Outbreaks	16	100	nPCR	6	94	68	Schreier *et al.*, 2000
National	Netherlands	96	Outbreaks	60	87	PCR	8	91	87	Vinjé *et al.*, 1997
National	Netherlands	94-95	Outbreaks	22	91	PCR	10	90	50	Vinjé *et al.*, 1996
National	Netherlands	96-01	Outbreaks	222	80	PCR	11	89	50*	Koopmans *et al.*, 2000, 2001 and unpublished data

*See Fig. 5 of this article.

Table 2

Summary table of community-based studies of (viral) gastroenteritis. (For abbreviations see legend of Table 1).

Study population	Country	Year	Design	n	% pos	Method	GGI	GGII	GGII.4	Reference
National	Finland	9/93-9/94	rotavirus vaccine trial, community	1447	22	PCR	8	92	2b: 47%	Pang *et al.*, 1999
City	Sweden	2/98-1/99	community survey	123	20	PCR	nr	nr	nr	Lindqvist *et al.*, 2001
National	Netherlands	99	community cohort	709	11	PCR	22	80	32	de Wit *et al.*, 2001a
National	Netherlands	96-99	physician-based study	857	5	PCR	14	86	29	de Wit *et al.*, 2001b
Regional	UK	8/93-1/96	physician-based study	2893	1,5	EM	nr	nr	nr	Tompkins *et al.*, 1999
Regional	UK	8/93-1/96	community cohort	715	8	EM	nr	nr	nr	Tompkins *et al.*, 1999
Regional	Canada	91-95	survey pediatric patients	768	18	PCR	3	97	2b: 65%	Levett *et al.*, 1996
Regional	Canada	11/97-7/98	pediatric patients	226	1	EM	nr	nr	nr	Waters *et al.*, 2000
Regional	Canada	11/97-7/98	day-care centers	211	1	EM	nr	nr	nr	Waters *et al.*, 2000
Regional	South Africa	10/91-10/95	survey pediatric patients	1296	2,5	EM, EIA, PCR	10	90	nr	Wolfaardt *et al.*, 1997
Regional	South Africa	89-91	community cases	299	1	EM, EIA, PCR	ins	ins	ins	Wolfaardt *et al.*, 1995
City	Brasil	8/88-91	community cohort, children	120	10	PCR	63	37	13	Parks *et al.*, 1999

Table 3

Summary of hospital-based studies of (viral) gastroenteritis. (For abbreviations see legend of Table 1).

Study population	Country	Year	Design	n	% pos	Method	GGI	GGII	GGII.4	Reference
Regional	Canada	11/97-7/98	hospitalised children	1638	<1	EM	nr	nr	nr	Waters *et al.*, 2000
City	France	12/95-2/98	hospitalised children	414	14	PCR	11	89	42	Bon *et al.*, 1999
Regional	France	96-98	hospitalised patients			PCR	8	79	nr	Traore *et al.*, 2000
City	Sweden	10/96-10/97	hospitalised patients, adults, case-control	851	3	EM	nr	nr	nr	Svenungsson *et al.*, 2000
City	Australia	95-96	children	360	2,5	PCR	89	11	0	Schnagl *et al.*, 2000
City	Australia	99	children	354	9	PCR	5	95	77	Kirkwood and Bishop, 2001
City	China	11/94-8/96	children	186	7,6	PCR	43	57	nr	Qiao *et al.*, 1999
Regional	India	11/88-1/99	hospital, all ages	80	15	EIA, PCR	30	70	nr	Kang *et al.*, 2000

Molecular detection techniques

The start of the molecular era for caliciviruses with the cloning of the Norwalk virus genome has greatly facilitated epidemiological studies addressing their role as causes of illness. Until then, the gold standard for calicivirus detection was visualisation of the particles directly in stools by electron microscopic (EM) examination of a preparation stained with electron-dense contrast fluids (Atmar and Estes, 2001). However, only few countries have sufficient and sufficiently experienced electron microscopists to be able to run a national reference service for caliciviruses based on EM detection. As a result, the importance of caliciviruses as a cause of illness remained obscure. Another disadvantage is the rather low sensitivity of EM. This may not be a problem in outbreak situations, when detecting the pathogens in a fraction of all cases may be diagnostic for a calicivirus outbreak. In the investigation of sporadic cases, however, diagnosis by EM will most likely underestimate the true number of positives (Wolfaardt *et al.*, 1995). These disadvantages have been overcome with the development of RT-PCR detection methods, following the successful cloning of the genome of Norwalk virus. Initially, RT-PCR assays were optimised using a limited range of variants, which resulted in highly sensitive detection but with a narrow diagnostic range (Jiang *et al.*, 1992a; Moe *et al.*, 1994; Ando *et al.*, 1995). With the finding that caliciviruses are genetically very diverse, several groups have tried successfully to develop broader, generic detection methods. At present, published protocols list a range of primer sets that have been developed and used for the detection of NLV and SLV (Ando *et al.*, 1995; Atmar and Estes 2001; Green *et al.*, 1993; Jiang *et al.*, 1999a; LeGuyader *et al.*, 1996a and b; Maunula *et al.*, 1999; Schreier *et al.*, 2000; Vinjé *et al.*, 1996 and 2000a, b; Wright *et al.*, 1998). Most assays target conserved regions within the RNA polymerase gene, which has also allowed the use of the sequence of the PCR products for typing (Fig. 2).

What is molecular epidemiology?

Molecular epidemiology is NOT drawing a set of samples from a historic collection, running some typing methods, and interpreting the findings. The real power of molecular epidemiology lies in the best of both approaches: a combination of sound epidemiological and molecular-diagnostic research. A range of strain characterisation methods can be used to provide possible explanations for the observations made through epidemiological studies. The laboratory methods used can be numerous, from biological assays (typing against a panel of neutralising monoclonal or polyclonal antibodies, looking at cytopathogenic effects in cell culture, effects on red blood cells etc.) to molecular analysis (such as sequence analysis of selected regions of the viral genome, mutational analysis, phylogeny). An example from the field of virology is the use of molecular typing to monitor progress in the global WHO campaign that aims at the eradication of poliovirus and the paralytic illness that it causes. Poliovirus has a positive sense RNA genome, that is genetically quite flexible in that the genome accu-

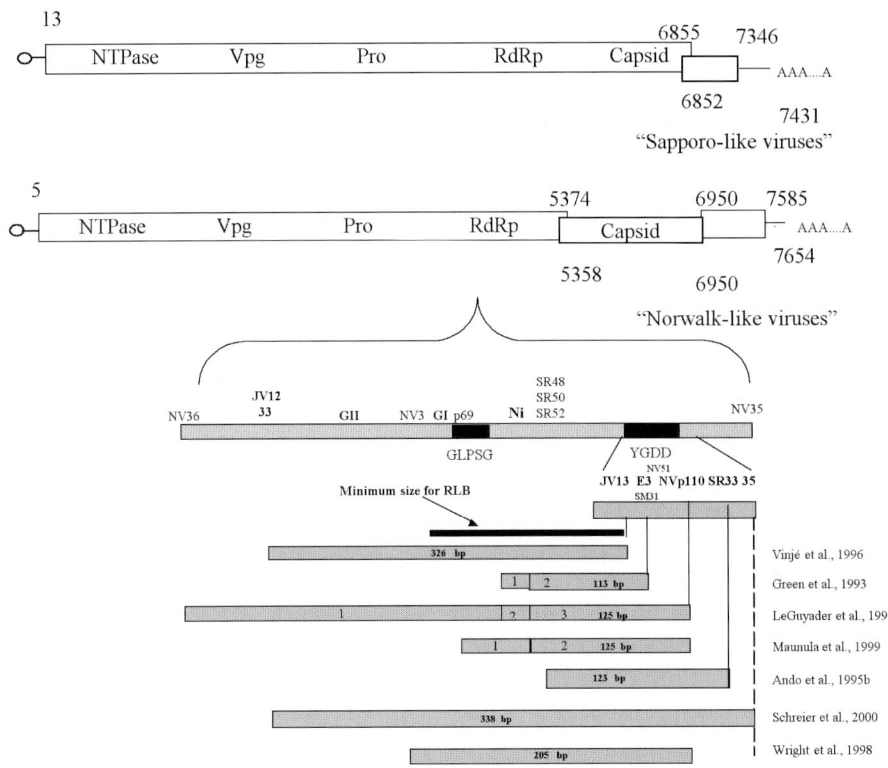

Fig. 2. Genome organisation of caliciviruses and localisation of assigned genes coding for the major capsid protein, an RNA-dependent RNA polymerase (RdRp), a 2C-like nucleoside triphosphatase (NTPase), a VPg, and a proteinase (Pro). Enlarged section represents target regions of commonly used RT-PCR assays for detection and strain typing of NLV all located within the RdRp. Conserved motivs are indicated (YGDD and GLPSG). Other abbreviations on the RdRp gene are primer codes. RLB is reverse line blot typing assay.

mulates mutations during replication of the virus in humans, at a rate of approximately 1% of nucleotides per year (Wimmer *et al.*, 1993). In addition to this, polioviruses can be divided genetically into lineages or genotypes that tend to cluster geographically (Lipskaya *et al.*, 1995). As a consequence, sequencing of a representative region of the poliovirus genome (e.g. the VP1-2a junction region) is used to map polioviruses imported into non-endemic countries to their probably region of origin (Mulders *et al.*, 1995a and b; van der Avoort *et al.*, 1995). Such importations may then point to an area, which requires intensified vaccination campaigns.

Molecular analysis of caliciviruses

A first question in molecular epidemiological studies is which methods to use for the laboratory investigations. The answer to this question depends on the issue that is addressed by the researcher (Table 4). For caliciviruses, most studies have focussed on understanding modes of transmission of viruses through the community. For this specific purpose, any method that would reliably group identical or highly similar viruses together, and separate them from less closely related relatives, would suffice. At present, sequence analysis of the viral genome suits this purpose. An important dilemma is which areas of the genome should be analysed. Relatively few calicivirus genomes have been completely sequenced, so that currently "the best genome region" for virus typing does not exist. An attempt at getting some more information was done by Vinjé *et al.* (2000b) and Green J. *et al.* (2000), who compiled a panel with 31 different NLVs from clinical specimens, and sequenced regions in the RdRp-gene, the complete capsid gene, and part of the 3' ORF. The panel was composed of a selection of stool samples that had been submitted for diagnostic tests to the two institutes over the past few years, and therefore were thought to represent the majority of the currently circulating strains in these countries (UK and The Netherlands). Based on their work, the viruses could be grouped into 15 distinct lineages, 7 belonging to genogroup

Table 4

Examples of questions for which molecular typing data can be used, and possible approaches to address these questions.

Question	Method
Taxonomic grouping	Mapping of signature sequences for calicivirus in non-structural genes, overall genome organisation
Is this a calicivirus?	Mapping of signature sequences for calicivirus in non-structural genes
Does this virus belong to a new calicivirus genotype?	Phylogeny based on complete capsid sequence, indicative grouping based on RdRp sequence type
Are these viruses from a common-source?	Sequence comparison of a representative genome region, e.g RdRp or capsid, preferably combined with epidemiological data
Are these viruses antigenically related?	Antigenic tests
Are there differences in virulence of viruses belonging to different genotypes?	Genotyping combined with clinical/ epidemiological data using standardised methods and well-designed epidemiological studies or a model system
Are these emerging viruses?	Systematic molecular typing data combined with epidemiological information

I and 8 belonging to genogroup II (Table 5; Fig. 3) (Koopmans *et al.*, 2001). Viruses belonging to one genotype had >80% amino-acid similarity across the capsid protein. By a similar approach, Ando *et al.* (2000) proposed a numerical system for genotypes based on phylogenetic grouping according to genetic relatedness in the major capsid protein. NLVs have been found that are almost equidistant from GGI and GGII, but were arbitrarily assigned to GGII because they had the highest similarity with the GGII viruses in a region at the start of the capsid gene (Fig. 3, strain designated GGNANA) (Vinjé and Koopmans, 2000). Ando *et al.* (2000) proposed to assign these viruses to a new genogroup. By analogy, the SLVs have been divided into 4 genotypes, belonging to 2 genogroups (Vinjé *et al.*, 2000b; Clarke and Lambden, 2000; Jiang *et al.*, 1997, 1999; Noel *et al.*, 1997; Liu *et al.*, 1995; Numata *et al.*, 1997).

Besides humans, animals may also carry caliciviruses related to the NLVs and SLVs. In cattle and pigs, viral sequences were found which cluster with GGI NLV (Jena virus,

Table 5

Classification of current genogroups and genotypes of NLV from humans (adapted from Green *et al.*, 2001).

GG	genotype	Reference virus	Examples
I	1	Norwalk/1968/US	KY/89/JPN
	2	Southampton/1991/UK	White Rose, Crawley
	3	Desert Shield/395/1990/SA	Birmingham291
	4	Chiba 407/1987/JP	Thistle Hall, Valetta, Malta
	5	Musgrove/1989/UK	Butlins
	6	Hesse 3/1997/GE	Sindlesham, Mikkeli, Lord Harris
	7	Winchester/1994/UK	Lwymontley
II	1	Hawaii/1971/US	Wortley, Girlington
	2	Melksham/1994/UK	Snow Mountain, Melksham
	3	Toronto 24/1991/CA	Mexico, Auckland, Rotterdam
	4	Bristol/1993/UK	Lordsdale, Camberwell, Pilgrim, SymGreen, Grimsby
	5	Hillingdon/1990/UK	White river, Welterhof
	6	Seacroft/1990/UK	
	7	Leeds/1990/UK	Gwynedd, Venlo, Creche
	8	Amsterdam/1998/NL	
III?	1	Alphatron/1998/NL	

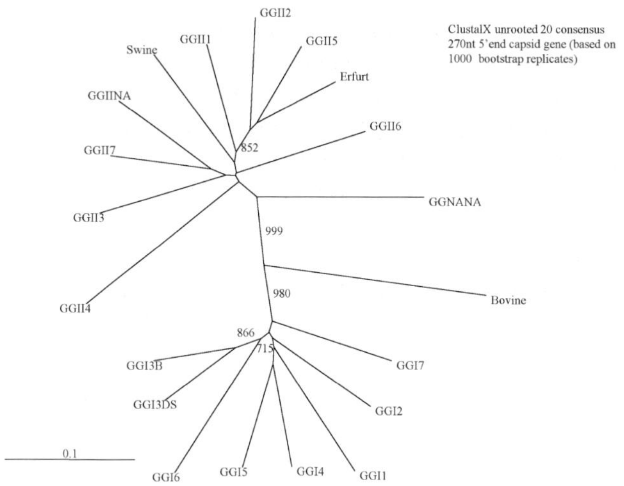

Fig. 3. Phylogenetic tree showing relationships between currently recognised clusters of NLV found in animals and humans, based on currently available NLV partial capsid gene sequences from Genbank or from a European database of published and unpublished NLV sequences. This includes published and unpublished sequences from France, Spain, the UK, The Netherlands, Germany, Finland, Sweden, and Denmark. This database will be described elsewhere. The alignments are based on sequences from the following variants: bovine NLV (n = 3); swine NLV (n = 2); GGI.1 Hawaii-like viruses (n = 38); GGII.2, Melksham-like viruses (n = 11); GGII.3 Mexico-like viruses (n = 50); GGII.4, Bristol-like viruses (n = 85); GGII.5 Hillingdon-like viruses (n = 16), GGII.6 Seacroft-like viruses (n = 23), GGII.7 Leeds-like viruses (n = 7); GGII.8 Amsterdam-like viruses (n = 8); GGNANA Alphatron-like viruses (n = 3); Erfurt-like viruses (n = 9); GGI.1 Norwalk-like viruses (n = 26); GGI.2 Southampton-like viruses (n = 10); GGI.3 Desert Shield-like viruses (n = 11) and Birmingham-like viruses (n = 13); GGI.4 Queens Arms-like viruses (n = 49); GGI.5 Musgrove-like viruses (n = 6); GGI.6 Sindlesham-like viruses (n = 3); and GGI.7 Winchester-like viruses (n = 3).

NA), GGII NLV (swine NLV) and SLV (swine SLV PEC) (Guo *et al.*, 1999; Dastjerdi *et al.*, 1999, 2000; Liu *et al.*, 1999; Sugieda *et al.*, 1998; van der Poel *et al.*, 2000). The genetic composition of the PEC virus suggests its placement in a separate GG within the genus SLV (39% aa identity with SLV). For the cattle NLVs, Ando *et al.* (2000) proposed to assign them to a distinct GG, although the reported sequences suggest that placement within the GGI viruses is more justified (Dastjerdi *et al.*, 2000).

What does this divergence mean for taxonomical grouping? Since caliciviruses have some similarities with viruses belonging to the family *Picornaviridae*, we could look at these viruses for comparison. The family *Picornaviridae* has been divided into 6 genera, with viruses as diverse as the poliovirus (genus *Enterovirus*), and foot-and-mouth-disease virus (genus *Aphthovirus*). Between genera, genomes differ by more than 55% in nts and 64% in aa (less than 36% aa identity), and have less than 33% aa homology across the VP1 protein sequence, which encodes the most external and immunodominant of the picornavirus proteins (Oberste *et al.*, 1998, 1999, 2002). Within a genus, different species of viruses share 44-58% nt in the VP1 sequence (aa identity 34-55%). This is in the range of differences seen between caliciviruses belonging to different genogroups (37-44% aa identity across the capsid gene). We have described a variant NLV (Hu/NLV/Alphatron/98/NL) which had 52% maximum sequence identity across the capsid with known NLVs and clustered almost equidistantly to GGI and GGII viruses (Vinjé and Koopmans, 2000). Picornavirus serotypes generally have >85% aa identity across the VP1 gene, which is in the range of the cut-off for calicivirus genotype (>80% aa identity) (Table 6) (Oberste *et al.*, 1999).

Table 6

Comparison of genetic similarities between viruses belonging to the family *Picornaviridae* and *Caliciviridae* in relation to taxonomic grouping. Indicated are % sequence similarities with other viruses in the family, and the ensuing taxonomic placement of a strain with that sequence.

Sequence similarity	Taxonomic grouping	
	Picornavirus	Calicivirus
45% nt, 34% aa	genus	genus
44-58% nt, 34-55% aa	species	genogroup
>85% aa	serotype	genotype

An important step in the use of sequence data for molecular epidemiological studies is the comparison of sequences and the subsequent analysis of their evolutionary relationships. This is done by making the best possible alignments. Aligning highly diverse sequences may be a challenging task. For caliciviruses, we have used the conserved motifs within the amplified diagnostic RT-PCR fragment (in the RdRp gene; Fig. 2) to check for the correctness of the alignments. After alignment, sequences are imported into one of several possible programmes for estimating relationships between viruses. Most commonly used are distance-based methods, in which the differences between sequences as expressed in percentage of nucleotide divergence are translated into an evolutionary distance. These methods assume that all positions across a sequence evolve at a constant rate, which clearly is not the case for viral genomes (Oberste *et al.*, 1999). Therefore, phylogenetic trees drawn need to be interpreted with caution. A modified distance based method designed by Van de Peer *et al.* (2000) may offer some advantages as it allows the calculation of position-specific evolutionary rates for a specified reference sequence database. A problem for the NLVs however remains the extreme diversity, even when focussing on the polymerase gene as one of the most conserved regions.

For the majority of caliciviruses studied by Vinjé *et al.* (2000b) and Green *et al.* (2000), the placement in phylogenetic trees was consistent, regardless which region of the genome or which clustering method was used for the phylogenetic inference. So far, most molecular epidemiology has been done on the basis of sequence analysis of a small part of the RdRp-gene. A pragmatic reason for this is that most diagnostic RT-PCR assays target this region (Fig. 2). Somewhat surprisingly, this work has shown that there is an extreme diversity at the level of these conserved regions, with as much as 40% nucleotide differences within the conserved RdRp fragment within a genogroup (Wang *et al.*, 1994; Ando *et al.*, 1995; Green *et al.*, 1994; Norcott *et al.*, 1994; Lew *et al.*, 1994). Since little sequence information is available for the other parts of the genome, it remains to be seen whether other regions are more conserved and therefore may be better targets for generic detection assays.

Antigenic diversity

The most biologically relevant antigenic typing is serotyping, in which viruses are divided into serogroups or serotypes on the basis of their reactivity with non-neutralising or neutralising antibodies. Human caliciviruses have not been adapted to cell culture, however, and this handicap precludes the use of virus neutralisation assays. Despite this, there is some evidence of antigenic diversity between caliciviruses, although the precise relationships are not defined (Lewis *et al.*, 1995). Sequence diversity across (parts of) the major capsid protein gene (ORF2) is informative, as the ORF2 product makes up most of the virion structure, and therefore is exposed to the host's immune system (Jiang *et al.*, 1992c). Therefore, one could speculate that diversity across the capsid gene would be correlated with antigenic diversity. Prior to molecular typing, comparison of viruses from stools of different sources by solid phase immuno-electron microscopy (SPIEM) with convalesent sera had already provided evidence for the presence of antigenic differences between variants. The viruses analysed by Vinjé *et al.* (2000b) and Green *et al.* (2000) belonged to 12 distinguishable SPIEM types that all segregated into different genetic clusters or genotypes based on sequence analysis of the complete capsid. Similarly, differential reactivity was observed of convalescent sera with recombinant baculovirus – expressed antigens assays (Green *et al.*, 1993; Gray *et al.*, 1994; Okhuysen *et al.*, 1995, Monroe *et al.*, 1993). Correlation between antigenic type and genotype can also be studied from work done with recombinant VLP ELISAs. Antigen detection ELISAs have been developed by using hyperimmune sera from laboratory animals (rabbits, guinea pigs, chicken) immunised with recombinant VLPs. The reported specificities vary: Kobayashi *et al.* (2000) tested all positive stool specimens detected by an ELISA using reagents specific for Chiba virus, a GGII virus, and found only viruses closely related to the Chiba strain by molecular typing. This suggested that rVLP-based ELISAs based on a single antigen have a specificity that is too narrow for their use as a diagnostic method for virus in stool. In contrast, Marks *et al.* (2000) reported the use of an ELISA based on recombinant Lordsdale virus (GGII.4) that detected some viruses from GGI outbreaks, suggesting the existence of one or more cross-reacting epitopes between viruses belonging to different genogroups.

Several groups have studied reactivity of convalescent sera collected from persons with recent infections. Jiang *et al.* (1992c) reported limited cross-reactivity in serologic responses in volunteers infected with different viruses using the rNV ELISA. Similarly, Myrmel and Rimstad (2000) cloned the capsid of a GGI virus, and found reactivities in sera from humans that had tested negative in the rNV ELISA for antibody detection. In contrast, Noel *et al.* (1999) found extensive serological cross-reactivity in convalescent sera for GGII viruses belonging to different genotypes within GGII. Others found occasional cross-reactivity between GG, although the homologous response was far greater (Hale *et al.*, 1998; Treanor *et al.*, 1993).

Combined, the data suggest that viruses belonging to different genotypes are antigenically distinct, but induce antibodies that cross-react with viruses belonging to the same genogroup. Low level cross-reactivity occurs with viruses belonging to a different GG. It is conceivable that the cross-reactive antibodies become detectable after repeated infections only, and that the lesser reported serological cross-reactivity reported for GGI results from the lower likelihood of repeated exposure to GGI viruses (Tables 1-3). Which epitopes are involved and how the observed antigenic differences correlate with serotypes remains unclear. The capsid gene of human caliciviruses can be divided in two domains: the N terminal, conserved, shell domain that encodes for a part of the capsid protein that constitutes the core of the particle, and the protruding domain, encoding parts of the protein that are surface exposed. Within the protruding domain, a highly variable region has been defined (P2 region, residues 281-404 for Norwalk virus), that probably plays a role in virus antigenicity (Green *et al.*, 2001) (See also Section IV, Chapter 1 of this book).

Consequences of molecular diversity for diagnostics

An issue that has been raised at the first Calicivirus Workshop in Atlanta in 1999, and that has not yet been addressed in comparative studies is how the data obtained with the molecular detection and typing assays can be compared. The most commonly used primers map to a region in the RdRp gene, which is slightly different for each of the RT-PCR assays used (Fig. 2). This raises several questions:

Can the data obtained with different assays be compared?

Comparative evaluation of assays is needed to study congruence between the tests. Since all published assays reportedly have a high sensitivity and specificity, this issue may seem trivial. However, the great diversity of caliciviruses makes it difficult to do a full evaluation of a diagnostic test. In fact, most assays have been optimised by use of RNA from the most commonly circulating variants as a reference template rather than specimens covering the full range of variants. Thus, the sensitivity of the assays in detecting less common variants is not known and may differ greatly between assays and variants. This information is crucial in order to be able to compare data from different studies.

In a comparative analysis of 4 currently used protocols, in which we compiled a panel of stool samples covering a wide range of variants, we found that the overall performance of the different assays was very similar. Overall sensitivity of the assays ranged from 82 to 100%. There were, however, significant differences when looking at sensitivity of detection of viruses belonging to different genogroups or genotypes, and assays were less than optimal for detecting the GGI viruses. The sensitivity of detection of GGI viruses ranged from 45 to 100%, and for GGII from 64 to 100% (Vinjé *et al.*, 2002). This stresses the importance of targeted evaluation of the currently used diagnostic tests. Special attention needs to be put into implementing generic tests in diagnostic laboratories. The optimised protocols result from a trade-off between broadness and sensitivity. For individual variants, lower detection limits may be reached with increased stringency of the assay (e.g. by raising the annealing temperature), and laboratory workers not familiar with generic tests often wish to redesign the protocols. That may, however, result in a decreased sensitivity of detection of other lineages of viruses.

Do data obtained with a specific primer set change over time?

RNA viruses have sloppy polymerases, which allow incorporation of mismatches during the replication of the genome, thereby introducing mutations. As a consequence, the progeny of one virus typically consists of a swarm of closely related sequences ("quasispecies"). Therefore, by analogy with other RNA viruses, calicivirus genomes may drift over time. Ando *et al.* (2000) have suggested that the positivity rate of PCR assays used for virus detection in outbreaks of gastroenteritis goes down over time as a result of accumulation of mutations in the primer binding sites. Clearly, this depends on the target region: the reverse primer of our standard RT-PCR for virus detection and typing anneals to the highly conserved YGDD motif of the RdRp. These amino acids are essential in the function of the RNA polymerase, and therefore mutant viruses will not survive competition with the existing virus population.

How informative is phylogeny based on a short RNA fragment?

Currently, the overlap between the amplified fragments for the assays developed by Ando *et al.* (1995), Vinjé and Koopmans (1996), Norcott *et al.* (1994), LeGuyader *et al.* (1996), and Maunula *et al.* (1999) is a marginal 60 nucleotides. It goes without saying that this fragment is too small for reliable clustering. This is a problem in the case of suspected international common source outbreaks: viruses that are grouped as identical based on the short 60 nt fragment may in fact be quite different. To demonstrate this, we aligned a panel of 71 different sequences from our collection, and clustered them by our routine method, after deleting increasing amounts of nucleotides. Of the sequences that were different based on the complete sequence in our collection, 65-75% were grouped as identical when the sequences were truncated to the 60 nt minimal fragment, depending on which segment of the amplicon was deleted (Fig. 4). This does not mean that these comparisons are useless. It does, however, imply that

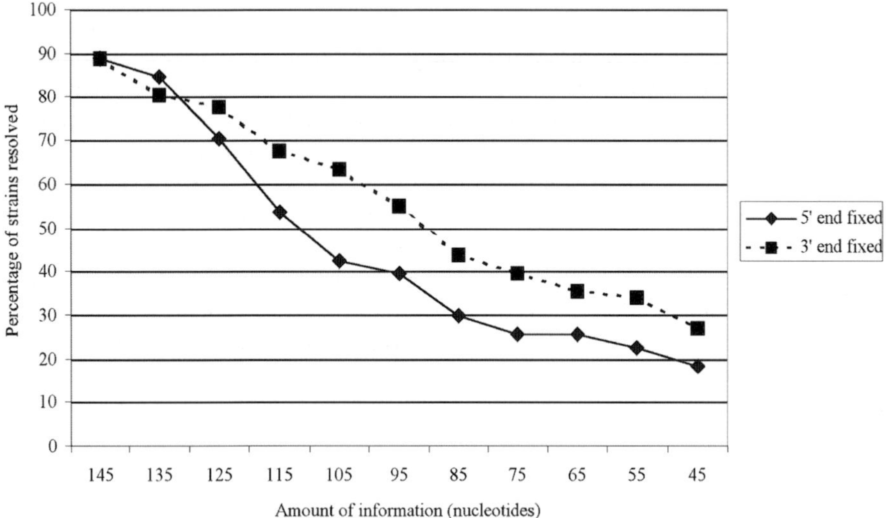

Fig. 4. Proportion of viral genomes matched as identical by fragment length of the amplified product in RT-PCR. Genome fragments in the RdRp gene (145 bp, Vinjé *et al.*, 1996) of 71 different sequences were aligned and clustered for phylogenetic analysis after increasing deletions of the fragment either from the 3'- (5'-fixed) or from the 5'-end (3'-fixed) of the PCR product.

viruses cannot be labelled as identical based on the small overlap alone, and that additional analysis will have to be done to confirm or disprove possible relatedness.

Use of molecular techniques to study calicivirus epidemiology

In recent years, a number of studies have been done in which molecular typing of caliciviruses was used (Tables 1-3). The studies addressed different segments of the population (Fig. 1). Overall, GGII viruses were detected most commonly, with the proportion of strains belonging to GGII ranging from 11-100%. There seems to be a difference in developing countries, where GGI viruses were reported more commonly (average 37%, range 10-63) as compared with data from countries with higher standards of living (average proportion GGI 15%, range 0-89). There are no data to explain this difference, although GGI viruses are more commonly found in water- or sewage-related contamination. Therefore, the higher rates of GGI viruses found in the studies in developing countries may result from exposure to contaminated water. The apparent association of GGI viruses with a waterborne mode of transmission suggests possible differences in stability of caliciviruses outside the host.

Molecular epidemiological studies of caliciviruses are still in their infancy. Very few community-based studies of gastroenteritis have been done, and even less have used molecular typing techniques. In our community studies, we found viruses belong-

540

ing to 9 different genotypes circulating within the same year in 1999 (de Wit *et al.*, 2001 a and b; Koopmans *et al.*, 2001). Viruses in genogroups II.4 (Bristol/Lordsdale-like viruses) and II.7 (Leeds-like viruses) were the most common. Outbreak surveys supported the notion that some genotypes are more common than others, and especially the GGII.4 viruses have been seen as predominant. The proportion of outbreaks attributed to viruses of this genotype ranges from 7 to 91% (Table 1). Based on this observation, recombinant GGII.4 (rGGII.4) antigen detection ELISA for screening stool samples has been introduced in some laboratories (Marks *et al.*, 2000). Similarly, typing methods selective for identification of viruses belonging to this genotype are used (Mattick *et al.*, 2000). Unfortunately, however, these assays do not provide insight into the molecular epidemiology until the antigenic relationships between viruses and their effect on sensitivity of antigen detection is resolved. Until then, we need molecular typing methods.

The dominance of some genotypes over others is not a constant feature, and several groups have reported seasonal fluctuation. In Fig. 5 the annual proportion of GGII.4 viruses is plotted for the passive outbreak surveillance system in The Netherlands. This shows that in this country the GGII.4 variants are common, but that the proportion of outbreaks attributed to this genotype fluctuates significantly.

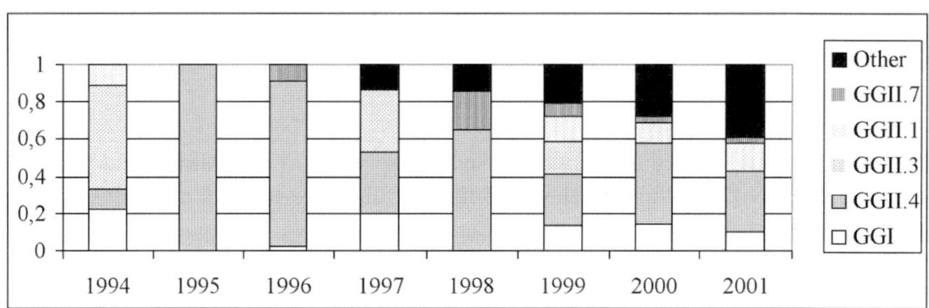

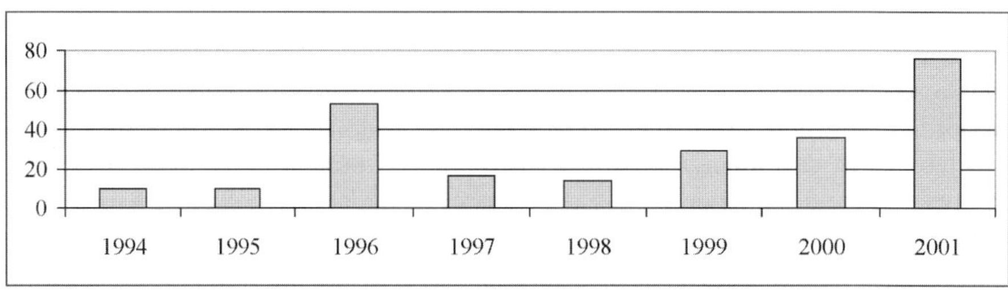

Fig. 5. Distribution of NLV genotypes in outbreaks of gastroenteritis in The Netherlands from 1994-2001 (top), and total number of outbreaks reported per year (bottom).

Emerging variants, mystery or misinterpretation?

In 1996, GGII.4 viruses were the dominant variant, as was observed in other countries. Based on this observation, Noel *et al.* (1999) suggested that this variant spread epidemically across the globe and might be a novel variant. Other studies, however, have shown that similar GGII.4 variants were present in the community long before the GGII.4 epidemic (Green *et al.*, 2002). Wright *et al.* (1998) found significant amino acid differences when comparing the genome of a common GGII.4 strain detected in 1988, with the 1996 "global strain". Could it be that GGII.4 viruses are the most common worldwide, and that the so-called epidemic resulted from a simultaneous flurry of studies which brought this to light? Or was there really something peculiar about the 1996 common strain, which might explain the increased number of outbreaks overall in The Netherlands (Fig. 5) and in the US when this virus was circulating (Vinjé *et al.*, 1997; Noel *et al.*, 1999)? When comparing distribution of genotypes for different age groups in a community-based case-control study of gastroenteritis in The Netherlands, NLVs of the genotype GII.4 were found more frequently in young children than other NLVs. In addition, GGII.4 viruses were not detected in persons without symptoms, whereas the other genotypes were (30% of GGII.3 positives; chi square, p = 0.03) (Koopmans *et al.*, unpublished observations). This is where the lack of a cell culture system and of an animal model is painfully clear: it would be quite interesting to compare biological behaviour of GGII.4 viruses from outbreaks during the "epidemic years" with those from historic collections.

Interestingly, through a collaborative approach to outbreak investigations across Europe, we saw another "epidemic" rise in the winter of 2000/2001 when a novel variant emerged almost simultaneously in 7 countries in Europe. In this case, the epidemic could be traced back to introductions of the virus by contaminated food and its subsequent spread (Koopmans *et al.*, manuscript in preparation). International food- or waterborne outbreaks have been described, and may be much more common than we think. Numerous outbreak reports illustrate the lack of awareness of the possibility of virus transmission via this route. In The Netherlands, we estimated that approximately 100.000 cases of NLV infection can be traced back to consumption of contaminated food, which is more than the number of cases of salmonellosis (de Wit *et al.*, submitted for publication). The CDC in Atlanta has published even higher estimates and suggests that 66% of all food-related illness in the US may be due to NLV infection (Mead *et al.*, 1999). Currently microbiological quality controls for food are not aimed at detection of viruses used and are known to be poor indicators of their presence.

Do animals constitute a reservoir for enteric caliciviruses?

As described above, viruses similar to the NLVs and SLVs in humans have been found in cattle and pigs. One of the popular hypothesis to explain the epidemic rise of a single variant was that these might be novel to the population and had recently been introduced from some animal reservoir(s). As attractive as it sounds, there are no

542

data to support this hypothesis at present. Despite an increasing number of studies in which all viruses found are subjected to genotyping, no "animal" caliciviruses have so far been found in humans. Nevertheless, this merely shows that animal-human transmissions are not the most common mode of transmission. It may still happen. Recent data from work with hepatitis E virus may be leading the way: HEV was considered to be an infection with a rather narrow host tropism, and a travel-associated disease in developed countries. Recently, however, viruses highly related to HEV were found in pigs, and people with professional exposure to pigs (vets, farmers) were shown to have increased prevalence of anti-HEV antibodies, suggesting possible zoonotic infections (Meng, 2000; Meng *et al.*, 1997, 1998). Indeed, a few human infections with viruses that group with the pig HEV variants have now been described, and there are indications of exposure in risk groups (Meng *et al.*, 2002). This could be a scenario for the caliciviruses: there is no evidence of very efficient zoonotic transmission, yet occasional species jumps may take place (in both directions). If so, chances are that – given the high incidence of calicivirus infections - dual infections occur, and recombinant viruses are generated which subsequently can spread through the community. Not surprisingly, recombination has been documented for caliciviruses (Vinjé *et al.*, 2000b). Additional studies, both serological and molecular studies are needed to clarify this interesting and important issue.

Use of molecular epidemiology in outbreak tracing

A promising application is the use of molecular typing to help establish modes of transmission in outbreak investigations. This has been used to support evidence for food-borne-, waterborne-, and environmental transmission routes (Table 7 and references therein). A more detailed review of the data in light of the issues described above shows that proving such transmission by molecular methods is not as easy as it may seem. The outbreaks in which the GGII.4–like viruses were detected may seem clear-cut. But given the high prevalence of these viruses in the community and in outbreak surveys, how can one disprove that related sequences are found by chance? Cheesebrough *et al.* (2000) reported that some sequence variation occurred in the cases in a protracted outbreak in a hotel in the UK that all had a GGII.4 virus. Isn't that remarkable in view of the highly conserved structure of the GGII.4 cluster? In 1996, in The Netherlands 60 consecutive seemingly unrelated outbreaks were caused by a virus in the GGII.4 cluster, and viruses with identical partial sequences of the polymerase gene were found in several countries all over the world in the same period (Fig. 6; Sweden, Japan, France, Australia, US) (Vinjé *et al.*, 1997; Noel *et al.*, 1999). The sequence variation described in the case strains by Cheesebrough *et al.* (2000) may therefore illustrate that this was not a protracted outbreak, but a series of outbreaks resulting from repeated introductions. Clearly, the hygienic conditions in the hotel were less than optimal, since widespread environmental contamination was demonstrated elegantly in this study. Parashar *et al.* (1998) linked cases to a food-handler based on identical sequence in a 81 nt region. However, the data presented in Fig. 4

Table 7

Outbreaks of gastroenteritis in which molecular typing has been used to build evidence (for abbreviations see legend of Table 1).

Setting	Country	n (sick)	Probable source	Level of evidence	Proposed explanation	Typing	Reference
University	US	125	Deli bar	identical virus in stool and food	asymptomatic food handler, sick child	GGII.4	Daniels *et al.*, 2000
Town	Finland	>1655	Drinking water	identical virus in stool and water	breakdown in chlorination	GGII.4 (stool, water) GGI (stool)	Kukkula *et al.*, 1999
Hospital	France	6	Drinking water	identical virus in stool and water	nr	GGII.2	Schvoerer *et al.*, 1999
Restaurant	UK	>52	Sick person	identical virus in all cases	aerosols from vomit	GGII.7	Marks *et al.*, 2000
Company	Finland	>509	Frozen raspberries	food-specifc odds ratio	spraying berries with surface water	GGII	Ponka *et al.*, 1999
Hotel	UK	850	Environmental contamination	virus in environmental swabs	contamination following an initial outbreak	GGII.4	Cheesebrough *et al.*, 2000
Hotel	UK	94	Kitchen sink	identical virus in food handler and case	food preparation in contaminated sink	GGII.4	Patterson *et al.*, 1997
Restaurant	Canada	66	Raspberries	identical virus in food and cases	contaminated food imported from Bosnia	GGI	Gaulin *et al.*, 1999
Company	US	>85	Sandwich	identical virus in food and foodhandler	asymptomatic food handler	GGII.2	Parashar *et al.*, 1998

544

show that this may be misclassification, since only 30-40% of different sequences were clustered as different based on a 81 bp region of the polymerase sequence. In the studies listed in Table 1 that reported GII.4 viruses and had typing information, identical viruses were found in a proportion of 7 out of 9 studies that had sequenced a fragment of less than 200 base pairs. In contrast, none of the studies in which a longer sequence was used reported identical viruses. These are examples to illustrate the importance of careful interpretation of the data: finding identical sequences is more convincing with a longer stretch of sequence, and if the viruses belong to clusters that are rare in the community. Background data on the diversity of data from viruses circulating in the community at the time of the outbreaks help support evidence. It remains essential that the findings are supported by epidemiological evidence, at least until sufficient outbreak reports have been done to know the reliability.

Special attention in outbreak tracing is needed for outbreaks in which the source of virus may be from sewage contamination. Typically, in such outbreaks mixtures of viruses may be detected within cases involved in a common-source outbreak. Some people may test positive for one virus, others for a different variant. It is clear from the outbreak surveys that this is a rule rather than an exception in the outbreaks in which sewage-contaminated oysters are involved (Ohyama *et al.*, 1999; Iritani *et al.*, 2000, 2002; Greening *et al.*, 2001; Maunula *et al.*, 1999). As a consequence, finding a virus in the food that differs from that in stools from cases does not necessarily disprove a causal link in these cases.

Surveys within a geographic region over several years

From the studies published so far, caliciviruses appear to have a worldwide distribution. In outbreak surveys, the GGII viruses are predominant. The most commonly detected variants overall are the GGII.4 viruses. An alignment of currently available partial GGII.4 sequences from Genbank, as well as unpublished sequences from a European database (Koopmans *et al.,* manuscript in preparation) is shown in Fig. 6. Variants detected in Europe and outside of Europe appear on every branch of the dendrogram, suggesting that there is no geographic clustering. Similarly, from data available over a 17-year period, there is no apparent drift in the sequences of this particular region. These data are consistent with worldwide endemic circulation of viruses belonging to this genotype. However, closer inspection of sub-branches shows that identical sequences often originate during the same season and from the same country. This is somewhat contrasting with the first observation, and suggests events that favour the distribution of progeny of a particular virus. That could happen through a seeding event (e.g. large common-source outbreak) or by selection of variants with increased fitness. For the latter explanation, one would expect that such variants would persist as for instance the 1995/6 " common strain" (Fig. 6) which was found world-wide in a relatively short span of time (Noel *et al.*, 1999). Only with prolonged, harmonised surveillance in which epidemiological and virological information is collected concurrently, and characterisation of carefully selected variants, we may learn whether these

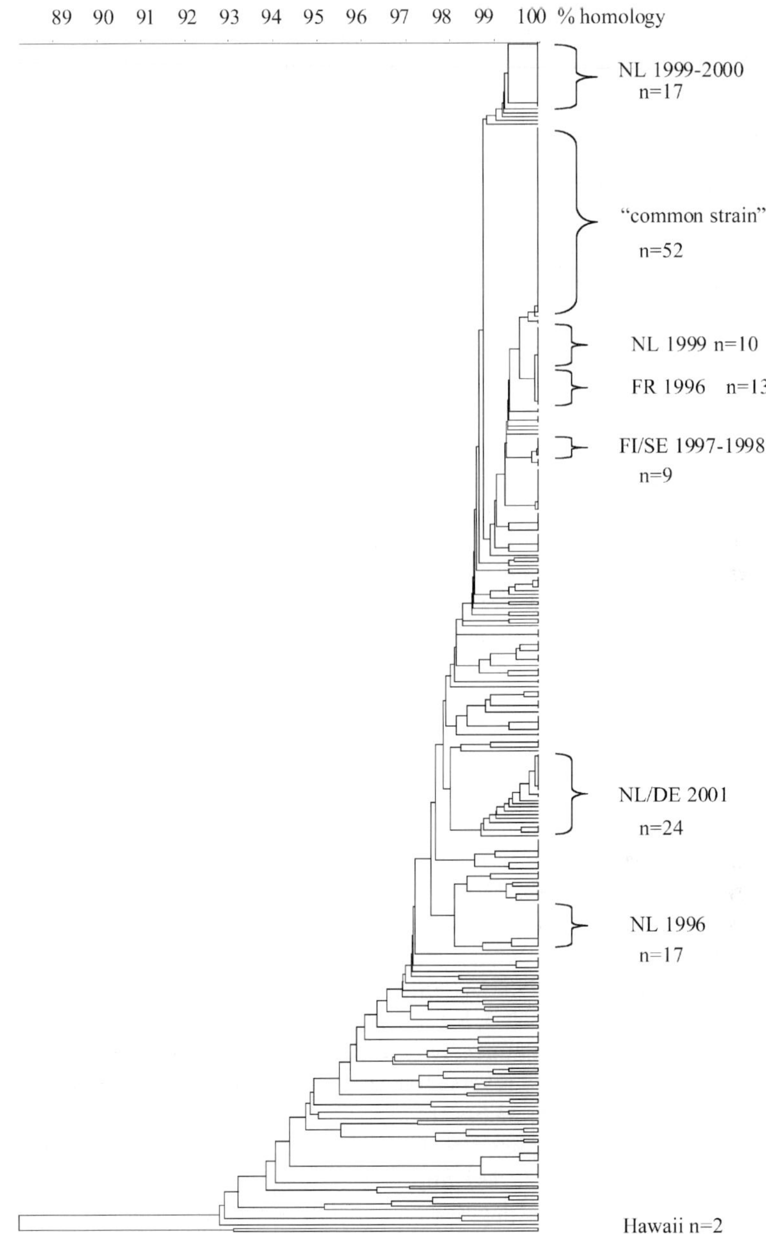

Fig. 6. Alignment of viruses within the GGII.4 cluster, based on nucleotide sequences of a fragment of the polymerase gene for 338 variants detected in association with outbreaks and sporadic cases of gastroenteritis worldwide. Sequences obtained were either downloaded from Genbank (all GGII.4 sequences, n = 55), or drawn from the European database of calicivirus sequences (see legend for Fig. 3). Sequences from GGII.1 (Hawaii-like viruses) were used as an outgroup (bottom of figure).

546

observations are correct. The answers to these questions are important, since they show us a lot about the modes of transmission of emerging enteric pathogens through the population.

Concluding remarks

Molecular epidemiological studies are increasingly popular in the field of clinical calicivirus research. Based on an (arbitrary) genetic typing system, caliciviruses are divided in genera, genogroups and genotypes. Most illness in humans appears to be associated with NLVs, although data on community-based studies are sparse. Therefore, the role of SLVs remains to be studied with similar approaches. The NLV GGII viruses dominate in all parts of the world. Several genotypes co-circulate, but the majority of infections is associated with only a few genotypes. The epidemiology of these genotypes is only beginning to be addressed. Future studies are needed in which systematic surveillance of outbreaks of gastroenteritis is done for prolonged periods of time and with harmonised methods in order to find explanations for the apparent emergence of calicivirus variants in populations, the fluctuations in their presence, virulence differences and modes of transmission including the possibility of zoonotic transmission. International collaborations coupled with rapid exchange of information on outbreak data are crucial if we want to look further than the horizons of our own countries. After all, with foodborne transmission being a major source of calicivirus infections, what use are borders?

References

Ando, T., Jin, Q., Gentsch, J. R., Monroe, S. S., Noel, J. S., Dowell, S. F., Cicirello, H. G., Kohn, M. A., and Glass, R. I. Epidemiologic applications of novel molecular methods to detect and differentiate small round structured viruses (Norwalk-like viruses). *J Med Virol*. 1995a; **47**:145-52.

Ando, T., Monroe, S. S., Gentsch, J. R., Jin, Q., Lewis, D. C., and Glass, R. I. Detection and differentiation of antigenically distinct small round- structured viruses (Norwalk-like viruses) by reverse transcription-PCR and southern hybridization. *J Clin Microbiol*. 1995b; **33**:64-71.

Ando, T., Noel, J. S., and Fankhauser, R. L. Genetic classification of "Norwalk-like viruses. *J Infect Dis*. 2000; **181 Suppl 2**:S336-48.

Atmar, R. L. and Estes, M. K. Diagnosis of noncultivatable gastroenteritis viruses, the human caliciviruses. *Clin Microbiol Rev*. 2001; **14**:15-37.

Bon, F., Fascia, P., Dauvergne, M., Tenenbaum, D., Planson, H., Petion, A. M., Pothier, P., and Kohli, E. Prevalence of group A rotavirus, human calicivirus, astrovirus, and adenovirus type 40 and 41 infections among children with acute gastroenteritis in Dijon, France. *J Clin Microbiol*. 1999; **37**:3055-8.

Cheesebrough, J. S., Green, J., Gallimore, C. I., Wright, P. A., and Brown, D. W.

Widespread environmental contamination with Norwalk-like viruses (NLV) detected in a prolonged hotel outbreak of gastroenteritis. *Epidemiol Infect.* 2000; **125**:93-8.

Clarke, I. N. and Lambden, P. R. Organization and expression of calicivirus genes. *J Infect Dis.* 2000; **181 Suppl 2**:S309-16.

Daniels, N. A., Bergmire-Sweat, D. A., Schwab, K. J., Hendricks, K. A., Reddy, S., Rowe, S. M., Fankhauser, R. L., Monroe, S. S., Atmar, R. L., Glass, R. I., and Mead, P. A foodborne outbreak of gastroenteritis associated with Norwalk-like viruses: first molecular traceback to deli sandwiches contaminated during preparation. *J Infect Dis.* 2000; **181**:1467-70.

Dastjerdi, A. M., Green, J., Gallimore, C. I., Brown, D. W., and Bridger, J. C. The bovine Newbury agent-2 is genetically more closely related to human SRSVs than to animal caliciviruses. *Virology* 1999; **254**:1-5.

Dastjerdi, A. M., Snodgrass, D. R., and Bridger, J. C. Characterisation of the bovine enteric calici-like virus, Newbury agent 1. *FEMS Microbiol Lett.* 2000; **192**: 125-31.

de Wit, M. A., Koopmans, M. P., Kortbeek, L. M., van Leeuwen, N. J., Vinjé, J., and van Duynhoven, Y. T. Etiology of gastroenteritis in sentinel general practices in the netherlands. *Clin Infect Dis.* 2001a; **33**:280-8.

de Wit, M. A., Koopmans, M. P., Kortbeek, L. M., Wannet, W. J., Vinjé, J., van Leusden, F., Bartelds, A. I., and van Duynhoven, Y. T. Sensor, a population-based cohort study on gastroenteritis in the Netherlands: incidence and etiology. *Am J Epidemiol.* 2001b; **154**:666-74.

Fankhauser, R. L., Noel, J. S., Monroe, S. S., Ando, T., and Glass, R. I. Molecular epidemiology of "Norwalk-like viruses" in outbreaks of gastroenteritis in the United States. *J Infect Dis.* 1998; **178**:1571-8.

Foley, B., O'Mahony, J., Hill, C., and Morgan, J. G. Molecular detection and sequencing of "Norwalk-like viruses" in outbreaks and sporadic cases of gastroenteritis in Ireland. *J Med Virol.* 2001; **65**:388-94.

Gaulin, C., Frigon, M., Poirier, D., and Fournier, C. Transmission of calicivirus by a foodhandler in the pre-symptomatic phase of illness. *Epidemiol. Infect.* 1999; **123**:475-8.

Gray, J. J., Cunliffe, C., Ball, J., Graham, D. Y., Desselberger, U., and Estes, M. K. Detection of immunoglobulin M (IgM), IgA, and IgG Norwalk virus- specific antibodies by indirect enzyme-linked immunosorbent assay with baculovirus-expressed Norwalk virus capsid antigen in adult volunteers challenged with Norwalk virus. *J Clin Microbiol.* 1994; **32**:3059-63.

Green, J., Vinjé, J., Gallimore, C. I., Koopmans, M., Hale, A., and Brown, D. W. Capsid protein diversity among Norwalk-like viruses. *Virus Genes* 2000; **20**: 227-36.

Green, K., Chanock, R., and Kapikian, A. Human caliciviruses. *In*: Knipe D. M., Howley P. M. *et al.* (Eds). Fields Virology, 4th Edition. pp 841-874. Lippincott Williams and Wilkins, Philadelphia, 2001.

Green, K. Y., Ando, T., Balayan, M. S., Berke, T., Clarke, I. N., Estes, M. K., Matson,

D. O., Nakata, S., Neill, J. D., Studdert, M. J., and Thiel, H. J. Taxonomy of the caliciviruses. *J Infect Dis*. 2000; **181 Suppl 2**:S322-30.

Green, K. Y., Belliot, G., Taylor, J. L., Valdesuso, J., Lew, J. F., Kapikian, A. Z., and Lin, F. Y. A predominant role for Norwalk-like viruses as agents of epidemic gastroenteritis in Maryland nursing homes for the elderly. *J Infect Dis*. 2002; **185**:133-46.

Green, K. Y., Lew, J. F., Jiang, X., Kapikian, A. Z., and Estes, M. K. Comparison of the reactivities of baculovirus-expressed recombinant Norwalk virus capsid antigen with those of the native Norwalk virus antigen in serologic assays and some epidemiologic observations. *J Clin Microbiol*. 1993; **31**:2185-91.

Green, S. M., Dingle, K. E., Lambden, P. R., Caul, E. O., Ashley, C. R., and Clarke, I. N. Human enteric Caliciviridae: a new prevalent small round-structured virus group defined by RNA-dependent RNA polymerase and capsid diversity. *J Gen Virol*. 1994; **75**:1883-8.Green, S. M., Lambden, P. R., Caul, E. O., and Clarke, I. N. Capsid sequence diversity in small round structured viruses from recent UK outbreaks of gastroenteritis. *J Med Virol*. 1997; **52**:14-9.

Greening, G. E., Mirams, M., and Berke, T. Molecular epidemiology of 'Norwalk-like viruses' associated with gastroenteritis outbreaks in New Zealand. *J Med Virol*. 2001; **64**:58-66.

Guo, M., Chang, K. O., Hardy, M. E., Zhang, Q., Parwani, A. V., and Saif, L. J. Molecular characterization of a porcine enteric calicivirus genetically related to Sapporo-like human caliciviruses. *J Virol*. 1999; **73**:9625-31.

Hale, A., Mattick, K., Lewis, D., Estes, M., Jiang, X., Green, J., Eglin, R., and Brown, D. Distinct epidemiological patterns of Norwalk-like virus infection. *J Med Virol*. 2000; **62**:99-103.

Hale, A. D., Lewis, D. C., Jiang, X., and Brown, D. W. Homotypic and heterotypic IgG and IgM antibody responses in adults infected with small round structured viruses. *J Med Virol*. 1998; **54**:305-12.

Iritani, N., Seto, Y., Haruki, K., Kimura, M., Ayata, M., and Ogura, H. Major change in the predominant type of "Norwalk-like viruses" in outbreaks of acute nonbacterial gastroenteritis in Osaka City, Japan, between April 1996 and March 1999. *J Clin Microbiol*. 2000; **38**:2649-54.

Iritani, N., Seto, Y., Kubo, H., Haruki, K., Ayata, M., and Ogura, H. Prevalence of "Norwalk-like virus" infections in outbreaks of acute nonbacterial gastroenteritis observed during the 1999-2000 season in Osaka City, Japan. *J Med Virol*. 2002; **66**:131-8.

Jiang, X., Cubitt, W. D., Berke, T., Zhong, W., Dai, X., Nakata, S., Pickering, L. K., and Matson, D. O. Sapporo-like human caliciviruses are genetically and antigenically diverse. *Arch Virol*. 1997; **142**:1813-27.

Jiang, X., Wang, J., Graham, D. Y., and Estes, M. K. Detection of Norwalk virus in stool by polymerase chain reaction. *J Clin Microbiol*. 1992a; **30**:2529-34.

Jiang, X., Wang, M., Graham, D. Y., and Estes, M. K. Expression, self-assembly, and antigenicity of the Norwalk virus capsid protein. *J Virol*. 1992b; **66**:6527-32.

Jiang, X., Wang, M., Wang, K., and Estes, M. K. Sequence and genomic organization of Norwalk virus. *Virology.* 1993; **195**:51-61.

Jiang, X., Zhong, W., Kaplan, M., Pickering, L. K., and Matson, D. O. Expression and characterization of Sapporo-like human calicivirus capsid proteins in baculovirus. *J Virol Methods.* 1999; **78**:81-91.

Kang, G., Hale, A. D., Richards, A. F., Jesudason, M. V., Estes, M. K., and Brown, D. W. Detection of 'Norwalk-like viruses' in Vellore, southern India. *Trans R Soc Trop Med Hyg.* 2000; **94**:681-3.

Kawamoto, H., Yamazaki, K., Utagawa, E., and Ohyama, T. Nucleotide sequence analysis and development of consensus primers of RT- PCR for detection of Norwalk-like viruses prevailing in Japan. *J Med Virol.* 2001; **64**:569-76.

Kirkwood, C. D. and Bishop, R. F. Molecular detection of human calicivirus in young children hospitalized with acute gastroenteritis in Melbourne, Australia, during 1999. *J Clin Microbiol.* 2001; **39**:2722-4.

Kobayashi, S., Sakae, K., Natori, K., Takeda, N., Miyamura, T., and Suzuki, Y. Serotype-specific antigen ELISA for detection of Chiba virus in stools. *J Med Virol.* 2000; **62**:233-8.

Koopmans, M., Vinjé, J., Duizer, E., de Wit, M., and van Duijnhoven, Y. Molecular epidemiology of human enteric caliciviruses in The Netherlands. *Novartis Found Symp.* 2001; **238**:197-214; discussion 214-8.

Kukkula, M., Maunula, L., Silvennoinen, E., and von Bonsdorff, C. H. Outbreak of viral gastroenteritis due to drinking water contaminated by Norwalk-like viruses. *J Infect Dis.* 1999; **180**:1771-6.

Lambden, P. R., Caul, E. O., Ashley, C. R., and Clarke, I. N. Sequence and genome organization of a human small round-structured (Norwalk-like) virus. *Science* 1993; **259**:516-9.

Le Guyader, F., Estes, M., Hardy, M., Neill, F., Green, J., Brown, D., and Atmar, R. Evaluation of a degenerate primer for the PCR detection of human caliciviruses. *Arch. Virol.* 1996a;2225-35.

Le Guyader, F., Neill, F. H., Estes, M. K., Monroe, S. S., Ando, T., and Atmar, R. L. Detection and analysis of a small round-structured virus strain in oysters implicated in an outbreak of acute gastroenteritis. *Appl Environ Microbiol.* 1996b; **62**:4268-72.

Levett, P. N., Gu, M., Luan, B., Fearon, M., Stubberfield, J., Jamieson, F., and Petric, M. Longitudinal study of molecular epidemiology of small round-structured viruses in a pediatric population. *J Clin Microbiol.* 1996; **34**:1497-501.

Lew, J. F., Kapikian, A. Z., Valdesuso, J., and Green, K. Y. Molecular characterization of Hawaii virus and other Norwalk-like viruses: evidence for genetic polymorphism among human caliciviruses. *J Infect Dis.* 1994; **170**:535-42.

Lewis, D., Ando, T., Humphrey, C. D., Monroe, S. S., and Glass, R. I. Use of solid-phase immune electron microscopy for classification of Norwalk-like viruses into six antigenic groups from 10 outbreaks of gastroenteritis in the United States. *J Clin Microbiol.* 1995; **33**:501-4.

Lindqvist, R., Andersson, Y., Lindback, J., Wegscheider, M., Eriksson, Y., Tidestrom, L., Lagerqvist-Widh, A., Hedlund, K. O., Lofdahl, S., Svensson, L., and Norinder,

550

A. A one-year study of foodborne illnesses in the municipality of Uppsala, Sweden. *Emerg Infect Dis.* 2001; **7 (Suppl)**:S88-92.

Lipskaya, G. Y., Chervonskaya, E. A., Belova, G. I., Maslova, S. V., Kutateladze, T. N., Drozdov, S. G., Mulders, M., Pallansch, M. A., Kew, O. M., and Agol, V. I. Geographical genotypes (geotypes) of poliovirus case isolates from the former Soviet Union: relatedness to other known poliovirus genotypes. *J Gen Virol.* 1995; **76**:1687-99.

Liu, B. L., Clarke, I. N., Caul, E. O., and Lambden, P. R. Human enteric caliciviruses have a unique genome structure and are distinct from the Norwalk-like viruses. *Arch Virol.* 1995; **140**:1345-56.

Liu, B. L., Lambden, P. R., Gunther, H., Otto, P., Elschner, M., and Clarke, I. N. Molecular characterization of a bovine enteric calicivirus: relationship to the Norwalk-like viruses. *J Virol.* 1999; **73**:819-25.

Maguire, A. J., Green, J., Brown, D. W., Desselberger, U., and Gray, J. J. Molecular epidemiology of outbreaks of gastroenteritis associated with small round-structured viruses in East Anglia, United Kingdom, during the 1996-1997 season. *J Clin Microbiol.* 1999; **37**:81-9.

Marks, P. J., Vipond, I. B., Carlisle, D., Deakin, D., Fey, R. E., and Caul, E. O. Evidence for airborne transmission of Norwalk-like virus (NLV) in a hotel restaurant. *Epidemiol Infect.* 2000; **124**:481-7.

Mattick, K. L., Green, J., Punia, P., Belda, F. J., Gallimore, C. I., and Brown, D. W. The heteroduplex mobility assay (HMA) as a pre-sequencing screen for Norwalk-like viruses. *J Virol Methods* 2000; **87**:161-9.

Maunula, L., Piiparinen, H., and von Bonsdorff, C. H. Confirmation of Norwalk-like virus amplicons after RT-PCR by microplate hybridization and direct sequencing. *J Virol Methods* 1999; **83**:125-34.

Mayo, M. A. Virus Taxonomy - Houston 2002. *Arch. Virol.* 2002; **147**:1071-6.

Mead, P. S., Slutsker, L., Dietz, V., McCaig, L. F., Bresee, J. S., Shapiro, C., Griffin, P. M., and Tauxe, R. V. Food-related illness and death in the United States. *Emerg Infect Dis.* 1999; **5**:607-25.

Meng, X. J. Novel strains of hepatitis E virus identified from humans and other animal species: is hepatitis E a zoonosis? *J Hepatol.* 2000; **33**:842-5.

Meng, X. J., Halbur, P. G., Shapiro, M. S., Govindarajan, S., Bruna, J. D., Mushahwar, I. K., Purcell, R. H., and Emerson, S. U. Genetic and experimental evidence for cross-species infection by swine hepatitis E virus. *J Virol.* 1998; **72**:9714-21.

Meng, X. J., Purcell, R. H., Halbur, P. G., Lehman, J. R., Webb, D. M., Tsareva, T. S., Haynes, J. S., Thacker, B. J., and Emerson, S. U. A novel virus in swine is closely related to the human hepatitis E virus. *Proc Natl Acad Sci USA* 1997; **94**:9860-5.

Meng, X. J., Wiseman, B., Elvinger, F., Guenette, D. K., Toth, T. E., Engle, R. E., Emerson, S. U., and Purcell, R. H. Prevalence of antibodies to hepatitis E virus in veterinarians working with swine and in normal blood donors in the United States and other countries. *J Clin Microbiol.* 2002; **40**:117-22.

Moe, C. L., Gentsch, J., Ando, T., Grohmann, G., Monroe, S. S., Jiang, X., Wang,

J., Estes, M. K., Seto, Y., Humphrey, C., *et al.* Application of PCR to detect Norwalk virus in fecal specimens from outbreaks of gastroenteritis. *J Clin Microbiol.* 1994; **32**:642-8.

Monroe, S. S., Stine, S. E., Jiang, X., Estes, M. K., and Glass, R. I. Detection of antibody to recombinant Norwalk virus antigen in specimens from outbreaks of gastroenteritis. *J Clin Microbiol.* 1993; **31**:2866-72.

Mulders, M. N., Lipskaya, G. Y., van der Avoort, H. G., Koopmans, M. P., Kew, O. M., and van Loon, A. M. Molecular epidemiology of wild poliovirus type 1 in Europe, the Middle East, and the Indian subcontinent. *J Infect Dis.* 1995; **171**: 1399-405.

Mulders, M. N., van Loon, A. M., van der Avoort, H. G., Reimerink, J. H., Ras, A., Bestebroer, T. M., Drebot, M. A., Kew, O. M., and Koopmans, M. P. Molecular characterization of a wild poliovirus type 3 epidemic in The Netherlands (1992 and 1993). *J Clin Microbiol.* 1995; **33**:3252-6.

Myrmel, M. and Rimstad, E. Antigenic diversity of Norwalk-like viruses: expression of the capsid protein of a genogroup I virus, distantly related to Norwalk virus. *Arch Virol.* 2000; **145**:711-23.

Nakata, S., Honma, S., Numata, K. K., Kogawa, K., Ukae, S., Morita, Y., Adachi, N., and Chiba, S. Members of the family Caliciviridae (Norwalk virus and Sapporo virus) are the most prevalent cause of gastroenteritis outbreaks among infants in Japan. *J Infect Dis.* 2000; **181**:2029-32.

Noel, J. S., Fankhauser, R. L., Ando, T., Monroe, S. S., and Glass, R. I. Identification of a distinct common strain of "Norwalk-like viruses" having a global distribution. *J Infect Dis.* 1999; **179**:1334-44.

Noel, J. S., Liu, B. L., Humphrey, C. D., Rodriguez, E. M., Lambden, P. R., Clarke, I. N., Dwyer, D. M., Ando, T., Glass, R. I., and Monroe, S. S. Parkville virus: a novel genetic variant of human calicivirus in the Sapporo virus clade, associated with an outbreak of gastroenteritis in adults. *J Med Virol.* 1997; **52**:173-8.

Norcott, J. P., Green, J., Lewis, D., Estes, M. K., Barlow, K. L., and Brown, D. W. Genomic diversity of small round structured viruses in the United Kingdom. *J Med Virol.* 1994; **44**:280-6.

Numata, K., Hardy, M. E., Nakata, S., Chiba, S., and Estes, M. K. Molecular characterization of morphologically typical human calicivirus Sapporo. *Arch Virol.* 1997; **142**:1537-52.

O'Ryan, M., Salinas, A. M., Mamani, N., Matson, D. O., Jiang, X., and Vial, P. Detection of Norwalk and Mexico viruses, two human Caliciviruses in stools of Chilean children. *Rev Med Chil.* 1999; **127**:523-31.

Oberste, M. S., Maher, K., Kilpatrick, D. R., and Pallansch, M. A. Molecular evolution of the human enteroviruses: correlation of serotype with VP1 sequence and application to picornavirus classification. *J Virol.* 1999; **73**:1941-8.

Oberste, M. S., Maher, K., and Pallansch, M. A. Molecular phylogeny of all human enterovirus serotypes based on comparison of sequences at the 5' end of the region encoding VP2. *Virus Res.* 1998; **58**:35-43.

Oberste, M. S., Maher, K., Pallansch, M. A. Molecular phylogeny and proposed

552

classification of the simian picornaviruses. *J Virol.* 2002; **76**:1244-51.

Ohyama, T., Yoshizumi, S., Sawada, H., Uchiyama, Y., Katoh, Y., Hamaoka, N., and Utagawa, E. Detection and nucleotide sequence analysis of human caliciviruses (HuCVs) from samples in non-bacterial gastroenteritis outbreaks in Hokkaido, Japan. *Microbiol Immunol.* 1999; **43**:543-50.

Okhuysen, P. C., Jiang, X., Ye, L., Johnson, P. C., and Estes, M. K. Viral shedding and fecal IgA response after Norwalk virus infection. *J Infect Dis.* 1995; **171**:566-9.

Pang, X. L., Joensuu, J., and Vesikari, T. Human calicivirus-associated sporadic gastroenteritis in Finnish children less than two years of age followed prospectively during a rotavirus vaccine trial. *Pediatr Infect Dis J.* 1999; **18**: 420-6.

Parashar, U. D., Dow, L., Fankhauser, R. L., Humphrey, C. D., Miller, J., Ando, T., Williams, K. S., Eddy, C. R., Noel, J. S., Ingram, T., Bresee, J. S., Monroe, S. S., and Glass, R. I. An outbreak of viral gastroenteritis associated with consumption of sandwiches: implications for the control of transmission by food handlers. *Epidemiol Infect.* 1998; **121**:615-21.

Parks, C. G., Moe, C. L., Rhodes, D., Lima, A., Barrett, L., Tseng, F., Baric, R., Talal, A., and Guerrant, R. Genomic diversity of "Norwalk like viruses" (NLVs): pediatric infections in a Brazilian shantytown. *J Med Virol.* 1999; **58**:426-34.

Patterson, W., Haswell, P., Fryers, P. T., and Green, J. Outbreak of small round structured virus gastroenteritis arose after kitchen assistant vomited. *Commun Dis Rep Rev.* 1997; **7**:R101-3.

Ponka, A., Maunula, L., von Bonsdorff, C. H., and Lyytikainen, O. An outbreak of calicivirus associated with consumption of frozen raspberries. *Epidemiol Infect.* 1999; **123**:469-74.

Prasad, B. V., Hardy, M. E., Jiang, X., and Estes, M. K. Structure of Norwalk virus. *Arch Virol.* 1996; **12(Suppl.)**:237-42.

Prasad, B. V., Rothnagel, R., Jiang, X., and Estes, M. K. Three-dimensional structure of baculovirus-expressed Norwalk virus capsids. *J Virol.* 1994; **68**:5117-25.

Qiao, H., Nilsson, M., Abreu, E. R., Hedlund, K. O., Johansen, K., Zaori, G., and Svensson, L. Viral diarrhea in children in Beijing, China. *J Med Virol.* 1999; **57**: 390-6.

Schnagl, R. D., Barton, N., Patrikis, M., Tizzard, J., Erlich, J., and Morey, F. Prevalence and genomic variation of Norwalk-like viruses in central Australia in 1995-1997. *Acta Virol.* 2000; **44**:265-71.

Schreier, E., Doring, F., and Kunkel, U. Molecular epidemiology of outbreaks of gastroenteritis associated with small round structured viruses in Germany in 1997/98. *Arch Virol.* 2000; **145**:443-53.

Schvoerer, E., Bonnet, F., Dubois, V., Rogues, A. M., Gachie, J. P., Lafon, M. E., and Fleury, H. J. A hospital outbreak of gastroenteritis possibly related to the contamination of tap water by a small round structured virus. *J Hosp Infect.* 1999; **43**:149-54.

Sethi, D., Cumberland, P., Hudson, M. J., Rodrigues, L. C., Wheeler, J. G., Roberts, J. A., Tompkins, D. S., Cowden, J. M., and Roderick, P. J. A study of infectious intestinal disease in England: risk factors associated with group A rotavirus in children. *Epidemiol Infect.* 2001; **126**:63-70.

Sugieda, M., Nagaoka, H., Kakishima, Y., Ohshita, T., Nakamura, S., and Nakajima, S. Detection of Norwalk-like virus genes in the caecum contents of pigs. *Arch Virol.* 1998; **143**:1215-21.

Svenungsson, B., Lagergren, A., Ekwall, E., Evengard, B., Hedlund, K. O., Karnell, A., Lofdahl, S., Svensson, L., and Weintraub, A. Enteropathogens in adult patients with diarrhea and healthy control subjects: a 1-year prospective study in a Swedish clinic for infectious diseases. *Clin Infect Dis.* 2000; **30**:770-8.

Tompkins, D. S., Hudson, M. J., Smith, H. R., Eglin, R. P., Wheeler, J. G., Brett, M. M., Owen, R. J., Brazier, J. S., Cumberland, P., King, V., and Cook, P. E. A study of infectious intestinal disease in England: microbiological findings in cases and controls. *Commun Dis Public Health* 1999; **2**:108-13.

Traore, O., Belliot, G., Mollat, C., Piloquet, H., Chamoux, C., Laveran, H., Monroe, S. S., and Billaudel, S. RT-PCR identification and typing of astroviruses and Norwalk-like viruses in hospitalized patients with gastroenteritis: evidence of nosocomial infections. *J Clin Virol.* 2000; **17**:151-8.

Treanor, J. J., Jiang, X., Madore, H. P., and Estes, M. K. Subclass-specific serum antibody responses to recombinant Norwalk virus capsid antigen (rNV) in adults infected with Norwalk, Snow Mountain, or Hawaii virus. *J Clin Microbiol.* 1993; **31**:1630-4.

Vainio, K., Stene-Johansen, K., Oystein Jonassen, T., Bruu, A. L., and Grinde, B. Molecular epidemiology of calicivirus infections in Norway. *J Med Virol.* 2001; **65**:309-14.

Van de Peer, Y., Baldauf, S. L., Doolittle, W. F., and Meyer, A. An updated and comprehensive rRNA phylogeny of (crown) eukaryotes based on rate-calibrated evolutionary distances. *J Mol Evol.* 2000; **51**:565-76.

van der Avoort, H. G., Reimerink, J. H., Ras, A., Mulders, M. N., and van Loon, A. M. Isolation of epidemic poliovirus from sewage during the 1992-3 type 3 outbreak in The Netherlands. *Epidemiol Infect.* 1995; **114**:481-91.

van der Poel, W. H., Vinjé, J., van Der Heide, R., Herrera, M. I., Vivo, A., and Koopmans, M. P. Norwalk-like calicivirus genes in farm animals. *Emerg Infect Dis.* 2000; **6**:36-41.

Vinjé, J., Altena, S. A., and Koopmans, M. P. The incidence and genetic variability of small round-structured viruses in outbreaks of gastroenteritis in The Netherlands. *J Infect Dis.* 1997; **176**:1374-8.

Vinjé, J., Deijl, H., van der Heide, R., Lewis, D., Hedlund, K. O., Svensson, L., and Koopmans, M. P. Molecular detection and epidemiology of Sapporo-like viruses. *J Clin Microbiol.* 2000a; **38**:530-6.

Vinjé, J., Green, J., Lewis, D. C., Gallimore, C. I., Brown, D. W., and Koopmans, M. P. Genetic polymorphism across regions of the three open reading frames of "Norwalk-like viruses". *Arch Virol.* 2000b; **145**:223-41.

Vinjé, J. and Koopmans, M. P. Molecular detection and epidemiology of small round-structured viruses in outbreaks of gastroenteritis in the Netherlands.

J Infect Dis. 1996; **174**:610-5.

Vinjé, J., and Koopmans, M. Simultaneous detection and genotyping of "Norwalk-like viruses" by oligonucleotide array in a reverse line blot hybridization format. *J Clin Microbiol.* 2000; **38**:2595-601.

Vinjé, J., Vennema, H., Maunula, L., Bonsdorff, C-H., Hoehne, M., Schreier, E., Richards, A., Green, J., Brown, D., Noel, J., Monroe, S., de Bruin, E., Svensson, L., and Koopmans, M. An international collaborative study to compare RT-PCR assays for the detection and genotyping of "Norwalk-like viruses". Submitted, 2002.

Wang, J., Jiang, X., Madore, H. P., Gray, J., Desselberger, U., Ando, T., Seto, Y., Oishi, I., Lew, J. F., Green, K. Y., *et al.* Sequence diversity of small, round-structured viruses in the Norwalk virus group. *J Virol.* 1994; **68**:5982-90.

Waters, V., Ford-Jones, E. L., Petric, M., Fearon, M., Corey, P., and Moineddein, R. Etiology of community-acquired pediatric viral diarrhea: a prospective longitudinal study in hospitals, emergency departments, pediatric practices and child care centers during the winter rotavirus outbreak, 1997 to 1998. The Pediatric Rotavirus Epidemiology Study for Immunization Study Group. *Pediatr Infect Dis J.* 2000; **19**:843-8.

Wimmer, E., Hellen, C. U., and Cao, X. Genetics of poliovirus. *Annu Rev Genet.* 1993; **27**:353-436.

Wolfaardt, M., Taylor, M. B., Grabow, W. O., Cubitt, W. D., and Jiang, X. Molecular characterisation of small round structured viruses associated with gastroenteritis in South Africa. *J Med Virol.* 1995; **47**:386-91.

Wright, P. J., Gunesekere, I. C., Doultree, J. C., and Marshall, J. A. Small round-structured (Norwalk-like) viruses and classical human caliciviruses in southeastern Australia, 1980-1996. *J Med Virol.* 1998; **55**:312-20.

IV, 6. Calicivirus RNA recombination

David O. Matson

Infectious Diseases Section, Center for Pediatric Research, Norfolk, Virginia 23510

Introduction

In this review, I will discuss evidence for the occurrence of RNA recombination in *Caliciviridae*, both within and outside the family. Constraints on recombination provided by the genomic diversity of caliciviruses (CVs), as well as implications of recombination on the natural diversity of CV strains and the clinical and biologic significance of RNA recombination, also will be considered. First, I will review some features of CVs that affect understanding of recombination.

Overview of genomic structure in CVs

The CV genome is a positive-sense, single-stranded, polyadenylated RNA molecule of about 7500 nucleotides in length. CVs fall into four genera that differ in their genomic organization (Green *et al.*, 2000a) (Fig. 1). Norwalk-like viruses (NLVs) have three open reading frames (ORFs). ORF1 encodes a polyprotein cleaved during replication into a set of nonstructural proteins, ORF2 encodes the capsid protein, and ORF3 encodes a protein that appears to be a minor structural protein (Sosnovtsev and Green, 2000). Where studied, CVs have been shown to synthesize a positive-sense subgenomic RNA that begins at the 5' of the capsid gene and that is co-terminus with the genome (Meyers *et al.*, 1991; Neill and Mengeling, 1988; Sosnovtser and Green, Section IV, Chapter 2 of this book). Vesiviruses differ from NLVs in having a longer genome that in some Vesiviruses (e.g., Pan-1; Rinehart-Kim *et al.*, 1999), but not others (e.g, feline CV; Carter *et al.*, 1992), includes a longer ORF1 with an additional predicted protein at its N-terminus. ORF2 of vesiviruses is longer than that of NLVs, with the extra nucleotides at the 5' end of ORF2. This extra sequence encodes a protein fragment that must be post-translationally cleaved to agree with experimental data of *Vesivirus* virion structure (Prasad *et al.*, 1994). ORF3 of vesiviruses is about one-half the size of that of NLVs (~120 amino acids vs. 250-275 amino acids, respectively). In lagoviruses and Sapporo-like viruses (SLVs), the genes that are in ORF1 and ORF2 of NLVs and vesiviruses are fused into one longer ORF1. A gene comparable to that of ORF3 of NLVs also is present. An ORF in another frame at the 5' end of the capsid gene occurs among SLVs, but not in all SLV strains (Liu *et al.*, 1999; Jiang *et al.*, 1997).

556

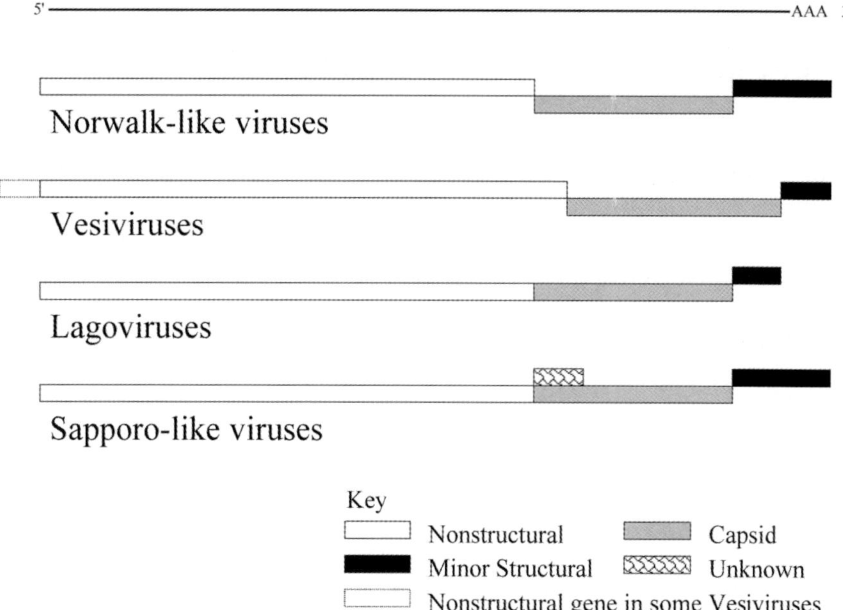

Fig. 1. Genome organizations of CV genera. The genome of Norwalk-like viruses (NLVs) has three open reading frames (ORFs), that 5' to 3' encode a nonstructural polypeptide (ORFI, white bars), the virion capsid gene (ORFII, grey bars), and a minor structural protein (ORFIII, black bars). The genomes of the other three genera differ from that of NLVs in the length of the ORFs, including an unique gene at the 5' end of ORF1 in at least one Vesivirus and a post-translationally cleaved N-terminus of the capsid protein in vesiviruses and Sapporo-like viruses (SLVs). Some sequence comparisons suggest that the longer ORF3 of NLV (and the comparable gene of SLVs) arose by intragenic recombination.

CV antigenic and genomic sequence diversity

The antigenic determinants (neutralization epitopes) that induce immunity against CVs presumably are located on the surface of the virion capsid. This capsid is composed of 180 copies of the capsid gene product, paired into 90 dimers (Prasad *et al.*, 1994; Prasad *et al.*, 1999). Despite the existence of just one capsid protein, CVs exhibit extensive antigenic diversity. In the best-characterized genus, *Vesivirus*, at least 40 distinct serotypes (neutralization types) exist, not including feline CVs and closely related strains, which among themselves are so diverse antigenically that definition of serotypes has been problematic (Lauritzen *et al.*, 1997; Hohdatsu *et al.*, 1999; Smith, 2000). The distinct *Vesivirus* serotypes are certainly determined by differences in nucleotide sequence of the capsid gene, resulting in differences in surface epitopes. The nucleotide differences sufficient to change the serotype are unknown, but likely to occur in a few distinct regions of the capsid gene (Neill, 1992; Rinehart-Kim *et al.*, 1999; Neill *et al.*, 2000).

When many capsid nucleotide sequences from different CV strains are simultaneously compared in phylogenetic analyses, the sequences within a genus fall into statistically significant clades (Berke *et al.*, 1997; Green *et al.*, 2000b). The biologic significance of these distinct clades is unknown. It is clear that such clades are related to differences in capsid gene sequences; sequence differences are less marked in the RNA polymerase gene: when RNA polymerase region sequences are analyzed in phylogenetic analyses, statistically significant differences similar to those observed among capsid gene sequences do not occur (Berke *et al.*, 1997). It is possible that separate capsid sequence clades within a genus indicate separate serotypes, but, even for *Vesivirus* capsid sequences, an insufficient number of strains have been analyzed to associate specific sequence differences with differences in serotype.

Evidence for recombination within CVs

With the description of statistically significant phylogenetic clades within CV genera, data were available to recognize strains that might be natural recombinants within CVs. Two examples are the well-characterized Argentine strain 320 (Arg320) and Snow Mountain virus (SMV), one of the prototype CVs, recognized to be recombinants when the RNA polymerase and capsid regions of these strains were characterized (Hardy *et al.*, 1997; Jiang *et al.*, 1999) (Fig. 2). At the time of publication, recombination was more certain for Arg320, because the sequence was derived from a single cDNA insert spanning the ~3.0 Kb at the 3' end of the genome, including the end of ORF1 and all of ORF2, ORF3, and the 3' non-coding region. In Arg320, the change of relative sequence identity occurred at the ORF1/capsid gene junction, indicating that the recombination occurred there. This site also was suggested (see below) to be the break-and-rejoin site for recombination between CVs and picornaviruses. For Arg320, the ORF1 sequence was closest to that of Lordsdale virus, among sequenced NLVs, and the capsid and ORF3 sequences were closest to those of Mexico virus. A similar change of relative sequence identity also occurred in SMV, when partial polymerase and capsid sequences were compared to reference Mexico and Melksham viruses. While SMV was likely also to be a recombinant virus, the capsid and RNA polymerase region amplicons of SMV were generated separately and that fact did not exclude the possibility of different sources of strains. Xi Jiang has confirmed the recombinant status of SMV by sequencing a single cDNA derived from a single RT-PCR amplicon (X Jiang, personal communication).

Potential origins of recombinants within CVs

Generation of recombinants within CVs requires biologic and molecular attributes of CVs. Outbreaks caused by multiple CV strains and co-infection by different HuCV strains occur (Matson *et al.*, 1995; Gray *et al.*, 1997; Reuter *et al.*, 2002). Infection of single cells simultaneously by two CVs implies absence of immune or molecular

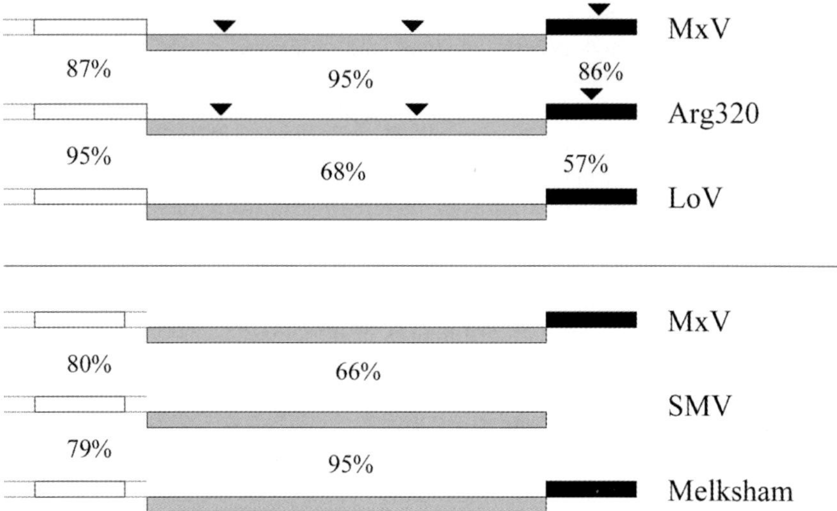

Fig. 2. Sequence comparisons of Argentine CV strain 320 (Arg320; top) and Snow Mountain virus (SMV; bottom), each with two other NLVs. Arg320 is significantly closer to Lordsdale virus (LoV) than Mexico virus (MxV) in the RNA polymerase genome region and significantly closer to MxV than LoV in the capsid and ORF3 genes. Deletions/insertions in the capsid and ORF3 genes of Arg320 are shared with MxV, but not LoV. SMV is equally close to MxV and Melksham virus in the known polymerase region, but much closer to Melksham than MxV in the capsid region. ORF3 sequence is not available for SMV. (Box shadings as in Fig. 1. Data from Jiang *et al.*, 1999; Hardy ME *et al.*, 1997; and DO Matson, unpublished.)

Fig. 3. Sequence comparisons of the first 5' genomic 40 nt of a CV strain (ID="A" sequence for this Fig.) and of 40 nt near the 5' end of that strain's capsid gene (ID="B" sequence for this Fig.). Each A or B sequence is from a CV with a known complete genome sequence, is on a single line, and is repeated in two columns. In the left-hand column, each A or B sequence is compared with the first 40 nt of the Norwalk virus genome, i.e., the Norwalk virus A sequence (Jiang *et al.*, 1993). In the second column, the A or B sequence is compared with the first 40 nt (A sequence) of a prototype sequence for that genus. Within a genus, the A sequences are listed first and the B sequences given next. "-" indicates the nt at that site is identical to that in the comparison sequence. For EBHSV, the "*" indicates a residue I inserted into the 5' noncoding region sequence to improve alignment with the comparison sequence, RHDV Meyers. For the Norwalk-like (NLV) sequences, the two columns are identical. For each CV strain evaluated, A and B sequences of that strain showed sequence conservation. These comparisons demonstrate that A and B sequences of NLVs are highly conserved, with the strongest conservation being in the first 20 nt of the two regions. "A" sequences of NLV strains are much less conserved with homologous A or B sequences of strains in the other CV genera. The observation that the AUG (□) of the A and B regions for a single strain fall at the same position in the

```
Genus                Genomic    RNA Sequence                              RNA Sequence
ID Host/Strain       Region     Hu/NV Jiang For Comparison                First Strain of Genus For Comparison

Norwalk-like
A  Hu/NV Jiang        1-40       GUGAAUGAUGAUGGCGUCAAAAGACGUCGUUCCUACUGCU  GUGAAUGAUGAUGGCGUCAAAAGACGUCGUUCCUACUGCU
A  Hu/NV Schreier     1-40       -----------------G-----------G-A---AAC    -----------------G-----------G-A---AAC
A  Hu/Southampton     1-40       -----------------G-----------G-A---AA-    -----------------G-----------G-A---AA-
A  Hu/Chiba           1-40       -----------------G-----------G-AG--A--    -----------------G-----------G-AG--A--
A  Hu/Lordsdale       1-40       -------A---------U--C----CUUCCG--G-C---   -------A---------U--C----CUUCCG--G-C---
A  Hu/Camberwell      1-40       -------A---------U--C----CUUCCG--G-C---   -------A---------U--C----CUUCCG--G-C---
A  Hu/GII Clarke      1-40       -------A---------U--C----CUUCCG--G-C---   -------A---------U--C----CUUCCG--G-C---
A  Hu GII Maryland    1-40       -------A---------U--C----CUUCCG--G-C---   -------A---------U--C----CUUCCG--G-C---
A  Hu/Hawaii          1-40       -------A---------U--C----CUUCCG--G-C---   -------A---------U--C----CUUCCG--G-C---
A  Bo/Jena            1-40       -------A--CUUU-A-G-U-UGGA-UC----A-GUUA    -------A--CUUU-A-G-U-UGGA-UC----A--GUUA
B  Hu/NV Jiang        5354-93    --A--------------U--G----CUACAU-A-CC-UC   --A--------------U--G----CUACAU-A-CC-UG
B  Hu/Chiba           5342-81    --A--------------U--G----CUACA--A-GC--A   --A--------------U--G----CUACA--A-GC--A
B  Hu/NV Schreier     5339-78    --A--------------U--G----C-CCA--AU-GC--  --A--------------U--G----C-CCA--AU-GC--
B  Hu/Southampton     5351-90    --A--------------U--G----C-CC--AA-GC---  --A--------------U--G----C-CC--AA-GC---
B  Hu/Lordsdale       5081-120   ---------A-------G--U----C-AAC--AU---A-   ---------A-------G--U----C-ACC--UCU-AU
B  Hu/Camberwell      5081-120   ---------A-------G-GU----C-AAC--AU---A-   ---------A-------G-GU----C-ACC--UCU-AU
B  Hu/GII Clarke      5081-120   ---------A-------G--U----C-AAC--AU---A-   ---------A-------G--U----C-ACC--UCU-AU
B  Hu/GII Maryland    5081-120   ---------A-------G-GU----C-AAC--AU---A-   ---------A-------G-GU----C-ACC--UCU-AU
B  Hu/Hawaii          5081-120   ---------A-------G-U-----C--CC--AU--AA-   ---------A-------G--U----C--CC--AU--AA-
B  Bo/Jena            5047-86    --A----A----A-UGAC-UGACAAGGA-GUGC-GGAAG   --A----A----A-UGAC-U-ACAAGG-UG-GC--GAA-

Sapporo-like
A  Hu/Manchester      1-40       ----U--G-U-GAUG--UUCCAAGCCAUUCAAGCCAAUAG   GUGAUUGGUUAGAUGGUUUCC AAGCCAUUCAAGCCAAUAG
A  Po/Saif            1-40       ----UC-UGAUG-CUAAUUGCCGUCCGUUGCCUAUUGGGC   -----C-UGAU-GCUAA--G-CGU--G--GCCUAUUGGGC
B  Hu/Manchester      5158-80    A-AGUGUUU--GAUG-AGGGC-AUG-CUCCAA-CCAGAGC   A--G-GUU-G------GGG---UGGC-C---C---GAGC
B  Po/Saif            5131-48    ---UUC-UGAUG-AG-CGCCUGCC-CAACCCGUUCGGUUG   ---U-C-UGAU-GA--CGC-UGCC---AC-CGUU-GG-UG

Lagovirus
A  RHDV Meyers        1-40       -----A-UUAUG-CG-CU-UGUCG--C-U-A-UGG-AUGA   GUGAAAGUUAUGGCGGCUAUGUCGCGCCUUACUGGCAUGA
A  RHDV Czech         1-40       -----A-U-AUG-CG-CU-UGUCG--C-U-A-UGG-AUGA   -----------------------------------------
A  RHDV Iowa 2000     1-40       -----A-U-AUG-CG--U-UGUCG--C-U-AUUGG-AUGA   ---------------U--------------U---------
A  RHDV Rossi         1-40       -----A-U-AUG-GG-CU-UGUCG--C-U-A-UGG-AUGA   -----------------------------------------
A  RHDV SD            1-40       -----A-U-AUG-CG-CU-UGUCG--C-U-A-UGG-AUGA   -----------------------------------------
A  EBHSV              1-40*      -----AUUAUGGC-GU-GCGUC-CGCC-U-G-GGCG--GC   -----*-------------U-GC--------C-UGC---G--G
B  RHDV Meyers        5296-335   --------U-AUG-AG-G----GCC---GCAG-GCCGCAAG   -----U----A--GC-AAG-C--UGCAG-GCCGCAAG
B  RHDV Czech         5296-335   --------A-AUG-AG-G----GCC---GCAG-GCCGCAAG   -----U----A--GC-AAG-C--UGCAG-GCCGCAAG
B  RHDV Iowa 2000     5296-335   --------U-AUG-AG-G----GCC--CACAG-GCCGCAAG   -----U----A--GC-AAG-C---ACAG-GCCGCAAG
B  RHDV Rossi         5296-335   --------U-AUG-AG-G----GCC---ACAG-GCCGCAAG   -----U----A--GC-AAG-C---ACAG-GCCGCAAG
B  RHDV SD            5296-335   --------U-AUG-AC-C----GCC--CACAG-GCCGCAAA   -----U----A--GC-AAC-C---ACAG-GCCGCAAC
B  EBHSV              5274-313   --------U-AUG-AG-GU--GCCU--GGC-GA-GC-CCUG   -----U----A--G--A-C-U--GGC-GAC-C-CCUG

Vesivirus
A  Fe/F4              1-40       --A--A--AAU-U-A-A--UGUCUCAAAC--UG-GCUUC    GUAAAGAAAUUUGAGACAUGUCUCAAACUCUGAGCUUC
A  Fe/F9              1-40       --A--A--AAU-U-A-A--UGUCUCAAAC--UG-GCUUC    -----------------------------------------
A  Fe/CFI             1-40       --A--A--AAU-U-A-A--UGUCUCAAAC--UG-GCUUC    -----------------------------------------
A  Fe/F65             1-40       --A--A--AAU-U-A-A--UGUCUCAAAC--UG-GCUUC    -----------------------------------------
A  PP/Pan-1           1-40       --A-----GA--UUGAG-U-UGGCUCAAACGCUCUCAAA    -----U--G-A--UGAG-U---G--------G--CUCAAAA
A  Po/A48             1-40       --A-----GA--UUGAG-U-UG-CUCAAACG-UCU-AAAA   -----U--C-A--UGAG-U---G--------C--CUCAAAA
A  Ca/Canine          1-40       --U----GA-AU-GC-UCUGCC-UCG-UC--UCGAGCUC    --U--U--G-AA--GCUUCUGCCA--GCU----CGAGC--
B  Fe/F4              5297-338   ---UUC--A-U-U-A-CAUGUGCU-AA-C-G-GCUAACG-   --GUUC---G-------CAUG--CUCA-CCUG-GCUAACGU
B  Fe/F9              5297-338   ---UUC--A-U-U-A-CAUGUGCU-AA-C-G-GCUAACG-   --GUUC---G-------CAUG--CUCA-CCUG-GCUAACGU
B  Fe/CFI             5294-335   ---UUC--A-U-U-A-CAUGUGCU-AA-C-G-GCUAACG-   --GUUC---G-------CAUG--CUCA-CCUG-GCUAACGU
B  Fe/F65             5297-336   --UUUCA-AAU-U-A-CAUGUGCU-AA-C-G-GCUAACG-   --UUUCA---------CAUG--CUCA-CCUG-GCUAACGU
B  PP/Pan-1           5649-88    UGUUUGAGAAUUA--CA-U-UG-CUACUAC--AC--GCU-   UGUUUGAG----A-CC--U---G--ACU----AC-CGC-U
B  Po/A48             5652-91    UGUUUGA-AAU-AA-CA-C-UG-CUACUAC--A---GCU-   UGUUUGA-----AACC--C---G--ACU----AU-CGC-U
B  Ca/Canine          5787-826   ---UU---CA--U-A-CU-UGGCU--CUA-CUUG-ACUCA   --GUUU--G-A-----CU-UG-CUCGCU-UCU---A--CA
```

homologous regions strengthens the conclusion that the existence of B regions is meaningful for CVs. These comparisons demonstrate that if sequence identity of A and B sequences is needed for (some types of) RNA recombination to occur, with the available sequence information, recombination would be more favored between NLV strains, between *Lagovirus* strains, or between feline *Vesivirus* strains, than between strains from different hosts within a CV genus or between strains representing different genera. Genbank accession numbers for the analyzed sequences are: Hu/NV Jiang (M87661), Hu/NV Schreier (AF093797), Hu/Southampton (L07418), Hu/Chiba (AB042808), Hu/Lordsdale (NC;001674), Hu/Camberwell (NC;002614), Hu/GII Clarke (X86557), Hu GII Maryland (AY032605), Hu/Hawaii (U07611), Bo/Jena (AJ011099), Hu/Manchester (X86560), Po/Saif (AF182760), RHDV Meyers (M67473), RHDV Czech (U54983), RHDV Iowa 2000 (AF258618), RHDV Rossi (X87607), RHDV SD (Z29514), EBHSV (NC;002615), Fe/F4 (D31836), Fe/F9 (M8679), Fe/CFI (U13992), Fe/F65 (AF109465), PP/Pan-1 (AF091736), Po/A48 (U76874), and Ca/Canine (AB070225). Abbreviations used include: Hu = human, NV = Norwalk virus, GII = genogroup II, Bo = bovine, Po = Porcine, RHDV = rabbit hemorrhagic disease virus, EBHSV = European brown hare syndrome virus, Fe = feline, PP = *Pan paniscus,* and Ca = canine.

interference. CV RNA also must have an attribute that permits/favors recombination, such as a site where errors in procession of RNA polymerase can occur. The subgenomic RNA is the most likely molecule to participate in recombination as noted above. The highly conserved 5' end sequence of the genome and at the 5' end of the capsid gene in NLVs is an obvious common target for CV RNA polymerase, for genomic and subgenomic RNA synthesis.

The sequence data indicated that recombination in strain Arg320 occurred at the ORF1/capsid gene junction where high sequence identity exists between the putative parent clades. The genomic sequence of 10 NLV strains has been determined. A comparison of sequence at the 5' genomic sequence with sequence near the 5' end of ORF2 shows a high degree of sequence identity (Fig. 3). No other region of an NLV strain genome shares this degree of identity, even closely, with the 5' end of the genome. In addition, sequence identity comparable to that shown between the NLV 5' end and near the 5' end of the NLV capsid may occur among CVs of a single genus from a single host, once enough strains are sequenced.

The "copy choice" model has been preferred for recombination of single-stranded RNA viruses, including picornaviruses and coronaviruses (Kirkegaard and Baltimore, 1986; Makino *et al.*, 1986; Lai and Cavanagh, 1997; Nagy and Simon, 1997). In the copy-choice model, recombination occurs during RNA replication when the viral RNA polymerase switches templates from the RNA derived from one strain (donor template) to the RNA derived from a second strain (acceptor template), at a highly conserved genome region, without releasing the nascent strand (Lai and Cavanagh, 1997). Models for RNA virus recombination have utilized two terminologies to describe the degree that features of the donor and acceptor templates are shared: homologous, aberrant homologous, and non-homologous types (Lai and Cavanagh, 1997) or sequence similarity-essential, similarity-assisted, and similarity-nonessential (Nagy and Simon, 1997). The putative parent clades of intrageneric CV recombinants have a long region of identical sequence and predicted stable hairpin structures at the proposed recombination site, which supports the classification of these recombinants as homologous or (at least) similarity-assisted. Interaction like that between genomic and subgenomic CV RNAs could occur by the same mechanisms for two genomic 5'ends, but the outcome of such recombination events would be hard to predict. Furthermore, if a virion contains genomic and subgenomic RNAs, recombination could occur in a generation after initial co-infection.

Evidence for recombination of CVs with other virus families

Evidence for recombination of CVs depends upon sequence comparisons. Upon sequencing a portion of a feline CV strain F9, Neill (1990) observed that (what later was designated) ORF1 contained significant sequence identity with picornaviruses. This significant identity was concentrated around certain amino acid motifs within ORF1 that are homologous to those within the non-structural region of picornaviruses, encoding, in order, 2C, 3C, and 3D genes. The order of these motifs and the approximate number of nucleotides between them were the same in both virus families (Fig. 4). The capsid gene of CVs also is homologous to the VP1 to VP4 capsid proteins of picornaviruses to the extent of a shared PPG amino acid motif in a relatively conserved 5' portion of the capsid gene(s), formation of capsomeres having polypeptide β-pleated sheets as a core structural element, and formation of a spherical virion capsid by the protein(s) (Prasad *et al.*, 1999). These findings led to the hypothesis that at some point in time CVs and picornaviruses were/are "recombination partners" (Dinulos and Matson, 1994).

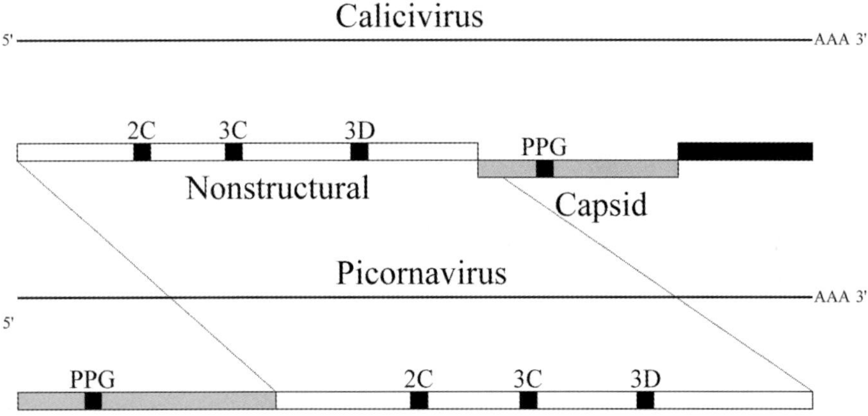

Fig. 4. Genome organization of CVs and picornaviruses showing the switch in order of the nonstructural and structural genes. In CVs, the nonstructural genes are encoded in ORF1; several of the genes were recognized because of the presence of significant sequence identity and consensus motifs like those known for picornaviruses (2C, 3C, and 3D). The capsid gene of CVs is encoded in ORF2 (NLVs and vesiviruses; *cf.* Fig. 1), which lies 3' to the nonstructural genes, and is marked by a "PPG" motif that signals a relatively conserved region between the families. In picornaviruses, this order of nonstructural-structural "gene cassettes" is reversed, with the order of motifs within the nonstructural peptide the same, and about the same number of nucleotides from each other in the genome. (Box shadings as in Fig. 1.)

In a recent report, Gibbs and Weiller (1999) suggested from sequence analyses that CVs (RNA genome, mostly in vertebrates) may have recombined with *Nanovirus* (ss DNA genome, plant virus) to generate (a) *Circovirus*(es) (Fig. 5). The sequence data, biologic information about host distribution of the putative parents, and pathways of DNA and RNA synthesis suggested to the authors that a *Nanovirus* crossed the invertebrate—vertebrate barrier, where, in a mixed infection, a 2C region cDNA of a CV recombined with the *Nanovirus*. A cDNA from a CV RNA limited to the 2C region was inserted into the *Nanovirus* DNA. The steps would not need to be contemporaneous, but would include: 1) *Nanovirus* infecting the vertebrate, 2) cDNA production from CV RNA, 3) excision of the 2C region, and 4) and fusion of the 2C region cDNA with the ligated

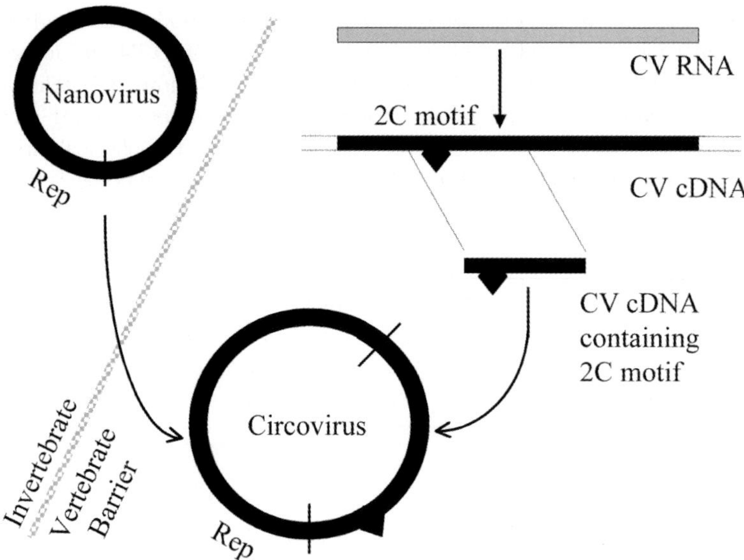

Fig. 5. Suggested origin of *Circovirus*(es) from recombination of *Nanovirus*(es) with a fragment of a CV genome. In this model, *Nanovirus* crossed the invertebrate-vertebrate barrier and infected a cell containing a CV 2C gene cDNA, which was inserted into the *Nanovirus* genome to derive a *Circovirus*. Therefore, reverse transcription of the CV RNA genome must have occurred and a fragment of the CV genome have been excised, either before or after fusion with the *Nanovirus* DNA genome. (Drawn from data in Gibbs and Weiller, 1999.)

Nanovirus DNA. That this set of steps occurred is suggested by significant sequence identities of two regions of circoviruses, one including the Rep (ligation initiation) gene of nanoviruses and the other 2C-like sequences closest to those of CVs. However, a reverse-transcriptase initiation site is not known in the CV genome. The possibility that recombination occurred in invertebrates is not excluded, given the existence of viruses in insects with close sequence identity to CVs (Govan *et al.*, 2000).

Genomic recombination as a mechanism for generating CV diversity

The differences in genome organization among CV genera imply different constraints on how RNA recombination might have occurred. For example, if the different genera are derived from a single "parental" genomic structure, different events must have occurred to generate the diversity of genome structures exhibited by the different CV genera—even within genera—for some genes are absent and others present. Alternatively, if, as discussed above, CVs are "recombination partners" with (an)other virus family(ies), then "convergent evolution" might explain the shared genomic features of CVs, despite multiple "parental" genomic structures.

Recombinants extend our knowledge of the genetic diversity within CVs. They also place constraints on methods that "genotype" CVs. If the RNA recombination of Arg320 and SMV is a common phenomenon, genotyping would be more difficult. For example, many reports of CV genotyping have been based upon sequence of the RNA polymerase region, due to its relativity high sequence conservation and relative ease of designing RT-PCR primers. In contrast, fewer capsid genes have been characterized (See also Jiang, Section IV, Chapter 4 of this book). The viral capsid protein is responsible for virion antigenicity and probably for inducing immunity. Genotyping of CVs based upon the RNA polymerase sequences clearly is not the best choice if recombination at the ORF1/capsid gene is common. In addition, it remains unclear whether additional recombination sites exist. Recombination in NLVs at the ORF1-ORF2 junction has been described upon the characterization of this genomic region for relatively few strains. Thus, one might discover other types of recombinants as more strains are characterized.

Clinical and biologic implications of recombination

One recombinant NLV, Arg320, was first recovered from ill children and adults in Argentina, the United States and the Netherlands (Jiang *et al.*, 1999; Jiang *et al.*, 2000; M Koopmans, personal communication). SMV (Morens *et al.*, 1979) and many very similar strains have been recovered from outbreaks of gastroenteritis worldwide. Many SMV-like NLVs have been characterized at the genomic level only in the RNA polymerase genome region. Each of these strains is a potential SMV-like recombinant, like the prototype, awaiting sufficient characterization of the capsid sequence to draw this conclusion. The possible widespread occurrence of recombinants in symptomatic persons suggests their ready infectivity in the host(s), their easy transmissibility, furthermore that recombination does not necessarily ablate virulence, and that recombinants are genetically and ecologically stable. Perhaps the most striking feature of Arg320 and SMV is that they and their associated illness were otherwise unremarkable. Their recombinant status was recognized only because their genomes were initially characterized in both the RNA polymerase region and capsid regions. Also, the two potential parental strains for each of Arg320 and SMV are within the range of genetic diversity of strains currently co-circulating. Therefore, the recombination event could have occurred recently, but not necessarily during the infection of the child from whom Arg320 was recovered. On the other hand, it would not be difficult to imagine that many CV strains currently co-circulating could have derived from remote recombination events in the past.

Recombination may permit CVs to escape host immunity quickly, analogous to antigenic shifts in influenza viruses, but by a different molecular mechanism. Recombination during calicivirus replication may be common or rare; it is possible to envision the generation of many non-viable or attenuated recombinants. The ORF1 polyprotein genes could persist in virus with a new capsid selected for by the host's immunity. Viable recombinants could be a model for laboratory manipulation of capsids (e.g., Neill *et al.*, 2000), including study of packaging constraints and antigenicity. In the two natural recombinants described above, ORF3, encoding a minor structural protein, segregated with the capsid gene in the recombinants. Whether recombination can occur with an ORF3 derived from another strain is unknown.

Summary

RNA recombination apparently contributed to the evolution of CVs. Nucleic acid sequence homology or identity and similar RNA secondary structure of CVs and non-CVs may provide a locus for recombination within CVs or with non-CVs should co-infections of the same cell occur. Natural recombinants have been demonstrated among other enteric viruses, including *Picornaviridae* (Kirkegaard and Baltimore, 1986; Furione *et al.*, 1993), *Astroviridae* (Walter *et al.*, 2001), and possibly rotaviruses (e.g., Desselberger, 1996; Suzuki *et al.*, 1998), augmenting the natural diversity of these pathogens and complicating viral gastroenteritis prevention strategies based upon traditional vaccines. Such is the case for CVs and *Astroviridae*, whose recombinant strains may be a common portion of naturally circulating strains. The taxonomic — and perhaps biologic — limits of recombination are defined by the suggested recombination of *Nanovirus* and CV, viruses from hosts of different biologic orders; the relationship of picornaviruses and CVs, viruses in different families, as recombination partners; and the intra-generic recombination between different clades of NLVs.

Acknowledgements

Thank you to my friend Xi Jiang for a special 10-year collaboration. I thank Tamas Berke for continued insights.

References

Berke T, Golding B, Jiang X, Cubitt WD, Wolfaardt M, Smith AW, Matson DO (1997). A phylogenetic analysis of the caliciviruses. *J. Med. Virol.* **52**: 419-424.

Carter MJ, Milton ID, Meanger J, Bennett M, Gaskell RM, Turner PC (1992). The complete nucleotide sequence of a feline calicivirus. *Virology* **190**: 443-448.

Desselberger U (1996). Genome rearrangements of rotaviruses. *Adv. Virus Res.* **46**:69-95.

Dinulos MB, Matson DO (1994). Recent developments with human caliciviruses. *Pediatr. Infect. Dis. J.* **13**:998-1003.

Furione M, Guillot S, Otelea D, Balanant J, Candrea A, Crainic R (1993). Polioviruses with natural recombinant genomes isolated from vaccine-associated paralytic poliomyelitis. *Virology* **196**: 199-208.

Gibbs MJ, Weiller GF (1999). Evidence that a plant virus switched hosts to infect a vertebrate and then recombined with a vertebrate-invecting virus. *Proc. Natl.*

Acad. Sci. U.S.A. **96**: 8022-8027.

Govan VA, Leat N, Allsopp M, Davison S (2000). Analysis of the complete genome sequence of acute bee paralysis virus shows that it belongs to the novel group of insect-infecting RNA viruses. *Virology* **277**: 457-463.

Gray JJ, Green J, Cunliffe C, Gallimore C, Lee JV, Neal K, Brown DW (1997). Mixed genogroup SRSV infection among a party of canoeists exposed to contaminated recreational water. *J. Med. Virol.* **52**: 425-429.

Green J, Vinjé J, Gallimore CI, Koopmans M, Hale A, Brown DW (2000). Capsid protein diversity among Norwalk-like viruses. *Virus Genes* **20**: 227-236.

Green KY, Ando T, Balayan MS, Clarke IN, Estes MK, Matson DO, Nakata S, Neill JD, Studdert MJ, Theil H-J (2000). Caliciviridae. In: van Regenmortel MHV, Fauquet CM, Bishop DHL, Carstens EB, Estes MK, Lemon SM, Maniloff J, Mayo MA, McGeoch DJ, Pringle CR, Wickner RB, eds. *Virus Taxonomy: Classification and Nomenclature of Viruses*. Seventh Report of the International Committee on Taxonomy of Viruses, pp. 725-735. San Diego: Academic Press.

Hardy ME, Kramer SF, Treanor JJ, Estes MK (1997). Human calicivirus genogroup II capsid sequence diversity revealed by analyses of the prototype Snow Mountain agent. *Arch. Virol.* **142**: 1469-1479.

Hohdatsu T, Sato K, Tajima T, Koyama H (1999). Neutralizing features of commercially available feline calicivirus (FCV) vaccine immune sera against FCV field isolates. *J. Vet. Med. Sci.* **61**: 299-301.

Jiang X, Berke T, Zhong W, Dai X, Dunham D, Pickering LK, Matson DO (1997). Sapporo-like human caliciviruses are genetically and antigenically diverse. *Arch. Virol.* **142**: 1813-1827.

Jiang X, Wang M, Wang K, Estes MK (1993). Sequence and genomic organization of Norwalk virus. *Virology* **195**:51-61.

Jiang X, Espul C, Zhong WM, Cuello H, Matson DO (1999). Characterization of a novel human calicivirus that may be a naturally occurring recombinant. *Arch. Virol.* **144**: 2377-2387.

Jiang X, Wilton N, Zhong WM, Farkas T, Huang PW, Barrett E, Guerrero M, Ruiz-Palacios G, Green KY, Green J, Hale AD, Estes MK, Pickering LK, Matson DO (2000). Diagnosis of human caliciviruses by use of enzyme immunoassays. *J. Infect. Dis.* **181**: S349-S359.

Kirkegaard K, Baltimore D (1986). The mechanism of RNA recombination in poliovirus. *Cell* **47**: 433-443.

Lai MM, Cavanagh D (1997). The molecular biology of coronaviruses. *Adv. Virus Res.* **48**: 1-100.

Lauritzen A, Jarrett O, Sabara M (1997). Serological analysis of feline calicivirus isolates from the United States and United Kingdom. *Vet. Microbiol.* **56**: 55-63.

Liu BL, Lambden PR, Gunther H, Otto P, Elschner M, Clarke IN (1999). Molecular characterization of a bovine enteric calicivirus: relationship to the Norwalk-like viruses. *J. Virol.* **73**:819-825.

Makino S, Keck JG, Stohlman SA, Lai MMC (1986). High-frequency RNA recombination of murine coronaviruses. *J. Virol.* **57**: 729-737.

Matson DO, Mitchell DK, Van R, Jiang X, Pickering LK (1995). Enteric viral pathogens as causes of outbreaks of diarrhea among children attending day care centers during one year of observation. *Pediatr. Res.* **37**:826.

Morens DM, Zweighaft RM, Vernon TM, Gary GW, Eslien JJ, Wood BT, Holman RC, Dolin R (1979). A waterborne outbreak of gastroenteritis with secondary person-to-person spread. Association with a viral agent. *Lancet* **1**:964-966.

Meyers G, Wirblich C, Thiel HJ (1991). Genomic and subgenomic RNAs of rabbit hemorrhagic disease virus are both protein-linked and packaged into particles. *Virology* **184**:677-686.

Nagy PD, Simon AE (1997). New insights into the mechanisms of RNA recombination. *Virology* **234**: 1-9.

Neill JD (1990). Nucleotide sequence of a region of the feline calicivirus genome which encodes picornavirus-like RNA-dependent RNA polymerase, cysteine protease and 2C polypeptides. *Virus Res.* **17**: 145-160.

Neill JD (1992). Nucleotide sequence of the capsid protein gene of two serotypes of San Miguel sea lion virus: identification of conserved and non-conserved amino acid sequences among calicivirus capsid proteins. *Virus Res.* **24**: 211-222.

Neill JD, Mengeling WL (1988). Further characterization of the virus-specific RNAs in feline calicivirus infected cells. *Virus Res.* **11**:59-72.

Neill JD, Sosnovtsev SV, Green KY (2000). Recovery and altered neutralization specificities of chimeric viruses containing capsid protein domain exchanges from antigenically distinct strains of feline calicivirus. *J. Virol.* **74**: 1079-1084.

Prasad BV, Hardy ME, Dokland T, Bella J, Rossmann MG, Estes MK (1999). X-ray crystallographic structure of the Norwalk virus capsid. *Science* **286**: 287-290.

Prasad V, Matson DO, Smith AW (1994). Three-dimensional structure of the primate calicivirus. *J. Mol. Biol.* **240**: 256-264.

Reuter G, Farkas T, Berke T, Jiang X, Szucs G, Matson DO (2003). Molecular epidemiology of human calicivirus gastroenteritis outbreaks in Hungary, 1998 to 2000. *J. Med. Virol.,* in press.

Rinehart-Kim J, Zhong W-M, Jiang X, Smith AW, Matson DO (1999). Complete nucleotide sequence and genomic organization of a primate calicivirus, Pan-1. *Arch. Virol.* **144**: 199-208.

Smith AW (2000). Virus cycles in aquatic mammals, poikilotherms, and invertebrates. In: Hurst CJ, ed. *Viral Ecology*, pp. 447-491. San Diego: Academic Press.

Sosnovtsev SV, Green KY (2000). Identification and genomic mapping of the ORF3 and VPg proteins in feline calicivirus virions. *Virology* **277**: 193-203.

Suzuki Y, Gojobori T, Nakagomi O (1998). Intragenic recombinations in rotaviruses. *FEBS Letters* **427**: 183-187.

Walter JE, Briggs J, Guerrero ML, Matson DO, Pickering LK, Ruiz-Palacios G, Berke T, Mitchell DK (2001). Molecular characterization of a novel recombinant strain of human astrovirus associated with gastroenteritis in children. *Arch. Virol.* **146**: 2357-2367.

Webster RG, Laver WG, Air GM, Schild GC (1982). Molecular mechanisms of variation in influenza viruses. *Nature* **296**: 115-121.

Viral Gastroenteritis
U. Desselberger and J. Gray (editors)
© 2003 Elsevier Science B.V. All rights reserved

SECTION V

Astroviruses

Introduction

Human astroviruses are members of the *Astroviridae* family. They are non-enveloped viruses possessing a single stranded RNA of positive polarity as their genome (Matsui, 1997; Matsui and Greenberg, 2000). Negatively stained astroviruses appear by EM as spherical particles of 35-40 nm diameter with a characteristic 5-6 pointed star surface configuration which has provided the name for these viruses. Astroviruses are relatively easy to detect in human faeces as they are produced in significant numbers at the onset of illness (by contrast to human caliciviruses, see Section IV), and they have been found to be cultivatable *in vitro* (Lee and Kurtz, 1981; Brinker *et al.*, 2000). This has allowed detailed analysis of the viral replication cycle (Matsui, 1997; Willcocks *et al.*, 1999; Matsui and Greenberg, 2000).

Cryo-electron microscopy and image reconstruction of astroviruses have identified spherical particles of uniform size (330Å diameter) with 30 diploid spikes extending about 50Å from the surface and leading to an external diameter of the particle of approximately 420Å (Yeager and Matsui, 2000; Matsui *et al.*, 2001). These investigations clarified and extended earlier EM data obtained from cell culture grown virus (Risco *et al.*, 1995).

U Geigenmüller *et al.* have reviewed astrovirus replication (Section V, Chapter 1). The genome of astroviruses measures approximately 6.8 kilobases in length, excluding the poly A tract at the 3' end, and encodes 3 open reading frames (ORFs; Jiang *et al.*, 1993; Willcocks *et al.*, 1994) (Fig. 2 in Section V, Chapter 1). ORFs 1A and 1B at the 5' end encode the viral protease and RNA-dependent RNA polymerase, respectively, and ORF2 at the 3' third of the genome encodes a structural protein that is a precursor of the structural proteins of the mature virus (Belliot *et al.*, 1997a; Bass and Qui, 2000). ORFs 1A and 1B are overlapping by 70 nucleotides. This area is highly conserved among human astrovirus serotypes and contains a frame-shifting signal of a "shifty" heptamer (A AAA AAC); downstream there are sequences which form a stem loop structure (Marczinke *et al.*, 1994; Lewis and Matsui 1995, 1996) essential for frameshifting, as part of a replication strategy seen with other viruses (retroviruses, Jacks *et al.*, 1988; coronaviruses, Brierley *et al.*, 1989). I Brierley and M Visakovic have reviewed and updated this topic (Section V, Chapter 2).

A further characteristic of astrovirus replication in cells infected *in vitro* is the production of full length genomic and ORF2-specific subgenomic RNAs; the latter is mainly used for the production of large amounts of structural proteins (Monroe *et al.*,

1993). More detailed investigation of expression products of the ORF1A and deletions thereof identified a protease active domain at the N terminus and a nuclear localization signal domain at the C terminus (Matsui, *et al.*, 2001).

As astroviruses have a genome of single-stranded RNA of positive polarity and as replication *in vitro* after infection with virus particles had been demonstrated, it was tempting to explore the infectivity of naked RNA. This was shown to be possible by transfection of viral RNA of human astrovirus type 1 into BHK cells (which were easier to transfect) and propagation of progeny virus obtained from these cells in Caco-2 cells (Geigenmüller *et al.*, 1997). In the same publication it was shown that infectious RNA could also be transcribed from a full length cDNA clone of human astrovirus serotype 1. This achievement has allowed to identify the functions of sequence regions necessary to produce fully infectious virus (Matsui *et al.*, 2001). A replication and transcription scheme for astroviruses was constructed similar to the one explored in more detail for the life cycles of alphaviruses such as Sindbis virus (Matsui *et al.*, 2001). U Geigenmüller *et al.* have presented recent studies based on the use of infectious cDNA and of mutagenizal derivatives thereof (Section V, Chapter 1). Recently, an infectious cDNA clone has also been obtained from another astrovirus, avian nephritis virus (Imada *et al.*, 2000).

The development of sensitive tests for the presence of astrovirus, e.g. using group reactive monoclonal antibodies (Herrmann *et al.*, 1988) has led to the conclusion that astroviruses are the cause of more cases of childhood diarrhoea than previously assumed. This was confirmed by the development of RT-PCR detection assays which also proved not infrequent asymptomatic shedding of astroviruses in older children (Jonassen *et al.*, 1995; Mitchell *et al.*, 1995). Astroviruses have also been identified as the cause of major outbreaks of diarrhoea and vomiting (e.g. Oishi *et al.*, 1994).

Different serotypes of human astrovirus have been defined on the basis of immune electron microscopy, neutralization tests and type-specific EIAs (Lee and Kurtz, 1994; Noel *et al.*, 1995). So far 8 different serotypes have been identified, and it has been shown that differences in the sequences of RT-PCR products from a region within ORF2 correlated precisely with antigenic types determined by type-specific EIA (Noel *et al.*, 1995).

The study of astrovirus serotypes in various populations has demonstrated that type 1 was the most commonly detected, whilst types 6-8 were rarely found and types 2-5 at intermediate frequencies (Lee and Kurtz, 1994; Noel and Cubitt, 1994; Noel *et al.*, 1995). Age-stratified seroprevalence studies of neutralising antibodies to astroviruses types 1-7 in humans in the Netherlands showed that subjects had been infected by several serotypes and that the antibody prevalence against several serotypes exceeded their relative isolation rate, suggesting that many infections are of milder clinical consequence or asymptomatic (Koopmans *et al.*, 1998). Human astroviruses were found to a very high percentage in shellfish and mussel (Le Guyader *et al.*, 2000).

Cocirculation of multiple astrovirus serotypes was shown by several groups in different locations (Noël *et al.*, 1995; Cunliffe *et al.*, 1998; Naficy *et al.*, 1999; Mustafa *et al.*, 2000; Monroe *et al.*, 2001), and the different types are seen to cluster in 2 genogroups (A: types 1-5; B: types 6, 7; Belliot *et al.*, 1997b). The assignment of type 8 to a genogroup seems to depend on the genome area used for evaluation and raised the possibility that RNA recombination might have occurred in astroviruses. (Belliot *et al.*, 1997b; Monroe *et al.*, 2001). S Monroe has described recent aspects of the molecular epidemiology of astroviruses (Section V, Chapter 3).

Progress in diagnosis of astrovirus infections with molecular techniques has established them as a not uncommon cause of acute gastroenteritis in infants and young children. Although enteric astrovirus strains have been isolated from animals, e.g. from turkeys (Koci *et al.*, 2000), there are so far no firm data supporting the possibility that human astroviruses are derived from an animal reservoir or that animal astroviruses infect humans on a larger scale.

References

Bass DM, Qiu S (2000). Proteolytic processing of the astrovirus capsid. *J. Virol.* **74:** 1810-1814.

Belliot G, Laveran H, Monroe SS (1997a). Capsid protein composition of reference strains and wild isolates of human astroviruses. *Virus Res.* **49**: 49-57.

Belliot G, Laveran H, Monroe SS *et al.* (1997b). Detection and genetic differentiation of human astroviruses: phylogenetic grouping varies by coding region. *Arch.Virol.* **142:**1323-1334.

Brierley I, Digard P, Inglis SC (1989). Characterization of an efficient coronavirus ribosomal frameshifting signal: requirement for an RNA pseudoknot. *Cell* **57:** 537-547.

Brinker JP, Blacklow NR, Herrmann JE (2000). Human astrovirus isolation and propagation in multiple cell lines. *Arch Virol.* **145:** 1847-1856.

Cunliffe NA, Bresee JS, Gondwe J, Hart CA (1998). The epidemiology of diarrhoeal disease in children at Queen Elizabeth Central Hospital, Blantyre, Malawi, 1994-1997. *Malawi Med.J.* **11:** 21-25.

Geigenmüller U, Ginzton NH, Matsui SM (1997). Construction of a genome-length cDNA clone for human astrovirus serotype 1 and synthesis of infectious RNA transcripts. *J. Virol.* **71:** 1713-1717.

Herrmann JE, Hudson RW, Perron-Henry DM, Kurtz JB, Blacklow NR (1988). Antigenic characterization of cell-cultivated astrovirus serotypes and development of astrovirus-specific monoclonal antibodies. *J. Infect. Dis.* **158:** 182-185.

Imada I, Yamaguchi S, Mase M, Tsukamoto K, Kubo M, Morooka A (2000). Avian nephritis virus (ANV) as a new member of the family *Astroviridae* and construction of infectious ANV cDNA. *J. Virol.* **74:** 8487-8493.

Jacks T, Power MD, Masiarz FR, Luciw PA, Barr PJ, Varmus HE (1988). Characterization of ribosomal frameshifting in HIV-1 gag-pol expression. *Nature* **331**: 280-283.

Jiang B, Monroe SS, Koonin EV, Stine SE, Glass RI (1993). RNA sequence of astrovirus: distinctive genomic organization and a putative retrovirus-like ribosomal frameshifting signal that directs the viral replicase synthesis. *Proc. Natl. Acad.*

Sci. USA **90:** 10539-10543.

Jonassen TO, Monceyron C, Lee TW, Kurtz JB, Grinde B (1995). Detection of all serotypes of human astrovirus by the polymerase chain reaction. *J. Virol Methods* **52:** 327-334.

Koopmans MP, Bijen MH, Monroe SS, Vinjé J (1998). Age stratified seroprevalence of neutralizing antibodies to astrovirus types 1 to 7 in humans in The Netherlands. *Clin Diagn. Lab Immunol.* **5:** 33-37.

Koci MD, Seal BS, Schultz-Cherry S (2000). Molecular characterization of an avian astrovirus. *J. Virol.* **74:** 6173-6177.

Lee TW, Kurtz JB (1981). Serial propagation of astrovirus in tissue culture with the aid of trypsin. *J. Gen. Virol.* **57:** 421-424.

Lee TW, Kurtz JB (1994). Prevalence of human astrovirus serotypes in the Oxford region 1976-1992, with evidence of two new serotypes. *Epidemiol. Infect.* **112:** 187-193.

Le Guyader F, Haugarreau L, Miossec L, Dubois E, Pommepuy M (2000). Three year study to assess human enteric viruses in shellfish. *Appl. Environm. Microbiol.* **66:** 3241-3248.

Lewis TL, Matsui SM (1995). An astrovirus frameshift signal induces ribosomal frameshifting *in vitro*. *Arch Virol.* **140:** 1127-1135.

Lewis TL, Matsui SM (1996). Astrovirus ribosomal frameshifting in an infection-transfection transient expression system. *J. Virol.* **70:** 2869-2875.

Marczinke B, Bloys AJ, Brown TDK, Willcocks MM, Carter MJ, Brierley I. (1994). The human astrovirus RNA-dependent RNA polymerase coding region is expressed by ribosomal frameshifting. *J. Virol.* **68:** 5588-5595.

Matsui SM (1997). Astrovirus. In: Richman DD, Whitley RJ, Hayden FG (eds) *Clinical Virology*. Churchill Livingstone, New York, pp. 1111-1121.

Matsui SM, Greenberg HB (2001). Astroviruses. In: Knipe DM, Howley PM *et al.* (eds) *Fields Virology,* 4th edition, Lippincott Williams and Wilkins, Philadelphia, pp. 875-891.

Matsui S, Kiang D, Ginzton N, Chew T, Geigermüller-Gnirke U (2001). Molecular biology of astroviruses: selected highlights. *Novartis Found Symp* **238:**219-233 (Discussion 233-236).

Mitchell DK, Monroe SS, Jiang X, Matson DO, Glass RI, Pickering LK (1995). Virologic features of an astrovirus diarrhea outbreak in a day care center revealed by reverse transcription-polymerase chain reaction. *J. Infect. Dis.* **172:** 1437-1444.

Monroe SS, Holmes JL and Belliot GM (2001). Molecular epidemiology of human astroviruses. *Novartis Found. Symp.* **238:**237-245 (Discussion 245-249).

Monroe SS, Jiang B, Stine SE, Koopmans M, Glass RI (1993). Subgenomic RNA sequence of human astrovirus supporos classification of *Astroviridae* as a new family of RNA viruses. *J. Virol.* **67:** 3611-3614.

Mustafa H, Palombo EA, Bishop RF (2000). Epidemiology of astrovirus infection in young children hospitalized with acute gastroenteritis in Melbourne, Australia, over a period of four consecutive years, 1995-1998. *J. Clin. Microbiol.* **38:** 1058-1062.

Naficy AB, Abu-Elyazeed R, Holmes JL *et al.* (1999). Epidemiology of rotavirus diarrhea in Egyptian children and implications for disease control. *Am. J. Epidemiol.* **150:** 770-777.

Noel JS, Cubitt D. (1994). Identification of astrovirus serotypes from children treated at the Hospitals for Sick Children, London, 1981 to 1993. *Epidemiol. Infect.* **113:** 1253-159.

Noel JS, Lee TW, Kurtz JB, Glass RI, Monroe SS (1995). Typing of human astroviruses from clinical isolates by enzyme immunoassay and nucleotide sequencing. *J. Clin. Microbiol.* **33:** 797-801.

Oishi I, Yamazaki K, Kimoto T, Minekawa Y, Utagawa E, Yamazaki S *et al.* (1994). A large outbreak of acute gastroenteritis associated with astrovirus among students and teachers in Osaka, Japan. *J. Infect. Dis.* **170:** 439-443.

Risco C, Carrascosa JL, Pedregosa AM, Humphrey CD, Sánchez-Fauquier A (1995). Ultrastructure of human astrovirus serotype 2. *J. Gen. Virol.* **76:** 2075-2080.

Willcocks MM, Brown TDK, Madeley CR, Carter MJ (1994). The complete sequence of a human astrovirus. *J. Gen. Virol.* **75:** 1785-1788.

Willcocks MM, Boxall AS, Carter MJ (1999). Processing and intracellular location of human astrovirus non-structural proteins. *J. Gen. Virol.* **80:** 2607-2611.

Yeager M, Tilova M, Novotry N, Matsui S (2002). Icosahedral design of human astrovirus, submitted.

Viral Gastroenteritis
U. Desselberger and J. Gray (editors)
© 2003 Elsevier Science B.V. All rights reserved

573

V, 1. Studies on the molecular biology of human astrovirus

Ute Geigenmüller[1], Ernesto Méndez[2] and Suzanne M. Matsui[1]

[1] *Department of Medicine, Division of Gastroenterology and Hepatology,*
Stanford University School of Medicine, Stanford, CA 94305, and Department of Medicine,
Gastroenterology Section, VA Palo Alto Health Care System, Palo Alto, CA 94304

[2] *Departamento de Genética y Fisiología Molecular, Instituto de Biotecnología,*
Universidad Nacional Autónoma de México. Apartado Postal 510-3, Colonia Miraval,
Cuernavaca, Morelos, 62250, México

Introduction

Our research has focused on the molecular characterization of human astrovirus serotypes 1 and 8 (HAstV-1 and HAstV-8, respectively). HAstV-1 is the most prevalent of the human astroviruses that cause acute gastroenteritis. A full-length cDNA copy of the genomic RNA of HAstV-1 (Oxford strain) was assembled, from which infectious viral RNA can be transcribed *in vitro* (pAVIC, Geigenmüller *et al.*, 1997). Since Caco-2 cells, which allow efficient propagation of HAstV (Willcocks, 1990), proved to be very poorly transfectable, we selected BHK cells for transfection with viral RNA that had been transcribed *in vitro*, achieving high rates of transfection. The transfected BHK cells produced astroviral particles that were capable of infecting Caco-2 cells at a titer of 2×10^7 infectious units per μg RNA used for transfection. The viral particles found in BHK cells transfected with AVIC RNA formed dense viral aggregates and were indistinguishable by electron microscopy from those seen in Caco-2 cells infected with wild type HAstV-1 (Fig. 1). Availability of the pAVIC clone and the sequential transfection-infection cell system allowed the introduction of targeted mutations into the astroviral genome and the assessment of their effect on the astroviral replication cycle. For HAstV-8, development of antibodies specific for different regions of the virally encoded proteins has allowed detailed analysis of those proteins in infected cells (Méndez *et al.*, manuscripts submitted and in preparation).

In the following we will first review briefly the genomic organization and proposed replication strategy of astrovirus, then discuss each of the known viral proteins in detail. We will focus on the data obtained for HAstV-1 and HAstV-8 through the use of the systems outlined above, and discuss them in the context of the general literature on the astroviral molecular biology. Since the sizes of the genome and the virally encoded proteins vary slightly between the eight known serotypes of HAstV, for reasons of simplicity the numbers given will always relate to HAstV-1 (Oxford strain), unless stated otherwise.

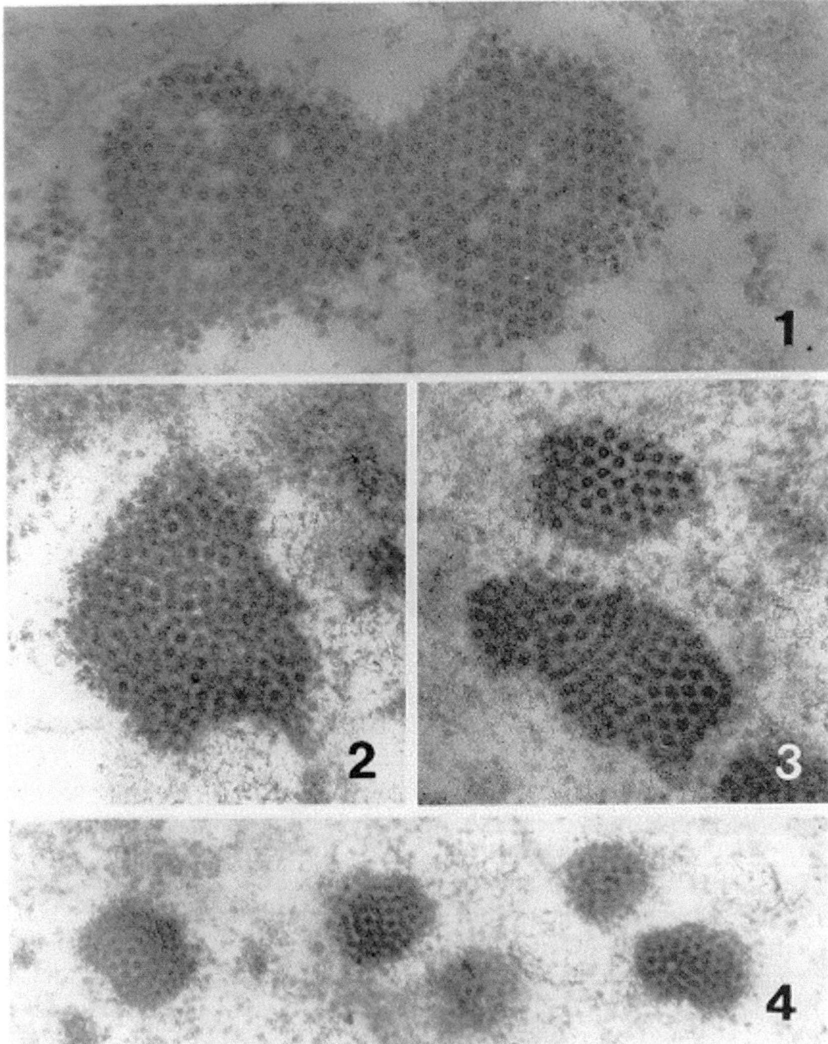

Fig. 1. HAstV-1 virion production in infected and transfected cells. Aggregates of viral particles are visible in the cytoplasm of: 1) Caco-2 cells infected with wild type HAstV-1 viral stock (x100,000), 2) BHK cells transfected with AVIC RNA (x90,000), 3) Caco-2 cells infected with the combined cell lysates and supernatants of BHK cells transfected with AVIC RNA (x90,000), and 4) Caco-2 cells infected with the combined cell lysates and supernatants of BHK cells transfected with wild type HAstV-1 RNA (x65,000). After overnight incubation in the absence of trypsin and presence of serum, infected or transfected cells were scraped off, pelleted, and fixed over night at 4°C in 2 % glutaraldehyde in 0.1 M phosphate buffer, pH 7.2. The fixed cell pellet was then treated like a tissue sample and prepared for electron microscopy as described in Glauert (1965). Electron micrographs are courtesy of Dr. L.S. Oshiro, California Department of Health Services, Berkeley, CA.

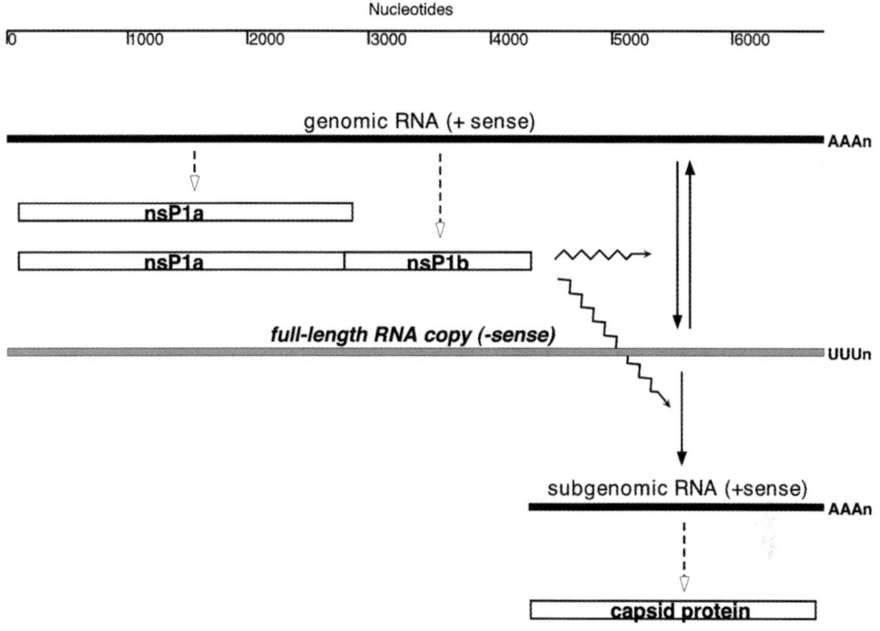

Fig. 2. Schematic representation of the astroviral replication strategy. Black lines stand for plus stranded RNA, shaded line for negative strand RNA. Open boxes represent viral proteins. Solid arrows show transcription, open arrows translation. The squiggly arrow indicates enzymatic function.

The RNA genome of HAstV-1 (Oxford strain) consists of 6771 nt excluding the 3′ terminal poly(A) tail. In infected cells, a polyadenylated subgenomic RNA co-linear with the 3′ 2455 nt of the genomic RNA is also found at an estimated 10-fold (or more) molar excess over full-length viral RNA (Monroe et al., 1993). The genomic RNA contains three long open reading frames, ORF1a, ORF1b, and ORF2 (Jiang et al., 1993; Lewis et al., 1994; Willcocks et al., 1994a). ORF1a and 1b span the 5′ two-thirds of the genome, encoding nonstructural proteins (nsP), while ORF2, which is contained in both the genomic and the subgenomic RNA, codes for the structural (capsid) protein (Monroe et al., 1993; Lewis et al., 1994).

The proposed replication strategy for astrovirus (Fig. 2) has in large part been inferred from the well-characterized life cycle of the enveloped alphaviruses, which show a genome organization and strategy for expression of the structural protein(s) similar to the astroviruses (reviewed in Schlesinger and Schlesinger, 1996). Specifically, upon infection, nsP1a and nsP1b are believed to be translated directly from the input positive strand viral genome. The newly synthesized nsPs then enable the transcription of a full-length negative strand RNA which subsequently serves as the template for the transcription of new plus strand genomic RNA and subgenomic RNA. The latter functions as messenger RNA for the translation of the capsid protein. Based on this replication strategy, and by analogy to the eukaryotic expression vectors that have been developed from several of the alphaviruses (Schlesinger, 2001), astrovirus can be used to express a foreign protein. When the region of ORF2 encoding aa 11 to 725 of the capsid protein was replaced by the gene for green fluorescent protein (GFP, Chalfie *et al.*, 1994) in pAVIC, efficient expression of a capsid protein-GFP fusion protein could be detected in BHK cells transfected with the vector RNA.

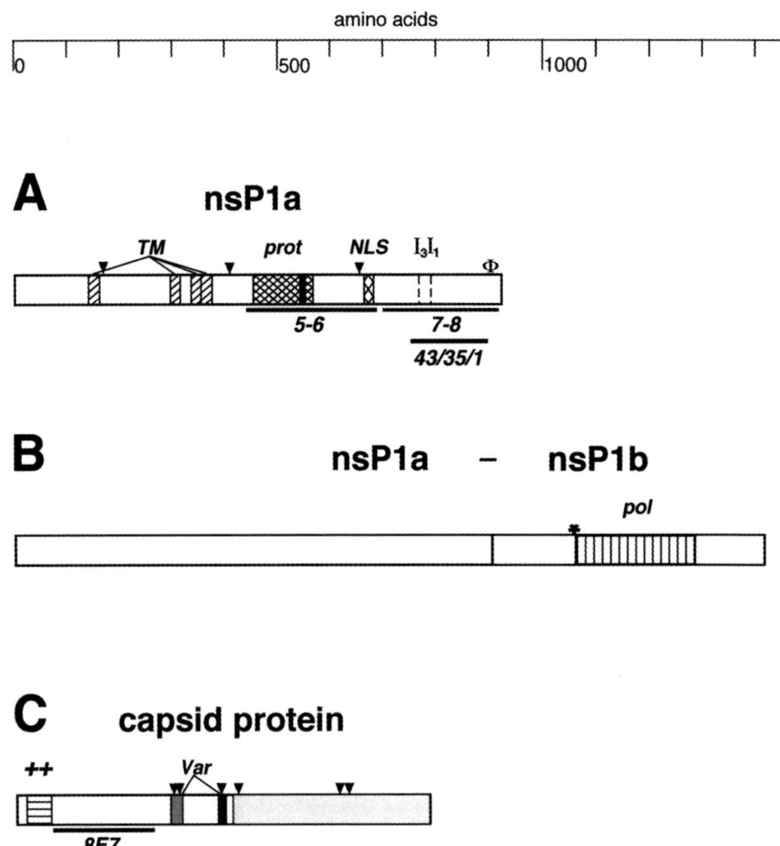

Fig. 3. Schematic representation of known features of the astroviral proteins. Open boxes show viral proteins. Specific regions within the proteins are indicated by hatching or shading. (A) TM, transmembrane regions; prot, protease motif; the black box within the protease motif marks the site of the 9 aa substitution (see text); NLS, nuclear localization signal; I_3, I_1, sites of insertions found in HAstV-3 (Oh and Schreier, 2001) and HAstV-1 (strain A2/88 Newcastle; Willcocks *et al.*, 1994b), respectively; Φ, frameshift site; (B) *, site of frameshift in nsP1b due to deletion of nt 3329 in ORF1b (ΔT3329); pol, RNA-dependent RNA polymerase motif; (C) ++, region rich in Arg and Lys; Var, variable regions. Shading outlines areas of the capsid protein with high sequence variability. (A, C) Black lines under the boxes outline epitopes for the antibodies indicated. Black triangles above the boxes mark protein cleavage sites.

ORF1a extends from nt 86 through to nt 2848 (2,763 nt) and is translated into a polypeptide of 920 amino acids (aa) (nsP1a, Fig. 3A). NsP1a contains a viral serine protease motif from aa 454-569 that can be aligned with the picornavirus 3C protease and other putative chymotrypsin-like proteases (Gorbalenya *et al.*, 1989). Surprisingly, no motifs suggestive of an RNA helicase or methyltransferase are found in ORF1a (or elsewhere in the HAstV genome), while all other positive strand RNA viruses with genomes larger than 5.8 kb encode at least one potential helicase (Kadaré and Haenni, 1997). A potential VPg-encoding region in ORF1a was identified by computer analysis (Jiang *et al.*, 1993), but it has not yet been determined whether HAstV indeed produces a VPg-like protein. Presence of an epitope in nsP1a that was identified by expression cloning and immunostaining with antiserum to purified viral particles (A43/35/1, aa 757-899, Matsui *et al.*, 1993) suggests that nsP1a or parts thereof may be incorporated into the viral capsid. Within the A43/35/1 epitope region there are two sites, where insertions or deletions occur among the different serotypes with known ORF1a sequence (HAstV-1, 2, 3, and 8). Compared to the other serotypes and the Oxford strain of HAstV-1, 7 additional aa are present in between aa 766/767 in HAstV-3 (Oh and Schreier, 2001), and 15 additional aa are found in between aa 790/791 in the A2/88 Newcastle strain of HAstV-1 (Willcocks *et al.*, 1994b). A sequence encoding a probable nuclear localization signal (NLS) from aa 666-682, downstream of the viral protease motif, has also been identified (Jiang *et al.*, 1993; Willcocks *et al.*, 1994a). Between aa 139 and 372 of nsP1a four transmembrane helices have been predicted (Jiang *et al.*, 1993).

The protease encoded in ORF1a has been shown to be functional and involved in processing of nsP1a. A single autocatalytic processing event was observed in nsP1a in a cell-free expression system (Kiang and Matsui, 2002). When nsP1a of HAstV-1 was expressed in BHK cells using a vaccinia virus-based expression system, multiple nsP1a-derived processing products were detected by immunoprecipitation with an nsP1a-specific antibody or an antibody specific for an N-terminally linked epitope tag (Geigenmüller *et al.*, 2002a). Mapping of two processing products, the N-terminal p20 and an internal p27, suggested cleavage sites near aa 170, and around aa 410 and aa 655 of nsP1a, respectively. The internal p27 processing product resulting from cleavage at approximately aa 410 and aa 655 was also found in Caco-2 cells that had been infected with HAstV-1. Cleavage at around aa 410 and aa 655, but not aa 170, was abolished when a 9 aa substitution was introduced into the protease motif in nsP1a, suggesting that processing at the former sites, but not at the latter position, occurs autocatalytically. Very similar nsP1a-derived processing products have been identified in HAstV-8 -infected Caco-2 cells, using a panel of nsP1a-specific antibodies (Méndez *et al.*, manuscript in preparation). During processing, both the proposed transmembrane helices on the N-terminal side of the predicted cleavage site near aa 410 and the NLS distal to the predicted cleavage site near aa 655 are separated from the ORF1a-encoded protease domain.

Data regarding the functionality of the nuclear localization signal (NLS) encoded in ORF1a are conflicting. While nuclear localization of the region of nsP1a containing the NLS motif has been reported by Willcocks *et al.* (1999), we found no evidence for

578

the presence of nsP1a in the nucleus. Specifically, Caco-2 cells infected with HAstV-1 were immunostained with antibodies produced in mice against either the region of nsP1a encompassing the protease motif and the NLS (aa 445-688; pAb 5-6) or the region on the C-terminal side of the NLS (aa 703-918, pAb 7-8). In both cases, nsP1a-specific staining was clearly confined to the cytoplasm (Fig. 4).

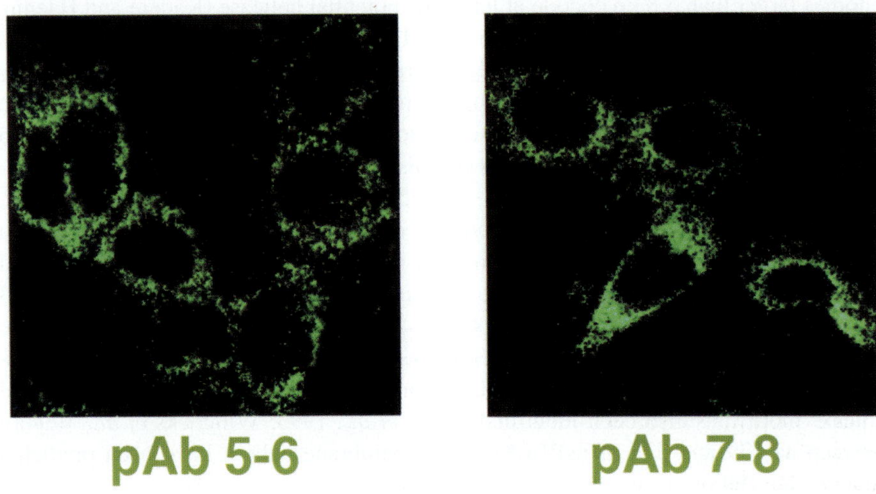

Fig. 4; Subcellular localization of nsP1a in Caco-2 cells infected with HAstV-1. Caco-2 cells were infected with HAstV-1, and after a 6 h incubation in the absence of trypsin, cells were immunostained with pAb 5-6 or pAb 7-8 and a secondary goat anti-mouse FITC-conjugated IgG (green). Cells were viewed under a confocal Laser scanning microcope (Nikon), using the Laser scanning confocal imaging system MRC-1024 (BioRad).

ORF1b is in the (-1) frame with respect to ORF1a and overlaps ORF1a by 70-nt (Lewis *et al.*, 1994). ORF1b (1,542 nt) encodes a potential product of 513 aa (nsP1b, Fig. 3B), and contains the highly conserved RNA-dependent RNA polymerase motif that can be aligned with RNA polymerases of supergroup I, including those of picornaviruses, caliciviruses and several positive strand RNA plant viruses (Koonin, 1991; Jiang *et al.*, 1993; Willcocks *et al.*, 1994a; Lewis *et al.*, 1994). The ORF1b product is translated by (-1) ribosomal frameshifting (Marczinke *et al.*, 1994; Lewis and Matsui, 1995). The 70-nt overlap region between ORF1a and ORF1b contains the frameshift signal, a heptamer with the sequence A AAA AAC which is common to all HAstV strains sequenced thus far. Nearby sequences (6-nt downstream), that could potentially form a stem-loop structure (14-nt GC-rich stem and 10-nt loop), also participate in the frameshifting process (See Brierley and Vidakovic, Section V, Chapter 2 of this book).

When a single nucleotide deletion was introduced into ORF1b in pAVIC (ΔT3329), resulting in a frameshift and subsequent premature termination during the translation of nsP1b, no production of infectious viral particles or expression of ORF2 could be detected in BHK cells transfected with ΔT3329 RNA. However, expression of ORF2, which requires replication of the genomic RNA as well as transcription of a subgenomic RNA, could be rescued, if cells were concomitantly transfected with infectious AVIC RNA (Matsui *et al.*, 2001). This finding proves, that the nsP1b function is crucial for expression of ORF2 and that it can be provided in *trans*.

When BHK cells were transfected with viral RNA that had been transcribed *in vitro* from a pAVIC derivative containing a 9 aa substitution within the protease motif (see above), neither production of infectious viral particles nor expression of the capsid protein could be detected. Since the transfected cells are expected to express a full-length nsP1a-1b fusion protein containing a defective protease but an intact polymerase, it seems likely that processing of the nsP1a-1b fusion protein by the ORF1a-encoded protease is necessary for activity of the ORF1b-encoded RNA-dependent RNA polymerase.

ORF2 (2,364 nt), which spans the 3′ one-third of the genome, overlaps ORF1b by 8 nt, and is in the same reading frame as ORF1a. It is followed by a 3′ UTR of 83 nt and the poly (A) tail, and encodes the viral structural (capsid) protein (Fig. 3C). The capsid protein consists of 787 aa and has a predicted molecular mass (MM) of 87 kDa (Lewis *et al.*, 1994). Recent data, however, indicate a higher apparent molecular mass of about 98 kDa for HAstV-1 (see below), suggesting post-translational modification of the capsid protein. The N-terminal 415 aa of the capsid protein are highly conserved among HAstV strains, while the sequence between aa 416 and 787 diverges considerably. Within the conserved N-terminal domain, two short variable regions exist (aa 292-319 and aa 387-399; Méndez-Toss *et al.*, 2000).

Near the N-terminus, a region that spans from aa 18 to aa 62 contains an unusually high proportion of basic amino acids (40%). This basic nature of the N-terminal region of the astroviral capsid protein is reminiscent of the capsid or coat proteins of other icosahedral RNA viruses, where a highly basic N-terminal region has been shown to be involved in packaging of the viral genomic RNA into the virion (Geigenmüller-Gnirke *et al.*, 1993; Baer *et al.*, 1994; Rao and Grantham, 1996). The RNA binding potential of the N-terminal region of the astroviral capsid protein is demonstrated by the ability of the 171 N-terminal aa of the capsid protein to direct a fused GFP to the nucleus and nucleoli of BHK cells expressing the 171-GFP fusion protein (Fig. 5). The green signal emanating from GFP is surrounded by a ring of red immunostaining specific for the nuclear pore complexes in the nuclear envelope (mAb 414, Davis and Blobel, 1986). Nuclear localization of the 171-GFP (predicted MM of 46 kDa) seems to be due to passive diffusion into the nucleus rather than active transport, since the N-terminal 171 aa of the capsid protein are unable to direct a capsid protein-GFP fusion protein retaining 725 capsid protein-derived aa at its N-terminus (725-GFP, predicted MM of 107 kDa, Fig. 5) or a full-length capsid protein (predicted MM of 87 kDa; data not shown) into the nucleus. It is likely that the 171-GFP fusion protein gets trapped in the nucleus by binding to nucleic acids, in particular the ribosomal RNA present in abundance in the nucleoli. If 15 positively charged aa are removed from the capsid-derived sequence in 171-GFP by deleting aa 11-54 (171Δ11-54-GFP), the fusion protein is no longer concentrated in the nucleus, but found throughout the entire cell. A similar distribution is seen for a fusion protein containing only the N-terminal 10 aa of the capsid protein linked to GFP (10-GFP).

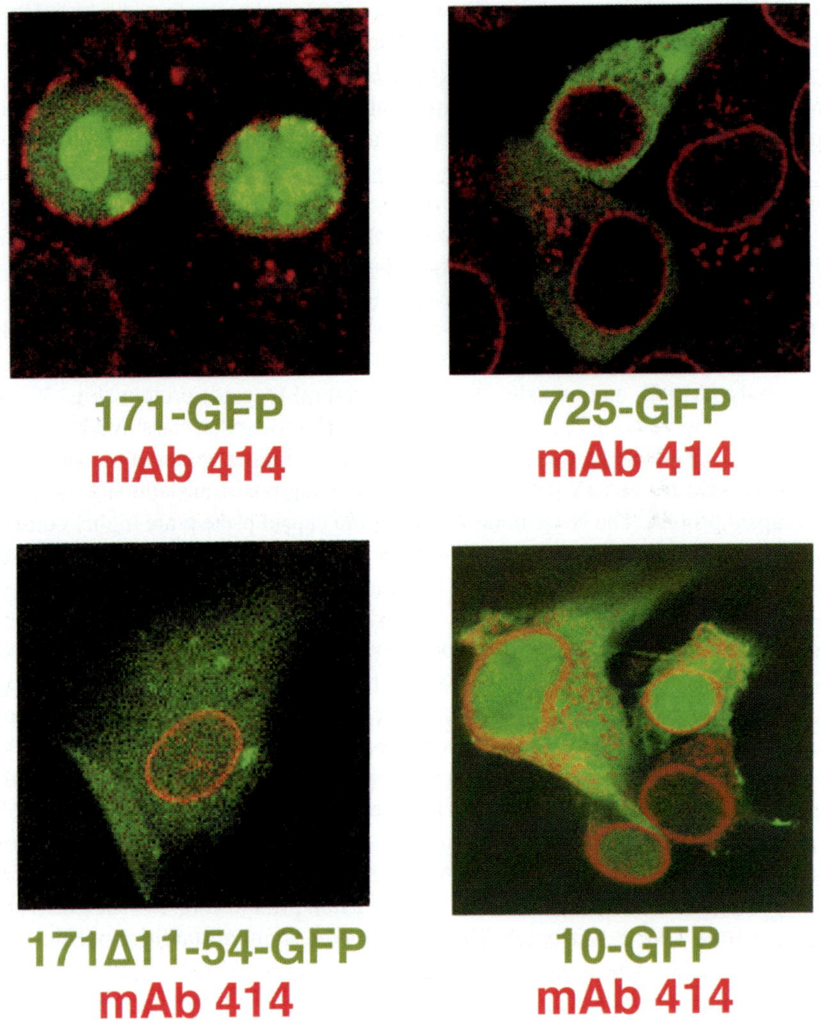

Fig. 5. Subcellular localization of capsid-protein GFP fusion proteins in transfected BHK cells. BHK cells were transfected with *in vitro* transcribed astroviral RNA encoding capsid protein-GFP fusion proteins retaining the indicated number of capsid-protein derived aa at their N-terminus. After overnight incubation, cells were fixed and immunostained with a monoclonal antibody against a nucleoporin in the nuclear pore complex (mAb414) and a secondary goat anti-mouse Texas Red-conjugated IgG. Cells were viewed under a confocal Laser scanning microcope (Nikon), using the Laser scanning confocal imaging system MRC-1024 (BioRad). Different fluorochromes were imaged sequentially to avoid bleed through of emission signals.

It has long been known that trypsin-dependent cleavage of the capsid protein is necessary for viral infectivity (Lee and Kurtz, 1981), but until recently few data were available on the precise nature of intracellular or extracellular processing of the capsid protein. Reports on intracellular processing of the capsid protein in infected cells have been controversial. Bass and Qiu (2000) identified an intracellular cleavage site between aa 70 and aa 71. They proposed that cleavage at this site is necessary for virion assembly, since only the C-terminal 79 kDa processing product, but not the full-length capsid protein, could be detected extracellularly. However, such a mechanism seems surprising given the potential role of the N-terminal region of the capsid protein in binding to the viral RNA for packaging. To re-examine intracellular processing of the capsid protein in infected Caco-2 cells, two astroviral mutants were constructed that coded for either an antigenic epitope tag inserted between aa 10 and aa 31 of the capsid protein or a strep-tag (Schmidt and Skerra, 1993) replacing aa 782 to aa787 of the capsid protein (Geigenmüller *et al.*, 2002b). RNA transcribed *in vitro* from these mutants was only 3- and 15-fold, respectively, less infective than wild type AVIC RNA. Apparently the deletion and insertion introduced between aa 10 and 31 of the capsid protein does not dramatically affect its ability to assemble into infectious viral particles. It is therefore likely that this part of the capsid protein is rather unstructured and not involved in conformation-dependent interactions with other capsid proteins. When intracellular processing of the wild type or tagged HAstV-1 capsid proteins in infected Caco-2 cells was analyzed by immunoprecipitation of cell lysates with an HAstV-specific monoclonal antibody (mAb 8E7, Herrmann *et al.*, 1991) that recognizes capsid protein sequences between aa 70 and 261, only one major band of about 98 kDa was detected. A protein of the same size was also precipitated with either an antibody directed against the N-proximal epitope tag, or with streptavidin directed against the C-terminal strep-tag, identifying the 98 kDa band as the full-length capsid protein. There was no indication of cleavage near the N-terminus of the capsid protein. Two minor bands of 82 kDa and 75 kDa detected with the capsid-specific mAb 8E7, in addition to the major 98 kDa band, were also immunoprecipitated by the antibody against the N-proximal tag, and therefore could not result from cleavage near the N-terminus. Instead the faint 82 kDa and 75 kDa bands may represent products of unspecific degradation or specific cleavage of a small proportion of the capsid protein near the C-terminus. Under these same conditions, where at best very limited processing of the capsid protein could be detected, paracrystalline arrays of astroviral particles could be readily observed in infected cells by electron microscopy (Fig. 1), suggesting that viral assembly does not depend on capsid protein processing. Recent studies on the processing of the capsid protein of HAstV-8 (Méndez *et al.*, submitted) confirmed the lack of processing near the N-terminus of the capsid protein. In these studies, antibodies raised against various regions of the capsid protein (see below) were used to detect capsid-derived proteins in lysates of HAstV-8-infected Caco-2 cells. In contrast to the results for HAstV-1 (Geigenmüller *et al.*, 2002b), Méndez *et al.* (submitted) observed slow intracellular cleavage of the 90 kDa capsid protein near the C-terminus, resulting in a 70 kDa N-terminal processing product. In this context it is noteworthy, that Caco-2 cells infected with HAstV-8 were grown for 3 days in the presence of low amounts of trypsin, while HAstV-1-infected Caco-2 cells were grown overnight in the presence of serum. C-proximal processing of the capsid protein in HAstV-8-infected cells could therefore be due to leakage of trypsin, which is known to be necessary for extracellular processing of the capsid protein (see below), into the damaged cells. In the case of HAstV-8, only the 70 kDa protein was detected in purified particles; it is not clear, however, if C-proximal processing of the capsid protein occurs before or after assembly into viral particles.

Several reports have been published on the extracellular processing of the astroviral capsid protein. At least three proteins have been found on the mature, trypsin-activated viral particle. The proposed sizes of these fragments range from 24 to 26 kDa, 29 to 31.5 kDa and 32 to 34 kDa (Willcocks *et al.*, 1990; Monroe *et al.*, 1991; Sánchez-Fauquier *et al.*, 1994; Belliot *et al.*, 1997; Bass and Qiu, 2000), depending on the serotype and the isolation method used. In one study (Sánchez-Fauquier *et al.*, 1994), N-terminal sequencing of the 26 and 29 kDa fragments derived from the capsid protein of HAstV-2, indicated cleavage after aa 361 and aa 395 of the capsid protein, respectively, identifying the 26 kDa protein as an N-terminally truncated version of the 29 kDa fragment. The most extensive analysis of capsid protein processing so far has been conducted for HAstV-8 (Méndez *et al.*, submitted). Using antibodies specific for aa 4-20, aa 3-208, aa 209-341 and aa 386-594 of the capsid protein, the authors determined the sequence of trypsin-dependent processing steps of the capsid-protein derived 70 kDa fragment found on purified viral particles, pinpointing some of the cleavage sites by N-terminal sequencing of the processing products (Fig. 6). Trypsin-treatment of purified viral particles led to cleavage of the 70 kDa fragment (VP70) at Arg393 to yield an N-terminal 41 kDa (VP41) and a C-terminal 28 kDa (VP28) fragment. VP41 was further processed at its C-terminus, giving rise to a 38.5 kDa (VP38.5), a 35 kDa (VP35) and a 34 kDa product (VP34). VP28 was processed either at its C-terminus, yielding a 27 kDa product (VP27), or at Arg423 near its N-terminus, giving rise to a 25 kDa fragment (VP25); a 26 kDa fragment (VP26) was also observed, and is believed to be an intermediary in the processing from VP28 to VP25. VP41, VP38.5, VP35, VP28, and VP26 constitute intermediary processing products that could only be detected after mild trypsin digestion, and did not correlate with viral infectivity. After addition of more trypsin, which led to a dramatic increase in infectivity, only VP34, VP27 and VP25 were detected. VP27 and VP25 probably correspond to the 29 and 26 kDa products identified for HAstV-2. It is noteworthy that Arg393 and Arg 423 are conserved among all known serotypes except serotype 2.

In summary, the following preliminary model for the replication cycle of astrovirus can be proposed: The infecting viral RNA is translated into nsP1a and an nsP1a-nsP1b fusion protein. Processing of the nsP1a and nsP1a-nsP1b fusion proteins depends on both a host protease and autocatalytic cleavage by the virally encoded protease contained in nsP1a. The resulting cleavage products allow replication of the viral genome and transcription of a subgenomic RNA leading to expression of ORF2. The unprocessed capsid protein assembles into viral particles, and viral genomic RNA is packaged by binding to the N-terminal region of the capsid protein. After release of the virions from the cell, exposure to trypsin results in a regulated cascade of cleavages

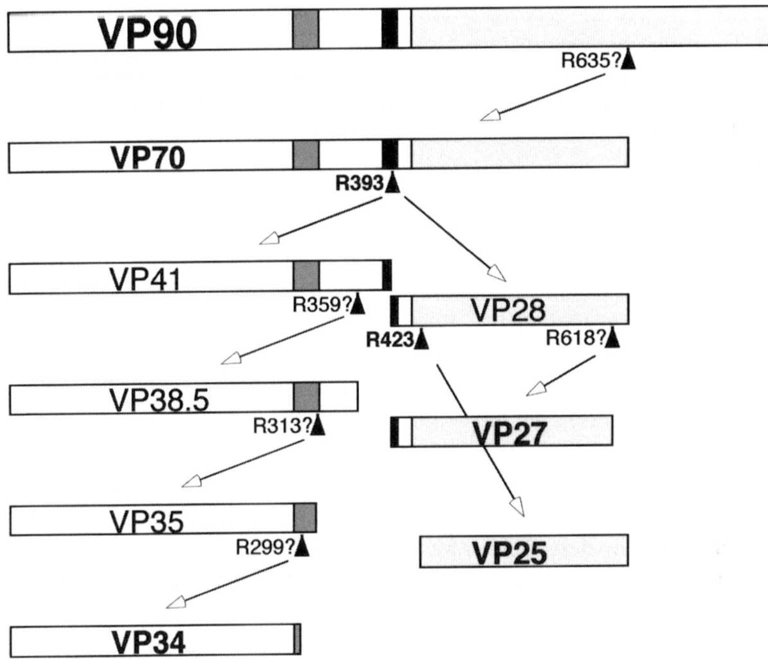

Fig. 6. Proposed trypsin processing pathway for the capsid protein of HAstV-8. Boxes represent the full-length translation product (VP90) or the various processing products. Stable intermediary (VP70) or end-products (VP34, VP27, VP25) of processing are indicated in bold print. Shading outlines regions with high sequence variability. Black triangles indicate potential (indicted by a question mark) or known (in bold print) cleavage sites.

in the capsid protein on the viral particle. Only viral particles composed of the final products of these processing events (VP34, VP25 and VP27, or similar sized proteins depending on the serotype) are able to infect cells to start a new replication cycle.

Acknowledgements

We thank Nancy Ginzton,Teri Chew, and Teresa Fernández for technical help and many helpful discussions. We are also indebted to Dr. Lyndon S. Oshiro for his help with the EM studies. Work on HAstV-1 was supported by the National Institutes of Health (R21 AI43513 to S.M.M. and P30 DK56339 to the Stanford Digestive Disease Center) and the Office of Research and Development, Department of Veterans Affairs (Merit Review to S.M.M.). Work on HAstV-8 was supported by grants MENSE31739 from the National Council for Sience and Technology - Mexico and IN200999 from DGAPA-UNAM.

References

Baer, M.L., Houser, F., Loesch-Fries, L.S. and Gehrke, L. (1994). Specific RNA binding by amino-terminal peptides of alfalfa mosaic virus coat protein. *EMBO J.* **13**, 727-735

Bass, D.M., Qiu, S. (2000). Proteolytic processing of the astrovirus capsid. *J. Virol.* **74**, 1810-1814

Belliot, G., Laveran, H., Monroe, S.S. (1997). Capsid protein composition of reference strains and wild isolates of human astrovirus. *Virus Res.* **49**, 49-57

Chalfie, M., Tu, Y., Euskirchen, G., Ward, W.W., Prasher, D.C. (1994) Green fluorescent protein as a marker for gene expression. *Science* **263**, 802-805

Davis, L.I., Blobel, G (1986). Identification and characterization of a nuclear pore complex protein. *Cell* **45**, 699-709

Geigenmüller-Gnirke, U., Nitschko, H., Schlesinger. S. (1993). Deletion analysis of the capsid protein of Sindbis virus: identification of the RNA binding region. *J. Virol.* **67**, 1620-1626

Geigenmüller, U., Ginzton, N.H, Matsui, S.M. (1997). Construction of a genome-length cDNA clone for human astrovirus serotype 1 and synthesis of infectious RNA transcripts. *J. Virol.* **71**,1713-1717

Geigenmüller, U., Chew, T., Ginzton, N., Matsui, S.M. (2002a). Processing of non-structural protein 1a of human astrovirus. *J. Virol.* 76, 2003-2008

Geigenmüller, U., Chew, T., Ginzton, N., Matsui, S.M. (2002b). Studies on intracellular processing of the capsid protein of human astrovirus serotype 1 in infected cells. *J. Gen. Virol.* **83**, 1691-1695

Glauert, A.M. (1965). The fixation and embedding of biological specimens. In: *Techniques for Electron Microscopy*, Edited by Kay, D.H., pp. 166-212. Philadelphia, F.A. Davis Company

Gorbalenya, A.E., Donchenko, A.P., Blinov, V.M., Koonin, E.V. (1989). Cysteine protease of positive strand RNA viruses and chymotrypsin-like serine proteases: a distinct protein superfamily with a common structural fold. *FEBS Lett.* **243**, 103-114

Herrmann, J.E., Taylor, D.N., Echeverria, P., Blacklow, N.R. (1991). Astroviruses as a cause of gastroenteritis in children. *N. Engl. J. Med.* **324**, 1757-1760

Jiang, B., Monroe, S.S., Koonin, E.V., Stine, S.E., Glass, R.I. (1993). RNA sequence of astrovirus: distinctive genomic organization and a putative retrovirus-like ribosomal frameshifting signal that directs the viral replicase synthesis. *Proc. Natl. Acad. Sci. USA* **90**, 10539-10543

Kadaré, G., Haenni, A.-L. (1997) Virus-encoded RNA helicases. *J. Virol.* 71, 2583-2590

Kiang, D., Matsui, S.M. (2002) Proteolytic processing of a human astrovirus nonstructural protein. *J. Gen. Virol.* **83**, 25-34

Koonin, E.V. (1991). The phylogeny of RNA-dependent RNA polymerases of positive-strand RNA viruses. *J. Gen.Virol.* **72**, 197-206

Lee, T.W., Kurtz, J.B. (1981). Serial propagation of astrovirus in tissue culture with the aid of trypsin. *J. Gen. Virol.* **57**, 421-424

Lewis, T.L., Greenberg, H.B., Herrmann, J.E., Smith, L.S., Matsui, S.M. (1994). Analysis of astrovirus serotype 1 RNA, identification of the viral RNA-dependent RNA polymerase motif, and expression of a viral structural protein. *J. Virol.* **68**, 77-83

Lewis, T.L., Matsui, S.M. (1995). An astrovirus frameshift signal induces ribosomal frameshifting *in vitro*. *Arch. Virol.* **140**, 1127-1135

Marczinke, B., Bloys, A.J., Brown, T.D.K., Willcocks, M.M., Carter, M.J., Brierley, I. (1994). The human astrovirus RNA-dependent RNA polymerase coding region is expressed by ribosomal frameshifting. *J. Virol.* **68**, 5588-5595

Matsui, S.M., Kim, J.P., Greenberg, H.B., Young, LaV.,M., Smith, L.S., Lewis, T.L., Herrmann, J.E., Blacklow, N.R., Dupuis, K., Reyes, G.R. (1993). Cloning and characterization of human astrovirus immunoreactive epitopes. *J. Virol.* **67**, 1712-1715

Matsui, S.M., Kiang, D., Ginzton, N., Chew, T., Geigenmüller-Gnirke, U. (2001). The molecular biology of astrovirus: selected highlights. *Novartis Found. Symp.* **238**, 219-233

Méndez-Toss, M., Romero-Guido, P., Munguía, M.E., Méndez, E., Arias, C.F. (2000). Molecular analysis of a serotype 8 human astrovirus genome. *J. Gen. Virol.* **81**, 2891-2897

Monroe, S.S., Stine, S.E., Gorelkin, L., Herrmann, J.E., Blacklow, N.R., Glass, R.I. (1991). Temporal synthesis of proteins and RNAs during human astrovirus infection of cultured cells. *J. Virol.* **65**, 641-648

Monroe, S.S., Jiang, B., Stine, S.E., Koopmans, M., Glass, R.I. (1993). Subgenomic RNA sequence of human astrovirus supports classification of *Astroviridae* as a new family of RNA viruses. *J. Virol.* **67**, 3611-3614

Oh, D., Schreier, E. (2001). Molecular characterization of human astroviruses in Germany. *Arch. Virol.* **146**, 443-455

Rao, A.L., Grantham, G.L. (1996). Molecular studies on bromovirus capsid protein. II. Functional analysis of the amino-terminal arginine-rich motif and its role in encapsidation, movement, and pathology. *Virology* **226**, 294-305

Sánchez-Fauquier, A., Carrascosa, A.L., Carrascosa, J.L., Otero, A., Glass, R.I., Lopez, J.A., San Martin, C., Melero, J.A. (1994). Characterization of a human astrovirus serotpe 2 structural protein (VP26) that contains an epitope involved in virus neutralization. *Virology* **201**, 312-320

Schlesinger, S. (2001). Alphavirus vectors: development and potential therapeutic applications. *Expert Opin. Biol. Ther.* **1**, 177-191

Schlesinger, S., Schlesinger, M.J. (1996) Togaviridae: the viruses and their replication. In: *Fields Virology*, third ed., pp. 825-841. Edited by Fields, B.N., Knipe, D.M., Howley, P.M. *et al.*, Philadelphia, Lippincott-Raven

Schmidt, T. G. and Skerra, A. (1993). The random peptide library-assisted engineering of a C-terminal affinity peptide, useful for the detection and purification of a functional Ig Fv fragment. *Protein Engineering* **6**, 109-122

Willcocks M.M., Carter, M.J., Laidler, F. R., Madeley, C.R. (1990). Growth and characterisation of human faecal astrovirus in a continuous cell line. *Arch. Virol.*

113, 73-81

Willcocks, M.M., Brown, T.D.K., Madeley, C.R., Carter, M.J. (1994a). The complete sequence of a human astrovirus. *J. Gen. Virol.* **75**, 1785-1788

Willcocks, M.M., Ashton, N., Kurtz, J.B., Cubitt, W.D., Carter, M.J. (1994b). Cell culture adaptation of astrovirus involves a deletion. *J. Virol.* **68**, 6057-6058

Willcocks, M. M., Boxall, A.S., Carter, M.J. (1999). Processing and intracellular location of human astrovirus non-structural proteins. *J. Gen. Virol.* **80**, 2607-2611

Viral Gastroenteritis
U. Desselberger and J. Gray (editors)

587

V, 2. Ribosomal frameshifting in astroviruses

Ian Brierley and Marijana Vidakovic

Division of Virology, Department of Pathology, University of Cambridge, Cambridge CB2 1QP, U.K.

Introduction

For the vast majority of eukaryotic mRNAs, translation initiation is a 5'-end dependent process beginning with recognition of the cap structure by the cap-binding complex eIF4F (Pestova *et al.*, 2001). Subsequently, a scanning ribosome complex is assembled which migrates along the mRNA until an AUG codon is encountered in a suitable context (Kozak, 1983). At this point, the elongation phase of protein synthesis begins and the ribosome translates along the mRNA in triplet steps, the reading frame being set by that of the initiator AUG. Upon encounter of an in-frame termination codon, the polypeptide is released and the ribosome dissociates from the mRNA. Thus for most mRNAs, only the 5'-most open reading frame (ORF) is translated and downstream ORFs, at least in principle, are not reached by the ribosome. This 5'-end dependence is a problem faced by many RNA viruses with polycistronic genomes and elaborate strategies have been developed to overcome the translational limitation. Some viruses produce subgenomic-length mRNAs in which the relevant downstream ORF is effectively moved to the 5'-end of the RNA from where it can be efficiently translated. Where the replication cycle involves a nuclear step, RNAs can be spliced by the cellular machinery, and many cytoplasmically-replicating viruses have evolved mechanisms to produce subgenomic mRNAs during transcription. In other viruses, the 5'-end problem is obviated simply by encoding all of the required information in a single ORF and subsequently processing the encoded polyprotein proteolytically.

However, viruses have also evolved a number of unconventional translation strategies to express distal ORFs. These include leaky scanning, where the AUG of the 5'-most ORF is inefficiently recognised and ribosomes scan on to initiate at a downstream ORF; ribosomal re-initiation, where a post-termination ribosome remains associated with the mRNA and reinitiates translation at a downstream ORF, and translational fusion, where two (or more) ORFs separated by a stop codon or in an overlapping configuration are translated as a single protein following termination codon suppression or programmed ribosomal frameshifting respectively (reviewed in Pe'ery and Mathews, 2000). Of these unconventional mechanisms, the best studied is programmed −1 ribosomal frameshifting, a process where specific signals in the mRNA instruct the ribosome to change reading frame from the zero to the −1 frame (movement 5'-wards) at a certain efficiency, and to continue translation in the new frame (see Chandler and Fayet, 1993; Brierley, 1995; Dinman, 1995; Futterer and Hohn, 1996; Farabaugh, 2000; Brierley and Pennell, 2001 for reviews). A growing number of viruses are found to employ frameshifting during their replication cycle, including many retroviruses, several eukaryotic positive-strand RNA viruses, double-stranded RNA viruses of yeast, some plant RNA viruses and certain bacteriophages. In most of the systems studied to date, frameshifting is involved in the expression of replicases. In retroviruses, it allows the synthesis of the Gag-Pol and Gag-Pro-Pol polyproteins from which reverse transcriptase is derived, and for most other viruses, frameshifting is required for expression of RNA-dependent RNA polymerases. In this article, we will review ribosomal frameshifting with an emphasis on the frameshifting process in astroviruses.

Programmed −1 ribosomal frameshifting signals

Eukaryotic ribosomal frameshift signals contain two essential mRNA elements: a "slippery" sequence, where the ribosome changes reading frame, and a stimulatory RNA secondary structure, usually an RNA "pseudoknot", located a few nucleotides downstream (Jacks *et al.*, 1988a; Brierley *et al.*, 1989; ten Dam *et al.*, 1990). A spacer region between the slippery sequence and the stimulatory RNA structure is also required, and a precise length of this spacer must be maintained for maximal frameshifting efficiency (Brierley *et al.*, 1989, 1992; Kollmus *et al.*, 1994). The slippery sequence is a heptanucleotide stretch that contains two homopolymeric triplets and conforms in the vast majority of cases to the motif XXXYYYZ (for example, UUUAAAC in the coronavirus infectious bronchitis virus (IBV) *1a/1b* signal). In eukaryotic systems, frameshifting at this sequence is thought to occur by "simultaneous-slippage" of two ribosome-bound tRNAs, presumably peptidyl and aminoacyl tRNAs, which are translocated from the zero (X XXY YYZ) to the -1 phase (XXX YYY) (Jacks *et al.*, 1988a; see Fig. 1). The homopolymeric nature of the slippery sequence seems to be required to allow the tRNAs to remain base-paired to the mRNA in at least two out of three anticodon positions following the slip. Frameshift assays, largely carried out *in vitro*, have revealed that the X triplet can be A, C, G or U, but the Y triplet must be A or U (Jacks *et al.*, 1988a, Dinman *et al.*, 1991; Brierley *et al.*, 1992). In addition to these restrictions, slippery sequences ending in G (XXXAAAG or XXXUUUG) do not function efficiently in *in vitro* translation systems (Brierley *et al.*, 1992), nor in yeast (Dinman *et al.*, 1991) or mammalian cells (Marczinke *et al.*, 2000). At naturally-occurring frameshift sites, of the possible codons which are decoded in the ribosomal aminoacyl (A)-site prior to tRNA slippage (XXXY<u>YYN</u>), only five are represented in eukaryotes, AAC, AAU, UUA, UUC, UUU (Farabaugh, 1996). Together with the *in vitro* data, it seems that the sequence restrictions observed are a manifestation of the need for the pre-slippage codon-anticodon complex in the A-site to be weak enough such that the tRNAs can detach from the codon during the process of frameshifting. G-C pairs are therefore avoided.

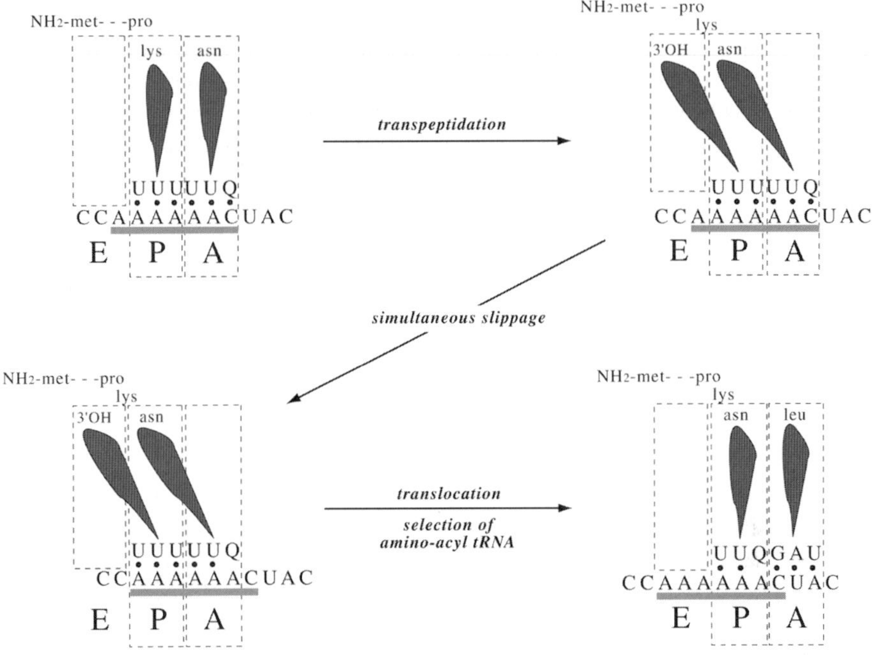

Fig. 1. The simultaneous slippage model of programmed −1 ribosomal frameshifting. The model shown is a variant of the original (Jacks *et al.*, 1988a) as proposed by Weiss *et al.* (1989). The first stage shows the amino-acyl- (A) and peptidyl- (P) tRNAs base-paired to the slippery sequence (AAAAAAC) in the zero frame before transpeptidation. In the second stage, the tRNAs are still in the zero frame but occupy hybrid sites (P-A and E (exit)-P), based on the displacement model for the peptidyl transfer reaction (Moazed and Noller, 1989). The third stage shows the tRNAs slipping back by one nucleotide, retaining (in the case of this slippery sequence) a total of five of six anticodon-codon base-pairs. In the fourth step, the incoming tRNA[Leu] decodes the −1 frame codon, completing the cycle. Q represents the base queuosine, the hypermodified derivative of guanosine present at the wobble position of tRNA[Asn] (Björk *et al.*, 1999). The specific modification of the wobble base of mammalian tRNA[Lys], 5-methylcarboxymethyl-2-thiouridine (mcm^{5}s^{2}U), is not shown.

Efficient frameshifting requires the presence of a stimulatory RNA structure located a few nucleotides downstream of the slippery sequence. In some cases this is a stem-loop, but more commonly, an RNA pseudoknot is present (an example from the coronavirus IBV is shown in Fig. 2, alongside a selection of stem-loop-containing signals). These structures are sometimes referred to as frameshifter stem-loops or frameshifter pseudoknots to distinguish them from related structures in cellular RNAs. Pseudoknots are formed when nucleotides in the loop of stem-loop base-pair with a region elsewhere in the mRNA to create a quasi-continuous double-helix joined by single-stranded connecting loops (Pleij *et al.*, 1985; Hilbers *et al.*, 1998; Giedroc *et al.*, 2000).

590

The interaction of the ribosome with the stimulatory RNA is thought to pause ribosomes in the act of decoding the slippery sequence, allowing more time for the tRNAs to realign in the -1 reading frame (Jacks *et al.*, 1988a). Our knowledge of the folding of these RNAs has been derived from site-specific mutagenesis, chemical and enzymatic structure probing and more recently from NMR and X-ray crystallography. The topic has recently been the subject of several reviews (Hilbers *et al.*, 1998; Giedroc *et al.*, 2000; Brierley and Pennell, 2001).

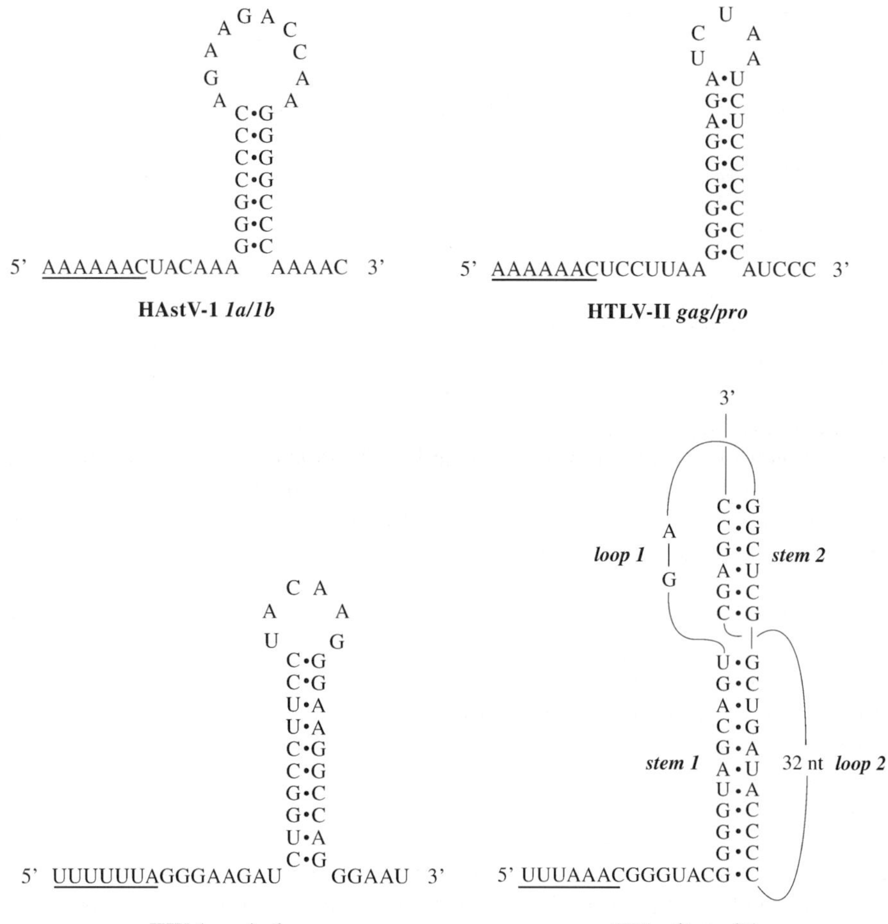

Fig. 2. Examples of stem-loop- and pseudoknot-containing ribosomal frameshift signals. Illustrated are the stem-loop signals of HAstV-1 *1a/1b* (Marczinke *et al.*, 1994), HTLV-II *gag/pro* (ten Dam *et al.*, 1990), HIV-1 *gag/pol* (Jacks *et al.*, 1988b; Kang, 1998) and the pseudoknot of the coronavirus IBV (Brierley *et al.*, 1991). The slippery sequences are underlined.

Astrovirus frameshifting signals

Astroviruses are small, non-enveloped viruses with a positive-sense, single-stranded RNA genome of about 7kb in length. Human astrovirus serotype 1 (HAstV-1; Willcocks *et al.*, 1994), like all astroviruses, contains three sequential open reading frames (ORFs) designated ORFs 1a, 1b and 2. The most 3' coding region, ORF 2, encodes the viral structural proteins (Matsui *et al.*, 1993; Lewis *et al.*, 1994) and is expressed from a sub-genomic mRNA (Monroe *et al.*, 1991; Matsui *et al.*, 1993; Geigenmüller *et al.*, Section V, Chapter 1 of this book). ORFs 1a and 1b contain amino-acid motifs indicative of non-structural proteins (reviewed in Cubitt, 1996), and a characteristic YGDD motif found in the RNA-dependent RNA polymerases of a variety of RNA viruses (Kamer and Argos, 1984) is located in ORF 1b. Initially, the mechanism of expression of this ORF was uncertain, since it overlapped ORF1a by 71 nucleotides and was in the −1 reading frame with respect to ORF1a. An examination of the sequence information present within the overlap region of the two ORFs suggested that ORF1b was expressed as a translational fusion with the upstream 1a ORF following a −1 ribosomal frameshift event within the ORF1a/1b overlap region (Jiang *et al.*, 1993; Lewis *et al.*, 1994; Willcocks *et al.* 1994). This was subsequently confirmed experimentally in *in vitro* translation systems; the HAstV signal was shown to cause some 5-7% of ribosomes to frameshift (Marczinke *et al.*, 1994; Lewis and Matsui, 1995). The frameshift signal of HAst-1 is one of the simplest described to date, comprising the slippery sequence AAAAAAC and a small GC-rich stem-loop located some six nucleotides (nts) downstream (see Fig. 2). Frameshifting at this signal was tested by cloning a region of the astrovirus genome containing the putative frameshift signal into a plasmid-borne reporter gene and performing *in vitro* transcription and translation studies using the rabbit reticulocyte lysate (RRL) system (Marczinke *et al.*, 1994). Site-specific mutagenesis confirmed the site of frameshifting as the AAAAAAC sequence, the presence of the stem region and also that the primary sequence of the loop nucleotides was unimportant, arguing against the presence of a pseudoknot. Further experimental evidence that the HAst-1 stimulatory RNA is a stem-loop and not a pseudoknot was obtained by RNA structure mapping of the *in vitro* transcripts, which indicated that the loop nucleotides were accessible to single-strand specific chemical and enzymatic probes (Marczinke *et al.*, 1994). A study of astrovirus frameshifting in an infection-transfection transient expression system also found that a minimal cassette of slippery sequence and downstream stem-loop was sufficient to induce efficient frameshifting both *in vitro* and *in vivo* arguing against the need for pseudoknot formation (Lewis and Matsui, 1996). The astrovirus signal closely resembles that present at the human T cell lymphotropic virus type II (HTLV-II) *gag/pro* junction and is similar to that used by human immundodeficiency virus type 1 (HIV-1) at the *gag/pol* junction, although the latter virus employs a U-rich rather than an A-rich slippery sequence (Fig. 2).

The complete genomic sequences of several human and animal astroviruses have been determined in the last few years. Fig. 3a shows an alignment of the sequences of the frameshift region of these astroviruses, and in Fig. 3b the predicted folding of

```
HAstV-1 [W]    ---GCAGGUUUGGAAGGUUUUCUCCAAAGGGUUAAAUCGAAAAACAAGGCCCCAAAAAAC 57
HAstV-1 [L]    ---GCAGGUUUGGAAGGUUUCCUCCAAAGGGUUAAAUCAAAAAACAAGGCCCCAAAAAAC 57
HAstV-2        ---GCAGGUCUAGAAGGUUUCCUCCAAAGAGUUAAAUCGAAAAACAAGGCCCCAAAAAAC 57
HAstV-5        ---GCAGGUCUAGAAGGUUUCCUCCAAAGAGUUAAAUCGAAAAAUAAGGCUCCAAAAAAC 57
HAstV-8        ---GCAGGUUUAGAAGGCUUCCUCCAAAGGGUUAAGUCAAAAAACAAGGCCCCAAAAAAC 57
HAstV-3        ---GCUGGUUUGGAAGGCUUUCUCCAGAGGGUCAAAUCAAAAAAUAAGGCUCCAAAAAAC 57
OAstV          -GAGCAUGGACUAAUGCCUUUCACUCAGCG--UAGGAAGCGUGUCCAGCAGCCAAAAAAC 57
TAstV          UUUUCAGGAAUUGAGAAGUUAG---AAGAUCAUGUGGUCAGUGGAGAGUGUCAAAAAAC 57
ANV            ---GAUUGUGGCGAGACUUUCGUUGAAAGGCAAGACUUCCACGUUUGUAAGUCAAAAAAC 57
                        *       *      **                                  *******
                                                                            slippery
                                                                            sequence

HAstV-1 [W]    UA-CAAAGGGCCCCAGAAGACCAAGGGGCCCAAAACUACCACUCAUUAGAUGCAUGGAAA 116
HAstV-1 [L]    UA-CAAAGGGCCCCAGAAGACCAAGGGGCCCAAAAUUACCACUCAUUAGAUGCGUGGAAA 116
HAstV-2        UA-CAAAGGGCCCCAGAAGACCAAGGGGCCCAAAAUUAUCACUCAUUAGAUGCAUGGAAA 116
HAstV-5        UA-CAAAGGGCCCCAGAAGACCAAGGGGCCCAAAACUACCACUCAUUAGAUGCAUGGAAA 116
HAstV-8        UA-CAAAGGGCCCCAGAAGACCAAGGGGCCCAAAACUACCACUCAUUAGAUGCAUGGAAG 116
HAstV-3        UA-CAAAGGGCCCCAGAAGACCAGGGGGCCCAAAACUACCACUCACUAGAUGCAUGGCAG 116
OAstV          UC-CAAAGGGGCCCUGAAGACCCGGGCCCCGAAGAGUGCAA---AUUAGACUACUGGGAG 113
TAstV          UAAUAGAGGGGCCUGUGACAACAAAGGCCCCUACCCCCGUACCAGAUUGGCUUAAAAUAU 117
ANV            UA--AAUGAGCCCCCUUCGG---GGGGCUACACACCUGUCCCUGACCAUCUUAGGUGGAA 112
                  *   *   *  *  **                    *                       *
                  spacer   5'-stem   loop   3'-stem

HAstV-1 [W]    UUGUUGCUAGAGCC-UCCGCGG--- 137
HAstV-1 [L]    UUGUUGCUAGAGCC-UCCGCGG--- 137
HAstV-2        UCAUUGCUAGAACC-UCCACGU--- 137
HAstV-5        UCUCUGCUAGAACC-UCCGCGG--- 137
HAstV-8        UCGCUACUAGAACC-UCCGCGG--- 137
HAstV-3        CUACUACUGGAGGC-CCCACGG--- 137
OAstV          CAGCUUGUUGAACCAUCUAAGGAA- 137
TAstV          UUGCAUGGGAAGAUGACAUA----- 137
ANV            CAACUGGCAAAUCUAUAUGGAACCU 137
                  *
```

Fig. 3. (A). Sequence alignment of the frameshift regions of various human and animal astroviruses. The alignment was performed using the ClustalW program within the facilities of the European Bioinformatics Institute (http://www.ebi.ac.uk/index.html). Asterisks indicate fully conserved residues. The slippery sequence AAAAAAC is such a fully conserved motif. The sequences studied were from HAstV-1 [W] (Willcocks *et al.*, 1994), HAstV-1 [L] (Lewis *et al.*, 1994), HAstV-2 (Jiang *et al.*, 1993), HAstV-3 (Oh and Schreier, 2001), HAstV-5 (Lewis *et al.*, 1994), HAst-8 (Mendez-Toss *et al.*, 2000), OAstV (Jonassen *et al.*, 2000), TAstV (Koci *et al.*, 2000) and ANV (Imada *et al.*, 2000).

the stem-loop stimulatory RNAs for the various serotypes and strains is indicated. All of the astroviruses employ the slippery sequence AAAAAAC and have the potential to form a hairpin precisely six nucleotides downstream of this sequence. Within the human astroviruses (HAstV-serotype 1, 2, 3, 5, 8) there is very little sequence variation. Indeed, in the region encompassing the slippery sequence and downstream hairpin, they are identical except for HAstV-3, where the ultimate loop nucleotide is G, not A. For the mammalian astrovirus of sheep (OAstV; Jonassen *et al.*, 2001), several changes are seen within the frameshift region, but they are all predicted to be silent with respect to the ability of the signal to promote frameshifting. Most of the changes are single nucleotide substitutions within the spacer and loop regions. There is also phylogenetic support for the formation of the hairpin structure in OAstV.

Two sequence differences are observed within the stem region, but together they maintain base-pairing, generating a G-C base pair at the equivalent position of the HAstV-1 C-G pair. When the comparison is extended to avian astroviruses there is increased sequence divergence. The frameshift region of turkey astrovirus (TAstV) is still recognisably similar to that of HAstV, despite the potential for an eight base-pair stem (as opposed to seven in HAstV) and a more variable loop sequence (Fig. 3b). The TAstV signal retains the G-rich 5'-arm of the stem and a 10 nucleotide loop as seen in most HAstV. However, the frameshift signal of avian nephritis virus (ANV), a proposed

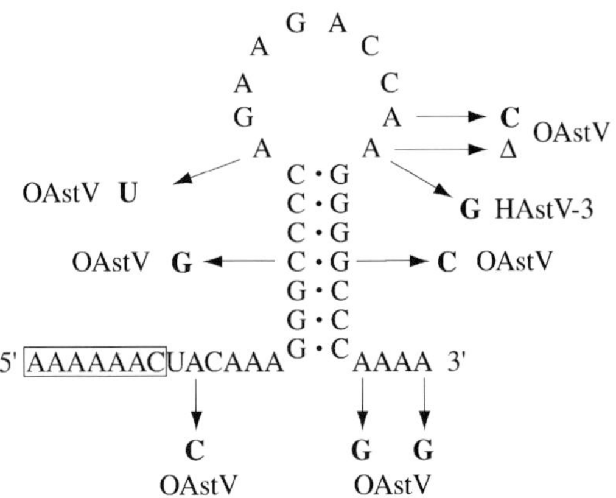

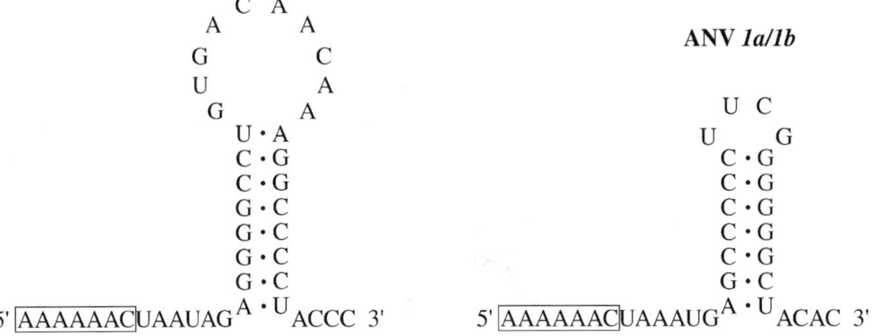

Fig. 3. (B). Secondary structure predictions of the stimulatory RNAs from the frameshift signals of HAstV-1, -2, -5, -8, TAstV and ANV. No sequence differences are present between HAstV -1, -2, -5 and -8 in the region studied.

594

astrovirus of chickens (Imada *et al.*, 2000), shows quite notable differences (Fig. 3b). The most obvious feature is the reduced loop size, to only four nucleotides, but also the stem has a C-rich 5'-arm as opposed to the G-rich stretch seen in other astroviruses. One interesting feature of the avian astroviruses is that the 1a termination codon (UAA) is located immediately downstream of the slippery sequence, whereas it is present after the frameshift signal in the HAstVs. There is no experimental evidence, however, to suggest that having a termination codon immediately downstream of the slippery sequence modifies the efficiency of the process (Brierley *et al.*, 1992).

The conservation of the slippery sequence in all astrovirus genomes sequenced to date suggests a preference for the use of lysyl- and asparaginyl-tRNAs in the frameshift process. There is considerable interest in whether the tRNAs involved in frameshifting are canonical tRNAs or special "shifty" tRNAs, more prone to frameshift than their "normal" counterparts (Jacks *et al.*, 1988a). In this respect, no novel tRNAs have been described as yet, but it has been noted that all the A-site tRNAs that function in frameshifting are decoded by tRNAs with a highly modified base in the anticodon loop (see Hatfield *et al.*, 1992 and references therein). In tRNA^{Asn} (decoding AAC and AAU), the wobble base is queuosine (Q), in tRNA^{Phe} (UUC, UUU), wyebutoxine (Y) is present just 3' of the anticodon, in tRNA^{Lys} (AAA, AAG), the wobble base is 5-methylcarboxymethyl-2-thiouridine (mcm^{5}s^{2}U) (eukaryotes), and in tRNA^{Leu} (UUA) 2-methyl-5-formylcytidine is present at the wobble position. Thus the tRNAs that decode the astrovirus frameshift signal (tRNA^{Lys}, tRNA^{Asn}) each have a modified wobble base. Hatfield *et al.* (1992) have raised the possibility that hypomodified variants (with the modification removed) of these tRNAs may exist which could act as specific "shifty" tRNAs, since such variants would have a considerably less bulky anticodon and be more free to move around at the decoding site. Support for this hypothesis comes from the observation that purified tRNA^{Phe} populations devoid of the Y modification can stimulate frameshifting at certain slippery sequences in RRL (Carlson *et al.*, 1999, 2001). In contrast, the frameshift capacity of tRNA^{Asn} appears to be uninfluenced by the absence of the Q modification in either prokaryotic (Brierley *et al.*, 1997) or eukaryotic cells (Marczinke *et al.*, 2000); thus, a simple lack of a bulky modification is not in itself sufficient to stimulate frameshifting. At present, therefore, the role of modified or hypomodified bases in frameshifting is uncertain. In some retroviruses, the relative proportions of hyper- and hypo-modified tRNAs varies in response to virus infection (Hatfield *et al.*, 1989). The possibility exists therefore that frameshift efficiencies could change during infection as a consequence of virus-induced changes in cellular tRNA modification levels. This could potentially allow the regulation of frameshifting (and hence downstream product levels) during the virus life cycle, although there is no evidence for this as yet.

Models for the ribosomal frameshifting process

Despite considerable study, the mechanism of programmed −1 ribosomal frameshifting has remained elusive. The central theory of frameshifting is that a specific ribosomal pause occurs upon encounter of the stimulatory RNA. In its simplest form, pausing will increase the time at which ribosomes are held over the slippery sequence, allowing more time for the tRNAs to realign in the -1 frame (Jacks *et al.*, 1988a). There is good experimental evidence that pausing occurs at frameshift signals. Polypeptide intermediates corresponding to ribosomes paused at RNA pseudoknots have been detected at the frameshift sites of IBV (Somogyi *et al.*, 1993) and of the yeast double-stranded RNA virus L-A (Lopinski *et al.*, 2000), and footprinting studies of elongating ribosomes have defined the site of pausing at the L-A signal (Tu *et al.*, 1992; Lopinski *et al.*, 2000) and more recently at the IBV and simian retrovirus-1 (SRV-1) signals (Kontos *et al.*, 2001). One of the great virtues of the pausing model is its ability to accommodate the variety of stimulatory RNAs that are present at -1 frameshifting signals, including stem-loops. Notwithstanding the range of secondary and tertiary features presented to the ribosome, as long as pausing occurs, frameshifting results. Unfortunately, the idea that a pause alone is sufficient to induce frameshifting is highly questionable. Simple provision of a roadblock to ribosomes in the form of stable RNA hairpins (Brierley *et al.*, 1991; Somogyi *et al.*, 1993), a tRNA (Chen *et al.*, 1995) or even different kinds of RNA pseudoknots (Napthine *et al.*, 1999; Liphardt *et al.*, 1999) is not sufficient to bring about frameshifting. Furthermore, non-frameshifting pseudoknots and stem-loops exist that can still pause ribosomes (Tu *et al.*, 1992; Somogyi *et al.*, 1993; Lopinski *et al.*, 2000; Kontos *et al.*, 2001). These experiments, of course, do not rule out a contribution of pausing to the mechanism of frameshifting since there are no documented examples where frameshifting has occurred in the absence of a detectable pause.

Regardless of whether or not pausing is the sole mediator of frameshifting, a contributor to the process, or simply a by-product of the stimulatory RNA-ribosome interaction, it is a matter of considerable interest how it is brought about. One possibility is that the stimulatory RNAs are especially resistant to unwinding by a ribosome-associated, hypothetical RNA helicase responsible for unwinding mRNA structures ahead of the decoding centre. RNA pseudoknots have been shown to possess unusual structural features which may be refractory to standard helix unwinding, for example triple helical regions formed between pseudoknot stem 1 and loop 2 regions (Le *et al.*, 1998; Su *et al.*, 1999; Michiels *et al.*, 2001). Modelling studies predict that such triplexes are likely to be one of the first features of the pseudoknot to be encountered by the ribosome (Giedroc *et al.*, 2000), where it could function to stabilise stem 1 and increase the time taken to unwind the structure. As noted several years ago (Draper, 1990) and re-iterated recently (Michiels *et al.*, 2001), another potential barrier to unwinding is the unusual topology of the pseudoknot at the beginning of stem 1, where in addition to the two base-paired strands, loop 2 adds a third strand in close proximity (Fig. 2). Perhaps the ribosome does not deal effectively with this kind of topological arrangement. In one form or another, it has been speculated that the resistance to unwinding creates a strain in the mRNA that can only be relieved by tRNA movement, i.e. by frameshifting. Although there is no direct experimental evidence to support this hypothesis, it is a compelling one.

When frameshifter pseudoknots were first described, it was proposed that they could act as binding sites for cellular or viral proteins responsible for promoting or

596

regulating the frameshift process (Jacks *et al.*, 1988a; Brierley et., 1989). Up to date, however, no such pseudoknot binding proteins have been unearthed. Certainly, if such factors exist, they are not easily titratable. The addition of a large molar excess of the SRV-1 pseudoknot to an *in vitro* translation reaction programmed with an SRV-1 frameshift reporter mRNA has no effect on frameshifting (ten Dam *et al.*, 1994), and the same is true for the IBV pseudoknot (Brierley and Inglis, unpublished observations). Furthermore, several laboratories have reported a failure to identify frameshifter pseudoknot binding factors in searches using traditional band-shift and UV crosslinking methods. These experiments do not rule out the involvement of a protein or protein complex that is tightly associated with the translation apparatus, perhaps even an integral ribosomal component, since the pseudoknot may be recognised only in the context of the elongating ribosome.

Astrovirus frameshifting: stem-loop versus pseudoknot?

The question of whether the stimulatory RNA of astroviruses is a stem-loop or a pseudoknot is an important one from the perspective of models for ribosomal frameshifting which, as seen above, are mainly tailored towards pseudoknot-containing sites and rely on pseudoknot-specific features. One way in which this can be rationalised is to propose that the stem-loop stimulators are in fact pseudoknots that have not been identified yet. Most studies of frameshifter stem-loops have been carried out on regions subcloned from the context of the natural mRNA, and it is possible that some long-range interactions have been overlooked. This is certainly true for astroviruses and needs to be addressed, although the present experimental and phylogenetic evidence supports the idea that the astrovirus signal is indeed a stem-loop. The availability of infectious astrovirus cDNA clones (Geigenmüller *et al.*, 1997; Imada *et al.*, 2000) should allow an inspection of the precise sequence requirements for astrovirus frameshifting in the context of the complete genome. A second possibility is that the stem-loops themselves possess hitherto uncharacterised, novel structural features that can promote frameshifting. A high resolution structure of a frameshifter stem-loop would be informative in this regard. Of course, it may be that our models of frameshifting are incomplete. Mutational analysis of such sites has already provided hints that the traditional combination of slippery sequence and hairpin may not be the sole defining feature of the signal and that other elements may contribute. Kim *et al.* (2001) have recently measured the frameshift efficiencies evoked *in vitro* by a series of human immunodeficiency virus type 1 (HIV-1) *gag/pol* - human T-cell lymphotropic virus type II (HTLV-II) *gag/pro* chimeras. They defined four elements, namely the slippery sequence, spacer, stem-loop *and* a region upstream of the slippery sequence and combined these in various ways to create a range of hybrid sites. Surprisingly, it was found that the regions flanking the slippery sequence and stem-loop could influence frameshifting quite dramatically, possibly by modulating stem-loop unfolding kinetics. It is not yet known whether the frameshifting efficiencies of the human and animal astroviruses differ. Certainly the spacer and flanking nucleotide sequences vary, and on the basis of the studies discussed above, this could affect the level of frameshifting. Another poorly studied area is how the translational environment influences frameshifting at stem-loop-containing sites. Experiments have revealed that under certain conditions, the HAstV-1 frameshift signal can produce frameshift efficiencies in the region of 25%, but the molecular basis for this enhanced frameshifting is not fully understood (Lewis and Matsui, 1996). Animal virus frameshift signals have largely been studied in *in vitro* translation systems, where the ribosomal load on the mRNA is thought to be low. To date, only a few groups have studied the process *in vivo,* where translation takes place on polysomes and the ribosomal load is probably higher. Under these circumstances, one would predict a reduced efficiency of frameshifting, since refolding of the stimulatory RNA unwound by the first ribosome in the polysome may not occur before the next ribosome is encountered. This topic is discussed in more detail elsewhere (Lewis and Matsui, 1996).

Frameshifting as a target for antiviral intervention

For most viruses that employ frameshifting during the replication cycle, its exact role is uncertain, but it is reasonably well understood for the retroviruses where it allows replicative enzymes (as part of the Pol polyprotein) to be synthesised as a C-terminal extension of the structural proteins (Gag). The product of ribosomal frameshifting, the Gag-Pol polyprotein, is incorporated into virions and this is an essential step in the virus life cycle, since the reverse transcriptase enzyme (encoded in *pol*) is required for subsequent events. There is growing evidence that modulation of frameshift efficiency can reduce retroviral infectivity, either by eliminating the replicase from virions, or by influencing particle assembly (Karacostas *et al.*, 1993; Hung *et al.*, 1998; Shehu-Xhilaga *et al.*, 2001). The same is true for the yeast double-stranded RNA virus L-A, which has Gag and Pol homologues (Dinman and Wickner, 1992). Modulation of frameshift efficiency may also diminish the infectivity of positive-strand RNA viruses, since it would alter the stoichiometry of non-structural proteins within the infected cell. Various antibiotics have been found to influence frameshifting efficiency at the L-A signal, and this has led to the working hypothesis that they could be used as antiviral drugs (Dinman *et al.*, 1997, 1998). High-throughput screening has also been employed in the search for candidate anti-frameshift drugs active against the HIV-1 stem-loop signal (Hung *et al.*, 1998). One such compound, RG501 {1,4-bis-[N-(3-N,N-dimethylpropyl) amidino] benzene tetrahydrochloride} was found to stimulate frameshifting at the HIV-1 signal about two-fold and inhibited HIV-1 replication in tissue culture. The drug was also able to stimulate frameshifting at the stem-loop-containing signals of HIV-2, simian immunodeficiency virus type 1 and HTLV I *gag/pro*, but not HTLV-1 *pro/pol*, which is thought to contain a pseudoknot. Recently, we have demonstrated that RG501 also stimulates frameshifting at the HAstV-1 signal *in vitro,* but not at pseudoknot containing sites (Vidakovic, Hamirally and Brierley; unpublished observations). It has been speculated (Hung *et al.*, 1998) that RG501 acts by binding to the loop region of hairpins (perhaps by intercalation), stabilising the structure and promoting frameshifting by increasing ribosomal pausing. We have sought evidence that RG501 binds to the HAstV-1 stimulatory stem-loop by probing the secondary structure of the HAstV-1 frameshift signal in the presence and absence of RG501, and the results are shown in Fig. 4. In this experiment, a short (109 nucleotides) transcript containing the HAstV-1 frameshift region was prepared, end-labelled with ^{33}P-γ-ATP and treated with limiting quantities of enzymatic or chemical structure-specific probes. The single-strand specific enzymatic probes employed were RNase T1, which cleaves 3' of unpaired G residues and RNase U2, which cleaves 3' of single-stranded A or G, with a preference for A. To probe double-stranded regions, RNase V1 was employed, which cleaves in helical regions. RNase V1 is not base-specific but cleaves RNA that is in helical conformation, whether base-paired (a minimum of 4-6bp are required) or single stranded and stacked. We also used the single-strand-specific chemical probes imidazole (Vlassov *et al.*, 1995) and lead acetate (Krzyzosiak *et al.*, 1988; Kolchanov *et al.*, 1996). In reactions with RG501, the compound was preincubated with the RNA for 10 minutes at 30°C prior to structure probing. The structure probing reactions were analysed on a 15% denaturing polyacrylamide gel and bands visualised by autoradiography (Fig. 4a). A summary of the structure probing data is shown in Fig. 4b. The reaction conditions are such that only a proportion of the input RNA is cleaved and the remainder appears as a major band at the top of the gel. The 5'-cleavage fragments migrate at a faster rate and are identified by their specific reactivities and by comparison to a marker ladder of bands differing by single base increments prepared by limited alkaline hydrolysis of the input RNA. The outcome of this experiment was unambiguous in that the regions of the frameshift signal proposed to be single- or double-stranded behaved exactly as predicted, confirming that when the HAstV-1 signal is expressed on a short transcript it folds into a stem-loop. However, we were unable to find any differences in the structure probing pattern when the RNA had been preincubated with RG501, even though the concentrations of RG501 were chosen to mimic those that had an effect on frameshifting *in vitro*. A related approach has been used to demonstrate the binding of Hoechst 33258 dye to the TAR RNA of HIV-1 (Dassonneville *et al.*, 1997), but in our case, we were unable to find any differences. It is possible that the RG501 did bind to the astrovirus transcript but the resulting complexes were not sufficiently altered in conformation to affect cleavage by the various enzymatic and chemical probes.

Fig. 4. (A). Secondary structure probing of the HAst-1 [W] ribosomal frameshifting signal in the presence and absence of compound RG501. A short transcript of 109 nucleotides derived from plasmid pAV1 (Marczinke *et al.*, 1994) was 5' end-labelled with γ-[^{33}P]-ATP and subjected to limited digestion by enzymatic and chemical probes (according to Napthine *et al.*, 1999) in the presence and absence of varying quantities of RG501. Sites of cleavage were identified by comparison with a ladder of bands created by limited alkaline hydrolysis of the RNA (OH⁻) and the position of known RNase U2 and T1 cuts, determined empirically. Products were analysed on a 15% acrylamide/7M urea gel. Enzymatic structure probing was with RNases U2, V1 and T1. Uniquely cleaved nucleotides were identified by their absence in untreated control lanes (0). The number of units of enzyme added to each reaction (final volume 50µl) is indicated. Chemical structure

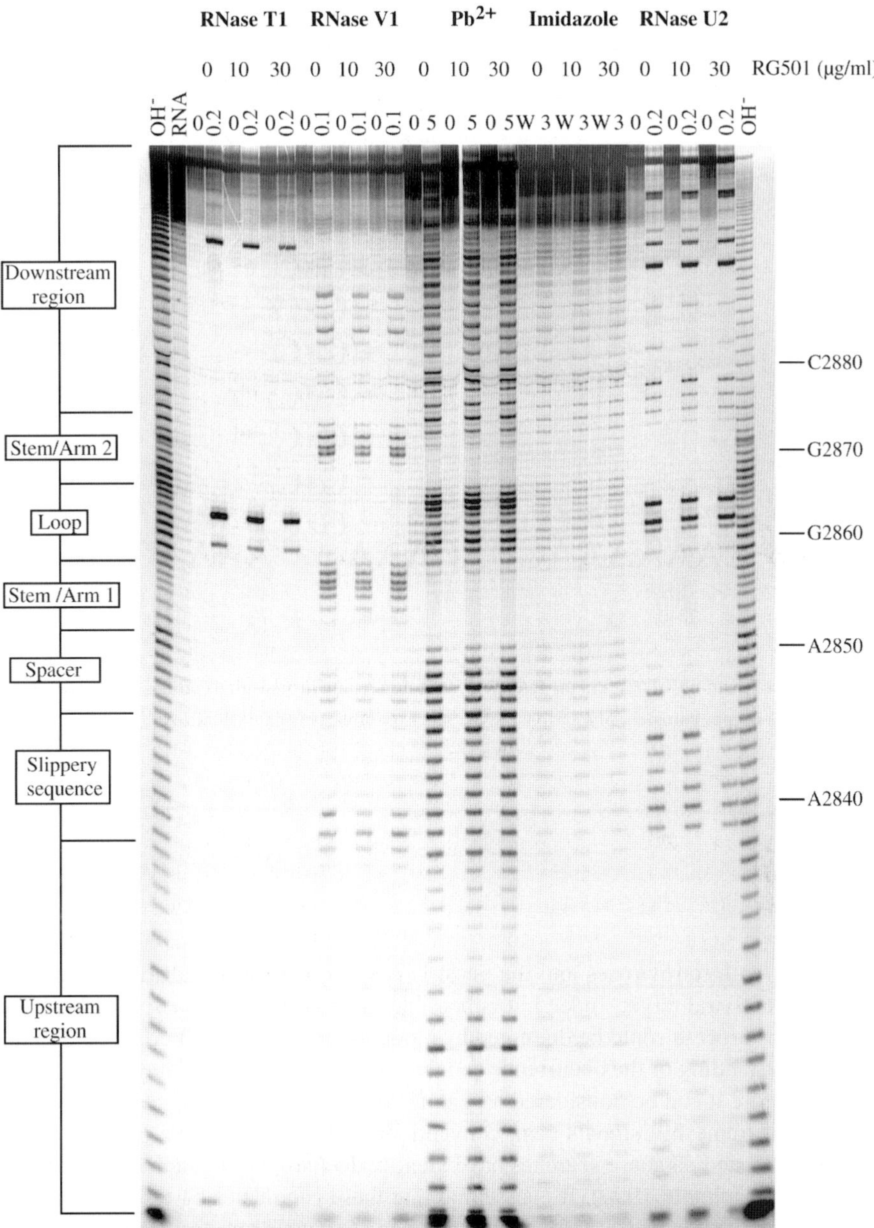

probing was with 2M imidazole (I, hours of reaction) or lead acetate (Pb^{2+}; mM concentration in reaction). The water lane (W) in the imidazole panel represents RNA which was dissolved in water, incubated for 3 hours and processed in parallel to the imidazole-treated samples. RNA represents an aliquot of the purified RNA loaded directly onto the gel without incubation in a reaction buffer.

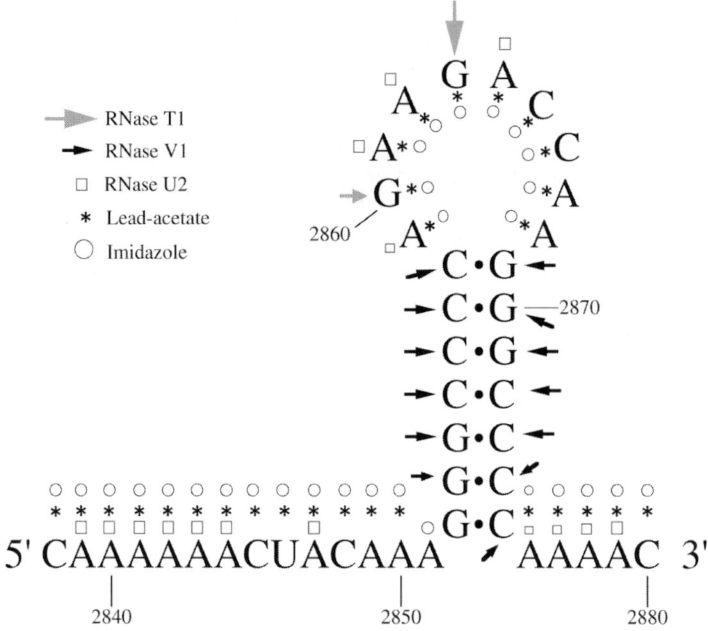

Fig. 4. (B). The sensitivity of bases in the HAstV-1 frameshift region to the various probes is shown. The size of the symbols is approximately proportional to the intensity of cleavage at that site.

Conclusions

From work with retroviruses and the yeast L-A virus, it is clear that frameshifting is a potential antiviral target. Indeed, it is possible that the replication cycle of any virus that uses this process could be disrupted by modulation of frameshift efficiencies, but a better understanding of the occurrence and the molecular basis of frameshifting will be required before it can be considered a genuine target. To date, there are no confirmed examples of frameshift signals from conventional eukaryotic cellular genes, although computer-assisted database searches have identified a number of candidates (Hammell et al., 1999; Liphardt, 1999). It is essential that these candidates be tested rigorously; although compounds like RG501 were developed as specific anti-frameshifting agents, the occurrence of cellular frameshifting signals would potentially preclude the use of such agents. For all positive-stranded RNA viruses, the exact role of frameshifting is unknown. Presumably, the frameshift allows the required ratio of viral proteins to be produced, but it may also serve to downregulate levels of viral replicases, which may be toxic in high amounts. Regarding astrovirus frameshifting, there is still much to learn. The key issue will be to determine the sequence requirements for frameshifting within the context of the complete virus genome, especially regarding the stimulatory RNA and whether it can form a pseudoknot. Infectious cDNA clones will be invaluable in this analysis and will also allow the study of the effect on virus replication of varying the frameshift efficiency. From our own perspective, with an interest in the frameshift mechanism, we are currently sequencing astrovirus clinical isolates to establish the level of natural sequence variation that is tolerated within the astrovirus frameshift region. It may be that we can identify conserved bases that may play a role in frameshifting, either through their primary sequence (binding factor recognition) or in long range, tertiary interactions with distal regions of the genome. Finally, it is important that stem-loop containing frameshift signals receive the same degree of experimental attention as pseudoknot-containing signals. It is very likely that such studies will reveal important clues to the mechanism of the frameshift process.

Acknowledgements

Research in the authors' laboratory is supported by the Medical Research Council UK and the Biotechnology and Biological Sciences Research Council, UK. MV would like to thank the Cambridge Overseas Trust for financial support during her PhD studies.

References

Björk G.R, Durand J.M., Hagervall T.G., Leipuviene R., Lundgren H.K., Nilsson K., Chen P., Qian Q., and Urbonavicius J. (1999). Transfer RNA modification: influence on translational frameshifting and metabolism. *FEBS Lett.* **452**:47-51.

Brierley I. 1995. Ribosomal frameshifting on viral RNAs. *J. Gen. Virol.* **76**: 1885-1892.

Brierley I., Digard P., and Inglis S.C. 1989. Characterisation of an efficient coronavirus ribosomal frameshifting signal: requirement for an RNA pseudoknot. *Cell* **57**: 537-547.

Brierley I., Jenner A.J., and Inglis S.C. 1992. Mutational analysis of the "slippery sequence" component of a coronavirus ribosomal frameshifting signal. *J. Mol. Biol.* **227**: 463-479.

Brierley I., Meredith M.R., Bloys A.J., and Hagervall T.G. 1997. Expression of a coronavirus ribosomal frameshift signal in *Escherichia coli*: influence of tRNA anticodon modification on frameshifting. *J. Mol. Biol.* **270**: 360-373.

Brierley I., and Pennell S.P. 2001. Structure and function of the stimulatory RNAs involved in programmed eukaryotic -1 ribosomal frameshifting. Cold Spring Harbor Symposia on Quantitative Biology, Vol. LXVI, Cold Spring Harbor, N.Y.

Brierley I., Rolley N.J., Jenner A.J., and Inglis S.C. 1991. Mutational analysis of the RNA pseudoknot component of a coronavirus ribosomal frameshifting signal. *J. Mol. Biol.* **220**: 889-902.

Carlson B.A., Kwon S.Y., Chamorro M., Oroszlan S., Hatfield D.L., and Lee B.J.

1999. Transfer RNA modification status influences retroviral frame-shifting. *Virology* **255**: 2-8.

Carlson B.A., Mushinski J.F., Henderson D.W., Kwon S.Y., Crain P.F., Lee B.J., and Hatfield D.L. 2001. 1-Methylguanosine in place of Y base at position 37 in phenylalanine tRNA is responsible for its shiftiness in retroviral ribosomal frameshifting. *Virology* **279**: 130-135.

Chandler M., and Fayet O. 1993. Translational frameshifting in the control of transposition in bacteria. *Mol. Microbiol.* **7**: 497-503.

Chen X., Chamorro M., Lee S.I., Shen L.X., Hines J.V., Tinoco I.Jr., and Varmus H.E. 1995. Structural and functional studies of retroviral RNA pseudoknots involved in ribosomal frameshifting: nucleotides at the junction of the two stems are important for efficient ribosomal frameshifting. *EMBO J.* **14**: 842-852.

Cubitt W.D. 1996. Historical background and classification of caliciviruses and astroviruses. *Arch. Virol. Suppl.* **12**: 225-235.

Dassonneville L., Hamy F., Colson P., Houssier C., and Bailly C. 1997. Binding of Hoechst 33258 to the TAR RNA of HIV-1. Recognition of a pyrimidine bulge-dependent structure. *Nucleic Acids Res.* **25**: 4487-4492.

Dinman J.D. 1995. Ribosomal frameshifting in yeast viruses. *Yeast* **11**: 1115-1127.

Dinman J.D., Icho T., and Wickner R.B. 1991. A -1 ribosomal frameshift in a double-stranded RNA virus of yeast forms a Gag-Pol fusion protein. *Proc. Natl. Acad. Sci. U.S.A.* **88**: 174-178.

Dinman J.D., Ruiz-Echevarria M.J., Czaplinski K., and Peltz S.W. 1997. Peptidyl-transferase inhibitors have antiviral properties by altering programmed -1 ribosomal frameshifting efficiencies: development of model systems. *Proc. Natl. Acad. Sci. U.S.A.* **94**: 6606-6611.

Dinman J.D., Ruiz-Echevarria M.J., and Peltz S.W. 1998. Translating old drugs into new treatments: ribosomal frameshifting as a target for antiviral agents. *Trends Biotechnol.* **16**: 190-196.

Dinman J.D., and Wickner R.B. 1992. Ribosomal frameshifting efficiency and *gag/gag-pol* ratio are critical for yeast M1 double-stranded RNA virus propagation. *J. Virol.* **66**: 3669-3676.

Draper D.E. 1990. Pseudoknots and the control of protein synthesis. *Curr. Opin. Cell Biol.* **2**: 1099-1103.

Farabaugh P.J. 1996. Programmed translational frameshifting. *Microb. Rev.* **60**: 103-134.

Farabaugh P.J. 2000. Translational frameshifting: implications for the mechanism of translational frame maintenance. *Prog. Nucleic Acids Res. Mol. Biol.* **64**: 131-170.

Futterer J., and Hohn T. 1996. Translation in plants--rules and exceptions. *Plant Mol. Biol.* **32**: 159-189.

Geigenmüler U., Ginzton N.H., and Matsui S.M. 1997. Construction of a genome-length cDNA clone for human astrovirus serotype 1 and synthesis of infectious RNA transcripts. *J. Virol.* **71**: 1713-1717.

Giedroc D.P., Theimer C.A., and Nixon P.L. 2000. Structure, stability and function of RNA pseudoknots involved in stimulating ribosomal frameshifting. *J. Mol. Biol.* **298**: 167-185.

Hammell A.B., Taylor R.C., Peltz S.W., and Dinman J.D. 1999. Identification of putative programmed -1 ribosomal frameshift signals in large DNA databases. *Genome Res.* **9:** 417-427.

Hatfield D., Feng Y.X., Lee B.J., Rein A., Levin J.G., and Oroszlan S. 1989. Chromatographic analysis of the aminoacyl-tRNAs which are required for translation of codons at and around the ribosomal frameshift sites of HIV, HTLV-1, and BLV. *Virology* **173**: 736-742.

Hatfield D., Levin J.G., Rein A., and Oroszlan S. 1992. Translational suppression in retroviral gene expression. *Adv. Virus Res.* **41:** 193-239.

Hilbers C.W., Michiels P.J., and Heus H.A. 1998. New developments in structure determination of pseudoknots. *Biopolymers* **48:** 137-153.

Hung M., Patel P., Davis S., and Green S.R. 1998. Importance of ribosomal frameshifting for human immunodeficiency virus type 1 particle assembly and replication. *J. Virol.* **72:** 4819-4824.

Imada T., Yamaguchi S., Mase M., Tsukamoto K., Kubo M., and Morooka A. 2000. Avian nephritis virus (ANV) as a new member of the family *Astroviridae* and construction of infectious ANV cDNA. *J. Virol.* **74**: 8487-8493.

Jacks T., Madhani H.D., Masiarz F.R., and Varmus H.E. 1988a. Signals for ribosomal frameshifting in the Rous sarcoma virus *gag-pol* region. *Cell* **55:** 447-458.

Jacks T., Power M.D., Masiarz F.R., Luciw P.A., Barr P.J., and Varmus H.E. 1988b. Characterization of ribosomal frameshifting in HIV-1 *gag-pol* expression. *Nature* **331:** 280-286.

Jiang B., Monroe S.S., Koonin E.V., Stine S.E., and Glass R.I. 1993. RNA sequence of astrovirus: distinctive genomic organization and a putative retrovirus-like ribosomal frameshifting signal that directs the viral replicase synthesis. *Proc. Natl. Acad. Sci. U.S.A.* **90:** 10539-10543.

Jonassen C.M., Jonassen T.O., Saif Y.M., Snodgrass D.R., Ushijima H., Shimizu M., and Grinde B. 2001. Comparison of capsid sequences from human and animal astroviruses. *J. Gen. Virol.* **82**: 1061-1067.

Kamer G., and Argos P. 1984. Primary structural comparison of RNA-dependent RNA polymerases from plant, animal and bacterial viruses. *Nucleic Acids Res.* **12**: 7269-7282.

Kang H. 1998. Direct structural evidence for formation of a stem-loop structure involved in ribosomal frameshifting in human immunodeficiency virus type 1. *Biochim. Biophys. Acta* **1397**: 73-78.

Karacostas V., Wolffe E.J., Nagashima K., Gond, M.A., and Moss B. 1993. Overexpression of the HIV-1 *gag-pol* polyprotein results in intracellular activation of HIV-1 protease and inhibition of assembly and budding of virus-like particles. *Virology* **193:** 661-671.

Kim Y.G., Maas S., and Rich A. 2001. Comparative mutational analysis of *cis*-acting RNA signals for translational frameshifting in HIV-1 and HTLV-2. *Nucleic Acids Res.* **29:** 1125-1131.

Koci M.D., Seal B.S., and Schultz-Cherry S. 2000. Molecular characterization of an avian astrovirus. *J. Virol.* **74**: 6173-6177.

Kollmus H., Honigman A., Panet A., and Hauser H. 1994. The sequences of and distance between two *cis*-acting signals determine the efficiency of ribosomal frameshifting in human immunodeficiency virus type 1 and human T-cell leukemia virus type II *in vivo*. *J. Virol.* **68:** 6087-6091.

Kontos H., Napthine S., and Brierley I. 2001. Ribosomal pausing at a frameshifter RNA pseudoknot is sensitive to reading phase but shows little correlation with frameshift efficiency. *Mol. Cell. Biol.* **21:** 8657-8670.

Kolchanov N.A., Titov I.I., Vlassova I.E., and Vlassov V.V. 1996. Chemical and computer probing of RNA structure. *Prog. Nucleic Acids Res. Mol. Biol.* **53:** 131-196.

Kozak M. 1983. Comparison of initiation of protein synthesis in procaryotes, eucaryotes, and organelles. *Microbiol. Rev.* **47:** 1-45.

Krzyzosiak W.J., Marciniec T., Wiewiorowski M., Romby P., Ebel J.-P., and Giege, R. 1988. Characterisation of the lead (II)-induced cleavages in tRNAs in solution and effect of the Y-base removal in yeast tRNA Phe. *Biochemistry* **27:** 5771-5777.

Le S.Y., Chen J.H., Pattabiraman N., and Maizel J.V. 1998. Ion-RNA interactions in the RNA pseudoknot of a ribosomal frameshifting site: molecular modeling studies. *J. Biomol. Struct. Dyn.* **16:** 1-11.

Lewis T. L., Greenberg H.B., Herrmann J.E., Smith L.S., and Matsui S.M. 1994. Analysis of astrovirus serotype 1 RNA, identification of the viral RNA-dependent RNA polymerase motif and expression of a viral structural protein. *J. Virol.* **68:** 77-83.

Lewis T.L., and Matsui S.M. 1995. An astrovirus frameshift signal induces ribosomal frameshifting *in vitro*. *Arch. Virol.* **140:** 1127-1135.

Lewis T.L, and Matsui S.M. 1996. Astrovirus ribosomal frameshifting in an infection-transfection transient expression system. *J. Virol.* **70:** 2869-2875.

Liphardt J.T. 1999. The mechanism of -1 ribosomal frameshifting: experimental and theoretical analysis. Ph.D Thesis, University of Cambridge, United Kingdom.

Liphardt J., Napthine S., Kontos H., and Brierley I. 1999. Evidence for an RNA pseudoknot loop-helix interaction essential for efficient -1 ribosomal frameshifting. *J. Mol. Biol.* **288:** 321-335.

Lopinski J.D., Dinman J.D., and Bruenn J.A. 2000. Kinetics of ribosomal pausing during programmed -1 ribosomal frameshifting. *Mol. Cell. Biol.* **20:** 1095-1103.

Marczinke B., Bloys A.J., Brown T.D., Willcocks M.M., Carter M.J., and Brierley I. 1994. The human astrovirus RNA-dependent RNA polymerase coding region is expressed by ribosomal frameshifting. *J. Virol.* **68:** 5588-5595.

Marczinke B., Hagervall T., and Brierley I. 2000. The Q-base of asparaginyl-tRNA is dispensible for efficient -1 ribosomal frameshifting in eukaryotes. *J. Mol. Biol.* **295:** 179-191.

Matsui S.M., Kim J.P., Greenberg H.B., Young L.M., Smith L.S., Lewis T.L., Herrmann J.E., Blacklow N.R., Dupuis K., and Reyes G.R. 1993. Cloning and characterization of human astrovirus immunoreactive epitopes. *J. Virol.* **67:** 1712-1715.

Mendez-Toss M., Romero-Guido P., Munguia M.E., Mendez E., and Arias C.F. 2000. Molecular analysis of a serotype 8 human astrovirus genome. *J. Gen. Virol.* **81**: 2891-2897.

Michiels P.J., Versleijen A.A., Verlaan P.W., Pleij C.W., Hilbers C.W., and Heus H.A. 2001. Solution structure of the pseudoknot of SRV-1 RNA, involved in ribosomal frameshifting. *J. Mol. Biol.* **310**: 1109-1123.

Moazed D., and Noller H.F. 1989. Intermediate states in the movement of transfer RNA in the ribosome. *Nature* **342**: 142-148.

Monroe S.S., Stine S.E., Gorelkin L., Herrmann J.E., Blacklow N.R., Glass R.I. 1991. Temporal synthesis of proteins and RNAs during human astrovirus infection of cultured cells. *J. Virol.* **65**: 641-648.

Napthine S., Liphardt J., Bloys A., Routledge S., and Brierley I. 1999. The role of RNA pseudoknot stem 1 length in the promotion of efficient -1 ribosomal frameshifting. *J. Mol. Biol.* **288**: 305-320.

Oh D., and Schreier E. 2001. Molecular characterization of human astroviruses in Germany. *Arch. Virol.* **146**: 443-455.

Pe'ery T., and Mathews M.B. 2000. Viral translational strategies and host defense mechanisms. In *"Translational control of gene expression"* (N. Sonenberg, J. Hershey and M. Mathews, eds.), p371. Cold Spring Harbor Laboratory Press, Cold Spring Harbor, New York.

Pestova T.V., Kolupaeva V.G., Lomakin I.B., Pilipenko E.V., Shatsky I.N., Agol V.I., and Hellen C.U. 2001. Molecular mechanisms of translation initiation in eukaryotes. *Proc. Natl. Acad. Sci. U S A.* **98**: 7029-7036.

Pleij C.W.A., Rietveld K., and Bosch L. 1985. A new principle of RNA folding based on pseudoknotting. *Nucleic Acids Res.* **13**: 1717-1731.

Shehu-Xhilaga M., Crowe S.M., and Mak J. 2001. Maintenance of the Gag/Gag-Pol ratio is important for human immunodeficiency virus type 1 RNA dimerization and viral infectivity. *J. Virol.* **75**: 1834-1841.

Somogyi P., Jenner A.J., Brierley I., and Inglis S.C. 1993. Ribosomal pausing during translation of an RNA pseudoknot. *Mol. Cell. Biol.* **13**: 6931-6940.

Su L., Chen L., Egli M., Berger J.M., and Rich A. 1999. Minor groove RNA triplex in the crystal structure of a ribosomal frameshifting viral pseudoknot. *Nature Struct. Biol.* **6**: 285-292.

Ten Dam E.B., Pleij C.W.A., and Bosch L. 1990. RNA pseudoknots: translational frameshifting and readthrough on viral RNAs. *Virus Genes* **4**: 121-136.

Ten Dam E., Brierley I., Inglis S., and Pleij C. 1994. Identification and analysis of the pseudoknot-containing *gag-pro* ribosomal frameshift signal of simian retrovirus-1. *Nucleic Acids Res.* **22**: 2304-2310.

Tu C., Tzeng T-H., and Bruenn J.A. 1992. Ribosomal movement impeded at a pseudoknot required for frameshifting. *Proc. Natl. Acad. Sci. U.S.A..* **89**: 8636-8640.

Vlassov V.V., Zuber G., Felden B., Behr J-P., and Giege R. 1995. Cleavage of tRNA with imidazole and spermine imidazole constructs: a new approach for probing RNA structure. *Nucleic Acids Res.* **23**: 3161-3167.

Weiss R.B., Dunn D.M., Shuh M., Atkins J.F., and Gesteland R.F. 1989. *E. coli* ribosomes re-phase on retroviral frameshift signals at rates ranging from 2 to 50 percent. *New Biol.* **1:** 159-169.

Willcocks M.M., Brown T.D., Madeley C.R., and Carter M.J. 1994. The complete sequence of a human astrovirus. *J. Gen. Virol.* **75**: 1785-1788.

Viral Gastroenteritis
U. Desselberger and J. Gray (editors)
© 2003 Elsevier Science B.V. All rights reserved

V, 3. Molecular epidemiology of human astroviruses

Stephan S. Monroe

Division of Viral and Rickettsial Diseases, National Center for Infectious Diseases,
Centers for Disease Control and Prevention, Atlanta, GA, 30333 USA,

Introduction

Astroviruses were first detected by electron microscopy (Appleton *et al.*, 1975; Madeley and Crosgrove, 1975), and the original assays for typing human astrovirus strains employed time-consuming immune electron microscopy or cell culture neutralization tests (Lee and Kurtz, 1994). The development of enzyme immunoassays for detecting (Herrmann *et al.*, 1990) and typing (Noel *et al.*, 1995) human astroviruses greatly simplified the characterization of strains and provided confirmation of the earlier studies indicating that serotype 1 (HAstV-1) is most common, accounting for over half of all strains detected (Noel *et al.*, 1995). The further development of RT-PCR assays for detecting human astroviruses allowed for more detailed characterization of astrovirus strains by analysis of nucleotide sequence information (Jonassen *et al.*, 1995; Mitchell *et al.*, 1995; Saito *et al.*, 1995). The usefulness of this approach was demonstrated when genotypes inferred by phylogenetic analysis of sequences from a region within the conserved P2 domain of open reading frame 2 (ORF2) were shown to correlate precisely with antigenic types determined by type-specific EIA (Noel *et al.*, 1995). The increasing application of genetic comparisons for characterization of astrovirus strains has provided interesting molecular insights into the epidemiology of astrovirus infection.

In the past few years, researchers in several countries have conducted studies on the prevalence of astrovirus in children with diarrhea, and many have employed a standard approach for detection and genotyping strains by RT-PCR. Altogether, 461 astrovirus strains were detected in 9624 samples from six continents with an overall detection rate of 4.8 percent. In all studies except for one conducted in Mexico, HAstV-1 was the most commonly detected genotype, accounting for over 56 percent of the typed strains. Types HAstV-2, HAstV-3, and HAstV-4 accounted for 9.4, 13.3, and 10.1 percent, respectively, while types HAstV-5, HAstV-6, and HAstV-7 each accounted for less than 4 percent of the strains. Interestingly, the recently described HAstV-8 while accounting for only 4.3 percent of strains overall, was relatively common in two recent studies in Egypt and Spain (Guix *et al.*, 2002; Naficy *et al.*, 2001), suggesting that it may be an emerging genotype.

608

Table 1

Studies on human gastroenteritis caused by astroviruses

	Astrovirus Detection					Astrovirus Typing							
Location	Period	Study Population	No. of Samples Tested	No. of Astrovirus Positive (%)	Typed (%)	T1	T2	T3	T4	T5	T6	T7	T8
Mexico[1]	1988-1991	Children	365	23 (6.3)	31[†] (100)	3	13	4	7	2	0	2	N/D
USA[2]	1993-1994	Child Care Centers	179	36 (20.1)	27 (75.0)	22	5	0	0	0	0	0	0
Japan[3]	1995-1998	Children	1382	82 (5.9)	82 (100.0)	66	0	14	2	0	0	0	0
Australia[4]	1995-1998	Hospitalized Children	1327	40 (3.0)	40 (100.0)	30	2	0	7	0	0	0	1
Egypt[5]	1995-1999	Children	3477	123 (3.5)	84[‡] (68.3)	36	3	10	4	13	7	0	11
USA[6]	1996-1997	Hospitalized Children	267	26 (9.7)	18 (69)	15	2	1	0	0	0	0	N/D
Spain[7]	1997-2000	Hospitalized Children	2347	116 (4.9)	49 (42)	18	3	10	12	0	0	0	6
Colombia, Venezuela[8]	1997-1999	Children	280	15 (5.4)	11 (73)	6	3	1	1	0	0	0	0

Table 1

Continued

	Astrovirus Detection					Astrovirus Typing							
Location	Period	Study Population	No. of Samples Tested	No. of Astrovirus Positive (%)	Typed (%)	T1	T2	T3	T4	T5	T6	T7	T8
Total Detected			9624	461 (4.8%)									
Various[9]	N/A	Various	N/A	57	57 (100)	32	7	10	6	1	1	0	N/D
Germany[10]	1997-1999	N/A	N/A	16	16 (100)	7	1	5	3	0	0	0	0
Total Typed				534	415 (78)	235	39	55	42	16	8	2	18
Percent					100.0	56.6	9.4	13.3	10.1	3.9	1.9	0.5	4.3

[1] Walter *et al.*, 2001
[2] Mitchell *et al.*, 1999
[3] Sakamoto *et al.*, 2000
[4] Mustafa *et al.*, 2000
[5] Naficy *et al.*, 2000
[6] Shastri *et al.*, 1998
[7] Guix *et al.*, 2002
[8] Medina *et al.*, 2000
[9] Noel *et al.*, 1995
[10] Oh and Schreier, 2001

[†] Includes eight strains from older subjects
[‡] Includes two mixed infections
N/D = not done
N/A = not available

Although the frequency of detection of astrovirus strains by genotype is relatively constant across these studies, distinct geographic patterns begin to emerge when one looks more closely using detailed comparisons of sequence information. Consistent with HAstV-1 strains being those most commonly detected, there is the most sequence data available from strains classified in this genogroup (Fig. 1). Within this genogroup, one can distinguish four distinct lineages, which have been denoted 1a-1d (Fig. 1). Lineage 1a, which includes the prototype strain, OXFORD_T1_GBR, also includes strains from South America (Columbia and Peru), North America (United States), Europe (Spain, Germany, United Kingdom), Asia (Korea), and Australia, with little apparent geographic linkage. Somewhat in contrast, the strains in lineages 1b and 1d appear to be relatively closely linked by location, with distinct clusters of strains from Spain, Columbia, Egypt, Australia, Malawi, and Germany. A Venezuelan strain, Ven-835, is the only characterized member of lineage 1c.

Although detection of HAstV-2 strains is fairly common, few sequences of strains from this genogroup are available for analysis. These strains segregate into two distinct lineages (2a and 2b), with a scattering of strains from several collections in each lineage (Fig. 2). On particular note in the comparison of HAstV-3 strains is the close similarity of two strains from Germany isolated in 1998/99 to two strains from Australia (Fig. 2).

One limitation of strain classification based on partial nucleotide sequence information from the relatively conserved P2 domain of ORF2 becomes apparent when one compares sequences from HAstV-4 and HAstV-8 strains. The between-type pairwise distances are much lower for genotypes 4 and 8 than they are for the other genotypes, raising concerns about the validity of considering them as distinct types (Belliot *et al.*, 1997). When one compares complete amino acid sequences from the variable P3 domain, however, all eight genotypes are nominally equidistant from one another, consistent with the current classification scheme (Monroe *et al.*, 2001). Despite their recent recognition, there is considerable sequence variation within HAstV-8 strains, with only those from the Spanish collection having identical sequences (Fig. 3). This suggests that these strains have been circulating undetected for some time. Twelve HAstV-4 strains have been characterized from a collection in Ehime, Japan, and they segregate into three distinct clusters (Fig. 3, arrows). One tightly clustered group of eight strains covers a period of seven years with identical sequences detected in strains from 1987 and 1993 (Fig. 3, middle arrow). Cocirculating in this community at the same time were two other clusters of HAstV-4 strains, one of which is quite similar to strains detected in Germany ten years later (Fig. 3, upper arrow).

Finally, for the rare genotypes, HAstV-5, -6, and -7, there are sequences of fewer strains to compare, but they do not appear to be restricted geographically (Fig. 4). The twelve strains classified as HAstV-5, for example, are from samples collected on five different continents.

The choice of which region of the genome is used for genetic comparison can influence the outcome of the analysis. In contrast to the close correlation between antigenic type and genotype based on sequences from ORF2, a similar analysis of sequences from within ORF1a resulted in a quite different grouping of strains (Belliot *et al.*, 1997).

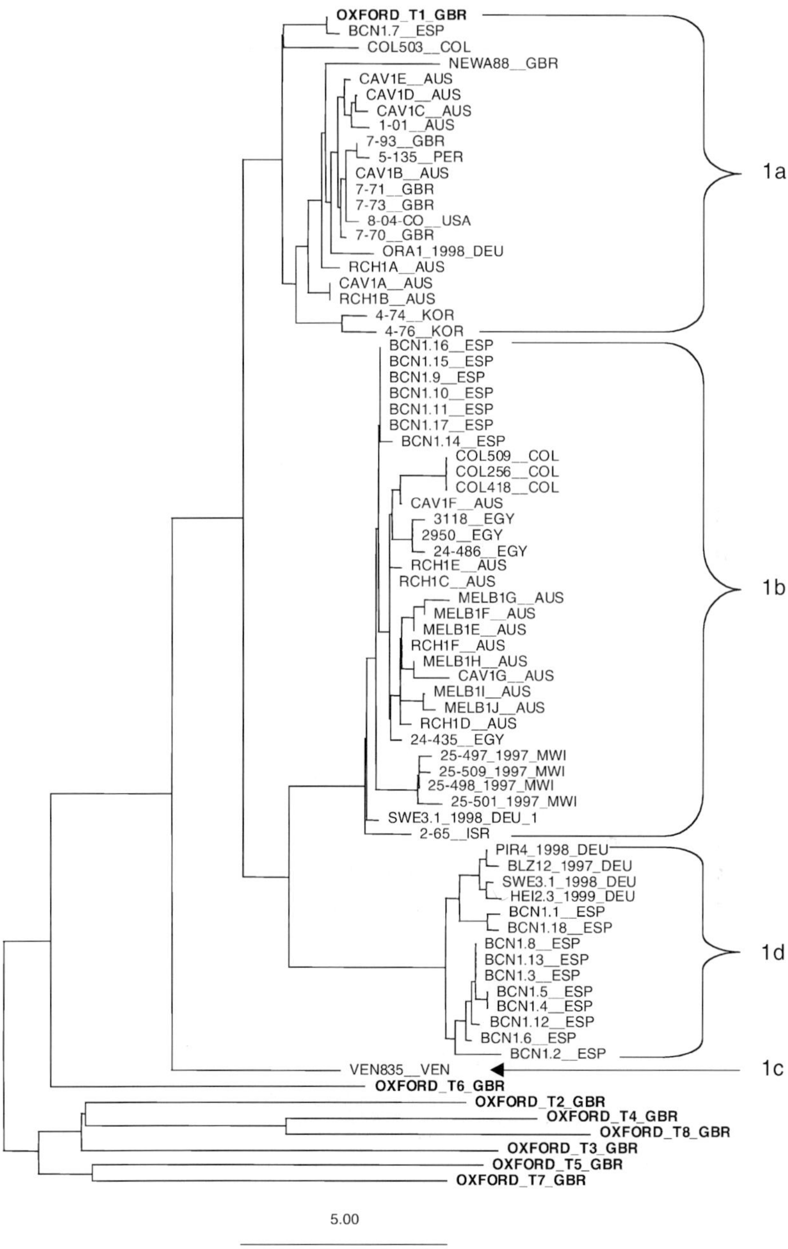

Fig. 1. Genetic relatedness of HAstV-1 strains based on 348-bp sequences from the conserved P2 region of the capsid protein. A neighbor-joining tree was constructed from uncorrected pairwise distances. Oxford reference strains are indicated in boldface. Other strains are indicated by sampleID_year(if known)_country(ISO3166codes). Braces indicate lineages within the HAstV-1 genotype.

612

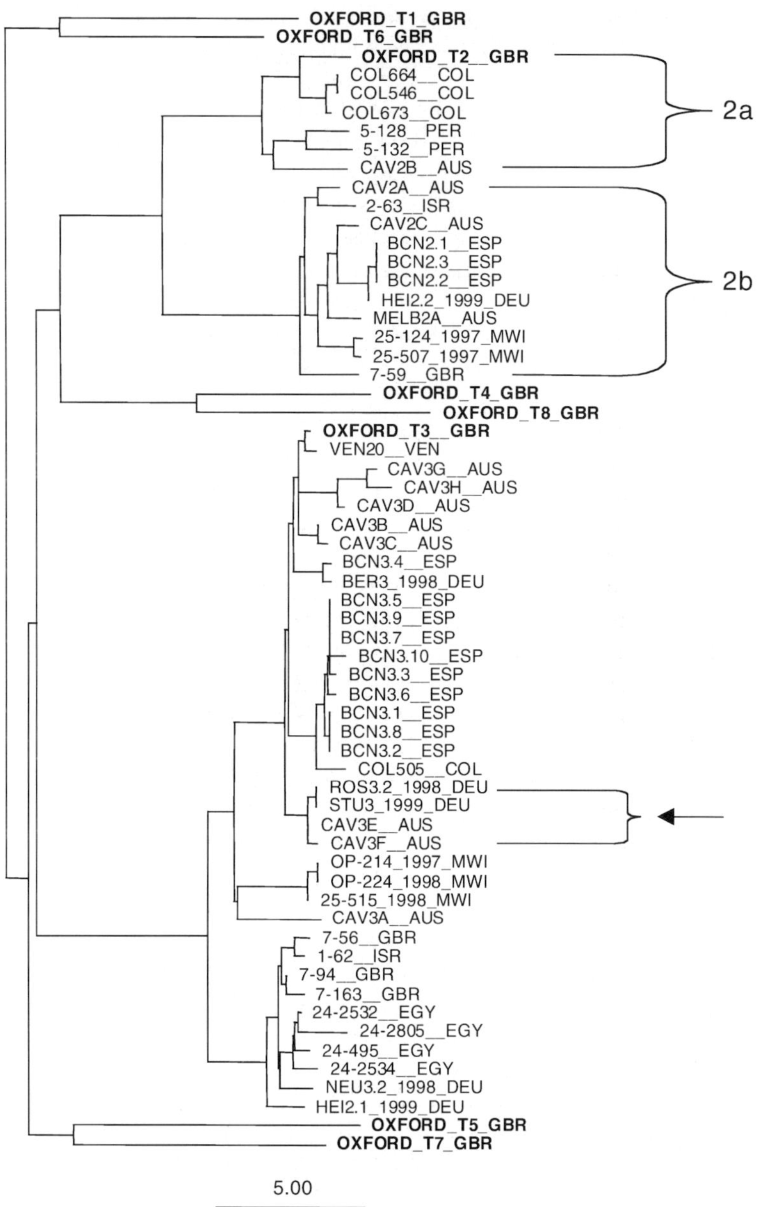

Fig. 2. Genetic relatedness of HAstV-2 and HAstV-3 strains based on 348-bp sequences from the conserved P2 region of the capsid protein. A neighbor-joining tree was constructed from uncorrected pairwise distances. Oxford reference strains are indicated in boldface. Other strains are indicated by sampleID_year(if known)_country(ISO3166codes). Arrowed brace indicates a cluster of closely related HAstV-3 strains from Australia and Germany.

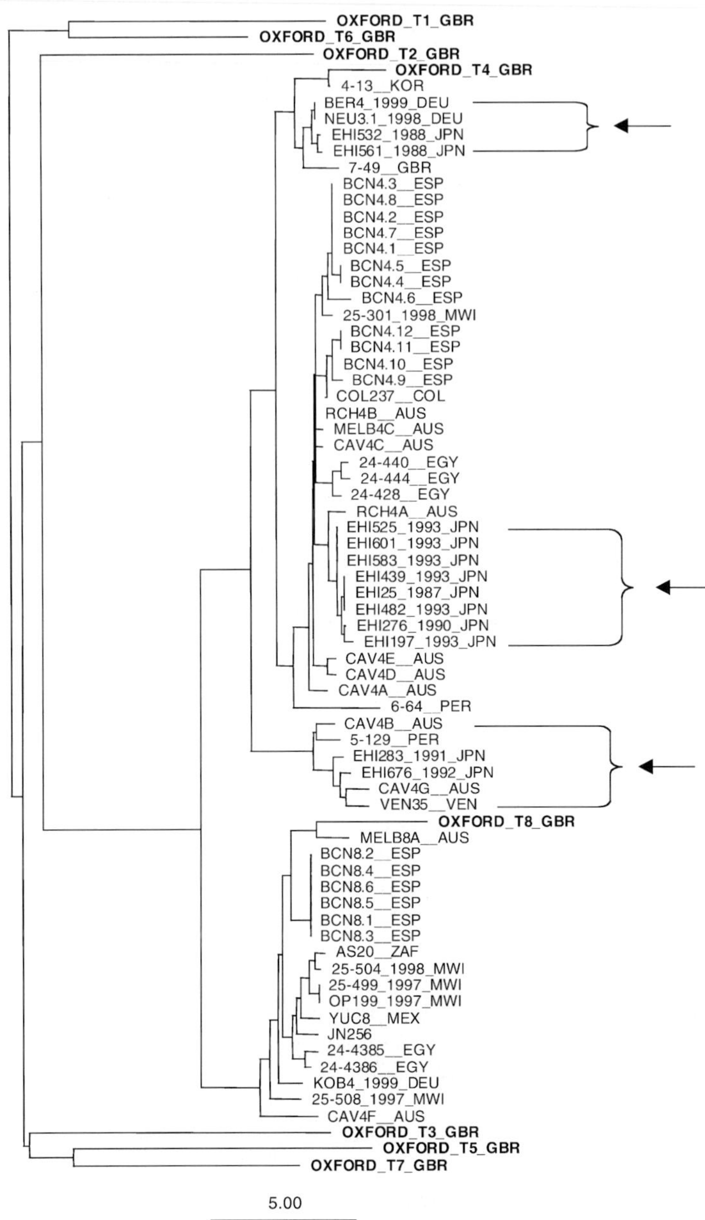

Fig. 3. Genetic relatedness of HAstV-4 and HAstV-8 strains based on 348-bp sequences from the conserved P2 region of the capsid protein. A neighbor-joining tree was constructed from uncorrected pairwise distances. Oxford reference strains are indicated in boldface. Other strains are indicated by sampleID_year(if known)_country(ISO3166codes). Arrowed braces indicate three different clusters of HAstV-4 strains from Japan. Two out of the three clusters have astrovirus representatives from other countries.

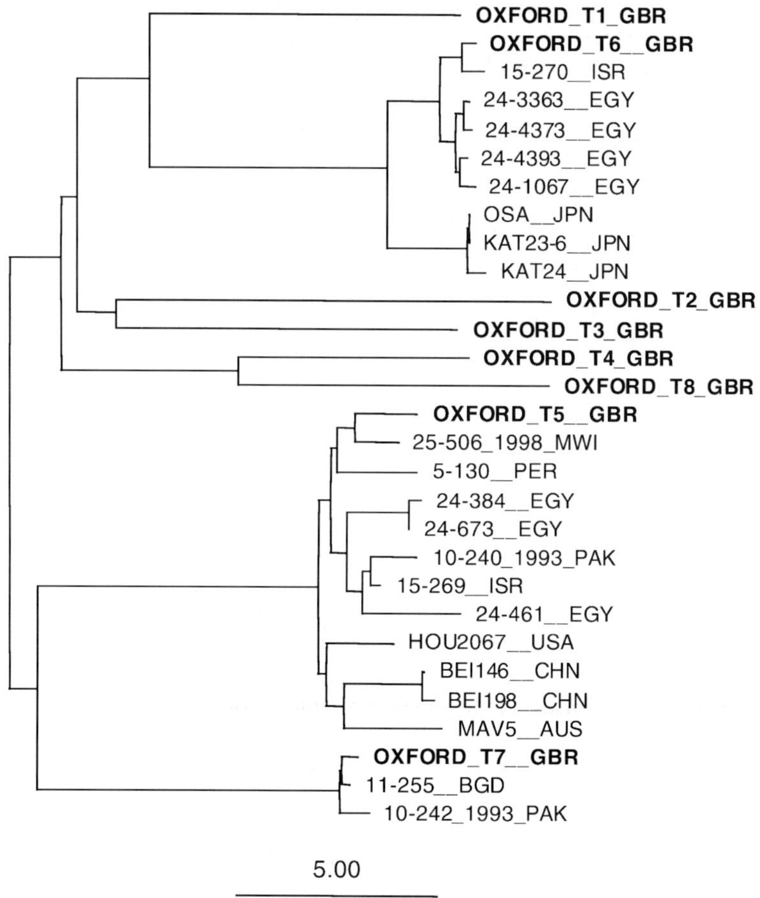

Fig. 4. Genetic relatedness of HAstV-5, HAstV-6, and HAstV-7 strains based on 348-bp sequences from the conserved P2 region of the capsid protein. A neighbor-joining tree was constructed from uncorrected pairwise distances. Oxford reference strains are indicated in boldface. Other strains are indicated by sampleID_year(if known)_country(ISO3166codes).

Strains of HAstV types 1-5 clustered in one distinct lineage, termed genogroup A, while those of types 6 and 7 clustered in genogroup B. While it was possible to clearly separate strains into these two well-resolved genogroups by using analysis of nucleotide sequence information, it was not possible to further separate strains within a genogroup into individual genotypes (Belliot *et al.* 1997). This may be an artifact of the high degree of sequence conservation in ORF1a, which makes this region a good target for broadly reactive detection assays, but limits its usefulness for phylogenetic analysis.

Overall, these results suggest a somewhat mixed conclusion about the molecular epidemiology of astroviruses. Astrovirus type 1 continues to be the predominant type detected in almost every study. There is evidence for the long-term circulation of astrovirus strains with little sequence variation in local variants in the same community, as in the case of HAstV-4 in Japan. Other strains appear to have a global distribution, with nearly identical sequences detected in widely separated locations. Finally, while some strains appear to predominate in a population, there are frequently four or five different astrovirus genotypes circulating simultaneously in the same community, as in Australia, Spain, Germany, Egypt, and Malawi.

The recent application of molecular diagnostic assays has firmly established that astroviruses are associated with five to eight percent of acute gastroenteritis in young children. It is not yet clear to what extent heterotypic protection is induced by a primary astrovirus infection. Continued monitoring of astrovirus strains will be useful to detect regional shifts in the predominant genotypes in circulation or the emergence of new genotypes.

References

Appleton, H., Buckley, M., Thom, B.T., Cotton, J.L. and Henderson, S. (1975) Virus-like particles in winter vomiting disease. *Lancet* **1**, 1297.

Belliot, G., Laveran, H. and Monroe, S.S. (1997) Detection and genetic differentiation of human astroviruses: phylogenetic grouping varies by coding region. *Archives of Virology* **142**, 1323-34.

Guix, S., Caballero, S., Villena, C., Bartolome, R., Latorre, C., Rabella, N., Simo, M., Bosch, A. and Pinto, R.M. (2002) Molecular epidemiology of astrovirus infection in Barcelona, Spain. *Journal of Clinical Microbiology* **40**, 133-9.

Herrmann, J.E., Nowak, N.A., Perron-Henry, D.M., Hudson, R.W., Cubitt, W.D. and Blacklow, N.R. (1990) Diagnosis of astrovirus gastroenteritis by antigen detection with monoclonal antibodies. *Journal of Infectious Diseases* **161**, 226-9.

Jonassen, T.O., Monceyron, C., Lee, T.W., Kurtz, J.B. and Grinde, B. (1995) Detection of all serotypes of human astrovirus by the polymerase chain reaction. *Journal of Virological Methods* **52**, 327-34.

Lee, T.W. and Kurtz, J.B. (1994) Prevalence of human astrovirus serotypes in the Oxford region 1976-92, with evidence for two new serotypes. *Epidemiology and Infection* **112**, 187-93.

Madeley, C.R. and Cosgrove, B.P. (1975) 28 nm particles in faeces in infantile gastroenteritis. *Lancet* **ii**, 451-452.

Medina, S.M., Gutierrez, M.F., Liprandi, F. and Ludert, J.E. (2000) Identification and type distribution of astroviruses among children with gastroenteritis in Colombia and Venezuela. *Journal of Clinical Microbiology* **38**, 3481-3.

Mitchell, D.K., Matson, D.O., Jiang, X., Berke, T., Monroe, S.S., Carter, M.J., Willcocks, M.M. and Pickering, L.K. (1999) Molecular epidemiology of childhood astrovirus infection in child care centers. *Journal of Infectious Diseases* **180**, 514-7.

Mitchell, D.K., Monroe, S.S., Jiang, X., Matson, D.O., Glass, R.I., and Pickering, L.K. (1995). Virologic features of an astrovirus diarrhea outbreak in a day care center revealed by reverse transcription – polymerase chain reaction. *Journal of Infectious Diseases* **172**:1437-1444.

Mitchell, D.K., Van, R., Morrow, A.L., Monroe, S.S., Glass, R.I. and Pickering, L.K. (1993) Outbreaks of astrovirus gastroenteritis in day care centers. *Journal of Pediatrics* **123**, 725-32.

Monroe, S.S., Holmes, J.L. and Belliot, G.M. (2001) Molecular epidemiology of human astroviruses. In: *Gastroenteritis Viruses, Novartis Foundation Symposium*, **238**, 237-249.

Mustafa, H., Palombo, E.A. and Bishop, R.F. (2000) Epidemiology of astrovirus infection in young children hospitalized with acute gastroenteritis in Melbourne, Australia, over a period of four consecutive years, 1995 to 1998. *Journal of Clinical Microbiology* **38**, 1058-62.

Naficy, A.B., Rao, M.R., Holmes, J.L., Abu-Elyazeed, R., Savarino, S.J., Wierzba, T.F., Frenck, R.W., Monroe, S.S., Glass, R.I. and Clemens, J.D. (2000) Astrovirus diarrhea in Egyptian children. *Journal of Infectious Diseases* **182**, 685-90.

Noel, J.S., Lee, T.W., Kurtz, J.B., Glass, R.I. and Monroe, S.S. (1995) Typing of human astroviruses from clinical isolates by enzyme immunoassay and nucleotide sequencing. *Journal of Clinical Microbiology* **33**, 797-801.

Oh, D. and Schreier, E. (2001) Molecular characterization of human astroviruses in Germany. *Archives of Virology* **146**, 443-55.

Saito, K., Ushijima, H., Nishio, O., Oseto, M., Motohiro, H., Ueda, Y., Takagi, M., Nakaya, S., Ando, T. and Glass, R. (1995) Detection of astroviruses from stool samples in Japan using reverse transcription and polymerase chain reaction amplification. *Microbiology and Immunology* **39**, 825-8.

Sakamoto, T., Negishi, H., Wang, Q.H., Akihara, S., Kim, B., Nishimura, S., Kaneshi, K., Nakaya, S., Ueda, Y., Sugita, K., Motohiro, T., Nishimura, T. and Ushijima, H. (2000) Molecular epidemiology of astroviruses in Japan from 1995 to 1998 by reverse transcription-polymerase chain reaction with serotype-specific primers (1 to 8). *Journal of Medical Virology* **61**, 326-31.

Shastri, S., Doane, A.M., Gonzales, J., Upadhyayula, U. and Bass, D.M. (1998) *Prevalence of astroviruses in a children's hospital. Journal of Clinical Microbiology* **36**, 2571-4.

Walter, J.E., Mitchell, D.K., Guerrero, M.L., Berke, T., Matson, D.O., Monroe, S.S., Pickering, L.K. and Ruiz-Palacios, G. (2001) Molecular epidemiology of human astrovirus diarrhea among children from a periurban community of Mexico City. *Journal of Infectious Diseases* **183**, 681-6.

Viral Gastroenteritis
U. Desselberger and J. Gray (editors)

SECTION VI

Other viruses causing gastroenteritis

Introduction

Besides the viruses producing the majority of human viral gastroenteritis (Sections II-V), other viruses infect more rarely, but are sometimes able to cause epidemics. In particular, they cause chronic infection in the immunocompromised.

Toroviruses

Toroviruses comprise a genus of the *Coronaviridae* family (Enjuanes *et al.*, 2000). They are enveloped and possess a genome of single stranded RNA of positive polarity which is approximately 20-25kb in size. The RNA is surrounded by nucleoprotein (N) in helical symmetry. The nucleocapsid has a toroid shape (inspiring the name from *lat.* torus = convex moulding of a column) and is enclosed in an envelope consisting of membrane protein (M) and a lipid bilayer. Inserted in the envelope are the surface proteins S (for *s*pike) and HE (for *h*aemagglutinin-*e*sterase). The HE protein has sequence similarities with corresponding proteins of coronaviruses and influenza C viruses. Toroviruses may have acquired this gene by a recombination event in the past. Replication is special in that mRNA is synthesized from 'core promoters' of a negative stranded RNA template and not by fusion of a common leader sequence with subgenomic transcripts as in the coronaviruses (Snijder and Horzinek, 1995).

Toroviruses are a well-described cause of diarrhoea in calves (BoTV; Breda V) and horses (EqTV, Berne virus) but may also infect sheep, goats and pigs. Humans seem to become infected by a closely related, but distinct virus (HuTV) (Koopmans and Horzinek, 1994; Enjuanes *et al.*, 2000). In cases of human diarrhoea toroviruses have been diagnosed by electron microscopy and EIA (Koopmans *et al.*, 1993). In a recent survey in Canada of 1365 faeces from children with diarrhoea, rotavirus was found in 32%, adenovirus in 4%, torovirus in 3%, Norwalk like viruses in 2%, and astroviruses and Sapporo-like viruses each in less than 1% (Waters *et al.*, 2000). This suggested that toroviruses are not a very frequent, but a consistent cause of diarrhoea in humans (Koopmans *et al.*, 1997; Waters *et al.*, 2000). M Petric has described the epidemiology of toroviruses (Section VI, Chapter 1). As molecular tests become available, the true extent of prevalence and incidence of human TV infections will become apparent.

618

Picobirnaviruses

Picobirnaviruses are related to, but not recognized yet as, members of the *Birnaviridae* family (Leong *et al.*, 2000). They comprise icosahedral, single shelled, non-enveloped particles which contain 2 segments of double stranded RNA as their genome. These viruses (diameter 60 nm) occur in three known genera: *Aquabirnavirus* (in fish; type species: infectious pancreatic necrosis virus, IPNV), *Avibirnavirus* (in birds; type species: infectious bursal disease virus, IBDV) and *Entomobirnaviruses* (in *Drosophila;* type species: Drosophila X virus (BXV); Leong *et al.*, 2000).

Picobirnaviruses are only 30-40 nm in diameter and have 2 (or sometimes 3) segments of dsRNA as their genome (Leite *et al.,* 1990). They were detected in children with diarrhoea and in several animal species (asymptomatic) (Pereira, 1991; Ludert *et al.*, 1991; Gallimore *et al.*, 1993, 1995). Recently they were found in the faeces of HIV-infected patients with diarrhoea more frequently than in HIV-infected patients without diarrhoea (Grohmann *et al.*, 1993; Giordano *et al.*, 1999), but a virus-specific immune response was not measurable (Grohmann *et al.*, 1993). The viral genome has recently been cloned and sequenced, and reagents derived from this will help to unravel the epidemiology of picorbirnavirus (Rosen *et al.*, 2000). B I Rosen has reviewed aspects of the molecular biology and epidemiology of these viruses (Section VI, Chapter 2).

Enteroviruses

The genus *Enterovirus* of the *Picornaviridae* family contains a large number of species [polioviruses of types 1-3, Coxsackieviruses of groups A (24 types) and B (6 types) and echoviruses (>70 types), Melnick, 1996]. Enteroviruses consist of an almost featureless capsid of 27-30 nm diameter containing 60 protomers, each possessing 3 surface proteins, VP1, VP2, VP3, and, in most viruses, an internal capsid protein, VP4. The capsid encloses a genome of single stranded RNA of positive polarity of 7-8.5 kb length. The RNA has a single open reading frame (ORF) encoding a polyprotein precursor which is co- and post-translationally cleaved in a complex cascade of events into the 3 to 4 structural and a number of non-structural proteins (for details see Racaniello, 2001).

All enteroviruses infect man via the gastrointestinal tract where they have their first site of replication, probably in lymphoid tissues of the pharynx and gut. Usually enteroviruses are excreted in the stool for several weeks after infection but can also be isolated easily from the pharynx. After primary replication viruses spread via blood to other organs (nerve, muscle, fatty tissue) where they replicate further. Most infections are asymptomatic, but enteroviruses can be the cause of meningitis, meningoencephalitis (poliomyelitis), myocarditis, pleurodynia, conjunctivitis and rashes (Melnick, 1996).

In most cases of enterovirus infection there is *no* diarrhoea. However, some diarrhoea outbreaks have been found to be associated with enterovirus infections;

Coxsackievirus A1 and echoviruses of types 4, 11, 14, 18, 19 and 22 have been involved (Townsend *et al.*, 1982; Patel *et al.* 1985; Melnick, 1996).

Recently, echoviruses of types 22 and 23 have been removed from the *Enterovirus* genus and, as they are very different in their sequence from all other *Picornaviridae*, been classified in a separate genus *Parechovirus* (King *et al.*, 2000).

The *Aichi*virus is an at present unassigned species in the *Picornaviridae* family (King *et al.*, 2000), but is proposed as a separate new genus (Yamashita *et al.* 1998). It is icosahedral, and the capsid has only 3 proteins. Virions are stable at pH 3.5. *Aichi*virus grows well in cell culture (BSC-1 and Vero cells) and has been shown to be a consistent cause of human gastroenteritis (Yamashita *et al.*, 1991, 1993). It has also been found as a cause of traveller's diarrhoea (Yamashita *et al.*, 1995). Its genome has recently been cloned and sequenced (Yamashita *et al.*, 1998), and it is hoped that the epidemiology of this virus will be further clarified with the availability of specific molecular reagents. T Yamashita and K Sakae have summarised our present knowledge of Aichivirus in Section VI, Chapter 3.

Human Immunodeficiency Virus (HIV)

HIV, the causative agent of the Acquired Immunodeficiency Syndrome (AIDS), is a member of the *Lentivirus* genus of the *Retroviridae* family. It infects the CD4 subset of lymphocytes overwhelmingly, but also macrophages, using different sets of receptors/coreceptors. Its replication (for details see Freed and Martin, 2001) damages the functions of infected cells before they are destroyed. The infection is symptomless for a long time, but then severe, generalised secondary infections appear (caused by other viruses, bacteria, fungi or protozoa) which are generally lethal.

There is evidence of extensive HIV infection in gut-associated lymphoid tissue (GALT) and also in enterocytes which contribute to the development of chronic diarrhoea in many AIDS patients (Nelson *et al.*, 1988; Heise *et al.*, 1991; Kotler *et al.*, 1991; Rabeneck, 1994).

Herpesviruses

Cytomegalovirus (CMV) and herpes simplex viruses, members of the *Herpesviridae* family, are found as the cause of colitis and oesophagitis, mainly in HIV-infected patients (Levinson and Bennets, 1985; Jacobson and Mills, 1988; Dieterich and Robinson, 1991; Theise *et al.* 1991; Mentec *et al.*, 1994; Cotte *et al.*, 1996). With the application of highly active antiretroviral therapy (HAART) it has become less urgent or indicated to commence and maintain a specific anti-CMV therapy with ganciclovir (Whitcup *et al.*, 1999; Pollok, 2001).

Coronaviruses

Coronavirus is another genus of the *Coronaviridae* family (Enjuanes *et al.*, 2000). The particles are 120-160 nm is diameter, enveloped and possess a genome of linear. positive-sense, single stranded RNA of 27 – 31 kb in size. The genome is surrounded by nucleocapsid protein (N; in helical symmetry) which in turn is contained in an envelope consisting of M protein and a host cell-derived lipid bilayer into which are inserted the surface proteins S (forming spikes), a small membrane protein (E) and the haemagglutinin esterase (HE) protein. Protein is expressed from RNA molecules which are in most cases subgenomic, and all mRNAs carry a leader sequence derived from the 5' end of the genome (for details see Lai and Holmes, 2001).

Coronaviruses infect the respiratory and gastrointestinal tract. They are a recognized cause of common cold in man (McIntosh, 1996). Whilst coronavirus infections are firmly associated with gastroenteritis in animals (e.g. Transmissible gastroenteritis virus (TGEV) of pigs; Kim *et al.*, 2001), their association with gastroenteritis in man has often been claimed, but its significance for human diarrhoeic disease has been controversial and not yet been firmly established (Gerna *et al.*, 1985; Zhang *et al.*, 1994; Siddell and Snijder, 1998; Holmes, 2001).

J Grant has produced a review of the histopathological findings of viral gastrointestinal tract infections, with particular emphasis on the data obtained from the immunocompromised (Section VI, Chapter 4).

References

Cotte L, Drouet E, Bailly F *et al.* (1996). Cytomegalovirus DNA level on biopsy specimens during treatment of cytomegalovirus gastrointestinal disease. *Gastroenterology* **111:** 439-444.

Dieterich DT, Rahmin M (1991). Cytomagalovirus colitis in AIDS: presentation in 44 patients and a review of the literature. *J. AIDS (Suppl.)* **4:** S29-S35.

Enjuanes L, Brian D, Cavanagh D *et al.* (2000). *Coronaviridae*. In: Van Regenmortel MHV *et al.* (eds) *Virus Taxonomy*. Seventh Report of the International Committee on Taxonomy of Viruses, pp835-849. Academic Press, San Diego.

Freed EO, MartinMA (2001). HIVs and their replication. In: *Fields Virology*, 4th edition (DM Knipe, PM Howley, D Griffin *et al.*, eds), pp. 1971-2041. Lippincott Williams and Wilkins, Philadelphia.

Gallimore CI, Appleton H, Lewis D, Green J, Brown DW (1995). Detection and characterization of bisegmented double stranded RNA viruses (Picobirnavirus) in human faecal specimes. *J. Med. Virol.* **45:** 135-140.

Gallimore CI, Lewis D, Brown DW (1993). Detection and characterization of a novel bisegmented double-stranded RNA virus (Picobirnavirus) from rabbit faeces. *Arch. Virol.* **133:** 63-73.

Gerna G, Passarini N, Battaglia M, Rondanelli EG (1985). Human enteric coronaviruses: antigenic relatedness to human coronavirus OC43 and possible etiologic role in viral gastroenteritis. *J. Infect. Dis.* **151:** 796-803.

Giordino MO, Martinez LC, Rinaldi D *et al.* (1999). Diarrhea and enteric emerging viruses in HIV-infected patients. *AIDS Res. Hum. Retroviruses* **15:** 1427-1432.

Grohmann GS, Glass RI, Pereira HG *et al.* (1993). Enteric viruses and diarrhoea in HIV-infected patients. *N. Engl. J. Med.* **329:** 14-20.

Heise C, Danderar S, Kumar C, Duplantier R, Donovan RM, Halstead CH (1991). Human immunodeficiency virus infection of enterocytes and mononuclear cells in human jejunal mucosa. *Gastroenterology* **100:** 1521-1527.

Holmes KV (2001). Coronaviruses. In: *Fields Virology*, 4th edition (DM Knipe, PM Howley, *et al.*, eds), pp. 1187-1203. Lippincott Williams and Wilkins, Philadelphia.

Jacobson MA, Mills J (1988). Serious cytomegalovirus disease in the acquired immunodeficiency syndrome (AIDS). Clinical findings, diagnosis and treatment. *Ann. Intern. Med.* **108:** 585-594.

Kim SJ, Song DS, Park BK (2001). Differential detection of transmissible gastroenteritis virus and epidemic diarrhea virus by duplex RT-PCR. *J. Vet. Diagn. Invest.* **13:** 516-520.

King AMQ, Brown F, Christian P *et al.* (2000). Picornaviridae. In: Van Regenmortel MHV *et al.* (eds). *Virus Taxonomy.* Seventh Report of the International Committee on Taxonomy of Viruses, pp 657-678. Academic Press, San Diego.

Koopmans MP, Goosen ES, Lima AA, McAuliffe IT *et al.* (1997). Association of torovirus with acute and persistent diarrhoea in children. *Pediatr. Infect. Dis. J.* **16:** 504-507.

Koopmans M, Horzinek MC (1994). Toroviruses of animals and humans: a review. *Adv. Virus. Res.* **43:** 233-273.

Koopmans M, Petric M, Glass RI, Monroe SS (1993). Enzyme-linked immunosorbent assay reactivity of torovirus-like particles in fecal specimens of humans with diarrhea. *J. Clin. Microbiol.* **31:** 2738-2744.

Kotler DP, Reka SA, Borcich A, Cronin WJ (1991). Detection, localisation and quantitation of HIV-associated antigens in intestinal biopsies from patients with AIDS. *Amer. J. Pathol.* **139:** 823-830.

Lai MMC, Holmes KV (2001). *Coronaviridae:* The viruses and their replication. In: *Fields Virology*, 4th edition (DM Knipe, PM Howley, D Griffin *et al.*, eds), pp. 1163-1185. Lippincott Williams and Williams, Philadelphia.

Leite JPG, Monteiro SP, Fialho AM, Pereira HG (1990). A novel avian virus with trisegmented double stranded RNA and further observations on previously described similar viruses with bisegmented genome. *Virus Res.* **16:** 119-126.

Leong JC, Brown D, Dobos P *et al.* (2000). Birnaviridae. In: *Virus Taxonomy.* Van Regenmortel MHV *et al.* (eds). Seventh Report of the International Committee on Taxonomy of Viruses, pp 481-490. Academic Press, San Diego.

Levinson W, Bennets RW (1985). Cytomegalovirus colitis in acquired immunodeficiency – a chronic disease with varying manifestations. *Amer. J. Gastroenterol.* **80:** 445-447.

Ludert JE, Hidalgo M, Gil F, Liprandi F (1991). Identification in porcine faeces of a novel virus with a bisegmented double stranded RNA genome. *Arch. Virol.* **117:** 97-107.

McIntosh K (1996). Coronaviruses. In: *Fields Virology,* 3rd Edition (Fields BN, Knipe DM, Howley PM *et al.*, eds.), pp 1095-1103. Lippincott Raven, Philadelphia.

Melnick JL. Enteroviruses: Polioviruses, Coxsackieviruses, echoviruses and newer enteroviruses. In: *Fields Virology,* 3rd edition (Fields BN, Knipe DM, Howley PM *et al.*, eds), pp 655-712. Lippincott-Raven, Philadelphia.

Mentec H, Laport C, Laport J *et al.* (1994). Cytomegalovirus colitis in HIV-1 infected patients: a prospective research in 55 patients. *AIDS* **8:** 461-467.

Nelson JA, Wiley CA, Reynolds-Kohler C, Reese CE, Margaretten W, Levy JA (1988). Human immunodeficiency virus detected in bowel epithelium from patients with gastrointestinal symptoms. *Lancet* **i:**259-262.

Patel JR, Daniel J, Mathan V (1985). An epidemic of acute diarrhoea in sural Southern India associated with echovirus type 11 infection. *J. Hyg. (Camb)* **95:** 483-492.

Pereira HG (1991). Double-stranded RNA viruses. *Semin. Virol.* **2:** 39-53.

Pollok RCG (2001). Viruses causing diarrhoea in AIDS. In: *Gastroenteritis viruses, Novartis Found. Symp.* 2001: **238:** 276-282 (Discussion 282-288).

Rabeneck L (1994). AIDS enteropathy: What's in the name? *J. Clin. Gastroenterol.* **19:** 154-157.

Racaniello VR (2001). *Picornaviridae*: The viruses and their replication. In: *Fields Virology*, 4th edition (DM Knipe, PM Howley *et al.*, eds), pp. 685-722. Lippincott Williams and Williams, Philadelphia.

Rosen BI, Fang ZY, Glass RI, Monroe SS (2000). Cloning of human picobirnavirus genomic segments and development of a RT-PCR assay. *Virology* **277:** 316-329.

Siddell SG, Snijder EJ (1998). Coronaviruses, toroviruses and arteriviruses. In: *Topley and Wilson's Microbiology and Microbial Infections,* Ninth Edition. Volume 1: *Virology* (Collier L, Mahy BWJ, eds), pp 463-484. E. Arnold, London.

Snijder EJ, Horzinek MC (1995). The molecular biology of toroviruses. In: *The Coronaviridae* (Siddell SG, ed). pp 219-238. Plenum Press, New York.

Theise ND, Rotterdam H, Dieterich D (1991). Cytomegalovirus esophagitis in AIDS diagnosis by endoscopic biopsy. *Amer. J. Gastroenterol.* **86:** 1123-1126.

Townsend TR, Bolyard EA, Yolken RH *et al.* (1982). Outbreak of Coxsackie A1 gastroenteritis; a complication of bone marrow transplantation. *Lancet* **i:** 820-823.

Waters V, Ford-Jones RL, Petric M, Fearon M, Corey P, Moineddein R (2000). Etiology of community-acquired pediatric viral diarrhea: a prospective longitudinal study in hospitals, emergency departments, pediatric practices and child care centres during the winter rotavirus outbreak, 1997 to 1998. The Pediatric Rotavirus Epidemiology Study for Immunization Study Group. *Pediatr. Infect. Dis. J.* **19:** 843-848.

Whitcup SM, Fortin E, Lindblad AS *et al.* (1999). Discontinuation of anticytomegalovirus therapy in patients with HIV infection and cytomegalovirus retinitis. *J. Amer. Med. Ass.* **282:** 1633-1637.

Yamashita T, Kobayashi S, Sakae K *et al.* (1991). Isolation of cytopathic small round viruses with BSC-1 cells from patients with gastroenteritis. *J. Infect. Dis.* **164:** 954-957.

Yamashita T, Sakae K, Ishihara Y, Isomura S, Utagawa E (1993). Prevalence of newly isolated cytopathic small round virus (*Aichi* strain) in Japan. *J. Clin. Microbiol.* **31:** 2938-2943.

Yamashita T, Sakae K, Kobayashi S *et al.* (1995). Isolation of cytopathic small round virus (*Aichi* virus) fro Pakistani children and Japanese travellers from South East Asia. *Microbiol. Immunol.* **39:** 433-435.

Yamashita T, Sakae K, Tsuzuki H *et al.* (1998). Complete nucleotide sequence and genetic organization of *Aichi* virus, a distinct member of the *Picornaviridae* associated with acute gastroenteritis in humans. *J. Virol.* **72:** 8408-8412.

Zhang XM, Herbst W, Kousoulas KG, Storz J (1994). Biological and genetic characterization of haemagglutinating coronavirus isolated from a diarrhoeic child. *J. Med. Virol.* **44:** 152-161.

VI, 1. Epidemiology of toroviruses

Martin Petric

*Virology Laboratory, The Hospital for Sick Children, 555 University Ave,
Toronto, Ontario, CANADA M1S 1C7*

The toroviruses

The discovery of the toroviruses, which comprise a genus of the family *Coronaviridae,*
began with the serendipitous isolation of the morphologically unique Berne virus
(ETV) in feces of a horse (Weiss *et al.*, 1983). This was followed by the discovery of
the Breda virus (BoTV), a morphologically similar enveloped virus particle, which was
detected in, but not isolated from the stools of diarrheic calves (Woode *et al.*, 1982).
Breda viruses were subsequently found to exist as two antigenically distinct serotypes,
referred to as BoTV-1 and BoTV-2 (Woode *et al.*, 1985). Morphologically similar
pleomorphic particles characterized by a well described fringe around the outer edges
were observed in human stool specimens (Beards *et al.*, 1884; Duckmanton *et al.*,
1997). These have been generally referred to as torovirus-like particles or coronavirus-
like particles (Mortensen *et al.*, 1985). More recently, a porcine torovirus (PoTV) has
been found and partially characterized (Kroneman *et al.*, 1998). There is serological
evidence that toroviruses are present in several other animal species (Weiss *et al.*,
1984).

Biological basis for torovirus-specific diagnostic assays

Investigations of the epidemiology of toroviruses necessitate having tests for the detection of the virus, its antigens and its nucleic acids in clinical specimens and having
well defined antigens for tests to detect the immune response to the virus. ETV was
first detected by the isolation of the virus in cell culture (Weiss *et al.*, 1983). Growth
of ETV in cell culture allowed for the characterization of the virus, in terms of its
morphology, morphogenesis, protein and nucleic acid composition, and facilitated
investigations into its replication strategy (Weiss and Horzinek 1986; Horzinek *et al.*,
1987; Snijder *et al.*, 1988). These studies demonstrated that the virus has a unique
morphology with a helical nucleocapsid in a torus configuration surrounded by an
envelope containing a spike protein, S, and a trans-membrane envelope protein, E.
The nucleocapsid is composed of the N protein and a single stranded positive sense

RNA genome. The sequence of the genome was shown to encode 3 structural proteins, N, M and S and to possess a pseudogene subsequently shown to be homologous to the gene which encodes the HE protein of BoTV (Snijder *et al.*, 1993). Studies on the viral replication in cell culture revealed that the virus proteins are translated from nested, 3'-coterminal mRNAs (Snijder *et al.*, 1988). This information, along with the viral structure and nucleic acid sequence established that toroviruses are a genus of the family *Coronaviridae*. (Cavanaugh, 1997)

After the description of the characteristic morphology of ETV, torovirus-like particles were readily identified by electron microscopy (EM). Unfortunately, the pleomorphic nature of the viruses makes EM for this virus less definitive than, for example, in the case of rotavirus or even coronavirus which have very well defined structural features. The major criteria for the EM diagnosis of torovirus include a characteristic fringe of spike proteins on the periphery of the particle, a diameter of approximately 100 nm and a subset of particles having a comma shaped appearance (Weiss *et al.*, 1983; Duckmanton *et al.*, 1997). Since stool specimens can contain a wide assortment of particulate matter, it is difficult to unambiguously identify torovirus-like particles in such preparations. Accordingly, for definitive EM diagnosis the presence of virus should be confirmed by IEM with a reference antibody to torovirus (Beards *et al.*, 1984; Duckmanton *et al.*, 1997).

The successful propagation of ETV in cell cultures facilitated the development of a virus neutralization assay, which made it possible to gain an insight into the sero-epidemiology of this virus in various animal populations (Weiss *et al.*, 1984). In contrast, no other torovirus has as yet been isolated in cell culture, which has been a significant impediment to the investigation of these agents. A bovine respiratory torovirus was reported to have been isolated from calves but has since been shown to be a coronavirus (Cornelissen *et al.*, 1998). ETV was further shown to hemagglutinate rabbit and guinea pig and human group O erythrocytes (Zanoni *et al.*, 1986). BoTV purified from calf stools was shown to agglutinate mouse and rat erythrocytes (Woode *et al.*, 1982). The hemagglutination is believed to be mediated by the S protein of ETV and the HE protein of the BoTV virus. These findings have allowed for the development of the hemagglutination inhibition assay as a more versatile serological test to follow the sero-epidemiology of these viruses (Jamieson *et al.*, 1998).

Specific antibodies to BoTV from infected gnotobiotic calves and purified BoTV were used in the development of ELISAs for the detection of viral antigens and antibodies, respectively (Brown *et al.*, 1987, Woode *et al.*, 1985, Koopmans *et al.*, 1993). The preparation of reference reagents for BoTV was facilitated by purification of virus from stool specimens of gnotobiotic and then infected calves (Woode *et al.*, 1985). Reference antisera to BoTV were also adapted to identify these viruses by immune electron microscopy (IEM) (Beards *et al.*, 1984). These assays facilitated studies on virus shedding and serological responses and also showed that antigenic cross-reactivity exists between these viruses. Finally, when the partial sequence of the genomes of ETV and BoTV were obtained (Snijder *et al.*, 1993; Cornelissen *et al.*, 1998; Duckmanton *et al.*, 1998b), it became possible to detect these viruses in stool specimens by hybridization or by RT-PCR (Koopmans *et al.*, 1991a; Duckmanton *et al.*, 1998b).

Epidemiology of ETV

Although ETV has been isolated only once from a horse, there is evidence that antigenically related viruses are widespread in the equine population (Weiss *et al.*, 1984). The sero-prevalence of ETV-specific antibody as measured by virus neutralization test is very high in the horse population with 35% of randomly collected sera in Germany (Liebermann, 1990) and 81% in Switzerland being positive (Weiss *et al.*, 1983, 1984). There is further evidence that the virus is actively circulating since 9% of 273 horses examined experienced sero-conversion when sequential sera separated by 3 to 45 days were tested for the presence of antibody (Weiss *et al.*, 1984). Neutralizing antibody was also found in sera from cattle, goats, sheep, pigs, rabbits and mice but not in veterinarians. These early studies provide an important insight into the prevalence of ETV or antigenically closely related viruses among different animal species.

Epidemiology of BoTV

Testing for antibody to BoTV has been performed by ELISA using virus purified from stools of infected calves as antigen (Woode *et al.*, 1985). Screening of sera from calves in the U.S.A. and Germany, showed that 88.5 and 75%, respectively, were found to be reactive for antibody to BoTV by this method (Woode *et al.*, 1985; Liebler *et al.*, 1992). Likewise, cattle in the UK and the Netherlands were found to be sero-positive for antibody to BoTV (Brown *et al.*, 1987; Koopmans *et al.*, 1989). Overall, it has been estimated that 90-95% of random serum samples from cattle are reactive for antibody to either ETV or BoTV (Koopmans and Horzinek, 1994). Electron microscopic surveys of stool specimens from diarrheic calves in France showed that 76% contained torovirus-like particles (Lamouliatte *et al.*, 1987). These early findings showed that BoTV is widespread in cattle over geographically diverse locations.

Evidence for the role of BoTV in gastroenteritis

The definitive etiological role of BoTV in diarrhea was established by the inoculation of germ-free calves with Breda virus (Pohlenz *et al.*, 1984). The calves became symptomatic and shed an abundant amount of virus. Studies on infected calf tissues by electron and immunofluorescence microscopy demonstrated that the sites of infection were predominantly in the crypt cell area of the gut. A study on sentinel calves showed that the animals readily became infected with BoTV from a contaminated environment and continued to shed the virus for a period of 4 months (Koopmans *et al.*, 1990). During this time virus was eliminated from a subset of calves but on re-introduction into the herd they became re-infected.

Studies on two populations of cattle, breeding calves up to 2 months of age and adult cattle experiencing dysentry, showed that virus shedding, measured by ELISA, occurred in 6% of symptomatic calves, and torovirus seroconversion was manifest in over 40% of adult cattle with dysentry (Koopmans *et al.*, 1991c). In a survey of stools from diarrheic calves collected from Canadian farms, BoTV was found by both EM and RT-PCR in 27.9% of symptomatic animals compared to 11.6% of asymptomatic controls (Duckmanton *et al.*, 1998a). These findings demonstrate that BoTV is associated with diarrhea in infected cattle and that such infections are common. To what extent toroviruses are associated with diarrhea in other animals is unknown. In the case of piglets, virus shedding can be detected within a month of age, but there is no firm evidence that this is associated with disease (Kroneman *et al.*, 1998). Likewise, foals experimentally inoculated with ETV remained asymptomatic despite manifesting a rise in neutralizing antibody (Weiss *et al.*, 1984).

Transmission of BoTV

It is generally believed that torovirus spreads in cattle by the oral-fecal route. This is supported by the high concentrations of virus (up to 10^{11} particles per mL) shed from experimentally infected calves (Koopmans and Horzinek, 1994). Hence in an outbreak situation, the environment is likely to contain substantial amounts of virus, which would place many animals at risk of infection. However, it has been reported that toroviruses may also spread by the airborne route (Vanopdenbosch *et al.*, 1992). Specifically, BoTV has been found in the respiratory tract of calves. Likewise, calves were successfully infected with BoTV by the intranasal route (Pohlenz *et al.*, 1984). Finally, young calves, which were shown to seroconvert also manifest respiratory symptoms (Koopmans *et al.*, 1989). In view of the report that some of these viruses from the respiratory tract which grow in cell culture are in fact coronaviruses, the respiratory route of transmission of torovirus may be open to question (Cornelissen *et al.*, 1998). This issue has to some extent been resolved by the recent molecular studies on BoTV as described below (Hoet *et al.*, 2002).

Epidemiology of human torovirus

Preliminary evidence of the existence of a human torovirus came from the reports of finding torovirus-like particles by EM in stool specimens of symptomatic patients which were shown to be reactive on IEM with BoTV specific antiserum and human convalescent serum (Beards *et al.*, 1984). These findings were augmented by the report that coronavirus-like particles were the sole identifiable entity found in a very high number of stool specimens of patients with gastroenteritis (Mortensen *et al.*, 1985; Payne *et al.*, 1986). Because of the pleomorphic appearance of these particles, when viewed by electron microscopy, there is justifiable concern whether these were indeed viruses as opposed to blebs of membrane from the mucosa or breakdown products of bacteria. Supportive evidence of their identity as torovirus was provided by the finding that 68% of stool specimens containing torovirus-like particles were reactive in an ELISA based on antibody to BoTV (Koopmans *et al.*, 1993). Additionally, RNA extracted from stool specimens in which torovirus-like particles were demonstrated by EM, was found to hybridize with cDNA probes derived from the 3' end of the ETV genome (Koopmans *et al.*, 1991b). Hence there is evidence at the morphological, antigenic and nucleic acid level that supports the existence of the human torovirus.

The prevalence of human torovirus remains controversial. Initial sero-surveys with ETV neutralization tests and BoTV ELISA failed to detect an appreciable presence of antibody to these viruses in human sera (Weiss *et al.*, 1984; Brown *et al.*, 1986). This observation is at odds with reports from laboratories in which viral gastroenteritis diagnosis is performed by EM that a very high number of specimens containing torovirus-like particles can be detected (Mortensen *et al.*, 1985). The presence of torovirus-like particles was shown to be associated with gastroenteritis in a case-control study in which 35% of symptomatic patients compared to 14.5% of matched controls were found to be shedding torovirus-like particles (Jamieson *et al.*, 1998). In this setting, excretion of torovirus-like particles was found more commonly in immunocompromised patients and those with nosocomially acquired gastroenteritis. Likewise, in a case control study on childhood diarrhea in Brazil, torovirus was detected by ELISA based on BoTV antiserum in 27% of children with acute diarrhea, in 27% of children with persistent diarrhea but not in controls (Koopmans *et al.*, 1997). While the former study was directed to hospitalized children, the latter addressed the role of virus in the community. A subsequent community and hospital based study showed that torovirus-like particles were detected in 6% of symptomatic patients (Waters *et al.*, 2000). These studies provide supportive evidence for the existence of human toroviruses and an insight into its prevalence in different populations. However, there remains a need to establish definitive nucleic acid based assays to augment the sensitivity and specificity of diagnosing these viruses in clinical specimens.

Molecular epidemiology of toroviruses

The sequences of approximately half of the genome of the ETV and the 3' end of the BoTV, which encode the structural proteins have been reported (Snijder *et al.*, 1993; Cornelissen *et al.*, 1997; Duckmanton *et al.*, 1998b). It is therefore possible to apply RT-PCR to detect these viruses in specimens from infected animals with enhanced sensitivity and specificity. This approach has been exploited in the testing of cattle for the prevalence of BoTV. In a study of 112 calves with diarrhea, 43% were positive for BoTV by RT-PCR compared to 30% when tested by EM (Duckmanton *et al.*, 1998). In a more recent study of feedlot cattle, BoTV was found to be shed in stools in 37% of 57 steers 6-7 month old when tested by ELISA and in 95% when tested by RT-PCR (Hoet *et al.*, 2002). Furthermore, it was demonstrated that nasal shedding occurred in 27% of the cattle when tested by ELISA and in 100% by RT-PCR. This study serves as an excellent model for the application of RT-PCR to the detection of toroviruses in symptomatic animals and has promise for similar investigations in humans. It also provides us with new insights on the sites of shedding of these viruses and their mode spread. Based on the published genome sequences it is now possible to express viral structural proteins of BoTV to produce well pedigreed supplies of antigens and antibodies for the design screening tests for these viruses (Cornelissen *et al.*, 1997; Duckmanton *et al.*, 1998b). The further development of RT-PCR assays is critical to define the epidemiology of these viruses, namely to establish their prevalence in different populations, define their mode of spread, explore the possibility of animals being a reservoir for human infection and definitively resolve their role as agents of human gastroenteritis.

References

Beards, G.M., Green, J., Hall, C., Flewett, T.H., Lamouliatte, F., Du Pasquier, P. 1984. An enveloped virus in stools of children and adults with gastroenteritis that resembles the Breda virus of calves. *Lancet* **1**, 1050-1052.

Brown, D.W., Beards, G.M., Flewett, T.H. 1987. Detection of Breda virus antigen and antibody in humans and animals by enzyme immunoassay. *J. Clin. Microbiol.* **25**, 637-640.

Cavanagh, D. 1997. Nidovirales: a new order comprising *Coronaviridae* and *Arteriviridae*. *Arch. Virol.* **142**, 629-633.

Cornelissen, L.A., Wierda, C.M., van der Meer, F.J., Herrewegh, A.A., Horzinek, M.C., Egberink, H.F., de Groot, R.J. 1997. Hemagglutinin-esterase, a novel structural protein of torovirus. *J. Virol.* **71**, 5277-5286.

Cornelissen, L.A., van Woensel, P.A., de Groot. R.J., Horzinek, M.C., Visser, N., Egberink, H.F. 1998. Cell culture-grown putative bovine respiratory torovirus identified as a coronavirus. *Vet. Rec.* 1998; **142**, 683-686.

Duckmanton, L., Luan, B., Devenish, J., Tellier, R., Petric, M. 1997. Characterization of torovirus from human fecal specimens. *Virology* **239**, 158-168.

Duckmanton, L., Carman, S., Nagy, E., Petric, M. 1998a. Detection of bovine torovirus in fecal specimens of calves with diarrhea from Ontario farms. *J. Clin. Microbiol.* **36**, 1266-1270.

Duckmanton, L.M, Tellier, R., Liu, P., Petric, M. 1998b. Bovine torovirus: sequencing of the structural genes and expression of the nucleocapsid protein of Breda virus. *Virus Res.* **58**, 83-96.

Hoet, A.E., Cho, K.O., Chang, K.O., Loerch, S.C., Wittum, T.E., Saif, L.J. 2002. Enteric and nasal shedding of bovine torovirus (Breda virus) in feedlot cattle. *Am. J. Vet. Res.* **63**, 342-348.

Jamieson, F.B., Wang, E.E., Bain, C., Good, J., Duckmanton, L., Petric, M. 1998. Human torovirus: a new nosocomial gastrointestinal pathogen. *J. Infect. Dis.* **178**,1263-1269.

Horzinek, M.C., Flewett, T.H., Saif, L.F., Spaan, W.J., Weiss, M., and Woode, G.N. 1987. A new family of vertebrate viruses: *Toroviridae. Intervirology* **27**, 17-24.

Koopmans, M., van den Boom, U., Woode, G.N., Horzinek, M.C. 1989. Seroepidemiology of Breda virus in cattle using ELISA. *Vet. Res.* **19**, 233-243.

Koopmans, M., Cremers, H., Woode, G., Horzinek, M.C. 1990. Breda virus (*Toroviridae*) infection and systemic antibody response in sentinel calves. *Am. J.*

Vet. Res. **51**, 1443-1448.

Koopmans, M., Snijder, E., Horzinek, M.C. 1991a. cDNA probes for the diagnosis of bovine torovirus (Breda virus) infection. *J. Clin. Microbiol.* **29**, 493-497.

Koopmans, M., Herrewegh, A., Horzinek, M.C. 1991b. Diagnosis of torovirus infection. *Lancet.* **337**, 859.

Koopmans, M., van Wuijckhuise-Sjouke, L., Schukken, Y.H., Cremers, H., Horzinek, M.C. 1991c. Association of diarrhea in cattle with torovirus infections on farms. *Am. J. Vet. Res.* **52**, 1769-73.

Koopmans, M., Petric, M., Glass, R., Monroe, S.S. 1993. Enzyme-linked immunosorbent assay reactivity of torovirus-like particles in fecal specimens form humans with diarrhea. *J. Clin. Microbiol.* **31**, 2738-2744.

Koopmans, M., Horzinek, M.C. 1994. Toroviruses of animals and humans: a review. *Adv. Virus Res.* **43**, 233-273.

Koopmans, M., Goosen, E.S., Lima, A.A., McAuliffe, I.T., Nataro, J.P., Barrett. L.J., Glass. R.I., Guerrant, R.L. 1997. Association of torovirus with acute and persistant diarrhea in children. *Pediatr. Infect. Dis. J.* **16**, 504-507.

Kroneman, A., Cornelissen, L.A., Horzinek, M.C., deGroot, R.J., Egberink, H.F. 1998. Identification and characterization of a porcine torovirus. *J. Virol.* **72**, 3507-3511.

Lamouliatte, F., du Pasquier, P., Rossi, F., Laporte, J., Loze, J,P. 1987. Studies on bovine Breda virus. *Vet. Microbiol.* **15**, 261-278.

Liebermann, H. 1990. New types of virus infections of domestic animals in the German Democratic Republic. 1. Serologic survey studies of the distribution of equine torovirus infections in the GDR. *Arch. Exp. Veterinärmed.* **44**, 251-253.

Liebler, E.M., Kluver, S., Pohlenz, J., Koopmans, M. 1992. The significance of bredavirus as a diarrhea agent in calf herds in Lower Saxony. *Dtsch. Tierärztl. Wochenschr.* **99**, 195-200.

Mortensen, M.L., Ray, C.G., Payne, C. Friedman, A.D., Minnich, L.L., Rousseau, C. 1985. Coronavirus-like particles in human gastrointestinal disease. *Am. J. Dis. Childh.* **139**, 928-934.

Payne, C., Ray, C.G., Borodouin, V., Minnich, L., Lebowitz, M.D. 1986. An eight year study of the viral agents of acute gastroenteritis in humans: ultrastructural observations and seasonal distribution with a major emphasis on coronavirus-like particles. *Diagn. Microbiol. Infect. Dis.* **5**, 39-54.

Pohlenz, J.F., Cheville, N.F., Woode, G.N., Mokresh, A.H. 1984. Cellular lesions in intestinal mucosa of gnotobiotic calves experimentally infected with a new unclassified bovine virus (Breda virus). *Vet. Pathol.* **21**, 407-417.

Snijder, E.J., Ederveen, J., Spaan, W.J., Weiss, M., Horzinek, M.C. 1988. Characterization of Berne virus genomic and messenger RNAs. *J. Gen. Virol.* **69**, 2135-2144.

Snijder E.J., Horzinek M.C., 1993. Toroviruses: replication, evolution and comparison with other members of the coronavirus-like superfamily. *J. Gen. Virol.* **74**, 2305-2316.

Vanopdenbosch, E., Wellmanns, G., Oudewater, J., Petroff, K. 1992. Prevalence of torovirus infections in Belgian cattle and theiur role in respiratory, digestive and reproductive disorders. *Vlaams Diergeneeskd. Tijdschr.* **61**, 187-191.

Weiss, M., Steck, F., Horzinek, M.C. 1983. Purification and partial characterization of a new enveloped RNA virus (Berne virus). *J. Gen. Virol.* **64**, 1849-1858.

Waters, V., Ford-Jones, E.L., Petric, M., Fearon, M., Corey, P., Moineddein, R. 2000. Etiology of community-acquired pediatric viral diarrhea: a prospective longitudinal study in hospitals, emergency departments, pediatric practices and child care centers during the winter rotavirus outbreak, 1997 to 1998. The Pediatric Rotavirus Epidemiology Study for Immunization Study Group. *Pediatr. Infect. Dis. J.* **19**, 843-848.

Weiss, M., Steck, F., Kaderli, R., Horzinek, M.C. 1984. Antibodies to Berne virus in horses and other animals. *Vet. Microbiol.* **9**, 523-531.

Weiss, M., and Horzinek, M.C. 1986. Morphogenesis of Berne virus (proposed family *Toroviridae*). *J. Gen. Virol.* **67**, 1305-1314.

Woode, G.N., Reed, D.E., Runnels, P.L., Herrig, M.A., Hill, T.H. 1982. Studies with an unclassified virus isolated from diarrheic calves. *Vet. Microbiol.* **7**, 221-240.

Woode, G.N., Saif, M., Quesada, M., Winnand, N.J., Pohlenz, J.F., Kelso Gourley, N. 1985. Comparative studies on three isolates of Breda virus of calves. *Am. J. Vet. Res.* **46**, 1003-1010.

Zanoni, R., Weiss, M., Peterhans, E. 1986. The haemagglutinating activity of Berne virus. *J. Gen. Virol.* **67**, 2485-2488.

VI, 2. Molecular characterization and epidemiology of picobirnaviruses

Blair I. Rosen

New York State Department of Health, Wadsworth Center, Griffin Laboratory, Albany, NY 12201-0509

Introduction

Picobirnavirus is the tentative name for a group of viruses and for a new virus genus of the same name in the family *Birnaviridae* (Leong *et al.*, 2000), that are characterized by a bisegmented, double-stranded RNA (dsRNA) genome. The virus was first reported in 1988 by two research groups involved in investigations of gastrointestinal infections of humans (Pereira *et al.*, 1988a) and animals (Alfieri *et al.*, 1988; Pereira *et al.*, 1988b). Pereira *et al.* (1988a) reported the discovery of a virus with two double-stranded RNA (dsRNA) segments during investigations of viruses in human stool samples. The virus differed from viruses in the *Birnaviridae* family, the only other known two-segmented dsRNA viruses to infect vertebrates, by the smaller size of the virus particles and of the genome. These characteristics led to the suggestion of the name picobirnavirus, referring to pico (small) and birna (bi-RNA) virus. In a subsequent study, Pereira *et al.* (1988b) reported the detection of a virus with similar physical characteristics in the intestinal contents of rats. A second research group investigating gastrointestinal viruses in chickens simultaneously reported the detection of a small virus that had characteristics consistent with those described in humans and rats (Alfieri *et al.*, 1988). Although the name picodiplornavirus was suggested for this newly discovered avian virus, the name *Picobirnavirus* has been used most often for members of this genus.

The role of picobirnavirus as a cause of gastroenteritis has not been clearly established. Epidemiologic studies have shown an association of the virus with gastroenteritis in humans and animals. Evidence supporting picobirnavirus infection of vertebrates includes: the detection of serologic responses to picobirnaviruses in pigs and rabbits (Gallimore *et al.*, 1993; Ludert *et al.*, 1991); limited replication of picobirnavirus strains from guinea pigs in first passage attempts in human and animal cell lines (Pereira *et al.*, 1989); prolonged periods of virus excretion in immunocompetent and HIV-infected people that suggests extended periods of infection and replication (Grohmann *et al.*, 1993; Gallimore *et al.*, 1995a; Giordano *et al.* 1999); and the detection of picobirnaviruses in the stools of suckling pigs and guinea pigs, which provides evidence against the ingestion of picobirnaviruses with food and passive excretion (Ludert *et al.*, 1991; Pereira *et al.*, 1989). Further investigations are needed to determine the location of the replication of picobirnaviruses and their role in gastroenteritis in humans and animals.

Morphology and virion particle characteristics

Electron microscopic examination of cesium chloride-purified picobirnaviruses from humans and animals has revealed nonenveloped virions with a round or hexagonal morphology that lack distinct surface structures (Pereira *et al.*, 1988a, 1988b; Gallimore *et al.*, 1993, 1995a; Ludert *et al.*, 1991, 1995). The surface morphology of virions was reported in one study to show symmetry along two fivefold axes, with a sixfold axis in the middle of the triangular icosahedrons (Ludert *et al.*, 1991). This pattern was suggested as evidence for an icosahedral surface lattice with a triangulation number equal to 3 (T=3). The buoyant density of picobirnavirus virions has been reported to be in the range of 1.38 to 1.42 g/ml (Pereira *et al.*, 1988b; Gallimore *et al.*, 1995; Ludert *et al.*, 1995).

Estimates of the diameter of the virions have ranged from 25 to 41 nm (Alfieri *et al.*, 1988; Pereira *et al.*, 1988b; Ludert *et al.*, 1991; Gallimore *et al.*, 1993, 1995; Haga *et al.*, 1999). However, virions with a diameter of approximately 35 nm have been cited most often (Pereira *et al.*, 1988b; Ludert *et al.*, 1991; Gallimore *et al.*, 1995a).

Genomic properties

The dsRNA nature of the picobirnavirus genome was confirmed by the sensitivity of the nucleic acid segments to degradation by RNaseA, and the resistance of the segments to digestion by RNase T1 and DNase (Alfieri *et al.*, 1988; Pereira *et al.*, 1988b; Ludert *et al.*, 1991; Gallimore *et al.*, 1995a). The association of the two dsRNA segments of picobirnavirus observed in gels after polyacrylamide electrophoresis (PAGE) with virions was initially confirmed by the detection of viral RNA with the greatest staining intensity in cesium chloride gradient fractions containing peak numbers of virus particles, as determined by electron microscopy (Pereira *et al.*, 1988a, 1988b; Ludert *et al.*, 1991; Gallimore *et al.*, 1993). Additional indirect evidence was provided by observations of equimolar amounts of the two genomic segments in stained polyacrylamide gels of nucleic acid extracted from picobirnavirus-positive fecal samples (Pereira *et al.*, 1988b; Gatti *et al.*, 1989; Green *et al.*, 1999).

Picobirnavirus genomic segments migrate in polyacrylamide gels within the migration range of the genomic segments of group A rotaviruses (Pereira *et al.*, 1988a; Ludert *et al.*, 1991; Gallimore *et al.*, 1993; Rosen *et al.*, 2000), a factor that led to the discovery of picobirnaviruses and that contributes to their continued detection. The larger picobirnavirus genomic segment, genomic segment 1, migrates within the range of rotavirus segments 2, 3, or 4. Genomic segment 2, the smaller picobirnavirus genomic segment, migrates in the region of rotavirus segment 5.

The electrophoretic migrations of the genomic segments of picobirnaviruses in poly-acrylamide gels vary among different virus strains (Gatti *et al.*, 1989; Ludert *et al.*, 1995, Gallimore *et al.*, 1993). The migration patterns have been described as wide, narrow, or intermediate (Giordano *et al.*, 1998; Gallimore *et al.*, 1995). Similar degrees of variability have been observed in other segmented dsRNA viruses (Christensen, 1989).

The length of picobirnavirus genomic segments visualized by PAGE has been estimated on the basis of comparisons with the electrophoretic migration of previously sequenced rotavirus genomic segments. The estimated lengths of picobirnavirus genomic segments 1 and 2 range from 2.2 to 2.7 kb and 1.2 to 1.95 kb, respectively (Gatti *et al.*, 1989; Gallimore *et al.*, 1995; Giordano *et al.*, 1998; Haga *et al.*, 1999). The variations in the size estimates by different research groups may be attributed to comparisons with rotavirus strains with different genomic segment mobilities, variations in the mobility of the segments of different picobirnavirus strains, and/or potential differences in the polyacrylamide gel systems of different laboratories. In group A rotaviruses, variations in the mobility of the genomic segments have been proposed to be a result of sequence heterogeneity and conformational folding of the RNA, rather than the result of differences in genome segment lengths (Estes, 1996). The factors responsible for the differences in the migration of picobirnavirus genomic segments have not been defined at the present time.

There is limited information available about the genetic sequence of picobirnaviruses. Currently, genomic segment 1 of a rabbit and a human picobirnavirus and segment 2 of two human picobirnaviruses have been cloned (Green *et al.*, 1999; Rosen *et al.*, 2000). Additional information has been provided by partial-length amplicons of the genomic segment 2 from several strains of human picobirnaviruses (Rosen *et al.*, 2000). All information pertaining to picobirnavirus proteins has been based on the analysis of the nucleic acid sequences of the genome. The small quantities of picobirnavirus in fecal samples, as well as the inability to propagate or passage the virus in cell culture or animals, respectively, have precluded direct analyses of the viral proteins. The consensus sequence of the genomic segment 1 of one strain of rabbit picobirnavirus was reported by Green *et al.* (1999). The genomic segment 1 consists of 2,362 bp and has a single major open reading frame (ORF1) from nucleotides 543 to 2315 that codes for a predicted protein of 591 amino acids in length, with a predicted molecular mass of 65.8 kDa. Two smaller ORFs were identified at nucleotides 51 to 215 (ORF2) and 212 to 532 (ORF3) with the potential to code for proteins of 55 and 155 amino acids in length, respectively, and with molecular masses of 6 kDa and 11.9 kDa. Although three separate proteins can potentially be expressed, it has been noted that two frame shifts during translation could result in the expression of a single putative protein coded by nucleotides 51 to 2313 (Green *et al.*, 1999).

A partial length clone of the genomic segment 1 of a human picobirnavirus, strain 3-GA-91, was reported by Rosen *et al.* (2000). The clone was prepared from picobirnavirus RNA extracted from a stool sample of an HIV-infected patient. Comparisons of the nucleic acid and predicted amino acid sequences of the 1572 bp clone with the genomic segment 1 of rabbit picobirnavirus revealed limited regions of homology. The greatest region of nucleic acid identity, 54.7%, occurred between nucleotides 670 to 797 and nucleotides 1523 to 1650 of the human and rabbit picobirnavirus strains, respectively. Alignments of the predicted amino acid sequences revealed an identity of 29.9% (42.9% similarity) between amino acids 19 to 172 and 310 to 465 of the human and rabbit picobirnavirus genomic segments 1, respectively.

Nearly full length consensus sequences of genomic segment 2 from two strains of human picobirnaviruses were derived from clones prepared from viral nucleic acid extracted from the stools of an HIV-infected person in Atlanta, Georgia (strain 4-GA-91) and a nonHIV-infected person in China (strain 1-CHN-97) (Rosen *et al.*, 2000). The respective sequences of strains 4-GA-91 and 1-CHN-97 consist of 1674 bp and 1696 bp. The genomic segment of each strain contains a single open reading frame. Strain 4-GA-91 contains an ORF from nucleotides 90 to 1643 that codes for a protein of predicted 517 amino acids in length, with a molecular mass of 59.6 kDa. The ORF of segment 2 of strain 1-CHN-97 spans nucleotides 58 to 1650 and codes for a predicted protein of 530 amino acids with a molecular mass of 60.3 kDa. The 5' noncoding regions of the genomic segment 2 of both picobirnavirus strains were AT rich, with 82.9 and 74.1% A and T nucleotide residues present in the 4-GA-91 and 1-CHN-97 strains, respectively.

The functions of the proteins encoded by the picobirnavirus genomic segments have been proposed on the basis of comparisons of the sequences of the genomic segments with sequences deposited in EMBL and GenBank, and searches for known amino acid motifs. Three amino acid motifs characteristic of the RNA-dependent RNA polymerases (RDRP) of dsRNA and some single-stranded RNA viruses were identified in the genomic segment 2 of the 4-GA-91 and 1-CHN-97 human picobirnavirus strains (Cohen *et al.*, 1989; Bruenn, 1991; Rosen *et al.*, 2000). The genomic segment 1 of rabbit and human picobirnavirus strains lack significant sequence similarities or identifiable motifs in common with other viruses (Green *et al.*, 1999; Rosen *et al.*, 2000). Based on these findings, the genomic segment 1 of picobirnaviruses was postulated to code for the capsid protein of the virus (Green *et al.*, 1999; Rosen *et al.*, 2000). Similar coding assignments have been determined for two-segmented dsRNA viruses in the *Birnaviridae* family.

Information on the genetic diversity of picobirnaviruses has been determined on the basis of comparisons of the sequences of the human picobirnavirus genomic segment 2 clones and partial-length amplicons produced by RT-PCR amplification of viral RNA from stool samples (Rosen *et al.*, 2000). The predicted phylogenetic relationships of nine picobirnavirus strains were determined for picobirnaviruses from HIV-infected persons in Atlanta, Georgia and Argentina, from a nonHIV-infected person from China, and from samples collected from two outbreaks of gastroenteritis in care of the elderly facilities in Florida. Eight of the picobirnavirus strains were related to the 1-CHN-97 strain and were considered to be representative of one genogroup. Viruses within this genogroup differed by 3 to 44% in nucleic acid sequence and by 2 to 46% in amino acid sequence. No picobirnaviruses with a specific genomic segment 2 sequence were detected exclusively in HIV-infected people or in people not infected with HIV. Picobirnavirus strain 4-GA-91, which could not be amplified with primers derived from the China strain of picobirnavirus, differed from all other picobirnavirus strains by 60 to 63% in nucleic acid sequence, and by 72 to 83% in amino acid sequence. These results suggest that picobirnavirus strain 4-GA-91 may be representative of a different genogroup.

The unique physical characteristics of the picobirnavirus particles and genome were reported during initial investigations of picobirnaviruses in humans and animals, and have led to suggestions for the classification of picobirnaviruses in a new virus taxon (Pereira *et al.*, 1988b; Ludert *et al.*, 1991, 1995). These observations have been supported by the lack of similarities between the nucleic acid and predicted amino acid sequences of rabbit and human picobirnavirus genomic segments with sequences deposited in GenBank and EMBL (Green *et al.*, 1999, Rosen *et al.*, 2000). Comparisons of the physical characteristics and the RDRP genomic coding assignment of picobirnaviruses with those of other one-, two-, and three-segmented dsRNA viruses were reported by Rosen *et al.* (2000). In that study, picobirnaviruses were shown to be distinct from viruses in other established families infecting bacteria, protozoa, fungi, plants, and animals. Picobirnaviruses were additionally shown to be distinct from the unclassified dsRNA viruses of *Cryptosporidium parvum*. Viruses with a three-segmented genome and tentatively named Picotrirnaviruses have been detected, similar to Picobirnaviruses, in the fecal samples of animals and humans (Leite *et al.*, 1990; Ludert and Liprandi, 1993). Picotrirnaviruses have been included in the genus Picobirnavirus (Leong *et al.*, 2000). The classification, relationships, and official species names of viruses in the genus Picobirnavirus await further consideration by the ICTV.

Epidemiology

Picobirnaviruses were initially discovered in stool samples during investigations of human gastroenteritis (Pereira *et al.*, 1988a). The viruses have subsequently been detected in fecal samples of many animal species including rats (Pereira *et al.*, 1988b), chickens (Alfieri *et al.*, 1988; Leite *et al.*, 1990), hamsters (Pereira *et al.*, 1988b); guinea pigs (Pereira *et al.*, 1989); pigs (Gatti *et al.*, 1989; Chasey, 1990; Ludert *et al.*, 1991; Pongsuwanna *et al.*, 1996), foals (Browning *et al.*, 1991), and giant anteaters (Haga *et al.*, 1999). Additional reports of picobirnavirus-like viruses have been reported in oocysts of *Cryptosporidium parvum* in human stool samples and in calves (Gallimore *et al.*, 1995b; Vanopdenbosch and Wellemans, 1990). However, differences in genomic segment size or sequence (Vanopdenbosch and Wellemans, 1990; Green *et al.*, 1999) and pathogenicity (Vanopdenbosch and Wellemans, 1990) have suggested or proven these viruses to be different from picobirnaviruses.

Information on the epidemiology of picobirnaviruses has been based primarily on the detection of viral dsRNA by PAGE followed by silver staining, and by electron microscopy. The detection of picobirnavirus by these assays has been difficult because of the low titers of the virus in many clinical samples (Grohmann *et al.*, 1993; Gallimore *et al.*, 1995a). Picobirnavirus RNA has been reported to be more readily detected by PAGE when RNA is extracted with guanidinium isothiocyanate and silica, as opposed to extractions with phenol and chloroform (Gallimore *et al.*, 1995a).

The frequency of detection of picobirnavirus in studies of gastroenteritis in humans and animals has varied in different investigations. In the initial report on human picobirnaviruses by Pereira and coworkers (1988a), the frequency of detection ranged from 2.2% to 20% of samples obtained from outbreaks of gastroenteritis in Brazil. Gallimore and coworkers (1995a) detected picobirnaviruses in 9 to 13% of 832 stool samples collected from patients with and without gastroenteritis in a study in the United Kingdom. The age range of patients in this study was from 3 to >65 years. During the course of this investigation, picobirnavirus excretion in one adult patient who had only mild diarrhea of 1-day duration was observed in seven consecutive stool samples collected over an interval of 107 days after the onset of symptoms. The dsRNA electrophoretic pattern of the picobirnavirus detected, as determined by analysis by PAGE, was consistent in all seven samples examined. The frequency of detection of picobirnaviruses in surveys conducted in children in Italy (Cascio *et al.*, 1996), Brazil (Pereira *et al.*, 1993), and Venezuela (Ludert and Liprandi, 1993) has been lower, with frequencies of 0.43, 0.5, and 0.5%, respectively.

A method for the detection of picobirnaviruses by reverse transcription-PCR (RT-PCR) was developed, following the determination of the sequence of the genomic segment 2 from two strains of human picobirnaviruses (Rosen *et al.*, 2000). The cloned genomic segments were prepared from picobirnavirus nucleic acid extracted from the stools of an HIV-infected person in Atlanta, Georgia (strain 4-GA-91) and a nonHIV-infected person in China (strain 1-CHN-97). Picobirnaviruses with a genomic segment 2 related to the China strain were the predominant viruses detected in stool samples collected from HIV-infected people in Argentina, Venezuela, and Atlanta, Georgia. In contrast, primers derived from strain 4-GA-91 produced amplicons of the expected size only with nucleic acid extracted from stool samples of the Atlanta patient from whom the clones were derived. During the course of investigations with the RT-PCR assay, picobirnaviruses were identified in stool samples collected from outbreaks of gastroenteritis in residents at two care of the elderly facilities in Florida. Picobirnaviruses were confirmed in one of five samples examined in one outbreak, and in three of five samples in the second outbreak. Amplicons of the expected size were produced only with primers derived from the 1-CHN-97 picobirnavirus strain. All of the samples that were positive for picobirnavirus in the two outbreaks were also positive for Norwalk-like virus, and thus a firm association of picobirnavirus infection with these outbreaks is not possible.

Associations of picobirnavirus with gastroenteritis

Picobirnaviruses have been detected in investigations of human gastroenteritis associated with infection with human immunodeficiency virus (HIV) or treatments with immunosuppressive drugs. An association of picobirnavirus with diarrhea in HIV-infected patients was first reported in a longitudinal study in the United States by Grohmann and coworkers (1993). In this study, picobirnavirus was detected in 9% of patients with diarrhea and in 2% of patients without diarrhea and was the second most commonly detected virus after astrovirus. Prolonged shedding of picobirnavirus was observed in one patient for up to seven months, and intermittent shedding in a second patient during three separate episodes of diarrhea. Giordano *et al.* (1998, 1999) detected picobirnavirus at frequencies of 8.8% and 14.6% in HIV-infected patients with diarrhea in two studies in Argentina, with no reports of picobirnaviruses in HIV-infected patients without diarrhea, or in nonHIV-infected patients. In contrast, the frequency of detection of picobirnaviruses in a study of HIV-infected patients in Venezuela (Gonzalez *et al.*, 1998) was lower, with virus detection in 2.3% of stool samples obtained only from HIV-infected patients without diarrhea. Valle *et al.* (2001) investigated the association of viral agents with diarrhea in 42 kidney transplant patients who had received immunosuppressive treatments. Picobirnavirus and group A rotavirus were detected in 1 and 2 non-coinfected patients, respectively, of 9 patients with severe diarrhea, with no viruses detected in patients without diarrhea. The viral agents were reported in patients who had high concentrations of the immunosuppressant cyclosporine in their blood (>290 ng/ml), or who had prolonged immunosuppressive treatments.

An association of picobirnaviruses with gastroenteritis in animals was reported by Gatti *et al.* (1989). In this study, fecal samples from 912 pigs collected from 28 municipalities in the state of Sao Paulo, Brazil were screened by PAGE. Picobirnaviruses were detected in samples from 18 of the 28 municipalities investigated. The virus was detected alone or together with rotavirus in 15.3% of animals with diarrhea, and in 9.6% of animals without diarrhea (Table 1). Samples containing both rotavirus and picobirnavirus were observed in 3.1% of pigs with diarrhea and in 1.9% of pigs without diarrhea. Investigations for picobirnavirus in swine in Venezuela by Ludert *et al.* (1991) failed to show an association of picobirnavirus infection with diarrhea. However, in contrast to the study of Gatti *et al.* (1989), the animal population studied was located on one farm. In the Venezuelan study, picobirnaviruses were detected in 11.1% of 244 fecal samples as determined by PAGE. The frequency of virus detection was 10.0% in pigs with diarrhea and 12.3% in pigs without diarrhea (Table 1). The highest frequency of detection, 16.9%, occurred in pigs 15 to 35 days old. A similar prevalence of picobirnaviruses was reported in chickens in a study in Brazil, with 17.1% of 257 chickens picobirnavirus-positive as determined by PAGE (Leite *et al.*, 1990) (Table 2). In this study, samples were collected from animals in a slaughterhouse and the small and large intestinal contents were analyzed separately. Picobirnaviruses were detected exclusively in the contents from the large intestine. Picobirnaviruses have been discovered in the feces of several animal species without signs of clinical gastroenteritis. Continued surveillance for picobirnaviruses is likely to find the virus widely dispersed in many domestic and wild animal species.

Table 1

Summary of picobirnavirus detection in pigs by PAGE.

No. of Fecal Samples	Epidemiology Data	Positive Samples					
		Picobirnavirus		Rotavirus		Picobirnavirus and Rotavirus	
		Animals with diarrhea	Animals without diarrhea	Animals with diarrhea	Animals without diarrhea	Animals with diarrhea	Animals without diarrhea
912[1]	Samples from pigs from 28 municipalities. Age: 9 to 61 days	49 (15.0%) (n = 321)	57 (9.6%) (n = 591)	121 (37.7%) (n = 321)	87 (14.7%) (n = 591)	10 (3.1%) (n = 321)	11 (1.9%) (n = 591)
244[2]	Samples from pigs on 1 farm. Age: 0 to >50 days	13 (10.0%) (n = 130)	14 (12.3%) (n = 114)	39 (30.0%) (n = 130)	16 (14.0%) (n = 114)	5 (3.8%) (n = 130)	3 (2.6%) (n = 114)

[1] Gatti *et al.*, 1989
[2] Ludert *et al.*, 1991

Table 2

Detection of picobirnavirus in chickens by PAGE.[1]

No. of Samples[2]	Epidemiology Data	Small Intestines – Positive Samples			
		Picobirnavirus	Picotrirnavirus	Reovirus	Rotavirus
250	Samples from animals in slaughterhouse - total from 3 separate collections. Age: 38 to 43 days	0 (0.0%)	0 (0.0%)	6 (2.4%)	1 (0.4%)

No. of Samples[2]		Large Intestines – Positive Samples				
		Picobirnavirus	Picotrirnavirus	Picobirnavirus and Picotrirnavirus	Picobirnavirus and Rotavirus	Picobirnavirus and Rotavirus and Reovirus
257	Same as above.	44 (17.1%)	11 (4.3%)	16 (6.2%)	1 (0.4%)	1 (0.4%)

[1] Leite *et al.*, 1990

[2] The intestinal contents of the small and large intestines were examined separately from the same 250 chickens. The contents of the large intestines were examined from 7 additional chickens.

642

References

Alfieri, A.F., Alfieri, A., Resende, J.S., and Resende, M. (1988). A new bisegmented double stranded RNA virus in avian feces. Arg. Bras. Med. Vet. Zootec. **40**: 437-440.

Browning, G.F., Chalmers, R.M., Snodgrass, D.R., Batt, R.M., Hart, C.A., Ormarod, S.E., Leadon, D., Stoneham, S.J., and Rossdale, P.D. (1991). The prevalence of enteric pathogens in diarrhoeic Thoroughbred foals in Britain and Ireland. Equine Vet. J. **23**:405-409.

Bruenn, J.A. (1991). Relationships among the positive strand and double-strand RNA viruses as viewed through their RNA-dependent RNA polymerases. Nucleic Acids Res. **19**:217-226.

Cascio, A., Bosco, M., Vizzi, E., Giammanco, A., Ferraro, D., and Arista, S. (1996). Identification of picobirnavirus from faeces of Italian children suffering from acute diarrhea. Eur. J. Epidemiol. **12**:545-547.

Chasey, D. (1990). Porcine picobirnavirus in UK? Vet. Rec. **126**:465.

Christensen, M.L. (1989). Human viral gastroenteritis. Clin. Microbiol. Rev. **2**:51-89.

Cohen, J., Charpilienne, A., Chilmonczyk, S., and Estes M.K. (1989). Nucleotide sequence of bovine rotavirus gene 1 and expression of the gene product in baculovirus. Virology **171**:131-140.

Estes, M.K. (1996). Rotaviruses and their replication. In: Fields Virology (Fields, B.N., Knipe, D.M., and Howley, P.M., Eds), 3rd edition, pp. 1625-1655. Lippincott-Raven, Philadelphia, PA.

Gallimore, C.I., Appleton, H., Lewis, D., Green, J., and Brown, D.W.G. (1995a). Detection and characterisation of bisegmented double-stranded RNA viruses (picobirnaviruses) in human faecal specimens. J. Med. Virol. **45**:135-140.

Gallimore, C.I., Green, J., Casemore, D.P., and Brown, D.W.G. (1995b). Detection of a picobirnavirus associated with *Cryptosporidium* positive stools from humans. Arch. Virol. **140**:1275-1278.

Gallimore, C., Lewis, D., and Brown, D. (1993). Detection and characterization of a novel bisegmented double-stranded RNA virus (picobirnavirus) from rabbit faeces. Arch. Virol. **133**:63-73.

Gatti, M.S.V., De Castro, A.F.P., Ferraz, M.M.G., Fialho, A.M., and Pereira, H.G. (1989). Viruses with bisegmented double-stranded RNA in pig faeces. Res. Vet. Sci. **47**:397-398.

Giordano, M.O., Martinez, L.C., Rinaldi, D., Espul, C., Martinez, N., Isa, M.B., Depetris, A.R., Medeot, S.I., and Nates, S.V. (1999). Diarrhea and enteric emerging viruses in HIV-infected patients. J. AIDS Res. Hum. Retroviruses **15**: 1427-1432.

Giordano, M.O., Martinez, L.C., Rinaldi, D., Guinard, S., Naretto, E., Casero, R., Yacci, M.R., Depetris, A.R., Medeot, S.I., and Nates, S.V. (1998). Detection of picobirnavirus in HIV-infected patients with diarrhea in Argentina. J. Acquir. Imm. Defic. Syndr. **18**:380-383.

Gonzalez, G.G., Pujol, F.H., Liprandi, F., Deibis, L., and Ludert, J.E. (1998).

Prevalence of enteric viruses in human immunodeficiency virus seropositive patients in Venezuela. J. Med. Virol. **55**:288-292.

Green, J., Gallimore, C.I., Clewley, J.P., and Brown, D.W.G. (1999). Genomic characterisation of the large segment of a rabbit picobirnavirus and comparison with the atypical picobirnavirus of *Cryptosporidium parvum*. Arch. Virol. **144**: 2457-2465.

Grohmann, G.S., Glass, R.I., Pereira, H.G., Monroe, S.S., Hightower, A.W., Weber, R., and Bryan, R.T. (1993). Enteric viruses and diarrhea in HIV-infected patients. N. Engl. J. Med. **329**:14-20.

Haga, I.R., Martins, S.S., Hosomi, S.T., Vicentini, F., Tanaka, H., and Gatti, M.S.V. (1999). Identification of a bisegmented double-stranded RNA virus (picobirnavirus) in faeces of giant anteaters (*Myrmecophaga tridactyla*). Vet. J. **158**:234-236.

Leite, J.P.G., Monteiro, S.P., Fialho, A.M., and Pereira, H.G. (1990). A novel avian virus with trisegmented double-stranded RNA and further observations on previously described similar viruses with bisegmented genome. Virus Res. **16**:119-126.

Leong, J.C., Brown, D., Dobos, P., Kibenge, F.S.B., Ludert, J.E., Muller, H., Mundt, E., and Nicholson, B. (2000). Family *Birnaviridae*. In: Virus Taxonomy, Classification and Nomenclature of Viruses. Seventh Report of the International Committee on Taxonomy of Viruses (van Regenmortel, M.H.V. *et al.*, Eds), pp.481-490. Academic Press, San Diego, CA.

Ludert, J.E., Abdul-Latiff, L., Liprandi, A., and Liprandi, F. (1995). Identification of picobirnavirus, viruses with bisegmented double stranded RNA, in rabbit faeces. Res. Vet. Sci. **59**:222-225.

Ludert, J.E., Hidalgo, M., Gil, F., and Liprandi, F. (1991). Identification in porcine faeces of a novel virus with a bisegmented double stranded RNA genome. Arch. Virol. **117**:97-107.

Ludert, J.E., and Liprandi, F. (1993). Identification of viruses with bi- and trisegmented double-stranded RNA genome in faeces of children with gastroenteritis. Res. Virol. **144**:219-224.

Pereira, H.G., De Araujo, H.P., Fialho, A.M., De Castro, L., and Monteiro, S.P. (1989). A virus with bi-segmented double-stranded RNA genome in guinea pig intestines. Mem. Inst. Oswaldo Cruz **84**:137-140.

Pereira, H.G., Fialho, A.M., Flewett, T.H., Teixeira, J.M.S., and Andrade, Z.P. (1988a). Novel viruses in human feces. Lancet **2**:103-104.

Pereira, H.G., Flewett, T.H., Candeias, J.A.N., and Barth, O.M. (1988b). A virus with a bisegmented double-stranded RNA genome in rat (*Oryzomys nigripes*) intestines. J. Gen. Virol. **69**:2749-2754.

Pereira, H.G., Linhares, A.C., Candeias, J.A.N., and Glass, R.I. (1993). National laboratory surveillance of viral agents of gastroenteritis in Brazil. Bull. PAHO **27**: 224-233.

Pongsuwanna, Y., Taniguchi, K., Chiwakul, M., Urasawa, T., Wakasugi, F., Jayavasu, C., and Urasawa, S. (1996). Serological and genomic characterization of porcine rotaviruses in Thailand: Detection of a G10 porcine rotavirus. J. Clin. Microbiol. **34**:1050-1057.

Rosen, B.I., Fang, Z.-Y., Glass, R.I., and Monroe, S.S. (2000). Cloning of human pico-birnavirus genomic segments and development of an RT-PCR detection assay. Virology **277**:316-329.
Valle, M.C., Martinez, L.C., Ferreyra, L.J., Giordano, M.O., Isa, M.B., Pavan, J.V., De Boccardo, G., Massari, P.U., and Nates, S.V. (2001). Viral agents related to diarrheic syndrome in kidney transplanted patients. Medicina **61**:179-182.
Vanopdenbosch, E., and Wellemans, G. (1990). Bovine birna type virus: A new etiological agent of neonatal calf diarrhoea? Vlaams Dierg. Tijdschr **59**:1-4.

Viral Gastroenteritis
U. Desselberger and J. Gray (editors)

VI, 3. Molecular biology and epidemiology of Aichi virus and other diarrhoeogenic enteroviruses

Teruo Yamashita and Kenji Sakae

Department of Microbiology, Aichi Prefectural Institute of Public Health, 7-6, Nagare, Tsujimachi, Kita-ku, Nagoya, Aichi 462-8576, Japan

Introduction

In March 1989, an outbreak of gastroenteritis occurred in Aichi Prefecture, Japan. Out of 29 people who participated in the farewell party of a company, 21 complained of gastroenteritis symptoms such as diarrhea, stomachache, vomiting, and fever approximately 63 hours later. The probable cause of this outbreak was traced to raw oyster in vinegar dressing. Because the search for a bacterial cause remained negative, the feces and paired sera of 5 patients were transported to our laboratory. Spherical virus-like particles of approximately 30nm in diameter were detected by electron microscopy. In addition, agents cytopathic for BS-C-1 cells were also detected in 2 of the 5 fecal samples. Furthermore, seroconversion against these agents was confirmed in 4 of 5 serum pairs, and the antibody titer rises were confirmed against both fecal isolates using a neutralization test. It was conceivable that the two agents were the same virus and the causative agents of this gastroenteritis outbreak (Yamashita *et al.*, 1991). Rotavirus, adenovirus, Norwalk-like virus and astrovirus are well known to be causative agents of gastroenteritis (Ashley *et al.*, 1978; Blacklow 1988; Blacklow *et al.*, 1991; Cruz *et al.*, 1992; Cubitt, 1987; Cukor *et al.*, 1984; Dolin *et al.*, 1987). Our isolates were not related to such gastroenteric viruses or to enteroviruses which are common cytopathic agents in humans. Therefore, we called this agent "Aichi virus", a new type of virus (Yamashita *et al.*, 1993). Genetic analyses performed on Aichi virus lead to the conclusion that it should be classified as a new genus of the *Picornaviridae* family rather than a member of the other genera such as *Enterovirus, Rhinovirus, Cardiovirus, Aphthovirus, Hepatovirus,* or *Parechovirus* (Yamashita *et al.*, 1998), and recently this virus has been proposed to belong to a new genus, *Kobuvirus*, in the *Picornaviridae* family (King *et al.*, 1999; King *et al.*, 2000). "Kobu" means bump or knob in Japanese, which is derived from the characteristic morphology of the virus particle.

Biophysical and biochemical properties

Aichi virus can be propagated in monkey kidney cells such as BS-C-1 and Vero cells (accompanied by a cytopathic effect characterized by detachment of the cells from the culture flask surface), but not in human cells such as HeLa, HEL, or RD cells nor in newborn mice. Aichi virus can be stabilized by the addition of IUDR to the culture medium. Its infectivity withstands treatment with a combination of chloroform and acid pH (3.5), and with heat at 50°C for 30 min; however, the virus is thermolabile at 60°C for 30 min. The buoyant density is 1.36 g/ml as determined by cesium chloride gradient ultracentrifugation (Yamashita *et al.*, 1991). Aichi virus, negatively stained with 2% phosphotungstic acid (pH7.2) for 2 min, is spherical, measuring approximately 30 nm in diameter, and has a rough surface which is different from that seen in an enterovirus, astrovirus, or calicivirus (Fig. 1). Three capsid proteins of 42, 30, and 22 kDa could be observed in SDS-PAGE. The 42-kDa protein corresponds to VP0, and no VP4-VP2 cleavage occurs. The 30- and 22-kDa proteins correspond to VP1 and VP3, respectively (Yamashita *et al.*, 1998).

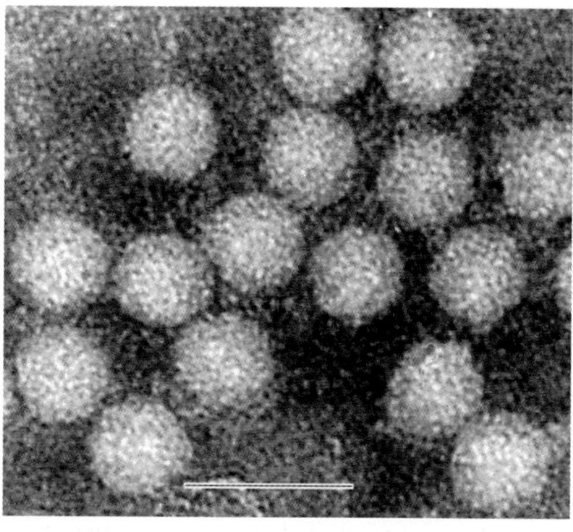

Fig. 1. Electron micrograph of Aichi virus after staining with 2% phosphotungstic acid (pH7.2) for 2 min. Calibration bar: 50 nm

Aichi virus genome

The virion of Aichi virus contains a single-stranded RNA molecule as the genome. The RNA genome consists of 8,280 nucleotides (nt), excluding a poly(A) tract. A large open reading frame with 7,299 nt that encodes a potential polyprotein precursor of 2,432 amino acid (aa) has been found and is preceded by 744 nt and followed by 237 nt and a poly(A). The 42 nt at the 5' end of the genome form a stable stem-loop structure and play an essential role in the formation of virus particles as well as in RNA replication (Sasaki *et al.*, 2001). Although the precise secondary structure of the 5'NTR has not been defined, the location of the pyrimidine tract and the initiator methionine suggests that the IRES of Aichi virus is similar to that of aphtho-, cardio- and hepato-viruses. The 3' NTR of the Aichi virus genome (237 nt) is by more than 100 nt longer than that of the encephalomyocarditis virus (EMCV) which until then possessed the longest 3'NTR in the *Picornaviridae* family. Whether the 3'NTR of Aichi virus consists of three double-stranded hairpin stems, as seen in EMCV, two stems, as seen in Poliovirus 1 and Foot-and-mouth disease virus (FMDV), or a single stem, as in human rhinovirus and hepatitis A virus (Auvinen and Hyypiä, 1990; Poyry *et al.*, 1996), has not yet been determined (Yamashita *et al.*, 1998).

The organization of the deduced amino acid sequence indicated in Fig. 2 is analogous to that of the other picornaviruses. A leader (L) protein, consisting of 170 aa, is present upstream of VP0. The length of the L protein is a little shorter than that of FMDV (217 aa) and more than twice as long as that of EMCV (67 aa). However, neither the catalytic dyad (Cys and His) conserved in a papain-like thiol protease and found in the FMDV L protein (Gorbalenya *et al.*, 1991; Piccone *et al.*, 1995) nor a putative zinc-binding motif, Cys-His-Cys-Cys, found in EMCV or Theiler's murine encephalo-myelitis virus (Chen *et al.*, 1995), could be identified. The function of the Aichi virus L protein is unknown at the moment. The homology of Aichi virus structural proteins (VP0, VP3, and VP1) with corresponding polypeptides of other picornaviruses varies between 19 and 32%. The dendrogram based on structural proteins is depicted in Fig. 3, indicating that the Aichi virus should be separated from the known genera of the *Picornaviridae* including entero-, rhino-, cardio-, aphtho-, hepato-, parecho-, erbo-, and teschoviruses. The nucleotide sequences compared in Fig. 3 have been obtained from DDBJ, EMBL, and GenBank: bovine enterovirus 1 (BEV-1), D00214; Coxsackievirus A16(CV-A16), U05876; Coxsackievirus B3 (CV-B3), M16572; Enterovirus 70 (EV-70), D00820; Poliovirus 1 (PV-1), J02281; Human rhinovirus 2 (HRV-2), X02316; Human rhinovirus 14(HRV-14), X01087; Encephalomyocarditisvirus (EMCV), M81861; Theiler's murine encephalomyelitis virus (ThV), M20301; Foot-and-mouth disease virus O (FMDV-O), X00871; Equine rhinitis A virus (ERAV), L43052; Equine rhinitis B virus (ERBV), X96871; Porcine teschovirus (PTV), AJ011380; Hepatitis A virus (HAV), M14707; Avian encephalomyelitis-like virus (AEV), AJ225173; Human parechovirus 1 (HPeV-1), L02971; and Aichi virus (AiV), AB 040749.

The 2A protein of picornaviruses is known to have a cis-acting proteolytic activity and has been classified into two types. In entero- and rhinoviruses, 2A protein functions auto-catalytically to cleave the P1 polyprotein at its own N-terminus and mediates the cleavage of the p220 component of the *cap*-binding complex eIF-4γ, leading to the shutoff of host cellular protein synthesis. Like in 3C protease, a catalytic triad conserved in trypsin-like proteases has been identified (Bazan and Fletterick, 1988; Yu and Lloyd, 1992). In aphtho- and cardioviruses, on the other hand, 2A mediates the cleavage at its own C-terminus and the autocleavage motif, NPEG, is conserved in the C-terminus of the 2A protein (Donnelly *et al.*, 1997; Ryan and Drew, 1994). In the Aichi virus 2A protein, neither the critical GXCG motif of trypsin-like protease nor the NPEG motif could be found.

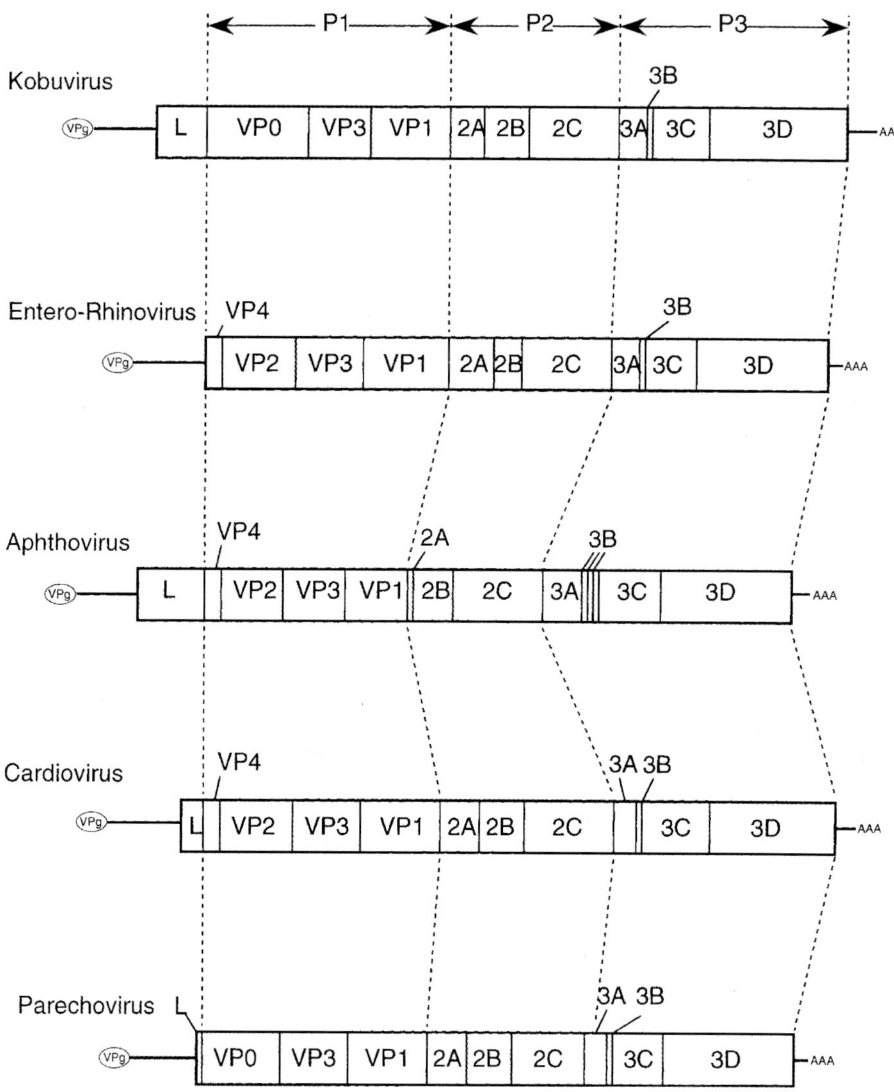

Fig. 2. Genome organization of Aichi virus and comparison of the structure among Picornaviruses. The genome organizations have been shown according to the L434 system. P1 represents viral structural proteins. P2 and P3 represent nonstructural proteins. P1 proteins of human parechovirus produce 12-aa leaders.

Recently, it has been reported that the 2A protein of Aichi virus as well as those of human parechoviruses and avian encephalomyelitis virus contain conserved motifs that are characteristic of a family of cellular proteins involved in the control of cell proliferation

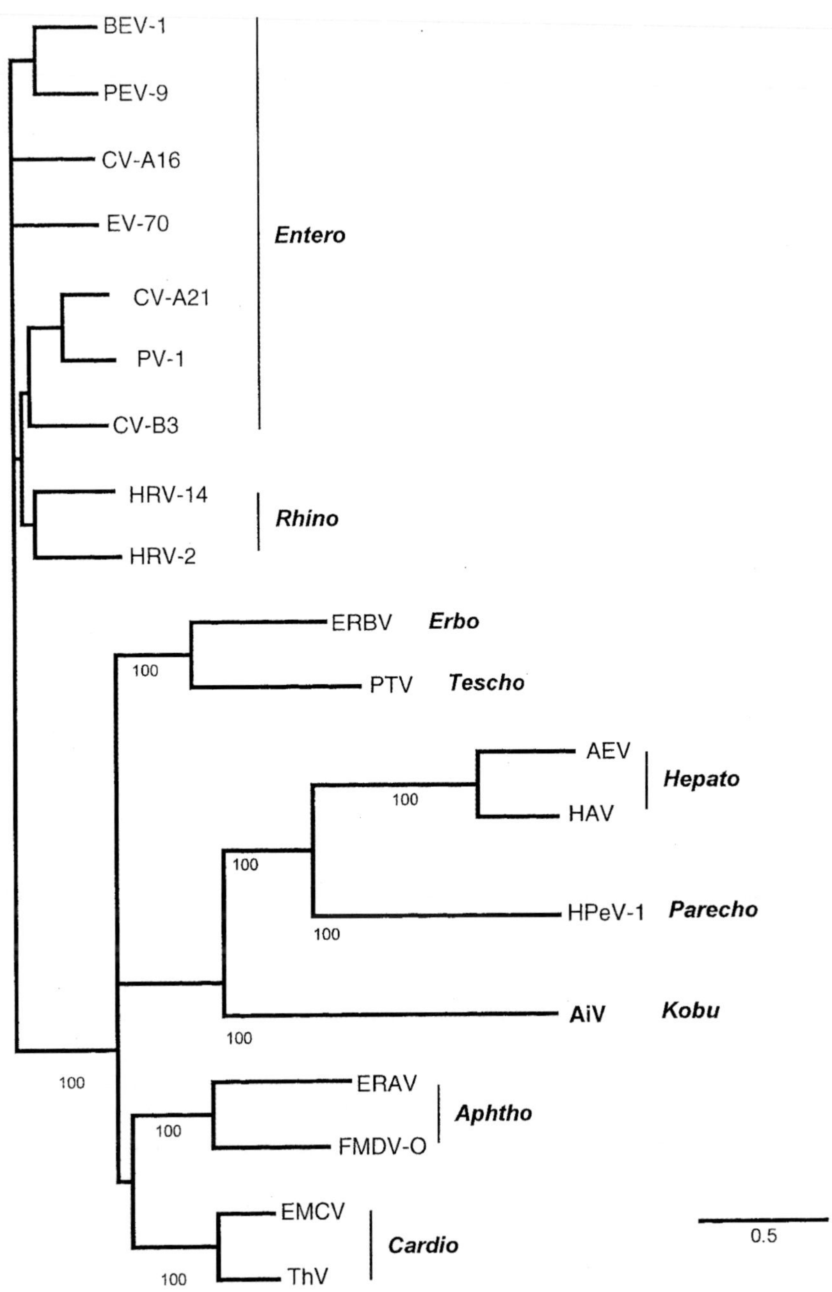

Fig. 3. Relationships between Aichi virus and other picornaviruses based on amino acid differences of structural proteins. The dendrogram was generated by evolutionary distances computed by UPGMA. The current picornavirus genera are indicated. Bootstrap values are indicated at branch roots.

(Hughes and Stanway, 2000). Amino acid sequences of the 2C, 3C, and 3D regions are well aligned with corresponding sequences of other picornaviruses. Although the function of 2C protein has not been completely elucidated, a highly conserved motif (GxxGXGKT, X: uncharged, x: nonconserved amino acid) in the nucleotide binding domain of the putative picornavirus helicase is found in the Aichi virus 2C protein. The 3B protein (VPg) of Aichi virus (27 aa) is by 3 to 7 aa longer than those of other picornaviruses. A tyrosine residue is conserved at the third N-terminal amino acid position as observed in other VPg proteins of picornaviruses. The 3C protein that participates in most of the cleavages of picornavirus polyproteins contains a catalytic triad formed by histidine, aspartate/glutamate and cysteine. These amino acids are conserved in all picornaviruses, and they are seen in Aichi virus 3C at positions 42, 84 and 143, respectively. A motif GXCGG conserved at the C terminus of enterovirus and rhinovirus 3C proteins is considered to form part of the active site, and this motif is present but altered to GxCGS in Aichi virus (x, nonconserved amino acid). An identical change has been observed in cardioviruses. A histidine residue that probably participates in the substrate binding pocket in the trypsin-like protease (Boniotti *et al.*, 1994; Bazan and Fletterick, 1988) is found at aa position 161 of the Aichi virus 3C protein. Highly conserved motifs (KDELR, YGDD and FLKR) in picornavirus 3D are found at aa residues 160-164, 328-331 and 377-380, respectively, in the Aichi virus 3D protein (Yamashita *et al.*, 1998).

Molecular epidemiology

The epidemiology of Aichi virus as a medically important pathogen has not been well defined. Aichi virus was isolated in Vero cells from 7 (12.5%) of 56 patients in 6 gastroenteritis outbreaks, 5 (0.7%) of 722 Japanese travelers returning from tours to Southeast Asian countries and complaining of gastrointestinal symptoms at the quarantine station of Nagoya International Airport in Japan, and 5(2.3%) of 222 Pakistani children with gastroenteritis (Yamashita *et al.*, 1993; Yamashita *et al.*, 1995). In the ELISA, 15 (26.8%) of 56 stool samples from adult patients in six oyster-associated gastroenteritis outbreaks were found to be positive for Aichi virus. Seroconversion against Aichi virus was also observed in 20 (47%) of 43 patients involved in these five outbreaks by neutralization test using paired sera (Yamashita *et al.*, 1993). Based on these results, Aichi virus was recognized to be one of the causative agents of human gastroenteritis.

Seventeen Aichi virus isolates, obtained in Vero cells as described above, were examined for variation, based on their reactivity with a monoclonal antibody raised against the standard strain (A486/88) and on reverse transcription (RT)-polymerase chain reactions (PCR) of three genomic regions. The RNA sequences of these 17 isolates were determined over 519 nucleotides at the putative junction between the C-terminus of 3C and the N-terminus of 3D. The analyses revealed an approximately 90% homology between these isolates that were then divided into two groups: group one (genotype A) included 6 isolates from the 4 outbreaks and one isolate from a traveler; group two (genotype B) included one isolate from the other outbreak, 4 isolates from returning travelers, and all of the isolates from the Pakistani children (Yamashita *et al.*, 2000).

Based on these results, a primer pair was devised for amplification and detection of Aichi virus RNA in fecal specimens. The Aichi virus RNA was detected in 54 (55%) of 99 fecal specimens from patients in 12 (32%) of 37 outbreaks of gastroenteritis in Japan. Of these 12 outbreaks, 11 were suspected to be due to infections with genotype A viruses (Table 1). Aichi virus RNA was also detected in 11 (1.9%) of 567 travelers returning with diarrhoea from India, Nepal, Thailand, Indonesia, Singapore, and Vietnam between 1996 and 1998 (Table 2). Of the 11 isolates, 7 were of genotype A, but slightly different from isolates identified in Japan between 1987 and 1991, as indicated in Fig. 4. The other 4 isolates were of type B, and were also different from isolates from Pakistani children and Japanese travelers identified between 1989 and 1992. These results indicated that circulating Aichi viruses differ, both geographically and temporally. The nucleotide sequences described above were deposited in the DDBJ, EMBL, and GenBank databases under accession numbers AB092824 to AB092834. For the comparison in Fig. 4, the following nucleotide sequences were obtained from DDBJ, EMBL, and GenBank: A846/88, AB040749; N128/91, AB034655; A1156/87, AB079268; N1277/91, AB034654; P803/90, AB0324659; A942/89, AB034653; M166, AB034657; P766/90, AB034658.

Table 1

Outbreaks of gastroenteritis positive for Aichi virus by RT-PCR in Japan, 1987-1998

Outbreak			RT-PCR	%	Genotype (n)[a]
No.	Yr	Likely source	No. of fecal specimens positive/no. tested		
1	1987	Oysters	5/9	55.6	A (3)
2	1988	Oysters	5/7	71.4	A (3)
3		Oysters	9/11	81.8	A (3)
4	1989	Oysters	4/5	80.0	A (2)
5		School excursion	9/14	64.3	A (3)
6	1990	Oysters	2/4	50.0	B (2)
7		Oysters	6/11	54.5	A (2)
8	1991	Oysters	3/6	50.0	A (2)
9	1994	Oysters	2/14	14.3	A (2)
10	1997	Oysters	5/8	62.5	A (3)
11	1998	Oysters	2/4	50.0	A (2)
12		Oysters	2/6	33.0	A (2)
Total			54/99	54.5	A (27), B (2)

[a] Genotypes were determined in a limited number of fecal samples.

652

Table 2

Detection of Aichi virus by RT-PCR in travelers returning to Japan, 1996-1998

Travel destination	No. of travelers tested	No. of travelers positive by RT-PCR	%
Indonesia	176	3	1.7
Thailand	145	2	1.4
Singapore	16	2	12.5
India	8	2	25
Vietnam	7	1	14.3
Nepal	4	1	25
Others*	211	0	0
Total	567	11	1.9

*:The Philippines, Malaysia, Hong Kong, Taiwan, South Korea, Maledives, China, Egypt, Cambodia, Mexico

Antibody prevalence levels by age in Japan are very low in childhood, and rapidly rise during adolescence and young adulthood, reaching about 80% of the population by middle age (Yamashita *et al.*, 1993). This age-related seroprevalence has not significantly changed during the last 25 years in Japan as shown in Table 3. This observation demonstrates that there are likely to be many asymptomatic infections and that this is not a new but a newly discovered virus. Antibody to Aichi virus could be detected using a neutralization test and an ELISA. These methods were used for identification of Aichi virus infection in paired serum samples. The diagnosis of IgM and IgA responses will be useful in a single serum sample of a patient infected with Aichi virus. The prevalence of fever in patients with IgM response is significantly higher than in patients without (Yamashita *et al.*, 2001). Antibody to Aichi virus has not been detected in monkey, cattle, horse, pig, dog, cat, and rat sera.

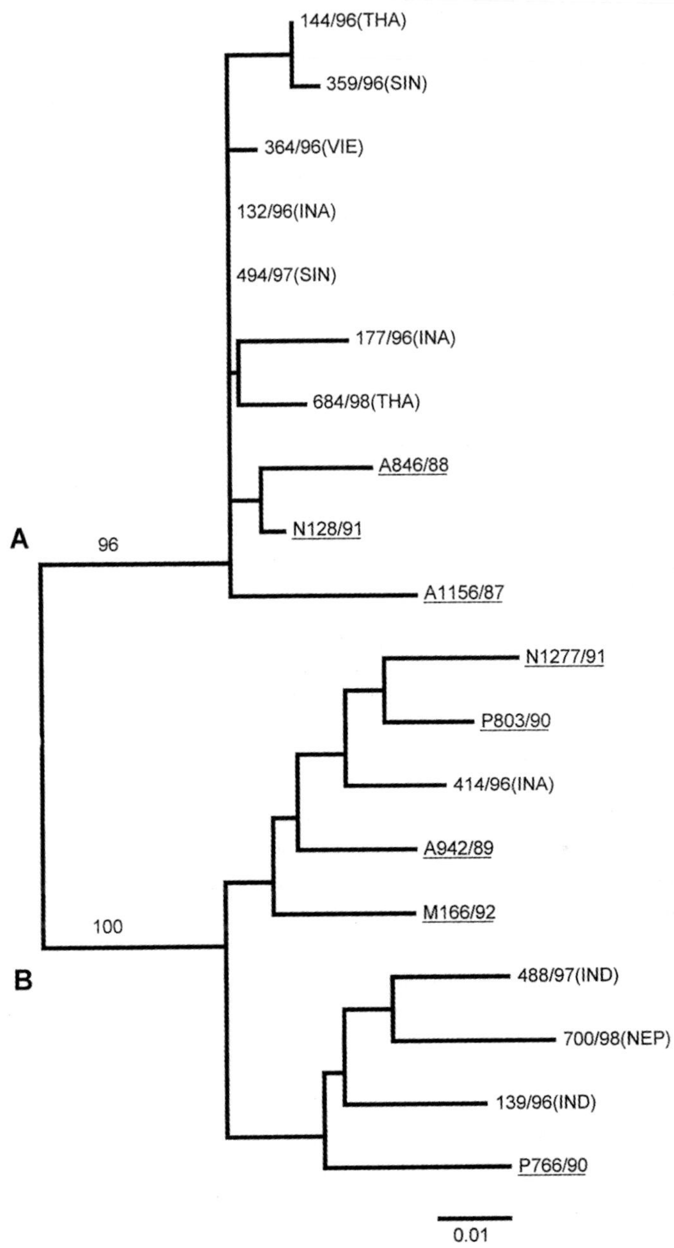

Fig. 4. Dendrogram showing the different sequences of PCR products in the 3CD regions between representative Aichi virus isolates (underlined; Yamashita *et al.* 2000) and stool samples of Japanese travelers returned from tours to India (IND), Nepal (NEP), Indonesia (IND), Singapore (SIN) Vietnam (VIE), and Thailand (THA) between 1996 and 1998. Bootstrap values are indicated at branch roots.

Table 3

Prevalence rates for antibody to Aichi virus in different age groups in Japan, 1973-1998

Age group (yr)	Positive rate (No. of serum samples positive / no. tested) in		
	1973	1989	1998
0-4	7.7 (4 /52)	7.2 (9 /125)	14.3 (3 /21)
5-9	35.7 (10/28)	17.8 (18 /101)	34.8 (8 /23)
10-14	55.0 (11/20)	31.9 (15 / 47)	41.2 (7 /17)
15-19	61.5 (8 /13)	50.0 (30 / 60)	65.2 (15/23)
20-29	67.6 (23/34)	62.5 (75 /120)	72.0 (18/25)
30-39	90.0 (9 /10)	79.2 (95 /120)	87.5 (21/24)
40-49	100 (4 / 4)	81.7 (98 /120)	100 (23/23)
50-59		85.5 (100/117)	100 (25/25)
60-64		87.0 (20 / 23)	100 (9 / 9)

Other diarrhoeogenic enteroviruses

Enteroviruses replicate in the gastrointestinal tract, and the resulting infection is usually symptomless. The clinical expressions of infection, when they do occur, range from paralysis to febrile illnesses. Gastrointestinal upset is commonly reported among associated symptoms in infections by a number of enteroviruses in which other clinical features predominate. In many outbreaks, the enterovirus reported might merely have been a passenger virus unrelated to the illness (Melnick, 1996). Several investigators have reviewed diarrhoeogenicity. There were some reports of enterovirus-associated gastroenteritis. However, these reports were not proof of enteroviruses diarrhoeogenicity (Madeley, 1990). Some enteroviruses have been documented in relation to outbreaks of diarrhea, but diarrhea was not invariably present in those infected. It is ironic that enteroviruses, named after the major site of replication, ie, the gastrointestinal tract, rarely cause gastroenteritis. Enterovirus-associated gastroenteritis may occur in the course of infection associated with another clinical picture (Moore and Morens, 1984).

In an epidemic outbreak of diarrhea in rural southern India, however, the peak of the epidemic curve was associated with echovirus 11 infections. The isolation of the echovirus 11 from 14 of 18 patients with diarrhea and the serological response established that an epidemic of echovirus 11 infections was present in the village at the same time as the wave of the epidemic of diarrhea (Patel *et al.*, 1985). In a bone-marrow transplantation unit of a hospital, 7 of 14 transplant recipients were infected with coxsackievirus A1 during a 3-week period. Diarrhea and mortality were significantly associated with infection (Townsend *et al.*, 1982). A further prospective study in the same unit yielded 37 enteric pathogens from 31 of 78 patients. Coxsackieviruses were found in the stools of four patients, occurring as the sole viral pathogen in two infections correlated with the occurrence of diarrhea and abdominal cramps (Yolken *et al.*, 1982).

The enteroviruses previously classified as echovirus 22 and 23, now constitute a new genus, *Parechovirus*, within the *Picornaviridae* family because of molecular characteristics considerably different from those of typical enteroviruses (Hyypiä *et al.*, 1992; King *et al.*, 2000). Human parechovirus 1 (echovirus 22) is often isolated from children with diarrhea and from patients during gastroenteritis outbreaks. A review of previously published studies concerning the epidemiology of echovirus 22 infections showed that gastroenteritis and respiratory infections are the most common symptoms observed in parechovirus 1 infection (Ehrnst and Eriksson, 1993; Joki-Korpeea and Hyypiä, 1998).

References

Ashley CR, Caul EO, Paver WK (1978). Astrovirus-associated gastroenteritis in children. *J. Clin. Pathol.* **31**: 939-943.

Auvinen P, Hyypiä T (1990). Echoviruses include genetically distinct serotypes. *J. Gen. Virol.* **71**:2133-2139.

Bazan JF, Fletterick RJ (1988). Viral cysteine proteases are homologous to the trypsin-like family of serine proteases: structural and functional implications. *Proc. Natl. Acad. Sci. USA* **85**: 7872-7876.

Blacklow NR (1988). Medical virology of small round gastroenteritis viruses. In: *Medical Virology 9.* (de la Maza LM, Peterson EM eds), pp111-128. Plenum Press, New York etc.

Blacklow NR, Greenberg HB (1991). Viral gastroenteritis. *N. Engl. J. Med.* **325**: 252-264.

Boniotti B, Wirblich C, Sibilia M, Meyers G, Thiel HJ, Rossi C (1994). Identification and characterization of a 3C-like protease from rabbit hemorrhagic disease virus, a calicivirus. *J. Virol.* **68**: 6487-6495.

Chen HH, Kong WP, Roos RP (1995).The leader peptide of Theiler's murine encephalomyelitis virus is a zinc-binding protein. *J. Virol.* **69**: 8076-8078.

Cruz JR, Bartlett AV, Herrmann JE, Caceres P, Blacklow NR, Cano F (1992). Astrovirus-associated diarrhea among Guatemalan ambulatory rural children. *J. Clin. Microbiol.* **30**: 1140-1144.

Cubitt WD (1987). The candidate caliciviruses. *Ciba. Found. Symp.* **128**: 126-143.

Cukor G, Blacklow NR (1984). Human viral gastroenteritis. *Microbiol. Rev.* **48**: 157-179.

Dolin R, Treanor J, Madore HP (1987). Novel agents of viral enteritis in humans. *J. Infect. Dis.* **155**: 365-376.

Donnelly ML, Gani D, Flint M, Monaghan S, Ryan MD (1997). The cleavage activities of aphthovirus and cardiovirus 2A proteins. *J. Gen. Virol.* **78**: 13-21.

Ehrnst A, Eriksson M (1993). Epidemiological features of type 22 echovirus infection. *Scand. J. Infect. Dis.* **25**:275-281.

Gorbalenya AE, Koonin EV, Lai MM (1991). Putative papain-related thiol proteases of positive-strand RNA viruses. Identification of rubi- and aphthovirus proteases and delineation of a novel conserved domain associated with proteases of rubi-, alpha- and coronaviruses. *FEBS Lett.* **288**: 201-205.

Hughes PJ, Stanway G (2000). The 2A proteins of three picornaviruses are related to each other and to the H-rev107 family of proteins involved in the control of cell proliferation. *J. Gen. Virol.* **81**: 201-207.

Hyypiä T, Horsnell C, Maaronen M *et al.* (1992). A distinct picornavirus group identified by sequence analysis. *Proc. Natl. Acad. Sci. USA* **89**: 8847-8851.

Joki-Korpeea P, Hyypiä T (1998). Diagnosis and epidemiology of echovirus 22 infection. *Clin. Infect. Dis.* **27**:129-136.

King, AMQ, Brown F, Christian P *et al.* (1999). Picornavirus taxonomy: a modified species definition and proposal for three new genera. *XIth International Congress of Virology.* Sydney.

King, AMQ, Brown F, Christian P *et al.* (2000). Picornaviridae. In: *Virus Taxonomy.* Seventh Report of the International Committee on Taxonomy of Viruses (van Regenmortel MHV, Fauquet CM, Bishop DHL *et al.*, eds), pp657-678. Academic Press, San Diego etc.

Madeley CR (1990). Viruses associated with acute diarrhoeal disease. In: *Principles and Practice of Clinical Virology*, 2nd edition (Zuckerman AJ, Banatvala JE, Pattison JR Eds), pp173-209. John Wiley and Sons, Chichester etc.

Melnick JL (1996). Enteroviruses: Polioviruses, coxsackieviruses, echoviruses, and newer enteroviruses. In: *Fields Virology,* 3rd edition (Fields BN, Knipe DM, Howley PM *et al.* eds), pp655-712. Lippincott-Raven, Philadelphia.

Moore M, Morens DM (1984). Enterovirus, including polioviruses. In *Textbook of Human Virology* (Belshe RB, ed), pp407-483. PSG Publishing Company, Inc. Littleton etc.

Patel JR, Daniel J, Mathan VI (1985). An epidemic of acute diarrhoea in rural southern India associated with echovirus type 11 infection. *J. Hyg. (Camb.)* **95**: 483-492.

Piccone ME, Zellner M, Kumosinski TF, Mason PW, Grubman MJ (1995). Identification of the active-site residues of the L proteinase of foot- and-mouth disease virus. *J. Virol.* **69**: 4950-4956.

Poyry T, Kinnunen L, Hyypiä T *et al.* (1996). Genetic and phylogenetic clustering of enteroviruses. *J. Gen. Virol.* **77**: 1699-1717.

Ryan MD, Drew J (1994). Foot-and-mouth disease virus 2A oligopeptide mediated cleavage of an artificial polyprotein. *EMBO J.* **13**: 928-933.

Sasaki J, Kusuhara Y, Maeno Y *et al.* (2001). Construction of an infectious cDNA clone of Aichi virus (a new member of the family Picornaviridae) and mutational analysis of a stem-loop structure at 5' end of the genome. *J. Virol.* **75**: 8021-8030.

Townsend TR, Bolyard EA, Yolken RH, Beschorner WE, Bishop CA, Burns WH, Santos GW, Saral R (1982). Outbreak of coxsackie A1 gastroenteritis: A complication of bone-marrow transplantation. *Lancet* **I**: 820-823.

Yamashita T, Ito M, Tsuzuki H, Sakae K (2001). Identification of Aichi virus infection by measurement of immunoglobulin responses using ELISA. *J. Clinic. Microbiol.* 39: 4178-4180.

Yamashita T, Kobayashi S, Sakae K *et al.* (1991). Isolation of cytopathic small round viruses with BS-C-1 cells from patients with gastroenteritis. *J. Infect. Dis.* **164**: 954-957.

Yamashita T, Sakae K, Ishihara Y, Isomura S, Utagawa E (1993). Prevalence of newly isolated, cytopathic small round virus (Aichi strain) in Japan. *J. Clin. Microbiol.* **31**: 2938-2943.

Yamashita T, Sakae K, Kobayashi S *et al.* (1995). Isolation of cytopathic small round virus (Aichi virus) from Pakistani children and Japanese travelers from Southeast Asia. *Microbiol. Immunol.* **39**: 433-435.

Yamashita T, Sakae K, Tsuzuki H *et al.* (1998). Complete nucleotide sequence and genetic organization of Aichi virus: a distinct member of the *Picornaviridae* associated with acute gastroenteritis in humans. *J. Virol.* **72**: 8408-8412.

Yamashita T, Sugiyama M, Tsuzuki H, Sakae K, Suzuki Y, Miyazaki Y (2000). Application of a polymerase chain reaction for identification and differentiation of Aichi virus, a new member of the picornavirus family associated with gastroenteritis in humans. *J. Clin. Microbiol.* **38**: 2955-2961.

Yolken RH, Bishop CA, Townsend TR, Bolyard EA, Bartlett J, Santos GW, Saral R (1982). Infectious gastroenteritis in bone-marrow-transplant recipients. *N. Engl. J. Med.* **306**: 1009-1012.

Yu SF, Lloyd RE (1992). Characterization of the roles of conserved cysteine and histidine residues in poliovirus 2A protease. *Virology* **186**: 725-735.

Viral Gastroenteritis
U. Desselberger and J. Gray (editors)
© 2003 Elsevier Science B.V. All rights reserved

VI, 4. Histopathology of viral gastrointestinal infections in the immunocompromised

John W. Grant

Department of Histopathology, Addenbrooke's Hospital, Cambridge CB2 2QQ

Introduction

Those with an impaired immune system are generally more susceptible to viral infections than the immunocompetent. Patients with T cell deficiencies, either congenital or acquired/iatrogenic, are more likely to experience viral infections, particularly with herpes viruses. Intestinal diseases, of which viral infections are a large component, are a major cause of morbidity and mortality in the immunocompromised. In AIDS, for example, gastrointestinal symptoms, many of which are of viral aetiology, affect 30-50% of North American and European and 90% of African patients (Fenoglio-Preiser *et al.*, 1999). This chapter presents an overview of viral infections of the gastrointestinal tract in the immunocompromised.

Mouth

Hairy leukoplakia is an uncommon oral lesion, which is seen almost exclusively in patients infected with the human immunodeficiency virus (HIV) (Neville *et al.*, 1995). It can sometimes be the presenting symptom of the infection, preceding other manifestations of AIDS by several years. White confluent patches of thickened hyperkeratotic "hairy" epithelium are present anywhere on the oral mucosa (Fig. 1). Microscopically there is evidence of mucosal acanthosis with hyperkeratosis (Fig. 2). Sometimes there is also koilocytosis of the superficial epidermal cells, which is indicative of human papilloma virus (HPV) infection. *In situ* hybridisation has revealed HPV, Epstein Barr virus (EBV) and sometimes also HIV in these lesions. EBV is now considered to be the major cause of this condition (Greenspan *et al.*, 1984; Greenspan *et al.*, 1985; Schiodt *et al.*, 1987; Sciubba, 1996) causing squamous cell proliferation of the oral mucosa. Sometimes there can also be an accompanying infection with *Candida* spp., which contributes to the macroscopic appearance of "hairiness".

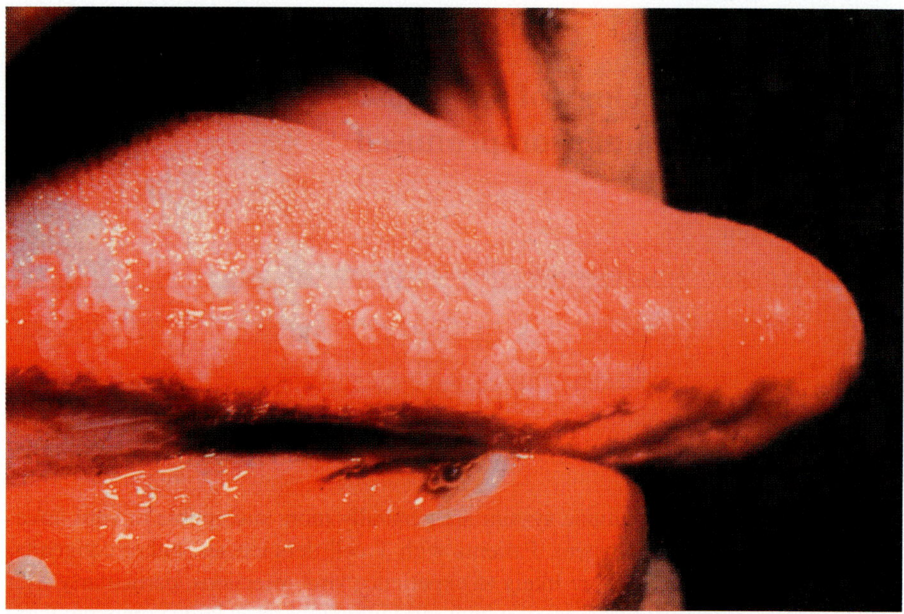

Fig. 1. Hairy leukoplakia on the tongue of an AIDS patient. Macroscopic appearance.

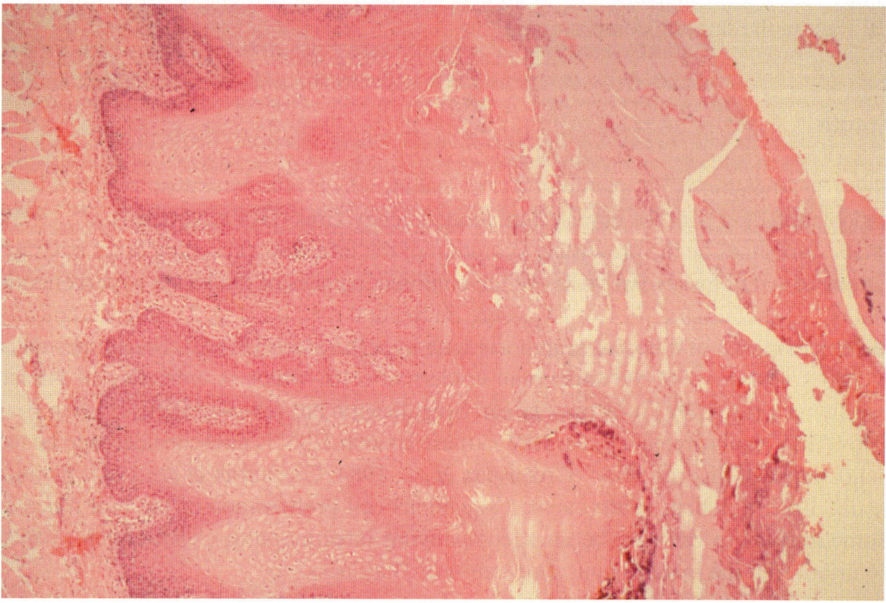

Fig. 2. Hairy leukoplakia on the tongue of an AIDS patient. Microscopic appearances showing acanthosis and hyperkeratosis. HE stain, ×4.

Oesophagus

Viral infection of the oesophageal mucosa is found particularly in the immunocompromised. Oesophagitis resulting from infection with herpes simplex virus (HSV) can occur in all age groups. Few cases of self-limited disease have also been reported in healthy immunocompetent adults (Springer *et al.*, 1979). Herpes oesophagitis most frequently affects immunosuppressed patients with malignant disease, particularly malignant lymphomas and leukaemias. It can also occur in a wide variety of other clinical settings such as following burns, in AIDS or after organ transplantation. It presents with the sudden onset of odynophagia, fever and retrosternal pain. Most frequently the lower third of the oesophagus is involved. Multiple ulcers are present, which coalesce and result in the formation of a pseudomembrane (McBane and Gross, 1991). If the ulceration is severe it can cause haemorrhage (Deshmukh *et al.*, 1984; Rattner *et al.*, 1985). There are occasional reports of spontaneous oesophageal perforation (Cronstedt *et al.*, 1992; Shintaku *et al.*, 1992). On biopsy the squamous epithelial cells at the margins of ulcers contain homogenous round or oval Cowdry type A intranuclear inclusions. There are also multinucleate syncytial squamous cells containing moulded nuclei and a typical "ground glass" appearance (Fig. 3) (Nash and Ross, 1974).

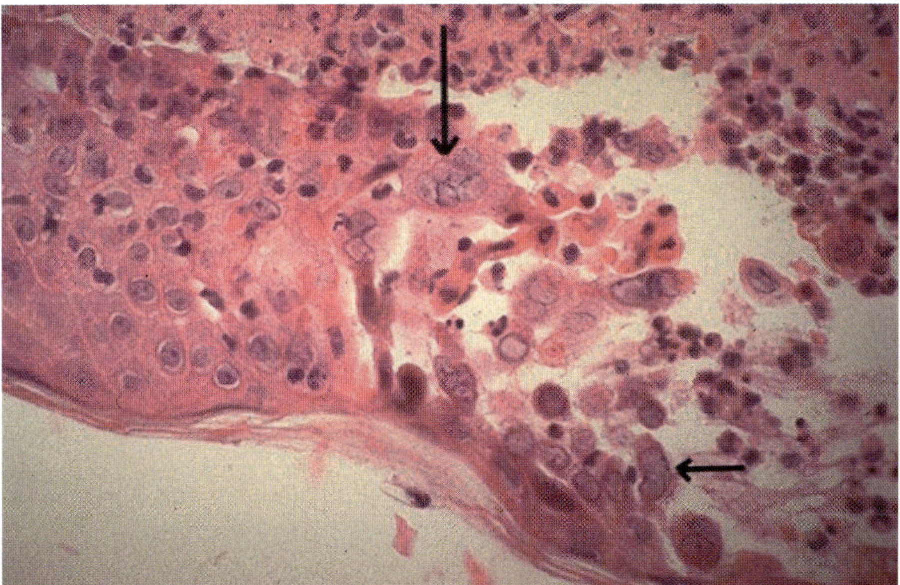

Fig. 3. Herpes simplex oesophagitis in an AIDS patient. Multinucleate syncytial squamous cells arrowed. HE stain, ×20

Varicella zoster virus (VZV) is a rare cause of oesophageal infection in children with disseminated chickenpox. Adults usually have a reactivation of a latent infection

662

acquired in childhood. Primary VZV infection is more severe in immunocompromised children, including those with AIDS, and is associated with visceral dissemination and increased mortality. Grossly, there is a necrotising oesophagitis with superficial and deep ulceration. The histological appearances are similar to those seen in HSV infection. Vesicles form, and there are foci of necrosis. Intranuclear eosinophilic inclusions are present in epithelial, endothelial and stromal cells (Fenoglio-Preiser *et al.,* 1999).

Cytomegalovirus (CMV) tends to affect the debilitated, elderly or immunocompromised. CMV is the most commonly recognised opportunistic infection in patients who are immunosuppressed as a result of malignancy or chemotherapy, as recipients of transplants, and as patients with AIDS. The infection may be acquired by reactivation of the virus or by acquisition from exogenous sources such as blood transfusions or grafts from seropositive donors. CMV oesophagitis presents with nausea, vomiting, fever, epigastric pain, diarrhoea and weight loss. Difficulty in swallowing and retrosternal pain are less common than with HSV infections (Weber *et al.,* 1987).

Infection with HPV may be implicated in some forms of oesophageal disease. It has been demonstrated in oesophageal papillomas (Fig. 4) and carcinomas and also in the pathological entity of clear cell acanthosis (Hille *et al.,* 1986; Morris and Price, 1986; Williamson *et al.,* 1991; Winkler *et al.,* 1985).

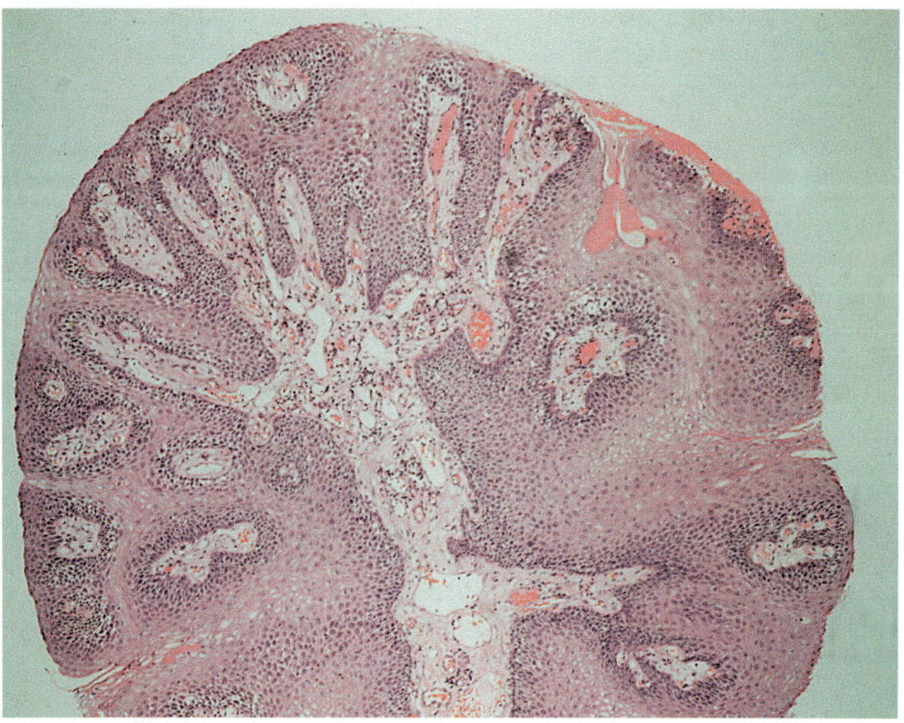

Fig. 4. Oesophageal squamous papilloma. HE stain, ×4

Acute HIV infection may be responsible for multiple acute oesophageal aphthous ulcers (Rabeneck *et al.,* 1990). They develop in the first few months following HIV infection. Patients present with malaise, fever, skin rash, dysphagia and severe odynophagia. The oesophageal epithelium is eroded or ulcerated. Sometimes so called "giant oesophageal ulcers" can form. These can progress to life threatening size, eroding vessels and compromising nutrition (Rabeneck *et al.,* 1990). In the submucosa there is a mixed inflammatory infiltrate consisting of neutrophils and mononuclear cells. Usually no infectious agent other than HIV can be detected. Its presence in mononuclear cells can be confirmed by *in situ* hybridisation.

Oesophageal ulceration has been described in cases in which the adjacent epithelium shows changes similar to those seen in oral leukoplakia (Kitchen *et al.,* 1990). In three cases, *in situ* hybridisation revealed the presence of the DNA of EBV and HPV with no evidence of CMV, HSV or any other potential cause. It is thought therefore that the pathogenesis is similar to that of the oral lesions (Allday and Crawford, 1988; Daniels *et al.,* 1987; Greenspan *et al.,* 1984; Greenspan *et al.,* 1985; Schiodt *et al.,* 1987). The ulcers of EBV infection differ from those of HSV in that they are deep, linear and located in the mid-oesophagus.

Stomach

Histologically recognisable infections of the stomach are rare. The principal cause is CMV infection, generally associated with AIDS or immunosuppression following transplantation or treatment for haematological malignancies (Minami *et al.,* 1990; Stenglein *et al.,* 1993). Gastric lesions of CMV can also be seen in neonates and occasionally in otherwise healthy individuals (Garcia *et al.,* 1987).

CMV gastritis presents with epigastric pain, nausea and vomiting. The condition can be complicated by haemorrhage, gastric outlet obstruction and perforation. In mild infections the lamina propria appears slightly hypocellular and only occasional characteristic inclusions are found (Fig. 5). Biopsies in severe infections show inflammation, ulceration and prominent viral intranuclear inclusions. Deep biopsies can show the presence of CMV in ganglion cells. In these patients there may be disordered gastric motility (Fenoglio-Preiser *et al.,* 1999). CMV can also induce diffuse antral thickening, which can simulate malignancy.

Other infections with viruses such as HSV and VZV are much less common. Grossly herpetic gastritis shows yellowish plaques or oedematous mucosal nodules separated by areas of ulceration. Biopsies from the vesicles and ulcer bases show eosinophilic intranuclear inclusions and ballooning cytoplasm.

Small intestine

Although viruses causing gastroenteritis are genetically and morphologically different they cause similar lesions in the gastrointestinal tract. In rotavirus infections there is

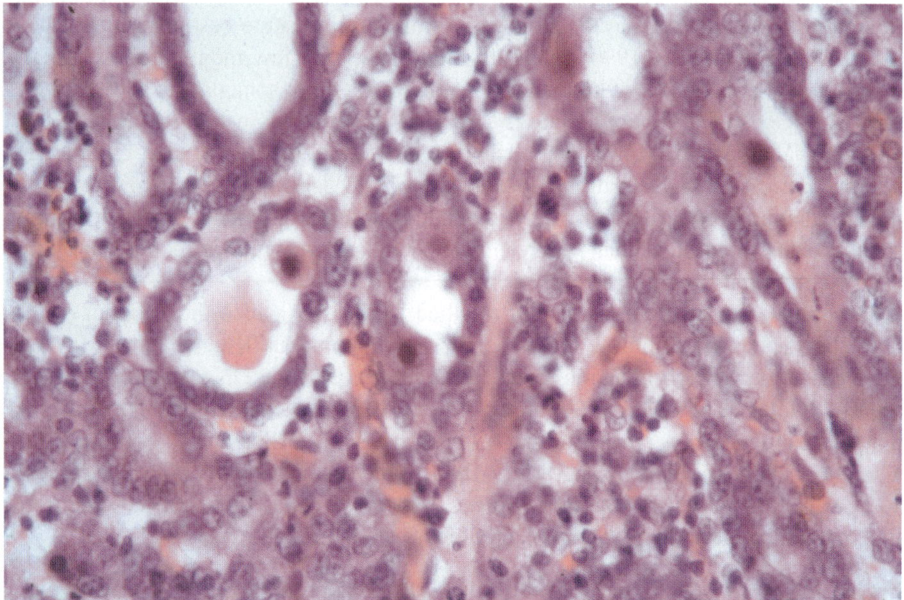

Fig. 5. CMV gastritis. Biopsy showing inflammation and prominent intranuclear inclusions. HE stain, ×20

blunting and destruction of the villous epithelial cells. Viral particles can be detected within them by electron microscopy and by immunofluorescence with virus-specific antibodies. There is secondary hyperplasia of the mucosal crypts and a mixed inflammatory infiltrate in the lamina propria (Burke and Desselberger, 1996).

Some adenoviruses can cause gastroenteritis (Blacklow and Greenberg, 1991). Besides rotaviruses, they are a frequent cause of diarrhoea amongst infants and young children. The infection typically lasts for 5-12 days, producing protracted watery diarrhoea. Whilst adenoviruses are an uncommon cause of intestinal illness in immunocompetent adults, they have been identified in 7% of HIV-infected homosexual men suffering from diarrhoea (Janoff *et al.*, 1991). They have also been reported to cause infection in other immunosuppressed patient groups including those with primary immunodeficiency, and in bone marrow transplant patients (Hierholzer, 1992). The macroscopic mucosal appearances were normal or near normal. Microscopy shows foci of mucosal necrosis and amphophilic nuclear inclusions within degenerating epithelial cells (Janoff *et al.*, 1991). EM shows characteristic adenovirus particles with hexagonal circumference within the inclusions.

A shorter survival has been demonstrated in HIV positive patients with diarrhoea who excrete adenovirus from the GI tract. This occurs at an advanced stage of HIV disease. Sabin *et al.* (1999) suggested that adenoviruses may contribute to mortality in this population.

HIV-induced changes play a central role in the pathogenesis of the gastrointestinal manifestations of HIV infection. In humans infected with HIV there is a more rapid, early and pronounced loss of CD4+ T cells in the mucosa than in peripheral blood. This may explain the frequent occurrence of opportunistic mucosal infections. There is evidence that small intestinal damage occurs independent of secondary infections resulting in an HIV enteropathy, although the mucosal response to HIV can be masked by other pathogens (Batman *et al.*, 1998). In late stages of HIV infection, there is evidence of villous atrophy with hyporegeneration and dysmaturation of the intestinal epithelial cells (Zeitz *et al.*, 1998). Infection of cells in the lamina propria of the jejunum with HIV stimulates crypt cell proliferation, whilst there is atrophy in the villous surface area. Early and late stage enteropathy in HIV infection may represent two types of immunologically mediated mucosal changes in which the number and state of activation of regulatory T cells determine whether hypo- or hyperproliferative villous atrophy occurs.

HIV-associated primary enteropathy is defined as the presence of chronic diarrhoea, malabsorption and wasting, without evidence of an enteric infection (Simon and Brandt, 1993). This may be caused by occult infections, indirect effects related to localised immunological dysfunction or to direct effects of HIV on the gastrointestinal epithelium. On histological examination (Fig. 6) the intestinal mucosa shows oedema, chronic inflammation, villous atrophy, crypt hyperplasia, nuclear enlargement, apoptosis and haemosiderin deposition (Heise *et al.*, 1991). Early in the AIDS epidemic a jejunal enteropathy characterised by villous atrophy and crypt hyperplasia was described in male homosexuals at different clinical stages of infection without associated secondary opportunistic infections (Batman *et al.*, 1989; Ullrich *et al.*, 1989). Many of the patients were found to have fat malabsorption and diarrhoea. The degree of steatorrhoea correlated with the degree of villous atrophy. When viewed by EM, the jejunal enterocytes show proliferation of the smooth endoplasmic reticulum. There was also a decrease in numbers of mitochondria. These changes were maximal in patients with late stage AIDS. Degenerative changes were seen in enterocytes, in enteric nerves and in the microvasculature of the lamina propria (Griffin, *et al.*, 1988; Mathan *et al.*, 1990). These observations suggest a primary HIV associated enteropathy. They were supported by the finding of the HIV genome in argentaffin cells and macrophages in the lamina propria (Fox *et al.*, 1989; Nelson *et al.*, 1988). HIV protein was also present in the epithelium and lamina propria cells (Jarry *et al.*, 1990; Kotler *et al.*, 1991). HIV particles have been demonstrated in duodenal biopsy specimens (Ehrenpreis *et al.*, 1992). More recently some authors (e.g. Pollok, 2001) have questioned the clinical significance of HIV associated enteropathy suggesting it is probably limited.

Nearly all patients at risk of HIV infection have serological evidence of previous infection with CMV (Mintz *et al.*, 1983). Reactivation of latent virus or reinfection causes a variety of clinical manifestations in the gastrointestinal tract and can lead to lesions occurring from the mouth to the anus. These may take the form of ulcers in the oesophagus, stomach or duodenum or more subtle appearances. Duodenitis due to CMV is seen in AIDS (Wilcox and Schwartz, 1992). It can also present with symptomatic ulcer disease. CMV infection may be associated with perforation in some cases.

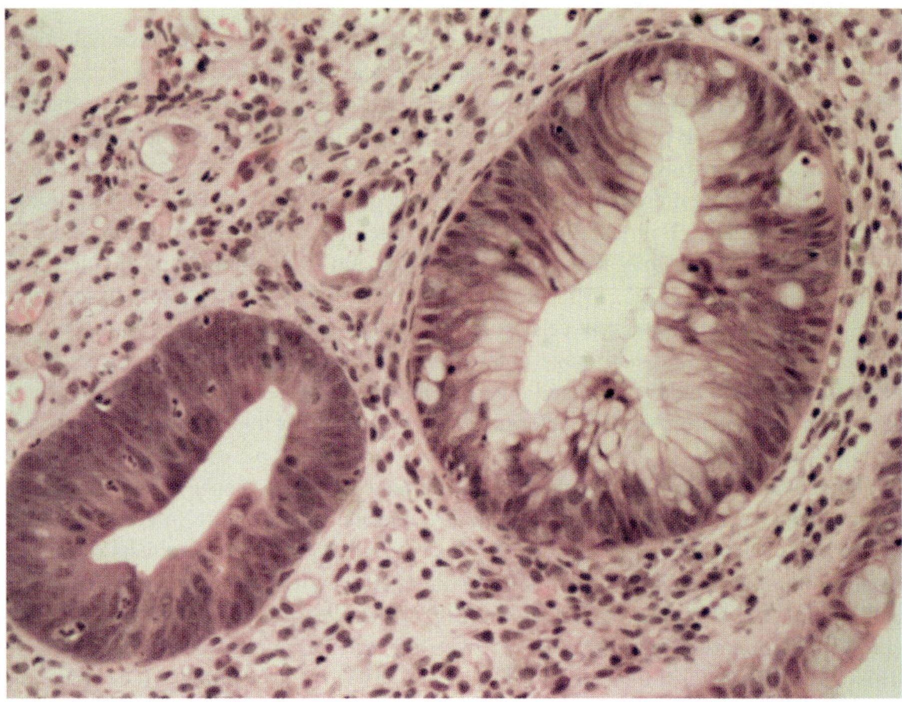

Fig. 6. HIV associated enteropathy. Biospy showing inflammation, oedema, nuclear enlargement and apoptosis. HE stain, x20.

There is an accompanying mixed inflammatory infiltrate, which includes plasma cells, lymphocytes, neutrophils and eosinophils. Occasionally an inflammatory mass can mimic a tumour (Rich *et al.*, 1992). Serious CMV disease has been described in up to 10% of HIV positive patients (Francis *et al.*, 1989). The cell types infected include endothelial cells, pericytes, muscle cells, macrophages in the lamina propria or the inflammatory infiltrate, and epithelial cells. Their involvement differs according to the site of the lesions in the gastrointestinal tract. Squamous epithelial cells of the oesophagus never show obvious cytomegalic infection. However, in the epithelial cells of the stomach and small intestine large numbers of inclusions are common, particularly in their deeper layers and in Brunner's glands (Francis, 1995). At all sites the cells of the lamina propria, small vessels and muscularis mucosae may be infected. Infection can extend across the muscularis mucosae (Aqel *et al.*, 1991), in which case muscle cells can also be seen to be infected with CMV. The endothelial cells of the submucosal vessels may also show inclusions. Immunohistochemistry and *in situ* hybridisation can increase the detectable number of infected cells. The pathogenesis of injury is unclear. In cases where there has been perforation, damage to muscle cells may be an important factor (Fernandes *et al.*, 1986). A necrotising vasculitis has been demonstrated in others (Frank and Raicht, 1984; Goodman and Porter, 1973). In some cases fibrin thrombi are seen in the lamina propria suggesting a possible ischaemic mechanism (Francis *et al.*, 1989).

Gastrointestinal viruses are detected in more than 50% of patients with symptomatic HIV infection and diarrhoea. However, they are also frequently found in those without diarrhoea (Cunningham *et al.*, 1988). These may represent passenger viruses, which have not been eliminated because of immunosuppression (Kaljot *et al.*, 1989). However, the detection of rotaviruses (Cunningham *et al.*, 1988) and adenoviruses (Pollok, 2001), even in the absence of cytopathic changes (Francis *et al.*, 1989), suggests that they may have a pathogenic role. It is possible that abortive viral infections modify the surface antigens of epithelial cells, providing a target for immune-mediated damage (Francis, 1995).

Large intestine

Although clinically significant CMV infection of the gastrointestinal tract has been described in normal subjects (Starr, 1979; Kinney *et al.*, 1985), it is much more common in immunocompromised patients, especially those with AIDS. The most common site of infection is the colon (Mentec *et al.*, 1994). It can cause a wide variety of symptoms of which chronic or intermittent diarrhoea and abdominal pain are the most common. It can also cause colonic ulceration, rectal bleeding, toxic megacolon and perforation (Dieterich and Rahmin, 1991; Foucar *et al.*, 1981; Framm and Soave, 1997; Frank and Raicht, 1984; Knapp *et al.*, 1983; Meiselman *et al.*, 1985; Underwood and Corbett, 1978). CMV causes colonic ulcers, particularly in the ileocaecal region (Fig. 7). Rarely patients can develop a CMV pancolitis (Foucar *et al.*, 1981). Acute lower gastrointestinal haemorrhage in HIV–infected patients is most commonly caused by CMV colitis and is associated with a high short-term morbidity and mortality (Bini *et al.* 1999).

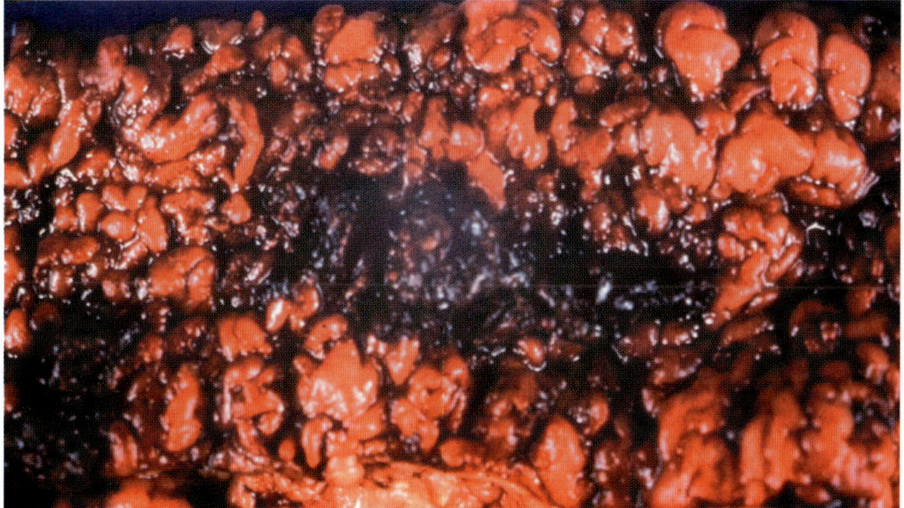

Fig. 7. CMV colitis with ulceration. Macroscopic appearance.

A definitive diagnosis of CMV enterocolitis requires an intestinal biopsy. Wilcox *et al.* (1998) prospectively examined 55 HIV patients with CMV infection by sigmoidoscopy and colonoscopy. Chronic diarrhoea occurred in 80% and abdominal pain in 50% of patients. 9% presented with lower GI bleeding with a previous history of diarrhoea. A variety of endoscopic appearances were observed: Colitis associated with ulceration was seen in 39%, ulceration alone in 38%, or colitis alone 20%. Subepithelial haemorrhage was seen commonly in all groups.

The pathogenesis of the mucosal disease in CMV is thought to be a virus-induced vasculitis with thrombosis and ischaemia. Haemorrhagic mucosal lesions can be seen on sigmoidoscopy. The characteristic histological changes are cytomegaly and acidophilic intranuclear inclusions with a clear halo, giving an "owl's eye" morphology associated with a granular basophilic cytoplasm (Fig. 8). The best yield of such cells is usually obtained from the base of CMV ulcers. The inclusions occur mainly in cells of mesenchymal origin, endothelial cells, fibroblasts, smooth muscle cells and macrophages. Colonic epithelial cells are occasionally involved and may have viral inclusions, but the cellular morphology is relatively well preserved compared to mesenchymal cells (Buckley *et al.,* 1990; Miller and Howell, 1988; Rotterdam, 1987). Classical inclusions are uncommon in AIDS. Morphologically atypical inclusions are present, which may be difficult to precisely identify as CMV-infected cells (Schwartz and Wilcox, 1992). Immunohistochemistry and *in situ* hybridisation can be used to confirm the diagnosis (Fig. 8, Inset) and increase the sensitivity of detection (Rasing *et al.,* 1990; Schwartz and Wilcox, 1992).

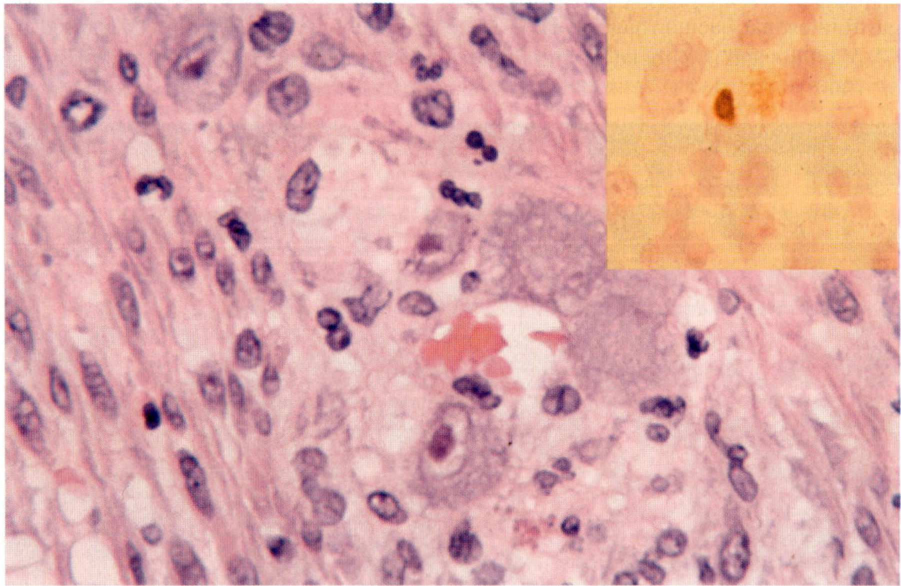

Fig. 8. CMV colitis showing inflamation and viral inclusions. HE stain, ×40.
Inset: immunohistochemical staining for CMV.

VZV infection is occasionally seen in the colon of patients infected with HIV (Pui *et al.*, 2001).

HSV is a common cause of non-gonococcal proctitis in homosexual men. In AIDS patients severe enlarging mucocutaneous ulcers can occur in the perianal or intergluteal skin, the anal canal or the rectum (Siegal *et al.*, 1981). Lesions are also found around the mouth or in the oesophagus. Histological examination of rectal biopsies may reveal intranuclear inclusion bodies, perivascular mononuclear cell infiltrates, giant cells and focal ulceration (Fig. 9) (Pavli and Doe, 1995). Areas of destructive ulceration will evolve in progressive mucocutaneous HSV infection. These lesions may respond to acyclovir therapy.

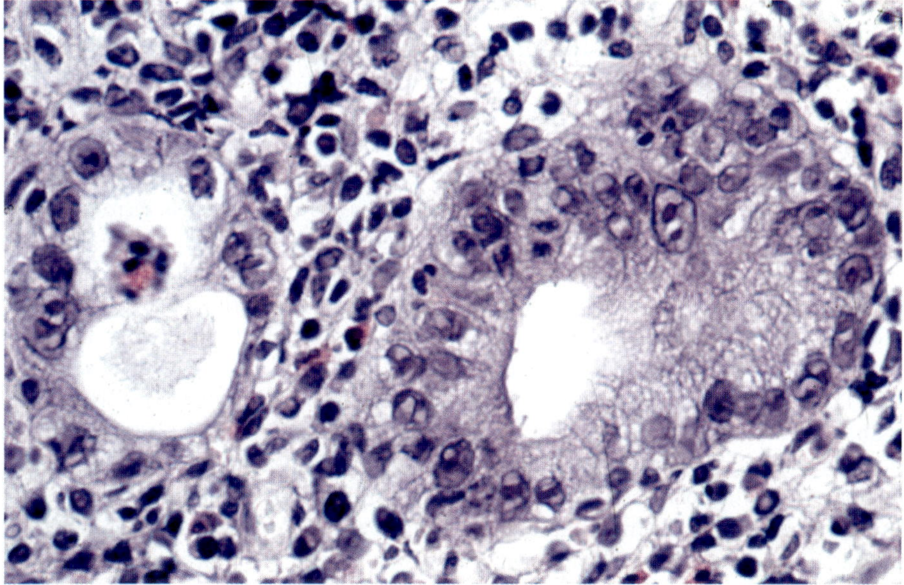

Fig. 9. Rectal biopsy in Herpes simplex virus type II infection showing inflammation and viral inclusions. HE stain, ×40

Since the introduction of highly active antiretroviral therapy (HAART) there has been a marked reduction in the morbidity and mortality associated with opportunistic GI infections in HIV infection and AIDS (Gazzard and Moyle, 1998; Monkemuller *et al.*, 2000). HAART can result in the remission of previously persistent opportunistic infections, including CMV. However the development of viral resistance and problems with compliance may lead to recurrent HIV viraemia and the re-emergence or re-acquisition of opportunistic GI infections.

Anus

Infections with HSV type 2 are seen in the anus. This presents initially with itching and soreness around the anus, which can develop into severe anorectal pain. Other symptoms include fever, inguinal lymphadenopathy, constipation, anorectal discharge and chronic perianal ulcers. Biopsies show multinucleated cells, epithelial intranuclear inclusions and perivascular lymphocytic infiltrates (Fenoglio-Preiser *et al.*, 1999).

CMV can also produce anal ulceration with histological appearances similar to those seen in the oesophagus. Inclusions are seen in endothelial cells, macrophages and epithelium (Fenoglio-Preiser *et al.*, 1999).

Gastrointestinal tumours in the immunocompromised

Post-transplant lymphoproliferative disorder (PTLD) occurs as a consequence of immunosuppression in recipients of solid organ or bone marrow allografts. The majority are associated with EBV infection and represent EBV induced monoclonal or less commonly polyclonal B-cell proliferations. They present as mass lesions with a tendency to involve extranodal sites and the allograft (Fig. 10) (Chan, 2000). Patients treated with cyclosporine or tacrolimus-based regimens develop PTLDs which tend to involve lymph nodes and the gastrointestinal tract (Penn, 1991). Bone marrow allograft recipients usually present with widespread disease involving nodal and extranodal sites including the gastrointestinal tract (Penn, 1991; Zutter *et al.*, 1988). A high incidence of PTLD was found in patients who developed GI bleeding or diarrhoea associated with EBV infection after paediatric liver transplantation (Cao *et al.*, 1998). In these patients endoscopy or biopsy may lead to early diagnosis of PTLD.

Some PTLD lesions show dramatic regression following reduction or cessation of immunosuppression. Fourteen percent of PTLDs are EBV-negative. These tend to appear after a longer interval following transplantation. Some of these also respond to a reduction in immunosuppression (Nelson *et al.*, 2000).

Although there is no evidence that HIV itself is capable of causing neoplastic transformation, patients with AIDS have a high incidence of certain tumours especially Kaposi's sarcoma, non-Hodgkin's lymphoma and cervical carcinoma. The basis for this increased risk is multifactorial and includes profound defects in T-cell immunity, dysregulated B-cell and monocyte function and infections with viruses such as human herpes virus type 8 (HHV8), EBV and HPV, respectively.

Most of the AIDS-associated lymphomas are high grade B-cell tumours (Fig. 11). Their pathogenesis probably involves polyclonal B-cell activation, followed by the emergence of monoclonal or oligoclonal B-cell populations. Some clones undergo somatic mutations and neoplastic transformation. EBV is a polyclonal mitogen for B-cells. Forty per cent of all AIDS lymphomas are thought to be associated with EBV (Chan, 2000).

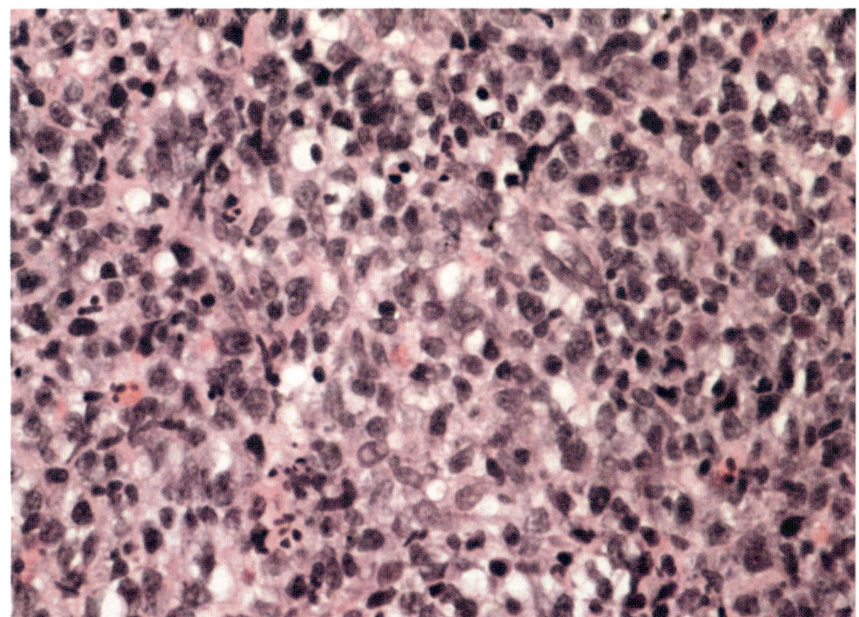

Fig. 10. The polymorphic infiltrate of a PTLD in a gastric biopsy of a post kidney transplant patient. HE stain, ×20.

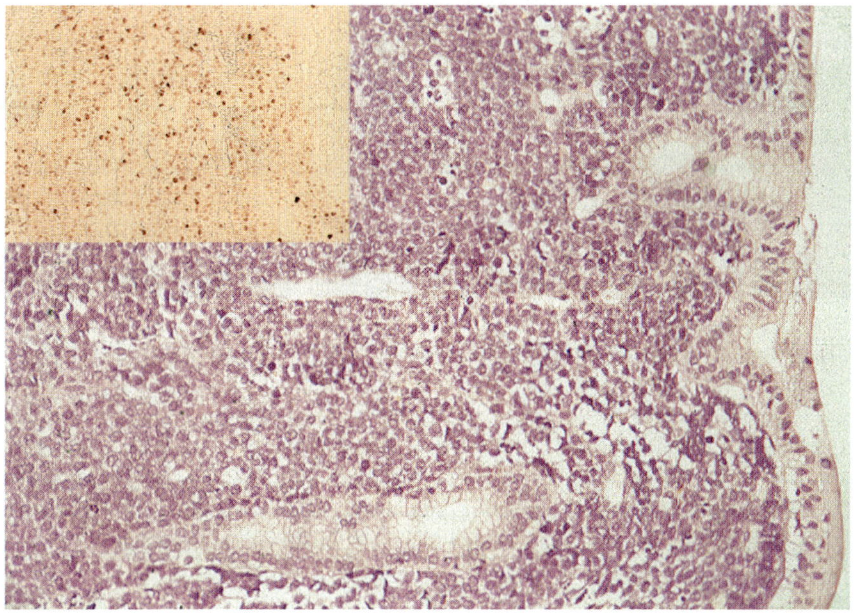

Fig. 11. High grade B cell intestinal lymphoma in an AIDS patient. HE stain, ×10. *In situ* hybridisation for EBER-2 in inset.

672

HHV8 has been identified in all types of Kaposi's sarcoma including both classic and endemic variants (Cesarman and Knowles, 1997; Kennedy *et al.*, 1997). The finding of monoclonality in multifocal lesions of Kaposi's sarcoma (Rabkin *et al.*, 1997) suggests that the condition may be a neoplastic process in which the virus may have an oncogenic role. Other studies have produced contradictory results suggesting that the condition may evolve through a polyclonal preneoplastic phase (Delabesse *et al.*, 1997; Gill *et al.*, 1998). In AIDS the condition is a disseminated aggressive process. In Western AIDS patients (as opposed to African), most cases occur in the homosexual risk group. The skin and gastrointestinal tract are commonly involved (Figs 12 and 13). Kaposi's sarcoma has also been seen in patients receiving immunosuppressive therapy, especially following kidney transplantation (Stribling *et al.*, 1978).

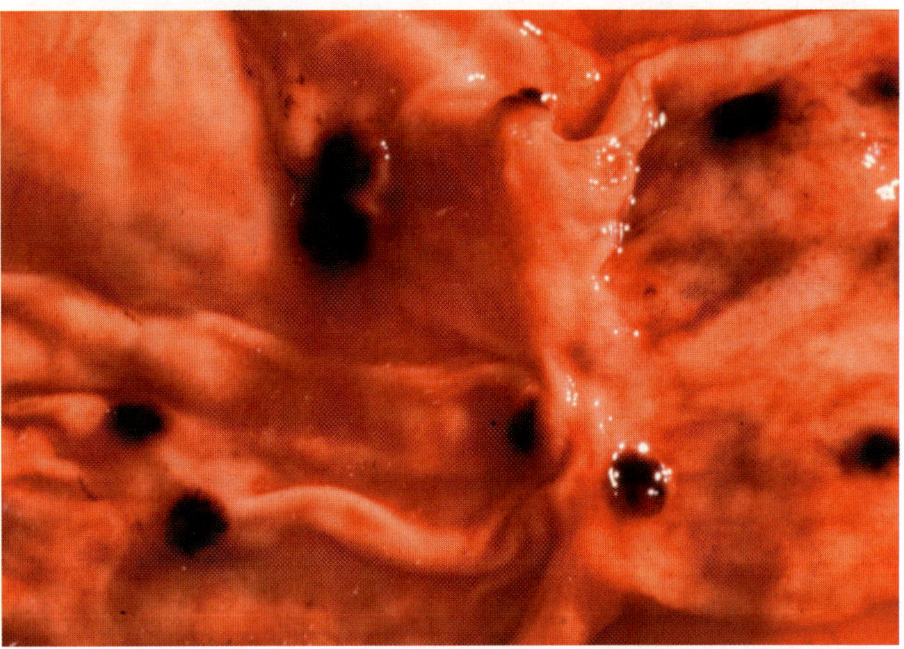

Fig. 12. Kaposi's sarcoma at gastro-oesophageal junction. Macroscopic appearance.

Infections with HPV produce koilocytosis, condylomas and squamous carcinomas of the lower anal canal (Figs 14 and 15). Although external anogenital lesions with HPV are rarer than cutaneous lesions in organ transplant recipients, their association with dysplastic changes and malignancy (Fig. 15) makes regular surveillance necessary (Euvrard *et al.*, 1997).

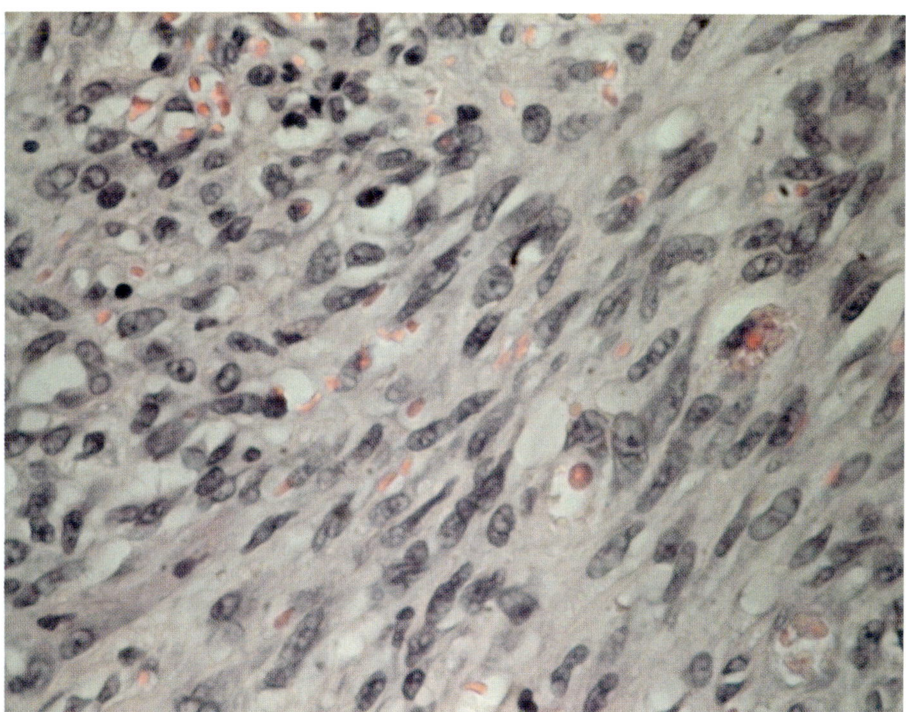

Fig. 13. Kaposi's sarcoma. Histology showing the characteristic spindle cell proliferation. HE stain, ×40

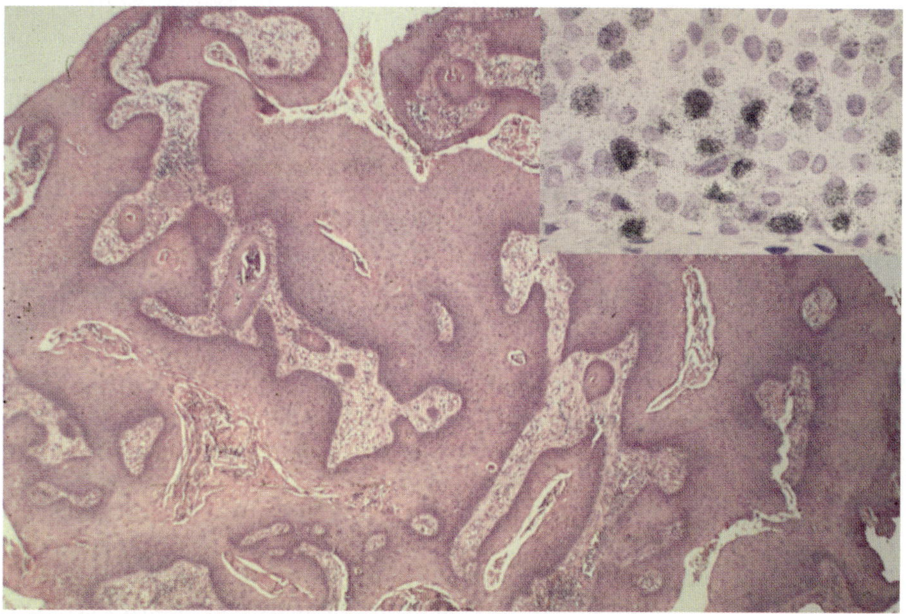

Fig. 14. Anal condyloma in AIDS. HE stain, ×2 Inset: *in situ* hybridisation for HPV.

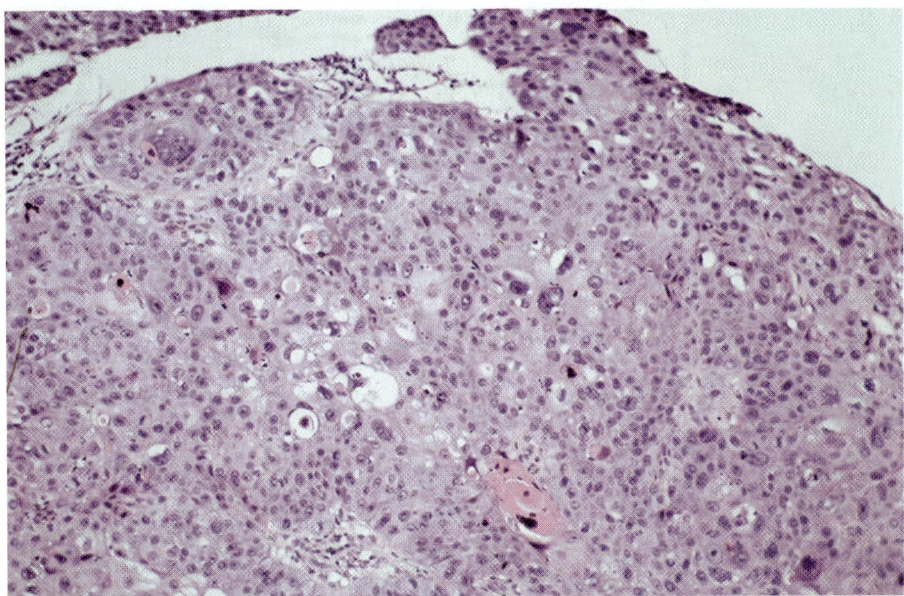

Fig. 15. Anal squamous carcinoma in AIDS. HE stain, ×4.

References

Allday, M., and Crawford, D. Role of epithelium in EBV persistence and pathogenesis of B cell tumours. *Lancet* 1988; **I:** 855-7.

Aqel, N. M., Tanner, P., Drury, A., Francis, N. D., and Henry, K. Cytomegalovirus gastritis with perforation and gastrocolic fistula formation. *Histopathology* 1991; **18:** 165-8.

Batman, P. A., Kapembwa, M. S., Miller, A. R., Sedgwick, P. M., Lucas, S., Sewankambo, N. K., Serwadda, D., Pudney, J., Moody, A., Harris, J. R., and Griffin, G. E. HIV enteropathy: comparative morphometry of the jejunal mucosa of HIV infected patients resident in the United Kingdom and Uganda. *Gut* 1998; **43:** 350-5.

Batman, P. A., Miller, A. R., Forster, S. M., Harris, J. R., Pinching, A. J., and Griffin, G. E. Jejunal enteropathy associated with human immunodeficiency virus infection: quantitative histology. *J Clin Pathol* 1989; **42:** 275-81.

Bini, E. J., Weinshel, E. H., and Falkenstein, D. B. Risk factors for recurrent bleeding and mortality in human immunodeficiency virus infected patients with acute lower GI hemorrhage. *Gastrointest Endosc* 1999; **49:** 748-53.

Blacklow, N. R., and Greenberg, H. B. Viral gastroenteritis. *N Engl J Med* 1991; **325:** 252-64.

Buckley, R. M., Braffman, M. N., and Stern, J. J. Opportunistic infections in the acquired immunodeficiency syndrome. *Semin Oncol* 1990; **17:** 335-49.

Burke, B., and Desselberger, U. Rotavirus pathogenicity. *Virology* 1996; **218:** 299-305.

Cao, S., Cox, K., Esquivel, C. O., Berquist, W., Concepcion, W., Ojogho, O., Monge, H., Krams, S., Martinez, O., and So, S. Posttransplant lymphoproliferative disorders and gastrointestinal manifestations of Epstein-Barr virus infection in children following liver transplantation. *Transplantation* 1998; **66:** 851-6.

Cesarman, E., and Knowles, D. M. Kaposi's sarcoma-associated herpes virus: a lymphotropic human herpes virus associated with Kaposi's sarcoma, primary effusion lymphoma, and multicentric Castleman's disease. *Semin Diagn Pathol* 1997; **14:** 54-66.

Chan, J. Tumors of the lymphoreticular system, including spleen and thymus. In: *Diagnostic histopathology of tumors* (CDM Fletcher, Ed.), pp. 1099-1315. Churchill Livingstone, London, 2000.

Cronstedt, J. L., Bouchama, A., Hainau, B., Halim, M., Khouqeer, F., and al Darsouny, T. Spontaneous esophageal perforation in herpes simplex esophagitis. *Am J Gastroenterol* 1992; **87:** 124-7.

Cunningham, A. L., Grohman, G. S., Harkness, J., Law, C., Marriott, D., Tindall, B., and Cooper, D. A. Gastrointestinal viral infections in homosexual men who were symptomatic and seropositive for human immunodeficiency virus. *J Infect Dis* 1988; **158:** 386-91.

Daniels, T. E., Greenspan, D., Greenspan, J. S., Lennette, E., Schiodt, M., Petersen, V., and de Souza, Y. Absence of Langerhans cells in oral hairy leukoplakia, an AIDS-associated lesion. *J Invest Dermatol* 1987; **89:** 178-82.

Delabesse, E., Oksenhendler, E., Lebbe, C., Verola, O., Varet, B., and Turhan, A. G. Molecular analysis of clonality in Kaposi's sarcoma. *J Clin Pathol* 1997; **50:** 664-8.

Deshmukh, M., Shah, R., and McCallum, R. W. (1984). Experience with herpes esophagitis in otherwise healthy patients. *Am J Gastroenterol* 1984; **79:** 173-6.

Dieterich, D. T., and Rahmin, M. Cytomegalovirus colitis in AIDS: presentation in 44 patients and a review of the literature. *J AIDS* 1991; **4 Suppl 1:** S29-S35.

Ehrenpreis, E. D., Patterson, B. K., Brainer, J. A., Yokoo, H., Rademaker, A. W., Glogowski, W., Noskin, G. A., and Craig, R. M. Histopathologic findings of duodenal biopsy specimens in HIV-infected patients with and without diarrhea and malabsorption. *Am J Clin Pathol* 1992; **97:** 21-8.

Euvrard, S., Kanitakis, J., Chardonnet, Y., Noble, C. P., Touraine, J. L., Faure, M., Thivolet, J., and Claudy, A. External anogenital lesions in organ transplant recipients. A clinicopathologic and virologic assessment. *Arch Dermatol* 1997; **133:** 175-8.

Fenoglio-Preiser, C., Lantz, P., Listrom, M., Davis, M., and Rilke, F. (Eds) "Gastrointestinal Pathology." 2nd edn. Lippincott-Raven, Philadelphia, 1999.

Fernandes, B., Brunton, J., and Koven, I. Ileal perforation due to cytomegaloviral enteritis. *Can J Surg* 1986; **29:** 453-6.

Foucar, E., Mukai, K., Foucar, K., Sutherland, D. E., and Van Buren, C. T. (1981). Colon ulceration in lethal cytomegalovirus infection. *Am J Clin Pathol* 1981; **76:** 788-801.

Fox, C. H., Kotler, D., Tierney, A., Wilson, C. S., and Fauci, A. S. Detection of HIV-1 RNA in the lamina propria of patients with AIDS and gastrointestinal disease. *J Infect Dis* 1989; **159:** 467-71.

Framm, S. R., and Soave, R. Agents of diarrhea. *Med Clin North Am* 1997; **81:** 427-47.

Francis, N. *Biopsy Pathology of the Oesophagus, Stomach and Duodenum.* 2nd edn (Day, D. W., and Dixon, M. F., Eds.). Chapman and Hall, London, 1995.

Francis, N. D., Boylston, A. W., Roberts, A. H., Parkin, J. M., and Pinching, A. J. Cytomegalovirus infection in gastrointestinal tracts of patients infected with HIV-1 or AIDS. *J Clin Pathol* 1989; **42:** 1055-64.

Frank, D., and Raicht, R. F. Intestinal perforation associated with cytomegalovirus infection in patients with acquired immune deficiency syndrome. *Am J Gastroenterol* 1984; **79:** 201-5.

Garcia, F., Garau, J., Sierra, M., and Marco, V. Cytomegalovirus mononucleosis-associated antral gastritis simulating malignancy. *Arch Intern Med* 1987; **147:** 787-8.

Gazzard, B., and Moyle, G. 1998 revision to the British HIV Association guidelines for antiretroviral treatment of HIV seropositive individuals. BHIVA Guidelines Writing Committee. *Lancet* 1998; **352:** 314-6.

Gill, P. S., Tsai, Y. C., Rao, A. P., Spruck, C. H., 3rd, Zheng, T., Harrington, W. A., Jr., Cheung, T., Nathwani, B., and Jones, P. A. Evidence for multiclonality in multicentric Kaposi's sarcoma. *Proc Natl Acad Sci USA* 1998; **95:** 8257-61.

Goodman, M. D., and Porter, D. D. Cytomegalovirus vasculitis with fatal colonic hemorrhage. *Arch Pathol* 1973 **96:** 281-4.

Greenspan, D., Greenspan, J. S., Conant, M., Petersen, V., Silverman, S., Jr., and de Souza, Y. Oral "hairy" leucoplakia in male homosexuals: evidence of association with both papillomavirus and a herpes-group virus. *Lancet* 1984; **2:** 831-4.

Greenspan, J. S., Greenspan, D., Lennette, E. T., Abrams, D. I., Conant, M. A., Petersen, V., and Freese, U. K. Replication of Epstein-Barr virus within the epithelial cells of oral "hairy" leukoplakia, an AIDS-associated lesion. *N Engl J Med* 1985; **313:** 1564-71.

Griffin, G. E., Miller, A., Batman, P., Forster, S. M., Pinching, A. J., Harris, J. R., and Mathan, M. M. Damage to jejunal intrinsic autonomic nerves in HIV infection. *AIDS* 1988; **2:** 379-82.

Heise, C., Dankekar, S., Kumar, P., Duplantier, R., Donovan, R.M., Halsted, C.H. Human immunodeficiency virus infection of enterocytes and mononuclear cells in human jejunal mucosa. *Gastroenterology* 1991; **100:** 1521-7

Hierholzer, J. C. Adenoviruses in the immunocompromised host. *Clin Microbiol Rev* 1992; **5:** 262-74.

Hille, J. J., Margolius, K. A., Markowitz, S., and Isaacson, C. Human papillomavirus infection related to oesophageal carcinoma in black South Africans. A preliminary study. *S Afr Med J* 1986; **69:** 417-20.

Janoff, E. N., Orenstein, J. M., Manischewitz, J. F., and Smith, P. D. Adenovirus colitis in the acquired immunodeficiency syndrome. *Gastroenterology* 1991; **100:** 976-9.

Jarry, A., Cortez, A., Rene, E., Muzeau, F., and Brousse, N. Infected cells and immune cells in the gastrointestinal tract of AIDS patients. An immunohistochemical study of 127 cases. *Histopathology* 1990; **16:** 133-40.

Kaljot, K. T., Ling, J. P., Gold, J. W., Laughon, B. E., Bartlett, J. G., Kotler, D. P., Oshiro, L. S., and Greenberg, H. B. Prevalence of acute enteric viral pathogens in acquired immunodeficiency syndrome patients with diarrhea. *Gastroenterology* 1989; **97:** 1031-2.

Kennedy, M. M., Lucas, S. B., Jones, R. R., Howells, D. D., Picton, S. J., Hanks, E. E., McGee, J. O., and O'Leary, J. J. HHV8 and Kaposi's sarcoma: a time cohort study. *Mol Pathol* 1997; **50:** 96-100.

Kinney, J. S., Onorato, I. M., Stewart, J. A., Pass, R. F., Stagno, S., Cheeseman, S. H., Chin, J., Kumar, M. L., Yaeger, A. S., Herrmann, K. L., *et al.* Cytomegaloviral infection and disease. *J Infect Dis* 1985; **151:** 772-4.

Kitchen, V. S., Helbert, M., Francis, N. D., Logan, R. P., Lewis, F. A., Boylston, A. W., Pinching, A. J., and Harris, J. R. Epstein-Barr virus associated oesophageal ulcers in AIDS. *Gut* 1990; **31:** 1223-5.

Knapp, A. B., Horst, D. A., Eliopoulos, G., Gramm, H. F., Gaber, L. W., Falchuk, K. R., Falchuk, Z. M., and Trey, C. Widespread cytomegalovirus gastroenterocolitis in a patient with acquired immunodeficiency syndrome. *Gastroenterology* 1983; **85:** 1399-402.

Kotler, D. P., Reka, S., Borcich, A., and Cronin, W. J. Detection, localization, and quantitation of HIV-associated antigens in intestinal biopsies from patients with HIV. *Am J Pathol* 1991; **139:** 823-30.

Mathan, M. M., Griffin, G. E., Miller, A., Batman, P., Forster, S., Pinching, A., and Harris, W. Ultrastructure of the jejunal mucosa in human immunodeficiency virus infection. *J Pathol* 1990; **161:** 119-27.

McBane, R. D., and Gross, J. B., Jr. Herpes esophagitis: clinical syndrome, endoscopic appearance, and diagnosis in 23 patients. *Gastrointest Endosc* 1991; **37:** 600-3.

Meiselman, M. S., Cello, J. P., and Margaretten, W. Cytomegalovirus colitis. Report of the clinical, endoscopic, and pathologic findings in two patients with the acquired immune deficiency syndrome. *Gastroenterology* 1985; **88:** 171-5.

Mentec, H., Leport, C., Leport, J., Marche, C., Harzic, M., and Vilde, J. L. Cytomegalovirus colitis in HIV-1-infected patients: a prospective research in 55 patients. *AIDS* 1994; **8:** 461-7.

Miller, S. E., and Howell, D. N. Viral infections in the acquired immunodeficiency syndrome. *J Electron Microsc Tech* 1988; **8:** 41-78.

Minami, H., Matsushita, T., Sugihara, T., Kodera, Y., Sakai, S., and Shimokata, K. Cytomegalovirus-induced gastritis in a bone marrow transplant patient. *Jpn J Med* 1990; **29:** 433-5.

Mintz, L., Drew, W. L., Miner, R. C., and Braff, E. H. Cytomegalovirus infections in homosexual men. An epidemiological study. *Ann Intern Med* 1983; **99:** 326-9.

Monkemuller, K. E., Call, S. A., Lazenby, A. J., and Wilcox, C. M. Declining prevalence of opportunistic gastrointestinal disease in the era of combination antiretroviral therapy. *Am J Gastroenterol* 2000; **95:** 457-62.

Morris, H., and Price, S. Langerhans' cells, papillomaviruses and oesophageal carcinoma. A hypothesis. *S Afr Med J* 1986; **69:** 413-7.

Nash, G., and Ross, J. S. Herpetic esophagitis. A common cause of esophageal ulceration. *Hum Pathol* 1974; **5:** 339-45.

Nelson, B. P., Nalesnik, M. A., Bahler, D. W., Locker, J., Fung, J. J., and Swerdlow, S. H. Epstein-Barr virus-negative post-transplant lymphoproliferative disorders: a distinct entity? *Am J Surg Pathol* 2000; **24:** 375-85.

Nelson, J. A., Wiley, C. A., Reynolds-Kohler, C., Reese, C. E., Margaretten, W., and Levy, J. A. Human immunodeficiency virus detected in bowel epithelium from patients with gastrointestinal symptoms. *Lancet* 1988; **1:** 259-62.

Neville, B. W., Waldron, C. A., and Herschraft, E. A. (Eds) *Oral and maxillofacial pathology.* Saunders, Philadelphia, 1995.

Pavli, P., and Doe, W. F. Gastrointestinal and oesophageal pathology (R, Whitehead, Ed.). Churchill Livingstone, Edinburgh, 1990.

Penn, I. The changing pattern of posttransplant malignancies. *Transplant Proc* 1991; **23:** 1101-3.

Pollok, R. Viruses causing diarrhoea in AIDS. *Novartis Found Symp* 2001; **238:** 276-88.

Pui, J. C., Furth, E. E., Minda, J., and Montone, K. T. Demonstration of varicella-zoster virus infection in the muscularis propria and myenteric plexi of the colon in an HIV-positive patient with herpes zoster and small bowel pseudo-obstruction (Ogilvie's syndrome). *Am J Gastroenterol* 2001; **96:** 1627-30.

Rabeneck, L., Popovic, M., Gartner, S., McLean, D. M., McLeod, W. A., Read, E., Wong, K. K., and Boyko, W. J. Acute HIV infection presenting with painful swallowing and esophageal ulcers. *J Am Med Ass* 1990; **263:** 2318-22.

Rabkin, C. S., Janz, S., Lash, A., Coleman, A. E., Musaba, E., Liotta, L., Biggar, R. J., and Zhuang, Z. Monoclonal origin of multicentric Kaposi's sarcoma lesions. *N Engl J Med* 1997; **336:** 988-93.

Rasing, L. A., De Weger, R. A., Verdonck, L. F., van der Bij, W., Compier-Spies, P. I., De Gast, G. C., Van Basten, C. D., and Schuurman, H. J. The value of immunohistochemistry and in situ hybridization in detecting cytomegalovirus in bone marrow transplant recipients. *APMIS* 1990; **98:** 479-88.

Rattner, H. M., Cooper, D. J., and Zaman, M. B. Severe bleeding from herpes esophagitis. *Am J Gastroenterol* 1985; **80:** 523-5.

Rich, J. D., Crawford, J. M., Kazanjian, S. N., and Kazanjian, P. H. Discrete gastrointestinal mass lesions caused by cytomegalovirus in patients with AIDS: report of three cases and review. *Clin Infect Dis* 1992; **15:** 609-14.

Rotterdam, H. Tissue diagnosis of selected AIDS-related opportunistic infections. *Am J Surg Pathol* 1987; **11 Suppl 1:** 3-15.

Sabin, C. A., Clewley, G. S., Deayton, J. R., Mocroft, A., Johnson, M. A., Lee, C. A., McLaughlin, J. E., and Griffiths, P. D. Shorter survival in HIV-positive patients with diarrhoea who excrete adenovirus from the GI tract. *J Med Virol* 1999; **58:**

280-5.

Schiodt, M., Greenspan, D., Daniels, T. E., and Greenspan, J. S. Clinical and histologic spectrum of oral hairy leukoplakia. *Oral Surg Oral Med Oral Pathol* 1987; **64:** 716-20.

Schwartz, D. A., and Wilcox, C. M. Atypical cytomegalovirus inclusions in gastrointestinal biopsy specimens from patients with the acquired immunodeficiency syndrome: diagnostic role of in situ nucleic acid hybridization. *Hum Pathol* 1992; **23:** 1019-26.

Sciubba, J. J. Opportunistic oral infections in the immunosuppressed patient: oral hairy leukoplakia and oral candidiasis. *Adv Dent Res* 1996; **10:** 69-72.

Shintaku, M., Hirai, T., and Kohno, K. Esophageal perforation in association with Herpesvirus infection: report of an autopsy case. *Am J Gastroenterol* 1992; **87:** 1524-5.

Siegal, F. P., Lopez, C., Hammer, G. S., Brown, A. E., Kornfeld, S. J., Gold, J., Hassett, J., Hirschman, S. Z., Cunningham-Rundles, C., Adelsberg, B. R., and *et al*. Severe acquired immunodeficiency in male homosexuals, manifested by chronic perianal ulcerative herpes simplex lesions. *N Engl J Med* 1981; **305:** 1439-44.

Simon, D., and Brandt, L. J. Diarrhea in patients with the acquired immunodeficiency syndrome. *Gastroenterology* 1993; **105:** 1238-42.

Springer, D. J., DaCosta, L. R., and Beck, I. T. A syndrome of acute self-limiting ulcerative esophagitis in young adults probably due to herpes simplex virus. *Dig Dis Sci* 1979; **24:** 535-9.

Starr, S. E. Cytomegalovirus. *Pediatr Clin North Am* 1979; **26:** 283-93.

Stenglein, S., Werner, M., Sinzger, C., Korn, K., Plachter, B., Beck, R., Neumayer, H. H., and Jahn, G. [Acute cytomegalovirus gastritis after kidney transplantation. Its diagnosis by the polymerase chain reaction, antigenemia assay and immunohistochemistry]. *Dtsch Med Wochenschr* 1993; **118:** 1597-602.

Stribling, J., Weitzner, S., and Smith, G. V. Kaposi's sarcoma in renal allograft recipients. *Cancer* 1978; **42:** 442-6.

Ullrich, R., Zeitz, M., Heise, W., L'Age, M., Hoffken, G., and Riecken, E. O. (1989). Small intestinal structure and function in patients infected with human immunodeficiency virus (HIV): evidence for HIV-induced enteropathy. *Ann Intern Med* 1989; **111:** 15-21.

Underwood, J. C., and Corbett, C. L. Persistent diarrhoea and hypoalbuminaemia associated with cytomegalovirus enteritis. *Br Med J* 1978; **1:** 1029-30.

Weber, J. N., Thom, S., Barrison, I., Unwin, R., Forster, S., Jeffries, D. J., Boylston, A., and Pinching, A. J. Cytomegalovirus colitis and oesophageal ulceration in the context of AIDS: clinical manifestations and preliminary report of treatment with Foscarnet (phosphonoformate). *Gut* 1987; **28:** 482-7.

Wilcox, C. M., Chalasani, N., Lazenby, A., and Schwartz, D. A. Cytomegalovirus colitis in acquired immunodeficiency syndrome: a clinical and endoscopic study. *Gastrointest Endosc* 1998; **48:** 39-43.

Wilcox, C. M., and Schwartz, D. A. Symptomatic CMV duodenitis. An important clinical problem in AIDS. *J Clin Gastroenterol* 1992; **14:** 293-7.

Williamson, A. L., Jaskiesicz, K., and Gunning, A. The detection of human papillomavirus in oesophageal lesions. *Anticancer Res* 1991; **11:** 263-5.

Winkler, B., Capo, V., Reumann, W., Ma, A., La Porta, R., Reilly, S., Green, P. M., Richart, R. M., and Crum, C. P. Human papillomavirus infection of the esophagus. A clinicopathologic study with demonstration of papillomavirus antigen by the immunoperoxidase technique. *Cancer* 1985; **55:** 149-55.

Zeitz, M., Ullrich, R., Schneider, T., Kewenig, S., Hohloch, K., and Riecken, E. O. HIV/SIV enteropathy. *Ann N Y Acad Sci* 1998; **859:** 139-48.

Zutter, M. M., Martin, P. J., Sale, G. E., Shulman, H. M., Fisher, L., Thomas, E. D., and Durnam, D. M. Epstein-Barr virus lymphoproliferation after bone marrow transplantation. *Blood* 1988; **72:** 520-9.

List of contributors

Dr Juana Angel
Instituto de Genética Humana
Pontificia Universidad Javeriana
Carrera 7 40-62
Bogotà COLOMBIA
Tel: 00 57 1 320 8320 ext 2790
Fax: 00 57 1 320 8320 ext 2793
Email: jangel@javeriana.edu.co

Dr Carlos F Arias
Departamento de Genética y Fisiologia Molecular
Instituto de Biotecnología
Universidad Nacional Autónoma de México
Av. Universidad 2001 Apartado Postal 510-3
Cuernavaca, Morelos, 62250 MEXICO
Tel: 0052 73 291615
Fax: 0052 73 172388
Email: arias@ibt.unam.mx

Dr Dorsey Bass
Department of Pediatrics Room G307
Stanford University Hospital
Stanford CA 94305-5208 USA
Tel: 001 650 725 1819
Fax: 001 650 725 3106
Email: dorseybass@hotmail.com

Dr A Richard Bellamy
School of Biological Sciences
University of Auckland
Private Bag 92019
Auckland NEW ZEALAND
Tel: 0064 9 373 7599 ext 7949
Fax: 0064 9 373 7414
Email: d.bellamy@auckland.ac.nz

Dr Andrea Bertolotti-Ciarlet
Department of Microbiology
302B Johnson Pavilion, 3610 Hamilton Walk,
Mailstop 6076, University of Pennsylvania,
Philadelphia, PA 19104 USA
Tel: 001 215 746 6757
Fax: 001 215 573 2883
Email: aciarlet@mail.med.upenn.edu

Dr Per Brandtzaeg
Laboratory of Immunohistochemistry
and Immunopathology (LIIPAT)
Institute of Pathology
University of Oslo
Rikshospitalet N-0027
Oslo NORWAY
Tel: 0047 2307 2743/1509/1507
Fax: 0047 2307 1511
Email: per.brandtzaeg@labmed.uio.no

Dr Ian Brierley
Division of Virology
Department of Pathology
University of Cambridge Level 5, Addenbrooke's Hospital
Cambridge CB2 2QQ UNITED KINGDOM
Tel: 0044 1223 336914
Fax: 0044 1223 336926
Email: ib103@mole.bio.cam.ac.uk

Dr Rong Chen
Verna and Marrs McLean Department of Biochemistry & Molecular Biology
Keir Center for Comparative Biology
Baylor College of Medicine
One Baylor Plaza
Houston TX 77030 USA
Tel: 001 713 798 3585
Fax: 001 713 798 3586

Dr Max Ciarlet
Merck & Co Inc Bioprocess and Bioanalytical Research
770 Sumneytown Pike
PO Box 4 - WP75-300
West Point PA 19486 USA
Tel: 001 215 652 1306
Fax: 001 215 652 9069
Email: max_ciarlet@merck.com

Dr Fred Clark
Section of Infectious Diseases
The Children's Hospital of Philadelphia
34th and Civic Center Blvd
Philadelphia PA 19104 USA
Tel: 001 215 590 2044
Fax: 001 215 590 2025
Email: Clarkf@email.chop.edu

Dr Jean Cohen
Virologie Molécularie et Structurale
UMR 2472 CNRS-INRA
1 avenue de la Terrasse, Bât. 14C
91198 Gif-sur-Yvette CEDEX / FRANCE
Tel: 0033 1 6982 3854
Fax: 0033 1 6982 4308
and
Virologie et Immunologie Moléculaires
INRA, Domaine de Vilvert
Jouy-en-Josas, Yvelines 78350 FRANCE
Tel: 0033 1 3465 2604
Fax: 0033 1 3465 2621
Email: cohen@jouy.inra.fr

Dr Margaret Conner
Baylor College of Medicine
Department of Molecular Virology
Mailstop BCM-385
One Baylor Plaza
Houston TX 77030 USA
Tel: 001 713 798 3590
Fax: 001 713 798 3586
Email: mconner@bcm.tmc.edu

Dr Mariela A Cuadras
Departments of Microbiology and
Immunology and of Medicine
Division of Gastroenterology
Stanford University School of Medicine
Stanford CA 94304 USA
Tel: 001 650 493 5000
Fax: 001 650 852 3259
Email: mariela@stanford.edu

Dr JC de Jong
Department of Virology
Erasmus University Rotterdam
Dr Molewaterplein 50
3015 GE Rotterdam THE NETHERLANDS
Tel: 0031 10 408 8067
Fax: 0031 10 408 9485
Email: jc.de.jong@wxs.nl

Dr Ulrich Desselberger
Clinical Microbiology and Public Health Laboratory
Level 6, Box 236
Addenbrooke's Hospital
Hills Road
Cambridge CB2 2QW UNITED KINGDOM
Present address:
Virologie Moléculaire et Structurale
UMR 2472, CNRS,
1 avenue de la Terrasse, Bât. 14B
91198 Gif-sur-Yvette Cedex, FRANCE
Tel: 0033 1 6982 3854
Fax: 0033 1 6982 4308
Email: ulrich.desselberger@gv.cnrs-gif.fr

Dr Mary Estes
Division of Molecular Virology
Baylor College of Medicine
One Baylor Plaza Room 923D
Houston TX 77030-3498 USA
Tel: 001 713 798 3585
Fax: 001 713 798 3586
Email: mestes@bcm.tmc.edu

Dr Dino A Feigelstock
Laboratory of Hepatitis and Related Emerging Agents
DETTD, OBRR, CBER, FDA
8800 Rockville Pike, Bldg 29, Room 231
National Institutes of Health
Bethesda MD 30892 USA
Tel: 001 301 480 4188
Fax: 001 301 480 7928
Email: dinoaf@hotmail.com

Dr Manuel A Franco
Instituto de Genética Humana
Pontificia Universidad Javeriana
Carrera 7 40-62
Bogotà COLOMBIA
Tel: 00 57 1 320 8320 ext 2790
Fax: 0057 1 320 8320 ext 2793
Email: mafranco@javeriana.edu.co

Dr Ute Geigenmüller
Department of Medicine
Division of Gastroenterology & Hepatology
Stanford University School of Medicine
Stanford CA 94305 USA
and
Department of Medicine
Gastroenterology Section (111-GI)
VA Palo Alto Healthcare System
3801 Miranda Avenue
Palo Alto CA 94304 USA
Tel: 001 650 493 5000 ext 63178
Fax: 001 650 852 3259
Email: ute@stanford.edu

Dr Roger Glass
Viral Gastroenteritis Section
Respiratory and Enteric Viruses Branch
Division of Viral and Rickettsial Diseases, CDC
1600 Clifton Road NE
Atlanta GA 30333 USA
Tel: 001 404 639 3577
Fax: 001 404 639 3645
Email: rglass@cdc.gov

Dr Ana Maria Gonzalez
Instituto de Genética Humana
Pontificia Universidad Javeriana
Carrera 7 40-62
Bogotà COLOMBIA
Present address:
FAHRP, Department of Veterinary Preventative Medicine
Ohio Agricultural Research and Development Center (OARDC)
Ohio State University
Wooster OH 44691-4096 USA
Tel: 00 57 1 320 8320 ext 2787
Tel: 001 330 263 3616 8041
Fax: 00 57 1 320 8320 ext 2793
Fax: 001 330 263 3677
Email: Amg21@yahoo.com
Email: Gonzalez.157@osu.edu

Dr John Grant
Department of Histopathology
Addenbrooke's Hospital
Cambridge CB2 2QQ UNITED KINGDOM
Tel: 0044 1223 216744
Fax: 0044 1223 216980
Email: jwg21@cam.ac.uk

Dr Jim Gray
Clinical Microbiology & Public Health Laboratory
Level 6, Box 236
Addenbrooke's Hospital
Cambridge CB2 2QW UNITED KINGDOM
Present Address:
Enteric Virus Unit
Enteric, Respiratory and Neurological Virus Laboratory
Central Public Health Laboratory
61 Colindale Avenue
London NW9 5DF UNITED KINGDOM
Tel: 0044 208 200 4400 ext 3211
Fax: 0044 208 205 8195
Email: Jgray1@phls.org.uk

Dr Kim Y Green
Laboratory of Infectious Diseases
National Institute of Allergy and Infectious Diseases
National Institutes of Health
50 South Drive MSC 8026
Bethesda MD 20892-8026 USA
Tel: 001 301 594 1666
Fax: 001 301 496 8312
Email: kgreen@niaid.nih.gov

Dr Harry B Greenberg
Medimmune Vaccines, Inc.
297 N. Bernardo Avenue
Mount View CA 94043 USA
and
VA Palo Alto Healthcare System
3801 Miranda Avenue, MC 154C
Palo Alto CA 94304 USA
Tel: 001 650 919 3711
Tel: 001 650 493 5000 ext 63121
Fax: 001 650 919 6686
Fax: 001 650 852 3259
Email: greenbergh@medimmune.com
Email: hbgreen@stanford.edu

Dr Ming Zhang Guo
Departments of Surgery and Medical Physiology
Cardiovascular Research Institute
Texas A&M University
System Health Science Center
702 Southwest HK Dodgen Loop
Temple TX 76504 USA
Tel: 001 254 742 7144
Fax: 001 254 742 7145
Email: Mzguo@neo.tamu.edu

Dr John E Herrmann
Division of Infectious Diseases
University of Massachusetts Medical School
364 Plantation Street
Worcester MA 01655-2324 USA
Tel: 001 508 856 2155
Fax: 001 508 856 8030
Email: john.herrmann@umassmed.edu

Dr Miren Iturriza-Gómara
Clinical Microbiology & Public Health Laboratory
Level 6 Addenbrooke's Hospital
Cambridge CB2 2QW UNITED KINGDOM
Present Address:
Enteric Virus Unit
Enteric, Respiratory and Neurological Virus Laboratory
Central Public Health Laboratory
61 Colindale Avenue
London NW9 5DF UNITED KINGDOM
Tel: 0044 208 200 4400
Fax: 0044 208 205 8195
Email: MIturriza@phls.org.uk

Dr Maria C Jaimes
Departments of Medicine, Microbiology & Immunology
Stanford University School of Medicine
300 Pasteur Drive - CCSR Bldg. Room 3115
Stanford CA 94305-5187 - USA
Tel: 001 650 493 5000 ext 63124
Fax: 001 650 852 3259
Email: mcjaimes@stanford.edu

Dr Xi Jiang
Cincinnati Children's Hospital Medical Center
Cincinnati - OH 45229-3039 USA
Tel: 001 513 636 0119
Fax: 001 513 636 6936
Email: jason.jiang@chmcc.org

Dr Finn-Erik Johansen
Laboratory of Immunohistochemistry and Immunopathology (LIIPAT)
Institute of Pathology
University of Oslo
Rikshospitalet N-0027 Oslo NORWAY
Tel: 0047 2307 2743
Fax: 0047 2307 1511
Email: Johansen@labmed.uio.no

Dr Karen Kearney
Laboratory of Infectious Diseases
National Institute of Allergy and Infectious Diseases
National Institutes of Health
50 South Drive MSC 8026, Room 6314
Bethesda MD 20892-8026 USA
Tel: 001 301 594 1615
Fax: 001 301 496 8312
Email: Kkearney@niaid.nih.gov

Dr Marion Koopmans
Diagnostic Laboratory for Infectious Diseases
Virology Division
National Institute for Public Health and the Environment
Antonie van Leeuwenhoeklaan 9
3720 BA Bilthoven THE NETHERLANDS
Tel: 0031 30 274 3945
Fax: 0031 30 274 4418
Email: marion.koopmans@rivm.nl

Dr Jean Lepault
Virologie Moléculaire et Structurale
UMR 2472 CNRS -INRA
1 Avenue de la Terrasse, Bât. 14C
91198 Gif-sur-Yvette CEDEX - FRANCE
Tel: 0033 1 6982 3163
Fax: 0033 1 6982 3150
Email: Jean.Lepault@cgm.cnrs-gif.fr

Dr Susana López
Departamento de Genética y Fisiología Molecular
Instituto de Biotecnología
Universidad Nacional Autónoma de México
Av. Universidad 2001
Apartado Postal 510-3
Cuernavaca, Morelos, 62250 MEXICO
Tel: 0052 73 291615
Fax: 0052 73 172388
Email: susana@ibt.unam.mx

Dr Ove Lundgren
Department of Physiology
Sahlgrenska Academy
Göteborg University, Box 432
S-40530 Göteborg SWEDEN
Tel: 0046 31 773 3522
Fax: 0046 31 773 3512
Email: Ove.lundgren@fysiologi.gu.se

Dr David O. Matson
Center for Pediatric Research
Eastern Virginia Medical School
855 W. Brambleton Ave.
Norfolk VA 23510-1001 USA
Tel: 001 757 668 6433
Fax: 001 757 668 6476
Email: dmatson@chkd.com

Dr Suzanne M Matsui
Division of Gastroenterology and Hepatology
Stanford University School of Medicine
Stanford CA 94305 USA
and
Department of Medicine
Gastroenterology Section (111-GI)
VA Palo Alto Healthcare System
3801 Miranda Avenue
Palo Alto CA 94304 USA
Tel: 001 650 493 5000 ext 63179
Fax: 001 650 852 3259
Email: sumatsui@stanford.edu

691

Dr Vivien Mautner
CRC Institute for Cancer Studies
University of Birmingham
Edgbaston
Birmingham B15 2TA UNITED KINGDOM
Tel: 0044 121 414 4484
Fax: 0044 121 414 3263
Email: v.mautner@bham.ac.uk

Dr Ernesto Méndez
Departamento de Genética y Fisiologia Molecular
Instituto de Biotecnología
Universidad Nacional Autónoma de México
Apartado Postal 510-3, Colonia Miraval
Cuernavaca, Morelos 62250 MEXICO
Tel: 0052 777 329 1612
Fax: 0052 777 317 2388
Email: ernesto@ibt.unam.mx

Dr Fabian Michelangeli
Laboratorio de Fisiología Gastrointestinal
Centro de Biofísica y Bioquímica
Instituto Venezolano de Investigacíones Científicas (IVIC)
PO Box 21827
Carácas 1020A VENEZUELA
Tel: 0058 212 504 1396
Fax: 0058 212 504 1093
Email: fabian@ivic.ve

Dr Stephan S Monroe
Division of Viral & Rickettsial Diseases
Centers for Disease Control & Prevention
Mailstop G-04
1600 Clifton Road N.E.
Atlanta GA 30333 USA
Tel: 001 404 639 2391
Fax: 001 404 639 3645
Email: stm2@cdc.gov

692

Dr Paul A Offit
Section of Infectious Diseases
The Children's Hospital of Philadelphia
34th and Civic Center Blvd
Philadelphia PA 19104 USA
Tel: 001 215 590 2020
Fax: 001 215 590 2025
Email: offit@email.chop.edu

Dr U Parashar
Viral Gastroenteritis Section
Division of Viral and Rickettsial Diseases
Centers for Disease Control & Prevention
Mailstop G-04
1600 Clifton Road NE
Atlanta GA 30333 USA
Tel: 001 404 639 3577
Fax: 001 404 639 3645
Email: uap2@cdc.gov

Dr John T Patton
Laboratory of Infectious Diseases
National Institute of Allergy and Infectious Diseases
National Institutes of Health
50 South Drive, MSC 8026, Room 6314
Bethesda MD 20892-0720 USA
Tel: 001 301 594 1615
Fax: 001 301 496 8312
Email: jpatton@niaid.nih.gov

Dr Joseph B Pesavento
Verna and Marrs McLean Department of Biochemistry & Molecular Biology
Baylor College of Medicine
One Baylor Plaza
Houston TX 77030 USA
Tel: 001 713 798 5686
Fax: 001 713 798 1625

Dr Martin Petric
Virology Laboratory
The Hospital for Sick Children
555 University Avenue
Toronto ON, M1S 1C7 CANADA
Tel: 001 416 813 6111
Fax: 001 416 813 6257
Email: martin.petric@sickkids.ca

Dr Didier Poncet
Virologie Moléculaire et Structurale
UMR 2472, CNRS-INRA
1 avenue de la Terrasse, Bât. 14B
91198 Gif-sur-Yvette Cedex, FRANCE
Tel: 0033 1 6982 3835
Fax: 0033 1 6982 4308
Email: poncet@gv.cnrs-gif.fr

Dr BV Venkataram Prasad
Verna and Marrs McLean Department of Biochemistry & Molecular Biology
Keir Center for Comparative Biology
Baylor College of Medicine
One Baylor Plaza
Houston TX 77030 USA
Tel: 001 713 798 5686
Fax: 001 713 798 1625
Email: vprasad@bcm.tmc.edu

Dr Félix A Rey
Virologie Moléculaire et Structurale
UMR 2472 CNRS
1 Avenue de la Terrasse, Bât. 14 C
91198 Gif-sur-Yvette CEDEX FRANCE
Tel: 0033 1 6982 3844
Fax: 0033 1 6982 4308
Email: rey@gv.cnrs-gif.fr

694

Dr Olga L Rojas
Instituto de Genética Humana
Pontificia Universidad Javeriana
Carrera 7 40-62
Bogotà COLOMBIA
Tel: 00 57 1 320 8320 ext 2790
Fax: 00 57 1 320 8320 ext 2793
Email: olga.rojas@javeriana.edu.co

Dr Blair I Rosen
Wadsworth Center
New York State Department of Health
Box 509
Albany, NY 12201-0509 USA
Tel: 001 518 869 4558
Fax: 001 518 869 6487
Email: bir01@health.state.ny.us

Dr Marie C Ruiz
Laboratorio de Fisiología Gastrointestinal
Centro de Biofísica y Bioquímica
Instituto Venezolano de Investigacíones Científicas (IVIC)
PO Box 21827
Carácas 1020A VENEZUELA
Tel: 0058 212 504 1164
Fax: 0058 212 504 1093
Email: mcir@ivic.ve

Dr Linda J Saif
Food Animal Health Research Program
Ohio Agricultural Research and Development Center
Veterinary Preventative Medicine Department
The Ohio State University
1680 Madison Avenue
Wooster OH 44691 USA
Tel: 001 330 263 3742 / 3744
Fax: 001 330 263 3677
Email: saif.2@osu.edu

Dr Kenji Sakae
Department of Microbiology
Aichi Prefectural Institute of Public Health
7-6 Nagare, Tsujimachi - Kita-ku
Nagoya Aichi 462-8576 JAPAN
Tel: 0081 52 910 5674
Fax: 0081 52 913 3641
Email: Kenji.sakae@nifty.com

Dr Alain L Servin
Unité 510 INSERM
Faculté de Pharmacie Paris XI
5, rue J B Clément
92296 Châtenay-Malabry CEDEX FRANCE
Tel: 0033 1 4683 5661
Fax: 0033 1 4683 5661
Email: alain.servin@cep.u-psud.fr

Dr Stanislav V Sosnovtsev
Laboratory of Infectious Diseases
National Institute of Allergy and Infectious Diseases
National Institutes of Health
50 South Drive MSC 8026
Bethesda MD 20892-8026 USA
Tel: 001 301 594 1666
Fax: 001 301 496 8312
Email: ss216m@nih.gov

Dr Fiona Stevenson
CRC Institute for Cancer Studies
University of Birmingham
Edgbaston
Birmingham B15 2TA UNITED KINGDOM
Tel: 0044 121 414 4484
Fax: 0044 121 414 3263
Email: FionaS@cancer.bham.ac.uk

Dr Lennart Svensson
Swedish Institute for Infectious Disease Control
Department of Virology
Nobelsväg 18 171 82
Solna SWEDEN
Tel: 0046 8 457 2696
Fax: 0046 8 301 635
Email: lensve@mbox.ki.se

Dr Zenobia Taraporewala
Laboratory of Infectious Diseases
National Institute of Allergy and Infectious Diseases
National Institutes of Health
50 South Drive Room 6314
Bethesda MD 20892-8026 USA
Tel: 001 301 594 1615
Fax: 001 301 496 8312
Email: ztarapore@niaid.nih.gov

Dr John A Taylor
School of Biological Sciences
University of Auckland
Private Bag 92019
Auckland NEW ZEALAND
Tel: 00649 373 7599 ext 2854
Fax: 00649 373 7414
Email: ja.taylor@auckland.ac.nz

Dr Elizabeth van Strien
Diagnostic Laboratory for Infectious Diseases
Virology Division
National Institute for Public Health and the Environment
Antonie van Leeuwenhoeklaan 9
3730 BA Bilthoven THE NETHERLANDS
Tel: 0031 30 274 3945
Fax: 0031 30 274 4449
Email: Elizabeth.vanStrien@rivm.nl

Dr Harry Vennema
Diagnostic Laboratory for Infectious Diseases
Virology Division
National Institute for Public Health and the Environment
Antonie van Leeuwenhoeklaan 9
3730 BA Bilthoven THE NETHERLANDS
Tel: 0031 30 274 3945
Fax: 0031 30 274 4449
Email: Harry.Vennema@rivm.nl

Dr Marijana Vidakovic
Division of Virology
Department of Pathology
Level 5 - Addenbrooke's Hospital
Cambridge CB2 2QW UNITED KINGDOM
Tel: 0044 1223 336914
Fax: 0044 1223 336926

Dr Richard Ward
Division of Infectious Diseases
Children's Hospital Medical Center
3333 Burnet Avenue
Cincinnati OH 45229-3039 USA
Tel: 001 513 636 7628
Fax: 001 513 636 7682
Email: Wardd0@chmcc.org

Dr Teruo Yamashita
Department of Microbiology
Aichi Prefectural Institute of Public Health
7-6 Nagare, Tsujimachi, Kita-ku
Nagoya Aichi 462-8576 JAPAN
Tel: 0081 52 910 5674
Fax: 0081 52 913 3641
Email: tyamashita@hi-ho.ne.jp

Dr Lijuan Yuan
Epidemiology Section
Laboratory of Infectious Diseases
National Institute of Allergy and Infectious Diseases
National Institutes of Health
50 South Drive Room 6310
Bethesda MD 20892-8026 USA
Tel: 001 301 594 1659
Fax: 001 301 496 8312
Email: lyuan@niaid.nih.gov

Subject Index

The main places where items are found are indicated in bold.
Cross references of key words have been made where appropriate